Test Bank
to Accompany

Calculus
with Analytic Geometry
FIFTH EDITION

HOWARD ANTON
DREXEL UNIVERSITY

Prepared by

Pasquale Condo
University of Lowell

Maureen Kelley
Catherine H. Pirri
Northern Essex Community College

Mark N. Pirri

JOHN WILEY & SONS, INC.
New York   Chichester   Brisbane   Toronto   Singapore

Contents

Contents

CHAPTER 1

Coordinates, Graphs, Lines

SECTION 1.1

1.1.1 Let $S = \{-3, 0, 2, -1/7, (4)^{1/3}, 8/7, 7.12, 10\ 8/9, \pi\}$

      (a) List the elements of S which are integers
      (b) List the elements of S which are rational numbers
      (c) List the elements of S which are irrational numbers.

1.1.2 In each part, sketch the set on a coordinate line.
      (a) $(-\infty, 2) \cap [1/2, +\infty)$ **(b)** $(-\infty, 2) \cup [1/2, +\infty)$.

1.1.3 Solve $2x + 4 < -x + 7$.   **1.1.4** Solve $\dfrac{3}{4}x \le \dfrac{2}{3}x - 4$.

1.1.5 Solve $-3 < \dfrac{3}{2}x + 4 < 2$.

1.1.6 Express $0.121121121\ldots$ as the ratio of two integers.

1.1.7 Express $1.921921921\ldots$ as the ratio of two integers.

1.1.8 Solve $3x + 2 \le 6 + 5x$.   **1.1.9** Solve $-\dfrac{2}{3}x \ge \dfrac{1}{6} - \dfrac{3}{4}x$.

1.1.10 Solve $8 - x^2 > 7x$.   **1.1.11** Solve $x^2 + 5x - 9 > 5$.

1.1.12 Solve $3x^2 - 1 \ge \dfrac{1}{2}(3 + x^2)$.   **1.1.13** Solve $6x^3 - 3x^2 > 2x - 1$.

1.1.14 Solve $4x^3 + x^2 - 4x - 1 \le 0$.   **1.1.15** Solve $\dfrac{3}{x+2} < \dfrac{2}{x-1}$.

1.1.16 Solve $\dfrac{2x - 3}{x + 2} \ge 1$.   **1.1.17** Solve $\dfrac{1}{x} < \dfrac{1}{x+1}$.

1.1.18 Solve $\dfrac{3x + 5}{x + 7} \ge \dfrac{1}{3}$.

SOLUTIONS

SECTION 1.1

1.1.1 **(a)** $\{-3, 0, 2\}$ **(b)** $\{-3, 0, 2, -1/7, 8/7, 7.12, 10\,8/9\}$ **(c)** $\{(4)^{1/3}, \pi\}$

1.1.2 **(a)**   **(b)**

1.1.3 $2x + 4 < -x + 7$
$x < 1$
$S = (-\infty, 1)$

1.1.4 $\dfrac{3}{4}x \le \dfrac{2}{3}x - 4$
$9x \le 8x - 48$
$x \le -48$
$S = (-\infty, -48]$

1.1.5 $-3 < \dfrac{3}{2}x + 4 < 2$
$-6 < 3x + 8 < 4$
$-14 < 3x < -4$
$\dfrac{14}{3} < x < -\dfrac{4}{3}$
$S = \left(-\dfrac{14}{3}, -\dfrac{4}{3}\right)$

1.1.6 $x = 0.121121121\ldots$
$1000x = 121.121121\ldots$
$999x = 121$
$x = \dfrac{121}{999}$

1.1.7 $x = 1.921921921\ldots$
$1000x = 1921.921921\ldots$
$999x = 1920$
$x = \dfrac{1920}{999} = \dfrac{640}{333}$

3

1.1.8 $3x + 2 \leq 6 + 5x$

$-2x \leq 4$

$x \geq -2$

$S = [-2, +\infty)$

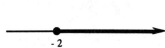

1.1.9 $-\dfrac{2}{3}x \geq \dfrac{1}{6} - \dfrac{3}{4}x$

$-8x \geq 2 - 9x$

$x \geq 2$

$S = [2, +\infty)$

1.1.10 $8 - x^2 > 7x$

$x^2 + 7x - 8 < 0$

$(x + 8)(x - 1) < 0$

$S = (-8, 1)$

Choose -9, 0, and 2 as test points within the intervals $(-\infty, -8)$, $(-8, 1)$, and $(1, +\infty)$ respectively.

Interval	Test Point	Sign of $(x+8)(x-1)$ at the Test Point
$(-\infty, -8)$	-9	$(-)(-) = +$
$(-8, 1)$	0	$(+)(-) = -$
$(1, +\infty)$	2	$(+)(+) = +$

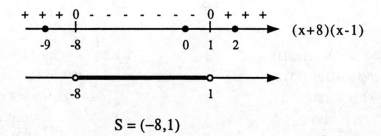

$$S = (-8, 1)$$

1.1.11 $x^2 + 5x - 9 > 5$

$x^2 + 5x - 14 > 0$

$(x + 7)(x - 2) > 0$

Choose -8, 0, and 3 as test points within the intervals $(-\infty, -7), (-7, 2)$, and $(2, +\infty)$ respectively.

Interval	Test Point	Sign of $(x + 7)(x - 2)$ at the Test Point
$(-\infty, -7)$	-8	$(-)(-) = +$
$(-7, 2)$	0	$(+)(-) = -$
$(2, +\infty)$	3	$(+)(+) = +$

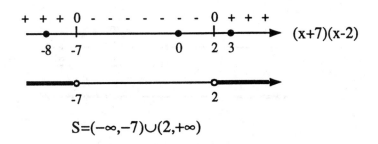

$S = (-\infty, -7) \cup (2, +\infty)$

1.1.12 $3x^2 - 1 \geq \dfrac{1}{2}(3 + x^2)$

$5x^2 - 5 \geq 0$

$(x + 1)(x - 1) \geq 0$

Choose -2, 0, and 2 as test points within the intervals $(-\infty, -1)$, $(-1, 1)$, and $(1, +\infty)$ respectively.

Interval	Test Point	Sign of $(x+1)(x-1)$ at the Test Point
$(-\infty, -1)$	-2	$(-)(-) = +$
$(-1, 1)$	0	$(+)(-) = -$
$(1, +\infty)$	2	$(+)(+) = +$

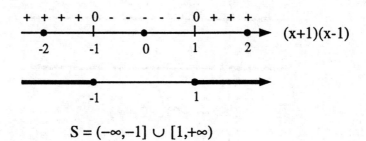

$$S = (-\infty, -1] \cup [1, +\infty)$$

1.1.13
$$6x^3 - 3x^2 > 2x - 1$$
$$6x^3 - 3x^2 - 2x + 1 > 0$$
$$3x^2(2x - 1) - (2x - 1) > 0$$
$$(\sqrt{3}\,x - 1)(2x - 1)(\sqrt{3}\,x + 1) > 0$$

Choose $-1, 0, 0.55$, and 1 as test points within the interval $\left(-\infty, -\dfrac{1}{\sqrt{3}}\right)$, $\left(-\dfrac{1}{\sqrt{3}}, \dfrac{1}{2}\right)$, $\left(\dfrac{1}{2}, \dfrac{1}{\sqrt{3}}\right)$, and $\left(\dfrac{1}{\sqrt{3}}, +\infty\right)$ respectively

Interval	Test Point	Sign of $(2x - 1)(\sqrt{3}\,x - 1)(\sqrt{3}\,x + 1)$ at the Test Point
$\left(-\infty, -\dfrac{1}{\sqrt{3}}\right)$	-1	$(-)(-)(-) = -$
$\left(-\dfrac{1}{\sqrt{3}}, \dfrac{1}{2}\right)$	0	$(-)(-)(+) = +$
$\left(\dfrac{1}{2}, \dfrac{1}{\sqrt{3}}\right)$	0.55	$(+)(-)(+) = -$
$\left(\dfrac{1}{\sqrt{3}}, +\infty\right)$	1	$(+)(+)(+) = +$

$$S = (-\tfrac{1}{\sqrt{3}}, \tfrac{1}{2}) \cup (\tfrac{1}{\sqrt{3}}, +\infty)$$

1.1.14　　$4x^3 + x^2 - 4x - 1 \leq 0$

$x^2(4x + 1) - (4x + 1) \leq 0$

$(x + 1)(4x + 1)(x - 1) \leq 0$

Choose $-2, -1/3, 0$, and 2 as test points within the intervals $(-\infty, -1)(-1, -1/4), (-1/4, 1)$, and $(1, +\infty)$ respectively.

Interval	Test Point	Sign of $(x + 1)(4x + 1)(x - 1)$ at the Test Point
$(-\infty, -1)$	-2	$(-)(-)(-) = -$
$(-1, -1/4)$	$-1/3$	$(+)(-)(-) = +$
$(-1/4, 1)$	0	$(+)(+)(-) = -$
$(1, +\infty)$	2	$(+)(+)(+) = +$

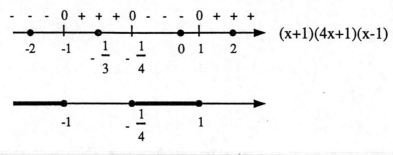

$$S = (-\infty, -1] \cup [-\tfrac{1}{4}, 1]$$

1.1.15
$$\frac{3}{x+2} < \frac{2}{x-1}$$

$$\frac{3}{x+2} - \frac{2}{x-1} < 0$$

$$\frac{x-7}{(x+2)(x-1)} < 0$$

Choose $-3, 0, 2,$ and 8 as test points within the intervals $(-\infty, -2), (-2, 1), (1, 7),$ and $(7, +\infty)$ respectively.

Interval	Test Point	Sign of $\dfrac{(x-7)}{(x+2)(x-1)}$ at the Test Point
$(-\infty, -2)$	-3	$(-)/(-)(-) = -$
$(-2, 1)$	0	$(-)/(+)(-) = +$
$(1, 7)$	2	$(-)/(+)(+) = -$
$(7, +\infty)$	8	$(+)/(+)(+) = +$

$$S = (-\infty, -2) \cup (1, 7)$$

1.1.16 $\dfrac{2x-3}{x+2} \geq 1$

$\dfrac{2x-3}{x+2} - 1 \geq 0$

$\dfrac{x-5}{x+2} \geq 0$

Choose -3, 0, and 6 as test points within the intervals $(-\infty, -2)$, $(-2, 5)$, and $(5, +\infty)$.

Interval	Test Point	Sign of $\dfrac{x-5}{x+2}$ at the Test Point
$(-\infty, -2)$	-3	$(-)/(-) = +$
$(-2, 5)$	0	$(-)/(+) = -$
$(5, +\infty)$	6	$(+)/(+) = +$

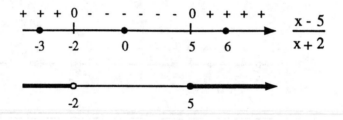

$S = (-\infty, -2) \cup [5, +\infty)$

1.1.17 $\dfrac{1}{x} < \dfrac{1}{x+1}$

$\dfrac{1}{x} - \dfrac{1}{x+1} < 0$

$\dfrac{1}{x(x+1)} < 0$

Choose -2, -0.5, 1 as test points within the intervals $(-\infty, -1)$, $(-1, 0)$, and $(0, +\infty)$ respectively

Interval	Test Point	Sign of $\dfrac{1}{x(x+1)}$ at the Test Point
$(-\infty, -1)$	-2	$1/((-)(-)) = +$
$(-1, 0)$	-0.5	$1/(+)(-) = -$
$(0, +\infty)$	1	$1/(+)(+) = +$

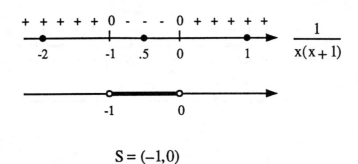

$S = (-1, 0)$

1.1.18 $$\frac{3x+5}{x+7} \geq \frac{1}{3}$$

$$\frac{3x+5}{x+7} - \frac{1}{3} \geq 0$$

$$\frac{x+1}{3(x+7)} \geq 0$$

Choose -8, -2, and 0 as test points within the intervals $(-\infty, -7)$, $(-7, -1)$, and $(-1, +\infty)$ respectively

Interval	Test Point	Sign of $\dfrac{x+1}{3(x+7)}$ at the Test Point
$(-\infty, -7)$	-8	$(-)/(-) = +$
$(-7, -1)$	-2	$(-)/(+) = -$
$(-1, +\infty)$	0	$(+)/(+) = +$

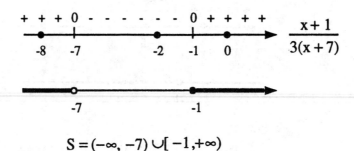

$$S = (-\infty,\ -7) \cup [\,-1, +\infty)$$

SECTION 1.2

1.2.1 Use the definition of the absolute value of a real number to show that $|-9| = 9$.

1.2.2 Solve $|2x - 3| = 5$.

1.2.3 Solve $|3x - 5| = \dfrac{2}{3}$.

1.2.4 Solve $|4x| - 9 = x$.

1.2.5 Solve $|2x| + 7 = 3x$.

1.2.6 Solve $|x + 2| = |3x - 5|$.

1.2.7 Solve $|2x + 1| = |3x - 7|$.

1.2.8 Solve $\left| \dfrac{x + 5}{x + 2} \right| = 4$.

1.2.9 Solve $|x + 1| > 3$ and sketch the solution on a coordinate line.

1.2.10 Solve $|x - 1| \leq 2$ and sketch the solution on a coordinate line.

1.2.11 Find the smallest value of M such that $|2x + 3| \leq M$ for all x in the interval $(-1, 5)$.

1.2.12 Find the smallest value of M such that $\left| \dfrac{1}{2x + 3} \right| \leq M$ for all x in the interval $(-1, 5)$.

SOLUTIONS

SECTION 1.2

1.2.1 $|a| = \begin{cases} a \text{ if } a \geq 0 \\ -a \text{ if } a < 0 \end{cases}$ identify a with -9
since $a < 0, |-9| = -(-9) = 9.$

1.2.2

$$|2x - 3| = 5$$

Case 1	Case 2
$2x - 3 = 5$	$2x - 3 = -5$
$2x = 8$	$2x = -2$
$x = 4$	$x = -1$

1.2.3

$$|3x - 5| = \frac{2}{3}$$

Case 1	Case 2
$3x - 5 = \dfrac{2}{3}$	$3x - 5 = -\dfrac{2}{3}$
$3x = \dfrac{17}{3}$	$3x = \dfrac{13}{3}$
$x = \dfrac{17}{9}$	$x = \dfrac{13}{9}$

1.2.4

$$|4x| - 9 = x$$

Case 1 : $x \geq 0$	Case 2 : $x < 0$
$4x - 9 = x$	$-4x - 9 = x$
$3x = 9$	$-5x = 9$
$x = 3$	$x = -9/5$

1.2.5

$$|2x| + 7 = 3x$$

Case 1 : $x \geq 0$	Case 2 : $x < 0$
$2x + 7 = 3x$	$-2x + 7 = 3x$
$-x = -7$	$-5x = -7$
$x = 7$	$x = \dfrac{7}{5}$; not a solution because x must also satisfy $x < 0.$

15

1.2.6 $$|x + 2| = |3x - 5|$$

Case 1	Case 2
$x + 2 = 3x - 5$	$x + 2 = -(3x - 5)$
$-2x = -7$	$x + 2 = -3x + 5$
$x = 7/2$	$4x = 3$
	$x = 3/4$

1.2.7 $$|2x + 1| = |3x - 7|$$

<u>Case 1</u>

$$2x + 1 = 3x - 7$$
$$-x = -8$$
$$x = 8$$

<u>Case 2</u>

$$2x + 1 = -(3x - 7)$$
$$2x + 1 = -3x + 7$$
$$5x = 6$$
$$x = \frac{6}{5}$$

1.2.8 $$\left|\frac{x + 5}{x + 2}\right| = 4$$

<u>Case 1</u>

$$x + 5 = 4(x + 2)$$
$$x + 5 = 4x + 8$$
$$-3x = 3$$
$$x = -1$$

<u>Case 2</u>

$$x + 5 = -4(x + 2)$$
$$x + 5 = -4x - 8$$
$$5x = -13$$
$$x = -\frac{13}{5}$$

1.2.9 $$|x + 1| > 3$$

<u>Case 1</u>

$$x + 1 > 3$$
$$x > 2$$
$$S = (-\infty, -4) \cup (2, +\infty).$$

<u>Case 2</u>

$$x + 1 < -3$$
$$x < -4$$

1.2.10 $|x - 1| \leq 2$

$$-2 \leq x - 1 \leq 2$$
$$-1 \leq x \leq 3$$
$$S = [-1, 3].$$

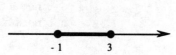

1.2.11 $-1 < x < 5$

$-2 < 2x < 10$

$1 < 2x + 3 < 13$

but $2x + 3 = |2x + 3|$ since $2x + 3 > 0$, thus, $|2x + 3| < 13$ and $M = 13$.

1.2.12 $-1 < x < 5$

$-2 < 2x < 10$

$1 < 2x + 3 < 13$

$1 > \dfrac{1}{2x + 3} > \dfrac{1}{13}$

but $\dfrac{1}{2x + 3} = \left| \dfrac{1}{2x + 3} \right|$ since $\dfrac{1}{2x + 3} > 0$, thus, $\left| \dfrac{1}{2x + 3} \right| < 1$ and $M = 1$.

SECTION 1.3

1.3.1 Determine whether the graph $y = 4x^2 - 2$ is symmetric about the x-axis, the y-axis, or the origin.

1.3.2 Determine whether the graph $y = 4x^3 + x$ is symmetric about the x-axis, the y-axis, or the origin.

1.3.3 Find all intercepts of $x^3 = 2y^3 - y$ and determine symmetry about the x-axis, the y-axis, or the origin.

1.3.4 Find all intercepts of $2x^2 - y^2 = 3$ and determine symmetry about the x-axis, the y-axis, or the origin.

1.3.5 Find all intercepts of $y = \dfrac{1}{3x + x^3}$ and determine symmetry about the x-axis, the y-axis, or the origin.

1.3.6 Find all intercepts of $x = y^4 - 3y^2$ and determine symmetry about the x-axis, the y-axis, or the origin.

1.3.7 Find all intercepts of $y^4 = |x| + 3$ and determine symmetry about the x-axis, the y-axis, or the origin.

1.3.8 Find all intercepts of $y^3 = |x| - 5$ and determine symmetry about the x-axis, the y-axis, or the origin.

1.3.9 Sketch $y = x^4 - x^2$ in the first quadrant and use symmetry to complete the rest of the graph.

1.3.10 Sketch $y = x^3 - x$ in the first quadrant and use symmetry to complete the rest of the graph.

1.3.11 Extend the graph of the figure given below so that it is symmetric about (a) the origin, (b) the x-axis, and (c) the y-axis.

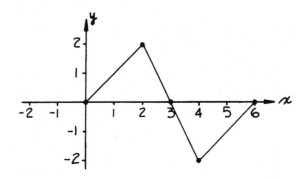

19

1.3.12   Extend the graph of the figure given below so that it is symmetric about (a) the origin, (b) the x-axis, and (c) the y-axis.

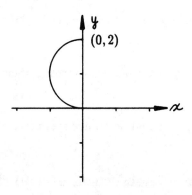

1.3.13   Show that $y = |x|$ is symmetric about the y-axis and sketch its graph.

1.3.14   Show that $y^2 = 4x + 4$ is symmetric about the x-axis and sketch its graph.

1.3.15   Show that $y = x^3$ is symmetric about the origin and sketch its graph.

1.3.16   Show that $xy = 4$ is symmetric about the origin and sketch its graph.

1.3.17   Match the given equations with its graph. [Equations are labeled (a)–(d), graphs are labeled (A)–(D).]

FUNCTION   GRAPH

(a) $y = \dfrac{1}{x^2 + 1}$ ———

(b) $y = \dfrac{x^2 - 1}{x^2 + 1}$ ———

(c) $y = \dfrac{1}{(x - 1)^2}$ ———

(d) $y = \dfrac{x}{x^2 + 1}$ ———

(A)

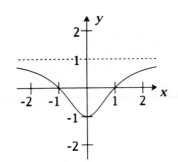

(B)

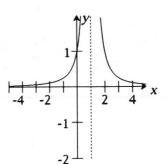

(C)

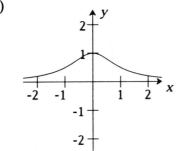

(D)

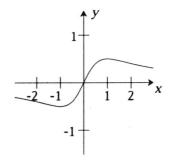

1.3.18   State which of the following statements are true and which are false.

(a) __ A graph which is symmetric about the x-axis and the y-axis must be symmetric about the origin.

(b) __ A graph which is symmetric about the origin must be symmetric about the x-axis and the y-axis.

(c) __ A graph which is symmetric about the origin and about the y-axis must be symmetric about the x-axis.

(d) __ A graph which is not symmetric about the origin is not symmetric about the x-axis and the y-axis.

SOLUTIONS

SECTION 1.3

1.3.1 $y = 4x^2 - 2$

Replace x with $(-x)$: $y = 4(-x)^2 - 2 = 4x^2 - 2$ thus the graph is symmetric about the y-axis.

1.3.2 $y = 4x^3 + x$

Replace x with $(-x)$ and y with $(-y)$:

$$-y = 4(-x)^3 + (-x)$$
$$-y = -(4x^3 + x)$$
$$y = 4x^3 + x$$

thus the graph is symmetric about the origin.

1.3.3 $x^3 = 2y^3 - y$

Set $y = 0$: $x^3 = 0$, (x-intercept) $x = 0$

Set $x = 0$: $0 = 2y^3 - y = y(2y^2 - 1)$ (y intercepts) $y = 0$ or $y = \pm\sqrt{2}/2$

Replace x with $(-x)$ and y with $(-y)$:

$$(-x)^3 = 2(-y)^3 - (-y)$$
$$-(x^3) = -(2y^3 - y)$$
$$x^3 = 2y^3 - y$$

thus the graph is symmetric about the origin.

1.3.4 $2x^2 - y^2 = 3$

Set $y = 0$: $2x^2 = 3$, $x = \pm\dfrac{\sqrt{6}}{2}$ (x-intercepts)

Set $x = 0$: $-y^2 = 3$ has no real solution so no y-intercept.

Replace x with $(-x)$

$$2(-x)^2 - y^2 = 3$$
$$2x^2 - y^2 = 3$$

thus the graph is symmetric about the y-axis. Replace y with $(-y)$

$$2x^2 - (-y)^2 = 3$$
$$2x^2 - y^2 = 3$$

thus the graph is symmetric about the y-axis. Since the graph is symmetric about both the x-axis and y-axis, it is symmetric about the origin.

1.3.5 $y = \dfrac{1}{3x + x^3}$

No x or y intercepts

Replace x with $(-x)$ and y with $(-y)$:

$$-y = \frac{1}{3(-x) + (-x)^3} = -\frac{1}{3x + x^3}$$

$$y = \frac{1}{3x + x^3}$$

thus the graph is symmetric about the origin.

1.3.6 $x = y^4 - 3y^2$

Set $x = 0$: $0 = y^4 - 3y^2 = y^2(y^2 - 3)$ (y-intercepts) $y = 0$ and $y = \pm\sqrt{3}$

Set $y = 0$: $x = 0$ (x-intercept)

Replace y with $(-y)$:

$x = (-y)^4 - 3(-y)^2 = y^4 - 3y^2$ thus the graph is symmetric about the x-axis.

1.3.7 $y^4 = |x| + 3$

Set $y = 0$: $0 = |x| + 3$, $|x| = -3$ no y-intercept

Set $x = 0$: $y^4 = 3$, $y = \pm\sqrt[4]{3}$ (y-intercepts)

Replace y with $(-y)$: $(-y)^4 = y^4 = |x| + 3$

Graph is symmetric about the x-axis

Replace x with $(-x)$: $y^4 = |-x| + 3 = |x| + 3$

Graph is symmetric about the y-axis

Since the graph is symmetric about both the x and y-axes the graph is symmetric about the origin.

1.3.8 $y^3 = |x| - 5$

Set $y = 0$: $0 = |x| - 5$ then $x = \pm 5$ (x-intercepts)

Set $x = 0$: $y^3 = -5$ then $y = \sqrt[3]{-5}$ (y-intercept)

Replace x with $(-x)$: $y^3 = |-x| - 5 = |x| - 5$

The graph is symmetric about the y-axis

1.3.9 $y = x^4 - x^2$

Set $y = 0$: $0 = x^2(x+1)(x-1)$, $x = 0$ and $x = \pm 1$ are x-intercepts

Set $x = 0$: $y = 0$ is a y-intercept

Replace x by $(-x)$: $y = (-x)^4 - (-x)^2 = x^4 - x^2$ thus the graph is symmetric about the y-axis

x	.5	1.5
y	-0.1875	2.8125

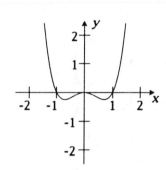

1.3.10 $y = x^3 - x$

Let $x = 0$: $y = 0$ is the y-intercept

Let $y = 0$: $0 = x^3 - x = x(x+1)(x-1)$,

$x = 0$, $x = \pm 1$ are the x-intercepts

Replace x by $(-x)$ and y by $(-y)$:

$-y = (-x)^3 - (-x)$

$-y = -(x^3 - x)$

$y = x^3 - x$.

The graph is symmetric about the y-axis

x	.5	1.5
y	-0.375	1.875

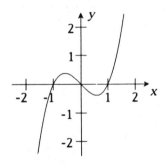

1.3.11 (a)

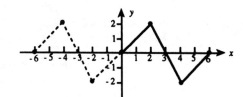

(b)

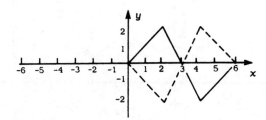

(c)

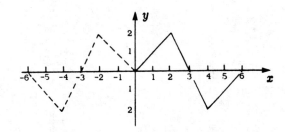

1.3.12 (a)

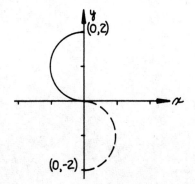

(b)

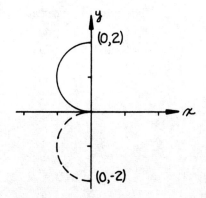

(c)

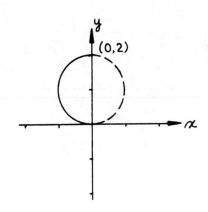

1.3.13 $y = |x|$

replace x by $-x$:
$$y = |-x| = |x|$$
Thus, the graph is symmetric
about the y-axis.

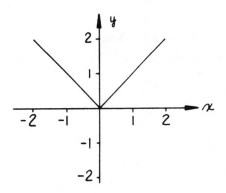

1.3.14 $y^2 = 4x + 4$

replace y with $(-y)$:
$$(-y)^2 = 4x + 4$$
$$y^2 = 4x + 4$$
Thus, the graph is symmetric
about the x-axis.

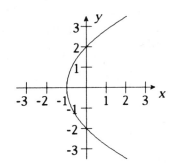

1.3.15 $y = x^3$

Replace x with $(-x)$ and y with $(-y)$:

$$(-y) = (-x)^3$$
$$-y = -x^3$$
$$y = x^3$$

thus, the graph is symmetric about the origin.

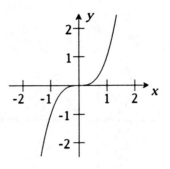

1.3.16   Replace x by $(-x)$ and (y) by $(-y)$:

$$(-x)(-y) = 4$$
$$xy = 4$$

Thus, the graph is symmetric about the origin.

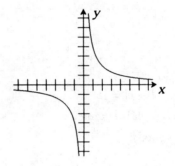

1.3.17   **(a)** (C)   **(b)** (A)   **(c)** (B)   **(d)** (D)

1.3.18   **(a)** T   **(b)** F   **(c)** T   **(d)** T

SECTION 1.4

1.4.1 Find the slope of the line through $(-4, 2)$ and $(3, 5)$.

1.4.2 Find the slope of the line through $(3, 4)$ and $(7, -4)$.

1.4.3 Find the slope of a line drawn perpendicular to the line through $(-2, -4)$ and $(3, 5)$.

1.4.4 Find the slope of a line drawn perpendicular to the line through $(3, 5)$ and $(6, -3)$.

1.4.5 Show that the line through $(-2, 14)$ and $(1, 8)$ is

    (a) parallel to the line through $(1, -2)$ and $(2, -4)$;
    (b) perpendicular to the line through $(1, 1)$ and $(3, 2)$.

1.4.6 Show that the line through $(3, -4)$ and $(7, 5)$ is

    (a) parallel to the line through $(1, -11)$ and $(5, -2)$;
    (b) perpendicular to the line through $(0, 0)$ and $(9, -4)$.

1.4.7 Use slopes to show that $(-2, 4)$, $(2, 0)$, and $(6, 4)$ are vertices of a right triangle.

1.4.8 Use slopes to show that $(9, -6)$, $(-3, 0)$, and $(0, 6)$ are vertices of a right triangle.

1.4.9 Use slopes to show that $(-1, -8)$, $(5, 0)$, and $(6, -7)$ are vertices of a right triangle.

1.4.10 Use slopes to find k so that $(k, -3)$, $(1, 1)$, and $(4, 3)$ lie on a straight line.

1.4.11 Use slopes to find k so that $(2, k)$, $(1, -1)$, and $(4, 4)$ lie on a straight line.

1.4.12 Use slopes to find k so that the line through $(k, -18)$ and $(1, -3)$ is perpendicular to the line through $(1, -3)$ and $(3, -13)$.

1.4.13 Use slopes to find k so that the line through $(-3, -2)$ and $(9, k)$ is perpendicular to the line through $(5, -3)$ and $(4, k)$.

1.4.14 Use slopes to show that $(-1, 1)$, $(4, 2)$, $(3, -2)$, and $(-2, -3)$ are vertices of a parallelogram.

1.4.15 Use slopes to show that $(-1, -3)$, $(8, 3)$, $(3, 4)$, and $(0, 2)$ are vertices of a trapezoid.

1.4.16   Show that $(-1, 3)$, $(6, 6)$, $(8, 2)$, and $(1, -1)$ are vertices of a parallelogram but not a rectangle.

1.4.17   Use slopes to show that $(3, -5)$, $(7, -2)$, $(2, -2)$ and $(-2, -5)$ are vertices of a rhombus.

1.4.18   Use slopes to show that $(-6, -1)$, $(-2, 5)$, $(1, 3)$ and $(-3, -3)$ are vertices of a rectangle.

1.4.19   Find the equation of the line through $(-1, 3)$ with slope $m = -2$.

1.4.20   Find the equation of the line through $(-3, -7)$ with slope $m = 3$.

1.4.21   Find the equation of the line through $(1, 3)$ and $(-2, 1)$.

1.4.22   Find the equation of the line through $(-2, -3)$ and $(5, -6)$.

1.4.23   Find the equation of the line through $(2, -1)$ and parallel to $3y + 5x - 6 = 0$.

1.4.24   Find the equation of the line through $(3, 4)$ and parallel to $4x + 3y + 7 = 0$.

1.4.25   Find the equation of the line through $(1, -1)$ and perpendicular to $2x - 3y - 8 = 0$.

1.4.26   Find the equation of the line through $(5, 2)$ and perpendicular to $4x - 7y - 10 = 0$.

1.4.27   Find the equation of the line through $(2, 2)$ and parallel to the line through $(3, 4)$ and $(6, 2)$.

1.4.28   Find the equation of the line that has an angle of inclination of $\phi = \frac{1}{4}\pi$ and passes through the point $(3, -2)$.

1.4.29   Find the equation of the line that passes through the point $(7, 3)$ and has an angle of inclination $\phi = \frac{1}{3}\pi$.

1.4.30   Find the point of intersection of the line through $(0, 2)$ and $(8, 8)$ with the line through $(1, 9)$ and $(7, 1)$.

1.4.31   Find the point of intersection of the line through $(-4, 1)$ that is perpendicular to the line through $(0, -6)$ and $(-2, -9)$.

1.4.32   Find the point of intersection of the line through $(0, 2)$ and $(4, -2)$ with the line through $(0, 0)$ and $(-2, 6)$.

1.4.33   Find k so that $(2,3)$ is on the line $3x - 5y + k = 0$.

1.4.34   Find k so that the line $2x + ky = 3$ is parallel to $3x + 4y = 1$.

1.4.35   Find k so that the line $3x + ky = 10$ is perpendicular to $3x + 4y = 1$.

SOLUTIONS

SECTION 1.4

1.4.1 $m = \dfrac{2-5}{-4-3} = \dfrac{3}{7}.$

1.4.2 $m = \dfrac{-4-4}{7-3} = -2.$

1.4.3 $m = \dfrac{5+4}{3+2} = \dfrac{9}{5}$, so any line with slope $-5/9$ will be perpendicular to the line through $(-2,-4)$ and $(3,5)$.

1.4.4 $m = \dfrac{-3-5}{6-3} = -\dfrac{8}{3}$, so any line with slope $3/8$ will be perpendicular to the line through $(3,5)$ and $6,-3)$.

1.4.5 Let m_1 be the slope of the line through $(-2,14)$ and $(1,8)$ and let m_2 and m_3 be the slope of the lines in parts (a) and (b), then

$$m_1 = \frac{8-14}{1+2} = -2;$$

    (a) $m_2 = \dfrac{-4+2}{2-1} = -2$, thus, $m_1 = m_2$ and the lines are parallel;

    (b) $m_3 = \dfrac{2-1}{3-1} = \dfrac{1}{2}$, thus, $m_1 m_3 = -1$ and the lines are perpendicular.

1.4.6 Let m_1 be the slope of the line through $(3,-4)$ and $(7,5)$ and let m_2 and m_3 be the slopes of the lines in parts (a) and (b), then

$$m_1 = \frac{5+4}{7-3} = \frac{9}{4};$$

    (a) $m_2 = \dfrac{-2+11}{5-1} = \dfrac{9}{4}$, thus, $m_1 = m_2$ and the lines are parallel;

    (b) $m_3 = \dfrac{-4-0}{9-0} = -\dfrac{4}{9}$, thus, $m_1 m_3 = -1$ and the lines are perpendicular.

1.4.7 Let $A(-2,4)$, $B(2,0)$, and $C(6,4)$ be the given vertices and let a, b, and c be the sides opposite the vertices, then

$$m_a = \frac{4-0}{6-2} = 1, \quad m_b = \frac{4-4}{6+2} = 0, \quad \text{and} \quad m_c = \frac{0-4}{2+2} = -1.$$

Since $m_a m_c = -1$, sides a and c are perpendicular thus ABC is a right triangle.

1.4.8   Let $A(9, -6)$, $B(-3, 0)$, and $C(0, 6)$ be the given vertices and let a, b, and c be the sides opposite the vertices, then

$$m_a = \frac{6-0}{0+3} = 2, \ m_b = \frac{-6-6}{9-0} = -\frac{4}{3}, \ \text{and} \ m_c = \frac{-6-0}{9+3} = -\frac{1}{2}.$$

Since $m_a m_c = -1$, sides a and c are perpendicular thus ABC is a right triangle.

1.4.9   Let $A(-1, -8)$, $B(5, 0)$, and $C(6, -7)$ be the given vertices and let a, b, and c be the sides opposite the vertices, then

$$m_a = \frac{-7-0}{6-5} = -7, \ m_b = \frac{-7+8}{6+1} = \frac{1}{7}, \ \text{and} \ m_c = \frac{0+8}{5+1} = \frac{4}{3}.$$

Since $m_a m_b = -1$, sides a and b are perpendicular thus, ABC is a right triangle.

1.4.10   Let m_1 be the slope of the line through $(k, -3)$ and $(1, 1)$, then $m_1 = \frac{-3-1}{k-1} = \frac{4}{1-k}$; let m_2 be the slope of the line through $(k, -3)$ and $(4, 3)$, then $m_2 = \frac{3+3}{4+k} = \frac{6}{4-k}$. The points will be on a line if $m_1 = m_2$, thus

$$\frac{4}{1-k} = \frac{6}{4-k}$$
$$4(4-k) = 6(1-k)$$
$$16 - 4k = 6 - 6k$$
$$-2k = 10$$
$$k = -5.$$

1.4.11   Let m_1 be the slope of the line through $(2, k)$ and $(1, -1)$, then, $m_1 = \frac{-1-k}{2-1} = 1+k$; let m_2 be the slope of the line through $(2, k)$ and $(4, 4)$, then $m_2 = \frac{4-k}{4-2} = \frac{4-k}{2}$. The points will be on a line if $m_1 = m_2$, thus, $1+k = \frac{4-k}{2}$, so $k = 2/3$.

1.4.12   Let m_1 be the slope of the line through $(k, -18)$ and $(1, -3)$, then $m_1 = \frac{-3+18}{1-k} = \frac{15}{1-k}$; let m_2 be the slope of the line through $(1, -3)$ and $(3, -13)$, then, $m_2 = \frac{-13+3}{3-1} = -5$. The lines will be perpendicular if $m_1 m_2 = -1$, thus,

$$\left(\frac{15}{1-k}\right)(-5) = -1$$
$$1 - k = 75$$
$$k = -74.$$

1.4.13 Let m_1 be the slope of the line through $(-3, -2)$ and $(9, k)$, then, $m_1 = \dfrac{k+2}{9+3} = \dfrac{k+2}{12}$; let m_2 be the slope of the line through $(5, -3)$ and $(4, k)$, then $m_2 = \dfrac{k+3}{4-5} = \dfrac{k+3}{-1}$. The lines will be perpendicular if $m_1 m_2 = -1$, thus,

$$\left(\frac{k+2}{12}\right)\left(\frac{k+3}{-1}\right) = -1$$
$$k^2 + 5k + 6 = 12$$
$$k^2 + 5k - 6 = 0$$
$$(k+6)(k-1) = 0, \text{ so } k = -6 \text{ or } k = 1.$$

1.4.14 The line through $(-1, 1)$ and $(4, 2)$ has slope $m_1 = 1/5$, the line through $(4, 2)$ and $(3, -2)$ has slope $m_2 = 4$, the line through $(3, -2)$ and $(-2, -3)$ has slope $m_3 = 1/5$, the line through $(-2, -3)$ and $(-1, 1)$ has slope $m_4 = 4$; since $m_1 = m_3$ and $m_2 = m_4$, opposite sides are parallel so the figure is a parallelogram.

1.4.15 The line through $(-1, -3)$ and $(8, 3)$ has slope $m_1 = 2/3$, the line through $(8, 3)$ and $(3, 4)$ has slope $m_2 = -1/5$, the line through $(3, 4)$ and $(0, 2)$ has slope $m_3 = 2/3$, the line through $(0, 2)$ and $(-1, -3)$ has slope $m_4 = 5$. So $m_1 = m_3$, the figure is a trapezoid since it has two parallel sides.

1.4.16 The line through $(-1, 3)$ and $(6, 6)$ has slope $m_1 = 3/7$, the line through $(6, 6)$ and $(8, 2)$ has slope $m_2 = -2$, the line through $(8, 2)$ and $(1, -1)$ has slope $m_3 = 3/7$, the line through $(1, -1)$ and $(-1, 3)$ has slope $m_4 = -2$; since $m_1 = m_3$ and $m_2 = m_4$, opposite sides are parallel so the figure is a parallelogram; since $m_1 m_2 \neq -1$, adjacent sides are not perpendicular and thus, the parallelogram is not a rectangle.

1.4.17 The line through $(3, -5)$ and $(7, -2)$ has slope $m_1 = \dfrac{-2+5}{7-3} = \dfrac{3}{4}$ the line through $(7, -2)$ and $(2, -2)$ has slope $m_2 = \dfrac{-2+2}{2-7} = 0$, the line through $(2, -2)$ and $(-2, -5)$ has slope $m_3 = \dfrac{-5+2}{-2-2} = \dfrac{3}{4}$ and the line through $(-2, -5)$ and $(3, -5)$ has slope $m_4 = \dfrac{-5+5}{3+2} = 0$. Since $m_1 = m_3$ and $m_2 = m_4$, opposite sides of the quadrilateral are parallel. The diagonal from $(3, -5)$ to $(2, -2)$ has slope $m_a = \dfrac{-2+5}{2-3} = -3$ and the diagonal from $(7, -2)$ to $(-2, -5)$ has slope $m_b = \dfrac{-5+2}{-2+7} = \dfrac{1}{3}$. Since $m_a m_b = -1$, the diagonals are perpendicular so the quadrilateral is a rhombus.

1.4.18 The line through $(-6, -1)$ and $(-2, 5)$ has slope $m_1 = \dfrac{5+1}{-2+6} = \dfrac{3}{2}$, the line through $(-2, 5)$ and $(1, 3)$ has slope $m_2 = \dfrac{3-5}{1+2} = -\dfrac{2}{3}$, the line through $(1, 3)$ and $(-3, -3)$ has slope $m_3 = \dfrac{-3-3}{-3-1} = \dfrac{3}{2}$, and the line through $(-3, -3)$ and $(-6, -1)$ has slope $m_4 = \dfrac{-1+3}{-6+3} = -\dfrac{2}{3}$. Since $m_1 = m_3$ and $m_2 = m_4$ opposite sides of the quadrilateral are parallel and since $m_1 m_2 = -1$, adjacent sides are perpendicular so the quadrilateral is a rectangle.

1.4.19 $y - 3 = (-2)(x + 1)$ **1.4.20** $y - (-7) = 3[x - (-3)]$
$y = -2x + 1$ $y = 3x + 2.$

1.4.21   The slope of the line through $(1, 3)$ and $(-2, 1)$ is $m = \dfrac{1 - 3}{-2 - 1} = \dfrac{2}{3}$; the required equation is
$y - 3 = \dfrac{2}{3}(x - 1)$ or $y = \dfrac{2}{3}x + \dfrac{7}{3}$.

1.4.22   The slope of the line through $(-2, -3)$ and $(5, -6)$ is $m = \dfrac{-6 - (-3)}{5 - (-2)} = -\dfrac{3}{7}$; the required
equation is $y - (-3) = -\dfrac{3}{7}[x - (-2)]$ or $y = -\dfrac{3}{7}x - \dfrac{27}{7}$.

1.4.23   Place $3y + 5x - 6 = 0$ into slope-intercept form to yield $y = -\dfrac{5}{3}x + 2$. The lines will be parallel
if the slope of the line $m = -5/3$, thus,

$$y - (-1) = -\frac{5}{3}(x - 2)$$

$$y = -\frac{5}{3}x + \frac{7}{3}.$$

1.4.24   Place $4x + 3y + 7 = 0$ into slope-intercept form to yield $y = -\dfrac{4}{3}x - \dfrac{7}{3}$. The lines will be
parallel if the slope of the line $m = 4/3$, thus,

$$y - 4 = -\frac{4}{3}(x - 3)$$

$$y = -\frac{4}{3}x + 8.$$

1.4.25   Place $2x - 3y - 8 = 0$ into slope-intercept form to yield $y = \dfrac{2}{3}x - \dfrac{8}{3}$. The lines will be
perpendicular if the slope of the line $m = -3/2$, thus,

$$y - (-1) = -\frac{3}{2}(x - 1)$$

$$y = -\frac{3}{2}x + \frac{1}{2}.$$

1.4.26   Place $4x - 7y - 10 = 0$ into slope-intercept form to yield $y = \dfrac{4}{7}x - \dfrac{10}{7}$. The lines will be
perpendicular if the slope of the line $m = -7/4$, thus,

$$y - 2 = -\frac{7}{4}(x - 5)$$

$$y = -\frac{7}{4}x + \frac{43}{4}.$$

1.4.27   The slope of the line through $(3,4)$ and $(6,2)$ is

$$\frac{2-4}{6-3} = -\frac{2}{3}.$$

The lines will be parallel if the slope of the line $m = -2/3$, thus,

$$y - 2 = -\frac{2}{3}(x-2)$$

$$y = -\frac{2}{3}x + \frac{10}{3}.$$

1.4.28 $m = \tan\dfrac{\pi}{4} = 1$ so

$$y - (-2) = x - 3$$
$$y + 2 = x - 3$$
$$y = x - 5$$

1.4.29 $m = \tan\dfrac{\pi}{3} = \sqrt{3}$

$$y - 3 = \sqrt{3}(x - 7)$$
$$y = \sqrt{3}\,x - 7\sqrt{3} + 3$$

1.4.30   The slope of the line through $(0,2)$ and $(8,8)$ is $m = \dfrac{8-2}{8-0} = \dfrac{3}{4}$ and the equation of the line through $(0,2)$ and $(8,8)$ is $y - 2 = \dfrac{3}{4}(x-0)$ or $3x - 4y = -8$; the slope of the line through $(1,9)$ and $(7,1)$ is $m = \dfrac{1-9}{7-1} = -\dfrac{4}{3}$ and the equation of the line through $(1,9)$ and $(7,1)$ is $y - 9 = -\dfrac{4}{3}(x-1)$ or $4x + 3y = 31$;

Solving $\left.\begin{array}{l} 3x - 4y = -8 \\ 4x + 3y = 31 \end{array}\right\}$ to get $x = 4$, $y = 5$, so the point is $(4,5)$.

1.4.31   The slope of the line through $(0,-6)$ and $(-2,-9)$ is

$$m = \frac{-9 - (-6)}{-2 - 0} = \frac{3}{2}$$

and the equation of the line through $(0,-6)$ and $(-2,-9)$ is

$$y - (-6) = \frac{3}{2}(x - 0) \quad \text{or} \quad 3x - 2y = 12$$

thus, the slope of the line through $(-4,1)$ that is perpendicular to the line through $(0,-6)$ and $(-2,-9)$ is $m = -\dfrac{2}{3}$. The equation of the line through $(-4,1)$ with slope $m = -\dfrac{2}{3}$ is

$$y - 1 = -\frac{2}{3}[x - (-4)]$$

$$y - \frac{2}{3}x - \frac{5}{3} \quad \text{or} \quad 2x + 3y = -5,$$

Solve $\left.\begin{array}{l}3x - 2y = 12 \\ 2x + 3y = -5\end{array}\right\}$ to get $x = 2$, $y = -3$, so the point is $(2, -3)$.

1.4.32 The slope of the line through $(0,2)$ and $(4,-2)$ is $m = \dfrac{-2-2}{4-0} = -1$ and the equation of the line through $(0,2)$ and $(4,-2)$ is $y - 2 = -(x - 0)$ or $x + y = 2$; the slope of the line through $(0,0)$ and $(-2,6)$ is $m = \dfrac{6-0}{-2-0} = -3$ and the equation of the line through $(0,0)$ and $(-2,6)$ is $y - 0 = -3(x - 0)$ or $y = -3x$;

Solving $\left.\begin{array}{l}x + y = 2 \\ 3x + y = 0\end{array}\right\}$ to get $x = -1$, $y = 3$, so the point is $(-1, 3)$.

1.4.33 If $(2,3)$ is on the line $3x - 5y + k = 0$, then $3(2) - 5(3) + k = 0$, thus, $k = 9$.

1.4.34 Place $2x + ky = 3$ into slope-intercept form to yield $y = -\dfrac{2x}{k} + \dfrac{3}{k}$, with $m_1 = -\dfrac{2}{k}$. Place $3x + 4y = 1$ into slope-intercept form to yield $y = -\dfrac{3x}{4} + \dfrac{1}{4}$, with $m_2 = -\dfrac{3}{4}$. The lines will be parallel if $m_1 = m_2$ thus

$$-\frac{2}{k} = -\frac{3}{4}$$
$$k = \frac{8}{3}$$

1.4.35 Place $3x + ky = 10$ into slope-intercept form to yield $y = -\dfrac{3x}{k} + \dfrac{10}{k}$, with $m_1 = -\dfrac{3}{k}$. Place $3x + 4y = 1$ into slope-intercept form to yield $y = -\dfrac{3}{4}x + \dfrac{1}{4}$, with $m_2 = -\dfrac{3}{4}$. The lines will be perpendicular if $m_1 m_2 = -1$ thus,

$$\left(-\frac{3}{k}\right)\left(-\frac{3}{4}\right) = -1$$
$$k = -\frac{9}{4}$$

SECTION 1.5

1.5.1 Find the distance between $A(-3, -6)$ and $B(5, -2)$, and find the coordinates of the midpoint of the line segment joining A and B.

1.5.2 Find the distance between $A(-1, -8)$ and $B(3, 15)$, and find the coordinates of the midpoint of the line segment joining A and B.

1.5.3 Show that the triangle with vertices $(-4, -1)$, $(2, 7)$, and $(3, 0)$ is isosceles and find its perimeter.

1.5.4 The ends of the base of an isosceles triangle are located at $(-2, 1)$ and $(8, 1)$. The altitude is 15 units. Find the coordinates of a third vertex (2 solutions possible).

1.5.5 Show that $(0, 6)$, $(-3, 0)$, and $(9, -6)$ are vertices of a right triangle and specify the vertex at which the right angle occurs.

1.5.6 Find the equation of the circle with center $(3, 2)$ and passing through $(6, 5)$.

1.5.7 Determine whether $2x^2 + 2y^2 - 8x - 12y + 24 = 0$ represents a circle, a point, or no graph. If the equation represents a circle, sketch it and find its center and radius.

1.5.8 Determine whether $x^2 + y^2 - 4x - 2y + 5 = 0$ represents a circle, a point, or no graph. If the equation represents a circle, sketch it and find its center and radius.

1.5.9 Find the equation of the two concentric circles with center at $(2, 6)$, one touching the x-axis and the other the y-axis.

1.5.10 Determine the value(s) for k for which the equation

$$x^2 + y^2 - 4x + 3y - k = 0$$

represents (a) a circle; (b) a point; and (c) no graph.

1.5.11 Find the equation of the circle which has as one of its diameters, the line segment through $(-2, -6)$ and $(6, 0)$.

1.5.12 Graph the parabola $y = x^2 + 4x + 2$. Label the coordinates of the vertex and intersections with the coordinate axes.

1.5.13 Graph the parabola $y = x^2 - 5x$. Label the coordinates of the vertex and intersections with the coordinate axes.

1.5.14 Graph the parabola $y^2 - 12 = x + y$. Label the coordinates of the vertex and intersections with the coordinate axes.

1.5.15 Find the equation of the circle whose diameter is the segment of the line $3x + 4y = 24$ included between the x and y-axes.

1.5.16 Determine the equation that the coordinates of (x, y) must satisfy if the distance between this point and the point $(2, -2)$ is twice the distance from the point to $(2, 1)$. Sketch the graph of the equation.

1.5.17 Find the equation of the line tangent to $x^2 + y^2 + 6x - 8y = 0$ at $(1, 7)$.

1.5.18 Find the equation of the circle with center $(-3, 8)$ that is tangent to $3y + 4x - 37 = 0$.

1.5.19 Solve $2x^2 + 5x - 3 \geq 0$

SOLUTIONS

SECTION 1.5

1.5.1 $d = \sqrt{(5+3)^2 + (-2+6)^2} = \sqrt{64 + 16} = \sqrt{80} = 4\sqrt{5};$

$\left(\dfrac{5-3}{2}, \dfrac{-2-6}{2}\right) = \left(\dfrac{2}{2}, -\dfrac{8}{2}\right) = (1, -4).$

1.5.2 $d = \sqrt{(3+1)^2 + (15+8)^2} = \sqrt{16 + 529} = \sqrt{545};$

$\left(\dfrac{3-1}{2}, \dfrac{15-8}{2}\right) = \left(\dfrac{2}{2}, \dfrac{7}{2}\right) = \left(1, \dfrac{7}{2}\right).$

1.5.3 Let $A(-4, -1)$, $B(2, 7)$, and $C(3, 0)$ be the given vertices and a, b, and c the lengths of the sides opposite the vertices, then

$$a = \sqrt{(3-2)^2 + (0-7)^2} = \sqrt{1 + 49} = \sqrt{50} = 5\sqrt{2}$$

$$b = \sqrt{(3+4)^2 + (0+1)^2} = \sqrt{49 + 1} = \sqrt{50} = 5\sqrt{2}$$

$$c = \sqrt{(2+4)^2 + (7+1)^2} = \sqrt{36 + 64} = \sqrt{100} = 10$$

thus, ABC is isosceles because it has two equal sides ($a = b$); the perimeter of $ABC = a + b + c = 5\sqrt{2} + 5\sqrt{2} + 10 = 10\sqrt{2} + 10.$

1.5.4 Let $A(-2, 1)$ and $B(8, 1)$ be the coordinates of endpoints of the base. The third vertex must lie on the perpendicular bisector of AB. The midpoint of AB is located at

$$\left(\dfrac{8-2}{2}, \dfrac{1+1}{2}\right) = (3, 1)$$

and hence, there are two points $(3, 1 + 15)$ or $(3, 16)$ and $(3, 1 - 15)$ or $(3, -14)$ either of which may serve as the third vertex.

1.5.5 Let $A(0, 6)$, $B(-3, 0)$, and $C(9, -6)$ be the given vertices and s_1, s_2, and s_3 be the lengths of the sides opposite these vertices. ABC is a right triangle if and only if the square of the longest side is equal to the sum of the squares of the other two sides (Pythagorean theorem), thus,

$$s_1^2 = (9+3)^2 + (-6-0)^2 = 144 + 36 = 180$$

$$s_2^2 = (9-0)^2 + (-6-6)^2 = 81 + 144 = 225$$

$$s_3^2 = (-3-0)^2 + (0-6)^2 = 9 + 36 = 45$$

and $s_2^2 = s_1^2 + s_3^2$, so ABC is a right triangle. The vertex at which the right angle occurs is located opposite the longest side at $B(-3, 0)$.

1.5.6 $r = \sqrt{(6-3)^2 + (5-2)^2} = \sqrt{18}$

$(x-3)^2 + (y-2)^2 = 18.$

1.5.7 $2(x^2 - 4x) + 2(y^2 - 6y) = -24$

$(x^2 - 4x + 4) + (y^2 - 6y + 9) = -12 + 4 + 9$

$(x-2)^2 + (y-3)^2 = 1$

circle, center $(2,3)$; $r = 1.$

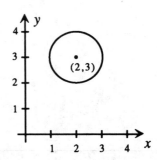

1.5.8 $(x^2 - 4x) + (y^2 - 2y) = -5$

$(x^2 - 4x + 4) + (y^2 - 2y + 1) = -5 + 4 + 1,$

$(x-2)^2 + (y-1)^2 = 0,$

the point, $(2,1).$

1.5.9 $(x-2)^2 + (y-6)^2 = 36$ touches the x-axis at $(2,0)$,

$(x-2)^2 + (y-6)^2 = 4$ touches the y-axis at $(0,6)$.

1.5.10 Place $x^2 + y^2 - 4x + 3y - k = 0$ into the standard form, thus,

$$(x^2 - 4x) + (y^2 + 3y) = k$$

$$(x^2 - 4x + 4) + (y^2 + 3y + 9/4) = k + 4 + \frac{9}{4}$$

$$(x-2)^2 + (y+3/2)^2 = k + \frac{25}{4}.$$

(a) The equation represents a circle if $k + \dfrac{25}{4} > 0$ or $k > -\dfrac{25}{4}$;

(b) the equation represents a point if $k + \dfrac{25}{4} = 0$ or $k = -\dfrac{25}{4}$;

(c) the equation represents no graph if $k + \dfrac{25}{4} < 0$ or $k < -\dfrac{25}{4}$.

1.5.11 The center is the midpoint of the line segment joining $(-2,-6)$ and $(6,0)$, thus, the center is $\left(\dfrac{-2+6}{2}, \dfrac{-6+0}{2}\right) = (2,-3)$; the radius is $\sqrt{(-2-2)^2 + (-6-(-3))^2} = \sqrt{25}$, so $(x-2)^2 + (y+3)^2 = 25.$

1.5.12

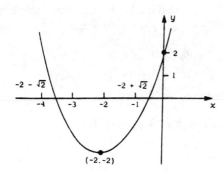

1.5.13

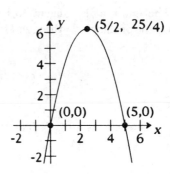

1.5.14

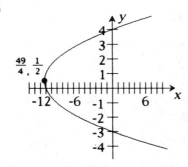

1.5.15 $3x+4y=24$ intercepts the coordinate axis at $(0,6)$ and $(8,0)$, thus, the center is the midpoint of the line segment joining $(0,6)$ and $(8,0)$. The center is

$$\left(\frac{0+8}{2}\right), \left(\frac{6+0}{2}\right) = (4,3);$$ the radius is $\sqrt{(0-4)^2+(6-3)^2} = \sqrt{25},$

so $(x-4)^2+(y-3)^2 = 25.$

1.5.16 Let d_1 be the distance from the point $(2,-2)$ to the point (x,y) and d_2 be the distance from the point (x,y) to $(2,1)$. $d_1 = 2d_2$ and

$$d_1 = \sqrt{(x-2)^2+(y+2)^2} = 2d_2 = 2\sqrt{(x-2)^2+(y-1)^2}$$

$$(x-2)^2+(y+2)^2 = 4[(x-2)^2+(y-1)^2]$$

$$(3x^2-12x)+(3y^2-12y) = -12$$

$$(x-2)^2+(y-2)^2 = 4.$$

Circle with center $(2,2)$ and radius 2.

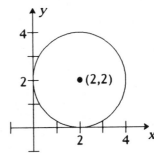

1.5.17 The line that is tangent to the circle at $(1,7)$ is perpendicular to the line drawn from the center of the circle through $(1,7)$. Place $x^2 + y^2 + 6x - 8y = 0$ into standard form, thus,

$$(x^2 + 6x) + (y^2 - 8y) = 0$$
$$(x^2 + 6x + 9) + (y^2 - 8y + 16) = 0 + 9 + 16$$
$$(x + 3)^2 + (y - 4)^2 = 25,$$

so the center is $(-3, 4)$. The slope m of a line through $(-3, 4)$ and $(1, 7)$ is

$$\frac{7 - 4}{1 + 3} = \frac{3}{4},$$

so $m_\perp = -4/3$. The equation of the tangent through $(1, 7)$ with slope $m_\perp = -4/3$ is

$$y - 7 = -\frac{4}{3}(x - 1)$$

$$y = -\frac{4}{3}x + \frac{25}{3}.$$

1.5.18 The radius of the circle is the distance from $(-3, 8)$ to the point of tangency. To find the point of tangency, first find m, the slope of the tangent, $3y + 4x - 37 = 0$, thus,

$y = -\frac{4}{3}x + \frac{37}{3}$, so $m = -4/3$. The slope of the perpendicular through $(-3, 8)$ to

$3y + 4x - 37 = 0$ is $m_\perp = 3/4$, so the equation of the perpendicular through $(-3, 8)$ with slope $m_\perp = 3/4$ is

$$y - 8 = \frac{3}{4}(x + 3)$$

$$y = \frac{3}{4}x + \frac{41}{4}.$$

Solve $\left.\begin{array}{l} 4x + 3y = 37 \\ 3x + 4y = -41 \end{array}\right\}$ to get $x = 1$, $y = 11$, so the point of tangency is $(1, 11)$ and the radius is $\sqrt{(1 - (-3))^2 + (11 - 8)^2} = \sqrt{25}$, so $(x + 3)^2 + (y - 8)^2 = 25$.

1.5.19 Solution

$$2x^2 + 5x - 3 = (x + 3)(2x - 1) \geq 0$$
$$S = (-\infty, -3] \cup [1/2, +\infty)$$

SUPPLEMENTARY EXERCISES, CHAPTER 1

In Exercises 1–5, use interval notation to describe the set of all values of x (if any) that satisfy the given inequalities.

1. (a) $-3 < x \leq 5$ (b) $-1 < x^2 \leq 9$ (c) $x^2 \geq \dfrac{1}{4}$.

2. (a) $|2x + 1| > 5$ (b) $|x^2 - 9| \geq 7$ (c) $1 \leq |x| \leq 3$.

3. (a) $2x^2 - 5x > 3$ (b) $x^2 - 5x + 4 \leq 0$.

4. (a) $\dfrac{x}{1 - x} \geq 3$.   (b) $\dfrac{2x + 3}{x} \geq x$.

5. (a) $\dfrac{|x| - 1}{|x| - 2} \leq 0$ (b) $|x - 1| \leq 2|x + 2|$.

6. Among the terms *integer*, *rational*, *irrational*, which ones apply to the given number?

    (a) $\sqrt{4/9}$ (b) 2^{-2} (c) $-4^{1/3}$

    (d) 0.87 (e) $-4^{1/2}$ (f) $0.1010010001\ldots$

    (g) 3.222 (h) $3/(-1)$.

7. (a) Find values of a and b such that $a < b$, but $a^2 > b^2$.
    (b) If $a < b$, what additional assumptions on a and b are required to ensure that $a^2 < b^2$?

8. Which of the following are true for all sets A and B?

    (a) $A \subset (A \cap B)$ (b) $(A \cap B) \subset A$

    (c) $\emptyset \subset A$ (d) $A \subset (A \cup B)$

    (e) $(A \cap B) \subset (A \cup B)$ (f) Either $A \subset B$ or $B \subset A$

    (g) $A \in B$

9. Prove $|x| \leq \sqrt{x^2 + y^2}$ and $|y| \leq \sqrt{x^2 + y^2}$. Interpret this geometrically.

In Exercises 10–14, draw a rectangular coordinate system and sketch the set of points whose coordinates (x, y) satisfy the given conditions.

10. (a) $y = 0$ and $x > 0$ (b) $2x - y \leq 3$.

11. (a) $xy = x^2$ (b) $y(x - 1) = x^2 - 1$.

12. (a) $y = (x^3 - 1)/(x - 1)$ (b) $y > x^2 - 9$.

13. (a) $y^2 - 6y + x^2 - 2x - 6 \geq 0$ (b) $x + |y - 2| = 1$.

14. (a) $|x| + |y| = 4$ (b) $|x| - |y| = 4$.

15. Where does the parabola $y = x^2$ intersect the line $y - 2 = x$?

In Exercises 16–19, sketch the graph of the given equation.

16. $xy + 4 = 0$.   17. $y = |x - 2|$.

18. $y = \sqrt{4 - x^2}$.   19. $y = x(x - 2)$.

In Exercises 20–24, find the standard equation for the circle satisfying the given conditions.

20.   The circle centered at $(3, -2)$ and tangent to the line $y = 1$.

21.   The circle centered at $(1, 2)$ and passing through the point $(4, -2)$.

22.   The circle centered on the line $x = 2$ and passing through the points $(1, 3)$ and $(3, -11)$.

23.   The circle of radius 5 tangent to the lines $y = 7$ and $x = 6$.

24.   The circle of radius 13 that passes through the origin and the point $(0, -24)$.

In Exercises 25–28, determine whether the equation represents a circle, a point, or has no graph. If it represents a circle, find the center and radius.

25. $x^2 + y^2 + 4x + 2y + 5 = 0$.   **26.** $4x^2 + 4y^2 - 4x + 8y + 1 = 0$.

27. $x^2 + y^2 - 3x + 2y + 4 = 0$.   **28.** $3x^2 + 3y^2 - 5x + 7y + 3 = 0$.

29.   In each part, find an equation for the line through A and B, the distance between A and B, and the coordinates of the midpoint of the line segment joining A and B.

   (a) $A(3, 4)$, $B(-3, -4)$ **(b)** $A(3, 4)$, $B(3, -4)$

   (c) $A(3, 4)$, $B(-3, 4)$ **(d)** $A(3, 4)$, $B(4, 3)$

30.   Show that the point $(8, 1)$ is not on the line through the points $(-3, -2)$ and $(1, -1)$.

31.   Where does the circle of radius 5 centered at the origin intersect the line of slope $-3/4$ through the origin?

32.   Find the slope of the line whose angle of inclination is

   (a) $30°$ **(b)** $120°$ **(c)** $90°$

In Exercises 33–35, find the slope-intercept form of the line satisfying the stated conditions.

33.   The line through $(2, -3)$ and $(4, -3)$.

34.   The line with x-intercept -2 and angle of inclination $\phi = 45°$.

35.   The line parallel to $x + 2y = 3$ that passes through the origin.

36.   Find an equation of the perpendicular bisector of the line segment joining $A(-2, -3)$ and $B(1, 1)$.

In Exercises 37–39, find equations of the lines L and L' and determine their point of intersection.

37. L passes through $(1, 0)$ and $(-1, 4)$.
         L' is perpendicular to L and has y-intercept -3.

38. L passes through $(-2, 0)$ and $(-2, 3)$.
         L' passes through $(-1, 4)$ and is perpendicular to L.

39. L has slope $2/5$ and passes through $(3, 1)$.
         L' has x-intercept $-8/3$ and y-intercept -4.

40.   Consider the triangle with vertices $A(5, 2)$, $B(1, -3)$, and $C(-3, 4)$. Find the point-slope form of the line containing

   (a) the median from C to AB **(b)** the altitude from C to AB.

41. Use slopes to show that the points $(5,6)$, $(-4,3)$, $(-3,-2)$, and $(6,1)$ are vertices of a parallelogram. Is it a rectangle?

42. For what value of k (if any) will the line $2x + ky = 3k$ satisfy the stated condition?

    (a) Have slope 3   **(b)** Have y-intercept 3.

    (c) Be parallel to the x-axis   **(d)** Pass through $(1,2)$

SOLUTIONS

SUPPLEMENTARY EXERCISES, CHAPTER 1

1. **(a)** $(-3, 5]$
   (b) $-1 < 0 \le x^2$ for all x, thus $-1 < x^2 \le 9$ is equivalent to $x^2 \le 9$, so $|x| \le 3$. The interval is $[-3, 3]$.
   (c) $x^2 \ge \dfrac{1}{4}$ is equivalent to $|x| \ge \dfrac{1}{2}$, which in interval notation is $(-\infty, -1/2] \cup [1/2, +\infty)$.

2. **(a)** $|2x + 1| > 5$; <u>Case 1</u>: $2x + 1 > 5$, $x > 2$,
   <u>Case 2</u>: $2x + 1 < -5$, $x < -3$ so $S = (-\infty, -3) \cup (2, +\infty)$
   (b) $|x^2 - 9| \ge 7$
   <u>Case 1</u>: $x^2 - 9 \ge 7$, $x^2 \ge 16$, $|x| \ge 4$; $S_1 = (-\infty, -4] \cup [4, +\infty)$
   <u>Case 2</u>: $x^2 - 9 \le -7$, $x^2 \le 2$, $|x| \le \sqrt{2}$; $S_2 = [-\sqrt{2}, \sqrt{2}]$
   so $S = S_1 \cup S_2 = (-\infty, -4] \cup [-\sqrt{2}, \sqrt{2}] \cup [4, +\infty)$
   (c) <u>Case 1</u>: If $x \ge 0$ then $1 \le x \le 3$,
   <u>Case 2</u>: If $x < 0$ then $1 \le -x \le 3$, $-1 \ge x \ge -3$ so $S = [-3, -1] \cup [1, 3]$

3. **(a)**
   $$2x^2 - 5x > 3$$
   $$2x^2 - 5x - 3 > 0$$
   $$(2x + 1)(x - 3) > 0$$
   $$(x + 1/2)(x - 3) > 0$$
   $$S = (-\infty, -1/2) \cup (3, +\infty)$$

   (b)
   $$x^2 - 5x + 4 \le 0$$
   $$(x - 1)(x - 4) \le 0$$
   $$S = [1, 4]$$

4. **(a)**
   $$\frac{x}{1 - x} \ge 3$$
   $$\frac{x - 3(1 - x)}{1 - x} \ge 0$$
   $$\frac{4x - 3}{1 - x} \ge 0$$
   $$\frac{x - 3/4}{1 - x} \ge 0$$
   $$S = [3/4, 1)$$

(b)
$$\frac{2x+3}{x} \geq x$$

$$\frac{2x+3-x^2}{x} \geq 0$$

$$\frac{x^2-2x-3}{x} \leq 0$$

$$\frac{(x+1)(x-3)}{x} \leq 0$$

$$S = (-\infty, -1] \cup (0, 3]$$

5. (a) $\dfrac{|x|-1}{|x|-2} \leq 0$

Case 1: $x \geq 0$,

$$\frac{x-1}{x-2} \leq 0$$

Case 2: $x < 0$,

$$\frac{-x-1}{-x-2} \leq 0; \quad \frac{x+1}{x+2} \leq 0$$

$S_1 = [1, 2)$ $S_2 = (-2, -1]$

$$S = S_1 \cup S_2 = (-2, -1] \cup [1, 2)$$

(b)
$$|x-1| \leq 2|x+2|$$
$$x^2 - 2x + 1 \leq 4(x^2 + 4x + 4)$$
$$-3x^2 - 18x - 15 \leq 0$$
$$x^2 + 6x + 5 \geq 0$$
$$(x+1)(x+5) \geq 0$$
$$S = (-\infty, -5] \cup [-1, +\infty)$$

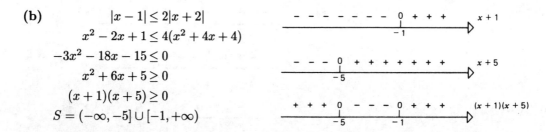

6. (a) rational   **(b)** rational   **(c)** irrational
    (d) rational   **(e)** integer, rational   **(f)** irrational
    (g) rational   **(h)** integer, rational

7. **(a)** $a = -2$ and $b = 1$

   (b) $a^2 < b^2$, $a^2 - b^2 < 0$, $(a+b)(a-b) < 0$; if $a < b$ then $a - b < 0$ so $(a+b)(a-b) < 0$ if $a + b > 0$

8. (b), (c), (d), (e)

9. $x^2 \leq x^2 + y^2$ because $y^2 \geq 0$, thus $\sqrt{x^2} \leq \sqrt{x^2 + y^2}$ and so $|x| \leq \sqrt{x^2 + y^2}$.

   Similarly, $|y| \leq \sqrt{x^2 + y^2}$. The right triangle with vertices $(0,0)$, (x,y), and $(x,0)$ has legs of lengths $|x|$ and $|y|$, and a hypotenuse of length $\sqrt{x^2 + y^2}$. The lengths of the legs cannot exceed the length of the hypotenuse.

10. **(a)**

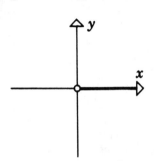

        (b)

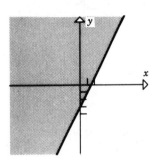

11. **(a)**
$$xy = x^2$$
$$xy - x^2 = 0$$
$$x(y - x) = 0,$$
so $x = 0$ or $y - x = 0$

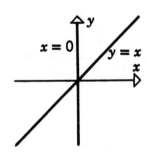

        (b)
$$y(x - 1) = x^2 - 1$$
$$y(x - 1) - (x^2 - 1) = 0$$
$$y(x - 1) - (x - 1)(x + 1) = 0$$
$$(x - 1)(y - x - 1) = 0,$$
so $x - 1 = 0$ or $y - x - 1 = 0$

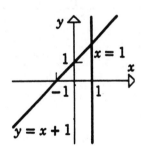

12. **(a)** $y = \dfrac{x^3 - 1}{x - 1}$

$\qquad = \dfrac{(x-1)(x^2 + x + 1)}{x - 1}$

$\qquad = x^2 + x + 1$ if $x \neq 1$

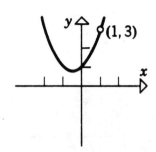

(b)

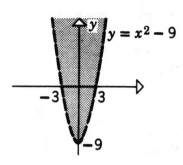

13. **(a)** Complete the square:

$(x^2 - 2x + 1) + (y^2 - 6y + 9) \geq 6 + 1 + 9$

$(x-1)^2 + (y-3)^2 \geq 16,$

all points on or outside the circle
with center $(1, 3)$ and radius 4.

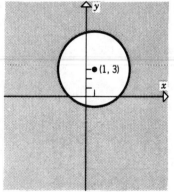

(b) $x + |y - 2| = 1$

If $y \geq 2$, then

$\qquad x + y - 2 = 1$

$\qquad\qquad y = -x + 3.$

If $y < 2$, then $x - (y - 2) = 1$

$\qquad\qquad\qquad y = x + 1.$

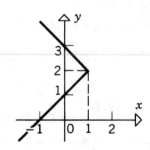

14. **(a)** $\quad x + y = 4$ if $x \geq 0$ and $y \geq 0,$

$\qquad -x + y = 4$ if $x < 0$ and $y \geq 0,$

$\qquad -x - y = 4$ if $x < 0$ and $y < 0,$

$\qquad x - y = 4$ if $x \geq 0$ and $y < 0.$

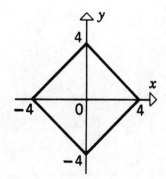

(b) $x - y = 4$ if $x \geq 0$ and $y \geq 0$,
         $-x - y = 4$ if $x < 0$ and $y \geq 0$,
         $-x + y = 4$ if $x < 0$ and $y < 0$,
         $x + y = 4$ if $x \geq 0$ and $y < 0$.

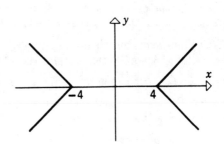

15. The curves intersect at (x, y) where $y = x^2$ and $y = x + 2$, so $x^2 = x + 2$, $x^2 - x - 2 = 0$, $(x + 1)(x - 2) = 0$, $x = -1$ or $x = 2$. The points of intersection are $(-1, 1)$ and $(2, 4)$.

16.

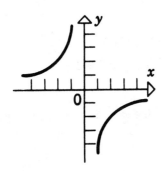

17. If $x \geq 2$, then $y = x - 2$;
     if $x < 2$, then $y = -x + 2$.

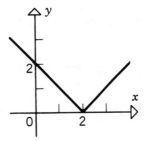

18.

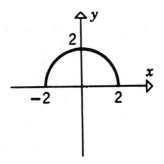

19.

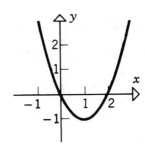

20. Because the point $(3, 1)$ is on the circle, the radius is $r = |1 - (-2)| = 3$, so the equation is $(x - 3)^2 + (y + 2)^2 = 9$.

21. The radius is $r = \sqrt{(4 - 1)^2 + (-2 - 2)^2} = 5$, so $(x - 1)^2 + (y - 2)^2 = 25$.

22. Let $(2, k)$ be the center, and r the radius. Then $(x - 2)^2 + (y - k)^2 = r^2$. But $(1, 3)$ is on the circle, so $(1 - 2)^2 + (3 - k)^2 = r^2$, $10 - 6k + k^2 = r^2$ (i), and $(3, -11)$ is on the circle so $(3 - 2)^2 + (-11 - k)^2 = r^2$, $122 + 22k + k^2 = r^2$ (ii). Eliminate r^2 from (i) and (ii) to get $122 + 22k + k^2 = 10 - 6k + k^2$, $28k = -112$, $k = -4$, thus, from (i), $r^2 = 10 - 6(-4) + (-4)^2 = 50$, so $(x - 2)^2 + (y + 4)^2 = 50$.

23. Let $C(h, k)$ be the center, then $|h - 6| = 5$ and $|k - 7| = 5$, so $h = 6 \pm 5$ and $k = 7 \pm 5$. There are four circles, $(x - h)^2 + (y - k)^2 = 25$, where $h = 1$ or 11 and $k = 2$ or 12.

24. Let $C(h, k)$ be the center, then $(x - h)^2 + (y - k)^2 = 169$. But $(0, 0)$ is on the circle, so $(0 - h)^2 + (0 - k)^2 = 169$, $h^2 + k^2 = 169$ (i), also $(0, -24)$ is on the circle so

    $(0 - h)^2 + (-24 - k)^2 = 169$, $h^2 + 576 + 48k + k^2 = 169$ (ii). Subtract (i) from (ii) to get

    $48k + 576 = 0$, $k = -12$, and then, from (i), $h^2 + (-12)^2 = 169$, $h^2 = 25$, $h = \pm 5$. There are two circles, $(x \pm 5)^2 + (y + 12)^2 = 169$.

25. $(x^2 + 4x + 4) + (y^2 + 2y + 1) = -5 + 4 + 1$, $(x + 2)^2 + (y + 1)^2 = 0$; the point $(-2, -1)$.

26. $4(x^2 - x + 1/4) + 4(y^2 + 2y + 1) = -1 + 1 + 4$,

    $(x - 1/2)^2 + (y + 1)^2 = 1$; circle, center $(1/2, -1)$, radius 1.

27. $(x^2 - 3x + 9/4) + (y^2 + 2y + 1) = -4 + 9/4 + 1$, $(x - 3/2)^2 + (y + 1)^2 = -3/4$; no graph

28. $3 \left(x^2 - \dfrac{5}{3}x + \dfrac{25}{36} \right) + 3 \left(y^2 + \dfrac{7}{3}y + \dfrac{49}{36} \right) = -3 + \dfrac{25}{12} + \dfrac{49}{12}$,

    $(x - 5/6)^2 + (y + 7/6)^2 = 19/18$; circle, center $(5/6, -7/6)$, radius $\sqrt{38}/6$

29. **(a)** $m = \dfrac{-4 - 4}{-3 - 3} = \dfrac{4}{3}$, so $y - 4 = \frac{4}{3}(x - 3)$, $y = \dfrac{4}{3}x$; $d = \sqrt{(-6)^2 + (-8)^2} = 10$;

    midpoint: $(0, 0)$.

    (b) $m = \dfrac{-4 - 4}{3 - 3} = \dfrac{-8}{0}$ which is not defined, so the vertical line $x = 3$; $d = |-4 - 4| = 8$;

    midpoint: $(3, 0)$.

    (c) $m = \dfrac{4 - 4}{-3 - 3} = 0$, so $y = 4$; $d = |-3 - 3| = 6$; midpoint: $(0, 4)$.

    (d) $m = \dfrac{3 - 4}{4 - 3} = -1$, so $y - 4 = -(x - 3)$, $y = -x + 7$; $d = \sqrt{(1)^2 + (-1)^2} = \sqrt{2}$;

    midpoint: $(7/2, 7/2)$.

30. $x - 4y = 5$ is an equation of the line through $(-3, -2)$ and $(1, -1)$, but $(8, 1)$ does not satisfy it.

31. Equation of circle is $x^2 + y^2 = 25$, equation of line is $y = -\dfrac{3}{4}x$.

    Eliminate y: $x^2 + \left(-\dfrac{3}{4}x \right)^2 = 25$, $x^2 + \dfrac{9}{16}x^2 = 25$, $\dfrac{25}{16}x^2 = 25$, $x^2 = 16$, so $x = \pm 4$.

    The points of intersection are $(-4, 3)$ and $(4, -3)$.

32. **(a)** $\tan 30° = 1/\sqrt{3}$ **(b)** $\tan 120° = -\sqrt{3}$ **(c)** $\tan 90°$ is not defined.

33. $m = \dfrac{-3 + 3}{4 - 2} = 0$, so $y = -3$.

34. $m = \tan 45° = 1$, and $(-2, 0)$ is on the line, so $y - 0 = (1)(x + 2)$, $y = x + 2$.

35. For $x + 2y = 3$, $m = -\dfrac{1}{2}$. A parallel line through the origin is $y - 0 = -\dfrac{1}{2}(x - 0)$, $y = -\dfrac{1}{2}x$.

36. The line segment joining $A(-2, -3)$ and $B(1, 1)$ has slope $m = \dfrac{4}{3}$ and midpoint $M \left(-\dfrac{1}{2}, -1 \right)$.

    The perpendicular bisector has slope $-\dfrac{3}{4}$ and goes through M, so $y + 1 = -\dfrac{3}{4} \left(x + \dfrac{1}{2} \right)$,

    $y = -\dfrac{3}{4}x - \dfrac{11}{8}$.

37. $L : m = -2$, so $y = -2(x - 1)$; $L' : m = \dfrac{1}{2}$ because L' is perpendicular to L, so $y = \dfrac{1}{2}x - 3$. L meets L' when $-2(x - 1) = \dfrac{1}{2}x - 3$, which gives $x = 2$. But $y = -2$ when $x = 2$, so the point is $(2, -2)$.

38. L : vertical line, so $x = -2$; L' : horizontal line because L is vertical, so $y = 4$. L meets L' at the point $(-2, 4)$.

39. $L : y - 1 = \dfrac{2}{5}(x - 3)$, $y = \dfrac{2}{5}x - \dfrac{1}{5}$; $L' : \left(-\dfrac{8}{3}, 0\right)$ and $(0, -4)$ are on the line, so $m = -\dfrac{3}{2}$ and therefore $y = -\dfrac{3}{2}\left(x + \dfrac{8}{3}\right) = -\dfrac{3}{2}x - 4$. L meets L' when $\dfrac{2}{5}x - \dfrac{1}{5} = -\dfrac{3}{2}x - 4$, which gives $x = -2$. But $y = -1$ when $x = -2$, so the point is $(-2, -1)$.

40. (a) The median from C to AB is the line segment joining C and the midpoint of AB. The midpoint of AB is $M(3, -1/2)$, thus the slope of the line through C and M is $-3/4$, so $y - 4 = (-3/4)(x + 3)$.

    (b) The altitude to AB is perpendicular to AB. The slope of AB is $5/4$, thus the slope of the line perpendicular to AB is $-4/5$, so $y - 4 = (-4/5)(x + 3)$.

41. Label the points as $A(5, 6)$, $B(-4, 3)$, $C(-3, -2)$, and $D(6, 1)$. Then $m_{AB} = 1/3$, $m_{BC} = -5$, $m_{CD} = -1/3$, and $m_{DA} = -5$, so $ABCD$ is a parallelogram because opposite sides are parallel ($m_{AB} = m_{CD}$, $m_{BC} = m_{DA}$). It is not a rectangle because sides AB and BC do not form a right angle ($m_{AB} \neq -1/m_{BC}$).

42. (a) $y = -2x/k + 3$, if $k \neq 0$; $m = -2/k = 3$ if $k = -2/3$.

    (b) $k \neq 0$ (if $k = 0$, then the line coincides with the y-axis and does not have a unique y-intercept).

    (c) $-2/k = 0$ is impossible for any real value of k.

    (d) $(1, 2)$ must satisfy $2x + ky = 3k$, so $2(1) + k(2) = 3k$ which gives $k = 2$.

CHAPTER 2

Functions and Limits

SECTION 2.1

2.1.1 If $h(x) = 3x^2 - 2$, find

    (a) $h(0)$ **(b)** $h(2a)$ **(c)** $h(a-4)$.

2.1.2 If $g(x) = \dfrac{x+1}{x}$, find

    (a) $g(1)$ **(b)** $g(0)$ **(c)** $g(-1)$ **(d)** $g(x-1)$.

2.1.3 If $f(\theta) = 2\sin\theta + \cos 2\theta$, find

    (a) $f(0)$ **(b)** $f(\pi/6)$ **(c)** $f(-\pi/3)$

2.1.4 If $\phi(x) = 2\sin 2x \cos 3x$, find

    (a) $\phi(\pi/6)$ **(b)** $\phi(-\pi/4)$ **(c)** $\phi(\pi/2)$.

2.1.5 Given that

$$s(x) = \begin{cases} -1, & x < 0 \\ 0, & x = 0 \\ 1, & x > 0 \end{cases}$$

    find $s(-1)$, $s(5)$, $s(0)$.

2.1.6 Given that

$$\phi(x) = \begin{cases} 1, & -1 \le x < 2 \\ 4-x, & 2 < x < 9 \\ x^2 - 4, & x \ge 9 \end{cases}$$

    find

    (a) $\phi(0)$ **(b)** $\phi(2)$ **(c)** $\phi(4)$

2.1.7 Express $f(x) = |x+4| - 6$ in piecewise form without using absolute values.

2.1.8 Express $g(y) = 2 - |5y - 10|$ in piecewise form without using absolute values.

2.1.9 Express $h(x) = |x - 2| + |x - 5|$ in piecewise form without using absolute values.

2.1.10 Find the domain and range for $f(x) = \sqrt{3x + 4}$.

2.1.11 Find the domain and range for $f(x) = \dfrac{2x - 5}{3x + 2}$.

2.1.12 Find the domain for $f(x) = \dfrac{1}{\sqrt{x - 1}}$.

2.1.13 Find the domain for $f(x) = \sqrt{\dfrac{x + 5}{x - 1}}$.

2.1.14 Find the domain and range for $g(x) = \dfrac{3x + 5}{2x + 3}$.

2.1.15 Find the domain for $h(x) = \sqrt{4 - 3x - x^2}$.

2.1.16 Find the domain for $f(x) = \sqrt{\dfrac{x}{x + 2}}$.

2.1.17 If $f(x) = \dfrac{x}{x + 2}$ and $g(x) = f\left(\dfrac{x + 2}{2}\right)$, write an expression for $g(x)$ and find its range and domain.

2.1.18 If $f(x) = \sqrt{1 - x^2}$ and $g(x) = f(2x)$, write an expression for $g(x)$ and find its range and domain.

2.1.19 If $h(x) = \dfrac{1}{|2 - x| + 4}$ and $g(x) = h(-2x)$, write an expression for $g(x)$ and find its domain and range.

SOLUTIONS

SECTION 2.1

2.1.1 **(a)** -2 **(b)** $12a^2 - 2$
       (c) $3(a-4)^2 - 2 = 3a^2 - 24a + 46.$

2.1.2 **(a)** 2 **(b)** not defined   **(c)** 0 **(d)** $\dfrac{x}{x-1}.$

2.1.3 **(a)** 1 **(b)** $3/2$ **(c)** $-\sqrt{3} - 1/2.$

2.1.4 **(a)** 0 **(b)** $\sqrt{2}$ **(c)** $0.$

2.1.5 **(a)** -1 **(b)** 1 **(c)** $0.$

2.1.6 **(a)** 1 **(b)** not defined   **(c)** $0.$

2.1.7 $f(x) = \begin{cases} x - 2, & x \geq -4 \\ -x - 10, & x < -4 \end{cases}.$ **2.1.8** $g(y) = \begin{cases} -5y + 12, & y \geq 2 \\ 5y - 8, & y < 2 \end{cases}.$

2.1.9 $f(x) = \begin{cases} -2x + 7, & x < 2 \\ 3, & 2 \leq x < 5 \\ 2x - 7, & x \geq 5 \end{cases}$

2.1.10 $3x + 4 \geq 0$ if $x \geq -4/3$, so the domain is $[-4/3, +\infty)$ and the range is $[0, +\infty)$.

2.1.11 $3x + 2 \neq 0$ so the domain is $(-\infty, -2/3) \cup (-2/3, +\infty)$. To get the range, let $y = \dfrac{2x - 5}{3x + 2}$ and solve for x, thus, $x = \dfrac{5 + 2y}{2 - 3y}$ so the range is $(-\infty, 2/3) \cup (2/3, +\infty)$.

2.1.12 $x \geq 0$ and $\sqrt{x} - 1 \neq 0$ so the domain is $[0, 1) \cup (1, +\infty)$.

2.1.13 $\dfrac{x + 5}{x - 1} \geq 0$ and $x - 1 \neq 0$ if $x \leq -5$ or $x > 1$ so the domain is $(-\infty, -5] \cup (1, +\infty)$.

2.1.14 $2x + 3 \neq 0$ so the domain is $(-\infty, -3/2) \cup (-3/2, +\infty)$. To get the range, let $y = \dfrac{3x + 5}{2x + 3}$ and solve for x, thus, $x = \dfrac{5 - 3y}{2y - 3}$ so the range is $(-\infty, 3/2) \cup (3/2, +\infty)$.

2.1.15 $4 - 3x - x^2 \geq 0$ if $-4 \leq x \leq 1$ so the domain is $[-4, 1]$.

2.1.16 $\dfrac{x}{x+2} \geq 0$ and $x \neq -2$ if $x < -2$ or $x \geq 0$ so the domain is $(-\infty, -2) \cup [0, +\infty)$.

2.1.17 $f(x) = \dfrac{x}{x+2}$, $g(x) = \dfrac{\dfrac{x+2}{2}}{\dfrac{x+2}{2} + 2} = \dfrac{x+2}{x+6}$. $x + 6 \neq 0$ so the domain is $(-\infty, -6) \cup (-6, +\infty)$.

To get the range, let $y = \dfrac{x+2}{x+6}$ and solve for x, thus, $x = \dfrac{2 - 6y}{y - 1}$ so the range is

$(-\infty, 1) \cup (1, +\infty)$.

2.1.18 $g(x) = \sqrt{1 - (2x)^2} = \sqrt{1 - 4x^2}$, $1 - 4x^2 \geq 0$ if $-1/2 \leq x \leq 1/2$ so the domain is

$[-1/2, 1/2]$ and the range is $[0, 1]$.

2.1.19 $g(x) = \dfrac{1}{|2 - (-2x)| + 4} = \dfrac{1}{|2 + 2x| + 4}$, $|2 + 2x| + 4 \neq 0$ so the domain is $(-\infty, +\infty)$ and

its range is $(0, 1/6]$.

SECTION 2.2

2.2.1   Let $f(x) = x^2 - x + 1$, find $\dfrac{f(2+h) - f(2)}{h}$.

2.2.2   Let $f(x) = \dfrac{x^2 - x - 6}{x}$ and $g(x) = x - 3$, find

    (a) $(f+g)(x)$ **(b)** $(f-g)(x)$ **(c)** $\left(\dfrac{f}{g}\right)(x)$.

2.2.3   Let $f(x) = \dfrac{x}{1+x^2}$ and $g(x) = \sqrt{x}$, find

    (a) $f(-2)$ **(b)** $g(x^2)$ **(c)** $f \circ g(x)$

2.2.4   Let $g(x) = 13x + 3$ and $h(x) = 2x - 1$, find

    (a) $g(t+2)$ **(b)** $g \circ h(x)$
    (c) $g \circ h(t+2)$ **(d)** the domain of $g \circ h(x)$

2.2.5   Let $f(x) = \sqrt{x-5}$ and $g(x) = \sqrt{x}$, find

    (a) $f \circ g(x)$ **(b)** the domain of $f \circ g(x)$
    (c) $g \circ f(x)$ **(d)** the domain of $g \circ f(x)$.

2.2.6   Let $f(x) = x^3$ and $g(x) = \dfrac{2}{\sqrt[3]{x}}$, find

    (a) $f \circ g(t+2)$ **(b)** $g \circ f(-x)$.

2.2.7   Let $f(x) = \dfrac{1}{x} + 1$ and $g(x) = 3x^2$, find

    (a) $f \circ g(x)$ **(b)** the domain of $f \circ g(x)$
    (c) $g \circ f(-x+1)$ **(d)** the domain of $g \circ f(x)$

2.2.8   Let $f(x) = \sqrt{4 - x^2}$ and $g(x) = \dfrac{2}{x}$, find

    (a) the domain of $f(x)$ **(b)** $g \circ f(x)$
    (c) $f \circ g(x)$ **(d)** the domain of $g \circ f(x)$

2.2.9   Let $f(x) = |x|$ and $g(x) = x^3 + 1$, find

    (a) $f \circ g(x)$ **(b)** $g \circ f(x)$
    (c) the domain of $f \circ g(x)$ **(d)** the domain of $g \circ f(x)$.

2.2.10   Let $f(x) = 2x - 1$ and $g(x) = \sqrt{x}$, find the domain for $f \circ g(x)$ and $g \circ f(x)$.

2.2.11   Express $h(x) = \sqrt{x^2 - 4}$ as the composition of two functions such that $h(x) = f \circ g(x)$.

2.2.12   Express $h(x) = |x^3 - 1|$ as the composition of two functions such that $h(x) = f \circ g(x)$.

2.2.13   Express $h(x) = \dfrac{3}{x - 4}$ as the composition of two functions such that $h(x) = f \circ g(x)$.

2.2.14   For what values of x does $f(x) = f(x + 1)$ and for what values of x does $f(x + 1) = f(x) + 1$ if $f(x) = x^2 - 2x + 1$?

2.2.15   For what values of x does $f(x) = f(x+3)$ and for what values of x does $f(x+3) = f(x) + f(3)$ if $f(x) = x^2 - 6x + 9$?

2.2.16   For what values of x does $f(x) = f(x + 1)$ if $f(x) = x^3 - x^2 - x + 1$?

2.2.17   Express $h(x) = \sin(x^2)$ as the composition of two functions such that $h(x) = f \circ g(x)$.

2.2.18   Express $h(x) = \cos(2x + \pi/3)$ as the composition of two functions such that $h(x) = f \circ g(x)$.

2.2.19   Let $f(x) = x^2 + 1$ and let h be any nonzero real number. Find $\dfrac{f(x + h) - f(x)}{h}$.

2.2.20   Let $f(x) = 3x - 1$ and let h be any nonzero real number. Find $\dfrac{f(x + h) - f(x)}{h}$.

SOLUTIONS

SECTION 2.2

2.2.1 $\dfrac{f(2+h) - f(2)}{h} = \dfrac{[(2+h)^2 - (2+h) + 1] - [(2)^2 - (2) + 1]}{h}$

$$= \dfrac{3h + h^2}{h} = \dfrac{h(3+h)}{h} = 3 + h, h \neq 0.$$

2.2.2 **(a)** $\dfrac{x^2 - x - 6}{x} + x - 3 = \dfrac{2x^2 - 4x - 6}{x}$

(b) $\dfrac{x^2 - x - 6}{x} - (x - 3) = \dfrac{2x - 6}{x}$

(c) $\dfrac{\frac{x^2 - x - 6}{x}}{x - 3} = \dfrac{x^2 - x - 6}{x(x-3)} = \dfrac{(x-3)(x+2)}{x(x-3)} = \dfrac{x+2}{x}, x \neq 3.$

2.2.3 **(a)** $-2/5$ **(b)** $\sqrt{x^2} = |x|$ **(c)** $\dfrac{\sqrt{x}}{1+x}.$

2.2.4 **(a)** $13t + 29$ **(b)** $26x - 10$
(c) $26t + 42$ **(d)** $(-\infty, +\infty)$

2.2.5 **(a)** $\sqrt{\sqrt{x} - 5}$ **(b)** $[25, +\infty)$
(c) $\sqrt{\sqrt{x - 5}} = \sqrt[4]{x - 5}$ **(d)** $[5, +\infty).$

2.2.6 **(a)** $f \circ g(x) = \dfrac{8}{x}$ so $f \circ g(t + 2) = \dfrac{8}{t+2}$

(b) $g \circ f(x) = \dfrac{2}{x}$ so $g \circ f(-x) = \dfrac{2}{-x}$ or $-\dfrac{2}{x}.$

2.2.7 **(a)** $f \circ g = \dfrac{1}{3x^2} + 1 = \dfrac{1 + 3x^2}{3x^2}$ **(b)** $(-\infty, 0) \cup (0, +\infty)$

(c) $3\left(\dfrac{1}{-1+x} + 1\right)^2 = 3\left(\dfrac{2-x}{1-x}\right)^2$ **(d)** $(-\infty, 0) \cup (0, +\infty)$

2.2.8 **(a)** $[-2, 2]$ **(b)** $\dfrac{2}{\sqrt{4 - x^2}}$
(c) $\sqrt{4 - \dfrac{4}{x^2}}$ **(d)** $(-2, 2)$

2.2.9   **(a)** $|x^3 + 1|$ **(b)** $|x|^3 + 1$ **(c)** $(-\infty, +\infty)$ **(d)** $(-\infty, +\infty)$

2.2.10 $f \circ g(x)$ is $2\sqrt{x} - 1$ so the domain of $f \circ g(x)$ is $[0, +\infty)$, $g \circ f(x)$ is $\sqrt{2x - 1}$ so the domain of $g \circ f(x)$ is $[1/2, +\infty)$.

2.2.11 $g(x) = x^2 - 4$, $f(x) = \sqrt{x}$.   **2.2.12** $g(x) = x^3 - 1$, $f(x) = |x|$.

2.2.13 $g(x) = x - 4$, $f(x) = \dfrac{3}{x}$.

2.2.14 $x^2 - 2x + 1 = (x+1)^2 - 2(x+1) + 1$ is true if $x = 1/2$ and $(x+1)^2 - 2(x+1) + 1 = (x^2 - 2x + 1) + 1$ is true if $x = 1$.

2.2.15 $x^2 - 6x + 9 = (x+3)^2 - 6(x+3) + 9$ is true only if $x = 3/2$ and
$(x+3)^2 - 6(x+3) + 9 = (x^2 - 6x + 9) + \left[(3)^2 - 6(3) + 9\right]$ is true only if $x = 3/2$.

2.2.16 $x^3 - x^2 - x + 1 = (x+1)^3 - (x+1)^2 - (x+1) + 1$
       $3x^2 + x - 1 = 0$

solve using the quadratic formula, thus the values of x are $\dfrac{-1 - \sqrt{13}}{6}$ and $\dfrac{-1 + \sqrt{13}}{6}$.

2.2.17 $g(x) = x^2$, $f(x) = \sin x$.

2.2.18 $g(x) = 2x + \pi/3$, $f(x) = \cos x$.

2.2.19 $\dfrac{(x+h)^2 + 1 - (x^2 + 1)}{h} = \dfrac{2xh + h^2}{h} = 2x + h$

2.2.20 $\dfrac{3(x+h) - 1 - (3x - 1)}{h} = \dfrac{3h}{h} = 3$

SECTION 2.3

2.3.1   Sketch the graph of $f(x) = 3 - 4x$, $[0, 2]$.

2.3.2   Use the graph of $f(x) = \sqrt{x}$ to sketch the graph of $f(x) = 1 + \sqrt{-x}$.

2.3.3   Sketch the graph of $f(x) = \sqrt{5 - 4x - x^2}$ by completing the square.

2.3.4   Sketch the graph of $g(x) = -\sqrt{6x - x^2}$.

2.3.5   Sketch the graph of $\phi(x) = \sin(-x/2)$.

2.3.6   Sketch the graph of $g(x) = 2 + \sin x$.

2.3.7   Sketch the graph of $f(x) = 2\sin x + \sin 2x$.

2.3.8   Express $f(x) = |x + 2| + 1$ in piecewise form without using absolute values and sketch its graph.

2.3.9   Express $g(x) = 7 - |2x - 4|$ in piecewise form without using absolute values and sketch its graph.

2.3.10   Sketch the graph of $\phi(x) = \begin{cases} x - 2, & x < 0 \\ x^2, & x \geq 0 \end{cases}$.

2.3.11   Sketch the graph of $h(x) = \dfrac{x}{|x|}$.

2.3.12   Use the graph of $f(x) = |x|$ to sketch the graph of $f(x) = 2 - |2 - x|$.

2.3.13   Sketch the graph of $f(x) = \begin{cases} 2x, & x \geq 1 \\ x^2, & x < 1 \end{cases}$.

2.3.14   Sketch the graph of $g(x) = \begin{cases} 3x - 1, & x > 1 \\ 3, & x = 1 \\ 2, & x < 1 \end{cases}$.

2.3.15   Sketch the graph of $h(x) = (x - 2)^3 - 1$.

2.3.16 Sketch the graph of $f(x) = \dfrac{x^2 - 2x - 3}{x - 3}$.

2.3.17 Sketch the graph of $x^2 + 2x - y - 3 = 0$.

2.3.18 Use the graph of $x = y^2$ to sketch the graph of $y^2 - 3y + \dfrac{5}{4} + x = 0$.

2.3.19 A function f with domain $[-2, 2]$ has the graph shown

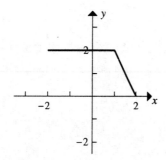

Use this graph to obtain the graphs of the equations

(a) $y - f(x) + 1$ (b) $y = f(x + 1)$ (c) $y = f(-x)$ (d) $y = -f(x)$

SOLUTIONS

SECTION 2.3

2.3.1

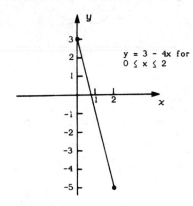

$y = 3 - 4x$ for
$0 \leq x \leq 2$

2.3.2

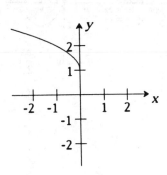

2.3.3 $f(x) = \sqrt{5 - 4x - x^2} = \sqrt{5 + 4 - (x^2 + 4x + 4)}$
$= \sqrt{9 - (x + 2)^2}$

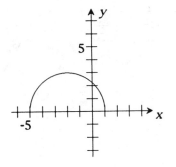

2.3.4
$$y = -\sqrt{6x - x^2}$$
$$y^2 = 6x - x^2$$
$$(x^2 - 6x) + y^2 = 0$$
$$(x^2 - 6x + 9) + y^2 = 9$$
$$(x - 3)^2 + y^2 = 9$$

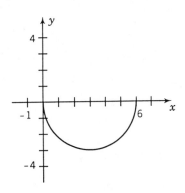

2.3.5

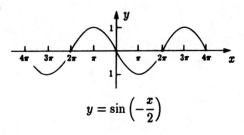

$$y = \sin\left(-\frac{x}{2}\right)$$

2.3.6

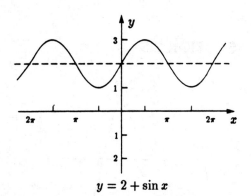

$$y = 2 + \sin x$$

2.3.7

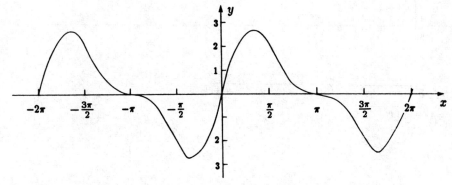

2.3.8 $f(x) = \begin{cases} x + 3, & x \geq -2 \\ -x - 1, & x < -2 \end{cases}$

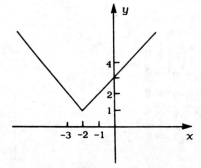

2.3.9 $\quad g(x) = \begin{cases} 11 - 2x, & x \geq 2 \\ 3 + 2x, & x < 2 \end{cases}$

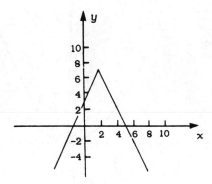

2.3.10 $\quad y = \begin{cases} x - 2, & x < 0 \\ x^2, & x \geq 0 \end{cases}$

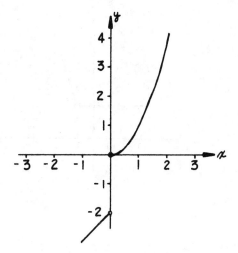

2.3.11

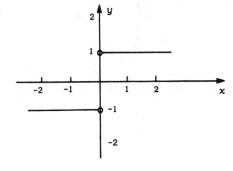

2.3.12

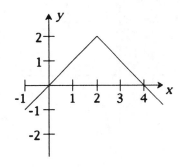

2.3.13 $y = \begin{cases} 2x, & x \geq 1 \\ x^2, & x < 1 \end{cases}$

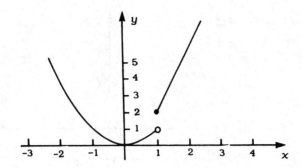

2.3.14 $y = \begin{cases} 3x - 1, & x > 1 \\ 3, & x = 1 \\ 2, & x < 1 \end{cases}$

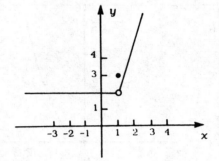

2.3.15

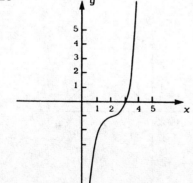

2.3.16

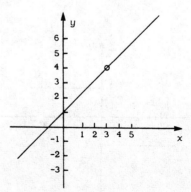

2.3.17 $x^2 + 2x - y - 3 = 0$

$$y = x^2 + 2x - 3$$
$$= (x^2 + 2x + 1) - 4$$
$$= (x + 1)^2 - 4$$

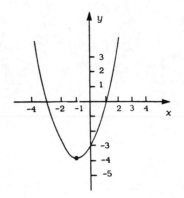

2.3.18

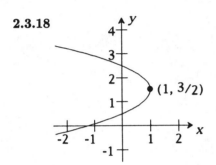

2.3.19   **(a)**   **(b)**

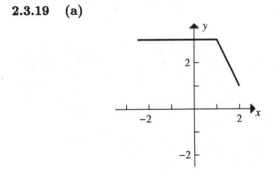   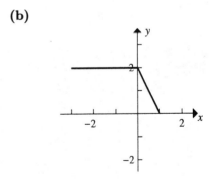

(c)   **(d)**

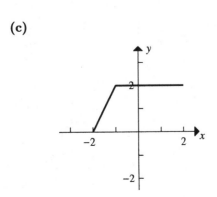   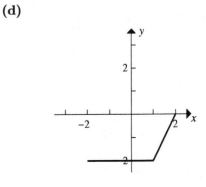

SECTION 2.4

2.4.1 For the function f graphed to the right, find

(a) $\lim\limits_{x \to 0^-} f(x)$

(b) $\lim\limits_{x \to 0^+} f(x)$

(c) $\lim\limits_{x \to 0} f(x)$

(d) $f(0)$

(e) $\lim\limits_{x \to -\infty} f(x)$

(f) $\lim\limits_{x \to +\infty} f(x)$

2.4.2 For the function f graphed to the right, find

(a) $\lim\limits_{x \to 2^-} f(x)$

(b) $\lim\limits_{x \to 2^+} f(x)$

(c) $\lim\limits_{x \to 2} f(x)$

(d) $f(2)$

(e) $\lim\limits_{x \to -\infty} f(x)$

(f) $\lim\limits_{x \to +\infty} f(x)$

2.4.3 For the function f graphed to the right, find

(a) $\lim\limits_{x \to -3^-} f(x)$

(b) $\lim\limits_{x \to -3^+} f(x)$

(c) $\lim\limits_{x \to -3} f(x)$

(d) $f(-3)$

(e) $f(0)$

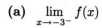

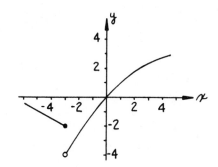

73

2.4.4 For the function f graphed to the right, find

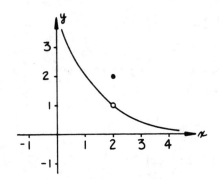

   (a) $\displaystyle\lim_{x\to 2^-} f(x)$

   (b) $\displaystyle\lim_{x\to 2^+} f(x)$

   (c) $\displaystyle\lim_{x\to 2} f(x)$

   (d) $f(2)$

   (e) $\displaystyle\lim_{x\to 0^+} f(x)$

   (f) $\displaystyle\lim_{x\to +\infty} f(x)$

2.4.5 For the function g graphed to the right, find

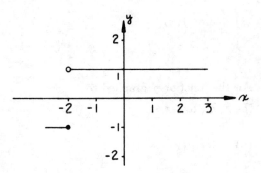

   (a) $\displaystyle\lim_{x\to -2^-} g(x)$

   (b) $\displaystyle\lim_{x\to -2^+} g(x)$

   (c) $\displaystyle\lim_{x\to -2} g(x)$

   (d) $g(-2)$

   (e) $\displaystyle\lim_{x\to +\infty} g(x)$

   (f) $\displaystyle\lim_{x\to -\infty} g(x)$

2.4.6 For the function f graphed to the right, find

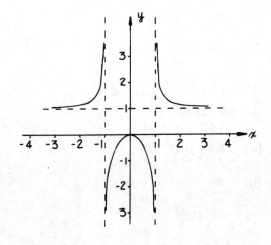

   (a) $\displaystyle\lim_{x\to -1^-} f(x)$

   (b) $\displaystyle\lim_{x\to -1^+} f(x)$

   (c) $\displaystyle\lim_{x\to -1} f(x)$

   (d) $f(-1)$

   (e) $\displaystyle\lim_{x\to +\infty} f(x)$

   (f) $\displaystyle\lim_{x\to -\infty} f(x)$

2.4.7

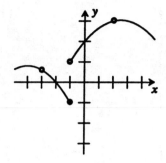

For the function h graphed above, find

(a) $h(-3)$ (b) $h(2)$ (c) $\lim\limits_{x \to -1^-} h(x)$

(d) $\lim\limits_{x \to -1^+} h(x)$ (e) $\lim\limits_{x \to -1} h(x)$ (f) $f(-1)$

2.4.8   For the function ϕ graphed to
        the right, find

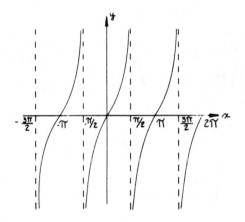

(a) $\lim\limits_{x \to \pi/2^-} \phi(x)$

(b) $\lim\limits_{x \to \pi/2^+} \phi(x)$

(c) $\lim\limits_{x \to \pi/2} \phi(x)$

(d) $\phi(\pi/2)$
(e) Can you identify this function?

2.4.9   For the function f graphed
        to the right, find

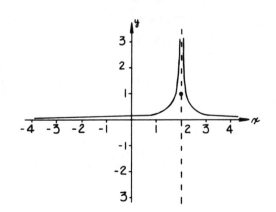

(a) $\lim\limits_{x \to 2^-} f(x)$

(b) $\lim\limits_{x \to 2^+} f(x)$

(c) $\lim\limits_{x \to 2} f(x)$

(d) $f(2)$
(e) $\lim\limits_{x \to -\infty} f(x)$

(f) $\lim\limits_{x \to +\infty} f(x)$

2.4.10 For the function f graphed to the right, find

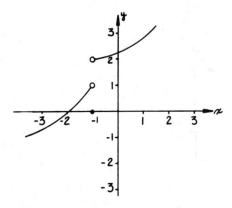

(a) $\displaystyle\lim_{x\to -1^-} f(x)$

(b) $\displaystyle\lim_{x\to -1^+} f(x)$

(c) $\displaystyle\lim_{x\to -1} f(x)$

(d) $f(-1)$

(e) $\displaystyle\lim_{x\to +\infty} f(x)$

(f) $\displaystyle\lim_{x\to -\infty} f(x)$

2.4.11 For the function f graphed to the right, find

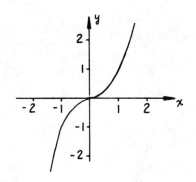

(a) $\displaystyle\lim_{x\to 1^-} f(x)$

(b) $\displaystyle\lim_{x\to 1^+} f(x)$

(c) $\displaystyle\lim_{x\to 1} f(x)$

(d) $f(1)$

(e) $\displaystyle\lim_{x\to +\infty} f(x)$

(f) $\displaystyle\lim_{x\to -\infty} f(x)$

2.4.12 Consider the function f graphed to the right. For what values of x_0 does $\displaystyle\lim_{x\to x_0} f(x)$ exist?

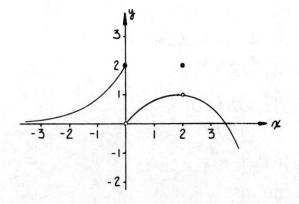

2.4.13 Consider the function
g graphed to the right.
For what values of x_0
does $\lim\limits_{x \to x_0} g(x)$ exist?

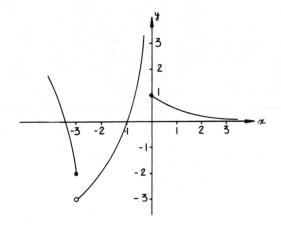

2.4.14 Consider the function g
graphed to the right. For what
values of x_0 does the $\lim\limits_{x \to x_0} g(x)$
exist?

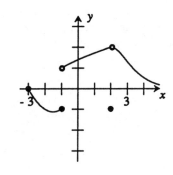

2.4.15 Consider the function f
graphed to the right. For what
values of x_0 does $\lim\limits_{x \to x_0} f(x)$
exist?

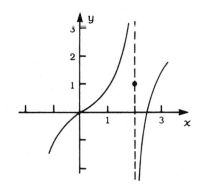

2.4.16 Consider the function f
graphed to the right. For what
values of x_0 does $\lim\limits_{x \to x_0} f(x)$
exist?

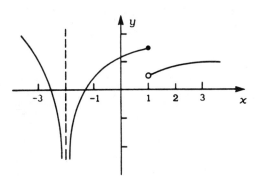

SOLUTIONS

SECTION 2.4

2.4.1   **(a)** 4   **(b)** $-\infty$ **(c)** does not exist
        (d) 0   **(e)** 1   **(f)** $+\infty$

2.4.2   **(a)** 3   **(b)** 3   **(c)** 3
        (d) 5   **(e)** -1 **(f)** does not exist

2.4.3   **(a)** -2 **(b)** -4 **(c)** does not exist
        (d) -2 **(e)** 0

2.4.4   **(a)** 1   **(b)** 1   **(c)** 1
        (d) 2   **(e)** $+\infty$ **(f)** 0

2.4.5   **(a)** -1 **(b)** 1   **(c)** does not exist
        (d) -1 **(e)** 1   **(f)** -1

2.4.6   **(a)** $+\infty$ **(b)** $-\infty$ **(c)** does not exist
        (d) does not exist   **(e)** 1   **(f)** 1

2.4.7   **(a)** 2   **(b)** does not exist   **(c)** -1
        (d) 1   **(e)** does not exist   **(f)** -1

2.4.8   **(a)** $+\infty$ **(b)** $-\infty$ **(c)** does not exist
        (d) does not exist   **(e)** $\phi(x) = \tan x$

2.4.9   **(a)** $+\infty$ **(b)** $+\infty$ **(c)** $+\infty$
        (d) 1   **(e)** 0   **(f)** 0

2.4.10   **(a)** 1   **(b)** 2   **(c)** does not exist
        (d) 0   **(e)** $+\infty$ **(f)** $-\infty$

2.4.11   **(a)** 1   **(b)** 1   **(c)** 1
        (d) 1   **(e)** $+\infty$ **(f)** $-\infty$

2.4.12 All values except 0.

2.4.13 All values except -3 and 0.

2.4.14 All values except -1.

2.4.15 All values except 2.

2.4.16 All values except 1 and -2.

SECTION 2.5

2.5.1 Find $\lim\limits_{x \to -2}(x^3 + 6x^2 - 16)$.

2.5.2 Find $\lim\limits_{x \to 0}\pi^2$.

2.5.3 Find $\lim\limits_{x \to 4}\dfrac{x^2 + 9}{x^2 - 1}$.

2.5.4 Find $\lim\limits_{x \to 4}\dfrac{x^2 - 16}{x^2 + x - 20}$.

2.5.5 Find $\lim\limits_{x \to 0}\dfrac{x^2 + 2x}{x - 2x^2}$.

2.5.6 Find $\lim\limits_{x \to 1}\dfrac{1 - x^2}{x^2 + 5x - 6}$.

2.5.7 Find $\lim\limits_{x \to 1}\dfrac{x^2 + x - 2}{x^2 - 4x + 3}$.

2.5.8 Find $\lim\limits_{x \to a}\dfrac{x^2 - a^2}{x - a}$.

2.5.9 Find $\lim\limits_{x \to 3}\dfrac{x^3 - 27}{x - 3}$.

2.5.10 Find $\lim\limits_{x \to 1}\dfrac{x^3 - 3x^2 + 2x}{x - 1}$.

2.5.11 Find $\lim\limits_{h \to 2}\dfrac{h^3 - 4h}{h^3 - 2h^2}$.

2.5.12 Find $\lim\limits_{x \to a}\dfrac{\dfrac{1}{x} - \dfrac{1}{a}}{x - a}$.

2.5.13 Find $\lim\limits_{h \to 0}\dfrac{\dfrac{1}{3 + h} - \dfrac{1}{3}}{h}$.

2.5.14 Find $\lim\limits_{x \to -a}\dfrac{x^3 + a^3}{x + a}$.

2.5.15 Find $\lim\limits_{x \to 3}\dfrac{x - 3}{x^3 - 27}$.

2.5.16 Find $\lim\limits_{x \to 2}\dfrac{1 - \dfrac{4}{x^2}}{1 - \dfrac{2}{x}}$.

2.5.17 Find $\lim\limits_{h \to 1}\dfrac{|h - 2| - 2}{h}$.

2.5.18 Find $\lim\limits_{x \to 4^-}\dfrac{x - 4}{|x - 4|}$.

2.5.19 Find $\lim\limits_{x \to 1^+}\dfrac{x - 1}{|x - 1|}$.

2.5.20 Find $\lim\limits_{x \to +\infty}\dfrac{2x^2 - 1}{x^2 + 1}$.

2.5.21 Find $\lim\limits_{x \to +\infty}\dfrac{x^3 + 2x}{3x^3 + 4x^2 + 5x}$.

2.5.22 Find $\lim\limits_{x \to +\infty}\dfrac{\sqrt{x^2 - 4}}{2x}$.

2.5.23 Find $\lim\limits_{x \to +\infty}\dfrac{\sqrt{3x^2 + 4x - 1}}{3 - x}$.

2.5.24 Find $\lim\limits_{x \to -\infty}\dfrac{\sqrt{x^2 - 4}}{x}$.

2.5.25 Find $\lim\limits_{x \to -\infty}\dfrac{\sqrt{3x^2 + 4x - 1}}{3 - x}$.

2.5.26 Find $\lim\limits_{x \to 3^-} f(x)$ where $f(x) = \begin{cases} \dfrac{|x - 3|}{x - 3}, & x < 3 \\ x, & x > 3 \end{cases}$.

2.5.27   Find $\lim\limits_{x \to 1} f(x)$ where $f(x) = \begin{cases} \dfrac{1}{x+2}, & x < 1 \\ 1 - 2x, & x > 1 \end{cases}$.

2.5.28   Find the right hand limit at $x = 1$ for $f(x) = \begin{cases} 1 - x, & x > 1 \\ 6, & x = 1 \\ 1 + x, & x < 1 \end{cases}$.

2.5.29   Find the left hand limit at $x = 0$ for $f(x) = \begin{cases} x^3 - 1, & x \geq 0 \\ x + 1, & x < 0 \end{cases}$.

2.5.30   Find $\lim\limits_{x \to 3} f(x)$ where $f(x) = \begin{cases} x^2 - 1, & x < 3 \\ (x-1)^3, & x > 3 \end{cases}$.

SOLUTIONS

SECTION 2.5

2.5.1 0.

2.5.2 π^2.

2.5.3 5/3.

2.5.4 $\displaystyle\lim_{x\to 4}\frac{(x+4)(x-4)}{(x+5)(x-4)}=\lim_{x\to 4}\frac{x+4}{x+5}=\frac{8}{9}$.

2.5.5 $\displaystyle\lim_{x\to 0}\frac{x(x+2)}{x(1-2x)}=\lim_{x\to 0}\frac{x+2}{1-2x}=2$.

2.5.6 $\displaystyle\lim_{x\to 1}\frac{(1+x)(1-x)}{(x+6)(x-1)}=\lim_{x\to 1}\frac{-(1+x)}{x+6}=-\frac{2}{7}$.

2.5.7 $\displaystyle\lim_{x\to 1}\frac{(x-1)(x+2)}{(x-1)(x-3)}=\lim_{x\to 1}\frac{x+2}{x-3}=-\frac{3}{2}$.

2.5.8 $\displaystyle\lim_{x\to a}\frac{(x-a)(x+a)}{x-a}=\lim_{x\to a}(x+a)=2a$.

2.5.9 $\displaystyle\lim_{x\to 3}\frac{(x-3)(x^2+3x+9)}{x-3}=\lim_{x\to 3}(x^2+3x+9)=27$.

2.5.10 $\displaystyle\lim_{x\to 1}\frac{x(x-1)(x-2)}{x-1}=\lim_{x\to 1}x(x-2)=-1$.

2.5.11 $\displaystyle\lim_{h\to 2}\frac{h(h-2)(h+2)}{h^2(h-2)}=\lim_{h\to 2}\frac{h+2}{h}=2$.

2.5.12 $\displaystyle\lim_{x\to a}\frac{\frac{1}{x}-\frac{1}{a}}{x-a}=\lim_{x\to a}\frac{a-x}{ax(x-a)}=\lim_{x\to a}\frac{-1}{ax}=-\frac{1}{a^2}$.

2.5.13 $\displaystyle\lim_{h\to 0}\frac{-h}{3h(3+h)}=\lim_{h\to 0}-\frac{1}{3(3+h)}=-\frac{1}{9}$.

2.5.14 $\displaystyle\lim_{x\to -a}\frac{(x+a)(x^2-ax+a^2)}{x+a}=\lim_{x\to -a}(x^2-ax+a^2)=3a^2$.

2.5.15 $\displaystyle\lim_{x\to 3}\frac{x-3}{(x-3)(x^2+3x+9)}=\lim_{x\to 3}\frac{1}{x^2+3x+9}=\frac{1}{27}$.

2.5.16 $\displaystyle\lim_{x\to 2}\frac{\left(1-\dfrac{2}{x}\right)\left(1+\dfrac{2}{x}\right)}{1-\dfrac{2}{x}}=\lim_{x\to 2}\left(1+\dfrac{2}{x}\right)=2.$

2.5.17 $-1.$

2.5.18 $\displaystyle\lim_{x\to 4^-}\frac{x-4}{-(x-4)}=-1.$ **2.5.19** $\displaystyle\lim_{x\to 1^+}\frac{x-1}{x-1}=1.$

2.5.20 $\displaystyle\lim_{x\to+\infty}\frac{2x^2}{x^2}=\lim_{x\to+\infty}2=2.$ **2.5.21** $\displaystyle\lim_{x\to+\infty}\frac{x^3}{3x^3}=\lim_{x\to+\infty}\frac{1}{3}=\frac{1}{3}.$

2.5.22 $\displaystyle\lim_{x\to+\infty}\frac{\dfrac{\sqrt{x^2-4}}{\sqrt{x^2}}}{\dfrac{2x}{x}}=\lim_{x\to+\infty}\frac{\sqrt{1-\dfrac{4}{x^2}}}{2}=\frac{1}{2}.$

2.5.23 $\displaystyle\lim_{x\to+\infty}\frac{\dfrac{\sqrt{3x^2+4x-1}}{\sqrt{x^2}}}{\dfrac{3-x}{x}}=\lim_{x\to+\infty}\frac{\sqrt{3+\dfrac{4}{x}-\dfrac{1}{x^2}}}{\dfrac{3}{x}-1}=-\sqrt{3}.$

2.5.24 $\displaystyle\lim_{x\to-\infty}\frac{\dfrac{\sqrt{x^2+4}}{-\sqrt{x^2}}}{\dfrac{x}{x}}=\lim_{x\to-\infty}\frac{-\sqrt{1-\dfrac{4}{x^2}}}{1}=-1.$

2.5.25 $\displaystyle\lim_{x\to-\infty}\frac{\dfrac{\sqrt{3x^2+4x-1}}{-\sqrt{x^2}}}{\dfrac{3-x}{x}}=\lim_{x\to-\infty}\frac{-\sqrt{3+\dfrac{4}{x}-\dfrac{1}{x^2}}}{\dfrac{3}{x}-1}=\sqrt{3}.$

2.5.26 $\displaystyle\lim_{x\to 3^-}\frac{-(x-3)}{x-3}=-1.$

2.5.27 $\displaystyle\lim_{x\to 1^-}\frac{1}{x+2}=\frac{1}{3};\ \lim_{x\to 1^+}x=-1,$ so the limit does not exist because $\displaystyle\lim_{x\to 1^-}f(x)\neq\lim_{x\to 1^+}f(x).$

2.5.28 $\displaystyle\lim_{x\to 1^+}(1-x)=0.$ **2.5.29** $\displaystyle\lim_{x\to 0^-}(x+1)=1.$

2.5.30 $\displaystyle\lim_{x\to 3^-}(x^2-1)=8;\ \lim_{x\to 3^+}(x-1)^3=8$ so $\displaystyle\lim_{x\to 3}f(x)=8.$

SECTION 2.7

2.7.1 Find any points of discontinuity for $f(x) = \dfrac{x-1}{x^2-1}$.

2.7.2 Find any points of discontinuity for $f(x) = \dfrac{x+1}{x^2+1}$.

2.7.3 Show that $f(x) = \dfrac{x^2-3}{x-\sqrt{3}}$ is not a continuous function.

2.7.4 Define $f(x) = \dfrac{x^3+1}{x+1}$ so that it will be continuous everywhere.

2.7.5 Define $g(x) = \dfrac{x^2+x-6}{x-2}$ so that it will be continuous everywhere.

2.7.6 Prove that $f(x) = \sqrt{x^2+x}$ is continuous on $[0,+\infty)$.

2.7.7 Assign a value to the constant k which will make g continuous.

$$g(x) = \begin{cases} \dfrac{x+2}{x^3+2x^2+x+2}, & x \neq -2 \\ k, & x = -2 \end{cases}$$

2.7.8 Assign a value to the constant k which will make h continuous.

$$h(x) = \begin{cases} \dfrac{x^3+3x^2+x+3}{x+3}, & x \neq -3 \\ k, & x = -3 \end{cases}$$

2.7.9 Assign a value to the constant k which will make f continuous.

$$f(x) = \begin{cases} \dfrac{x^2-4x+3}{x-1}, & x \neq 1 \\ k, & x = 1 \end{cases}$$

2.7.10 Show that $f(x) = \begin{cases} \dfrac{x^2-x-2}{x+1}, & x < -1 \\ 2x+2, & x \geq -1 \end{cases}$ is not continuous at $x = -1$ but is continuous from the right at $x = -1$.

2.7.11 Examine $h(x) = \begin{cases} \dfrac{2x^2 + 3x + 1}{x + 1}, & x < -1 \\ \dfrac{|x|}{x}, & -1 \le x < 0 \\ 2x, & x \ge 0 \end{cases}$ and determine if h is **(a)** continuous at $x = -1$, **(b)** continuous at $x = 0$, and **(c)** continuous from the right at $x = 0$.

2.7.12 Examine $g(x) = \begin{cases} \sqrt{\dfrac{2x + 3}{2 + x + x^2}}, & x < -1 \\ 2 - x^2, & x \ge -1 \end{cases}$ and determine if g is **(a)** continuous at $x = -1$, **(b)** continuous from the right at $x = -1$, and **(c)** continuous from the left at $x = -1$.

2.7.13 Let $g(x) = \begin{cases} |x + 1|, & x \le -2 \\ x + 1, & -2 < x < 1 \\ \sqrt{x + 3}, & 1 \le x \le 6 \\ \dfrac{6}{8 - x}, & 6 < x \le 7 \\ 6, & 7 < x \le 10 \end{cases}$.

(a) Determine if g is continuous from the right at $x = -2$.
(b) Determine if g is continuous from the left at $x = 1$.
(c) Determine if g is continuous at $x = 7$.
(d) Determine if g is continuous at $x = 9$.

2.7.14 Show that $f(x) = \dfrac{x - 1}{x(x + 1)}$ is not continuous at $x = 0$ or $x = -1$ and show also that the discontinuities at $x = 0$ and $x = -1$ are nonremovable.

2.7.15 Show that $f(x) = \dfrac{1}{(x - 1)^3}$ is not continuous at $x = 1$ and that the discontinuity at $x = 1$ is nonremovable.

2.7.16 Show that the equation $f(x) = x^3 + x + 6$ has at least one solution in the interval $[-3, 0]$.

2.7.17 Show that the equation $f(x) = x^3 + 3x + 1$ has at least one solution in the interval $[-1, 2]$.

2.7.18 Determine the interval for which $f(x) = \dfrac{1}{\sqrt{3 - x}}$ is a continuous function.

2.7.19 Show that $f(x) = \begin{cases} \dfrac{|x|}{x}, & x \ne 0 \\ k, & x = 0 \end{cases}$ cannot be made continuous for any assigned value of the constant k.

SOLUTIONS

SECTION 2.7

2.7.1 f is discontinuous at x if $x^2 - 1 = 0$, $x = \pm 1$.

2.7.2 f is continuous everywhere since $x^2 + 1 \neq 0$.

2.7.3 f is not continuous at $x = \sqrt{3}$ since $f(\sqrt{3})$ is not defined.

2.7.4 f is continuous everywhere except at $x = -1$, however,

$$\lim_{x \to -1} \frac{x^3 + 1}{x + 1} = \lim_{x \to -1} \frac{(x+1)(x^2 - x + 1)}{x + 1} = \lim_{x \to -1} (x^2 - x + 1) = 3, \text{ so let}$$

$$f(x) = \begin{cases} \dfrac{x^3 + 1}{x + 1}, & x \neq -1 \\ 3, & x = -1 \end{cases} \quad \text{thus } f \text{ is continuous at } x = -1 \text{ since } \lim_{x \to -1} f(x) = f(-1).$$

2.7.5 g is continuous everywhere except at $x = 2$, however,

$$\lim_{x \to 2} \frac{x^2 + x - 6}{x - 2} = \lim_{x \to 2} \frac{(x-2)(x+3)}{x - 2} = \lim_{x \to 2} (x + 3) = 5, \text{ so let}$$

$$g(x) = \begin{cases} \dfrac{x^2 + x - 6}{x - 2}, & x \neq 2 \\ 5, & x = 2 \end{cases}, \text{ thus } g \text{ is continuous at } x = 2 \text{ since } \lim_{x \to 2} g(x) = g(2).$$

2.7.6 For c in the interval $(0, \infty)$, $\lim_{x \to c} f(x) = \lim_{x \to c} \sqrt{x^2 + x} = \sqrt{\lim_{x \to c} (x^2 + x)} = \sqrt{c^2 + c} = f(c)$ so f is continuous on $(0, \infty)$. Also $\lim_{x \to 0^+} f(x) = \lim_{x \to 0^+} \sqrt{x^2 + x} = 0 = f(0)$. So f is continuous on $[0, +\infty)$.

2.7.7 g is continuous everywhere except, perhaps, at $x = -2$, however,

$$\lim_{x \to -2} \frac{x + 2}{x^3 + 2x^2 + x + 2} = \lim_{x \to -2} \frac{x + 2}{(x+2)(x^2 + 1)} = \lim_{x \to -2} \frac{1}{x^2 + 1} = \frac{1}{5} \text{ so let } k = 1/5, \text{ thus, } g \text{ is}$$
continuous at $x = -2$ since $\lim_{x \to -2} g(x) = g(-2)$.

2.7.8 h is continuous everywhere except, perhaps, at $x = -3$, however,

$$\lim_{x \to -3} \frac{x^3 + 3x^2 + x + 3}{x + 3} = \lim_{x \to -3} \frac{(x+3)(x^2 + 1)}{x + 3} = \lim_{x \to -3} (x^2 + 1) = 10 \text{ so let } k = 10, \text{ thus, } h \text{ is}$$
continuous at $x = -3$ since $\lim_{x \to -3} h(x) = h(-3)$.

2.7.9 f is continuous everywhere except, perhaps, at $x = 1$, however,

$$\lim_{x \to 1} \frac{x^2 - 4x + 3}{x - 1} = \lim_{x \to 1} \frac{(x-1)(x-3)}{x - 1} = \lim_{x \to 1} (x - 3) = -2, \text{ so let } k = -2, \text{ thus, } f \text{ is continuous}$$
at $x = 1$ since $\lim_{x \to 1} f(x) = f(1)$.

2.7.10 $\displaystyle\lim_{x\to-1^-} f(x) = \lim_{x\to-1^-} \frac{x^2-x-2}{x+1} = \lim_{x\to-1^-} \frac{(x+1)(x-2)}{x+1} = \lim_{x\to-1^-} (x-2) = -3;$

$\displaystyle\lim_{x\to-1^+} f(x) = \lim_{x\to-1^+} (2x+2) = 0$ so f is not continuous at $x=-1$ since $\displaystyle\lim_{x\to-1} f(x)$ does not exist, however, f is continuous from the right since $\displaystyle\lim_{x\to-1^+} f(x) = f(-1)$.

2.7.11 **(a)** $\displaystyle\lim_{x\to-1^-} f(x) = \lim_{x\to-1^-} \frac{2x^2+3x+1}{x+1} = \lim_{x\to-1^-} \frac{(x+1)(2x+1)}{x+1} = \lim_{x\to-1^-} (2x+1) = -1;$

$\displaystyle\lim_{x\to-1^+} f(x) = \lim_{x\to-1^+} \frac{|x|}{x} = \lim_{x\to-1^+} \frac{-x}{x} = -1$ and $f(-1) = -1$ so f is continuous at $x=-1$.

(b) $\displaystyle\lim_{x\to0^-} f(x) = \lim_{x\to0^-} \frac{|x|}{x} = \lim_{x\to0^-} \frac{-x}{x} = -1;\ \lim_{x\to0^+} f(x) = \lim_{x\to0^+} 2x = 0$ so f is not continuous at $x=0$ since $\displaystyle\lim_{x\to0} f(x)$ does not exist.

(c) f is continuous from the right at $x=0$ since $\displaystyle\lim_{x\to0^+} f(x) = f(0)$.

2.7.12 **(a)** $\displaystyle\lim_{x\to-1^-} g(x) = \lim_{x\to-1^-} \sqrt{\frac{2x+3}{2+x+x^2}} = \sqrt{\frac{1}{2}};$

$\displaystyle\lim_{x\to-1^+} g(x) = \lim_{x\to-1^+} (2-x^2) = 1$ so g is not continuous at $x=-1$ since $\displaystyle\lim_{x\to-1} g(x)$ does not exist.

(b) $\displaystyle\lim_{x\to-1^+} g(x) = g(-1) = 1$ so g is continuous from the right at $x=-1$.

(c) g is not continuous from the left at $x=-1$ since $\displaystyle\lim_{x\to-1^-} g(x) \neq g(-1)$.

2.7.13 **(a)** $\displaystyle\lim_{x\to-2^+} g(x) = \lim_{x\to-2^+} (x+1) = -1$ and $g(-2) = 1$ so g is not continuous from the right at $x=-2$ since $\displaystyle\lim_{x\to2^+} g(x) \neq g(-2)$.

(b) $\displaystyle\lim_{x\to1^-} g(x) = \lim_{x\to1^-} (x+1) = 2,\ g(1) = \sqrt{1+3} = 2$ so g is continuous from the left at $x=1$ since $\displaystyle\lim_{x\to1^-} g(x) = g(1)$.

(c) $\displaystyle\lim_{x\to7^-} g(x) = \lim_{x\to7^-} \frac{6}{8-x} = 6;\ \lim_{x\to7^+} g(x) = \lim_{x\to7^+} 6 = 6$ and $g(7) = 6$ so g is continuous at $x=7$.

(d) $\displaystyle\lim_{x\to9} g(x) = \lim_{x\to9} 6 = 6,\ g(9) = 6$ so g is continuous at $x=9$ since $\displaystyle\lim_{x\to9} g(x) = g(9)$.

2.7.14 $f(-1)$ and $f(0)$ are not defined, thus f is not continuous at $x=-1$ or $x=0$, moreover, $\displaystyle\lim_{x\to-1} f(x)$ and $\displaystyle\lim_{x\to0}$ do not exist thus the discontinuities at $x=-1$ and $x=0$ are nonremovable.

2.7.15 $f(1)$ is not defined, thus f is not continuous at $x=1$, moreover, $\displaystyle\lim_{x\to1} f(x)$ does not exist thus the discontinuity at $x=1$ is nonremovable.

2.7.16 $f(x) = x^3 + x + 6$ is continuous on $[-3,0]$, $f(-2) = -4$ and $f(-1) = 4$ have opposite signs so Theorem 2.7.10 applies.

2.7.17 $f(x) = x^3 + 3x + 1$ is continuous on $[-2, 2]$, $f(-1) = -3$ and $f(1) = 5$ have opposite signs so Theorem 2.7.10 applies.

2.7.18 $(-\infty, 3)$

2.7.19 $\lim\limits_{x \to 0^-} \dfrac{|x|}{x} = \lim\limits_{x \to 0^-} \dfrac{-x}{x} = -1$, $\lim\limits_{x \to 0^+} \dfrac{|x|}{x} = \lim\limits_{x \to 0^+} \dfrac{x}{x} = 1$, thus $\lim\limits_{x \to 0} f(x)$ does not exist and $f(x)$ is not continuous for any k.

SECTION 2.8

2.8.1 Find $\lim\limits_{\theta\to 0}\dfrac{\tan\theta}{\theta}$.

2.8.2 Find $\lim\limits_{\theta\to 0}\dfrac{\sin 2\theta}{\tan\theta}$.

2.8.3 Find $\lim\limits_{\alpha\to 0}\dfrac{\sin\alpha-\tan\alpha}{\sin^3\alpha}$.

2.8.4 Find $\lim\limits_{\theta\to 0}\theta\cot 4\theta$.

2.8.5 Find $\lim\limits_{\theta\to 0}\dfrac{\sin\sqrt{2\theta}}{\sqrt{\theta}}$.

2.8.6 Find $\lim\limits_{\phi\to 0}\dfrac{\phi^2}{\sin 3\phi^2}$.

2.8.7 Find $\lim\limits_{\theta\to 0}\dfrac{3}{\theta\csc\theta}$.

2.8.8 Find $\lim\limits_{\phi\to 0}\dfrac{\sin 3\phi}{\sin 2\phi}$.

2.8.9 Find $\lim\limits_{\alpha\to 0}\dfrac{\alpha}{\cos\alpha}$.

2.8.10 Find $\lim\limits_{t\to 0}\dfrac{t^2}{1-\cos^2 t}$.

2.8.11 Find $\lim\limits_{\phi\to 0}\dfrac{3\phi}{\cos 2\phi}$.

2.8.12 Find $\lim\limits_{\theta\to 0}\dfrac{\sin^2\theta}{\tan\theta}$.

2.8.13 Find $\lim\limits_{t\to 0}\dfrac{\sin t}{t^2+5t}$.

2.8.14 Find $\lim\limits_{\alpha\to 0}\dfrac{3\alpha^2+\sin 4\alpha}{\alpha}$.

2.8.15 Find $\lim\limits_{\theta\to 0}\dfrac{\sin^2\dfrac{\theta}{2}}{\theta^2}$.

2.8.16 Find $\lim\limits_{x\to 0}\dfrac{\cos\left(\dfrac{\pi}{2}+x\right)}{x}$.

2.8.17 Find a value for the constant k so that

$$f(\theta)=\begin{cases}\dfrac{\theta}{\sin 2\theta}, & \theta\neq 0\\[2mm] k, & \theta=0\end{cases}$$

will be continuous at $\theta=0$.

2.8.18 Find a value for the constant k so that

$$f(\theta)=\begin{cases}\dfrac{\sin 3\theta}{2\theta}, & \theta\neq 0\\[2mm] k, & \theta=0\end{cases}$$

will be continuous at $\theta=0$.

2.8.19 Find a value for the constant k so that

$$f(\theta)=\begin{cases}\dfrac{\tan\theta}{\theta}, & \theta\neq 0\\[2mm] k, & \theta=0\end{cases}$$

will be continuous at $\theta=0$.

SOLUTIONS

SECTION 2.8

2.8.1 $\displaystyle \lim_{\theta \to 0} \frac{\tan \theta}{\theta} = \lim_{\theta \to 0} \frac{\frac{\sin \theta}{\cos \theta}}{\theta} = \lim_{\theta \to 0} \frac{\sin \theta}{\theta \cos \theta} = \left(\lim_{\theta \to 0} \frac{\sin \theta}{\theta} \right) \left(\lim_{\theta \to 0} \frac{1}{\cos \theta} \right) = (1)(1) = 1.$

2.8.2 $\displaystyle \lim_{\theta \to 0} \frac{\sin 2\theta}{\tan \theta} = \lim_{\theta \to 0} \frac{\frac{\sin 2\theta}{\theta}}{\frac{\tan \theta}{\theta}} = \lim_{\theta \to 0} \frac{\frac{2 \sin 2\theta}{2\theta}}{\frac{\sin \theta}{\theta \cos \theta}}$

$$= \frac{2 \lim\limits_{\theta \to 0} \dfrac{\sin 2\theta}{2\theta}}{\left(\lim\limits_{\theta \to 0} \dfrac{\sin \theta}{\theta} \right) \left(\lim\limits_{\theta \to 0} \dfrac{1}{\cos \theta} \right)} = \frac{2(1)}{(1)(1)} = 2$$

2.8.3 $\displaystyle \lim_{\alpha \to 0} \frac{\sin \alpha - \tan \alpha}{\sin^3 \alpha} = \lim_{\alpha \to 0} \frac{\sin \alpha - \frac{\sin \alpha}{\cos \alpha}}{\sin^3 \alpha} = \lim_{\alpha \to 0} \frac{\sin \alpha (\cos \alpha - 1)}{\cos \alpha \sin^3 \alpha} = \lim_{\alpha \to 0} \frac{\cos \alpha - 1}{\cos \alpha \sin^2 \alpha}$

$$= \lim_{\alpha \to 0} \frac{\cos \alpha - 1}{\cos \alpha (1 - \cos^2 \alpha)} = \lim_{\alpha \to 0} \frac{\cos \alpha - 1}{\cos \alpha (1 - \cos \alpha)(1 + \cos \alpha)}$$

$$= \lim_{\alpha \to 0} \frac{-1}{\cos \alpha (1 + \cos \alpha)} = \frac{-1}{(1)(2)} = -\frac{1}{2}.$$

2.8.4 $\displaystyle \lim_{\theta \to 0} \theta \cot 4\theta = \lim_{\theta \to 0} \frac{\frac{\cos 4\theta}{\sin 4\theta}}{\theta} = \lim_{\theta \to 0} \frac{\cos 4\theta}{\frac{4 \sin 4\theta}{4\theta}}$

$$= \frac{\lim\limits_{\theta \to 0} \cos 4\theta}{4 \lim\limits_{\theta \to 0} \dfrac{\sin 4\theta}{4\theta}} = \frac{1}{4(1)} = \frac{1}{4}$$

2.8.5 $\displaystyle \lim_{\theta \to 0} \frac{\sin \sqrt{2\theta}}{\sqrt{\theta}} = \lim_{\theta \to 0} \frac{\sqrt{2} \sin \sqrt{2\theta}}{\sqrt{2\theta}}$

$$= \sqrt{2} \lim_{\theta \to 0} \frac{\sin \sqrt{2\theta}}{\sqrt{2\theta}} = \sqrt{2}(1) = \sqrt{2}$$

2.8.6 $\displaystyle \lim_{\phi \to 0} \frac{\phi^2}{\sin 3\phi^2} = \lim_{\phi \to 0} \frac{1}{\frac{\sin 3\phi^2}{\phi^2}} = \lim_{\phi \to 0} \frac{1}{\frac{3 \sin 3\phi^2}{3\phi^2}}$

$$= \frac{1}{3 \lim\limits_{\phi \to 0} \dfrac{\sin 3\phi^2}{3\phi^2}} = \frac{1}{(3)(1)} = \frac{1}{3}$$

2.8.7 $\displaystyle \lim_{\theta \to 0} \frac{3}{\theta \csc \theta} = \lim_{\theta \to 0} \frac{3 \sin \theta}{\theta} = 3 \lim_{\theta \to 0} \frac{\sin \theta}{\theta} = 3(1) = 3.$

2.8.8 $\displaystyle \lim_{\phi \to 0} \frac{\sin 3\phi}{\sin 2\phi} = \lim_{\phi \to 0} \frac{\dfrac{\sin 3\phi}{\phi}}{\dfrac{\sin 2\phi}{\phi}} = \lim_{\phi \to 0} \frac{\dfrac{3 \sin 3\phi}{3\phi}}{\dfrac{2 \sin 2\phi}{2\phi}}$

$$= \frac{3 \displaystyle\lim_{\phi \to 0} \dfrac{\sin 3\phi}{3\phi}}{2 \displaystyle\lim_{\phi \to 0} \dfrac{\sin 2\phi}{2\phi}} = \frac{3(1)}{2(1)} = \frac{3}{2}$$

2.8.9 $0.$

2.8.10 $\displaystyle \lim_{t \to 0} \frac{t^2}{1 - \cos^2 t} = \lim_{t \to 0} \frac{t^2}{\sin^2 t} = \lim_{t \to 0} \frac{1}{\dfrac{\sin^2 t}{t^2}} = \frac{1}{\left(\displaystyle\lim_{t \to 0} \dfrac{\sin t}{t}\right)^2} = \frac{1}{(1)^2} = 1.$

2.8.11 $0.$

2.8.12 $\displaystyle \lim_{\theta \to 0} \frac{\sin^2 \theta}{\tan \theta} = \lim_{\theta \to 0} \frac{\sin^2 \theta}{\dfrac{\sin \theta}{\cos \theta}} = \lim_{\theta \to 0} \cos \theta \sin \theta = \left(\lim_{\theta \to 0} \cos \theta\right)\left(\lim_{\theta \to 0} \sin \theta\right) = (1)(0) = 0.$

2.8.13 $\displaystyle \lim_{t \to 0} \frac{\sin t}{t^2 + 5t} = \lim_{t \to 0} \frac{\sin t}{t(t + 5)} = \left(\lim_{t \to 0} \frac{\sin t}{t}\right)\left(\lim_{t \to 0} \frac{1}{t + 5}\right) = (1)\left(\frac{1}{5}\right) = \frac{1}{5}.$

2.8.14 $\displaystyle \lim_{\alpha \to 0} \frac{3\alpha^2 + \sin 4\alpha}{\alpha} = \lim_{\alpha \to 0} \left(\frac{3\alpha^2}{\alpha} + \frac{\sin 4\alpha}{\alpha}\right) = \lim_{\alpha \to 0} \left(3\alpha + \frac{4 \sin 4\alpha}{4\alpha}\right)$

$$= \lim_{\alpha \to 0} 3\alpha + 4 \lim_{\alpha \to 0} \frac{\sin 4\alpha}{4\alpha} = 0 + 4(1) = 4.$$

2.8.15 $\displaystyle \lim_{\theta \to 0} \frac{\sin^2 \dfrac{\theta}{2}}{\theta^2} = \lim_{\theta \to 0} \frac{\sin^2 \dfrac{\theta}{2}}{4 \cdot \dfrac{\theta^2}{4}} = \frac{1}{4}\left(\lim_{\theta \to 0} \frac{\sin \dfrac{\theta}{2}}{\dfrac{\theta}{2}}\right)^2 = \frac{1}{4}(1)^2 = \frac{1}{4}.$

2.8.16 $\displaystyle \lim_{x \to 0} \frac{\cos\left(\dfrac{\pi}{2} + x\right)}{x} = \lim_{x \to 0} \left(-\frac{\sin x}{x}\right) = \lim_{x \to 0} \frac{\sin x}{x} = -1.$

2.8.17 $\displaystyle \lim_{\theta \to 0} \frac{\theta}{\sin 2\theta} = \lim_{\theta \to 0} \frac{1}{\dfrac{2 \sin 2\theta}{2\theta}} = \frac{1}{2 \displaystyle\lim_{\theta \to 0} \dfrac{\sin 2\theta}{2\theta}} = \frac{1}{2} \text{ so } k = 1/2.$

2.8.18 $\quad \lim\limits_{\theta \to 0} \dfrac{\sin 3\theta}{2\theta} = \dfrac{1}{2} \lim\limits_{\theta \to 0} \dfrac{3 \sin 3\theta}{3\theta} = \dfrac{3}{2} \lim\limits_{\theta \to 0} \dfrac{\sin 3\theta}{3\theta} = \dfrac{3}{2}$ so $k = 3/2$.

2.8.19 $\quad \lim\limits_{\theta \to 0} \dfrac{\tan \theta}{\theta} = \lim\limits_{\theta \to 0} \dfrac{\frac{\sin \theta}{\cos \theta}}{\theta} = \lim\limits_{\theta \to 0} \dfrac{\sin \theta}{\theta \cos \theta} = \left(\lim\limits_{\theta \to 0} \dfrac{\sin \theta}{\theta} \right) \left(\lim\limits_{\theta \to 0} \dfrac{1}{\cos \theta} \right) = (1)(1) = 1$ so $k = 1$.

SUPPLEMENTARY EXERCISES, CHAPTER 2

In Exercises 1–5, find the natural domain of f and then evaluate f (if defined) at the given values of x.

1. $f(x) = \sqrt{4 - x^2}; x = -\sqrt{2}, 0, \sqrt{3}$.

2. $f(x) = 1/\sqrt{(x-1)^3}; x = 0, 1, 2$.

3. $f(x) = (x-1)/(x^2 + x - 2); x = 0, 1, 2$.

4. $f(x) = \sqrt{|x| - 2}; x = -3, 0, 2$.

5. $f(x) = \begin{cases} x^2 - 1, & x \le 2 \\ \sqrt{x - 1}, & x > 2 \end{cases}; \quad x = 0, 2, 4$.

In Exercises 6 and 7, find

    (a) $f(x^2) - (f(x))^2$

    (c) $f(1/x) - 1/f(x)$

    (b) $f(x+3) - [f(x) + f(3)]$

    (d) $(f \circ f)(x)$.

6. $f(x) = \sqrt{3 - x}$.

7. $f(x) = \dfrac{3 - x}{x}$.

In Exercises 8–15, sketch the graph of f and find its domain and range.

8. $f(x) = (x - 2)^2$.

9. $f(x) = -\pi$.

10. $f(x) = |2 - 4x|$.

11. $f(x) = \dfrac{x^2 - 4}{2x + 4}$.

12. $f(x) = \sqrt{-2x}$

13. $f(x) = -\sqrt{3x + 1}$.

14. $f(x) = 2 - |x|$.

15. $f(x) = \dfrac{2x - 4}{x^2 - 4}$

16. In each part, complete the square, and then find the range of f.

    (a) $f(x) = x^2 - 5x + 6$

    (b) $f(x) = -3x^2 + 12x - 7$.

17. Express $f(x)$ as a composite function $(g \circ h)(x)$ in two different ways.

    (a) $f(x) = x^6 + 3$

    (b) $f(x) = \sqrt{x^2 + 1}$

    (c) $f(x) = \sin(3x + 2)$.

18. Find $\displaystyle\lim_{x \to k} \dfrac{x^3 - kx^2}{x^2 - k^2}$, where k is a constant.

In Exercises 19 and 20, sketch the graph of f and find the indicated limits of $f(x)$ (if they exist).

19. $f(x) = \begin{cases} 1/x, & x < 0 \\ x^2, & 0 \le x < 1 \\ 2, & x = 1 \\ 2 - x, & x > 1 \end{cases}$

    (a) as $x \to -1$

    (b) as $x \to 0$

    (c) as $x \to 1$

    (d) as $x \to 0^+$

    (e) as $x \to 0^-$

    (f) as $x \to 2^+$

    (g) as $x \to -\infty$

    (h) as $x \to +\infty$.

20. $f(x) = \begin{cases} 2, & x \le -1 \\ -x, & -1 < x < 0 \\ x/(2-x), & 0 < x < 2 \\ 1, & x \ge 2 \end{cases}$

(a) as $x \to -1^+$ (b) as $x \to -1^-$ (c) as $x \to -1$ (d) as $x \to 0$

(e) as $x \to 2^+$ (f) as $x \to 2^-$ (g) as $x \to 2$ (h) as $x \to -\infty$.

In Exercises 21–24, find $\lim\limits_{x \to a} f(x)$ (if it exists).

21. $f(x) = \sqrt{2 - x}$;
$a = -2, 1, 2^-, 2^+, -\infty, +\infty$.

22. $f(x) = \begin{cases} (x-2)/|x-2|, & x \ne 2 \\ 0, & x = 2 \end{cases}$
$a = 0, 2^-, 2^+, 2, -\infty, +\infty$.

23. $f(x) = (x^2 - 25)/(x - 5)$;
$a = 0, 5^+, -5^-, 5, -5, -\infty, +\infty$.

24. $f(x) = (x + 5)/(x^2 - 25)$;
$a = 0, 5^+, -5^-, -5, 5, -\infty, +\infty$.

In Exercises 25–32, find the indicated limit if it exists.

25. $\lim\limits_{x \to 0} \dfrac{\tan ax}{\sin bx}$ $(a \ne 0, b \ne 0)$.

26. $\lim\limits_{x \to 0} \dfrac{\sin 3x}{\tan 3x}$,

27. $\lim\limits_{\theta \to 0} \dfrac{\sin 2\theta}{\theta^2}$.

28. $\lim\limits_{x \to 0} \dfrac{x \sin x}{1 - \cos x}$.

29. $\lim\limits_{x \to 0^+} \dfrac{\sin x}{\sqrt{x}}$.

30. $\lim\limits_{x \to 0} \dfrac{\sin^2(kx)}{x^2}$, $k \ne 0$.

31. $\lim\limits_{x \to 0} \dfrac{3x \; \sin(kx)}{x}$, $k \ne 0$.

32. $\lim\limits_{x \to +\infty} \dfrac{2x + x \sin 3x}{5x^2 - 2x + 1}$.

SOLUTIONS

SUPPLEMENTARY EXERCISES, CHAPTER 2

1. $\sqrt{4-x^2}$ is real if and only if $4-x^2 \geq 0$, thus $4 \geq x^2$, so the domain is $|x| \leq 2$; $f(-\sqrt{2}) = \sqrt{2}$, $f(0) = 2$, $f(\sqrt{3}) = 1$.

2. domain: $x > 1$; $f(0)$ and $f(1)$ are not defined, $f(2) = 1$.

3. $f(x) = \dfrac{(x-1)}{(x+2)(x-1)}$, domain: all x except -2 and 1; $f(0) = 1/2$, $f(1)$ is not defined, $f(2) = 1/4$.

4. domain: $|x| \geq 2$; $f(-3) = 1$, $f(0)$ is not a real number, $f(2) = 0$.

5. domain: all x; $f(0) = -1$, $f(2) = 3$, $f(4) = \sqrt{3}$.

6. **(a)** $f(x^2) - (f(x))^2 = \sqrt{3 - x^2} - (3 - x)$

   (b) $f(x+3) - [f(x) + f(3)] = \sqrt{3 - (x+3)} - [\sqrt{3-x} + \sqrt{3-3}] = \sqrt{-x} - \sqrt{3-x}$

   (c) $f(1/x) - 1/f(x) = \sqrt{3 - 1/x} - 1/\sqrt{3-x}$

   (d) $f(f(x)) = \sqrt{3 - \sqrt{3-x}}$

7. **(a)** $f(x^2) - (f(x))^2 = \dfrac{3 - x^2}{x^2} - \left(\dfrac{3-x}{x}\right)^2 = \dfrac{3-x^2}{x^2} - \dfrac{9 - 6x + x^2}{x^2} = \dfrac{-2x^2 + 6x - 6}{x^2}$

   (b) $f(x+3) - [f(x) + f(3)] = \dfrac{3 - (x+3)}{x+3} - \left[\dfrac{3-x}{x} + \dfrac{3-3}{3}\right] = -\dfrac{9}{x(x+3)}$

   (c) $f(1/x) - 1/f(x) = \dfrac{3 - 1/x}{1/x} - \dfrac{x}{3-x} = 3x - 1 - \dfrac{x}{3-x} = \dfrac{3x^2 - 9x + 3}{x - 3}$

   (d) $f(f(x)) = f\left(\dfrac{3-x}{x}\right) = \dfrac{3 - \dfrac{3-x}{x}}{\dfrac{3-x}{x}} = \dfrac{4x - 3}{3 - x}$

8.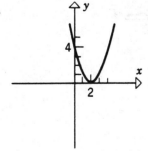

domain: all x
range: $y \geq 0$

9.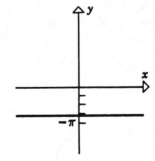

domain: all x
range: $y = -\pi$

10.

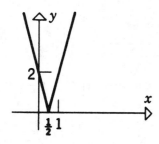

domain: all x
range: $y \geq 0$

11.

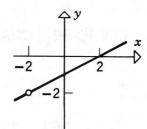

$$f(x) = \frac{x^2 - 4}{2x + 4} = \frac{1}{2}(x - 2).$$

$x \neq -2$
domain: all x except -2
range: all y except -2

12.

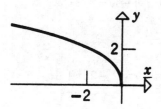

domain: $x \leq 0$
range: $y \geq 0$

13.

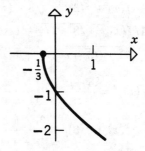

domain: $x \geq -1/3$
range: $y \leq 0$

14.

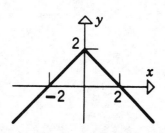

domain: all x
range: $y \leq 2$

15.

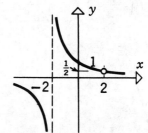

$$f(x) = \frac{2x - 4}{x^2 - 4} = \frac{2}{x + 2}, x \neq 2$$

domain: all x except $-2, 2$
range: all y except $0, 1/2$

16. **(a)** $y = f(x) = \left(x^2 - 5x + \frac{25}{4}\right) + 6 - \frac{25}{4} = \left(x - \frac{5}{2}\right)^2 - \frac{1}{4}$; range: $y \geq -\frac{1}{4}$.

    (b) $y = f(x) = -3(x^2 - 4x + 4) - 7 + 12 = -3(x - 2)^2 + 5$; range: $y \leq 5$.

17. Some possible answers are:

    (a) $h(x) = x^3$, $g(x) = x^2 + 3$; $h(x) = x^6$, $g(x) = x + 3$

    (b) $h(x) = x^2 + 1$, $g(x) = \sqrt{x}$; $h(x) = x^2$, $g(x) = \sqrt{x+1}$

    (c) $h(x) = 3x + 2$, $g(x) = \sin x$; $h(x) = 3x$, $g(x) = \sin(x + 2)$

18. $\displaystyle\lim_{x \to k} \frac{x^3 - kx^2}{x^2 - k^2} = \lim_{x \to k} \frac{x^2(x - k)}{(x + k)(x - k)} = \lim_{x \to k} \frac{x^2}{x + k} = \frac{1}{2}k$

19. **(a)** -1

    (b) does not exist

    (c) 1

    (d) 0

    (e) $-\infty$ (does not exist)

    (f) 0

    (g) 0

    (h) $-\infty$ (does not exist)

20. **(a)** 1

    (b) 2

    (c) does not exist

    (d) 0

    (e) 1

    (f) $+\infty$ (does not exist)

    (g) does not exist

    (h) 2

21. $f(x) = \sqrt{2 - x}$ is defined for $x \le 2$ and $\displaystyle\lim_{x \to a} f(x) = \sqrt{2 - a}$ if $a < 2$, so $\displaystyle\lim_{x \to a} f(x) = 2, 1, 0$ for $a = -2, 1, 2^-$. Because $f(x)$ is not defined for $x > 2$, $\displaystyle\lim_{x \to 2^+} f(x)$ and $\displaystyle\lim_{x \to +\infty} f(x)$ do not exist. Finally, $\displaystyle\lim_{x \to -\infty} f(x) = +\infty$, so this limit does not exist.

22. If $x \ne 2$, $f(x) = \dfrac{x - 2}{|x - 2|} = \begin{cases} 1, & x > 2 \\ -1, & x < 2 \end{cases}$, so $\displaystyle\lim_{x \to a} f(x) = \lim_{x \to a}(1) = 1$ for $a = 2^+$, $+\infty$ and $\displaystyle\lim_{x \to a} f(x) = \lim_{x \to a}(-1) = -1$ for $a = 0$, 2^-, $-\infty$. Because $\displaystyle\lim_{x \to 2^-} f(x) \ne \lim_{x \to 2^+} f(x)$, $\displaystyle\lim_{x \to 2} f(x)$ does not exist.

23. $f(x) = \dfrac{x^2 - 25}{x - 5} = x + 5$, $x \ne 5$, so $\displaystyle\lim_{x \to a} f(x) = \lim_{x \to a}(x + 5) = a + 5 = 5, 10, 0, 10, 0$ for $a = 0, 5^+, -5^-, 5, -5$. Also, $\displaystyle\lim_{x \to -\infty} f(x) = -\infty$ and $\displaystyle\lim_{x \to +\infty} f(x) = +\infty$, so neither of these limits exist.

24. $f(x) = \dfrac{x + 5}{x^2 - 25} = \dfrac{1}{x - 5}$, $x \ne -5$, so $\displaystyle\lim_{x \to a} f(x) = \lim_{x \to a} \dfrac{1}{x - 5} = \dfrac{1}{a - 5} = -\dfrac{1}{5}, -\dfrac{1}{10}, -\dfrac{1}{10}$ for $a = 0, -5^-, -5$. Also, $\displaystyle\lim_{x \to 5^+} f(x) = +\infty$ and $\displaystyle\lim_{x \to 5^-} f(x) = -\infty$, so $\displaystyle\lim_{x \to 5^+} f(x)$ and $\displaystyle\lim_{x \to 5} f(x)$ do not exist. Finally, $\displaystyle\lim_{x \to a} f(x) = 0$ for $a = -\infty, +\infty$.

25. $\displaystyle\lim_{x \to 0} \frac{\tan ax}{\sin bx} = \lim_{x \to 0} \frac{\sin ax}{\sin bx} \frac{1}{\cos ax} = \lim_{x \to 0} \frac{a[(\sin ax)/(ax)]}{b[(\sin bx)/(bx)]} \frac{1}{\cos ax} = \frac{a}{b}.$

26. $\displaystyle\lim_{x \to 0} \frac{\sin 3x}{\tan 3x} = \lim_{x \to 0} \cos 3x = 1$

27. $\displaystyle\lim_{\theta \to 0} \frac{\sin 2\theta}{\theta^2} = \lim_{\theta \to 0} \frac{\sin 2\theta}{2\theta} \frac{2}{\theta}$, but $\dfrac{\sin 2\theta}{2\theta} \to 1$ as $\theta \to 0$ and $\left|\dfrac{2}{\theta}\right| \to +\infty$ as $\theta \to 0$ so the limit does not exist.

28. $\displaystyle\lim_{x \to 0} \frac{x \sin x}{1 - \cos x} = \lim_{x \to 0} \frac{x \sin x}{1 - \cos x} \cdot \frac{1 + \cos x}{1 + \cos x} = \lim_{x \to 0} \frac{x \sin x (1 + \cos x)}{1 - \cos^2 x}$

$$= \lim_{x \to 0} \frac{x \sin x (1 + \cos x)}{\sin^2 x} = \lim_{x \to 0} \frac{1 + \cos x}{[(\sin x)/x]} = \frac{1 + 1}{1} = 2.$$

29. $\displaystyle\lim_{x \to 0^+} \frac{\sin x}{\sqrt{x}} = \lim_{x \to 0^+} \sqrt{x}\left(\frac{\sin x}{x}\right) = (0)(1) = 0$

30. $\displaystyle\lim_{x \to 0} \frac{\sin^2(kx)}{x^2} = \lim_{x \to 0} k^2 \left[\frac{\sin(kx)}{kx}\right]^2 = k^2$

31. $\displaystyle\lim_{x \to 0} \frac{3x - \sin(kx)}{x} = \lim_{x \to 0} \left[3 - k\frac{\sin(kx)}{kx}\right] = 3 - k$

32. $\displaystyle\lim_{x \to +\infty} \frac{2x + x \sin 3x}{5x^2 - 2x + 1} = \lim_{x \to +\infty} \frac{2 + \sin 3x}{5x - 2 + 1/x} = 0.$

CHAPTER 3
Differentiation

SECTION 3.1

3.1.1 Let $f(x) = \dfrac{1}{x^2}$;

    (a) Find the average rate of change of y with respect to x over the interval $[2, 3]$.
    (b) Find the instantaneous rate of change of y with respect to x at the point $x = 2$.
    (c) Find the instantaneous rate of change of y with respect to x at a general point x_0.
    (d) Sketch the graph of $y = f(x)$ together with the secant and tangent lines whose slopes are given by the results in parts (a) and (b).

3.1.2 Let $f(x) = x^2 + 1$.

    (a) Find the average rate of change of y with respect to x over the interval $[-2, -1]$.
    (b) Find the instantaneous rate of change of y with respect to x at the point $x = -2$.
    (c) Find the instantaneous rate of change of y with respect to x at a general point x_0.
    (d) Sketch the graph of $y = f(x)$ together with the secant and tangent lines whose slopes are given by the results in parts (a) and (b).

3.1.3 Let $f(x) = \dfrac{1}{x - 2}$.

    (a) Find the average rate of change of y with respect to x over the interval $[3, 5]$.
    (b) Find the instantaneous rate of change of y with respect to x at the point $x = 3$.
    (c) Find the instantaneous rate of change of y with respect to x at a general point x.
    (d) Sketch the graph of $y = f(x)$ together with the secant and tangent lines whose slopes are given by the results in parts (a) and (b).

3.1.4 Let $f(x) = \dfrac{1}{x + 1}$.

    (a) Find the average rate of change of y with respect to x over the given interval $[1, 3]$.
    (b) Find the instantaneous rate of change of y with respect to x at the point $x = 1$.
    (c) Find the instantaneous rate of change of y with respect to x at the general point x_0.
    (d) Sketch the graph of $y = f(x)$ together with the secant and tangent lines whose slopes are given by the results in parts (a) and (b).

3.1.5 Let $f(x) = \dfrac{2}{3-x}$.

(a) Find the slope of the tangent to the graph of f at a general point x_0 using the method of Section 3.1

(b) Use the result in part (a) to find the slope of the tangent at $x_0 = 1$.

3.1.6 Let $f(x) = \dfrac{3}{x-1}$.

(a) Find the slope of the tangent to the graph of f at a general point x_0 using the method of Section 3.1.

(b) Use the result in part (a) to find the slope of the tangent at $x_0 = 4$.

3.1.7 Let $f(x) = \dfrac{1}{x^2}$.

(a) Find the slope of the tangent to the graph of f at a general point x_0 using the method of section 3.1.

(b) Use the result in part (a) to find the slope of the tangent at $x_0 = -2$.

3.1.8 Let $f(x) = 3x^2$.

(a) Find the slope of the tangent to the graph of f at a general point x_0 using the method of section 3.1.

(b) Use the result in part (a) to find the slope of the tangent at $x = 3$.

3.1.9 A rock is dropped from a height of 144 feet and falls toward the earth in a straight line. In t seconds, the rock drops a distance of $s = 16t^2$ feet.

(a) What is the average velocity of the rock while it is falling?

(b) Use the method of 3.1 to find the instantaneous velocity of the rock when it hits the ground.

3.1.10 A rock is dropped from a height of 64 feet and falls toward the earth in a straight line. In t seconds, the rock drops a distance of $s = 16t^2$ feet.

(a) What is the average velocity of the rock while it is falling?

(b) Use the method of Section 3.1 to find the instantaneous velocity of the rock when it hits the ground.

3.1.11   A particle moves in a straight line from its initial position so that after t seconds, its distance is given by $s = t^2 + t$ feet from its initial position.

(a)   Find the average velocity of the particle over the interval $[1,3]$ seconds.
(b)   Use the method of Section 3.1 to find the instantaneous velocity of the particle at $t = 1$ second.

3.1.12   A particle moves in a straight line from its initial position so that after t seconds, its distance is given by $s = \dfrac{t}{t+2}$ feet from its initial position.

(a)   Find the average velocity of the particle over the interval $[2,3]$ seconds.
(b)   Use the method of Section 3.1 to find the instantaneous velocity of the particle at $t = 2$ seconds.

3.1.13   Let $f(x) = x^2$.
Use the method of Section 3.1 to show that the slope of the tangent to the graph of f at $x = x_0$ is $2x_0$.

3.1.14   Let $f(x) = ax^2 + b$, where a and b are constants. Use the method of Section 3.1 to show that the slope of the tangent to the graph of f at $x = x_0$ is $2ax_0$.

3.1.15   Let $f(x) = ax^3 + b$, where a and b are constants. Use the method of Section 3.1 to show that the slope of the tangent to the graph of f at $x = x_0$ is $3ax_0^2$.

3.1.16   A particle moves in a straight line from its initial position so that after t seconds, its distance is given by $s = 16t^2$ feet. Use the method of Section 3.1 to show that the instantaneous velocity of the particle at $t = t_0$ seconds is $32t_0$.

3.1.17   A particle moves in a straight line from its initial position so that after t seconds, its distance is given by $s = 4 - 16t^2$ feet. Use the method of Section 3.1 to show that the instantaneous velocity of the particle at $t = t_0$ seconds is $v = -32t_0$.

3.1.18 The figure shows the position versus time curves of four different particles moving on a straight line. For each particle, determine if its instantaneous velocity is increasing or decreasing with time.

(a)

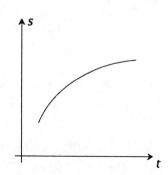

(b)

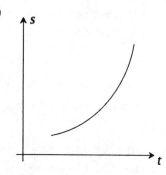

(c)

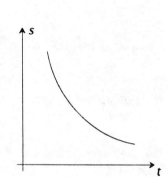

(d)

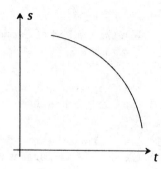

3.1.19 The figure shows the position versus time curve for a certain particle moving along a straight line. Estimate each of the following from the graph.

(a) The average velocity over the interval $0 \leq t \leq 4.6$
(b) The values of t at which the instantaneous velocity is zero
(c) The values of t at which the instantaneous velocity is maximum; minimum
(d) The instantaneous velocity when $t = 5$ seconds

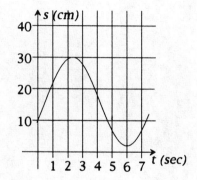

SOLUTIONS

SECTION 3.1

3.1.1 **(a)** $m_{\sec} = \dfrac{f(3) - f(2)}{3 - 2} = \dfrac{\frac{1}{(3)^2} - \frac{1}{(2)^2}}{1} = -\dfrac{5}{36}$.

Thus, on the average, y decreases 5 units per 36 units increase in x over the interval $[2, 3]$.

(b) $m_{\tan} = \lim\limits_{x_1 \to 2} \dfrac{\frac{1}{x_1^2} - \frac{1}{2^2}}{x_1 - 2} == \lim\limits_{x_1 \to 2} \dfrac{\frac{1}{x_1^2} - \frac{1}{4}}{x_1 - 2}$

$= \lim\limits_{x_1 \to 2} \dfrac{4 - x_1^2}{4x_1^2(x_1 - 2)} = \lim\limits_{x_1 \to 2} -\dfrac{(x_1 - 2)(x_1 + 2)}{4x_1^2(x_1 - 2)} = \lim\limits_{x_1 \to 2} -\dfrac{(x_1 + 2)}{4x_1 - 2} = -\dfrac{1}{4}$

Thus, y is decreasing at the point $x = 2$ at a rate of 1 unit per 4 units increase in x.

(c) $m_{\tan} = \lim\limits_{x_1 \to x_0} \dfrac{\frac{1}{(x_1)^2} - \frac{1}{(x_0)^2}}{x_1 - x_0}$

$= \lim\limits_{x_1 \to x_0} \dfrac{x_0^2 - x_1^2}{x_1^2 x_0^2(x_1 - x_0)} = \lim\limits_{x_1 \to x_0} \dfrac{-(x_1 + x_0)}{x_1^2 x_0^2} = -\dfrac{2}{x_0^3}$

Thus the instantaneous rate of change of y with respect to x at $x = x_0$ is $-\dfrac{2}{x_0^3}$.

(d)

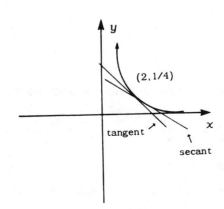

3.1.2 **(a)** $m_{\sec} = \dfrac{f(-1) - f(-2)}{-1 - (-2)} = \dfrac{\left[(-1)^2 + (-1)\right] - \left[(-2)^2 + 1\right]}{1} = -3$.

Thus, on the average, y decreases 3 units per unit increase in x over the interval $[-2, -1]$.

107

(b) $m_{\tan} = \lim\limits_{x_1 \to -2} \dfrac{f(x_1) - f(-2)}{x_1 - (-2)} = \lim\limits_{x_1 \to -2} \dfrac{[x_1^2 + 1] - [(-2)^2 + 1]}{x_1 + 2}$

$\quad\quad = \lim\limits_{x_1 \to -2} \dfrac{x_1^2 - 4}{x_1 + 2} = \lim\limits_{x_1 \to -2} x_1 - 2 = -4,$

Thus, y is decreasing at the point $x = -2$ at a rate of 4 units per unit increase in x.

(c) $m_{\tan} = \lim\limits_{x_1 \to x_0} \dfrac{[x_1^2 + 1] - [x_0^2 + 1]}{x_1 - x_0} = \lim\limits_{x_1 \to x_0} \dfrac{x_1^2 - x_0^2}{x_1 - x_0}$

$\quad\quad = \lim\limits_{x_1 \to x_0} (x_1 + x_0) = 2x_0$

Thus, the instantaneous rate of change of y with respect to x at $x = x_0$ is $2x_0$.

(d)

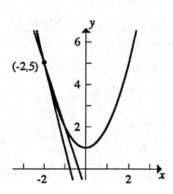

3.1.3 **(a)** $m_{\text{sec}} = \dfrac{f(5) - f(3)}{5 - 3} = \dfrac{\frac{1}{5-2} - \frac{1}{3-2}}{5 - 3} = \dfrac{\frac{1}{3} - 1}{2} = -\dfrac{1}{3}$

Thus, on the average, y decreases 1 unit per 3 units increase in x over the interval $[3, 5]$.

(b) $m_{\tan} = \lim\limits_{x_1 \to 3} \dfrac{\frac{1}{x_1-2} - \frac{1}{3-2}}{x_1 - 3} = \lim\limits_{x_1 \to 3} \dfrac{\frac{1}{x_1-2} - 1}{x_1 - 3} = \lim\limits_{x_1 \to 3} \dfrac{1 - (x_1 - 2)}{(x_1 - 2)(x_1 - 3)}$

$\quad\quad = \lim\limits_{x_1 \to 3} \dfrac{-1}{x_1 - 2} = -1$

Thus, y is decreasing at the point $x = 3$ at a rate of 1 unit per unit increase in x.

(c) $m_{\tan} = \lim\limits_{x_1 \to x_0} \dfrac{\frac{1}{x_1-2} - \frac{1}{x_0-2}}{x_1 - x_0} = \lim\limits_{x_1 \to x_0} \dfrac{(x_0 - 2)(x_1 - 2)}{(x_1 - 2)(x_0 - 2)(x_1 - x_0)}$

$\quad\quad = \lim\limits_{x_1 \to x_0} \dfrac{-1}{(x_1 - 2)(x_0 - 2)} = \dfrac{-1}{(x_0 - 2)^2}$

Thus, the instantaneous rate of change of y with respect to x at $x = x_0$ is $-\dfrac{1}{(x_0-2)^2}$.

(d)

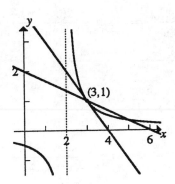

3.1.4 **(a)** $m_{\text{sec}} = \dfrac{f(2) - f(1)}{2 - 1} = \dfrac{\dfrac{1}{2+1} - \dfrac{1}{1+1}}{1} = -\dfrac{1}{6}$

Thus, on the average, y decreases one unit per six units increase in x over the interval $[1, 2]$.

(b) $m_{\text{tan}} = \lim\limits_{x_1 \to 1} \dfrac{f(x_1) - f(1)}{x_1 - 1} = \lim\limits_{x_1 \to 1} \dfrac{\dfrac{1}{(x_1 + 1)} - \dfrac{1}{1+1}}{x_1 - 1} = \lim\limits_{x_1 \to 1} \dfrac{2 - (x_1 + 1)}{2(x_1 + 1)(x_1 - 1)}$

$= \lim\limits_{x_1 \to 1} -\dfrac{(x_1 - 1)}{2(x_1 + 1)(x_1 - 1)} = \lim\limits_{x_1 \to 1} -\dfrac{-1}{2(x_1 + 1)} = -\dfrac{1}{4},$

Thus, y is decreasing at the point $x = 1$ at a rate of 1 unit per 4 units increase in x.

(c) $m_{\text{tan}} = \lim\limits_{x_1 \to x_0} \dfrac{\dfrac{1}{x_1 + 1} - \dfrac{1}{x_0 + 1}}{x_1 - x_0} == \lim\limits_{x_1 \to x_0} \dfrac{(x_0 + 1) - (x_1 + 1)}{(x_1 + 1)(x_0 + 1)(x_1 - x_0)}$

$= \lim\limits_{x_1 \to x_0} \dfrac{-1}{(x_1 + 1)(x_0 + 1)} = -\dfrac{1}{(x_0 + 1)^2}$

Thus, the instantaneous rate of change of y with respect to x at $x = x_0$ is $\dfrac{-1}{(x_0 + 1)^2}$

(d)

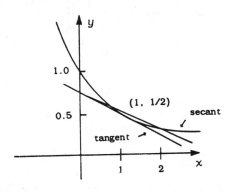

3.1.5 **(a)** $m_{\tan} = \lim\limits_{x_1 \to x_0} \dfrac{\frac{2}{3-x_1} - \frac{2}{3-x_0}}{x_1 - x_0} == \lim\limits_{x_1 \to x_0} \dfrac{2(3 - x_0) - 2(3 - x_1)}{(3 - x_0)(3 - x_1)(x_1 - x_0)}$

$= \lim\limits_{x_1 \to x_0} \dfrac{2}{(3 - x_0)(3 - x_1)} = \dfrac{2}{(3 - x_0)^2}$

(b) $m_{\tan}$, when $x_0 = 1$, is $\dfrac{2}{(3 - 1)} = 1$

3.1.6 **(a)** $m_{\tan} = \lim\limits_{x_1 \to x_0} \dfrac{\frac{3}{x_1-1} - \frac{3}{x_0-1}}{x_1 - x_0} == \lim\limits_{x_1 \to x_0} \dfrac{3(x_0 - 1) - 3(x_1 - 1)}{(x_1 - 1)(x_1 - x_0)(x_0 - 1)}$

$= \lim\limits_{x_1 \to x_0} \dfrac{-3}{(x_1 - 1)(x_0 - 1)} = \dfrac{-3}{(x_0 - 1)^2}$

(b) $m_{\tan}$, when $x_0 = 4$, is $\dfrac{-3}{(4 - 1)^2} = -\dfrac{1}{3}$

3.1.7 **(a)** $m_{\tan} = \lim\limits_{x_1 - x_0} \dfrac{\frac{1}{x_0^2} - \frac{1}{x_0^2}}{x_1 - x_0} = \lim\limits_{x_1 - x_0} \dfrac{x_0^2 - x_1^2}{x_1^2 x_0^2 (x_1 - x_0)}$

$= \lim\limits_{x_1 - x_0} \dfrac{-(x_1 + x_0)}{x_1^2 x_0^2} = -\dfrac{2}{x_0^3}$

(b) $m_{\tan}$ when $x_0 = -2$, is $-\dfrac{2}{(-2)^3} = -\dfrac{1}{4}$

3.1.8 **(a)** $m_{\tan} = \lim\limits_{x_1 \to x_0} \dfrac{3(x_1)^2 - 3(x_0)^2}{x_1 - x_0} = \lim\limits_{x_1 \to x_0} 3(x_1 + x_0) = 3x_0^2$

(b) $m_{\tan}$, when $x_0 - 3$, is $3(3)^2 = 27$.

3.1.9 **(a)** The rock will hit the ground when $16t^2 = 144$, $t = 3$ seconds, so the average velocity is

$$\dfrac{16(3)^2 - 16(0)^2}{3 - 0} = 48 \text{ feet per second}$$

(b) The instantaneous velocity $= \lim\limits_{t_1 \to 3} \dfrac{f(t_1) - f(3)}{t_1 - 3}$

$= \lim\limits_{t_1 \to 3} \dfrac{16t_1^2 - 16(3)^2}{t_1 - 3}$

$= \lim\limits_{t_1 \to 3} 16(t_1 + 3) = 16(6) = 96 \text{ feet per second}$

3.1.10 **(a)** The rock will hit the ground when $16t^2 = 64$, $t = 2$ seconds so the average velocity is
$\dfrac{16(2)^2 - 16(0)^2}{2 - 0} = 32 \text{ feet per second}$.

(b) The instantaneous velocity $= \lim\limits_{t_1 \to 2} \dfrac{f(t_1) - f(2)}{t_1 - 2} = \lim\limits_{t_1 \to 2} \dfrac{16t_1^2 - 16(2)^2}{t_1 - 2}$

$= \lim\limits_{t_1 \to 2} 16(t_1 + 2) = 16(4) = 64 \text{ feet per second}$.

3.1.11   **(a)**   average velocity $= \dfrac{f(3) - f(1)}{3 - 1} = \dfrac{[(3)^2 + (3)] - [(1)^2 + (1)]}{2}$

$$= 5 \text{ feet per second.}$$

(b)   The instantaneous velocity at $t = 1$ second is

$$\lim_{t_1 \to 1} \frac{f(t_1) - f(1)}{t_1 - 1} = \lim_{t_1 \to 1} \frac{(t_1^2 + t_1) - (1^2 + 1)}{t_1 - 1}$$

$$= \lim_{t_1 \to 1} \frac{t_1^2 + t_1 - 2}{t_1 - 1} = \lim_{t_1 \to 1} (t_1 + 2) = 3 \text{ feet per second.}$$

3.1.12   **(a)**   average velocity $= \dfrac{f(3) - f(2)}{3 - 2} = \dfrac{\dfrac{3}{3+2} - \dfrac{2}{2+2}}{1}$

$$= \frac{1}{10} \text{ feet per second.}$$

(b)   The instantaneous velocity at $t = 2$ seconds is

$$\lim_{t_1 \to 2} \frac{f(t_1) - f(2)}{t_1 - 2} = \lim_{t_1 \to 2} \frac{\dfrac{t_1}{t_1 + 2} - \dfrac{1}{2}}{t_1 - 2} = \lim_{t_1 \to 2} \frac{2t_1 - (t_1 + 2)}{2(t_1 + 2)(t_1 - 2)}$$

$$= \lim_{t_1 \to 2} \frac{1}{2(t_1 + 2)} = \frac{1}{8} \text{ feet per second.}$$

3.1.13   **(a)** $m_{\tan} = \lim_{x_1 \to x_0} \dfrac{x_1^2 - x_0^2}{x_1 - x_0} = \lim_{x_1 \to x_0} (x_1 + x_0) = 2x_0$

3.1.14   **(a)** $m_{\tan} = \lim_{x_1 \to x_0} \dfrac{(ax_1^2 + b) - (ax_0^2 + b)}{x_1 - x_0} = \lim_{x_1 \to x_0} \dfrac{a(x_1^2 - x_0^2)}{x_1 - x_0}$

$$= \lim_{x_1 \to x_0} a(x_1 + x_0) = 2ax_0$$

3.1.15   **(a)** $m_{\tan} = \lim_{x_1 \to x_0} \dfrac{(a(x_1)^3 + b) - (a(x_0)^3 + b)}{x_1 - x_0} = \lim_{x_1 \to x_0} \dfrac{a\left[(x_1)^3 - (x_0)^3\right]}{x_1 - x_0}$

$$= \lim_{x_1 \to x_0} a(x_1^2 + x_1 x_0 + x_0^2) = 3ax_0^2$$

3.1.16   Instantaneous velocity at $t = t_0$ seconds is

$$\lim_{t_1 \to t_0} \frac{16t_1^2 - 16t_0^2}{t_1 - t_0} = \lim_{t_1 \to t_0} 16(t_1 + t_0)$$

$$= 32t_0 \text{ feet per second}$$

3.1.17   Instantaneous velocity at $t = t_0$ seconds is

$$\lim_{t_1 \to t_0} \frac{4 - 16t_1^2 - (4 - 16t_0^2)}{t_1 - t_0} = \lim_{t_1 \to t_0} \frac{-16(t_1^2 - t_0^2)}{t_1 - t_0}$$

$$= \lim_{t_1 \to t_0} -16(t_1 + t_0) = 32t_0 \text{ feet per second}$$

3.1.18   **(a)** decreasing   **(b)** increasing   **(c)** increasing   **(d)** decreasing

3.1.19   **(a)** $\dfrac{f(4.6) - f(0)}{4.6 - 0} = \dfrac{10 - 10}{4.6} = 0$

(b)   The instantaneous velocity is zero when the slope of the tangent line to the curve is zero. The values of t at which the tangent line is horizontal are $t \approx 2.5, 6$.

(c)   The velocity is a maximum when $t \approx 2.25$. The velocity is a minimum when $t \approx 6$.

(d)   When $t = 5$, the instantaneous velocity ≈ -1.

SECTION 3.2

3.2.1 Use the definition of the derivative to calculate $f'(x)$ if $f(x) = 3x^2 - x$ and find the equation of the tangent to the graph of f at $x = 1$.

3.2.2 Use the definition of the derivative to calculate $f'(x)$ if $f(x) = 2x^2 - x + 1$.

3.2.3 Use the definition of the derivative to calculate $f'(x)$ if $f(x) = 2x^3 + 1$ and find the equation of the tangent line and the normal line to the graph of f at $x = 1$.

3.2.4 Use the definition of the derivative to calculate $f'(x)$ if $f(x) = x^3 - 3x$ and find the equation of the tangent line and the normal line to the graph of f at $x = 2$.

3.2.5 Use the definition of the derivative to calculate $f'(x)$ if $f(x) = \sqrt{2x}$ and find the equation of the tangent line and the normal line to the graph of f at $x = 2$.

3.2.6 Let $y = \sqrt{3x + 1}$. Use the definition of the derivative to find $\dfrac{dy}{dx}$.

3.2.7 Let $y = \frac{1}{x+2}$. Use the definition of the derivative to find $\dfrac{dy}{dx}$.

3.2.8 Use the definition of the derivative to calculate $f'(x)$ if $f(x) = \dfrac{2}{3 - x}$.

3.2.9 Given that $f(0) = 4$ and $f'(0) = -1$, find an equation for the tangent line to the graph of $y = f(x)$ at the point where $x = 0$.

3.2.10 Given that $f(2) = -1$ and $f'(2) = 5$, find an equation for the tangent line to the graph of $y = f(x)$ at the point where $x = 2$.

3.2.11 Use the definition of the derivative to calculate $f'(x)$ if $f(x) = \dfrac{1}{\sqrt{2x}}$ and find the equation of the tangent line and the normal the line to the graph of f at $x = 2$.

3.2.12 The volume of a sphere is given by $\dfrac{4}{3}\pi r^3$ where r is the radius of the sphere. Use the method of Section 3.2 to find the instantaneous rate of change of V with respect to r when $r = 4$.

3.2.13 The surface area of a sphere is given by $S = 4\pi r^2$ where r is the radius of the sphere. Use the method of Section 3.2 to find the instantaneous rate of change of S with respect to r when $r = 4$.

3.2.14 The volume of a sphere is given by $V = \dfrac{\pi}{6}D^3$ where D is the diameter of the sphere. Use the method of Section 3.2 to find the instantaneous rate of change of V with respect to D when $D = 2$.

3.2.15 Show that $f(x) = \begin{cases} x^2 - 5 & x \le 1 \\ x - 5 & x > 1 \end{cases}$ is continuous but not differentiable at $x = 1$. Sketch the graph of f.

3.2.16 Sketch the graph of the derivative of the function whose graph is shown.

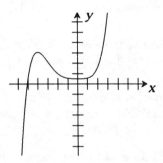

3.2.17 It has been observed that some large colonies of bacteria tend to grow at a rate proportional to the number of bacteria present. The graph shows bacteria count P (in thousands) versus time t (in seconds)

(a) Estimate P and $\dfrac{dP}{dt}$ when $t = 2$ sec

(b) This model for bacterial growth can be expressed as $\dfrac{dP}{dt} = kP$ where k is the constant of proportionality. Use the results in part (a) to estimate the value of k.

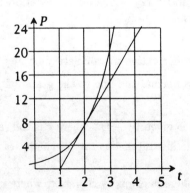

SOLUTIONS

SECTION 3.2

3.2.1 $f'(x) = \lim_{h \to 0} \dfrac{[3(x+h)^2 - (x+h)] - (3x^2 - x)}{h}$

$\qquad = \lim_{h \to 0} \dfrac{6xh + 3h^2 - h}{h} = \lim_{h \to 0} (6x + 3h - 1) = 6x - 1$

so the slope of the tangent at $(1,2)$ is $f'(1) = 6(1) - 1 = 5$, thus, the equation of the tangent to f at $(1,2)$ is $y - 2 = 5(x - 1)$ or $y = 5x - 3$.

3.2.2 $f'(x) = \lim_{h \to 0} \dfrac{[2(x+h)^2 - (x+h) + 1] - (2x^2 - x + 1)}{h}$

$\qquad = \lim_{h \to 0} \dfrac{4xh + 2h^2 - h}{h} = \lim_{h \to 0} (4x + 2h - 1) = 4x - 1.$

3.2.3 $f'(x) = \lim_{h \to 0} \dfrac{[2(x+h)^3 + 1] - (2x^3 + 1)}{h})$

$\qquad = \lim_{h \to 0} \dfrac{6x^2h + 6xh^2 + 2h^3}{h} = \lim_{h \to 0} (6x^2 + 6xh + 2h^2) = 6x^2,$

so the slope of the tangent at $(1,3)$ is $f'(1) = 6$, thus, the equation of the tangent to the graph of f at $(1,3)$ is $y - 3 = 6(x - 1)$ or $y = 6x - 3$; the slope of the normal at $(1,3)$ is $-\dfrac{1}{6}$ so the equation of the normal to the graph of f at $(1,3)$ is $y - 3 = -\dfrac{1}{6}(x - 1)$ or $y = -\dfrac{1}{6}x + \dfrac{10}{6}$.

3.2.4 $f'(x) = \lim_{h \to 0} \dfrac{[(x+h)^3 - 3(x+h)] - (x^3 - 3x)}{h}$

$\qquad = \lim_{h \to 0} \dfrac{3x^2h + 3xh^2 + h^3 - 3h}{h} = \lim_{h \to 0} (3x^2 + 3xh + h^2 - 3)$

$\qquad = 3x^2 - 3,$

so the slope of the tangent at $(2,2)$ is $f'(2) = 3(2)^2 - 3 = 9$, thus, the equation of the tangent to the graph of f at $(2,2)$ is $y - 2 = 9(x - 2)$ or $y = 9x - 16$; the slope of the normal at $(2,2)$ is $-\dfrac{1}{9}$ so the equation of the normal to the graph of f at $(2,2)$ is $y - 2 = -\dfrac{1}{9}(x - 2)$ or $y = -\dfrac{1}{9}x + \dfrac{20}{9}.$

3.2.5 $f'(x) = \lim\limits_{h \to 0} \dfrac{\sqrt{2(x+h)} - \sqrt{2x}}{h} = \lim\limits_{h \to 0} \left(\dfrac{\sqrt{2(x+h)} - \sqrt{2x}}{h} \right) \left(\dfrac{\sqrt{2(x+h)} + \sqrt{2x}}{\sqrt{2(x+h)} + \sqrt{2x}} \right)$

$\qquad = \lim\limits_{h \to 0} \dfrac{2(x+h) - 2x}{h\left(\sqrt{2(x+h)} + \sqrt{2x} \right)} = \lim\limits_{h \to 0} \dfrac{2h}{h\left(\sqrt{2(x+h)} + \sqrt{2x} \right)}$

$\qquad = \lim\limits_{h \to 0} \dfrac{2}{\sqrt{2(x+h)} + \sqrt{2x}} = \dfrac{1}{\sqrt{2x}},$

so the slope of the tangent at $(2,2)$ is $f'(2) = \dfrac{1}{\sqrt{2(2)}} = \dfrac{1}{2}$, thus, the equation of the tangent to the graph of f at $(2,2)$ is $y - 2 = \dfrac{1}{2}(x - 2)$ or $y = \dfrac{1}{2}x + 1$; the slope of the normal at $(2,2)$ is $-\dfrac{1}{(1/2)} = -2$ so the equation of the normal to the graph of f at $(2,2)$ is $y - 2 = -2(x - 2)$ or $y = -2x + 6$.

3.2.6 $\dfrac{dy}{dx} = \lim\limits_{h \to 0} \dfrac{\sqrt{3(x+h)+1} - \sqrt{3x+1}}{h}$

$\qquad = \lim\limits_{h \to 0} \left(\dfrac{\sqrt{3(x+h)+1} - \sqrt{3x+1}}{h} \right) \left(\dfrac{\sqrt{3(x+h)+1} + \sqrt{3x+1}}{\sqrt{3(x+h)+1} + \sqrt{3x+1}} \right)$

$\qquad = \lim\limits_{h \to 0} \dfrac{[3(x+h)+1] - (3x+1)}{h(\sqrt{3(x+h)+1} + \sqrt{3x+1})}$

$\qquad = \lim\limits_{h \to 0} \dfrac{3h}{h(\sqrt{3(x+h)+1} + \sqrt{3x+1})}$

$\qquad = \lim\limits_{h \to 0} \dfrac{3}{\sqrt{3(x+h)+1} + \sqrt{3x+1}} = \dfrac{3}{2\sqrt{3x+1}}.$

3.2.7 $\dfrac{dy}{dx} = \lim\limits_{h \to 0} \dfrac{\dfrac{1}{(x+h)+2} - \dfrac{1}{x+2}}{h}$

$\qquad = \lim\limits_{h \to 0} -\dfrac{h}{h[(x+h)+2][x+2]}$

$\qquad = \lim\limits_{h \to 0} -\dfrac{1}{[(x+h)+2][x+2]} = -\dfrac{1}{(x+1)^2}.$

3.2.8 $f'(x) = \lim\limits_{h \to 0} \dfrac{\dfrac{2}{3-(x+h)} - \dfrac{2}{3-x}}{h} = \lim\limits_{h \to 0} \dfrac{2h}{h(3-x)(3-x-h)}$

$\qquad = \lim\limits_{h \to 0} \dfrac{2}{(3-x)(3-x-h)} = \dfrac{2}{(3-x)^2}.$

3.2.9 The slope of the tangent line at $(0,4)$ is -1, thus the equation of the tangent to the graph of f at $(0,4)$ is $y - 4 = -1(x - 0)$ or $y = -x + 4$.

3.2.10 The slope of the tangent line at $(2, -1)$ is 5, thus the equation of the tangent to the graph of f at $(2, -1)$ is $y - (-1) = 5(x - 2)$ or $y = 5x - 11$.

3.2.11 $f'(x) = \lim_{h \to 0} \dfrac{\dfrac{1}{\sqrt{2(x+h)}} - \dfrac{1}{\sqrt{2x}}}{h} = \lim_{h \to 0} \left(\dfrac{\sqrt{2x} - \sqrt{2(x+h)}}{h\sqrt{2x}\sqrt{2(x+h)}} \right) \left(\dfrac{\sqrt{2x} + \sqrt{2(x+h)}}{\sqrt{2x} + \sqrt{2(x+h)}} \right)$

$= \lim_{h \to 0} \dfrac{-2h}{h\sqrt{2x}\sqrt{2(x+h)}(\sqrt{2x} + \sqrt{2(x+h)})}$

$= \lim_{h \to 0} \dfrac{-2}{\sqrt{2x}\sqrt{2(x+h)}\left(\sqrt{2x} + \sqrt{2(x+h)}\right)} = -\dfrac{1}{(2x)^{3/2}},$

so, the slope of the tangent at $(2, 1/2)$ is $f'(2) = -\dfrac{1}{(4)^{3/2}} = -\dfrac{1}{8}$, thus, the equation of the tangent to the graph of f at $(2, 1/2)$ is $y - \dfrac{1}{2} = -\dfrac{1}{8}(x - 2)$ or $y = -\dfrac{1}{8}x + \dfrac{3}{4}$; the slope of the normal at $(2, 1/2)$ is $-\dfrac{1}{-\frac{1}{8}} = 8$ so the equation of the normal to the graph of f at $(2, 1/2)$ is $y - \dfrac{1}{2} = 8(x - 2)$ or $y = 8x - \dfrac{31}{2}$.

3.2.12 $f'(r) = \lim_{h \to 0} \dfrac{\dfrac{4}{3}\pi(r+h)^3 - \dfrac{4}{3}\pi r^3}{h} = \lim_{h \to 0} \dfrac{\dfrac{4\pi}{3}\left(3r^2 h + 3rh^2 + h^3\right)}{h}$

$= \lim_{h \to 0} \dfrac{4\pi\left(3r^2 + 3rh + h^2\right)}{3} = 4\pi r^2,$

so the instantaneous rate of change of V with respect to r is $f'(4) = 4\pi(4)^2 = 64\pi$.

3.2.13 $f'(r) = \lim_{h \to 0} \dfrac{4\pi(r+h)^2 - 4\pi r^2}{h} = \lim_{h \to 0} \dfrac{4\pi(2rh + h^2)}{h}$

$= \lim_{h \to 0} 4\pi(2r + h) = 8\pi r,$

so the instantaneous rate of change of S with respect to r at $r = 4$ is $f'(4) = 8\pi(4) = 32\pi$.

3.2.14 $f'(D) = \lim_{h \to 0} \dfrac{\dfrac{\pi}{6}(D+h)^3 - \dfrac{\pi}{6}D^3}{h} = \lim_{h \to 0} \dfrac{\dfrac{\pi}{6}\left(3D^2 h + 3Dh^2 + h^3\right)}{h}$

$= \lim_{h \to 0} \dfrac{\pi\left(3D^2 + 3Dh + h^2\right)}{6} = \dfrac{\pi}{2}D^2,$

thus, the instantaneous rate at which V changes with respect to D when $D = 2$ is

$f'(2) = \dfrac{\pi}{2}(2)^2 = 2\pi.$

3.2.15 $\lim\limits_{x \to 1^-} f(x) = \lim\limits_{x \to 1^+} f(x) = f(1) = -4$ so f is continuous at $x = 1$

$$\lim_{h \to 0^-} \frac{f(1+h) - f(1)}{h} = \lim_{h \to 0^-} \frac{[(1+h)^2 - 5] + 4}{h}$$

$$= \lim_{h \to 0^-} \frac{2h + h^2}{h} = \lim_{h \to 0^-} (2 + h) = 2$$

$$\lim_{h \to 0^+} \frac{f(1+h) - f(1)}{h} \quad \lim_{h \to 0^+} \frac{[(1+h) - 5] + 4}{h}$$

$$= \lim_{h \to 0^+} \frac{h}{h} = 1$$

so $f'(1)$ does not exist.

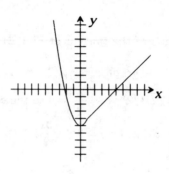

3.2.16

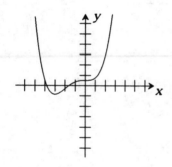

3.2.17 P is approximately 7.4 thousand

$\dfrac{dP}{dt}$ is approximately 7.4

$\dfrac{dP}{dt} = kP$, $7.4 = k7.4$ so $k = 1$.

SECTION 3.3

3.3.1 Find $\dfrac{dy}{dx}$ if $y = \dfrac{3x^3 + 5x^2 + \sqrt{x}}{x}$.

3.3.2 Find $\dfrac{dy}{dx}$ if $y = \dfrac{x^2 + 3x}{7 - 2x}$.

3.3.3 Find $f''(2)$ if $f(x) = \dfrac{-8}{x^2} + \dfrac{1}{5}x^5$.

3.3.4 Find $\dfrac{dy}{dx}$ if $y = -2\left(x^2 - 5x\right)\left(3 + x^7\right)$.

3.3.5 Find $f'(s)$ if $f(s) = (3s^2 + 4)(s^2 - 9s)$.

3.3.6 Find $f'(x)$ if $f(x) = \dfrac{2x + 1}{x^2 + 3x}$.

3.3.7 If $f(3) = 2, f'(3) = -1, g(3) = 3, g'(3) = 0$, find $F'(3)$

(a) $F(x) = 2f(x) - g(x)$
(b) $F(x) = \frac{1}{2}f(x)g(x)$
(c) $F(x) = \frac{1}{3}\dfrac{f(x)}{g(x)}$

3.3.8 Find $\dfrac{d^2y}{dt^2}$ if $y = -\dfrac{1}{t} - \dfrac{5}{t^2}$.

3.3.9 Find $f'(u)$ is $f(u) = \dfrac{u^2 - 5}{3u^2 - 1}$.

3.3.10 Find $\dfrac{dy}{dx}$ if $y = \left(x^2 - 2\right)\left(x^3 + 5x\right)$.

3.3.11 Find $\dfrac{dv}{dh}$ if $v = n\left(uh^2 - \dfrac{1}{3}h^3\right)$, a is a constant.

3.3.12 Find $f'(x)$ if $f(x) = \left(x^2 + 1\right)\left(x^3 - 2x^2 + x\right)$.

3.3.13 Find equations for the tangents and normals to the graph of $y = 4 - 3x - x^2$ at those points where the curve intersects the x-axis.

3.3.14 Find equations for the tangents and normals to the graph of $y = 6 - x - x^2$ at the points where the curve intersects the x-axis.

3.3.15 Find the points on the graph of $y = 2x^3 - 3x^2 - 12x + 20$ at which the tangent is parallel to the x-axis.

3.3.16 Show that the parabola $y = -x^2$ and the line $x - 4y - 18 = 0$ intersect at right angles at one of their points of intersection.

3.3.17 Find the equation of the tangents and normals to the graph of $y = \dfrac{x + 1}{x - 1}$ at $x = 2$.

3.3.18   Find the equation of the tangent and normal to the graph of $y = 10 - 3x - x^2$ at the point where the curve intersects the x-axis.

3.3.19   Show that the parabola $y = x^2$ and the line $x + 2y - 3 = 0$ intersect at right angles at one of their points of intersection.

SOLUTIONS

SECTION 3.3

3.3.1 $y = 3x^2 + 5x + x^{-1/2}$ so $\dfrac{dy}{dx} = 6x + 5 - \dfrac{1}{2}x^{-3/2}$.

3.3.2 $\dfrac{dy}{dx} = \dfrac{(7 - 2x)\dfrac{d}{dx}\left[x^2 + 3x\right] - \left(x^2 + 3x\right)\dfrac{d}{dx}[7 - 2x]}{(7 - 2x)^2}$

$\qquad = \dfrac{(7 - 2x)(2x + 3) - \left(x^2 + 3x\right)(-2)}{(7 - 2x)^2} = \dfrac{21 + 14x - 2x^2}{(7 - 2x)^2}$.

3.3.3 $f(x) = -8x^{-2} + \dfrac{1}{5}x^5$ so $f'(x) = 16x^{-3} + x^4$ and $f''(x) = -48x^{-4} + 4x^3$, then

$\qquad f''(2) = -48(2)^{-4} + 4(2)^3 = 29$.

3.3.4 $\dfrac{dy}{dx} = -2\left[\left(x^2 - 5x\right)\dfrac{d}{dx}\left[3 + x^7\right] + \left(3 + x^7\right)\dfrac{d}{dx}\left[x^2 - 5x\right]\right]$

$\qquad = -2\left[\left(x^2 - 5x\right)\left(7x^6\right) + \left(3 + x^7\right)(2x - 5)\right] = -18x^8 + 80x^7 - 12x + 30$.

3.3.5 $f'(s) = (3s^2 + 4)\dfrac{d}{ds}[s^2 - 9s] + (s^2 - 9s)\dfrac{d}{ds}[3s^2 + 4]$

$\qquad = (3s^2 + 4)(2s - 9) + (s^2 - 9s)(6s) = 12s^3 - 27s^2 - 46s - 36$.

3.3.6 $f'(x) = \dfrac{(x^2 + 3x)\dfrac{d}{dx}[2x + 1] - (2x + 1)\dfrac{d}{dx}\left[x^2 + 3x\right]}{(x^2 + 3x)^2}$

$\qquad = \dfrac{\left(x^2 + 3x\right)(2) - (2x + 1)(2x + 3)}{\left(x^2 + 3x\right)^2} = \dfrac{-2x^2 - 2x - 3}{\left(x^2 + 3x\right)^2}$.

3.3.7 **(a)** $F'(x) = 2f'(x) - g'(x)$;

$\qquad\qquad F'(3) = 2f'(3) - g'(3) = 2(-1) - 0 = -2$

$\qquad$ **(b)** $F'(x) = \dfrac{1}{2}\left[f(x)g'(x) + g(x)f'(x)\right]$;

$\qquad\qquad F'(3) = \dfrac{1}{2}\left[f(3)g'(3) + g(3)f'(3)\right] = \dfrac{1}{2}\left[(2)(0) + (3)(-1)\right] = -\dfrac{3}{2}$

$\qquad$ **(c)** $F'(x) = \dfrac{1}{3}\left[\dfrac{g(x)f'(x) - f(x)g'(x)}{\lfloor g(x)\rfloor^2}\right]$;

$\qquad$ **(d)** $F'(x) = \dfrac{1}{3}\left[\dfrac{g(3)f'(3) - f(3)g'(3)}{\lfloor g(3)\rfloor^2}\right] = \dfrac{1}{3}\left[\dfrac{(3)(-1) - (2)(0)}{(3)^2}\right] = -\dfrac{1}{9}$

3.3.8 $y = -t^{-1} - 5t^{-2}$ so $\dfrac{dy}{dt} = t^{-2} + 10t^{-3}$ and $\dfrac{d^2y}{dt^2} = -2t^{-3} - 30t^{-4}$.

3.3.9 $f'(u) = \dfrac{(3u^2 - 1)\dfrac{d}{du}[u^2 - 5] - (u^2 - 5)\dfrac{d}{du}[3u^2 - 1]}{(3u^2 - 1)^2}$

$\qquad = \dfrac{(3u^2 - 1)(2u) - (u^2 - 5)(6u)}{(3u^2 - 1)^2} = \dfrac{28u}{(3u^2 - 1)^2}$

3.3.10 $\dfrac{dy}{dx} = (x^2 - 2)\dfrac{d}{dx}[x^3 + 5x] + (x^3 + 5x)\dfrac{d}{dx}[x^2 - 2]$

$\qquad = (x^2 - 2)(3x^2 + 5) + (x^3 + 5x)(2x) = 5x^4 + 9x^2 - 10.$

3.3.11 $\dfrac{dv}{dh} = \pi(2ah - h^2).$

3.3.12 $f'(x) = (x^2 + 1)\dfrac{d}{dx}[x^3 - 2x^2 + x] + (x^3 - 2x^2 + x)\dfrac{d}{dx}[x^2 + 1]$

$\qquad = (x^2 + 1)(3x^2 - 4x + 1) + (x^3 - 2x^2 + x)(2x)$

$\qquad = 5x^4 - 8x^3 + 6x^2 - 4x + 1.$

3.3.13 The curve intersects the x−axis when $4 - 3x - x^2 = 0$; $x = -4$ or $x = 1$. $f'(x) = -3 - 2x$ so the slope of the tangent at $x = -4$ is $f'(-4) = -3 - 2(-4) = 5$ and at $x = 1$ is $f'(1) = -3 - 2(1) = -5$. The equation of the tangent to y at $(-4, 0)$ is $y - 0 = 5(x + 4)$ or $y = 5x + 20$ and at $(1, 0)$ is $y - 0 = -5(x - 1)$ or $y = -5x + 5$; the slope of the normal at $x = -4$ is $-\dfrac{1}{5}$ and at $x = 1$ is $-\dfrac{1}{-5} = \dfrac{1}{5}$ so the equation of the normal at $(-4, 0)$ is $y - 0 = -\dfrac{1}{5}(x + 4)$ or $y = -\dfrac{1}{5}x - \dfrac{4}{5}$ and at $(1, 0)$ is $y - 0 = \dfrac{1}{5}(x - 1)$ or $y = \dfrac{1}{5}x - \dfrac{1}{5}.$

3.3.14 The curve intersects the x-axis when $6 - x - x^2 = 0$; $x = -3$ or $x = 2$. $f'(x) = -1 - 2x$ so the slope of the tangent at $x = -3$ is $f'(-3) = -1 - 2(-3) = 5$ and at $x = 2$ is $f'(2) = -1 - 2(2) = -5$. The equation of the tangent to y at $(-3, 0)$ is $y - 0 = 5(x + 3)$ or $y = 5x + 15$ and at $(2, 0)$ is $y - 0 = -5(x - 2)$ or $y = -5x + 10$; the slope of the normal at $x = -3$ is $-\dfrac{1}{5}$ and at $x = 2$ is $-\dfrac{1}{-5} = \dfrac{1}{5}$ so the equation of the normal at $(-3, 0)$ is $y - 0 = -\dfrac{1}{5}(x + 3)$ or $y = -\dfrac{1}{5}x - \dfrac{3}{5}$ and at $(2, 0)$ is $y - 0 = \dfrac{1}{5}(x - 2)$ or $y = \dfrac{1}{5}x - \dfrac{2}{5}.$

3.3.15 The tangent is parallel to the x−axis when $f'(x) = 6x^2 - 6x - 12 = 0$, thus,

$6(x - 2)(x + 1) = 0$, $x = 2$ or $x = -1$ so the points on the graph of y are $(-1, 27)$ and $(2, 0)$.

3.3.16 Substitute $y = -x^2$ into $x - 4y - 18 = 0$ to get $4x^2 + x - 18 = 0$. Solve for x to get $x = 2$ or $x = -\dfrac{9}{4}$. Place $x - 4y - 18 = 0$ into the slope intercept form to get $y = \dfrac{1}{4}x - \dfrac{9}{2}$, thus the slope of the line is $m_1 = 1/4$. Differentiate $y = -x^2$ to get $\dfrac{dy}{dx} = -2x$ so that the slope of a line drawn tangent to $y = -x^2$ at $x = x_0$ is $m_2 = -2x_0$. When $x = 2$, $m_1 = 1/4$ and $m_2 = -4$, so the graphs intersect at right angles since $m_1 m_2 = (1/4)(-4) = -1$.

3.3.17   The slope of the tangent to $y = \dfrac{x+1}{x-1}$ is $m_1 = \dfrac{d}{dx}\left[\dfrac{x+1}{x-1}\right]_{x=2}$;

$\dfrac{d}{dx}\left[\dfrac{x+1}{x-1}\right] = \dfrac{(x-1)-(x+1)}{(x-1)^2} = \dfrac{-2}{(x-1)^2}$ so $m_1 = \dfrac{d}{dx}\left[\dfrac{x+1}{x-1}\right]_{x=2} = -2$. When $x = 2$,

$y = 3$ so the equation of the tangent line is $y - 3 = -2(x-2)$ or $y = -2x + 7$. The slope

of the normal to $y = \dfrac{x+1}{x-1}$ is $m_2 = -\dfrac{1}{m_1} = -\dfrac{1}{-2} = \dfrac{1}{2}$, so the equation of the normal is

$y - 3 = \dfrac{1}{2}(x-2)$ or $y = \dfrac{1}{2}x + 2$.

3.3.18   The curve intersects the x axis when $10 - 3x - x^2 = 0$; $x = -5$ or $x = 2$. $f'(x) = -3 - 2x$ so the slope of the tangent at $x = -5$ is $f'(-5) = 7$ and at $x = 2$, is $f'(2) = -7$. The equation of the tangent at $x = -5$ is $y = 7x + 35$ and at $x = 2$ is $y = -7x + 14$. The slope of the normal at $x = -5$ is $\dfrac{-1}{f'(-5)} = -\dfrac{1}{7}$ and the slope of the normal at $x = 2$ is $\dfrac{-1}{f'(2)} = \dfrac{1}{7}$. The equation of the normal at $x = -5$ is $y = -\dfrac{1}{7}x - \dfrac{5}{7}$ and at $x = 2$ is $y = \dfrac{1}{7}x - \dfrac{2}{7}$.

3.3.19   Substitute $y = x^2$ into $x + 2y - 3 = 0$ to get $2x^2 + x - 3 = 0$. Solve for x to get $x = 1$ or $x = -\dfrac{3}{2}$. Place $x + 2y - 3 = 0$ into the slope intercept form to get $y = -\dfrac{1}{2}x + \dfrac{3}{2}$, thus, the slope of the line is $m_1 = -\dfrac{1}{2}$.

Differentiate $y = x^2$ to get $\dfrac{dy}{dx} = 2x$ so that the slope of a line drawn tangent to $y = x^2$ at $x = x_0$ is $m_2 = 2x_0$. When $x_0 = 1$, $m_1 = -\dfrac{1}{2}$ and $m_2 = 2$, so the graphs intersect at right angles since $m_1 m_2 = \left(-\dfrac{1}{2}\right)(2) = -1$.

SECTION 3.4

3.4.1 Find $f'(x)$ if $f(x) = x \tan x$.

3.4.2 Find $f''(x)$ if $f(x) = x \sin x$.

3.4.3 Find $\dfrac{dy}{dx}$ if $y = \dfrac{\sin x}{x^2}$.

3.4.4 Find $\dfrac{dy}{dx}$ if $y = \sec x \tan x$.

3.4.5 Find $f'(x)$ if $f(x) = \dfrac{\cot x}{1 + \csc x}$.

3.4.6 Find $f'(x)$ if $f(x) = (5x^2 + 7)\cos x$.

3.4.7 Differentiate $y = \dfrac{\csc x}{\sqrt{x}}$.

3.4.8 Find $\dfrac{dy}{dx}$ if $y = \dfrac{\cos x}{1 - \sin x}$.

3.4.9 Find $\dfrac{dy}{dx}$ if $y = (x^3 + 7x)\tan x$.

3.4.10 Find $y''(x)$ if $y = 12\sin x + 5\cos x + \dfrac{x^4}{4}$.

3.4.11 Find $f'(\theta)$ if $f(\theta) = \dfrac{1}{1 - 2\cos\theta}$.

3.4.12 Find $\dfrac{dy}{dx}$ if $y = 2x\sin x - 2\cos x + x^2 \cos x$.

3.4.13 Find $f'(\theta)$ if $f(\theta) = \dfrac{1 + \sin\theta}{1 - \sin\theta}$.

3.4.14 Find $\dfrac{dy}{dt}$ if $y = \dfrac{1 + \tan t}{1 - \tan t}$.

3.4.15 Show by use of a trigonometric identity that $\dfrac{d}{dx}[\tan x - x] = \tan^2 x$.

3.4.16 Show by use of a trigonometric identity that

$$\frac{d}{dx}[x + \cot x] = -\cot^2 x$$

3.4.17 A 12 foot long ladder leans against a wall at an angle θ with the horizontal as shown in the figure. The top of the ladder is x feet above the ground. If the bottom of the ladder is pushed toward the wall, find the rate at which x changes with θ when $\theta = 60°$. Express the answer in units of feet/degree.

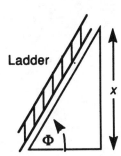

125

3.4.18 An airplane is flying on a horizontal path at a height of 4500 ft, as shown in the figure. At what rate is the distance s between the airplane and the fixed point P changing with θ when $\theta = 30°$. Express the answer in units of feet/degree.

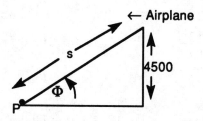

SOLUTIONS

SECTION 3.4

3.4.1 $f'(x) = x\left(\sec^2 x\right) + \tan x(1) = x\sec^2 x + \tan x.$

3.4.2 $f'(x) = x(\cos x) + \sin x(1) = x\cos x + \sin x;$
$f''(x) = x(-\sin x) + \cos x(1) + \cos x = 2\cos x - x\sin x.$

3.4.3 $\dfrac{dy}{dx} = \dfrac{x^2(\cos x) - \sin x(2x)}{x^4} = \dfrac{x^2\cos x - 2x\sin x}{x^4} = \dfrac{x\cos x - 2\sin x}{x^3}.$

3.4.4 $\dfrac{dy}{dx} = \sec x(\sec^2 x) + \tan x(\sec x\tan x) = \sec^3 x + \sec x\tan^2 x.$

3.4.5 $f'(x) = \dfrac{(1 + \csc x)(-\csc^2 x) - \cot x(-csc x\cot x)}{(1 + \csc x)^2}$

$= \dfrac{-\csc^2 x - \csc^3 x + \csc x\cot^2 x}{(1 + \csc x)^2} = \dfrac{\csc x(-\csc x - \csc^2 x + \cot^2 x)}{(1 + \csc x)^2}$

$= \dfrac{\csc x(-\csc x - 1)}{(1 + \csc x)^2} = \dfrac{-\csc x}{1 + \csc x}.$

3.4.6 $f'(x) = (5x^2 + 7)(-\sin x) + \cos x(10x) = -(5x^2 + 7)\sin x + 10x\cos x$

3.4.7 $\dfrac{dy}{dx} = \dfrac{x^{1/2}(-\csc x\cot x) - \csc x\left(\dfrac{1}{2}x^{-1/2}\right)}{x}$

$= \dfrac{-x^{1/2}\csc x\cot x - \dfrac{1}{2}x^{-1/2}\csc x}{x} = \dfrac{-2x\csc x\cot x - \csc x}{2x^{3/2}}.$

3.4.8 $\dfrac{dy}{dx} = \dfrac{(1 - \sin x)(-\sin x) - \cos x(-\cos x)}{(1 - \sin x)^2} = \dfrac{-\sin x + \sin^2 x + \cos^2 x}{(1 - \sin x)^2}$

$= \dfrac{-\sin x + 1}{(1 - \sin x)^2} = \dfrac{1}{1 - \sin x}.$

3.4.9 $\dfrac{dy}{dx} = \left(x^3 + 7x\right)\left(\sec^2 x\right) + \tan x\left(3x^2 + 7\right)$

$= \left(x^3 + 7x\right)\sec^2 x + \left(3x^2 + 7\right)\tan x.$

3.4.10 $y' = 12\cos x - 5\sin x + x^3,\ y'' = -12\sin x - 5\cos x + 3x^2$

3.4.11 $f'(\theta) = -\dfrac{2\sin\theta}{(1 - 2\cos\theta)^2}$ (reciprocal rule).

3.4.12 $\dfrac{dy}{dx} = 2x(\cos x) + 2\sin x(1) - 2(-\sin x) + x^2(-\sin x) + \cos x(2x)$

$\qquad = 4x\cos x + 4\sin x - x^2\sin x.$

3.4.13 $f'(\theta) = \dfrac{(1 - \sin\theta)(\cos\theta) - (1 + \sin\theta)(-\cos\theta)}{(1 - \sin\theta)^2} = \dfrac{2\cos\theta}{(1 - \sin\theta)^2}.$

3.4.14 $\dfrac{dy}{dt} = \dfrac{(1 - \tan t)\,(\sec^2 t) - (1 + \tan t)\,(-\sec^2 t)}{(1 - \tan t)^2} = \dfrac{2\sec^2 t}{(1 - \tan t)^2}.$

3.4.15 $\dfrac{d}{dx}[\tan x - x] = \sec^2 x - 1 = \tan^2 x.$

3.4.16 $\dfrac{d}{dx}[x + \cot x] = 1 - \csc^2 x = -\cot^2 x$

3.4.17
$$\sin\theta = \frac{x}{12}$$
$$x = 12\sin\theta$$
$$\frac{dx}{d\theta} = 12\cos\theta$$

$\theta = 60°\qquad \left.\dfrac{dx}{d\theta}\right|_{\theta=60°} = 12\cos 60° = 12\left(\dfrac{1}{2}\right) = 6$ ft/degree

3.4.18
$$\csc\theta = \frac{s}{4500}$$
$$s = 4500\csc\theta$$
$$\frac{ds}{d\theta} = -4500\csc\theta\cot\theta$$

$\theta = 30°\qquad \left.\dfrac{ds}{d\theta}\right|_{\theta=30°} = -45000\csc 30°\cot 30° = -4500(2)(\sqrt{3})$

$\qquad\qquad\qquad\qquad = -9000\sqrt{3}$ ft/degree

SECTION 3.5

3.5.1 Find $f'(x)$ where $f(x) = x^2(\sin 2x)^3$.

3.5.2 Find $f'(x)$ where $f(x) = \dfrac{3}{\left(x^2 - 2x + 2\right)^3}$.

3.5.3 Find $f'(x)$ where $f(x) = \sin(\tan 2x)$.

3.5.4 Find $f'(\theta)$ where $f(\theta) = (\theta + \sin 2\theta)^2$.

3.5.5 Find $f'(\theta)$ where $f(\theta) = \sin^2\left(2\theta^2 - \theta\right)^3$.

3.5.6 Find $f'\left(\dfrac{\pi}{12}\right)$ where $f(x) = \cos^3 2x$.

3.5.7 Find $f'\left(\dfrac{\pi}{8}\right)$ where $f(x) = \sin^2 2x$.

3.5.8 Find $f'(x)$ where $f(x) = \csc^3 4x$.

3.5.9 Find $f'(x)$ where $f(x) = \sec^2\left(3x - x^2\right)$.

3.5.10 Find $f'(x)$ where $f(x) = (x^2 - 3)^3(x^2 + 1)^2$.

3.5.11 Find $\dfrac{dy}{dx}$ where $y = (x + 4)^4 (3x + 2)^3$.

3.5.12 Find $\dfrac{dy}{dx}$ where $y = \left(\dfrac{x + 1}{x - 1}\right)^2$.

3.5.13 Find $y'(\pi)$ where $y = \left(\dfrac{1}{x} + \sin x\right)^{-1}$.

3.5.14 Find $f'(t)$ where $f(t) = \left(\dfrac{1}{t} + \dfrac{1}{t^2}\right)^4$.

3.5.15 Find equations for the tangent and normal lines to the graph of $f(x) = \sin\left(4 - x^2\right)$ at $x = 2$.

3.5.16 Find equations for the tangent and normal lines to the graph of $f(x) = x \cos 4x$ at $x = \pi/4$.

3.5.17 Find $f'(x)$ where $f(x) = (x^4 + 3x)^{52}$

3.5.18 Find $f'(x)$ where $f(x) = \sqrt{x^5 + 2x + 3}$

3.5.19 Find $\dfrac{d}{dx}\left[x^2y^3 - \dfrac{x}{y^2}\right]$ in terms of x, y and $\dfrac{dy}{dx}$ assuming that y is a differentiable function of x.

3.5.20 Find $\dfrac{d}{dx}\left[\sin\sqrt{x^2 + y^2}\right]$ in terms of x, y and $\dfrac{dy}{dx}$ assuming that y is a differentiable function of x.

3.5.21 Find $\dfrac{d}{dt}\left[\tan(x^2\sqrt{y})\right]$ in terms of x, y, $\dfrac{dx}{dt}$ and $\dfrac{dy}{dt}$ assuming x and y are differentiable functions of t.

3.5.22 Given that $f(1) = 2$, $f'(1) = 4$ and $g(x) = (f(x))^{-3}$, find $g'(1)$.

3.5.23 Find $(f \circ g)'(0)$ if $f'(0) = 4$, $g(0) = 0$ and $g'(0) = 2$.

SOLUTIONS

SECTION 3.5

3.5.1 $f'(x) = x^2 \dfrac{d}{dx}\left[(\sin 2x)^3\right] + (\sin 2x)^3 \dfrac{d}{dx}\left[x^2\right]$

$\qquad = x^2(3)(\sin 2x)^2 \dfrac{d}{dx}[\sin 2x] + (\sin 2x)^3(2x)$

$\qquad = 3x^2(\sin 2x)^2 \cos 2x \dfrac{d}{dx}[2x] + 2x(\sin 2x)^3$

$\qquad = 6x^2 \cos 2x(\sin 2x)^2 + 2x(\sin 2x)^3$

$\qquad = 2x(\sin 2x)^2(3x \cos 2x + \sin 2x).$

3.5.2 $f'(x) = 3(-3)\left(x^2 - 2x + 2\right)^{-4} \dfrac{d}{dx}\left[x^2 - 2x + 2\right]$

$\qquad = -9\left(x^2 - 2x + 2\right)^{-4}(2x - 2) = \dfrac{18(1 - x)}{\left(x^2 - 2x + 2\right)^4}.$

3.5.3 $f'(x) = \cos(\tan 2x)\dfrac{d}{dx}[\tan 2x]$

$\qquad = \cos(\tan 2x)(\sec^2 2x)\dfrac{d}{dx}[2x]$

$\qquad = \sec^2 2x \cos(\tan 2x)(2)$

$\qquad = 2\sec^2 2x \cos(\tan 2x).$

3.5.4 $f'(\theta) = 2(\theta + \sin 2\theta)\dfrac{d}{d\theta}[\theta + \sin 2\theta]$

$\qquad = 2(\theta + \sin 2\theta)(1 + \cos 2\theta \dfrac{d}{d\theta}[2\theta])$

$\qquad = 2(\theta + \sin 2\theta)(1 + 2\cos 2\theta).$

3.5.5 $f'(\theta) = 2\sin(2\theta^2 - \theta)^3 \dfrac{d}{d\theta}\left[\sin(2\theta^2 - \theta)^3\right]$

$\qquad = 2\sin(2\theta^2 - \theta)^3 \cos(2\theta^2 - \theta)^3 \dfrac{d}{d\theta}\left[(2\theta^2 - \theta)^3\right]$

$\qquad = 2\sin(2\theta^2 - \theta)^3 \cos(2\theta^2 - \theta)^3(3)(2\theta^2 - \theta)^2 \dfrac{d}{d\theta}\left[2\theta^2 - \theta\right]$

$\qquad = 6(2\theta^2 - \theta)^2 \sin(2\theta^2 - \theta)^3 \cos(2\theta^2 - \theta)^3(4\theta - 1)$

$\qquad = 6(4\theta - 1)(2\theta^2 - \theta)^2 \sin(2\theta^2 - \theta)^3 \cos(2\theta^2 - \theta)^3$

or $3(4\theta - 1)\left(2\theta^2 - \theta\right)^2 \sin 2\left(2\theta^2 - \theta\right)^3.$

3.5.6 $f'(x) = 3\cos^2 2x \dfrac{d}{dx}[\cos 2x]$

$$= 3\cos^2 2x (-\sin 2x) \frac{d}{dx}[2x]$$

$$= -3\cos^2 2x \sin 2x (2)$$

$$= -6\cos^2 2x \sin 2x, \ \text{so} \ f'\left(\frac{\pi}{12}\right) = -\frac{9}{4}.$$

3.5.7 $f'(x) = 2\sin 2x \dfrac{d}{dx}[\sin 2x]$

$$= 2\sin 2x \cos 2x \frac{d}{dx}[2x]$$

$$= 4\sin 2x \cos 2x, \ \text{so} \ f'\left(\frac{\pi}{8}\right) = 2.$$

3.5.8 $f'(x) = 3\csc^2 4x \dfrac{d}{dx}[\csc 4x]$

$$= 3\csc^2 4x (-\csc 4x \cot 4x) \frac{d}{dx}[4x]$$

$$= -3\csc^3 4x \cot 4x (4)$$

$$= -12\csc^3 4x \cot 4x.$$

3.5.9 $f'(x) = 2\sec\left(3x - x^2\right) \dfrac{d}{dx}\left[\sec\left(3x - x^2\right)\right]$

$$= 2\sec^2\left(3x - x^2\right) \tan\left(3x - x^2\right) \frac{d}{dx}\left[3x - x^2\right]$$

$$= 2\sec^2\left(3x - x^2\right) \tan\left(3x - x^2\right)(3 - 2x)$$

$$= 2(3 - 2x)\sec^2\left(3x - x^2\right) \tan\left(3x - x^2\right).$$

3.5.10 $f'(x) = \left(x^2 - 3\right)^3 \dfrac{d}{dx}\left[\left(x^2 + 1\right)^2\right] + \left(x^2 + 1\right)^2 \dfrac{d}{dx}\left[\left(x^2 - 3\right)^3\right]$

$$= \left(x^2 - 3\right)^3 (2)\left(x^2 + 1\right) \frac{d}{dx}\left[x^2 + 1\right] + \left(x^2 + 1\right)^2 (3)\left(x^2 - 3\right)^2 \frac{d}{dx}\left[x^2 - 3\right]$$

$$= 2\left(x^2 + 1\right)\left(x^2 - 3\right)^3 (2x) + 3\left(x^2 + 1\right)^2 \left(x^2 - 3\right)^2 (2x)$$

$$= 4x\left(x^2 + 1\right)\left(x^2 - 3\right)^3 + 6x\left(x^2 + 1\right)^2 \left(x^2 - 3\right)^2$$

$$= 2x\left(x^2 + 1\right)\left(x^2 - 3\right)^2 \left(5x^2 - 3\right).$$

3.5.11 $\dfrac{dy}{dx} = (x + 4)^4 (3)(3x + 2)^2 \dfrac{d}{dx}[3x + 2] + (3x + 2)^3 (4)(x + 4)^3 \dfrac{d}{dx}[x + 4]$

$$= 3(x + 4)^4 (3x + 2)(3) + 4(3x + 2)^3 (x + 4)^3 (1)$$

$$= (x + 4)^3 (3x + 2)^2 (21x + 44).$$

3.5.12 $\dfrac{dy}{dx} = 2\left(\dfrac{x + 1}{x - 1}\right) \dfrac{d}{dx}\left[\dfrac{x + 1}{x - 1}\right] = 2\left(\dfrac{x + 1}{x - 1}\right)\left[\dfrac{(x - 1)(1) - (x + 1)(1)}{(x - 1)^2}\right]$

$$= \frac{-4(x + 1)}{(x - 1)^3} = \frac{4(x + 1)}{(1 - x)^3}.$$

3.5.13 $y'(x) = -\left(x^{-1} + \sin x\right)^{-2} \dfrac{d}{dx}\left[x^{-1} + \sin x\right]$

$\qquad\qquad = -\left(x^{-1} + \sin x\right)^{-2}\left(-x^{-2} + \cos x\right)$

$\qquad$ so $y'(\pi) = -\left(\dfrac{1}{\pi} + \sin \pi\right)^{-2}\left(-\dfrac{1}{\pi^2} + \cos \pi\right) = \pi^2 + 1.$

3.5.14 $f'(t) = 4\left(t^{-1} + t^{-2}\right)^3 \dfrac{d}{dt}\left[t^{-1} + t^{-2}\right]$

$\qquad\qquad = 4\left(t^{-1} + t^{-2}\right)^3\left(-t^{-2} - 2t^{-3}\right) = -4\left(\dfrac{1}{t} + \dfrac{1}{t^2}\right)^3\left(\dfrac{1}{t^2} + \dfrac{2}{t^3}\right).$

3.5.15 $f'(x) = \cos\left(4 - x^2\right)\dfrac{d}{dx}\left[4 - x^2\right]$

$\qquad\qquad = \cos\left(4 - x^2\right)(-2x) = -2x\cos\left(4 - x^2\right),$

$\qquad$ so the slope of the tangent to the graph of f at $x = 2$ is $f'(2) = -4\cos 0 = -4$, thus, the
equation of the tangent to f at $(2,0)$ is $y - 0 = -4(x - 2)$ or $y = -4x + 8$; the slope of the
normal to f at $x = 2$ is $-\dfrac{1}{-4} = \dfrac{1}{4}$ so the equation of the normal to f at $(2,0)$ is $y - 0 = \dfrac{1}{4}(x-2)$
or $y = \dfrac{1}{4}x - \dfrac{1}{2}.$

3.5.16 $f'(x) = x\dfrac{d}{dx}[\cos 4x] + \cos 4x\dfrac{d}{dx}[x]$

$\qquad\qquad = x(-\sin 4x)\dfrac{d}{dx}[4x] + \cos 4x(1)$

$\qquad\qquad = -4x\sin 4x + \cos 4x,$

$\qquad$ so the slope of the tangent to the graph of f at $x = \dfrac{\pi}{4}$ is $f'\left(\dfrac{\pi}{4}\right) = -\dfrac{4\pi}{4}\sin\dfrac{4\pi}{4} + \cos\dfrac{4\pi}{4} = -1,$
thus the equation of the tangent to f at $\left(\dfrac{\pi}{4}, -\dfrac{\pi}{4}\right)$ is $y - \left(-\dfrac{\pi}{4}\right) = -\left(x - \dfrac{\pi}{4}\right)$ or $y = -x;$
the slope of the normal to f at $x = \pi/4$ is $-\dfrac{1}{-1} = 1$ so the equation of the normal to f at
$\left(\dfrac{\pi}{4}, -\dfrac{\pi}{4}\right)$ is $y - \left(-\dfrac{\pi}{4}\right) = \left(x - \dfrac{\pi}{4}\right)$ or $y = x - \dfrac{\pi}{2}.$

3.5.17 $f'(x) = 52(x^4 + 3x)^{51}\dfrac{d}{dx}(x^4 + 3x)$

$\qquad\qquad = 52(x^4 + 3x)^{51}(4x^3 + 3)$

3.5.18 $f'(x) = \dfrac{1}{2\sqrt{x^5 + 2x + 3}}\dfrac{d}{dx}(x^5 + 2x + 3)$

$\qquad\qquad = \dfrac{5x^4 + 2}{2\sqrt{x^5 + 2x + 3}}$

3.5.19 $\dfrac{d}{dx}\left[x^2y^3 - \dfrac{x}{y^2}\right] = x^2\dfrac{d}{dx}[y^3] + y^3\dfrac{d}{dx}[x^2] - \dfrac{\left[y^2\dfrac{d}{dx}[x] - x\dfrac{d}{dx}[y^2]\right]}{(y^2)^2}$

$$= x^2\left(3y^2\dfrac{dy}{dx}\right) + y^3(2x) - \dfrac{\left[y^2(1) - x(2y)\dfrac{dy}{dx}\right]}{y^4}$$

$$= 3x^2y^2\dfrac{dy}{dx} + 2xy^3 - \dfrac{\left[y - 2x\dfrac{dy}{dx}\right]}{y^3}$$

$$= y^{-3}\left[3x^2y^5\dfrac{dy}{dx} + 2xy^6 - y + 2x\dfrac{dy}{dx}\right]$$

$$= y^{-3}\left[x(3xy^5 + 2)\dfrac{dy}{dx} + y(2xy^5 - 1)\right]$$

3.5.20 $\dfrac{d}{dx}\left[\sin\sqrt{x^2+y^2}\right] = \cos\sqrt{x^2+y^2}\,\dfrac{d}{dx}\left[\sqrt{x^2+y^2}\right]$

$$= \cos\sqrt{x^2+y^2}\left(\dfrac{1}{2}(x^2+y^2)^{-1/2}\dfrac{d}{dx}[x^2+y^2]\right)$$

$$= \cos\sqrt{x^2+y^2}\left(\dfrac{1}{2}(x^2+y^2)^{-1/2}\left(2x + 2y\dfrac{dy}{dx}\right)\right)$$

$$= \dfrac{\cos\sqrt{x^2+y^2}\left(x + y\dfrac{dy}{dx}\right)}{\sqrt{x^2+y^2}}$$

3.5.21 $\dfrac{d}{dt}[\tan(x^2\sqrt{y})] = [\sec^2(x^2\sqrt{y})]\dfrac{d}{dt}[x^2\sqrt{y}]$

$$= [\sec^2(x^2\sqrt{y})]\left(x^2\dfrac{d}{dt}[\sqrt{y}] + \sqrt{y}\dfrac{d}{dt}[x^2]\right)$$

$$= [\sec^2(x^2\sqrt{y})]\left[x^2\left(1/2\,y^{-1/2}\dfrac{dy}{dt}\right) + \sqrt{y}\left(2x\dfrac{dx}{dt}\right)\right]$$

$$= [\sec^2(x^2\sqrt{y})]\left[\dfrac{x^2}{2\sqrt{y}}\dfrac{dy}{dt} + 2x\sqrt{y}\dfrac{dx}{dt}\right]$$

$$= \dfrac{x\left[x\dfrac{dy}{dt} + 4y\dfrac{dx}{dt}\right]}{2\sqrt{y}}\sec^2(x^2\sqrt{y})$$

3.5.22 $g(x) = (f(x))^{-3}$,

$g'(x) = -3\,(f(x))^{-4}\dfrac{d}{dx}f(x) = -3\,(f(x))^{-4}\,f'(x)$ so

$g'(1) = -3\,(f(1))^{-4}\,f'(1) = -3(2)^{-4}(4) = -\dfrac{3}{4}$

3.5.23 $(f \circ g)'(x) = f'\,(g(x))\,g'(x)$ so

$(f \circ g)'(0) = f'\,(g(0))\,g'(0) = f'(0)g'(0) = 8$

SECTION 3.6

3.6.1   Find $f'(x)$ if $f(x) = x^2\sqrt{x^2 + a^2}$, $a = $ constant.

3.6.2   Find $f'(x)$ if $f(x) = (2 + \cos 2x)^{1/2}$.

3.6.3   Find $\dfrac{dy}{dx}$ if $y = (x + 4)^{1/4}(3x + 2)^{1/3}$.

3.6.4   Find $\dfrac{dy}{dx}$ if $y = (2x + 4)^4(3x - 2)^{7/3}$.

3.6.5   Find $\dfrac{dy}{dx}$ if $y = \left(\dfrac{a^2 - x^2}{a^2 + x^2}\right)^{2/3}$; $a = $ constant.

3.6.6   Find $\dfrac{dy}{dx}$ if $\sin(x + y) = \tan xy$.

3.6.7   Find $\dfrac{dy}{dx}$ by implicit differentiation if $xy^2 + \sqrt{xy} = 2$.

3.6.8   Find $\dfrac{dy}{dx}$ by implicit differentiation if $x \sin y = y \cos 2x$.

3.6.9   Find $\dfrac{dy}{dx}$ by implicit differentiation if $a^2 x^{3/4} + b^2 y^{2/3} = c^2$; a, b, c are constants.

3.6.10   Use implicit differentiation to find $\dfrac{dy}{dx}$ if $\sin^2 xy \cos xy = 1$.

3.6.11   Find $\dfrac{dy}{dx}$ by implicit differentiation if $(x - y)^2 + 4x - 5y - 1 = 0$.

3.6.12   Find $\dfrac{dy}{dx}$ by implicit differentiation if $x^{-1/3} + y^{-1/3} = 1$.

3.6.13   Use implicit differentiation to find $\dfrac{dy}{dx}$ if $\tan^2(x^2 y) = y$.

3.6.14   Find $\dfrac{d^2y}{dx^2}$ by implicit differentiation if $x^2 + 3y^2 = 10$.

3.6.15   Find $\dfrac{d^2y}{dx^2}$ by implicit differentiation if $x^2 + 2xy - y^2 + 8 = 0$.

3.6.16   Find the equation of the tangent and normal lines to $2x^2 - 3xy + 3y^2 = 2$ at $(1, 1)$.

3.6.17   Use implicit differentiation to find the equations of the tangent and normal lines to the ellipse $3x^2 + y^2 = 4$ at $(1, 1)$.

3.6.18   Use implicit differentiation to find the equations of the tangent and normal lines to the hyperbola $5x^2 - y^2 = 4$ at $(1, 1)$.

3.6.19   Use implicit differentiation to show that for any constants a and b, the hyperbolas $xy = a$ and $x^2 - y^2 = b$ intersect at right angles at the point (x_0, y_0).

SOLUTIONS

SECTION 3.6

3.6.1 $f'(x) = x^2 \left(\dfrac{1}{2}\right)(x^2+a^2)^{-1/2}(2x) + (x^2+a^2)^{1/2}(2x) = \dfrac{3x^3 + 2a^2 x}{\sqrt{x^2+a^2}}.$

3.6.2 $f'(x) = \dfrac{1}{2}(2+\cos 2x)^{-1/2}(-\sin 2x)(2) = -\dfrac{\sin 2x}{\sqrt{2+\cos 2x}}.$

3.6.3 $\dfrac{dy}{dx} = (x+4)^{1/3}\left(\dfrac{1}{3}\right)(3x+2)^{-2/3}(3) + (3x+2)^{1/3}\left(\dfrac{1}{4}\right)(x+4)^{-3/4}(1)$

$\qquad = \dfrac{1}{4}(3x+2)^{-2/3}(x+4)^{-3/4}(7x+18).$

3.6.4 $\dfrac{dy}{dx} = (2x+4)^4\left(\dfrac{7}{3}\right)(3x-2)^{4/3}(3) + (3x-2)^{7/3}(4)(2x+4)^3(2)$

$\qquad = (2x+4)^3(3x-2)^{4/3}(38x+12).$

3.6.5 $\dfrac{dy}{dx} = \dfrac{2}{3}\left(\dfrac{a^2-x^2}{a^2+x^2}\right)^{-1/3}\left[\dfrac{\left(a^2+x^2\right)(-2x) - \left(a^2-x^2\right)(2x)}{\left(a^2+x^2\right)^2}\right]$

$\qquad = \dfrac{8a^2 x}{3\left(x^2-a^2\right)^{1/3}\left(a^2+x^2\right)^{5/3}}.$

3.6.6 $\cos(x+y)\left(1+\dfrac{dy}{dx}\right) = \sec^2 xy\left(x\dfrac{dy}{dx}+y\right)$

$\qquad \dfrac{dy}{dx} = \dfrac{y\sec^2 xy - \cos(x+y)}{\cos(x+y) - x\sec^2 xy}$

3.6.7 $x\left(2y\dfrac{dy}{dx}\right) + y^2(1) + \left(\dfrac{1}{2}\right)(xy)^{-1/2}\left(x\dfrac{dy}{dx}+y\right) = 0,$ so $\dfrac{dy}{dx} = -\dfrac{y+2x^{1/2}y^{5/2}}{4x^{3/2}y^{3/2}+x}.$

3.6.8 $x\cos y\dfrac{dy}{dx} + \sin y(1) = y(-\sin 2x)(2) + \cos 2x\dfrac{dy}{dx},$ so $\dfrac{dy}{dx} = \dfrac{2y\sin 2x + \sin y}{\cos 2x - x\cos y}.$

3.6.9 $a^2\left(\dfrac{3}{4}\right)x^{-1/4} + b^2\left(\dfrac{2}{3}\right)y^{-1/3}\dfrac{dy}{dx} = 0,$ so $\dfrac{dy}{dx} = -\dfrac{9a^2 y^{1/3}}{8b^2 x^{1/4}}.$

3.6.10 $-x\sin^3(xy)\dfrac{dy}{dx} - y\sin^3(xy) + 2x\cos^2(xy)\sin(xy)\dfrac{dy}{dx} + 2y\cos^2(xy)\sin(xy) = 0,$ so

$\qquad \dfrac{dy}{dx} = \dfrac{y\sin^3(xy) - 2y\cos^2(xy)\sin(xy)}{2x\cos^2(xy)\sin(xy) - x\sin^3(xy)}$

3.6.11 $2(x-y)\left(1-\dfrac{dy}{dx}\right)+4-5\dfrac{dy}{dx}=0$, so $\dfrac{dy}{dx}=\dfrac{2y-2x-4}{2y-2x-5}$.

3.6.12 $-\dfrac{1}{3}x^{-4/3}-\dfrac{1}{3}y^{-4/3}\dfrac{dy}{dx}=0$, so $\dfrac{dy}{dx}=-\left(\dfrac{y}{x}\right)^{4/3}$.

3.6.13 $2\tan(x^2y)\sec^2(x^2y)\left(x^2\dfrac{dy}{dx}+2xy\right)=\dfrac{dy}{dx}$, so $\dfrac{dy}{dx}=\dfrac{4xy\tan^2(x^2y)\sec^2(x^2y)}{1-2x^2\tan(x^2y)\sec^2(x^2y)}$

3.6.14 $2x+6y\dfrac{dy}{dx}=0$, $\dfrac{dy}{dx}=-\dfrac{x}{3y}$; $\dfrac{d^2y}{dx^2}=-\dfrac{1}{3}\left[\dfrac{y(1)-x\dfrac{dy}{dx}}{y^2}\right]$, but $\dfrac{dy}{dx}=-\dfrac{x}{3y}$,

so $\dfrac{d^2y}{dx^2}=-\dfrac{1}{3}\left[\dfrac{y-x\left(-\dfrac{x}{3y}\right)}{y^2}\right]=-\dfrac{3y^2+x^2}{9y^3}=-\dfrac{10}{9y^3}$ since $x^2+3y^2=10$.

3.6.15 $2x+2\left(x\dfrac{dy}{dx}+y\right)-2y\dfrac{dy}{dx}=0$, $\dfrac{dy}{dx}=\dfrac{y+x}{y-x}$;

$\dfrac{d^2y}{dx^2}=\dfrac{(y-x)\left(\dfrac{dy}{dx}+1\right)-(y+x)\left(\dfrac{dy}{dx}-1\right)}{(y-x)^2}=\dfrac{2y-2x\dfrac{dy}{dx}}{(y-x)^2}$, but $\dfrac{dy}{dx}=\dfrac{y+x}{y-x}$

so $\dfrac{d^2y}{dx^2}=\dfrac{2y-2x\left(\dfrac{y+x}{y-x}\right)}{(y-x)^2}=\dfrac{-2\left(x^2+2xy-y^2\right)}{(y-x)^3}=\dfrac{-2(8)}{(y-x)^3}=\dfrac{16}{(x-y)^3}$

since $x^2+2xy-y^2+8=0$.

3.6.16 $4x-3x\dfrac{dy}{dx}-3y+6y\dfrac{dy}{dx}=0$, so $\dfrac{dy}{dx}=\dfrac{3y-4x}{6y-3x}$. At $(1,1)$,

$m_{\text{tan}}=\dfrac{dy}{dx}\bigg|_{\substack{x=1\\y=1}}=\dfrac{3y-4x}{6y-3x}\bigg|_{\substack{x=1\\y=1}}=-\dfrac{1}{3}$ and $m_{\text{normal}}=3$ so the equation of the tangent line is

$y-1=-\dfrac{1}{3}(x-1)$ or $y=-\dfrac{1}{3}x+\dfrac{4}{3}$ and the equation of the normal line is $y-1=3(x-1)$ or $y=3x-2$.

3.6.17 $6x+2y\dfrac{dy}{dx}=0$, so $\dfrac{dy}{dx}=-\dfrac{3x}{y}$. At $(1,1)$, $m_{\text{tan}}=\dfrac{dy}{dx}\bigg|_{\substack{x=1\\y=1}}=-\dfrac{3x}{y}\bigg|_{1,1}=-3$ and

$m_{\text{normal}}=\dfrac{1}{3}$ so the equation of the tangent line is $y-1=-3(x-1)$ or $y=-3x+4$ and the

equation of the normal line is $y=\dfrac{1}{3}x+\dfrac{2}{3}$.

3.6.18 $10x - 2y\dfrac{dy}{dx} = 0$, so $\dfrac{dy}{dx} = \dfrac{5x}{y}$. At $(1,1)$, $m_{\text{tan}} = \dfrac{dy}{dx}\bigg|_{\substack{x=1\\y=1}} = \dfrac{5x}{y}\bigg|_{\substack{x=1\\y=1}} = 5$ and $m_{\text{normal}} = -\dfrac{1}{5}$

so the equation of the tangent line is $y - 1 = 5(x - 1)$ or $y = 5x - 4$ and the equation of the normal line is $y = -\dfrac{1}{5}x + \dfrac{6}{5}$.

3.6.19 $x\dfrac{dy}{dx} + y = 0$ so $\dfrac{dy}{dx} = -\dfrac{y}{x}$, similarly, $2x - 2y\dfrac{dy}{dx} = 0$ and $\dfrac{dy}{dx} = \dfrac{x}{y}$. At (x_0, y_0), let

$m_1 = -\dfrac{y}{x}\bigg|_{\substack{x=x_0\\y=y_0}} = -\dfrac{y_0}{x_0}$ and $m_2 = \dfrac{x}{y}\bigg|_{\substack{x=x_0\\y=y_0}} = \dfrac{x_0}{y_0}$, then $m_1 m_2 = \left(-\dfrac{y_0}{x_0}\right)\left(\dfrac{x_0}{y_0}\right) = -1$,

thus, the tangent lines are perpendicular to each other at (x_0, y_0) so the curves intersect at right angles.

SECTION 3.7

3.7.1   Let $y = x^3 - 1$.

   (a)   Find Δy if $\Delta x = 1$ and the initial value of x is $x = 1$.
   (b)   Find dy if $dx = 1$ and the initial value of x is $x = 1$.
   (c)   Make a sketch of $y = x^3 - 1$ and show Δy and dy in the picture.

3.7.2   Let $y = \dfrac{1}{2}x^2 + 1$.

   (a)   Find Δy if $\Delta x = 1$ and the initial value of x is $x = 1$.
   (b)   Find dy if $dx = 1$ and the initial value of x is $x = 1$.
   (c)   Make a sketch of $y = \dfrac{1}{2}x^2 + 1$ and show Δy and dy in the picture.

3.7.3   Use a differential to approximate $\sqrt[4]{14}$.

3.7.4   Use a differential to approximate $\sqrt[3]{9}$.

3.7.5   Use a differential to approximate $\sqrt[5]{29}$.

3.7.6   Use a differential to approximate $\sqrt[3]{10}$.

3.7.7   Use a differential to approximate $(1.98)^4$.

3.7.8   Use a differential to approximate $\cos 58°$.

3.7.9   Use a differential to approximate $\sin 31°$.

3.7.10   Use a differential to approximate $\tan 43°$.

3.7.11   The surface area of a sphere is given by $S = 4\pi r^2$ where r is the radius of the sphere. The radius is measured to be 3 cm with an error of ± 0.1 cm.

   (a)   Use differentials to estimate the error in the calculated surface area.
   (b)   Estimate the percentage error in the radius and surface area.

3.7.12   The surface area S of a cube is to be computed from a measured value of its side x. Estimate the maximum permissible percentage error in the side measurement if the percentage error in the surface area must be kept to within $\pm 4\%$.

3.7.13   A circular hole 6 inches in diameter and 10 feet deep is to be drilled out of a glacier. The diameter of the hole is exact but the depth of the hole is measured with an error of $\pm 1\%$. Estimate the percentage error in the volume of ice removed. ($V = \frac{\pi}{4} d^2 h$ is the volume of a cylinder of diameter d and height h.)

3.7.14   The pressure P, the volume V, and the temperature T of an enclosed gas are related by the Ideal Gas Law, $PV = kT$ where k is a constant. With the temperature held constant, the volume of the gas is calculated from a measured value of its pressure. Estimate the maximum permissible error in the pressure measurement if the percentage error in the volume must be kept to within $\pm 2\%$.

3.7.15   The magnetic force F acting on a particle is given by $F = \frac{k}{r^2}$, where r is the distance from the magnetic source and k is a constant. r is measured to be 3 cm with a possible error of $\pm 6\%$.

   (a)   Use differentials to estimate the error in the calculated value of F.
   (b)   Estimate the percentage error in F and r.

3.7.16   When a cubical block of metal is heated, each edge increases by 0.1% per degree increase in temperature. Use differentials to estimate the percentage increase in the surface area and volume of the block per degree increase in temperature.

3.7.17   When a spherical ball of metal is heated, the radius of the sphere increases by 0.1% per degree increase in temperature. Use differentials to estimate the percentage increase in the surface area and volume of the ball per degree increase in temperature.

$$\left(S = 4\pi r^2 \text{ and } V = \frac{4}{3}\pi r^3. \right)$$

3.7.18   The area of a circle is to be computed from a measured value of its diameter. Estimate the maximum permissible percentage error in the measurement if the percentage error in the area must be kept within 0.5%.

SOLUTIONS

SECTION 3.7

3.7.1 (a) $\Delta y = (x + \Delta x)^3 - 1 - (x^3 - 1) = (1+1)^3 - 1 - (1^3 - 1) = 7$

(b) $du = 3x^2 dx = 3(1)^2(1) = 3$

(c)

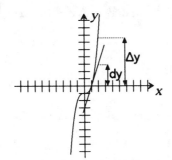

3.7.2 (a) $\Delta y = \left[\dfrac{1}{2}(x + \Delta x)^2 + 1 \right] - \left[\dfrac{1}{2}x^2 + 1 \right]$

$$= \left[\dfrac{1}{2}(1+1)^2 + 1 \right] - \left[\dfrac{1}{2}(1)^2 + 1 \right] = \dfrac{3}{2}$$

(b) $dy = x dx = (1)(1) = 1$

(c)

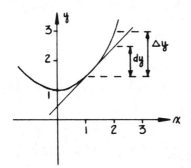

3.7.3 Let $f(x) = \sqrt[4]{x}$, $x_0 = 16$, $\Delta x = -2$, then $f'(x) = \dfrac{1}{4}x^{-3/4}$ and $x_0 + \Delta x = 14$ so

$$f(14) \approx f(16) + f'(16)(-2)$$
$$\approx \sqrt[4]{16} + \dfrac{1}{4(16)^{3/4}}(-2) = 2 - \dfrac{1}{16} = \dfrac{31}{16}.$$

3.7.4 Let $f(x) = \sqrt[3]{x}$, $x_0 = 8$, $\Delta x = 1$, then, $f'(x) = \dfrac{1}{3}x^{-2/3}$ and $x_0 + \Delta x = 9$, so

$$f(9) \approx f(8) + f'(8)(1)$$
$$\approx \sqrt[3]{8} + \dfrac{1}{3(8)^{2/3}}(1) = 2 + \dfrac{1}{12} = \dfrac{25}{12}.$$

3.7.5 Let $f(x) = \sqrt[5]{x}$, $x_0 = 32$, $\Delta x = -3$, then $f'(x) = \dfrac{1}{5}x^{-4/5}$ and $x_0 + \Delta x = 29$, so

$$f(29) \approx f(32) + f'(32)(-3)$$

$$\approx \sqrt[5]{32} + \frac{1}{5(32)^{4/5}}(-3) = 2 - \frac{3}{80} = \frac{157}{80}.$$

3.7.6 Let $f(x) = \sqrt[3]{x}$, $x_0 = 8$, $\Delta x = 2$, then $f'(x) = \dfrac{1}{3}x^{-2/3}$ and $x_0 + \Delta x = 10$, so

$$f(10) \approx f(8) + f'(8)(2)$$

$$\approx \sqrt[3]{8} + \frac{1}{3(8)^{2/3}}(2) = 2 + \frac{1}{6} = \frac{13}{6}.$$

3.7.7 Let $f(x) = x^4$, so $x_0 = 2$, $\Delta x = -0.02$, then $f'(x) = 4x^3$ and $x_0 + \Delta x = 1.98$, so

$$f(1.98) \approx f(2) + f'(2)(-0.02)$$

$$\approx (2)^4 + 4(2)^3(-0.02) = 16 - 0.64 = 15.36.$$

3.7.8 Let $f(x) = \cos x$, $x_0 = 60° = \dfrac{\pi}{3}$ radians, $\Delta x = -2° = \dfrac{-\pi}{90}$ radians, then, $f'(x) = -\sin x$ and $x_0 + \Delta x = \dfrac{29\pi}{90}$ radians, so

$$f\left(\frac{29\pi}{90}\right) = f\left(\frac{\pi}{3}\right) + f'\left(\frac{\pi}{3}\right)\left(\frac{-\pi}{90}\right) = \cos\frac{\pi}{3} - \sin\frac{\pi}{3}\left(\frac{-\pi}{90}\right)$$

$$= \frac{1}{2} + \frac{\sqrt{3}}{2}\left(\frac{\pi}{90}\right) \approx 0.5302.$$

3.7.9 Let $f(x) = \sin x$, $x_0 = 30° = \dfrac{\pi}{6}$ radians, $\Delta x = 1° = \dfrac{\pi}{180}$, then $f'(x) = \cos x$ and $x_0 + \Delta x = \dfrac{31\pi}{180}$, so

$$f\left(\frac{31\pi}{180}\right) \approx f\left(\frac{\pi}{6}\right) + f'\left(\frac{\pi}{6}\right)\left(\frac{\pi}{180}\right) \approx \sin\frac{\pi}{6} + \cos\frac{\pi}{6}\left(\frac{\pi}{180}\right)$$

$$\approx \frac{1}{2} + \frac{\sqrt{3}}{2}\left(\frac{\pi}{180}\right) \approx 0.515.$$

3.7.10 Let $f(x) = \tan x$, $x_0 = 45° = \dfrac{\pi}{4}$ radians, $\Delta x = -2° = \dfrac{-\pi}{90}$ radians, then $f'(x) = \sec^2 x$ and $x_0 + \Delta x = \dfrac{43\pi}{180}$, so

$$f\left(\frac{43\pi}{180}\right) \approx f\left(\frac{\pi}{4}\right) + f'\left(\frac{\pi}{4}\right)\left(\frac{-\pi}{90}\right) \approx \tan\frac{\pi}{4} + \sec^2\frac{\pi}{4}\left(\frac{-\pi}{90}\right)$$

$$\approx 1 + (2)\left(\frac{-\pi}{90}\right) \approx 0.930.$$

3.7.11 **(a)** $ds = 8\pi r \, dr = 8\pi(3)(\pm 0.1) = \pm 2.4\pi$

(b) The relative error in the radius is $\approx \dfrac{dr}{r} = \dfrac{\pm 0.1}{3} = \pm 0.033$ so the percentage error is $\approx \pm 3.3\%$; the relative error in the surface area is

$\approx \dfrac{ds}{s} = \dfrac{8\pi r \, dr}{4\pi r^2} = 2\dfrac{dr}{r} = 2(\pm 0.033) = \pm 0.066$ so the percentage error in the surface area is $\approx \pm 6.6\%$.

3.7.12 The relative error in S is $\approx \dfrac{dS}{S}$ where $S = 6x^2$ and $dS = 12x \, dx$, thus

$\dfrac{dS}{S} = \dfrac{12x \, dx}{6x^2} = 2\dfrac{dx}{x}$ so $\dfrac{dx}{x} = \left(\dfrac{1}{2}\right)\left(\dfrac{dS}{S}\right)$ and the percentage error is $\approx \dfrac{1}{2}(\pm 4\%) = \pm 2\%$.

3.7.13 The relative error in V is $\approx \dfrac{dV}{V}$ where $V = \dfrac{\pi}{4}d^2 h$, but d is exactly 6 inches or $\dfrac{1}{2}$ foot so

$V = \dfrac{\pi}{4}\left(\dfrac{1}{2}\right)^2 h = \dfrac{\pi}{16}h$ and $dV = \dfrac{\pi}{16}dh$, so $\dfrac{dV}{V} = \dfrac{\frac{\pi}{16}dh}{\frac{\pi}{16}h} = \dfrac{dh}{h} = \dfrac{\pm 0.01}{10} = \pm 0.001$, thus the

percentage error is $V \approx \pm 0.1\%$.

3.7.14 If $V = \dfrac{kT}{P}$, then $dV = -\dfrac{kT}{P^2}dP$ (T held constant) so $\dfrac{dV}{V} = \dfrac{-\frac{kT}{P^2}dP}{\frac{kT}{P}} = -\dfrac{dP}{P} \approx$ relative error

in P, thus $\dfrac{dP}{P} = -\dfrac{dV}{V} = -(\pm 2\%) = \pm 2\%$.

3.7.15 **(a)** $dF = -\dfrac{2k}{r^3}dr = \dfrac{-2k(\pm 0.06)}{(3)^3} = \dfrac{\pm 0.04k}{9} = \pm 0.0044k$.

(b) The relative error in $r \approx \dfrac{dr}{r} = \pm\dfrac{0.06}{3} = \pm 0.02$ so the percentage error in r is $\pm 2\%$; the

relative error in $F \approx \dfrac{dF}{F} = \dfrac{\frac{-2k}{r^3}dr}{\frac{k}{r^2}} = -2\dfrac{dr}{r} = -2(\pm 0.02) = \pm 0.04$ so the percentage

error in F is $\pm 4\%$.

3.7.16 The surface area of the block is $S = 6x^2$ and the relative error in the measurement of the surface area is approximately $\dfrac{dS}{S} = \dfrac{12x\,dx}{6x^2} = \dfrac{2dx}{x} = 2(\pm 0.001) = \pm 0.002$ so the percentage error is $S \approx \pm 0.2\%$; the volume of the block is $V = x^3$ and the relative error in the measurement of the volume is $\approx \dfrac{dV}{V} = \dfrac{3x^2\,dx}{x^3} = 3\dfrac{dx}{x} = 3(\pm 0.001) = \pm 0.003$ so the percentage error in $V \approx \pm 0.3\%$.

3.7.17   The relative increase in the surface area of the sphere is

$$\approx \frac{dS}{S} = \frac{4\pi(2r\,dr)}{4\pi r^2} = 2\frac{dr}{r} = 2(\pm 0.001) = \pm 0.002,$$ so the percentage error is $\pm 0.2\%$ where $\frac{dr}{r}$ is the relative increase in radius of the sphere; the relative increase in the volume of the

sphere is approximately $\dfrac{dV}{V} = \dfrac{\frac{4}{3}\pi(3r^2 dr)}{\frac{4}{3}\pi r^3} = 3\dfrac{dr}{r} = 3(\pm 0.001) = \pm 0.003,$ so the percentage

error is $\pm 0.3\%$.

3.7.18 $A = \dfrac{1}{4}\pi D^2$ where D is the diameter of the circle; $\dfrac{dA}{A} = \dfrac{(\pi D/2)dD}{\pi D^2/4} = 2\dfrac{dD}{D}$; $\dfrac{dD}{D} = \dfrac{1}{2}\dfrac{dA}{A}$ but

$\dfrac{dA}{A} \approx \pm 0.005$ so $\dfrac{dD}{D} \approx \pm\dfrac{0.005}{2}$, $\dfrac{dD}{D} \approx \pm 0.0025$; maximum permissible percentage error in $D \approx \pm 0.25\%$.

SUPPLEMENTARY EXERCISES, CHAPTER 3

In Exercises 1–4, use Definition 3.2.1 to find $f'(x)$.

1. $f(x) = kx$ (k constant).

2. $f(x) = (x - a)^2$ (a constant).

3. $f(x) = \sqrt{9 - 4x}$.

4. $f(x) = \dfrac{x}{x + 1}$.

5. Use Definition 3.2.1 to find $\dfrac{d}{dx}[|x|^3]\Big|_{x=0}$

6. Suppose $f(x) = \begin{cases} x^2 - 1, & x \leq 1 \\ k(x - 1), & x > 1. \end{cases}$

For what values of k is f

   (a) continuous

   (b) differentiable?

7. Suppose $f(3) = -1$ and $f'(3) = 5$. Find an equation for the tangent line to the graph of f at $x = 3$.

8. Let $f(x) = x^2$. Show that for any distinct values of a and b, the slope of the tangent line to $y = f(x)$ at $x = \frac{1}{2}(a + b)$ is equal to the slope of the secant line through the points (a, a^2) and (b, b^2).

9. Given the following table of values at $x = 1$ and $x = -2$, find the indicated derivatives in parts (a)-(1).

x	$f(x)$	$f'(x)$	$g(x)$	$g'(x)$
1	1	3	-2	-1
-2	-2	-5	1	7

(a) $\dfrac{d}{dx}[f^2(x) - 3g(xx^2)]\Big|_{x=1}$

(b) $\dfrac{d}{dx}[f(x)g(x)]\Big|_{x=1}$

(c) $\dfrac{d}{dx}\left[\dfrac{f(x)}{g(x)}\right]\Big|_{x=-2}$

(d) $\dfrac{d}{dx}\left[\dfrac{g(x)}{f(x)}\right]\Big|_{x=-2}$

(e) $\dfrac{d}{dx}[fg(x)]\Big|_{x=1}$

(f) $\dfrac{d}{dx}[f(g(x))]\Big|_{x=-2}$

(g) $\dfrac{d}{dx}[g(f(x))]\Big|_{x=-2}$

(h) $\dfrac{d}{dx}[g(g(x))]\Big|_{x=-2}$

(i) $\dfrac{d}{dx}[f(g(4 - 6x))]\Big|_{x=1}$

(j) $\dfrac{d}{dx}[g^3(x)]\Big|_{x=1}$

(k) $\dfrac{d}{dx}[\sqrt{f(x)}]\Big|_{x=1}$

(l) $\dfrac{d}{dx}[f(-\frac{1}{2}x)]\Big|_{x=-2}$

In Exercises 10–15, find $f'(x)$ and determine those values of x for which $f'(x) = 0$.

10. $f(x) = (2x + 7)^6(x - 2)^5$.

11 $f(x) = \dfrac{(x - 3)^4}{x^2 + 2x}$.

12. $f(x = \sqrt{3x + 1}(x - 1)^2$.

13. $f(x) = \left(\dfrac{3x + 1}{x^2}\right)^3$.

14. $f(x) = \dfrac{3(5x - 1)^{1/3}}{3x - 5}$.

15. $f(x) = \sqrt{x}\sqrt[3]{x^2 + x + 1}$.

16. Suppose that $f'(x) = 1/x$ for all $x \neq 0$.

    (a) Use the chain rule to show that for any nonzero constant a, $d(f(ax))/dx = d(f(x))/dx$.

    (b) If $y = f(\sin x)$ and $v = f(1/x)$, find dy/dx and dv/dx.

In Exercises 17–26, find the indicated derivatives.

17. $\dfrac{d}{dx}\left(\dfrac{\sqrt{2}}{x^2} - \dfrac{2}{5x}\right)$.

18. $\dfrac{dy}{dx}$ if $y = \dfrac{3x^2 + 7}{x^2 - 1}$.

19. $\dfrac{dz}{dr}\Big|_{r=\pi/6}$ if $z = 4\sin^2 r \cos^2 r$.

20. $g'(2)$ if $g(x) = 1/\sqrt{2x}$.

21. $\dfrac{du}{dx}$ if $u = \left(\dfrac{x}{x - 1}\right)^{-2}$.

22. $\dfrac{dw}{dv}$ if $w = \sqrt[5]{v^3} - \sqrt[4]{v}$.

23. $d(\sec^2 x - \tan^2 x)/dx$.

24. $\dfrac{dy}{dx}\Big|_{x=\pi/4}$ if $y = \tan t$ and $t = \cos(2x)$.

25. $F'(x)$ if $F(x) = \dfrac{(1/x) + 2x}{\frac{1}{2}(1/x^2) + 1}$.

26. $\Phi'(x)$ if $\Phi(x) = \dfrac{x^2 - 4x}{5\sqrt{x}}$.

27. Find all values of x for which the tangent to $y = x - (1/x)$ is parallel to the line $2x - y = 5$.

28. Find all values of x for which the tangent to $y = 2x^3 - x^2$ is perpendicular to the line $x + 4y = 10$.

29. Find all values of x for which the tangent to $y = (x + 2)^2$ passes through the origin.

30. Find all values of x for which the tangent to $y = x - \sin 2x$ is horizontal.

31. Find all values of x for which the tangent to $y = 3x - \tan x$ is parallel to the line $y - x = 2$.

In Exercises 32–34, find Δx, Δy, and dy.

32. $y = 1/(x - 1)$; x decreases from 2 to 1.5.

33. $y = \tan x$; x increases from $-\pi/4$ to 0.

34. $y = \sqrt{25 - x^2}$; x increases from 0 to 3.

35. Use a differential to approximate

(a) $\sqrt[3]{-8.25}$ (b) $\cot 46°$.

36. Let V and S denote the volume and surface area of a cube. Find the rate of change of V with respect to S.

37. The amount of water in a tank t minutes after it has started to drain is given by $W = 100(t-15)^2$ gal.

(a) At what rate is the water running out at the end of 5 min?

(b) What is the average rate at which the water flows out during the first 5 min?

In Exercises 40–42, find dy/dx by implicit differentiation and use it to find the equation of the tangent line at the indicated points.

38. $(x+y)^3 + 3xy = -7; (-2,1)$.   **39.** $xy^2 = \sin(x+2y); (0,0)$.

40. $(x+y^3) - 5x + y = 1$; at the point where the curve intersects the line $x+y = 1$.

41. Show that the curves whose equations are $y^2 = x^3$ and $2x^2 + 3y^2 = 5$ intersect at the point $(1,1)$ and that their tangent lines are perpendicular there.

42. Show that for any point $P_)(x_0, y_0)$ on the circle $x^2+y^2 = r^2$, the tangent line at P_0 is perpendicular to the radial line from the origin to P_0.

43. Verify that the function $y = \cos x - 3\sin x$ satisfies $y''' + y'' + y' + y = 0$.

44. Find d^2y/dx^2 implicitly if

(a) $y^3 + 3x^2 = 4y$ (b) $\sin y + \cos x = 1$.

SOLUTIONS

SUPPLEMENTARY EXERCISES, CHAPTER 3

1. $f'(x) = \lim\limits_{h \to 0} \dfrac{k(x+h) - kx}{h} = \lim\limits_{h \to 0} k = k$

2. $f'(x) = \lim\limits_{h \to 0} \dfrac{(x+h-a)^2 - (x-a)^2}{h} = \lim\limits_{h \to 0}[2(x-a) + h] = 2(x-a)$

3. $f'(x) = \lim\limits_{h \to 0} \dfrac{\sqrt{9 - 4(x+h)} - \sqrt{9-4x}}{h} = \lim\limits_{h \to 0} \dfrac{[9 - 4(x+h)] - [9-4x]}{h(\sqrt{9-4(x+h)} + \sqrt{9-4x})}$

$= \lim\limits_{h \to 0} \dfrac{-4}{\sqrt{9-4(x+h)} + \sqrt{9-4x}} = -\dfrac{2}{\sqrt{9-4x}}$

4. $f'(x) = \lim\limits_{h \to 0} \dfrac{\dfrac{x+h}{x+h+1} - \dfrac{x}{x+1}}{h} = \lim\limits_{h \to 0} \dfrac{(x+h)(x+1) - x(x+h+1)}{h(x+1)(x+h+1)}$

$= \lim\limits_{h \to 0} \dfrac{1}{(x+1)(x+h+1)} = \dfrac{1}{(x+1)^2}$

5. $\dfrac{d}{dx}\left(|x|^3\right)\Big|_{x=0} = \lim\limits_{h \to 0} \dfrac{|0+h|^3 - |0|^3}{h} = \lim\limits_{h \to 0} \dfrac{|h|^3}{h} = \lim\limits_{h \to 0} h|h| = 0$

6. **(a)** f is continuous everywhere for all k, except perhaps at $x = 1$;

$\lim\limits_{x \to 1^-} f(x) = \lim\limits_{x \to 1^-}(x^2 - 1) = 0$, $\lim\limits_{x \to 1^+} f(x) = \lim\limits_{x \to 1^+} k(x-1) = 0$, and $f(1) = 0$ thus $\lim\limits_{x \to 1} f(x) = f(1)$ for all k, so f is continuous for all k.

(b) f is differentiable everywhere for all k, except perhaps at $x = 1$. Using the theorem that precedes Exercise 71, Section 3.3, $\lim\limits_{x \to 1^-} f'(x) = \lim\limits_{x \to 1^-} 2x = 2$ and

$\lim\limits_{x \to 1^+} f'(x) = \lim\limits_{x \to 1^+} k = k$; these limits are equal if $k = 2$, so f is differentiable if $k = 2$.

7. $y - (-1) = 5(x - 3)$, $y = 5x - 16$.

8. $f'(x) = 2x$ so $m_{\text{tan}} = f'\left(\dfrac{a+b}{2}\right) = a + b$, but $m_{\text{sec}} = \dfrac{b^2 - a^2}{b - a} = b + a$ if $a \ne b$ so $m_{\text{tan}} = m_{\text{sec}}$.

9. **(a)** $2f(x)f'(x) - 3g'(x^2)(2x)\big|_{x=1} = 12$ **(b)** $f(x)g'(x) + f'(x)g(x)\big|_{x=1} = -7$

(c) $\dfrac{g(x)f'(x) - f(x)g'(x)}{g^2(x)}\bigg|_{x=-2} = 9$ **(d)** $\dfrac{f(x)g'(x) - g(x)f'(x)}{f^2(x)}\bigg|_{x=-2} = -\dfrac{9}{4}$

(e) $f'(g(x))g'(x)\big|_{x=1} = f'(g(1))g'(1) = f'(-2)(-1) = 5$

(f) $f'(g(x))g'(x)\big|_{x=-2} = f'(g(-2))g'(-2) = f'(1)(7) = 21$

(g) $g'(f(x))f'(x)\big|_{x=-2} = g'(f(-2))f'(-2) = g'(-2)(-5) = -35$

(h) $g'(g(x))g'(x)\big|_{x=-2} = g'(g(-2))g'(-2) = g'(1)(7) = -7$

(i) $f'(g(4 - 6x)g'(4-6x)(-6)\big|_{x=1} = f'(g(-2))g'(-2)(-6) = f'(1)(7)(-6) = -126$

(j) $3g^2(x)g'(x)\big|_{x=1} = 3(-2)^2(-1) = -12$

(k) $\dfrac{1}{2}[f(x)]^{-1/2}f'(x)\bigg|_{x=1} = \dfrac{1}{2}(1)^{-1/2}(3) = \dfrac{3}{2}$

(l) $f'(-x/2)(-1/2)\big|_{x=-2} = -\dfrac{3}{2}$

10. $f'(x) = (2x+7)^6 5(x-2)^4 + (x-2)^5 6(2x+7)^5 (2)$
$\qquad\quad = (2x+7)^5 (x-2)^4 [5(2x+7) + 12(x-2)]$
$\qquad\quad = (2x+7)^5 (x-2)^4 (22x+11) = 11(2x+7)^5 (x-2)^4 (2x+1)$

so $f'(x) = 0$ if $x = -7/2, 2, -1/2$.

11. $f'(x) = \dfrac{(x^2+2x)4(x-3)^3 - (x-3)^4(2x+2)}{(x^2+2x)^2}$

$\qquad = \dfrac{(x-3)^3 [4(x^2+2x) - (x-3)(2x+2)]}{(x^2+2x)^2}$

$\qquad = \dfrac{(x-3)^3(2x^2+12x+6)}{(x^2+2x)^2} = \dfrac{2(x-3)^3(x^2+6x+3)}{(x^2+2x)^2}$

so $f'(x) = 0$ if $x - 3 = 0$ or if $x^2 + 6x + 3 = 0$; the solution of $x - 3 = 0$ is $x = 3$, and the solution of $x^2 + 6x + 3 = 0$ is $x = -3 \pm \sqrt{6}$.

12. $f'(x) = (3x+1)^{1/2} 2(x-1) + (x-1)^2 \dfrac{1}{2}(3x+1)^{-1/2}(3)$

$\qquad = \dfrac{1}{2}(3x+1)^{-1/2}(x-1)[4(3x+1) + 3(x-1)] = \dfrac{(x-1)(15x+1)}{2\sqrt{3x+1}}$

so $f'(x) = 0$ if $x = -1/15, 1$.

13. $f'(x) = 3\left[\dfrac{3x+1}{x^2}\right]^2 \dfrac{x^2(3) - (3x+1)(2x)}{x^4} = -\dfrac{3(3x+2)(3x+1)^2}{x^7}$ so $f'(x) = 0$ if $x = -2/3, -1/3$.

14. $f'(x) = 3 \cdot \dfrac{(3x-5)(1/3)(5x-1)^{-2/3}(5) - (5x-1)^{1/3}(3)}{(3x-5)^2}$

$\qquad = \dfrac{(5x-1)^{-2/3}[5(3x-5) - 9(5x-1)]}{(3x-5)^2} = \dfrac{-2(15x+8)}{(3x-5)^2(5x-1)^{2/3}}$

so $f'(x) = 0$ if $x = -8/15$.

15. $f'(x) = x^{1/2}(1/3)(x^2+x+1)^{-2/3}(2x+1) + (x^2+x+1)^{1/3}(1/2)x^{-1/2}$

$\qquad = \dfrac{1}{6}x^{-1/2}(x^2+x+1)^{-2/3}[2x(2x+1) + 3(x^2+x+1)]$

$\qquad = \dfrac{7x^2+5x+3}{6x^{1/2}(x^2+x+1)^{2/3}}$

but $7x^2 + 5x + 3 = 0$ has no real solutions so there are no values of x for which $f'(x) = 0$.

16. **(a)** by the chain rule;

$\qquad \dfrac{d}{dx}[f(ax)] = f'(ax)\dfrac{d}{dx}(ax) = \dfrac{1}{ax}(a) = \dfrac{1}{x} = \dfrac{d}{dx}[f(x)]$

$\qquad$ **(b)** $\dfrac{dy}{dx} = f'(\sin x)\dfrac{d}{dx}(\sin x) = \dfrac{1}{\sin x}(\cos x) = \cot x;$

$\qquad\qquad \dfrac{dv}{dx} = f'(1/x)\dfrac{d}{dx}(1/x) = \dfrac{1}{(1/x)}(-1/x^2) = -1/x.$

17. $\dfrac{d}{dx}(\sqrt{2}x^{-2} - \dfrac{2}{5}x^{-1}) = -2\sqrt{2}x^{-3} + \dfrac{2}{5}x^{-2}$

18. $\dfrac{dy}{dx} = \dfrac{(x^2-1)(6x) - (3x^2+7)(2x)}{(x^2-1)^2} = -\dfrac{20x}{(x^2-1)^2}$

19. $z = (2\sin r\cos r)^2 = \sin^2 2r$ so $\dfrac{dz}{dr} = 2(\sin 2r)(\cos 2r)(2) = 2\sin 4r$

and $\dfrac{dz}{dr}\bigg|_{r=\pi/6} = 2\sin(2\pi/3) = \sqrt{3}$

20. $g(x) = (2x)^{-1/2}$ so $g'(x) = -\dfrac{1}{2}(2x)^{-3/2}(2) = -1/(2x)^{3/2}$ and $g'(2) = -1/4^{3/2} = -1/8$

21. $u = \left[\dfrac{x-1}{x}\right]^2 = (1-x^{-1})^2$ so $\dfrac{du}{dx} = 2(1-x^{-1})(x^{-2}) = 2(x-1)/x^3$.

22. $w = \left(v^3 - v^{1/4}\right)^{1/5}$ so $\dfrac{dw}{dv} = \dfrac{1}{5}\left(v^3 - v^{1/4}\right)^{-4/5}\left(3v^2 - \dfrac{1}{4}v^{-3/4}\right)$

23. $\dfrac{d}{dx}(\sec^2 x - \tan^2 x) = \dfrac{d}{dx}(1) = 0$

24. $\dfrac{dy}{dx} = \dfrac{dy}{dt}\dfrac{dt}{dx} = \sec^2 t(-2\sin 2x)$, if $x = \pi/4$ then $t = \cos(\pi/2) = 0$

so $\dfrac{dy}{dx}\bigg|_{x=\pi/4} = \sec^2(0)(-2\sin(\pi/2)) = -2$

25. $F(x) = \dfrac{2x + 4x^3}{1 + 2x^2} = \dfrac{2x(1 + 2x^2)}{1 + 2x^2} = 2x$ so $F'(x) = 2$.

26. $\Phi(x) = \left(x^{3/2} - 4x^{1/2}\right)/5$, $x \neq 0$ so $\Phi'(x) = \dfrac{1}{5}\left(\dfrac{3}{2}x^{1/2} - 2x^{-1/2}\right) = (1/10)(3x-4)/\sqrt{x}$

27. $y' = 1 + x^{-2}$, and the slope of $2x - y = 5$ is 2 so we want $1 + x^{-2} = 2$ which gives $x^2 = 1$, $x = \pm 1$.

28. $y' = 6x^2 - 2x$, and the slope of $x + 4y = 10$ is $-1/4$ so we want $6x^2 - 2x = 4$ which results in $x = -2/3, 1$.

29. $y' = 2(x+2)$ so at $(x_0, f(x_0))$ the tangent line is $y - f(x_0) = 2(x_0 + 2)(x - x_0)$, or
$y - (x_0+2)^2 = 2(x_0+2)(x-x_0)$. But if the line passes through the origin then $x = 0$, $y = 0$ must satisfy the latter equation thus $-(x_0 + 2)^2 = -2x_0(x_0 + 2)$ which leads to $(x_0+2)(x_0 - 2) = 0$ so $x_0 = -2, 2$.

30. $y' = 1 - 2\cos 2x$; the tangent is horizontal where $1 - 2\cos 2x = 0$ so $\cos 2x = 1/2$,
$2x = \pm\pi/3 + 2k\pi$, $x = \pm\pi/6 + k\pi$ where $k = 0, \pm 1, \pm 2 \cdots$.

31. $y' = 3 - \sec^2 x$, and the slope of $y - x = 2$ is 1 so we want $3 - \sec^2 x = 1$ which gives $\sec^2 x = 2$, $\sec x = \pm\sqrt{2}$, $x = \pi/4 + k\pi/2$ where $k = 0, \pm 1, \pm 2, \cdots$.

32. $\Delta x = 1.5 - 2 = -0.5$, $\Delta y = y|_{x=1.5} - y|_{x=2} = 2 - 1 = 1$,

$dy = \dfrac{dy}{dx}\bigg|_{x=2} dx = -\dfrac{1}{(2-1)^2}(-0.5) = 0.5$.

33. $\Delta x = 0 - (-\pi/4) = \pi/4$, $\Delta y = y|_{x=0} - y|_{x=-\pi/4} = 0 - (-1) = 1$,

$dy = \sec^2(-\pi/4)(\pi/4) = \pi/2$.

34. $\Delta x = 3 - 0 = 3$, $\Delta y = y|_{x=3} - y|_{x=0} = \sqrt{16} - \sqrt{25} = -1$,

$dy = \dfrac{dy}{dx}\bigg|_{x=0} dx = -\dfrac{0}{\sqrt{25 - 0^2}}(3) = 0$.

35. **(a)** Consider $y = f(x) = \sqrt[3]{x}$ with $x = -8$ and $dx = -0.25 = -1/4$, then
$$f(-8.25) \approx f(-8) + dy, \quad \sqrt[3]{-8.25} \approx \sqrt[3]{-8} + \frac{1}{3}(-8)^{-2/3}(-1/4) = -2 - 1/48 = -97/48.$$

(b) Consider $y = f(x) = \cot x$ (x in radians) with $x = 45° = \pi/4$ radians and
$dx = 1° = \pi/180$ radians, then $f(\pi/4 + 180/\pi) \approx f(\pi/4) + dy$,
$\cot 46° \approx \cot 45° + (-\csc^2 45°)(\pi/180) = 1 - \pi/90.$

36. $V = x^3$ and $S = 6x^2$ where x is the length of an edge thus $x = (S/6)^{1/2}$ so $V = (S/6)^{3/2}$ and
$dV/dS = (3/2)(S/6)^{1/2}(1/6) = \sqrt{S/6}/4.$

37. **(a)** $dW/dt|_{t=5} = 200(t-15)|_{t=5} = -2000$ so water is running out at the rate of
2000 gal/min.

(b) average rate of change of $W = (W|_{t=5} - W|_{t=0})/5 = (10,000 - 22,500)/5 = -2500$ so water
flows out at an average rate of 2500 gal/min during the first 5 minutes.

38. $3(x+y)^2 \left(1 + \dfrac{dy}{dx}\right) + 3\left(x\dfrac{dy}{dx} + y\right) = 0$ so $\dfrac{dy}{dx} = -\dfrac{y + (x+y)^2}{x + (x+y)^2}.$

$\left.\dfrac{dy}{dx}\right|_{(-2,1)} = 2$, the tangent line is $y - 1 = 2(x+2)$, $y = 2x + 5.$

39. $2xy\dfrac{dy}{dx} + y^2 = \cos(x+2y)\left(1 + 2\dfrac{dy}{dx}\right)$ so $\dfrac{dy}{dx} = \dfrac{\cos(x+2y) - y^2}{2xy - 2\cos(x+2y)}.$

$\left.\dfrac{dy}{dx}\right|_{(0,0)} = -\dfrac{1}{2}$, the tangent line is $y = -x/2.$

40. $3(x+y)^2\left(1 + \dfrac{dy}{dx}\right) - 5 + \dfrac{dy}{dx} = 0$ so $\dfrac{dy}{dx} = \dfrac{5 - 3(x+y)^2}{1 + 3(x+y)^2}.$ To find the points of intersection,
replace $x+y$ by 1, and y by $1-x$ in $(x+y)^3 - 5x + y = 1$ to get $1 - 5x + 1 - x = 1$, so $x = 1/6$
and $y = 1 - 1/6 = 5/6$. $\left.\dfrac{dy}{dx}\right|_{(1/6, 5/6)} - \dfrac{1}{2}$, the tangent line is $y - \dfrac{5}{6} = \dfrac{1}{2}\left(x - \dfrac{1}{6}\right).$

41. $x = y = 1$ satisfies both equations so they intersect at the point $(1,1)$. For $2x^2 + 3y^2 = 5$,
$\dfrac{dy}{dx} = -\dfrac{2x}{3y}$ so $\left.\dfrac{dy}{dx}\right|_{(1,1)} = -\dfrac{2}{3}$. For $y^2 = x^3$, $\dfrac{dy}{dx} = \dfrac{3x^2}{2y}$ so $\left.\dfrac{dy}{dx}\right|_{(1,1)} = \dfrac{3}{2}$. The tangent lines are
perpendicular at $(1,1)$ because the slope of one curve is the negative reciprocal of the slope of the
other curve.

42. If $x^2 + y^2 = r^2$, then $\dfrac{dy}{dx} = -\dfrac{x}{y}$ so $\left.\dfrac{dy}{dx}\right|_{(x_0,y_0)} = -\dfrac{x_0}{y_0}$, $y_0 \neq 0$. The radius from the origin to P_0
has slope $\dfrac{y_0}{x_0}$, $x_0 \neq 0$. Thus, if $x_0 \neq 0$ and $y_0 \neq 0$, the slope of the tangent to the circle at P_0
is the negative reciprocal of the slope of the radius from the origin to P_0 so the tangent line is
perpendicular to the radius at P_0. If $x_0 = 0$, then the tangent line is horizontal and the radius
to P_0 is vertical. If $y_0 = 0$, then the circle has a vertical tangent at P_0 and the radius to P_0 is
horizontal. Thus the tangent line at P_0 is perpendicular to the radius from the origin at P_0 for
any point $P_0(x_0, y_0)$ on the circle.

43. $y = \cos x - 3\sin x$, $y' = -\sin x - 3\cos x$, $y'' = -\cos x + 3\sin x$, $y''' = \sin x + 3\cos x$
so $y''' + y'' + y' + y = (-3 - 1 + 3 + 1)\sin x + (1 - 3 - 1 + 3)\cos x = 0.$

44.   **(a)** $3y^2 \dfrac{dy}{dx} + 6x = 4\dfrac{dy}{dx}, \ \dfrac{dy}{dx} = \dfrac{6x}{4 - 3y^2}$

$$\dfrac{d^2y}{dx^2} = 6\dfrac{(4 - 3y^2)(1) - x(-6y \, dy/dx)}{(4 - 3y^2)^2} = 6\dfrac{4 - 3y^2 + 6xy[6x/(4 - 3y^2)]}{(4 - 3y^2)^2}$$

$$= 6\left[(4 - 3y^2)^2 + 36x^2 y\right]/(4 - 3y^2)^3.$$

(b) $\cos y \dfrac{dy}{dx} - \sin x = 0, \ \dfrac{dy}{dx} = \dfrac{\sin x}{\cos y}$

$$\dfrac{d^2y}{dx^2} = \dfrac{\cos y \cos x - \sin x(-\sin y)dy/dx}{\cos^2 y} = (\cos^2 y \cos x + \sin^2 x \sin y)/\cos^3 y.$$

CHAPTER 4

Applications of Differentiation

SECTION 4.1

4.1.1   A shark, looking for dinner, is swimming parallel to a straight beach and 90 feet offshore. The shark is swimming at the constant speed of 30 feet per second. At time $t = 0$, the shark is directly opposite a lifeguard station. How fast is the shark moving away from the lifeguard station when the distance between them is 150 feet?

4.1.2   A ladder 13 feet long is leaning against a wall. If the base of the ladder is moving away from the wall at the rate of 1/2 foot per second, at what rate will the top of the ladder be moving when the base of the ladder is 5 feet from the wall?

4.1.3   A spherical balloon is inflated so that its volume is increasing at the rate of 3 cubic feet per minute. How fast is the radius of the balloon increasing at the instant the radius is 1/2 foot?

$$\left[V = \frac{4}{3}\pi r^3 \right]$$

4.1.4   Sand is falling into a conical pile so that the radius of the base of the pile is always equal to one half its altitude. If the sand is falling at the rate of 10 cubic feet per minute, how fast is the altitude of the pile increasing when the pile is 5 feet deep?

$$\left[V = \frac{1}{3}\pi r^2 h \right]$$

4.1.5   A metal cone contracts as it cools. Assume the height of the cone is 16 cm and the radius at the base of the cone is 4 cm. If the height of the cone is decreasing at 4.0×10^{-5} cm per second, at what rate is the volume of the cone decreasing when its height is 15 cm?

$$\left[V = \frac{1}{3}\pi r^2 h \right]$$

4.1.6   A spherical balloon is inflated so that its volume is increasing at the rate of 20 cubic feet per minute. How fast is the surface area of the balloon increasing when the radius is 4 feet? [Use $V = \frac{4}{3}\pi r^3$ and $S = 4\pi r^2$.]

4.1.7   Two ships leave port at noon. One ship sails north at 6 miles per hour and the other sails east at 8 miles per hour. At what rate are the two ships separating 2 hours later?

4.1.8 A conical funnel is 14 inches in diameter and 12 inches deep. A liquid is flowing out at the rate of 40 cubic inches per second. How fast is the depth of the liquid falling when the level is 6 inches deep?

$$\left[V = \frac{1}{3}\pi r^2 h \right]$$

4.1.9 A baseball diamond is a square 90 feet on each side. A player is running from home to first base at the rate of 25 feet per second. At what rate is his distance from second base changing when he has run half way to first base?

4.1.10 A ship, proceeding southward on a straight course at the rate of 12 miles/hr is, at noon, 40 miles due north of a second ship, which is sailing west at 15 miles/hr.

  (a) How fast are the ships approaching each other 1 hour later?
  (b) Are the ships approaching each other or are they receding from each other at 2 o'clock and at what rate?

4.1.11 An angler has a fish at the end of his line, which is being reeled in at the rate of 2 feet per second from a bridge 30 feet above the water. At what speed is the fish moving through the water towards the bridge when the amount of line out is 50 feet? (Assume the fish is at the surface of the water and that there is no sag in the line.)

4.1.12 A kite is 150 feet high and is moving horizontally away from a boy at the rate of 20 feet per second. How fast is the string being paid out when the kite is 250 feet from him?

4.1.13 A kite is flying horizontally at a constant height of 250 feet above the girl flying the kite. At a certain instant, the angle which the string makes with the girl is 30° and decreasing. If the string is paying out at 16 feet per second, how fast is the angle decreasing? Express your answer in degrees per second.

4.1.14 Consider a rectangle where the sides are changing but the area is always 100 square inches. If one side changes at the rate of 3 inches per second, when it is 20 inches long, how fast is the other side changing?

4.1.15 The sides of an equilateral triangle are increasing at the rate of 5 centimeters per hour. At what rate is the area increasing when the side is 10 centimeters?

4.1.16 A circular cylinder has a radius r and a height h feet. If the height and radius both increase at the constant rate of 10 feet per minute, at what rate is the lateral surface area increasing?

$$(S = 2\pi r h)$$

4.1.17   A straw is used to drink soda from the bottom of a cylindrical shaped cup. The diameter of the cup is 3 inches. The liquid is being consumed at the rate of 3 cubic inches per second. How fast is the level of the soda dropping?

$$[V = \pi r^2 h]$$

4.1.18   The edge of a cube of side x is contracting. At a certain instant, the rate of change of the surface area is equal to 6 times the rate of change of its edge. Find the length of the edge.

4.1.19   An aircraft is climbing at a 30° angle to the horizontal. Find the aircraft's speed if it is gaining altitude at the rate of 200 miles per hour.

SOLUTIONS

SECTION 4.1

4.1.1 Let x and r be the distances shown on the diagram. We want to find $\left.\dfrac{dr}{dt}\right|_{r=150}$ given that $\dfrac{dx}{dt} = 30$. From $r^2 = (90)^2 + x^2$,

we get $2r\dfrac{dr}{dt} = 2x\dfrac{dx}{dt}$, or

$\dfrac{dr}{dt} = \dfrac{x}{r}\dfrac{dx}{dt}$. When $r = 150$,

$x^2 = (150)^2 - (90)^2 = 22500 - 8100 = 14400$

and $x = 120$, thus

$\left.\dfrac{dr}{dt}\right|_{r=150} = \dfrac{120}{150}(30) = 24$ ft/sec.

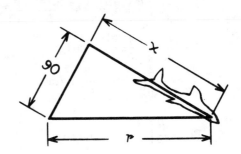

4.1.2 Let x and y be as shown on the diagram. We want to find

$\left.\dfrac{dy}{dt}\right|_{x=5}$ given that $\dfrac{dx}{dt} = \dfrac{1}{2}$.

From $y^2 + x^2 = (13)^2$, we get

$2y\dfrac{dy}{dt} + 2x\dfrac{dx}{dt} = 0$, or,

$\dfrac{dy}{dt} = -\dfrac{x}{y}\dfrac{dx}{dt}$. When $x = 5$,

$y^2 = (13)^2 - (5)^2 = 169 - 25 = 144$,

so $y = 12$ and $\left.\dfrac{dy}{dt}\right|_{x=5} = -\dfrac{5}{12}\left(\dfrac{1}{2}\right) = -\dfrac{5}{24}$ ft/sec.,

i.e., the top of the ladder is moving down

at the rate of $\dfrac{5}{24}$ ft/sec.

4.1.3 We want to find $\left.\dfrac{dr}{dt}\right|_{r=1/2}$ given that $\dfrac{dV}{dt} = 3$. So, from $V = \dfrac{4}{3}\pi r^3$, we get $\dfrac{dV}{dt} = \dfrac{4}{3}\pi(3)r^2\dfrac{dr}{dt}$

or $\dfrac{dr}{dt} = \dfrac{1}{4\pi r^2}\dfrac{dV}{dt}$ and $\left.\dfrac{dr}{dt}\right|_{r=1/2} = \dfrac{1}{4\pi\left(\dfrac{1}{2}\right)^2}(3) = \dfrac{3}{\pi}$ ft/min.

4.1.4   Find $\dfrac{dh}{dt}\Big|_{h=5}$ given that $\dfrac{dV}{dt} = 10$.

Since $V = \dfrac{1}{3}\pi r^2 h$ and $r = \dfrac{h}{2}$, then

$$V = \frac{1}{3}\pi\left(\frac{h}{2}\right)^2 h = \frac{\pi}{12}h^3,\quad \frac{dV}{dt} = \frac{\pi}{12}(3)h^2\frac{dh}{dt},$$

and $\dfrac{dh}{dt} = \dfrac{4}{\pi h^2}\dfrac{dV}{dt}$, thus

$$\frac{dh}{dt} = \frac{4}{\pi(5)^2}(10) = \frac{8}{5\pi}\ \text{ft/min.}$$

4.1.5   Find $\dfrac{dv}{dt}\Big|_{h=15\text{cm}}$ given that

$\dfrac{dh}{dt} = -4.0\times 10^{-5}$ cm/sec. Since

$$V = \frac{1}{3}\pi r^2 h \text{ and } r = \frac{1}{4}h,$$

then $V = \dfrac{1}{3}\pi\left(\dfrac{h}{4}\right)^2 h = \dfrac{\pi h^3}{48}$,

$$\frac{dV}{dt} = \frac{\pi}{48}(3)h^2\frac{dh}{dt} = \frac{\pi}{16}h^2\frac{dh}{dt}$$

$$\frac{dV}{dt}\Big|_{n=15} = \frac{\pi}{16}(15)^2(-4.0\times 10^{-5}) = -1.8\times 10^{-3}\ \text{cm}^3/\text{sec,}$$

i.e., the volume of the cone is

decreasing at the rate of

1.8×10^{-3} cm^3/sec.

4.1.6   Find $\dfrac{dS}{dt}\Big|_{r=4}$ given that $\dfrac{dV}{dt} = 20$. From $V = \dfrac{4}{3}\pi r^3$ and $S = 4\pi r^2$, we get

$$\frac{dV}{dt} = \frac{4}{3}\pi(3)r^2\frac{dr}{dt} \text{ and } \frac{dS}{dt} = 4\pi(2)r\frac{dr}{dt}. \text{ Then, solving for } \frac{dr}{dt},\ \frac{dr}{dt} = \frac{1}{4\pi r^2}\frac{dV}{dt} = \frac{1}{8\pi r}\frac{dS}{dt}$$

or $\dfrac{dS}{dt} = \dfrac{2}{r}\dfrac{dV}{dt}$ so, $\dfrac{dS}{dt}\Big|_{r=4} = \dfrac{2}{4}(20) = 10$, thus, the surface area is increasing at the rate of

10 sq. ft/min.

4.1.7   Let A and B be the two ships and
x, y, and r be their distances
as shown on the diagram. We
want to find $\left.\dfrac{dr}{dt}\right|_{t=2\text{ hrs}}$, given

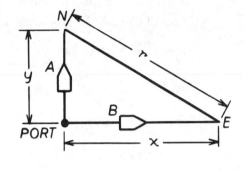

that $\dfrac{dy}{dt} = 6$ and $\dfrac{dx}{dt} = 8$. From

$r^2 = x^2 + y^2$, we get $2r\dfrac{dr}{dt} = 2x\dfrac{dx}{dt} + 2y\dfrac{dy}{dt}$ or

$\dfrac{dr}{dt} = \dfrac{1}{r}\left[x\dfrac{dx}{dt} + y\dfrac{dy}{dt}\right]$.

When $t = 2$ hours, B will have sailed 16 miles east and A will have sailed 12 miles north of
port, thus, $r^2 = (16)^2 + (12)^2 = 400$, $r = 20$. $\left.\dfrac{dr}{dt}\right|_{t=2} = \dfrac{1}{20}[16(8)+12(6)] = 10$, thus, the ships
are separating at the rate of 10 miles per hour.

4.1.8   Let r and h be the dimensions
shown on the diagram. We want to
find $\left.\dfrac{dh}{dt}\right|_{h=6}$ given that $\dfrac{dV}{dt} = -40$.

By similar triangles (see figure),

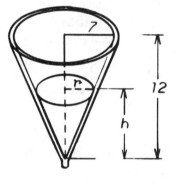

$\dfrac{r}{h} = \dfrac{7}{12}$ or $r = \dfrac{7h}{12}$, thus,

$V = \dfrac{\pi}{3}\left(\dfrac{7h}{12}\right)^2 h = \dfrac{49\pi}{432}h^3$.

$\dfrac{dV}{dt} = \dfrac{49\pi}{432}(3)h^2\dfrac{dh}{dt}$ or $\dfrac{dh}{dt} = \dfrac{144}{49\pi h^2}\dfrac{dV}{dt}$,

thus, $\left.\dfrac{dh}{dt}\right|_{h=6} = \dfrac{144}{49\pi(6)^2}(-40) = -\dfrac{160}{49\pi}$.

The depth of the liquid is falling at the rate of $-\dfrac{160}{49\pi}$ inches per second.

4.1.9 Let x, y, and r be the distances as shown in the diagram and let the baseball diamond be positioned as shown. We want to find $\left.\dfrac{dr}{dt}\right|_{x=45}$

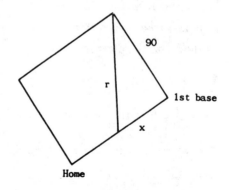

given that $\dfrac{dx}{dt} = -25$. Thus

$x^2 + y^2 = r^2$. $2x\dfrac{dx}{dt} = 2r\dfrac{dr}{dt}$

[y is constant] so $\dfrac{dr}{dt} = \dfrac{x}{r}\dfrac{dx}{dt}$

when $x = 45$, $r = \sqrt{(45)^2 + (90)^2} = \sqrt{2025 + 8100} = \sqrt{10125} = 45\sqrt{5}$ feet, so,

$\dfrac{dr}{dt} = \dfrac{45}{45\sqrt{5}}(-25) = -5\sqrt{5}$ feet per second.

4.1.10 **(a)** Let A be one ship and B be the other. Let x, y, and r be the dimensions shown in the figure at a certain instant of time. We want to find $\left.\dfrac{dr}{dt}\right|_{t=1}$, since x and y are both functions of time, let $x = 15t$ and $y = 40 - 12t$, then, $r^2 = x^2 + y^2 = (15t)^2 + (40 - 12t)^2$;

$2r\dfrac{dr}{dt} = 2(15t)(15) + 2(40 - 12t)(-12)$

or $\dfrac{dr}{dt} = \dfrac{1}{r}[369t - 480]$.

When $t = 1$, $r^2 = [15(1)]^2 + [40 - 12(1)]^2 = 1009$, $r = \sqrt{1009}$ and

$\dfrac{dr}{dt} = \dfrac{1}{\sqrt{1009}}[369(1) - 480] = -\dfrac{111}{\sqrt{1009}} \approx -3.49$ miles/hr. The ships are approaching

each other at the rate of $\dfrac{111}{\sqrt{1009}} \approx 3.49$ miles/hr 1 hour later.

(b) At 2 o'clock, $t = 2$ hours, so, $r^2 = [15(2)]^2 + [40 - 12(2)]^2 = 1156$,

$r = 34$. $\left.\dfrac{dr}{dt}\right|_{t=2} = \dfrac{1}{34}[369(2) - 480] = \dfrac{258}{34}$ or $\dfrac{129}{17}$. At 2 o'clock, the ships are receding

at the rate of $\dfrac{129}{17}$ miles/hour.

4.1.11   Let x and r be as shown on the diagram. We want to find

$$\left.\frac{dx}{dt}\right|_{r=50} \quad \text{given that} \quad \frac{dr}{dt} = -2.$$

From $x^2 + (30)^2 = r^2$,

$$2x\frac{dx}{dt} = 2r\frac{dr}{dt} \quad \text{and} \quad \frac{dx}{dt} = \frac{r}{x}\frac{dr}{dt}.$$

When $r = 50$,

$$x^2 = (50)^2 - (30)^2$$
$$= 2500 - 900 = 1600, x = 40$$

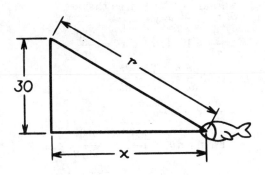

and $\left.\dfrac{dx}{dt}\right|_{r=50} = \dfrac{50}{40}(-2)$

$$= -\frac{100}{40} = -\frac{5}{2}.$$

The fish is moving towards the bridge at the rate of $\dfrac{5}{2}$ ft/sec when the amount of line out is 50 feet.

4.1.12   Let x and r be as shown on the diagram. We want to find $\left.\dfrac{dr}{dt}\right|_{r=250}$

given that $\dfrac{dx}{dt} = 20.$ From

$r^2 = (150)^2 + x^2$, we get

$$2r\frac{dr}{dt} = 2x\frac{dx}{dt} \quad \text{or} \quad \frac{dr}{dt} = \frac{x}{r}\frac{dx}{dt}.$$

When $r = 250$,

$$x^2 = (250)^2 - (150)^2$$
$$= 62500 - 22500 = 40000;$$

$x = 200$ and $\left.\dfrac{dr}{dt}\right|_{r=250} = \dfrac{200}{250}(20) = 16.$

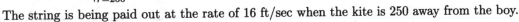

The string is being paid out at the rate of 16 ft/sec when the kite is 250 away from the boy.

4.1.13   Let θ and r be as shown in the diagram.

Find $\dfrac{d\theta}{dt}\bigg|_{\theta=30°}$ given that

$\dfrac{dr}{dt} = 16$. From $\sin\theta = \dfrac{250}{r}$,

$\cos\theta\dfrac{d\theta}{dt} = -\dfrac{250}{r^2}\dfrac{dr}{dt}$,

$\dfrac{d\theta}{dt} = -250\dfrac{\sec\theta}{r^2}\dfrac{dr}{dt}$.

From $\sin\theta = \dfrac{250}{r}$, $r = \dfrac{250}{\sin\theta}$.

When $\theta = 30°$, $r = \dfrac{250}{\sin 30} = 500$. So

$\dfrac{d\theta}{dt} = -250\dfrac{\sec 30°}{(500)^2}(16) = .0185$ deg/sec.,

i.e., the angle of the string is

decreasing at the rate of 0.0185 deg/sec.

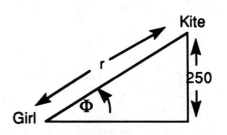

4.1.14   Let $xy = 100$. We want to find, say, $\dfrac{dy}{dt}\bigg|_{x=20}$ given that $\dfrac{dx}{dt} = 3$; $x\dfrac{dy}{dt} + y\dfrac{dx}{dt} = 0$ and

$\dfrac{dy}{dt} = -\dfrac{y}{x}\dfrac{dx}{dt}$. From $20y = 100$, $y = 5$, and $\dfrac{dy}{dt} = -\dfrac{5}{20}(3) = -\dfrac{3}{4}$ inch/sec, thus, the rate of

change of the second side is opposite to that of the first side and is at the rate of $\dfrac{3}{4}$ inch/sec.

4.1.15   Let x = side of the triangle.

The area is $A = \dfrac{1}{2}(x)\left(\dfrac{\sqrt{3}}{2}x\right) = \dfrac{\sqrt{3}}{4}x^2$.

We want to find $\dfrac{dA}{dt}\bigg|_{x=10}$ given that

$\dfrac{dx}{dt} = 5$, $\dfrac{dA}{dt} = \dfrac{\sqrt{3}}{4}(2)x\dfrac{dx}{dt}$. Thus,

$\dfrac{dA}{dt}\bigg|_{x=10} = \dfrac{\sqrt{3}}{2}(10)(5) = 25\sqrt{3}$ or

the area is increasing at the rate of

$25\sqrt{3}$ cm^2/hr when the side is 10 cm.

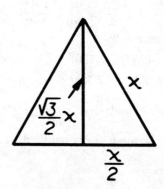

4.1.16   We want to find $\dfrac{dS}{dt}$ given that $\dfrac{dh}{dt} = \dfrac{dr}{dt} = 10$, from $S = 2\pi rh$, $\dfrac{dS}{dt} = 2\pi\left[r\dfrac{dh}{dt} + h\dfrac{dr}{dt}\right]$ or

$\dfrac{dS}{dt} = 2\pi[r(10) + h(10)] = 20\pi(r + h)$ ft^2 min.

4.1.17 With d and h as shown in the figure find $\dfrac{dh}{dt}$ given that $\dfrac{dV}{dt} = -3$ in^3/sec.

From $V = \pi r^2 h = \pi \left(\dfrac{d}{2}\right)^2 h = \dfrac{\pi}{4} d^2 h$

$\dfrac{dV}{dt} = \dfrac{\pi}{4} d^2 \dfrac{dh}{dt}$ and

$\dfrac{dh}{dt} = \dfrac{4}{\pi d^2} \dfrac{dV}{dt} = \dfrac{4}{\pi (3)^2}(-3) = \dfrac{4}{3\pi}$

or $-.42$ in/sec., i.e., the level of the soda is dropping at the rate of $\dfrac{4}{3\pi}$ in/sec. or 0.42 in/sec.

4.1.18 We want to find x given that $\dfrac{dS}{dt} = 6\dfrac{dx}{dt}$. From $S = 6x^2$, we get $\dfrac{dS}{dt} = 6(2)x\dfrac{dx}{dt}$, but $\dfrac{dS}{dt} = 6\dfrac{dx}{dt}$, so $6\dfrac{dx}{dt} = 12x\dfrac{dx}{dt}$, $6\dfrac{dx}{dt}(2x - 1) = 0$; $x = 1/2$.

4.1.19 With s and h as shown in the figure, we want to find $\dfrac{ds}{dt}$ given that $\dfrac{dh}{dt} = 200$ mph.

From the figure, $s = \dfrac{h}{\sin 30°} = 2h$ so

$\dfrac{ds}{dt} = 2\dfrac{dh}{dt} = 2(200) = 400$ mph.

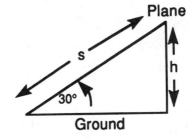

SECTION 4.2

4.2.1 $f(x) = x^4 - 24x^2$

  (a) Find the largest intervals where f is increasing and where f is decreasing.
  (b) Find the largest intervals where f is concave up and where f is concave down.
  (c) Find the location of any inflection points.

4.2.2 $f(x) = x^4 - 4x^3$

  (a) Find the largest intervals where f is increasing and where f is decreasing.
  (b) Find the largest intervals where f is concave up and where f is concave down.
  (c) Find the location of any inflection points.

4.2.3 $f(x) = x^4 + 8x^3 + 24$

  (a) Find the largest intervals where f is increasing and where f is decreasing.
  (b) Find the largest intervals where f is concave up and where f is concave down.
  (c) Find the location of any inflection points.

4.2.4 $f(x) = 5x^4 - x^5$

  (a) Find the largest intervals where f is increasing and where f is decreasing.
  (b) Find the largest intervals where f is concave up and where f is concave down.
  (c) Find the location of any inflection points.

4.2.5 $f(x) = 4x^3 - 15x^2 - 18x + 10$

  (a) Find the largest intervals where f is increasing and where f is decreasing.
  (b) Find the largest intervals where f is concave up and where f is concave down.
  (c) Find the location of any inflection points.

4.2.6 $f(x) = x(x - 6)^2$

  (a) Find the largest intervals where f is increasing and where f is decreasing.
  (b) Find the largest intervals where f is concave up and where f is concave down.
  (c) Find the location of any inflection points.

4.2.7 $f(x) = x^3 - 5x^2 + 3x + 1$

    (a) Find the largest intervals where f is increasing and where f is decreasing.
    (b) Find the largest intervals where f is concave up and where f is concave down.
    (c) Find the location of any inflection points.

4.2.8 $f(x) = 3x^4 - 4x^3 + 1$

    (a) Find the largest intervals where f is increasing and where f is decreasing.
    (b) Find the largest intervals where f is concave up and where f is concave down.
    (c) Find the location of any inflection points.

4.2.9 $f(x) = x(x + 4)^3$

    (a) Find the largest intervals where f is increasing and where f is decreasing.
    (b) Find the largest intervals where f is concave up and where f is concave down.
    (c) Find the location of any inflection points.

4.2.10 $f(x) = (x - 4)^4 + 4$

    (a) Find the largest intervals where f is increasing and where f is decreasing.
    (b) Find the largest intervals where f is concave up and where f is concave down.
    (c) Find the location of any inflection points.

4.2.11 $f(x) = x(x - 3)^5$

    (a) Find the largest intervals where f is increasing and where f is decreasing.
    (b) Find the largest intervals where f is concave up and where f is concave down.
    (c) Find the location of any inflection points.

4.2.12 $f(x) = \sin 2x \, (0, \pi)$

    (a) Find the largest intervals where f is increasing and where f is decreasing.
    (b) Find the largest intervals where f is concave up and where f is concave down.
    (c) Find the location of any inflection points.

4.2.13   Are the following true or false?

   (a) If $f''(x) > 0$ on the open interval (a, b) then $f'(x)$ is increasing on (a, b).
   (b) If $f''(x) > 0$ on the open interval (a, b) then $f(x)$ is increasing on (a, b).
   (c) If $f''(x) = 0$, then x is a point of inflection.
   (d) If x_0 is a point of inflection, then $f''(x_0) = 0$.
   (e) If $f'(x)$ is decreasing on (a, b), then $f(x)$ is concave down on (a, b).

4.2.14   Which of the following is correct if $f'(x) < 0$ and $f''(x) > 0$ on (a, b):

   (a) $f(x)$ is increasing and concave up.
   (b) $f(x)$ is decreasing and concave up.
   (c) $f(x)$ is increasing and concave down.
   (d) $f(x)$ is decreasing and concave down.

4.2.15   Sketch a continuous curve having the following properties:
   $f(-3) = 27$, $f(0) = 27/2$, $f(3) = 0$, $f'(x) > 0$ for $|x| > 3$
   $f'(-3) = f'(3) = 0$, $f''(x) < 0$ for $x < 0$, $f''x) > 0$ for $x > 0$.

4.2.16   Sketch a continuous curve $y = f(x)$ for $x > 0$ if $f(1) = 0$, and $f'(x) = 1/x$ for all $x > 0$. Is the curve concave up or concave down?

4.2.17   Sketch a continuous curve having the following properties:
   $f(0) = 4$, $f(-2) = f(2) = 0$; $f'(x) > 0$ for $(-\infty, 0)$ and
   $f'(x) < 0$ for $(0, +\infty)$, $f''(x) < 0$ for $(-\infty, +\infty)$.

SOLUTIONS

SECTION 4.2

4.2.1 $f'(x) = 4x^3 - 48x$, $f''(x) = 12x^2 - 48$

    (a) Increasing $[-2\sqrt{3}, 0]$, $[2\sqrt{3}, +\infty)$ decreasing $(-\infty, -2\sqrt{3}]$, $[0, 2\sqrt{3}]$

    (b) Concave up $(-\infty, -2)$, $(2, +\infty)$; concave down $(-2, 2)$

    (c) $(-2, -80)$ and $(2, -80)$

4.2.2 $f'(x) = 4x^3 - 12x^2$, $f''(x) = 12x^2 - 24x$

    (a) Increasing $[3, +\infty)$; decreasing $(-\infty, 3]$

    (b) Concave up $(-\infty, 0)$, $(2, +\infty)$; concave down $(0, 2)$

    (c) $(0, 0)$ and $(2, -16)$

4.2.3 $f'(x) = 4x^3 + 24x^2$, $f''(x) = 12x^2 + 48x$

    (a) Increasing $[-6, +\infty)$; decreasing $(-\infty, -6]$

    (b) Concave up $(-\infty, -4)$, $(0, +\infty)$; concave down $(-4, 0)$

    (c) $(-4, -232)$ and $(0, 24)$

4.2.4 $f'(x) = 20x^3 - 5x^4$, $f''(x) = 60x^2 - 20x^3$

    (a) Increasing $[0, 4]$; decreasing $(-\infty, 0]$, $[4, +\infty)$

    (b) Concave up $(-\infty, 0)$, $(0, 3)$; concave down $(3, +\infty)$

    (c) $(3, 162)$

4.2.5 $f'(x) = 12x^2 - 30x - 18$, $f''(x) = 24x - 30$

    (a) Increasing $(-\infty, -1/2]$, $[3, +\infty)$; decreasing $[-1/2, 3]$

    (b) Concave up $(5/4, +\infty)$; concave down $(-\infty, 5/4)$

    (c) $\left(\dfrac{5}{4}, -\dfrac{225}{16}\right)$

4.2.6 $f'(x) = (x - 6)(3x - 6)$, $f''(x) = 6x - 24$

    (a) Increasing $(-\infty, 2]$, $[6, +\infty)$; decreasing $[2, 6]$

    (b) Concave up $(4, +\infty)$; concave down $(-\infty, 4)$

    (c) $(4, 16)$

4.2.7 $f'(x) = 3x^2 - 10x + 3$, $f''(x) = 6x - 10$

    (a) Increasing $(-\infty, 1/3]$, $[3, +\infty)$; decreasing $[1/3, 3]$

    (b) Concave up $(5/3, +\infty)$; concave down $(-\infty, 5/3)$

    (c) $\left(\dfrac{5}{3}, -\dfrac{88}{27} \right)$

4.2.8 $f'(x) = 12x^3 - 12x^2$, $f''(x) = 36x^2 - 24x$

    (a) Increasing $[1, +\infty)$; decreasing $(-\infty, 1]$

    (b) Concave up $(-\infty, 0)$, $(2/3, +\infty)$; concave down $(0, 2/3)$

    (c) $(0, 1)$ and $(2/3, 11/27)$

4.2.9 $f'(x) = (x + 4)^2(4x + 4)$, $f''(x) = 12(x + 4)(x + 2)$

    (a) Increasing $[-1, +\infty)$; decreasing $(-\infty, -1]$

    (b) Concave up $(-\infty, -4)$, $(-2, +\infty)$; concave down $(-4, -2)$

    (c) $(-4, 0)$ and $(-2 - 16)$

4.2.10 $f'(x) = 4(x - 4)^3$, $f''(x) = 12(x - 4)^2$

    (a) Increasing $[4, +\infty)$; decreasing $(-\infty, 4]$

    (b) Concave up $(-\infty, +\infty)$

    (c) no inflection points

4.2.11 $f'(x) = (x - 3)^4(6x - 3)$, $f''(x) = 30(x - 3)^3(x - 1)$

    (a) Increasing $[1/2, +\infty)$; decreasing $(-\infty, 1/2]$

    (b) Concave up $(-\infty, 1)$, $(3, +\infty)$; concave down $(1, 3)$

    (c) $(1, -32)$ and $(3, 0)$

4.2.12 $f'(x) = 2\cos 2x$; $f''(x) = -4\sin 2x$

    (a) Increasing $(0, \pi/4]$, $[3\pi/4, \pi)$; decreasing $[\pi/4, 3\pi/4]$

    (b) Concave up $[\pi/2, \pi)$; concave down $(0, \pi/2]$

    (c) $(\pi/2, 0)$

4.2.13   **(a)** True   **(b)** True   **(c)** False

           (d) True   **(e)** True

4.2.14   (b) only

4.2.15

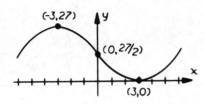

4.2.16 $f''(x) = -\dfrac{1}{x^2}$
Concave down for $(0, +\infty)$

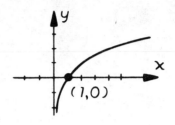

4.2.17

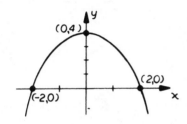

SECTION 4.3

4.3.1 Find the relative extrema for $f(x) = 3x^5 - 5x^4$.

4.3.2 Find the relative extrema for $f(x) = 12x^{2/3} - 16x$.

4.3.3 Find the relative extrema for $f(x) = x^{2/3}(5-x)$.

4.3.4 Find the relative extrema for $f(x) = \frac{2}{5}x^{5/3} + 8x^{2/3}$.

4.3.5 Find the relative extrema for $f(x) = \frac{1}{3}x^{4/3} - \frac{4}{3}x^{1/3}$.

4.3.6 Find the relative extrema for $f(x) = \frac{x^4}{4} - 2x^2 + 1$.

4.3.7 The derivative of a continuous function is $f'(x) = 2(x-1)^2(2x+1)$. Find all critical points and determine whether a relative maximum, relative minimum or neither occurs there.

4.3.8 The derivative of a continuous function is $f'(x) = \frac{2}{3}x^{\frac{1}{3}} - \frac{2}{3}x^{-\frac{2}{3}}$. Find all critical points and determine whether a relative maximum, relative minimum or neither occurs there.

4.3.9 Find the relative extrema for $f(x) = 2x + 2x^{2/3}$.

4.3.10 Find the relative extrema for $f(x) = \frac{1}{x} - \frac{1}{3x^3}$.

4.3.11 Find the relative extrema for $f(x) = x^{4/3} - 4x^{-1/3}$.

4.3.12 Find the relative extrema for $f(x) = 6x^2 - 9x + 5$.

4.3.13 Find the relative extrema for $f(x) = x^4 - 6x^2 + 17$.

4.3.14 Find the relative extrema for $f(x) = (x+1)^{-\frac{1}{3}}$.

4.3.15 Find the relative extrema for $f(x) = x + \cos 2x$, $0 < x < \pi$.

4.3.16 Find the relative extrema for $f(x) = x - \sin 2x$, $0 < x < \pi$.

4.3.17 Which of the following statements is correct if $f'(x_0) = 0$ and $f''(x_0) = 0$:

    (a) x_0 is a local minimum   (b) x_0 is a local maximum

    (c) x_0 is a point of inflection   (d) Any one of (a), (b), (c) may happen.

4.3.18 Which of the following statements about the graph of $f(x) = 2x^4 + x + 1$ is correct:

    (a) There is a relative minimum at $x = -\dfrac{1}{2}$ and a point of inflection at $x = 0$.

    (b) There is a relative maximum at $x = -\dfrac{1}{2}$ and a point of inflection at $x = 0$.

    (c) There are no relative extrema, but there is a point of inflection at $x = 0$.

    (d) There is a relative minimum at $x = -\dfrac{1}{2}$, but there is no point of inflection.

    (e) There are no local extrema and no points of inflection.

4.3.19 Which of the following statements about the graph of $g(x) = (x^2 - 1)^3$ is correct:

    (a) There are three relative minima and two points of inflection.

    (b) There are two relative minima and three points of inflection.

    (c) There is one local minimum and four points of inflection.

    (d) There are no local minima and five points of inflection.

    (e) There are two relative minima and two points of inflection.

SOLUTIONS

SECTION 4.3

4.3.1 $f'(x) = 5x^3(3x - 4)$;

critical points $x = 0, 4/3$

relative maximum of 0 at $x = 0$; $f'(x)$:

relative minimum of $\dfrac{-256}{81}$ at

$x = 4/3$ by first derivative test.

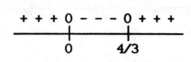

4.3.2 $f'(x) = 8x^{-1/3} - 16$;

critical points $x = 0, 1/8$

relative minimum of 0 at $x = 0$; $f'(x)$:

relative maximum of 1 at $x = 1/8$

by first derivative test.

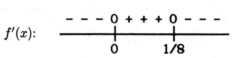

4.3.3 $f'(x) = \dfrac{10}{3}x^{-1/3} - \dfrac{5}{3}x^{2/3}$;

critical points, $x = 0, 2$

relative minimum of 0 at $x = 0$; $f'(x)$:

relative maximum of $3(2)^{2/3}$

at $x = 2$ by first derivative test.

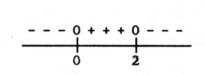

4.3.4 $f'(x) = \dfrac{2}{5}\dfrac{5}{3}x^{2/3} + 8\left(\dfrac{2}{3}\right)x^{-1/3}$;

critical points, $x = -8, 0$ $f'(x)$:

relative maximum of $\dfrac{96}{5}$ at $x = -8$;

relative minimum of 0 at $x = 0$

by first derivative test.

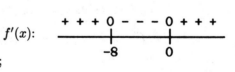

4.3.5 $f'(x) = \dfrac{4}{9}x^{1/3} - \dfrac{4}{9}x^{-2/3}$;

critical points, $x = 0, 1$ $f'(x)$:

relative minimum of -1 at

$x = 1$ by first derivative test.

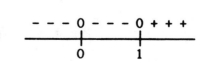

179

4.3.6 $f'(x) = x(x-2)(x+2)$; critical points $x = -2, 0, 2$. $f''(x) = 3x^2 - 4$; $f''(-2) > 0$, $f''(0) < 0$, $f''(2) > 0$. Relative minimum of -3 at $x = -2$, relative maximum of 1 at $x = 0$, relative minimum of -3 at $x = 2$.

4.3.7 $f'(x) = 2(x-1)^2(2x+1)$;

critical points $x = -1/2, 1$ $\qquad$ $f'(x)$:

relative minimum of $-\dfrac{27}{16}$ at

$x = -1/2$ by first derivative test.

$$- - - \; 0 \; + + + \; 0 \; + + +$$
$$-1/2 \qquad 1$$

4.3.8 $f'(x) = \dfrac{2}{3}x^{1/3} - \dfrac{2}{3}x^{-2/3}$;

critical points $x = 0, 1$; $\qquad$ $f'(x)$:

relative minimum of $-\dfrac{3}{2}$ at

$x = 1$ by first derivative test.

$$- - - \; 0 \; - - - \; 0 \; + + +$$
$$0 \qquad 1$$

4.3.9 $f'(x) = 2 + \dfrac{4}{3}x^{-1/3}$;

critical points $x = \dfrac{-8}{27}, 0$; $\qquad$ $f'(x)$:

relative maximum of $8/27$

at $x = \dfrac{-8}{27}$; relative minimum of

0 at $x = 0$ by first derivative test.

$$+ + + \; 0 \; - - - \; 0 \; + + +$$
$$-\frac{8}{27} \qquad 0$$

4.3.10 $f'(x) = -\dfrac{1}{x^2} + \dfrac{1}{x^4}$; critical points $x = -1, 1$ ($x = 0$ is not a critical point since $x = 0$ is not in the domain of f). $f''(x) = -\dfrac{2}{x^3} - \dfrac{4}{x^5}$; $f''(-1) > 0$, $f''(1) < 0$. Relative minimum of $-2/3$ at $x = -1$, relative maximum of $2/3$ at $x = 1$.

4.3.11 $f'(x) = \dfrac{4}{3}x^{1/3} + \dfrac{4}{3}x^{-4/3}$; critical points $x = -1$, ($x = 0$ is not a critical point since $x = 0$ is not in the domain of f). $f''(x) = \dfrac{4}{9}x^{-2/3} - \dfrac{16}{9}x^{-7/3}$; $f''(-1) > 0$, a relative minimum of 5 at $x = -1$.

4.3.12 $f'(x) = 12x - 9$; critical point $x = \dfrac{3}{4}$. $f''(x) = 12$, $f''(3/4) > 0$, relative minimum of $\dfrac{13}{8}$ at $x = \dfrac{3}{4}$.

4.3.13 $f'(x) = 4x\left(x^2 - 3\right)$; critical points $x = -\sqrt{3}, 0, \sqrt{3}$. $f''(x) = 12\left(x^2 - 1\right)$; $f''\left(-\sqrt{3}\right) > 0$, $f''(0) < 0$, $f''\left(\sqrt{3}\right) > 0$, relative minimum of 8 at $x = -\sqrt{3}$, relative maximum of 17 at $x = 0$, relative minimum of 8 at $x = \sqrt{3}$.

4.3.14 $f'(x) = -\dfrac{1}{3(x+1)^{4/3}}$; no critical points, ($x = -1$ is not in the domain of f)

4.3.15 $f'(x) = 1 - 2\sin 2x$; critical points $\dfrac{\pi}{12}, \dfrac{5\pi}{12}$

$f''(x) = -4\cos 2x$; $f''\left(\dfrac{\pi}{12}\right) < 0$; $f''\left(\dfrac{5\pi}{12}\right) > 0$;

relative maximum of $\dfrac{\pi}{12} + \dfrac{\sqrt{3}}{2}$ at $x = \dfrac{\pi}{12}$;

relative minimum of $\dfrac{5\pi}{12} - \dfrac{\sqrt{3}}{2}$ at $x = \dfrac{5\pi}{12}$

4.3.16 $f'(x) = 1 - 2\cos 2x$; critical points $x = \dfrac{\pi}{6}, 5\dfrac{\pi}{6}$. $f''(x) = 4\sin 2x$; $f''\left(\dfrac{\pi}{6}\right) > 0$;

$f''\left(\dfrac{5\pi}{6}\right) < 0$; relative minimum of $\dfrac{\pi}{6} - \dfrac{\sqrt{3}}{2}$ at $x = \dfrac{\pi}{6}$, relative maximum of $\dfrac{5\pi}{6} + \dfrac{\sqrt{3}}{2}$ at $x = \dfrac{5\pi}{6}$.

4.3.17 (d) **4.3.18** (d) **4.3.19** (c)

SECTION 4.4

4.4.1 Sketch the graph of $y = 5 - 2x - x^2$. Plot any stationary points and any points of inflection.

4.4.2 Sketch the graph of $y = x^3 - 9x^2 + 24x - 7$. Plot any stationary points and any points of inflection.

4.4.3 Sketch the graph of $y = x^3 + 6x^2$. Plot any stationary points and any points of inflection.

4.4.4 Sketch the graph of $y = x^3 - 5x^2 + 8x - 4$. Plot any stationary points and any points of inflection.

4.4.5 Sketch the graph of $y = x^3 - 12x + 6$. Plot any stationary points and any points of inflection.

4.4.6 Sketch the graph of $y = x^3 - 6x^2 + 9x + 6$. Plot any stationary points and any points of inflection.

4.4.7 Sketch the graph of $y = 3x^4 - 4x^3 + 1$. Plot any stationary points and any points of inflection.

4.4.8 Sketch the graph of $y = x^2(9 - x^2)$. Plot any stationary points and any points of inflection.

4.4.9 Sketch the graph of $y = x^4 - 2x^2 + 7$. Plot any stationary points and any points of inflection.

4.4.10 Sketch the graph of $y = x^3 + \frac{3}{2}x^2 - 6x + 12$. Plot any stationary points and any points of inflection.

4.4.11 Sketch the graph of $y = \left(\dfrac{x-3}{x-1}\right)^2$. Plot any stationary points and any points of inflection. Show any horizontal and vertical asymptotes.

4.4.12 Sketch the graph of $y = \dfrac{x^2}{x^2+1}$. Plot any stationary points and any points of inflection. Show any horizontal and vertical asymptotes.

4.4.13 Sketch the graph of $y = \dfrac{x^2-x}{(x+1)^2}$. Plot any stationary points and any points of inflection. Show any horizontal and vertical asymptotes.

4.4.14 Sketch the graph of $y = \dfrac{3x^2}{x^2-4}$. Plot any stationary points and any points of inflection. Show any horizontal and vertical asymptotes.

4.4.15   Sketch the graph of $y = \dfrac{8}{4 - x^2}$. Plot any stationary points and any points of inflection. Show any horizontal and vertical asymptotes.

4.4.16   Sketch the graph of $y = \dfrac{x^2}{x^2 - 9}$. Plot any stationary points and any points of inflection. Show any vertical and horizontal asymptotes.

4.4.17   Sketch the graph of $y = \dfrac{1}{x - 3} + 1$. Plot any stationary points and any points of inflection. Show any vertical and horizontal asymptotes.

4.4.18   Sketch the graph of $y = 2 - \dfrac{3}{x} - \dfrac{3}{x^2}$. Plot any stationary points and any points of inflection. Show any vertical and horizontal asymptotes.

4.4.19   Sketch $y = 1 + \dfrac{2}{x} - \dfrac{1}{x^2}$. Plot any stationary points and any points of inflection. Show any vertical and horizontal asymptotes.

4.4.20   Sketch the graph of $y = \dfrac{x^2 - 3}{x}$. Show all vertical, horizontal, and oblique asymptotes.

4.4.21   Sketch the graph of $y = \dfrac{x^2 - 2x - 2}{x + 1}$. Show all vertical, horizontal and oblique asymptotes.

SOLUTIONS

SECTION 4.4

4.4.1 $y = 5 - 2x - x^2$

$y' = -2 - 2x$

$y'' = -2$

Relative max at $(-1, 6)$

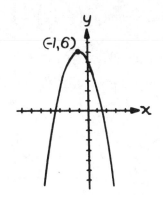

4.4.2 $y = x^3 - 9x^2 + 24x - 7$

$y' = 3x^2 - 18x + 24$

$y'' = 6x - 18$

Relative maximum at $(2, 13)$

Inflection point at $(3, 11)$

Relative minimum at $(4, 9)$

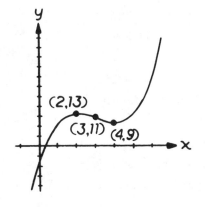

4.4.3 $y = x^3 + 6x^2$

$y' = 3x^2 + 12x$

$y'' = 6x + 12$

Relative maximum at $(-4, 32)$

Inflection point at $(-2, 16)$

Relative minimum at $(0, 0)$

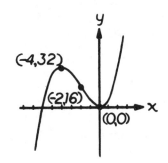

4.4.4 $y = x^3 - 5x^2 + 8x - 4$

$y' = 3x^2 - 10x + 8$

$y'' = 6x - 10$

Relative maximum at $\left(\dfrac{4}{3}, \dfrac{4}{27}\right)$

Inflection point at $\left(\dfrac{5}{3}, \dfrac{2}{27}\right)$

Relative minimum at $(2, 0)$

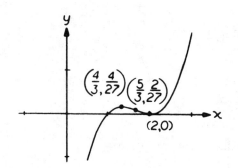

4.4.5 $y = x^3 - 12x + 6$

$y' = 3x^2 - 12$

$y'' = 6x$

Relative maximum at $(-2, 22)$

Inflection point at $(0, 6)$

Relative minimum at $(2, -10)$

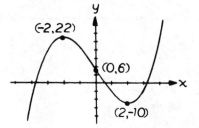

4.4.6 $y = x^3 - 6x^2 + 9x + 6$

$y' = 3x^2 - 12x + 9$

$y'' = 6x - 12$

Relative maximum at $(1, 10)$

Inflection point at $(2, 8)$

Relative minimum at $(3, 6)$

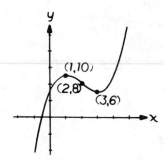

4.4.7 $y = 3x^4 - 4x^3 + 1$

$y' = 12x^3 - 12x^2$

$y'' = 36x^2 - 24x$

Inflection points at $(0, 1)$ and $\left(\dfrac{2}{3}, \dfrac{11}{27}\right)$

Relative minimum at $(1, 0)$

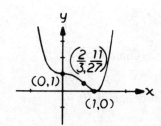

4.4.8 $y = 9x^2 - x^4$

$y' = 18x - 4x^3$

$y'' = 18 - 12x^2$

Relative maxima at $\left(\pm\dfrac{3}{\sqrt{2}}, \dfrac{81}{4}\right)$

Relative minimum at $(0,0)$

Inflection points at $\left(\pm\sqrt{\dfrac{3}{2}}, \dfrac{45}{4}\right)$

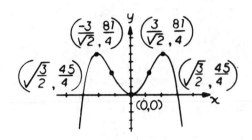

4.4.9 $y = x^4 - 2x^2 + 7$

$y' = 4x^3 - 4x$

$y'' = 12x^2 - 4$

Relative maximum $(0,7)$

Relative minima $(\pm 1, 6)$

Inflection points $\left(\pm\dfrac{1}{\sqrt{3}}, \dfrac{58}{9}\right)$

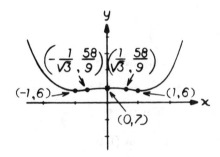

4.4.10 $y = x^3 + \frac{3}{2}x^2 - 6x + 12$

$y' = 3x^2 + 3x - 6$

$y'' = 6x + 3$

Relative maximum at $(-2, 22)$

Inflection point at $\left(-\dfrac{1}{2}, \dfrac{61}{4}\right)$

Relative minimum at $\left(1, \dfrac{17}{2}\right)$

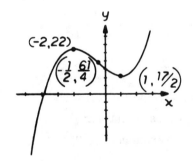

4.4.11 $y = \left(\dfrac{x-3}{x-1}\right)^2$

$y' = \dfrac{4(x-3)}{(x-1)^3}$, $y'' = \dfrac{8(4-x)}{(x-1)^4}$

Vertical asymptote at $x = 1$

Horizontal asymptote at $y = 1$

Relative minimum at $(3, 0)$

Inflection point at $\left(4, \dfrac{1}{9}\right)$

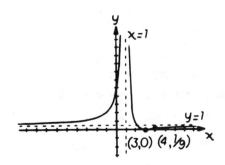

4.4.12 $y = \dfrac{x^2}{x^2 + 1}$, $y' = \dfrac{2x}{(x^2 + 1)^2}$,

$y'' = \dfrac{2\left(1 - 3x^2\right)}{\left(x^2 + 1\right)^3}$

Horizontal asymptote at $y = 1$

Inflection points at $\left(\pm\dfrac{1}{\sqrt{3}}, \dfrac{1}{4}\right)$

Relative minimum at $(0, 0)$

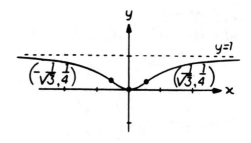

4.4.13 $y = \dfrac{x^2 - x}{(x + 1)^2}$, $y' = \dfrac{3x - 1}{(x + 1)^3}$

$y'' = \dfrac{6(1 - x)}{(x + 1)^4}$

Vertical asymptote at $x = -1$

Horizontal asymptote at $y = 1$

Relative minimum at $\left(\dfrac{1}{3}, -\dfrac{1}{8}\right)$

Inflection point at $(1, 0)$

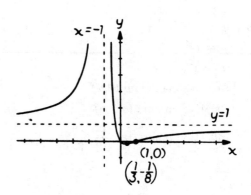

4.4.14 $y = \dfrac{3x^2}{x^2 - 4}$, $y' = -\dfrac{24x}{\left(x^2 - 4\right)^2}$

$y'' = \dfrac{24\left(3x^2 + 4\right)}{\left(x^2 - 4\right)^3}$

Vertical asymptotes at $x = \pm 2$

Horizontal asymptotes at $y = 3$

Relative maximum at $(0, 0)$

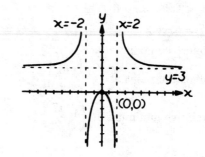

4.4.15 $y = \dfrac{8}{4 - x^2}$, $y' = \dfrac{16x}{\left(4 - x^2\right)^2}$

$y'' = \dfrac{16\left(4 + 3x^2\right)}{\left(4 - x^2\right)^3}$

Vertical asymptotes at $x = \pm 2$

Horizontal asymptotes at $y = 0$

Relative minimum at $(0, 2)$

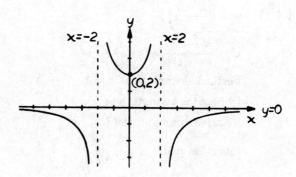

4.4.16 $y = \dfrac{x^2}{x^2 - 9},\ y' = -\dfrac{18x}{(x^2 - 9)^2}$

$y'' = \dfrac{54\left(x^2 + 3\right)}{\left(x^2 - 9\right)^3}$

Vertical asymptotes at $x = \pm 3$

Horizontal asymptotes at $y = 1$

Relative maximum at $(0,0)$

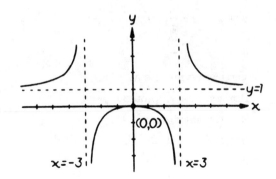

4.4.17 $y = \dfrac{1}{x - 3} + 1,\ y' = -\dfrac{1}{(x - 3)^2}$

$y'' = \dfrac{2}{(x - 3)^3}$

Vertical asymptote at $x = 3$

Horizontal asymptote at $y = 1$.

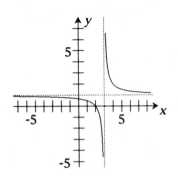

4.4.18 $y = 2 - \dfrac{3}{x} - \dfrac{3}{x^2},\ y' = \dfrac{3(x + 2)}{x^3}$

$y'' = \dfrac{-6(x + 3)}{x^4}$

Vertical asymptote at $x = 0$

Horizontal asymptote at $y = 2$

Relative maximum at $(-2, 11/4)$

Inflection point at $(-3, 8/3)$

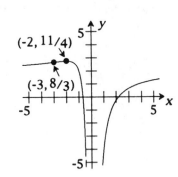

4.4.19 $y = 1 + \dfrac{2}{x} - \dfrac{1}{x^2}$, $y' = \dfrac{2(1-x)}{x^3}$

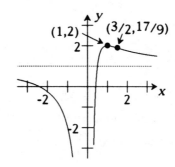

$y'' = \dfrac{2(2x-3)}{x^4}$

Vertical asymptote at $x = 0$

Horizontal asymptote at $y = 1$

Relative maximum at $(1, 2)$

Inflection point at $(3/2, 17/9)$

4.4.20 $y = \dfrac{x^2 - 3}{x} = x - \dfrac{3}{x}$

so $y = x$ is an oblique asymptote

$y' = \dfrac{x^2 + 3}{x^2}$

$y'' = -\dfrac{6}{x^3}$

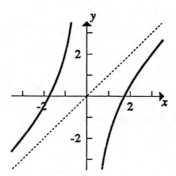

4.4.21 $y = \dfrac{x^2 - 2x - 2}{x + 1} = x - 3 + \dfrac{1}{x+1}$

so $y = x - 3$ is an oblique asymptote

$y' = \dfrac{x(x+2)}{(x+1)^2}$

$y'' = \dfrac{2}{(x+1)^3}$

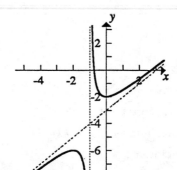

SECTION 4.5

4.5.1 Sketch the graph of $y = 1 + (x - 2)^{1/3}$. Plot any stationary points and any inflections points.

4.5.2 Sketch the graph of $y = x^{1/3}(x + 4)$. Plot any stationary points and any inflections points.

4.5.3 Sketch the graph of $y = (x + 1)^{1/3}(x - 4)$. Plot any stationary points and any inflections points.

4.5.4 Sketch the graph of $y = (x + 1)^{2/3}$. Plot any stationary points, inflections points, and cusps which may or may not exist.

4.5.5 Sketch the graph of $y = x^{2/3}(x + 5)$. Plot any stationary points, inflections points, and cusps which may or may not exist.

4.5.6 Sketch the graph of $y = x(x - 3)^{2/3}$. Plot any stationary points, inflections points, and cusps which may or may not exist.

4.5.7 Sketch the graph of $y = (x - 2)^{2/3} - 1$. Plot any stationary points, inflections points, and cusps which may or may not exist.

4.5.8 Sketch the graph of $y = x^{2/3}(x - 3)^2$. Plot any stationary points, inflection points, and cusps which may or may not exist.

4.5.9 Sketch the graph of $y = (x - 1)^{4/5}$. Plot any stationary points, inflection points, and cusps which may or may not exist.

4.5.10 Sketch the graph of $y = \sqrt{4 - x^2}$. Plot any stationary points.

4.5.11 Sketch the graph of $y = \sqrt{\dfrac{x}{4 - x}}$.

4.5.12 Sketch the graph of $y = \sqrt{x}(x - 2)$. Plot any stationary points and any inflection points.

4.5.13 Sketch the graph of $y = x - 2\sqrt{x}$. Plot any stationary points and any inflection points.

4.5.14 Sketch the graph of $y = \dfrac{1}{5}x^{5/2} - x^{3/2}$. Plot any stationary and any inflection points.

4.5.15 Sketch the graph of $y = \dfrac{1}{2}x^{4/3} - 2x^{1/3}$. Plot any stationary points and points of inflection.

SOLUTIONS

SECTION 4.5

4.5.1 $y = 1 + (x-2)^{1/3}$

$y' = \dfrac{1}{3(x-2)^{2/3}}$

$y'' = -\dfrac{2}{9(x-2)^{5/3}}$

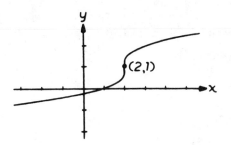

Inflection point at $(2,1)$

Vertical tangent at $(2,1)$

4.5.2 $y = x^{1/3}(x+4)$

$y' = \dfrac{4(x+1)}{3x^{2/3}}$

$y'' = \dfrac{4(x-2)}{9x^{5/3}}$

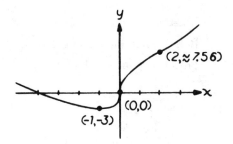

Relative minimum at $(-1,-3)$

Inflection points at $(0,0)$ and $(2, \approx 7.56)$

Vertical tangent at $(0,0)$

4.5.3 $y = (x+1)^{1/3}(x-4)$

$y' = \dfrac{4x-1}{3(x+1)^{2/3}}$

$y'' = \dfrac{4x+14}{9(x+1)^{5/3}}$

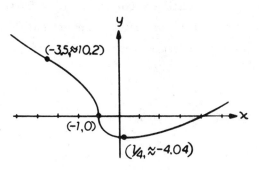

Relative minimum at $\left(\dfrac{1}{4}, \approx -4.04\right)$

Inflection points at $(-3.5, \approx 10.2)$ and $(-1,0)$

Vertical tangent $(-1,0)$

4.5.4 $y = (x+1)^{2/3}$

$y' = \dfrac{2}{3}(x+1)^{-1/3}$

$y'' = -\dfrac{2}{9}(x+1)^{-4/3}$

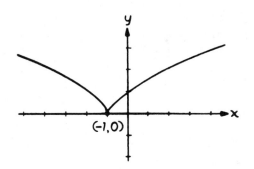

Relative minimum at $(-1, 0)$

Cusp at $(-1, 0)$

4.5.5 $y = x^{2/3}(x+5)$

$y' = \dfrac{5(x+2)}{3x^{1/3}}$

$y'' = \dfrac{10(x-1)}{9x^{4/3}}$

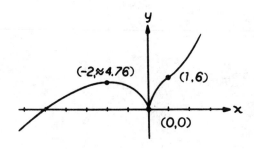

Relative maximum at $(-2, \approx 4.76)$

Relative minimum and cusps at $(0, 0)$

Inflection point at $(1, 6)$

4.5.6 $y = x(x-3)^{2/3}$

$y' = \dfrac{5x-9}{3(x-3)^{1/3}}$

$y'' = \dfrac{10x-36}{9(x-3)^{4/3}}$

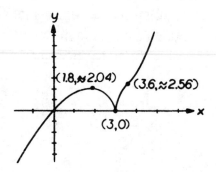

Relative maximum at $(1.8, \approx 2.04)$

Cusp and relative minimum at $(3, 0)$

Inflection point at $(3.6, \approx 2.56)$

4.5.7 $y = (x-2)^{2/3} - 1$

$y' = \dfrac{2}{3(x-2)^{1/3}}$

$y'' = -\dfrac{2}{9(x-2)^{4/3}}$

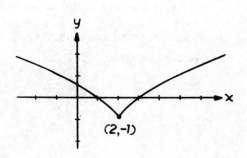

Relative minimum and cusp at $(2, -1)$

4.5.8 $y = x^{2/3}(x-3)^2$

$$y' = \frac{2(x-3)(4x-3)}{3x^{1/3}}$$

$$y'' = \frac{2(20x^2 - 30x - 9)}{9x^{4/3}}$$

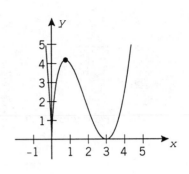

Relative maximum $(3/4, \approx 4.18)$

Relative minimum and cusp $(0,0)$,

relative minimum $(3,0)$

Inflection points $(\approx 1.76, \approx 2.24)$, $(\approx -0.26, \approx 4.33)$

4.5.9 $y = (x-1)^{4/5}$

$$y' = \frac{4}{5}(x-1)^{-1/5}$$

$$y'' = -\frac{4}{25}(x-1)^{-6/5}$$

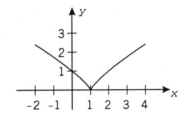

Relative minimum and cusp at $(1,0)$

4.5.10 $y = \sqrt{4-x^2}$

$$y' = -\frac{x}{\sqrt{4-x^2}}$$

$$y'' = -\frac{4}{(4-x^2)^{3/2}}$$

Relative maximum at $(0,2)$

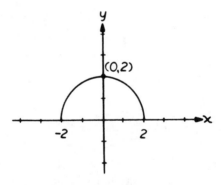

4.5.11 $y = \sqrt{\dfrac{x}{4-x}}$

$$y' = \frac{2}{x^{1/2}(4-x)^{3/2}}$$

$$y'' = \frac{4x-4}{x^{3/2}(4-x)^{5/2}}$$

Inflection point at $\left(1, \dfrac{\sqrt{3}}{3}\right)$

$x = 4$ is a vertical asymptote.

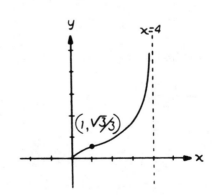

4.5.12 $y = \sqrt{x}(x - 2)$

$y' = \dfrac{3x - 2}{2x^{1/2}}$

$y'' = \dfrac{3x + 2}{4x^{3/2}}$

Relative minimum at $\left(\dfrac{2}{3}, -\dfrac{4\sqrt{6}}{9}\right)$

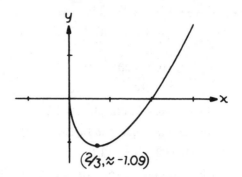

4.5.13 $y = x - 2\sqrt{x}$

$y' = \dfrac{\sqrt{x} - 1}{\sqrt{x}}$

$y'' = \dfrac{1}{2x^{3/2}}$

Relative minimum at $(1, -1)$

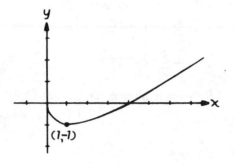

4.5.14 $y = \dfrac{1}{5}x^{5/2} - x^{3/2}$

$y' = \dfrac{\sqrt{x}(x - 3)}{2}$

$y'' = \dfrac{3(x - 1)}{4\sqrt{x}}$

Inflection point at $\left(1, -\dfrac{4}{5}\right)$

Relative minimum at $(3, \approx -2.08)$

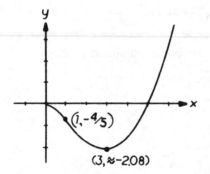

4.5.15 $y = \dfrac{1}{2}x^{4/3} - 2x^{1/3}$

$y' = \dfrac{2(x - 1)}{3x^{2/3}}$

$y'' = \dfrac{2(x + 2)}{9x^{5/3}}$

Inflection points at $(-2, \approx 3.8)$ and $(0, 0)$

Relative minimum at $(1, -1.5)$

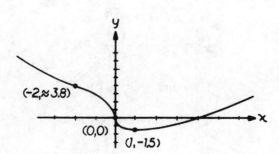

SECTION 4.6

4.6.1 Find the extreme values for $f(x) = \dfrac{x}{2} + 2$ on the interval $[0, 100]$ and determine where those occur.

4.6.2 Find the extreme values for $f(x) = 2x^3 - 3x^2 - 12x + 8$ on the interval $[-2, 2]$ and determine where those values occur.

4.6.3 Find the extreme values for $f(x) = \dfrac{x^3}{3} - x^2 - 3x + 1$ on the interval $[-1, 2]$ and determine where those values occur.

4.6.4 Find the extreme values for $f(x) = 2x^3 - 3x^2 - 12x + 5$ on the interval $[0, 4]$ and determine where those values occur.

4.6.5 Find the extreme values for $f(x) = x^3 - 6x^2 + 5$ on the interval $[-1, 5]$ and determine where those values occur.

4.6.6 Find the extreme values for $f(x) = x^3 + \dfrac{3}{2}x^2 - 18x + 4$ on the interval $[0, 4]$ and determine where those values occur.

4.6.7 Find the extreme values for $f(x) = x - \sin x$ on the interval $\left[-\dfrac{\pi}{2}, \dfrac{\pi}{2}\right]$ and determine where those values occur.

4.6.8 Find the extreme values for $f(x) = 1 - x^{2/3}$ on the interval $[-1, 1]$ and determine where those values occur.

4.6.9 Find the extreme values for $f(x) = 2\sec x - \tan x$ on the interval $\left[-\dfrac{\pi}{4}, \dfrac{\pi}{4}\right]$ and determine where those values occur.

4.6.10 Find the extreme values for $f(x) = x^{4/3} - 3x^{1/3}$ on the interval $[-1, 8]$ and determine where those values occur.

4.6.11 Find the extreme values for $f(x) = \dfrac{\sqrt{x}}{x^2 + 3}$ on the interval $(0, +\infty)$ and determine where those values occur.

4.6.12 Find the extreme values for $f(x) = \dfrac{x}{x^2 + 1}$ on the interval $[0, 2]$ and determine where those values occur.

4.6.13 Find the extreme values for $f(x) = \dfrac{|x|}{1 + |x|}$ on the interval $[0, +\infty)$ and determine where those values occur.

4.6.14 Find the extreme values for $f(x) = \dfrac{1}{x - x^2}$ on the interval $(0, 1)$ and determine where those values occur.

4.6.15 Find the extreme values for $f(x) = \begin{cases} x^2 & x < 0 \\ x^3 & x \geq 0 \end{cases}$ $(-\infty, +\infty)$ and determine where those values occur.

4.6.16 Find the extreme values for $f(x) = \begin{cases} -x - 1 & x < -1 \\ 1 - x^2 & -1 \leq x \leq 1 \\ x - 1 & x < 1 \end{cases}$ on the interval $[-2, +2]$ and determine where those values occur.

4.6.17 Find the extreme values for $f(x) = \begin{cases} 1 - x^2 & x < 0 \\ x^3 - 1 & x \geq 0 \end{cases}$ on the interval $[-2, 1]$ and determine where those values occur.

4.6.18 Find the extreme values for $f(x) = |3 - 2x|$ on the interval $[-2, 2]$ and determine where those values occur.

SOLUTIONS

SECTION 4.6

4.6.1 $f(x) = \dfrac{x}{2} + 2$; $f'(x) = \dfrac{1}{2}$, no critical points. $f(0) = 2$ and $f(100) = 52$ so f has a maximum of 52 at $x = 100$ and a minimum of 2 at $x = 2$.

4.6.2 $f(x) = 2x^3 - 3x^2 - 12x + 8$; $f'(x) = 6x^2 - 6x - 12 = 6(x-2)(x+1)$. $f'(x) = 0$ for $x = -1$ and $x = 2$. $f(-2) = 4$; $f(-1) = 15$; $f(2) = -12$ so f has a maximum of 15 at $x = -1$ and a minimum of -12 at $x = 2$.

4.6.3 $f(x) = \dfrac{x^3}{3} - x^2 - 3x + 1$; $f'(x) = x^2 - 2x - 3 = (x-3)(x+1)$. $f'(x) = 0$ for $x = -1$ and $x = 3$, but $x = 3$ is outside the interval. So $f(-1) = \dfrac{8}{3}$, $f(2) = -\dfrac{19}{3}$, thus f has a maximum of $\dfrac{8}{3}$ at $x = -1$ and a minimum of $-\dfrac{19}{3}$ at $x = 2$.

4.6.4 $f(x) = 2x^3 - 3x^2 - 12x + 5$; $f'(x) = 6x^2 - 6x - 12 = 6(x-2)(x+1)$. $f'(x) = 0$ when $x = -1$ and $x = 2$, however, $x = -1$ is outside the interval. $f(0) = 5$, $f(2) = -15$, and $f(4) = 37$, thus f has a maximum of 37 at $x = 4$ and a minimum of -15 at $x = 2$.

4.6.5 $f(x) = x^3 - 6x^2 + 5$; $f'(x) = 3x^2 - 12x = 3x(x-4)$. $f'(x) = 0$ when $x = 0$ and $x = 4$. $f(-1) = -2$; $f(0) = 5$; $f(4) = -27$; and $f(5) = -20$, thus, f has a maximum of 5 at $x = 0$ and a minimum of -27 at $x = 4$.

4.6.6 $f(x) = x^3 + \dfrac{3}{2}x^2 - 18x + 4$; $f'(x) = 3x^2 + 3x - 18 = 3(x-2)(x+3)$. $f'(x) = 0$ for $x = 2$ and $x = -3$. $f(0) = 4$; $f(2) = -18$; $f(4) = -20$, so f has a maximum of 4 at $x = 0$ and a minimum of -20 at $x = 4$.

4.6.7 $f(x) = x - \sin x$; $f'(x) = 1 - \cos x$. $f'(x) = 0$ when $x = 0$. $f\left(-\dfrac{\pi}{2}\right) = -\dfrac{\pi}{2} + 1$; $f(0) = 0$; $f\left(\dfrac{\pi}{2}\right) = \dfrac{\pi}{2} - 1$, so f has a maximum of $\dfrac{\pi - 2}{2}$ at $x = \dfrac{\pi}{2}$ and a minimum of $-\dfrac{(\pi - 2)}{2}$ at $x = -\dfrac{\pi}{2}$.

4.6.8 $f(x) = 1 - x^{2/3}$; $f'(x) = -\dfrac{2}{3x^{1/3}}$. $f'(x)$ does not exist at $x = 0$. $f(-1) = 0$, $f(0) = 1$, and $f(1) = 0$, thus, f has a maximum of 1 at $x = 0$ and a minimum of 0 which occurs at $x = -1$ and $x = 1$.

4.6.9 $f(x) = 2\sec x + \tan x$; $f'(x) = 2\sec x \tan x + \sec^2 x$. $f'(x) = 0$ for $x = -\dfrac{\pi}{6}$. $f\left(-\dfrac{\pi}{4}\right) = 2\sqrt{2} - 1$; $f\left(-\dfrac{\pi}{6}\right) = \dfrac{\sqrt{3}}{3}$; $f\left(\dfrac{\pi}{4}\right) = 2\sqrt{2} + 1$, so f has a maximum of $2\sqrt{2} + 1$ at $x = \dfrac{\pi}{4}$ and a minimum of $\dfrac{\sqrt{3}}{3}$ at $x = -\dfrac{\pi}{6}$.

4.6.10 $f(x) = x^{4/3} - 3x^{1/3}$; $f'(x) = \dfrac{4x - 3}{3x^{2/3}}$. $f'(x) = 0$ when $x = 3/4$ and $f'(x)$ does not exist when $x = 0$. $f(-1) = 4$; $f(0) = 0$; $f(3/4) = -\dfrac{9}{4}\left(\dfrac{3}{4}\right)^{1/3} \approx -2.04$; $f(8) = 10$. Thus, the maximum value is 10 at $x = 8$ and the minimum value is $-\dfrac{9}{4}\left(\dfrac{3}{4}\right)^{1/3} \approx -2.04$ at $x = \dfrac{3}{4}$.

4.6.11 $f(x) = \dfrac{\sqrt{x}}{x^2 + 3}$, $f'(x) = \dfrac{3 - 3x^2}{2\sqrt{x}\,(x^2 + 3)^2}$. $f'(x) = 0$ for $x = -1$, $x = 0$ and $x = 1$, however, $x = 0$ and $x = -1$ are outside the interval, thus f has a maximum of $1/4$ at $x = 1$ (first derivative test). There is no minimum.

4.6.12 $(x) = \dfrac{x}{x^2 + 1}$; $f'(x) = \dfrac{1 - x^2}{(x^2 + 1)^2}$. $f'(x) = 0$ when $x = -1$ and $x = 1$, however, $x = -1$ is outside the interval. $f(0) = 0$; $f(1) = 1/2$; and $f(2) = 2/5$, so f has a maximum of $1/2$ at $x = 1$ and a minimum of 0 at $x = 0$.

4.6.13 $f(x) = \dfrac{x}{1 + x}$ for x in $[0, +\infty)$; $f'(x) = \dfrac{1}{(1 + x)^2}$ so there are no critical points. $f(0) = 0$, thus, f has a minimum of 0 at $x = 0$. There is no maximum.

4.6.14 $f(x) = \dfrac{1}{x - x^2}$, $f'(x) = \dfrac{2x - 1}{(x - x^2)^2}$. $f'(x) = 0$ when $x = 1/2$; $f'(x)$ does not exist at $x = 0$ or $x = 1$, however, both of these values are outside the interval. f has a minimum of $+4$ at $x = 1/2$, there is no maximum.

4.6.15 $f(x) = \begin{cases} x^2 & x < 0 \\ x^3 & x \geq 0 \end{cases}$; $f'(x) = \begin{cases} 2x & x < 0 \\ 3x^2 & x > 0 \end{cases}$.

$f'(x) = 0$ when $x = 0$ which corresponds to a minimum value (first derivative test), there is no maximum.

4.6.16 $f(x) = \begin{cases} x - 1, & x < -1 \\ 1 - x^2, & -1 \leq x \leq 1 \\ x - 1, & x > 1 \end{cases}$; $f'(x) = \begin{cases} -1, & x < -1 \\ -2x, & -1 < x < 1 \\ 1, & x > 1 \end{cases}$.

$f'(x) = 0$ when $x = 0$, $f'(x)$ does not exist at $x = -1$ or $x = 1$. $f(-2) = 1$; $f(-1) = 0$; $f(0) = 1$; $f(1) = 0$; $f(2) = 1$, thus, f has a maximum of 1 at $x = -2$, $x = 0$, and $x = 2$; f has a minimum of 0 at $x = -1$ and $x = 1$.

4.6.17 $f(x) = \begin{cases} 1 - x^2, & x < 0 \\ x^3 - 1, & x \geq 0 \end{cases}$; $f'(x) = \begin{cases} -2x, & x < 0 \\ 3x^2, & x > 0 \end{cases}$.

$f'(x) = 0$ when $x = 0$. $f(-2) = -3$, $f(0) = -1$, $f(1) = 0$, thus, f has a maximum of 0 at $x = 1$ and a minimum of -3 at $x = -2$.

4.6.18 $f(x) = 3 - 2x = \begin{cases} 3 - 2x, & x \leq 3/2 \\ -3 + 2x, & x > 3/2 \end{cases}$; $f'(x) = \begin{cases} -2, & x < 3/2 \\ 2, & x > 3/2 \end{cases}$.

$f'(x)$ does not exist at $x = 3/2$. $f(-2) = 7$, $f\left(\dfrac{3}{2}\right) = 0$, $f(2) = 1$, thus, f has a maximum of 7 at $x = -2$ and a minimum of 0 at $x = 3/2$.

SECTION 4.7

4.7.1 Express the number 25 as a sum of two nonnegative terms whose product is as large as possible.

4.7.2 A rectangular lot is to be bounded by a fence on three sides and by a wall on the fourth side. Two kinds of fencing will be used with heavy duty fencing selling for $4 a foot on the side opposite the wall. The two remaining sides will use standard fencing selling for $3 a foot. What are the dimensions of the rectangular plot of greatest area that can be fenced in at a cost of $6600?

4.7.3 A sheet of cardboard 18 in square is used to make an open box by cutting squares of equal size from the corners and folding up the sides. What size squares should be cut to obtain a box with largest possible volume?

4.7.4 Prove that $(2,0)$ is the closest point on the curve $x^2 + y^2 = 4$ to $(4,0)$.

4.7.5 Find the dimensions of the rectangle of greatest area that can be inscribed in a circle of radius a.

4.7.6 A divided field is to be constructed with 4000 feet of fence as shown. For what value of x will the area be a maximum?

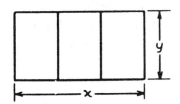

4.7.7 Find the dimensions of the rectangle of maximum area which may be embedded in a right triangle with sides of length 12, 16, and 20 feet as shown in the figure.

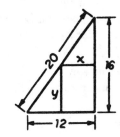

4.7.8   The infield of a 440 yard track
consists of a rectangle and 2
semicircles. To what dimensions
should the track be built in
order to maximize the area of
the rectangle?

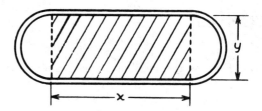

4.7.9   A long strip of copper 8 inches wide is to be made into a rain gutter by turning up the sides
to form a trough with a rectangular cross section. Find the dimensions of the cross section if
the carrying capacity of the trough is to be a maximum.

4.7.10   An isosceles triangle is drawn with its vertex at the origin and its base parallel to the x axis.
The vertices of the base are on the curve $5y = 25 - x^2$. Find the area of the largest such
triangle.

4.7.11   The strength of a beam with a
rectangular cross section varies
directly as x and as the square
of y. What are the dimensions
of the strongest beam that can
be sawed out of a round log whose
diameter is d? See figure on
right.

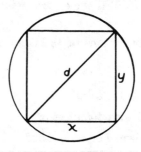

4.7.12   Find the area of the largest possible isosceles triangle with 2 sides equal to 6.

4.7.13   A lighthouse is 8 miles off a straight coast and a town is located 18 miles down the seacoast.
Supplies are to be moved from the town to the lighthouse on a regular basis and at a minimum
time. If the supplies can be moved at the rate of 7 miles/hour on water and 25 miles/hour
over land, how far from the town should a dock be constructed for shipment of supplies?

4.7.14   Find the circular cylinder of largest lateral area which can be inscribed in a sphere of radius
4 feet. [Surface area of a cylinder, $S = 2\pi rh$, where r = radius, h = height.]

4.7.15   If three sides of a trapezoid are 10 inches long, how long should the fourth side be if the area
is a maximum? [Area of a trapezoid $= \dfrac{1}{2}(a+b)h$ where a and b are the lengths of the parallel
sides and h = height.]

4.7.16 The stiffness of a beam of rectangular cross section is proportional to the product xy^3. Find the stiffest beam which can be cut from a log 2 feet in diameter. See figure on right.

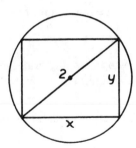

4.7.17 Find the dimensions of the rectangular area of maximum area which can be laid out within a triangle of base 12 and altitude 4 if one side of the rectangle lies on the base of the triangle.

4.7.18 Find the dimensions of the rectangle of greatest area with base on the x-axis and its other two vertices above the x-axis and on $4y = 16 - x^2$.

4.7.19 Find the dimensions of the trapezoid of greatest area with longer base on the x-axis and its other two vertices above the x-axis on $4y = 16 - x^2$. [Area of a trapezoid $= \frac{1}{2}(a + b)h$ where a and b are the length of the bases and h is the height.]

4.7.20 A poster is to contain 50 in^2 of printed matter with margins of 4 in. each at top and bottom and 2 in. at each side. Find the overall dimensions if the total area of the poster is to be a minimum.

4.7.21 A rancher is going to build a 3-sided cattle enclosure with a divider down the middle as shown to the right. The cost per foot of the 3 side walls will be \$6/foot, while the back wall, being taller, will be \$10/foot. If the rancher wishes to enclose an area of 180 ft^2, what dimensions of the enclosure will minimize his cost?

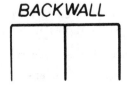

4.7.22 A can containing 16 in^3 of tuna and water is to be made in the form of a circular cylinder. What dimensions of the can will require the least amount of material? ($V = \pi r^2 h$, $S = 2\pi rh$, $A = \pi r^2$)

4.7.23 Find the maximum sum of 2 numbers given that the first plus the square of the second is equal to 30?

4.7.24 An open-top shipping crate with square bottom and rectangular sides is to hold 32 in^3 and requires a minimum amount of cardboard. Find the most economical dimensions.

4.7.25 Find the minimum distance from the point $(3,0)$ to $y = \sqrt{x}$.

4.7.26 The product of 2 positive numbers is 48. Find the numbers, if the sum of one number and the cube of the other is to be minimum.

4.7.27 Find values for x and y such that their product is a minimum if $y = 2x - 10$.

4.7.28 A container with a square base, vertical sides and open top is to be made from 192 ft^2 of material. Find the dimensions of the container with greatest volume.

4.7.29 The cost of fuel used in propelling a dirigible varies as the square of its speed and is \$200/hour when the speed is 100 miles/hour. Other expenses amount to \$300/hour. Find the most economical speed for a voyage of 1000 miles.

4.7.30 A rectangular garden is to be laid out with one side adjoining a neighbor's lot and is to contain 675 ft^2. If the neighbor agrees to pay for half the dividing fence, what should the dimensions of the garden be to insure a minimum cost of enclosure?

4.7.31 A rectangle is to have an area of 32 in^2. What should be its dimensions if the distance from one corner to the mid-point of a nonadjacent side is to be a minimum?

4.7.32 A slice of pizza, in the form of a sector of a circle, is to have a perimeter of 24 inches. What should be the radius of the pan to make the slice of pizza largest. (Hint: the area of a sector of a circle, $A = \dfrac{r^2}{2}\theta$ where θ is the central angle in radian and the arc length along a circle is $S = r\theta$ with θ in radians.)

4.7.33 Find the point on the parabola $2y = x^2$ which is closest to $(4,1)$.

4.7.34 A line is drawn through the point $P(3,4)$ so that it intersects the y−axis at $A(0,y)$ and the x−axis at $B(x,0)$. Find the smallest triangle formed if x and y are positive.

4.7.35 An open cylindrical trash can is to hold 6 cubic feet of material. What should be its dimensions if the cost of material used is to be a minimum? [Surface Area, $S = 2\pi rh$ where r = radius and h = height.]

4.7.36 Two fences, 16 feet apart are to be constructed so that the first fence is 2 feet high and the second fence is higher than the first. What is the length of the shortest pole that has one end on the ground, passing over the first fence and reaches the second fence? See figure.

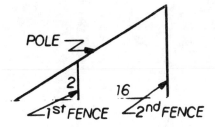

4.7.37 A line is drawn through the point $P(3,4)$ so that it intersects the y−axis at $A(0,y)$ and the x−axis at $B(x,0)$. Find the equation of the line through AB if the triangle formed is to have a minimum area and both x and y are positive.

SOLUTIONS

SECTION 4.7

4.7.1 Let $x =$ one number, $y =$ the other number, and $P = xy$ where $x + y = 25$ thus $y = 25 - x$ so $P = x(25 - x) = 25x - x^2$ for x in $[0, 25]$. $\frac{dP}{dx} = 25 - 2x$, $\frac{dP}{dx} = 0$ when $x = 12.5$. If $x = 0, 12.5, 25$ then $P = 0, 156.25, 0$ so P is maximum when $x = 12.5$ and $y = 12.5$.

4.7.2 $A = xy$ is subject to the cost condition $4x + 3(2y) = 6600$ or

$y = 1100 - \frac{2}{3}x$. Thus

$A = x\left(1100 - \frac{2}{3}x\right) = 1100x - \frac{2x^2}{3}$

for x in $[0, 1650]$. $\frac{dA}{dx} = 1100 - \frac{4x}{3}$,

$\frac{dA}{dx} = 0$ when $x = 825$.

If $x = 0, 825, 1650$ then

$A = 0, 453, 750, 0$. So the area is

greatest when $x = 825$ feet and

$y = 550$ feet.

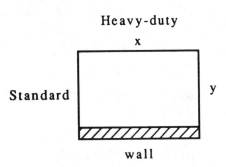

Heavy-duty

x

Standard   y

wall

4.7.3 $v = x(18 - 2x)^2$ for $0 \le x \le 9$

$\frac{dv}{dx} = 12(x - 9)(x - 3)$, $\frac{dv}{dx} = 0$ when

$x = 3$ for $0 < x < 9$. If $x = 0, 3, 9$ then

$v = 0, 432, 0$. So the volume is

largest when $x = 3$ in.

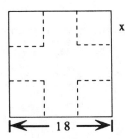

x

18

4.7.4 Let $P(x, y)$ be a point on the curve $x^2 + y^2 = 4$. The distance between $P(x, y)$ and $P_0(4, 0)$ is $D = \sqrt{(x - 4)^2 + y^2}$ but $y^2 = 4 - x^2$ so $D = \sqrt{(x - 4)^2 + (4 - x^2)} = 2\sqrt{5 - 2x}$ for $-2 \le x \le 2$. $\frac{dD}{dx} = \frac{-2}{\sqrt{5 - 2x}}$ which has no critical points for $-2 < x < 2$. If $x = -2, 2$ then $D = 6, 2$ so the closest point occurs when $x = 2$ and $y = 0$.

4.7.5 $A = xy$ and $x^2 + y^2 = 4a^2$, thus
$y^2 = 4a^2 - x^2$ and $y = \sqrt{4a^2 - x^2}$.
$A(x) = x\sqrt{4a^2 - x^2}$ for [0,2a];
$A'(x) = \dfrac{4a^2 - 2x^2}{\sqrt{4a^2 - 2x^2}}; A'(x) = 0$

for x in $[0, 2a]$ when $x = \sqrt{2}a$,
thus, $A(0) = 0; A\left(\sqrt{2}a\right) = 2a^2$;
$A(2a) = 0$; so, the area is maximum
when $x = \sqrt{2}a$ and $y = \sqrt{4a^2 - 2a^2} = \sqrt{2}a$.

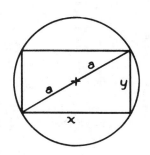

4.7.6 $A = xy$ and $2x + 4y = 4000$
$x = 2000 - 2y$ and
$A(y) = (2000 - 2y)(y) = 2\left(1000y - y^2\right)$
for y in $[0, 1000]$.
$A'(y) = 2(1000 - 2y); A'(y) = 0$
for y in $[0, 1000]$ when $y = 500$,
thus, $A(0) = 0$, $A(500) = 500000$,
$A(1000) = 0$, so the area is maximum
when $y = 500$ and $x = 1000$.

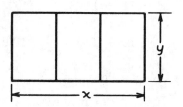

4.7.7 Let x and y be the dimensions as
shown in the figure, then $A = xy$,
and by similar triangles,

$\dfrac{x}{12} = \dfrac{16 - y}{16}, y = \dfrac{48 - 4x}{3}$ so

$A(x) = \dfrac{x(48 - 4x)}{3}$ for x in $[0, 12]$.

$A'(x) = \dfrac{48 - 8x}{3}$ and $A'(x) = 0$

when $x = 6$. Thus, $A(0) = 0; A(6) = 48$;

$A(12) = 0$ so the area is a maximum when $x = 6$ and $y = \dfrac{48 - 4 \cdot 6}{3} = 8$.

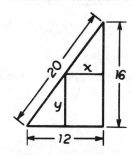

4.7.8 Let x and y be the dimensions as shown in the figure, then $A = xy$, and, $2x + \pi y = 440$, (radius of semicircle is $\dfrac{y}{2}$), $x = \dfrac{440 - \pi y}{2}$

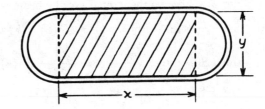

so that $A(y) = \dfrac{y(440 - \pi y)}{2}$

$A(y) = \dfrac{440y - \pi y^2}{2}$ for y in $\left[0, \dfrac{440}{\pi}\right]$. $A'(y) = \dfrac{440 - 2\pi y}{2}$ and $A'(y) = 0$ when $y = \dfrac{220}{\pi}$.

$A(0) = 0$, $A\left(\dfrac{220}{\pi}\right) = \dfrac{24200}{\pi}$, $A\left(\dfrac{440}{\pi}\right) = 0$, so the maximum area of the rectangle is

$\dfrac{24200}{\pi}$ when $y = \dfrac{220}{\pi}$ and $x = 110$.

4.7.9 Let x be as shown in the figure. Then the area of the cross section is $A(x) = x(8 - 2x)$ or $A(x) = 8x - 2x^2$ for x in $[0, 4]$. $\dfrac{dA}{dx} = 8 - 4x$ and

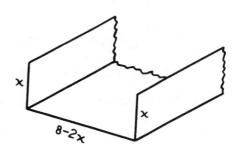

$\dfrac{dA}{dx} = 0$ when $x = 2$. $A(0) = 0$; $A(2) = 8$; $A(4) = 0$, thus, the carrying capacity is a maximum when the cross sectional area is 8. This occurs when $x = 2$.

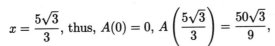

4.7.10 Let x and y be as shown in the figure. The area of the triangle is $A = \dfrac{1}{2}(2xy) = xy$ and $y = \dfrac{25 - x^2}{5}$

so $A(x) = \dfrac{x(25 - x^2)}{5} = \dfrac{25x - x^3}{5}$

for $[0, 5]$. $A'(x) = \dfrac{25 - 3x^2}{5}$.

$A'(x) = 0$ for x in $[0, 5]$ when

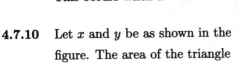

$x = \dfrac{5\sqrt{3}}{3}$, thus, $A(0) = 0$, $A\left(\dfrac{5\sqrt{3}}{3}\right) = \dfrac{50\sqrt{3}}{9}$,

$A(5) = 0$ so the maximum area of $\dfrac{50\sqrt{3}}{9}$ occurs when $x = \dfrac{5\sqrt{3}}{3}$ and $y = \dfrac{10}{3}$.

4.7.11   Let $S = kxy^2$ be the strength of the beam where k is a constant. $x^2 + y^2 = d^2$, $y^2 = d^2 - x^2$ and $S(x) = kx\left(d^2 - x^2\right)$ for $[0, d]$. $S'(x) = k\left(d^2 - 3x^2\right)$; $S'(x) = 0$ for x in $[0, d]$ when $x = \dfrac{\sqrt{3}d}{3}$, thus, $S(0) = 0$, $S\left(\dfrac{\sqrt{3}d}{3}\right) = \dfrac{2\sqrt{3}kd^3}{9}$, $S(d) = 0$ so the strength of the beam is a maximum of $\dfrac{2\sqrt{3}kd^3}{9}$ when $x = \dfrac{\sqrt{3}d}{3}$ and $y = \dfrac{\sqrt{6}d}{3}$.

4.7.12   Let x and y be as shown in the figure. $A = \dfrac{1}{2}xy$ and

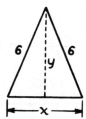

$\dfrac{x^2}{4} + y^2 = 36$ so

$y^2 = 36 - \dfrac{x^2}{4} = \dfrac{144 - x^2}{4}$ thus

$A(x) = \dfrac{1}{2}x\sqrt{\dfrac{144 - x^2}{4}} = \dfrac{x\sqrt{144 - x^2}}{4}$ for x in $[0, 12]$.

$A'(x) = \dfrac{144 - 2x^2}{4\sqrt{144 - x^2}}$, $A'(x) = 0$ for x in $[0, 12]$ when $x = 6\sqrt{2}$, thus, $A(0) = 0$, $A\left(6\sqrt{2}\right) = 18$, $A(12) = 0$ so the largest possible isosceles triangle with 2 sides equal to 6 has an area of 18 when $x = 6\sqrt{2}$ and $y = 3\sqrt{2}$.

4.7.13   Refer to the figure on the right. The distance from the lighthouse to the dock, then to town is $\sqrt{64 + x^2} + 18 - x$. The time required to move supplies is

$T(x) = \dfrac{\sqrt{64 + x^2}}{7} + \dfrac{18 - x}{25}$ for x

in $[0, 18]$. $T'(x) = \dfrac{x}{7\sqrt{64 + x^2}} - \dfrac{1}{25}$.

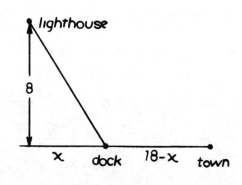

$T'(x) = 0$ when $x = \dfrac{7}{3}$, thus,

$T(0) = 1.86$, $T\left(\dfrac{7}{3}\right) = 1.82$, and

$T(18) = 2.81$, so the minimum time for shipment is 1.82 hours when the dock is located $18 - \dfrac{7}{3} = 15\dfrac{2}{3}$ miles from town.

4.7.14 $S = 2\pi rh$ and $\dfrac{h^2}{4} + r^2 = 16$, so,

$r^2 = 16 - \dfrac{h^2}{4}$ and

$S(h) = 2\pi h\sqrt{16 - \dfrac{h^2}{4}} = \pi h\sqrt{64 - h^2}$

for h in $[0, 8]$. $S'(h) = 0$ for

h in $[0, 8]$ when $h = 4\sqrt{2}$, thus,

$S(0) = 0$, $S(4\sqrt{2}) = 32\pi$, and $S(8) = 0$,

so, the cylinder of largest lateral

area is 32π with $h = 4\sqrt{2}$ and $r = 2\sqrt{2}$.

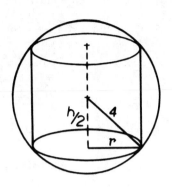

4.7.15   Let x be as shown in the figure.

Then $A(x) = \dfrac{1}{2}(10 + 10 + 2x)\sqrt{100 - x^2}$

$= (10 + x)\sqrt{100 - x^2}$ for x in $[0, 10]$.

$A'(x) = \dfrac{(10 + x)(-x)}{\sqrt{100 - x^2}} + \sqrt{100 - x^2}$ or

$A'(x) = \dfrac{100 - 10x - 2x^2}{\sqrt{100 - x^2}}$. $A'(x) = 0$

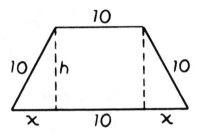

for x in $[0, 10]$ when $x = 5$, so $A(0) = 100$, $A(5) = 75\sqrt{3}$, $A(10) = 0$, thus the maximum area of the trapezoid is $75\sqrt{3}$ when $x = 5$ and the 4th side is 20.

4.7.16 $S = kxy^3$ and from the figure, $x^2 + y^2 = 4$, so $y = \sqrt{4 - x^2}$ and $S(x) = kx(4 - x^2)^{3/2}$ for x in $[0, 2]$. $S'(x) = k\left[-3x^2\left(4 - x^2\right)^{1/2} + \left(4 - x^2\right)3/2\right]$ or $S'(x) = 4k\left(1 - x^2\right)\sqrt{4 - x^2}$.

$S'(x) = 0$ for x in $[0, 2]$ when $x = 1$ and $x = 2$. $S(0) = 0$, $S(1) = k3^{3/2}$, $S(2) = 0$ so the stiffest beam is $k3^{3/2}$ when $x = 1$ and $y = \sqrt{3}$.

4.7.17   Let x and y be as shown in the

figure. $A = xy$ and by similar

triangles, $\dfrac{4 - y}{4} = \dfrac{x}{12}$ so $y = \dfrac{12 - x}{3}$

and $A(x) = x\left(\dfrac{12 - x}{3}\right) = \dfrac{12x - x^2}{3}$ for

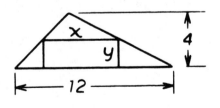

x in $[0, 12]$. $A'(x) = \dfrac{12 - 2x}{3}$ and $A'(x) = 0$ when $x = 6$, thus, $A(0) = 0$, $A(6) = 12$, $A(12) = 0$ so the maximum area of the rectangle is 12 when $x = 6$ and $y = 2$.

4.7.18 Refer to the figure on the right.

$A = 2xy$ and $y = 4 - \dfrac{x^2}{4}$ so

$A(x) = 2x\left(\dfrac{16 - x^2}{4}\right) = \dfrac{16x - x^3}{2}$ for

x in $[0, 4]$. $A'(x) = \dfrac{16 - 3x^2}{2}$

and $A'(x) = 0$ for x in $[0, 4]$

when $x = \dfrac{4\sqrt{3}}{3}$, thus, $A(0) = 0$,

$A\left(\dfrac{4\sqrt{3}}{3}\right) = \dfrac{64\sqrt{3}}{9}$, $A(4) = 0$,

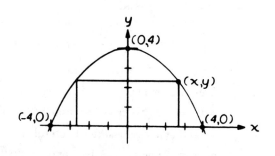

so, the maximum area of the rectangle is $\dfrac{64\sqrt{3}}{9}$ when $x = \dfrac{4\sqrt{3}}{3}$ and $y = \dfrac{8}{3}$.

4.7.19 Refer to the figure on right.

$A = \dfrac{1}{2}(8 + 2x)y$ and $y = \dfrac{16 - x^2}{4}$

so $A(x) = \dfrac{1}{2}(8 + 2x)\left(\dfrac{16 - x^2}{4}\right)$ or

$A(x) = \dfrac{64 + 16x - 4x^2 - x^3}{4}$ for x in

$[0, 4]$. $A'(x) = \dfrac{16 - 8x - 3x^2}{4}$ or

$A'(x) = \dfrac{(4 - 3x)(4 + x)}{4}$ and $A'(x) = 0$

for x in $[0, 4]$ when $x = \dfrac{4}{3}$, thus,

$A(0) = 16$, $A\left(\dfrac{4}{3}\right) = \dfrac{512}{27}$, $A(4) = 0$,

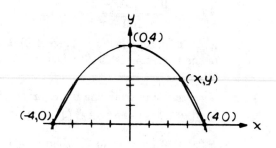

thus, the maximum area of the trapezoid is $\dfrac{512}{27}$ when $x = \dfrac{4}{3}$ and $y = \dfrac{32}{9}$.

4.7.20   Let the area of the printed matter
be xy. The area of the poster is
$A = (x + 4)(y + 8)$, thus,
$A = xy + 4y + 8x + 32$; since $xy = 50$,
substituting, $A = 82 + \dfrac{200}{x} + 8x$ for
$0 < x < +\infty$. $\dfrac{dA}{dx} = \dfrac{-200}{x^2} + 8$, $\dfrac{dA}{dx} = 0$
for $x = 5$. By second derivative test,
the minimum area of the poster is 162
square inches when $x = 5$ and $y = 10$.

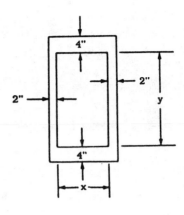

4.7.21   The area of the enclosure is
$A = xy = 180$ so $y = \dfrac{180}{x}$.
The cost of the walls is
$C = 10x + 3(6)y = 10x + 18\left(\dfrac{180}{x}\right)$

for $0 < x < +\infty$; $\dfrac{dC}{dx} = 10 - \dfrac{3240}{x^2}$.

$\dfrac{dC}{dx} = 0$ when $x = 18$. By second derivative test,

the minimum cost is \$360 when $x = 18$ ft and $y = 10$ ft.

4.7.22 $V = \pi r^2 h = 16$, so $h = \dfrac{16}{\pi r^2}$ and the total surface area of the can is

$S = 2\pi r^2 + 2\pi rh = 2\pi r^2 + (2\pi r)\left(\dfrac{16}{\pi r^2}\right) = 2\pi r^2 + \dfrac{32}{r}$, for $0 < r < +\infty$, $\dfrac{dS}{dr} = 4\pi r - \dfrac{32}{r^2}$.

$\dfrac{dS}{dr} = 0$ when $r = \dfrac{2}{\sqrt[3]{\pi}}$ in. By the second derivative test, the minimum surface area is $24\sqrt[3]{\pi}$ in^2

when $r = \dfrac{2}{\sqrt[3]{\pi}}$ in. and $h = \dfrac{4}{\sqrt[3]{\pi}}$ in.

4.7.23 Let $x =$ one number, $y =$ the other number; $S = x + y$ given that $x + y^2 = 30$. Thus, $x = 30 - y^2$ and $S = 30 - y^2 + y$ for $-\infty < y < +\infty$. $\dfrac{dS}{dy} = -2y + 1$, $\dfrac{dS}{dy} = 0$ when $y = \dfrac{1}{2}$. By second derivative test, S has a maximum of $\dfrac{121}{4}$ when $y = \dfrac{1}{2}$ and $x = \dfrac{119}{4}$.

4.7.24 Let x and y be the dimensions shown in the figure. The surface area $S = x^2 + 4xy$ and $V = x^2 y = 32$, thus, $y = \dfrac{32}{x^2}$ and $S = x^2 + \dfrac{128}{x}$ for $0 < x < +\infty$, $\dfrac{dS}{dx} = 2x - \dfrac{128}{x^2}$. $\dfrac{dS}{dx} = 0$ when $x = 4$ and $\dfrac{d^2 S}{dx^2} = 2 + \dfrac{256}{x^2} > 0$ so, S has a minimum of 48 in^2 when $x = 4$ and $y = 2$.

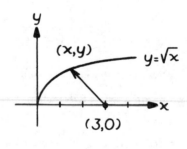

4.7.25 Let $d = \sqrt{(x - 3)^2 + (y - 0)^2} = \sqrt{(x - 3)^2 + y^2}$ but $y = \sqrt{x}$ so $d = \sqrt{(x - 3)^2 + x}$ for $0 \le x < +\infty$.

Let $S = d^2 = (x - 3)^2 + x$,

$\dfrac{dS}{dx} = 2(x - 3) + 1$ and $\dfrac{dS}{dx} = 0$ when $x = \dfrac{5}{2}$. When $x = 0$, $d = 3$ and when $x = \dfrac{5}{2}$, $d = \dfrac{\sqrt{11}}{2}$ so the minimum distance is $\dfrac{\sqrt{11}}{2}$ which occurs when $x = \dfrac{5}{2}$ and $y = \sqrt{\dfrac{5}{2}}$.

4.7.26 Let $x =$ one positive number and $y =$ the other positive number, then $S = x + y^3$ given that $xy = 48$, so, $x = \dfrac{48}{y}$ and $S = \dfrac{48}{y} + y^3$ for $0 < y < +\infty$. $\dfrac{dS}{dy} = \dfrac{-48}{y^2} + 3y^2$ so $\dfrac{dS}{dy} = 0$ when $y = 2$. $\dfrac{d^2 S}{dy^2} > 0$ for $y = 2$ so the minimum sum of the two numbers is 32 when $x = 24$ and $y = 2$.

4.7.27 Let $x =$ one number and $y =$ other number. $P = xy$ and $y = 2x - 10$ so $P = x(2x - 10)$ for x in $(-\infty, +\infty)$. $\dfrac{dP}{dx} = 4x - 10$ so, $\dfrac{dP}{dx} = 0$ when $x = \dfrac{5}{2}$. $\dfrac{d^2 P}{dx^2} > 0$ so the minimum product is $-\dfrac{25}{2}$ when $x = \dfrac{5}{2}$ and $y = -5$.

4.7.28 Let x and y be as shown in the
figure. $V = x^2 y$ and $x^2 + 4xy = 192$,

so $y = \dfrac{192 - x^2}{4x}$ and $V = x^2 \left(\dfrac{192 - x^2}{4x} \right)$

or $V = \dfrac{192x - x^3}{4}$ for $0 < x \le 8\sqrt{3}$.

$\dfrac{dV}{dx} = \dfrac{192 - 3x^2}{4}$ and $\dfrac{dV}{dx} = 0$ when $x = 8$.

When $x = 8$, $V = 256$ and when $x = 8\sqrt{3}$,
$V = 0$, so the maximum volume is 256
which occurs when $x = 8$ and $y = 4$.

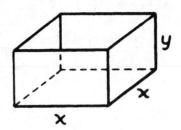

4.7.29 Let x be the speed of the dirigible and $t = \dfrac{1000}{x}$ be the length of time of the voyage. Let
$F = kx^2$ be the cost of fuel, then $k = \dfrac{F}{x^2} = \dfrac{200}{(100)^2} = \dfrac{1}{50}$ so that $F = \dfrac{x^2}{50}$. The total cost of the voyage is $C = \left(\dfrac{x^2}{50} + 300 \right) \left(\dfrac{1000}{x} \right) = 20x + \dfrac{300(1000)}{x}$ for $0 < x < +\infty$.
$\dfrac{dC}{dx} = 20 - \dfrac{300(1000)}{x^2}$; $\dfrac{dC}{dx} = 0$ for $x = \sqrt{15000} = 50\sqrt{6}$ and $\dfrac{d^2C}{dx^2} > 0$ for $x > 0$ so the cost is
a minimum when $x = 50\sqrt{6}$ miles/hour.

4.7.30 Let x and y be the dimensions
shown in the figure. The cost
of fencing is $C = 2x + \dfrac{3y}{2}$, then

$xy = 675$ and $x = \dfrac{675}{y}$ so that

$C = 2 \cdot \dfrac{675}{y} + \dfrac{3y}{2}$ for $0 < y < +\infty$.

$\dfrac{dC}{dy} = -\dfrac{1350}{y^2} + \dfrac{3}{2}$; $\dfrac{dC}{dy} = 0$

for $y = 30$, $\dfrac{d^2C}{dy^2} > 0$ so that

the cost of fencing is a minimum

when $y = 30$ and $x = \dfrac{45}{2}$.

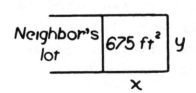

4.7.31 Let x, y, and D be as shown in the figure. Let $L = D^2 = x^2 + \dfrac{y^2}{4}$.

The area $= xy = 32$ so $x = \dfrac{32}{y}$ and

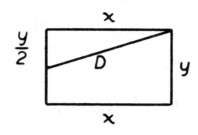

$L = \dfrac{32}{y}^2 + \dfrac{y^2}{4} = \dfrac{1024}{y^2} + \dfrac{y^2}{4}$ for

$0 < y < +\infty$. $\dfrac{dL}{dy} = -\dfrac{2048}{y^3} + \dfrac{2y}{4}$ and

$\dfrac{dL}{dy} = 0$ for $y > 0$ when $y = 8$.

$\dfrac{d^2 L}{dy^2} > 0$ when $y = 8$ so that the

minimum distance is $4\sqrt{2}$

when $x = 4$ and $y = 8$.

4.7.32 Let r and θ be as shown in the diagram. The perimeter of the slice is $P = 2r + r\theta = 24$ and the area of the slice is $A = \dfrac{1}{2}r^2\theta$. $\theta = \dfrac{24 - 2r}{r}$

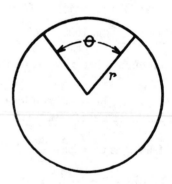

and $A = \dfrac{1}{2}r^2\left(\dfrac{24 - 2r}{r}\right) = r(12 - r) = 12r - r^2$

for $0 < r \le 12$. $\dfrac{dA}{dr} = 12 - 2r$, $\dfrac{dA}{dr} = 0$

when $r = 6$. When $r = 6$, $A = 36$

and when $r = 12$, $A = 0$, so the

slice of pizza is largest when the

radius of the pan is 6 inches.

4.7.33 Let $d = \sqrt{(x - 4)^2 + (y - 1)^2}$ be the distance from $(4, 1)$ to any point on the parabola. Substitute $2y = x^2$ into d to get $d = \sqrt{(x - 4)^2 + \left(\dfrac{1}{2}x^2 - 1\right)^2}$, then, let $S = d^2 = (x - 4)^2 + \left(\dfrac{1}{2}x^2 - 1\right)^2$ for x in $(-\infty, +\infty)$. $\dfrac{dS}{dx} = x^3 - 8$ which is zero when $x = 2$. Since $\dfrac{d^2 s}{dx^2} = 3x^2 > 0$, S has a minimum when $x = 2$ and the closest point on the parabola $2y = x^2$ to $(4, 1)$ is $(2, 2)$.

4.7.34 $A = \frac{1}{2}xy$. The slope of the line

through $(3,4)$ may be expressed as

$\frac{4-y}{3-0} = \frac{0-4}{x-3}$ or $\frac{4-y}{3} = \frac{-4}{x-3}$; thus

$y = \frac{4x}{x-3}$ and $A = \frac{1}{2}(x)\left(\frac{4x}{x-3}\right) = \frac{2x^2}{x-3}$

for $(3, +\infty)$. $\frac{dA}{dx} = \frac{2x^2 - 12x}{(x-3)^2}$, $\frac{dA}{dx} = 0$

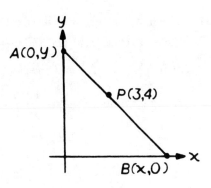

when $x = 6$. By the first derivative test, the area of the triangle is a minimum when $x = 6$
and $y = 8$.

4.7.35 Let r be the radius of the can and h be its height. Then $S = \pi r^2 + 2\pi rh$. $V = 6 = \pi r^2 h$

so $h = \frac{6}{\pi r^2}$ and $S = \pi r^2 + \frac{12}{r}$ for $0 < r < +\infty$. $\frac{dS}{dr} = 2\pi r - \frac{12}{r^2}$, $\frac{dS}{dr} = 0$ when $r = \sqrt[3]{\frac{6}{\pi}}$.

$\frac{d^2 S}{dr^2} > 0$ so the surface area S is a minimum when $r = \sqrt[3]{\frac{6}{\pi}}$ and $h = \sqrt[3]{\frac{6}{\pi}}$.

4.7.36 Let $x = AB + BC$ so that

$x = 2\csc\theta + 16\sec\theta$ for θ in $(0, \pi/2)$.

$\frac{dx}{d\theta} = -2\csc\theta\cot\theta + 16\sec\theta\tan\theta$,

$\frac{dx}{d\theta} = 0$ for $\tan^3\theta = \frac{1}{8}$; thus,

$\tan\theta = \frac{1}{2}$; $\theta \approx 26.6°$. By first

derivative test, x is a minimum

when $\theta \approx 26.6°$. To find x,

construct a triangle such that

$\tan\theta = \frac{1}{2}$ as follows

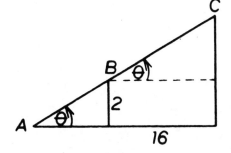

$x = 2\csc\theta + 16\sec\theta = 2\cdot\sqrt{5} + 16\cdot\frac{\sqrt{5}}{2} = 10\sqrt{5}$.

4.7.37 The line through $(3,4)$ intersects the $x-$axis at $(x,0)$ and the $y-$axis at $(0,y)$. $A = \frac{1}{2}xy$ and the slope of the line through $(3,4)$ is $\frac{4-y}{3-0} = \frac{0-4}{x-3}$ or $\frac{4-y}{3} = \frac{4}{x-3}$, thus, $y = \frac{4x}{x-3}$ and

$$A = \frac{1}{2}(x)\left(\frac{4x}{x-3}\right) = \frac{2x^2}{x-3} \text{ for } (3,+\infty).$$

$$\frac{dA}{dx} = \frac{2x^2 - 12x}{(x-3)^2}, \frac{dA}{dx} = 0 \text{ for } x = 6.$$

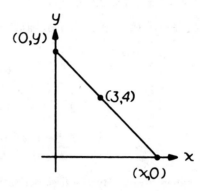

By first derivative test, the area is a minimum when $x = 6$ and $y = 8$. The slope of the line drawn through $(3,4)$ with intercepts at $(0,8)$ and $(6,0)$ is $m = -\frac{4}{3}$ and the equation of the line is $4x + 3y - 24 = 0$.

SECTION 4.8

4.8.1 Approximate $\sqrt{3}$ by applying Newton's Method to the equation $x^2 - 3 = 0$.

4.8.2 Approximate $\sqrt{11}$ by applying Newton's Method to the equation $x^2 - 11 = 0$.

4.8.3 Approximate $\sqrt{84}$ by applying Newton's Method to the equation $x^2 - 84 = 0$.

4.8.4 Approximate $\sqrt{66}$ by applying Newton's Method to the equation $x^2 - 66 = 0$.

4.8.5 Approximate $\sqrt{97}$ by applying Newton's Method to the equation $x^2 - 97 = 0$.

4.8.6 Approximate $\sqrt[3]{10}$ by applying Newton's Method to the equation $x^3 - 10 = 0$.

4.8.7 Approximate $\sqrt[3]{25}$ by applying Newton's Method to the equation $x^3 - 25 = 0$.

4.8.8 Approximate $-\sqrt[3]{72}$ by applying Newton's Method to the equation $x^3 + 72 = 0$.

4.8.9 Approximate $\sqrt[4]{36}$ by applying Newton's Method to the equation $x^4 - 36 = 0$.

4.8.10 Approximate $-\sqrt[5]{34}$ by applying Newton's Method to the equation $x^5 + 34 = 0$.

4.8.11 The equation, $x^3 - x - 2 = 0$ has one real solution for $1 < x < 2$. Approximate it by Newton's Method.

4.8.12 The equation, $x^3 - 3x + 1 = 0$ has one real solution for $0 < x < 1$. Approximate it by Newton's Method.

4.8.13 The equation, $x^3 + x^2 - 3x - 3 = 0$ has one real solution for $x > 1$. Approximate it by Newton's Method.

4.8.14 The equation, $x^3 + x^2 - 3x - 3 = 0$ has one real solution for $-2 < x < -1$. Approximate it by Newton's Method.

4.8.15 The equation, $x^3 - x^2 - 2x + 1 = 0$ has one real solution for $1 < x < 2$. Approximate it by Newton's Method.

4.8.16 The equation, $\sin x = x/3$ has one real solution for $\dfrac{\pi}{2} < x < \pi$. Approximate it by Newton's Method.

SOLUTIONS

SECTION 4.8

4.8.1 $f(x) = x^2 - 3$
$f'(x) = 2x$
$$x_{n+1} = \frac{x_n^2 + 3}{2x_n}$$
$x_1 = 1$
$x_2 = 2$
$x_3 = 1.75$
$x_4 = 1.7321429$
$x_5 = 1.7320508$
$x_6 = 1.7320508$

4.8.2 $f(x) = x^2 - 11$
$f'(x) = 2x$
$$x_{n+1} = \frac{x_n^2 + 11}{2x_n}$$
$x_1 = 3$
$x_2 = 3.3333333$
$x_3 = 3.3166667$
$x_4 = 3.3166248$
$x_5 = 3.3166248$

4.8.3 $f(x) = x^2 - 84$
$f'(x) = 2x$
$$x_{n+1} = \frac{x_n^2 + 84}{2x_n}$$
$x_1 = 9$
$x_2 = 9.1666667$
$x_3 = 9.1651515$
$x_4 = 9.1651514$
$x_5 = 9.1651514$

4.8.4 $f(x) = x^2 - 66$
$f'(x) = 2x$
$$x_{n+1} = \frac{x_n^2 + 66}{2x_n}$$
$x_1 = 8$
$x_2 = 8.125$
$x_3 = 8.1240385$
$x_4 = 8.1230384$
$x_5 = 8.1240384$

4.8.5 $f(x) = x^2 - 97$
$f'(x) = 2x$
$$x_{n+1} = \frac{x_n^2 + 97}{2x_n}$$
$x_1 = 10$
$x_2 = 9.95$
$x_3 = 9.8488579$
$x_4 = 9.8488578$
$x_5 = 9.8488578$

4.8.6 $f(x) = x^3 - 10$
$f'(x) = 3x^2$
$$x_{n+1} = \frac{2x_n^3 + 10}{3x_n^2}$$
$x_1 = 2$
$x_2 = 2.1666667$
$x_3 = 2.1545036$
$x_4 = 2.1544347$
$x_5 = 2.1544347$

4.8.7 $f(x) = x^3 - 25$
$f'(x) = 3x^2$

$$x_{n+1} = x_n - \frac{x_n^3 - 25}{3x_n^2}$$

$x_1 = 3$
$x_2 = 2.9259259$
$x_3 = 2.924019$
$x_4 = 2.9240177$
$x_5 = 2.9240177$

4.8.8 $f(x) = x^3 + 72$
$f'(x) = 3x^2$

$$x_{n+1} = x_n - \frac{x_n^3 + 72}{3x_n^2}$$

$x_1 = -4$
$x_2 = -4.1666667$
$x_3 = -4.1601678$
$x_4 = -4.1601677$
$x_5 = -4.1601677$

4.8.9 $f(x) = x^4 - 36$
$f'(x) = 4x^3$

$$x_{n+1} = x_n - \frac{x_n^4 - 36}{4x_n^3}$$

$x_1 = 2$
$x_2 = 2.625$
$x_3 = 2.4663205$
$x_4 = 2.4496612$
$x_5 = 2.4494898$
$x_6 = 2.4494897$
$x_7 = 2.4494897$

4.8.10 $f(x) = x^5 + 34$
$f'(x) = 5x^4$

$$x_{n+1} = x_n - \frac{x_n^5 + 34}{5x_n^4}$$

$x_1 = -2$
$x_2 = -2.025$
$x_3 = -2.0243978$
$x_4 = -2.0243975$
$x_5 = -2.0243975$

4.8.11 $f(x) = x^3 - x - 2$
$f'(x) = 3x^2 - 1$

$$x_{n+1} = x_n - \frac{x_n^3 - x_n - 2}{3x_n^2 - 1}$$

$x_1 = 1.5$
$x_2 = 1.5217391$
$x_3 = 1.5213798$
$x_4 = 1.5213797$
$x_5 = 1.5213797$

4.8.12 $f(x) = x^3 - 3x + 1$
$f'(x) = 3x^2 - 3$

$$x_{n+1} = x_n - \frac{x_n^3 - 3x_n + 1}{3x_n^2 - 3}$$

$x_1 = 0.5$
$x_2 = 0.3333333$
$x_3 = 0.3472222$
$x_4 = 0.3472964$
$x_5 = 0.3472964$

4.8.13 $f(x) = x^3 + x^2 - 3x - 3$
$f'(x) = 3x^2 + 2x - 3$

$$x_{n+1} = x_n - \frac{x_n^3 + x_n^2 - 3x_n - 3}{3x_n^2 + 2x_n - 3}$$

$x_1 = 1$
$x_2 = 3$
$x_3 = 2.2$
$x_4 = 1.8301508$
$x_5 = 1.7377955$
$x_6 = 1.7320723$
$x_7 = 1.7320508$
$x_8 = 1.7320508$

4.8.14 $f(x) = x^3 + x^2 - 3x - 3$
$f'(x) = 3x^2 + 2x - 3$

$$x_{n+1} = x_n - \frac{x_n^3 + x_n^2 - 3x_n - 3}{3x_n^2 + 2x_n - 3}$$

$x_1 = -1$
$x_2 = -2$
$x_3 = -1.8$
$x_4 = -1.7384619$
$x_5 = -1.7321176$
$x_6 = -1.7320508$
$x_7 = -1.7320508$

4.8.15 $f(x) = x^3 - x^2 - 2x + 1$
$f'(x) = 3x^2 - 2x - 2$

$$x_{n+1} = x_n - \frac{x_n^3 - x_n^2 - 2x_n + 1}{3x_n^2 - 2x_n - 2}$$

$x_1 = 1.5$
$x_2 = 2$
$x_3 = 1.8333333$
$x_4 = 1.801935$
$x_5 = 1.8019388$
$x_6 = 1.8019377$
$x_7 = 1.8019377$

4.8.16 $f(x) = \sin x - x/3$
$f'(x) = \cos x - 1/3$

$$x_{n+1} = x_n - \frac{\sin x_n - \frac{x_n}{3}}{\cos x_n - \frac{1}{3}}$$

$x_1 = 1.5$
$x_2 = 3.3945252$
$x_3 = 2.3328766$
$x_4 = 2.2799109$
$x_5 = 2.2788631$
$x_6 = 2.2788627$
$x_7 = 2.2788627$

SECTION 4.9

4.9.1   Verify that $f(x) = x^3 - x$ satisfies the hypothesis of Rolle's Theorem on the interval $[-1, 1]$ and find all values of C in $(-1, 1)$ such that $f'(C) = 0$.

4.9.2   Verify that $f(x) = x^3 - 3x + 2$ satisfies the hypothesis of the Mean-Value Theorem over the interval $[-2, 3]$ and find all values of C that satisfy the conclusion of the theorem.

4.9.3   Verify that $f(x) = x^2 + 2x - 1$ satisfies the hypothesis of the Mean-Value Theorem over the interval $[0, 1]$ and find all values of C that satisfy the conclusion of the theorem.

4.9.4   Verify that $f(x) = x^3 - 4x$ satisfies the hypothesis of Rolle's Theorem on the interval $[-2, 2]$ and find all values of C that satisfy the conclusion of the theorem.

4.9.5   Does $f(x) = \dfrac{1}{x^2}$ satisfy the hypothesis of the Mean-Value Theorem over the interval $[-1, 1]$? If so, find all values of C that satisfy the conclusion.

4.9.6   Verify that $f(x) = x^2 + 4$ satisfies the hypothesis of the Mean-Value Theorem on the interval $[0, 2]$ and find all values of C that satisfy the conclusion of the theorem.

4.9.7   Verify that $f(x) = x^3 - 3x + 1$ satisfies the hypothesis of the Mean-Value Theorem on the interval $[-2, 2]$ and find all values of C that satisfy the conclusion of the theorem.

4.9.8   Verify that $f(x) = \dfrac{4x}{4 - x}$ satisfies the hypothesis of the Mean-Value Theorem over the interval $[1, 3]$ and find all values of C that satisfy the conclusion of the theorem.

4.9.9   Use Rolle's Theorem to prove that the equation $7x^6 - 9x^2 + 2 = 0$ has at least one solution in the interval $(0, 1)$.

4.9.10   Verify that $f(x) = x^3 - 3x^2 - 3x + 1$ satisfies the hypothesis of the Mean-Value Theorem over the interval $[0, 2]$ and find all values of C that satisfy the conclusion of the theorem.

4.9.11   Use Rolle's Theorem to show that $f(x) = x^3 + x - 2$ does not have more than one real root.

4.9.12   Does $f(x) = \sqrt{x}$ satisfy the hypothesis of the Mean-Value Theorem over the interval $[0, 4]$? If so, find all values of C that satisfy the conclusion of the theorem.

4.9.13   Does $f(x) = \sqrt[3]{x}$ satisfy the hypothesis of the Mean-Value Theorem over the interval $[-1, 1]$? If so, find all values of C that satisfy the conclusion of the theorem.

4.9.14   An automobile starts from rest and travels 3 miles along a straight road in 4 minutes. Use the Mean-Value Theorem to show that at some instant during the trip its velocity was exactly 45 miles per hour.

4.9.15   Does $f(x) = \dfrac{x}{x-1}$ satisfy the hypothesis of the Mean-Value Theorem over the interval $[0, 2]$? If so, find all values of C that satisfy the conclusion of the theorem.

4.9.16   Does $f(x) = \sqrt[3]{x}$ satisfy the hypothesis of the Mean-Value Theorem over the interval $[0, 1]$? If so, find all values of C that satisfy the conclusion.

4.9.17   Use Rolle's Theorem to show that $f(x) = x^3 + ax + b$, where $a > 0$, cannot have more than one real root.

4.9.18   A cyclist starts from rest and travels 4 miles along a straight road in 20 minutes. Use the Mean-Value Theorem to show that at some instant during the trip his velocity was exactly 12 miles per hour.

SOLUTIONS

SECTION 4.9

4.9.1
$$f(-1) = f(1) = 0$$
$$f'(x) = 3x^2 - 1$$
$$3C^2 - 1 = 0$$
$$C = \pm\frac{\sqrt{3}}{3}$$

4.9.2
$$f(-2) = 0; \ f(3) = 20$$
$$f'(x) = 3x^2 - 3$$
$$3C^2 - 3 = \frac{20 - 0}{3 - (-2)} = 4$$
$$C^2 = \frac{7}{3}, \quad C = \pm\sqrt{\frac{7}{3}}$$

4.9.3
$$f(0) = -1; \ f(1) = 2$$
$$f'(x) = 2x + 2$$
$$2C^2 + 2 = \frac{2 - (-1)}{1 - 0} = 3$$
$$C = \frac{1}{2}$$

4.9.4
$$f(-2) = f(2) = 0$$
$$f'(x) = 3x^2 - 4$$
$$3C^2 - 4 = 0$$
$$C = \pm\frac{2\sqrt{3}}{3}$$

4.9.5 No, since f is not differentiable at $x = 0$ which is in $(-1, 1)$.

4.9.6
$$f(0) = 4; \ f(2) = 8$$
$$f'(x) = 2x$$
$$2C = \frac{8 - 4}{2 - 0} = 2$$
$$C = 1$$

4.9.7
$$f(-2) = -1; \ f(2) = 3$$
$$f'(x) = 3x^2 - 3$$
$$3C^2 - 3 = \frac{3 - (-1)}{2 - (-2)}$$
$$C^2 = \frac{4}{3}, \quad C = \pm\frac{2\sqrt{3}}{3}$$

4.9.8
$$f(1) = 4/3; \ f(3) = 12$$
$$f'(x) = \frac{16}{(4 - x)^2}$$
$$\frac{16}{(4 - C)^2} = \frac{12 - 4/3}{3 - 1} = \frac{16}{3}$$
$$(4 - C)^2 = 3; \ C = 4 \pm \sqrt{3} \text{ of which only } C = 4 - \sqrt{3} \text{ is in } (1, 3)$$

4.9.9 If $f(x) = x^7 - 3x^3 + 2x$, $f(0) = f(1) = 0$ and $f'(x) = 7x^6 - 9x^2 + 2$ there is at least one number c in $(0, 1)$ where $f'(c) = 0$.

4.9.10
$$f(0) = 1; \; f(2) = -9$$
$$f'(x) = 3x^2 - 6x - 3$$
$$3C^2 - 6C - 3 = \frac{-9-1}{2-0} = -5$$
$$3C^2 - 6C + 2 = 0$$
$$C = \frac{6 \pm \sqrt{36 - 24}}{6} = \frac{3 \pm \sqrt{3}}{3}$$

4.9.11 Suppose f has more than one real root. Let r_1 and r_2 be any two of these roots, then $f(r_1) = f(r_2) = 0$. By Rolle's Theorem, $f'(C) = 0$ for some C in (r_1, r_2), but,

$f'(x) = 3x^2 + 1$ and $3C^2 + 1 = 0$ has no real solution, so f cannot have more than one real root.

4.9.12 Yes, since f is continuous over $[0, 4]$ and differentiable over $(0, 4)$, thus $f(0) = 0$; $f(4) = 2$; and $f'(x) = \dfrac{1}{2\sqrt{x}}$

$$\frac{1}{2\sqrt{C}} = \frac{2-0}{4-0} = \frac{1}{2}$$
$$\sqrt{C} = 1, \; C = 1$$

4.9.13 No, since f is not differentiable at $x = 0$ which is in $(-1, 1)$.

4.9.14 Let $s = f(t)$ be the position versus time curve for the automobile moving in the positive direction along the straight road. Then f satisfies the hypothesis of the Mean-Value Theorem on the time interval $[0, 4]$ and there will be an instant t_0 where the instantaneous velocity at t_0 equals the average velocity over $[0, 4]$. Instantaneous velocity = 45 miles per hour at t_0.

Average velocity = $\dfrac{s(4) - s(0)}{4 - 0} = \dfrac{3}{4}$ miles per minute = 45 miles per hour. Thus at some t_0 in $[0, 4]$ the car's instantaneous velocity is equal to its average velocity.

4.9.15 No, f is not continuous at $x = 1$ which is in $[0, 2]$.

4.9.16 Yes, since f is continuous over $[0, 1]$ and differentiable over $(0, 1)$, thus, $f(0) = 0$; $f(1) = 1$; and $f'(x) = \dfrac{1}{3x^{2/3}}$.

$$\frac{1}{3C^{2/3}} = \frac{1-0}{1-0} = 1$$

$$3C^{2/3} = 1, \; C = \pm\frac{\sqrt{3}}{9} \text{ of which only } C = \frac{\sqrt{3}}{9} \text{ lies in } (0, 1).$$

4.9.17   Suppose f has more than one real root. Let r_1 and r_2 be any two of those roots, then, $f(r_1) = f(r_2) = 0$. By Rolle's Theorem, $f'(C) = 0$ for some C in (r_1, r_2), but, $f'(x) = 3x^2 + a$ and $3C^2 + a = 0$ has no real solution for $a > 0$, so, f cannot have more than one real root.

4.9.18   Let $s = f(t)$ be the position versus time curve for the cyclist moving in the positive direction along the straight road. Then f satisfies the hypothesis of the Mean-Value Theorem on the time interval $[0, 20]$ and there will be an instant t_0 where the instantaneous velocity at t_0 equals the average velocity over $[0, 20]$. Instantaneous velocity $= 12$ miles per hour at t_0. Average velocity $= \dfrac{s(20) - s(0)}{20 - 0} = \dfrac{1}{5}$ miles per minute $= 12$ miles per hour. Thus at some t_0 in $[0, 20]$ the cyclist's instantaneous velocity is equal to its average velocity.

SECTION 4.10

4.10.1 The graph below depicts the position function of a particle moving on a coordinate line at three different times. For each time specify whether the particle is moving to the left or right and whether or not it is speeding up or slowing

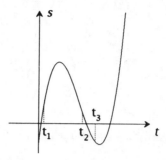

4.10.2 The graph below depicts the position function of a particle moving on a coordinate line at three different times. For each time, specify whether the particle is moving to the left or right and whether or not it is speeding up or slowing down.

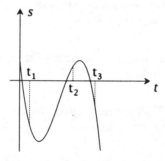

4.10.3 The graph below depicts the position function of a particle moving on a coordinate line at three different times. For each time, specify whether the particle is moving to the left or right and whether or not it is speeding up or slowing down.

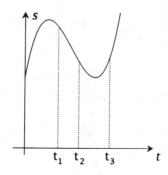

4.10.4   The graph below depicts the position function of a particle moving on a coordinate line at three different times. For each time, specify whether the particle is moving to the left or right and whether or not it is speeding up or slowing down.

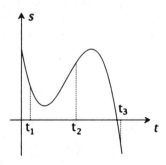

4.10.5   Let $s = 2t^3 - 12t^2 + 4t + 9$; find s and v when $a = 0$.

4.10.6   Let $s = 4t^3 - 12t^2$; find s and v when $a = 0$.

4.10.7   Let $s = 3t^3 - 9t^2 - 5t + 2$; find s and v when $a = 0$.

4.10.8   Let $s = 2t^2 - 6t - 9$ be the position function of a particle. Find the maximum speed of the particle during the time interval $1 \leq t \leq 4$.

4.10.9   Let $s = t^2 - 5t - 6$ be the position function of a particle. Find the maximum speed of the particle during the time interval $0 \leq t \leq 6$.

4.10.10   The position function of a particle is given by $s = 3t^2 - 4t + 1$ for $t \geq 0$. Describe the motion of the particle and make a sketch.

4.10.11   The position function of a particle is given by $s = 2t^3 - 9t^2 + 12t + 5$ for $t \geq 0$. Describe the motion of the particle and make a sketch.

4.10.12   The position function of a particle is given by $s = 4t^3 - 12t^2 + 9t - 1$ for $t \geq 0$. Describe the motion of the particle and make a sketch.

4.10.13   The position function of a particle is given by $s = t(t - 6)^2$ for $t \geq 0$. Describe the motion of the particle and make a sketch.

4.10.14   The position function of a particle is given by $s = t^3 - 3t^2 - 9t$ for $t \geq 0$. Describe the motion of the particle and make a sketch.

4.10.15   The position function of a particle is given by $s = t^3 - 5t^2 + 3t$ for $t \geq 0$. Describe the motion of the particle and make a sketch.

4.10.16   The position function of a particle is given by $s = \frac{1}{3}t^3 - 3t^2 + 8t + 1$ for $t \geq 0$. Describe the motion of the particle and make a sketch.

4.10.17   The position function of a particle is given by $s = 2t^3 - 5t^2 + 4t - 3$ for $t \geq 0$. Describe the motion of the particle and make a sketch.

SOLUTIONS

SECTION 4.10

4.10.1 At $t = t_1$, $v = \dfrac{ds}{dt} > 0$, $a = \dfrac{d^2s}{dt^2} < 0$ so the particle is moving to the right and slowing down; at $t = t_2$, $v < 0$, $a < 0$ so the particle is moving left and speeding up; at $t = t_3$, $v < 0$, $a > 0$ so the particle is moving to the left and slowing down.

4.10.2 At $t = t_1$, $v = \dfrac{ds}{dt} < 0$, $a = \dfrac{d^2s}{dt^2} > 0$ so the particle is moving to the left and slowing down; at $t = t_2$, $v > 0$, $a < 0$ so the particle is moving to the right and slowing down; at $t = t_3$, $v < 0$, $a < 0$ so the particle is moving to the left and speeding up.

4.10.3 At $t = t_1$, $v = \dfrac{ds}{dt} < 0$, $a = \dfrac{d^2s}{dt^2} < 0$ so the particle is moving to the left and speeding up; at $t = t_2$, $v < 0$ and $a > 0$ so the particle is moving to the left and slowing down; at $t = t_3$, $v > 0$, $a > 0$ so the particle is moving to the right and speeding up.

4.10.4 At $t = t_1$, $v = \dfrac{ds}{dt} < 0$, $a = \dfrac{d^2s}{dt^2} > 0$ so the particle is moving to the left and slowing down; at $t = t_2$, $v > 0$, $a < 0$ so the particle is moving to the right and slowing down; at $t = t_3$, $v < 0$, $a < 0$ so the particle is moving to the left and speeding up.

4.10.5 $v = 3(2t^2) - 12(2t) + 4(1) = 6t^2 - 24t + 4$; $a = 6(2t) - 24(1) = 12t - 24$, when $a = 0$, $t = 2$ and $s = 2(2)^3 - 12(2)^2 + 4(2) + 9 = -15$ and $v = 6(2)^2 - 24(2) + 4 = -20$.

4.10.6 $v = 3(4t^2) - 12(2t) = 12t^2 - 24t$; $a = 12(2t) - 24(1) = 24t - 24$, when $a = 0$, $t = 1$ and $s = 4(1)^3 - 12(1)^2 = -8$ and $v = 12(1)^2 - 24(1) = -12$.

4.10.7 $v = 3(3t^2) - 9(2t) - 5(1) = 9t^2 - 18t - 5$; $a = 9(2t) - 18(1) = 18t - 18$, when $a = 0$, $t = 1$ and $s = 3(1)^3 - 9(1)^2 - 5(1) + 2 = -9$ and $v = 9(1)^2 - 18(1) - 5 = -14$.

4.10.8 $v = 4t - 6$, speed $= |v| = |4t - 6|$. $\dfrac{dv}{dt}$ does not exist at $t = 3/2$ which is the only critical point, thus

t	1	$\dfrac{3}{2}$	4
$|v|$	2	0	10

, so, the maximum speed is 10.

4.10.9 $v = 2t - 5$, speed $= |v| = |2t - 5|$. $\dfrac{dv}{dt}$ does not exist at $t = 5/2$ which is the only critical point, thus

t	0	5/2	6
$|v|$	5	0	7

, so, the maximum speed is 7.

4.10.10 $s = 3t^2 - 4t + 1$

$v = 6t - 4$

$a = 6$

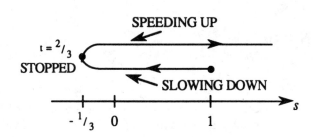

4.10.11 $s = 2t^3 - 9t^2 + 12t + 5$

$v = 6t^2 - 18t + 12 = 6(t-2)(t-1)$

$a = 12t - 18 = 6(2t-3)$

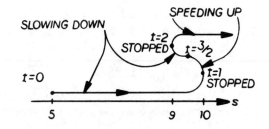

4.10.12 $s = 4t^3 - 12t^2 + 9t - 1$

$v = 12t^2 - 24t + 9 = 3(2t-1)(2t-3)$

$a = 24t - 24 = 24(t-1)$

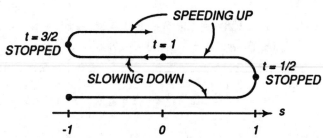

4.10.13 $s = t(t-6)^2$

$v = 3(t-6)(t-2)$

$a = 6(t-4)$

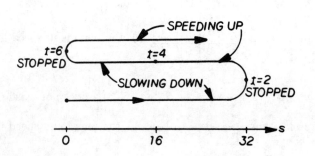

4.10.14 $s = t^3 - 3t^2 - 9t$

$v = 3t^2 - 6t - 9 = 3(t-3)(t+1)$

$a = 6t - 6 = 6(t-1)$

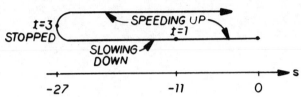

4.10.15 $s = t^3 - 5t^2 + 3t$

$v = 3t^2 - 10t + 3 = (3t-1)(t-3)$

$a = 6t - 10$

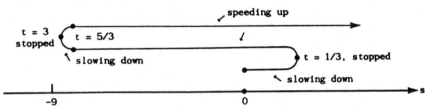

4.10.16 $s = 1/3 t^3 - 3t^2 + 8t + 1$

$v = t^2 - 6t + 8$

$a = 2t - 6.$

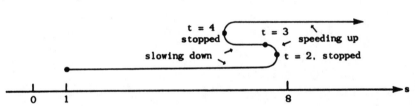

4.10.17 $s = 2t^3 - 5t^2 + 4t - 3$

$v = 6t^2 - 10t + 4$

$a = 12t - 10$

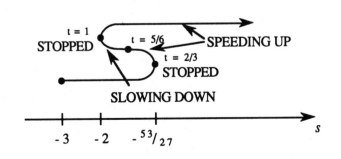

SUPPLEMENTARY EXERCISES, CHAPTER 4

1. For the hollow cylinder shown, assume that R and r are increasing at a rate of 2 m/sec, and h is decreasing at a rate of 3 m/sec. At what rate is the volume changing at the instant when $R = 7$ m. $r = 4$ m, and $h = 5$ m?

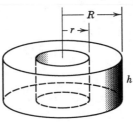

2. The vessel shown is filled at the rate of 4 ft^3/min. How fast is the fluid level rising at the instant when the level is 1 ft?

3. A ball is dropped from a point 10 ft away from a light at the top of a 48-ft pole as shown. When the ball has dropped 16 ft, its velocity (downward) is 32 ft/sec. At what rate is its shadow moving along the ground at that instant?

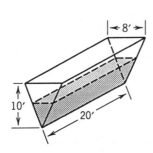

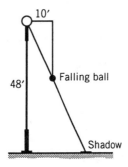

In Exercises 4–9, find the minimum value m and the maximum value M of f on the indicated interval (if they exist) and state where these extreme values occur.

4. $f(x) = 1/x; [-2, -1]$.

5. $f(x) = x^3 - x^4; \left[-1, \dfrac{3}{2}\right]$.

6. $f(x) = x^2(x - 2)^{1/3}; (0, 3]$.

7. $f(x) = 2x/(x^2 + 3); (0, 2]$.

8. $f(x) = 2x^5 - 5x^4 + 7; (-1, 3)$.

9. $f(x) = -|x^2 - 2x|; [1, 3]$.

10. Use Newton's Method to approximate the smallest positive solution of $\sin x + \cos x = 0$.

11. Use Newton's Method to approximate all three solutions of $x^3 - 4x + 1 = 0$.

In Exercises 12–19, sketch the graph of f. Use symmetry, where possible, and show all relative extrema, inflection points, and asymptotes.

12. $f(x) = (x^2 - 3)^2$.

13. $f(x) = \dfrac{1}{1 + x^2}$.

14. $f(x) = \dfrac{2x}{1 + x}$

15. $f(x) = \dfrac{x^3 - 2}{x}$.

16. $f(x) = (1 + x)^{2/3}(3 - x)^{1/3}$.

17. $f(x) = 2\cos^2 x, 0 \le x \le \pi$.

18. $f(x) = x - \tan x, 0 \le x \le 2\pi$

19. $f(x) = \dfrac{3x}{(x + 8)^2}$.

20. Use implicit differentiation to show that a function defined implicitly by $\sin x + \cos y = 2y$ has a critical point wherever $\cos x = 0$. Then use either the first or second derivative test to classify these critical points as relative maxima or minima.

21. Find the equations of the tangent lines at all inflection points of the graph of

$$f(x) = x^4 - 6x^3 + 12x^2 - 8x + 3$$

In Exercises 22–24, find all critical points and use the first derivative test to classify them.

22. $f(x) = x^{1/3}(x - 7)^2$.

23. $f(x) = 2\sin x - \cos 2x, 0 \le x \le 2\pi$.

24. $f(x) = 3x - (x - 1)^{3/2}$.

In Exercises 25–27, find all critical points and use the second derivative test (if possible) to classify them.

25. $f(x) = x^{-1/2} + \dfrac{1}{9}x^{1/2}$.

26. $f(x) = x^2 + 8/x$.

27. $f(x) = \sin^2 x - \cos x, 0 \le x \le 2\pi$.

28. Find two nonnegative numbers whose sum is 20 and such that (a) the sum of their squares is a maximum, and (b) the product of the square of one and the cube of the other is a maximum.

29. Find the dimensions of the rectangle of maximum area that can be inscribed inside the ellipse $(x/4)^2 + (y/3)^2 = 1$.

30. Find the coordinates of the point on the curve $2y^2 = 5(x + 1)$ that is nearest to the origin.

[*Note:* All points $P(x, y)$ on the curve satisfy $x \ge -1$.]

31. If a calculator factory produces x calculators per day, the total daily cost (in dollars) incurred is $0.25x^2 + 35x + 25$. If they are sold for $50 - \dfrac{1}{2}x$ dollars each, find the value of x that maximizes the daily profit.

In Exercises 35–37, determine if all hypotheses of Rolle's Theorem are satisfied on the stated interval. If not state which hypotheses fail; if so, find all values of c guaranteed in the conclusion of the theorem.

32. $f(x) = \sqrt{4 - x^2}$ on $[-2, 2]$.

33. $f(x) = x^{2/3} - 1$ on $[-1, 1]$.

34. $f(x) = \sin(x^2)$ on $0, \sqrt{\pi}\,]$.

In Exercises 38–41, determine if all hypotheses of the Mean-Value Theorem are satisfied on the stated interval. If not, state which hypotheses fail; if so, find all values of c guaranteed in the conclusion of the theorem.

35. $f(x) = |x - 1|$ on $[-2, 2]$.

36. $f(x) = \sqrt{x}$ on $[0, 4]$.

37. $f(x) = \dfrac{x + 1}{x - 1}$ on $[2, 3]$.

38. $f(x) = \begin{cases} 3 - x^2, & x \le 1 \\ 2/x, & x > 1 \end{cases}$ on $[0, 2]$.

SOLUTIONS

SUPPLEMENTARY EXERCISES, CHAPTER 4

1. $V = \pi R^2 h - \pi r^2 h = \pi(R^2 - r^2)h$, $dV/dt = \pi[(R^2 - r^2)dh/dt + h(2R\ dR/dt - 2r\ dr/dt)]$.
   But $dR/dt = dr/dt = 2$, $dh/dt = -3$ so for $R = 7$, $r = 4$, and $h = 5$
   $dV/dt = \pi[(49 - 16)(-3) + 5(14(2) - 8(2))] = -39\pi$. The volume is decreasing at the rate of 39π m^3/sec.

2. At any instant of time the volume of
   fluid is $V = \frac{1}{2}xy(20) = 10xy$. By
   similar triangles $y/x = 8/10$,
   $y = 8x/10$ so $V = 8x^2$ and
   $dV/dt = 16x\ dx/dt$. But $dV/dt = 4$
   so when $x = 1$ we get $4 = 16\ dx/dt$,
   $dx/dt = 1/4$ ft/min.

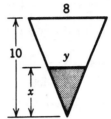

3. By similar triangles
   $x/48 = 10/y$, $x = 480/y$ so
   $dx/dt = -(480/y^2)dy/dt$. But
   $dy/dt = 32$ when $y = 16$ thus
   $dx/dt = -(480/16^2)(32) = -60$.
   The shadow is moving toward the pole
   at the rate of 60 ft/sec.

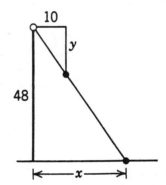

4. $f'(x) = -1/x^2$, no critical points in $(-2, -1)$; $f(-2) = -1/2$, $f(-1) = -1$ so $m = -1$ at $x = -1$ and $M = -1/2$ at $x = -2$.

5. $f'(x) = x^2(3 - 4x)$, critical points $x = 0, 3/4$; $f(-1) = -2$, $f(0) = 0$, $f(3/4) = 27/256$, $f(3/2) = -27/16$. $m = -2$ at $x = -1$, $M = 27/256$ at $x = 3/4$.

6. $f'(x) = \dfrac{x(7x - 12)}{3(x - 2)^{2/3}}$, critical points $x = 2, 12/7$; $f(2) = 0$, $f(12/7) = -\dfrac{144}{49}^3\sqrt{2/7} \approx -1.9$,
   $f(3) = 9$, $\lim\limits_{x \to 0^+} f(x) = 0$. $m \approx -1.9$ at $x = 12/7$, $M = 9$ at $x = 3$.

7. $f'(x) = 2(3 - x^2)/(x^2 + 3)^2$, critical point $x = \sqrt{3}$; $f(\sqrt{3}) = \sqrt{3}/3$, $f(2) = 4/7$, $\lim\limits_{x \to 0^+} f(x) = 0$.
   No minimum on $(0, 2]$, $M = \sqrt{3}/3$ at $x = \sqrt{3}$.

8. $f'(x) = 10x^3(x - 2)$, critical points $x = 0, 2$; $f(0) = 7$, $f(2) = -9$, $\lim\limits_{x \to -1^+} f(x) = 0$,
   $\lim\limits_{x \to 3^-} f(x) = 88$. $m = -9$ at $x = 2$, no maximum.

9. $x^2 - 2x \geq 0$ when $x \leq 0$ or $x \geq 2$, $x^2 - 2x < 0$ when $0 < x < 2$

$$f'(x) = \begin{cases} -2x + 2, & x < 0 \text{ or } x > 2 \\ 2x - 2, & 0 < x < 2 \end{cases}$$

and $f'(x)$ does not exist when $x = 0, 2$. The only critical point in $(1, 3)$ is $x = 2$; $f(1) = -1$, $f(2) = 0$, $f(3) = -3$, $m = -3$ at $x = 3$, $M = 0$ at $x = 2$.

10. $f(x) = \sin x + \cos x$, $f'(x) = \cos x - \sin x$, $x_{n+1} = x_n - \dfrac{\sin x_n + \cos x_n}{\cos x_n - \sin x_n} = x_n - \dfrac{\tan x_n + 1}{1 - \tan x_n}$

$x_1 = 2$, $x_2 = 2.372064374$, $x_3 = 2.356193158$, $x_4 = x_5 = 2.356194490$.

11. $f(x) = x^3 - 4x + 1$, $f'(x) = 3x^2 - 4$, $x_{n+1} = x_n - \dfrac{x_n^3 - 4x_n + 1}{3x_n^2 - 4}$

$x_1 = -2$, $x_2 = -2.125$, $x_3 = -2.114975450, \cdots, x_5 = x_6 = -2.114907541$

$x_1 = 0$, $x_2 = 0.25$, $x_3 = 0.254098361$, $x_4 = x_5 = 0.254101688$

$x_1 = 2$, $x_2 = 1.875$, $x_3 = 1.860978520, \cdots, x_5 = x_6 = 1.860805853$.

12. $f'(x) = 4x(x^2 - 3)$

$f''(x) = 12(x^2 - 1)$

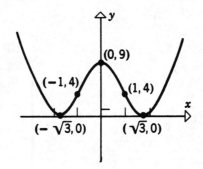

13. $f'(x) = -\dfrac{2x}{(1 + x^2)^2}$

$f''(x) = \dfrac{2(3x^2 - 1)}{(1 + x^2)^3}$

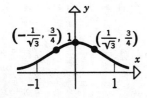

14. $f'(x) = 2/(1+x)^2$

$f''(x) = -4/(1+x)^3$

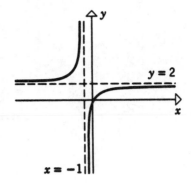

15. $f'(x) = \dfrac{2(x^3+1)}{x^2}$

$f''(x) = \dfrac{2(x^3-2)}{x^3}$

$f(x) = x^2 - \dfrac{2}{x}$ so $f(x)$ is

asymptotic to $y = x^2$ for $|x|$ large.

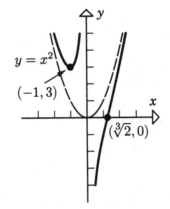

16. $f'(x) = \dfrac{5-3x}{3(1+x)^{1/3}(3-x)^{2/3}}$

$f''(x) = -\dfrac{32}{9(1+x)^{4/3}(3-x)^{5/3}}$

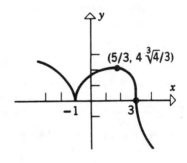

17. $f'(x) = -4\cos x \sin x$

$\quad\quad = -2\sin 2x$

$f''(x) = -4\cos 2x$

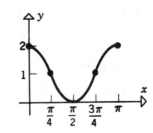

18. $f'(x) = 1 - \sec^2 x$

$f''(x) = -2\sec^2 x \tan x$

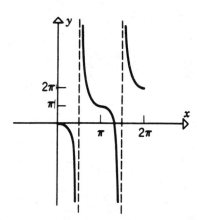

19. $f'(x) = \dfrac{3(8-x)}{(x+8)^3}$

$f''(x) = \dfrac{6(x-16)}{(x+8)^4}$

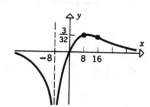

20. $\dfrac{dy}{dx} = \dfrac{\cos x}{2 + \sin y}$, $\dfrac{dy}{dx} = 0$ when $\cos x = 0$. Using the first derivative test, if x_0 is a critical point then $\cos x$ changes sign from $+$ to $-$ or from $-$ to $+$ as x increases through x_0 while $2 + \sin y$ remains $+$ so there is a relative extremum at each critical point.

$\dfrac{d^2y}{dx^2} = -\dfrac{(2 + \sin y)\sin x + \cos x \cos y (dy/dx)}{(2 + \sin y)^2}$. Using the second derivative test, when

$dy/dx = 0$ the critical points satisfy $\cos x = 0$ but $\sin x = \pm 1$ whenever $\cos x = 0$ so

$\dfrac{d^2y}{dx^2} = -\dfrac{(2 + \sin y)(\pm 1) + 0}{(2 + \sin y)^2} = \pm 1/(2 + \sin y)$ which is either $+$ or $-$ at a critical point so there is a relative extremum at each critical point.

21. $f'(x) = 4x^3 - 18x^2 + 24x - 8$

$f''(x) = 12x^2 - 36x + 24 = 12(x-1)(x-2)$

$f''(x) = 0$ when $x = 1, 2$; $f(1) = 2$, $f(2) = 3$. The inflection points are $(1,2)$ and $(2,3)$ because the concavity changes at these points.

$f'(1) = 2$ so the tangent line at $(1,2)$ is $y - 2 = 2(x-1)$, $y = 2x$.

$f'(2) = 0$ so the tangent line at $(2,3)$ is $y = 3$.

22. $f'(x) = \dfrac{7(x-7)(x-1)}{x^{2/3}}$; critical points $x = 0, 1, 7$ relative max at $x = 1$, relative min at $x = 7$, vertical tangent at $x = 0$.

23. $f'(x) = 2\cos x + 2\sin 2x = 2\cos x + 4\sin x \cos x = 2\cos x(1 + 2\sin x)$;
$f'(x) = 0$ when $\cos x = 0$ or $\sin x = -1/2$; critical points $x = \pi/2, 3\pi/2, 7\pi/6, 11\pi/6$
relative max at $x = \pi/2, 3\pi/2$, relative min at $x = 7\pi/6, 11\pi/6$

24. $f'(x) = \dfrac{3}{2}(2 - \sqrt{x-1})$; $f'(x) = 0$ when $\sqrt{x-1} = 2$; critical point $x = 5$, relative max at $x = 5$

25. $f'(x) = \dfrac{x-9}{18x^{3/2}}$; critical point $x = 9$ (0 is not a critical point, it is not in the domain of f)

$f''(x) = \dfrac{27 - x}{36x^{5/2}}$; $f''(9) > 0$, relative min at $x = 9$

26. $f'(x) = 2(x^3 - 4)/x^2$; critical point $x = \sqrt[3]{4}$
$f''(x) = 2 + 16/x^3$; $f''(\sqrt[3]{4}) > 0$, relative min at $x = \sqrt[3]{4}$

27. $f'(x) = \sin x(2\cos x + 1)$; $f'(x) = 0$ when $\sin x = 0$ or $\cos x = -1/2$, in $(0, 2\pi)$ the critical points are $x = \pi, 2\pi/3, 4\pi/3$
$f''(x) = 2\cos 2x + \cos x$; $f''(\pi) > 0$, $f''(2\pi/3) < 0$, $f''(4\pi/3) < 0$
relative max at $x = 2\pi/3, 4\pi/3$, relative min at $x = \pi$

28. Let x and y be the numbers, then $x + y = 20$ thus $y = 20 - x$ for $0 \le x \le 20$.
    (a) $S = x^2 + y^2 = x^2 + (20 - x)^2 = 2x^2 - 40x + 400$, $dS/dx = 4x - 40$, critical point at $x = 10$.
        If $x = 0, 10, 20$ then $S = 400, 200, 400$. S is a maximum for the numbers 0 and 20.
    (b) $P = x^2 y^3 = x^2(20 - x)^3$, $dP/dx = 5x(8 - x)(20 - x)^2$, critical point at $x = 8$. P is maximum for $0 \le x \le 20$ when $x = 8$, $y = 12$.

29. Let (x, y) be a point in the first quadrant that is on the ellipse, then $A = (2x)(2y) = 4xy$. But, from the equation of the ellipse, $y^2 = \dfrac{9}{16}(16 - x^2)$ so with

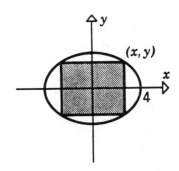

$S = A^2 = 16x^2 y^2$
$S = 9x^2(16 - x^2) = 9(16x^2 - x^4)$ for $0 < x < 4$,
$dS/dx = 36x(8 - x^2)$, critical point at $x = \sqrt{8} = 2\sqrt{2}$.
$d^2 S/dx^2 > 0$ at $x = 2\sqrt{2}$ thus S and hence A is maximum there. If $x = 2\sqrt{2}$ then $y = 3\sqrt{2}/2$.
The dimensions of the rectangle are $4\sqrt{2}$ by $3\sqrt{2}$.

30. If (x, y) is a point on the curve, then its distance L from the origin is $L = \sqrt{x^2 + y^2}$ where $y^2 = \frac{5}{2}(x+1)$ so with $S = L^2 = x^2 + \frac{5}{2}(x+1)$ for $x \geq -1$, $dS/dx = 2x + 5/2$, $dS/dx = 0$ when $x = -5/4$ so there are no critical points for $x > -1$. If $x = -1$ then $S = 1$. $\lim\limits_{x \to +\infty} S = +\infty$. The point nearest the origin occurs when $x = -1$, $y = 0$.

31. $P = $ (total daily sales) $-$ (total daily cost)
$= x(50 - 0.5x) - (0.25x^2 + 35x + 25) = -0.75x^2 + 15x - 25$ for

$0 < x < 100$, $dP/dx = -1.5x + 15$, critical point $x = 10$. $d^2P/dx^2 < 0$ so the profit is maximum when $x = 10$.

32. f is continuous on $[-2, 2]$, $f'(x) = -x/\sqrt{4 - x^2}$ so f is differentiable on $(-2, 2)$, $f(-2) = f(2) = 0$; hypotheses are satisfied. $f'(c) = 0$ for $c = 0$.

33. f is continuous on $[-1, 1]$, $f'(x) = \frac{2}{3}x^{-1/3}$ and $f'(0)$ does not exist, $f(-1) = f(1) = 0$; all hypotheses are not satisfied.

34. f is continuous on $[0, \sqrt{\pi}]$, $f'(x) = 2x\cos(x^2)$ so f is differentiable on $(0, \sqrt{\pi})$, $f(0) = f(\sqrt{\pi}) = 0$; hypotheses are satisfied. $f'(c) = 0$ when $2c\cos(c^2) = 0$ which yields $c = 0, \pm\sqrt{\pi/2}$ of which only $c = \sqrt{\pi/2}$ is in $(0, \sqrt{\pi})$.

35. f is continuous on $[-2, 2]$ but f does not have a derivative at $x = 1$ so all hypotheses are not satisfied.

36. f is continuous on $[0, 4]$ and differentiable on $(0, 4)$. $f'(c) = \dfrac{f(4) - f(0)}{4 - 0}$, $\dfrac{1}{2\sqrt{c}} = \dfrac{1}{2}$, $c = 1$

37. f is continuous on $[2, 3]$, $f'(x) = -2/(x-1)^2$ so f is differentiable on $(2, 3)$.
$f'(c) = \dfrac{f(3) - f(2)}{3 - 2}$, $-\dfrac{2}{(c-1)^2} = -1$, $(c-1)^2 = 2$, $c = 1 \pm \sqrt{2}$ of which only

$c = 1 + \sqrt{2}$ is in $(2, 3)$.

38. By inspection, f is continuous on $[0, 2]$ and differentiable on $(0, 2)$ except perhaps at $x = 1$. For $x = 1$, $\lim\limits_{x \to 1^-} f(x) = \lim\limits_{x \to 1^+} f(x) = f(1)$ so f is continuous at $x = 1$.

$\lim\limits_{x \to 1^-} f'(x) = \lim\limits_{x \to 1^-}(-2x) = -2$, $\lim\limits_{x \to 1^+} f'(x) = \lim\limits_{x \to 1^+}(-2/x^2) = -2$ so f is differentiable at $x = 1$ (see theorem preceding Exercise 71, Section 3.3). $f'(c) = \dfrac{f(2) - f(0)}{2 - 0} = \dfrac{1 - 3}{2} = -1$ so $c \neq 1$. If $x < 1$ then $f'(x) = -2x$ thus $f'(c) = -1$ for $c = 1/2$. If $x > 1$ then $f'(x) = -2/x^2$ thus $f'(c) = -1$ for $c = \sqrt{2}$. The values of c are $1/2$, $\sqrt{2}$.

CHAPTER 5

Integration

SECTION 5.2

5.2.1 Evaluate $\int \dfrac{(1+x)^2}{x^{1/2}} dx$.

5.2.2 Evaluate $\int (x^2 + 2x + 5)dx$.

5.2.3 Evaluate $\int \left(x^2 + 8x + \dfrac{3}{x^2} \right) dx$.

5.2.4 Evaluate $\int \dfrac{7\,dx}{\sqrt{x}}$.

5.2.5 Evaluate $\int (x^2 + 1)^2 dx$.

5.2.6 Evaluate $\int (3\sqrt{x} + 1)dx$.

5.2.7 Evaluate $\int \dfrac{x^4 + 1}{x^3} dx$.

5.2.8 Evaluate $\int (x^3 + 2)^2 dx$.

5.2.9 Evaluate $\int \dfrac{dx}{\cos^2 x}$.

5.2.10 Evaluate $\int (x^3 - x + 5)dx$.

5.2.11 Evaluate $\int \left(x^2 - \dfrac{3}{x^4} \right) dx$.

5.2.12 Evaluate $\int \dfrac{x^2 - 4}{\sqrt[3]{x^2}} dx$.

5.2.13 Evaluate $\int \left(\sqrt{x} + \dfrac{1}{\sqrt{x}} \right) dx$.

5.2.14 Evaluate $\int (x + 1)\sqrt{x}\,dx$.

5.2.15 Evaluate $\int \dfrac{7}{t^4} dt$.

5.2.16 Evaluate $\int (\sqrt{x} + 2)^2 dx$.

5.2.17 Evaluate $\int \left(\dfrac{x^3}{4} + \cos x \right) dx$.

5.2.18 Evaluate $\int (x - 2)^2 x\,dx$.

5.2.19 Evaluate $\int (x^{-2} + \sec^2 x + 3)dx$.

5.2.20 $\int (x^3 - \csc x \cot x + 7)\,dx$

SOLUTIONS

SECTION 5.2

5.2.1 $\displaystyle \int \frac{(1+x)^2}{x^{1/2}}dx = \int \frac{(1+2x+x^2)}{x^{1/2}}dx = \int (x^{-1/2} + 2x^{1/2} + x^{3/2})dx$

$$= 2x^{1/2} + \frac{4}{3}x^{3/2} + \frac{2}{5}x^{5/2} + C.$$

5.2.2 $\displaystyle \frac{x^3}{3} + x^2 + 5x + C.$ 　　　　　　　　　　**5.2.3** $\displaystyle \frac{x^3}{3} + 4x^2 - \frac{3}{x} + C.$

5.2.4 $14\sqrt{x} + C.$

5.2.5 $\displaystyle \int (x^4 + 2x^2 + 1)dx = \frac{x^5}{5} + \frac{2}{3}x^3 + x + C.$

5.2.6 $2x^{3/2} + x + C.$

5.2.7 $\displaystyle \int (x + x^{-3})dx = \frac{x^2}{2} - \frac{x^{-2}}{2} + C = \frac{x^2}{2} - \frac{1}{2x^2} + C.$

5.2.8 $\displaystyle \int (x^6 + 4x^3 + 4)dx = \frac{x^7}{7} + x^4 + 4x + C.$

5.2.9 $\displaystyle \int \sec^2 x\, dx = \tan x + C.$ 　　　　　**5.2.10** $\displaystyle \frac{x^4}{4} - \frac{x^2}{2} + 5x + C.$

5.2.11 $\displaystyle \int (x^2 - 3x^{-4})dx = \frac{x^3}{3} + \frac{1}{x^3} + C.$

5.2.12 $\displaystyle \int (x^{4/3} - 4x^{-2/3})dx = \frac{3}{7}x^{7/3} - 12x^{1/3} + C.$

5.2.13 $\displaystyle \frac{2}{3}x^{3/2} + 2x^{1/2} + C.$

5.2.14 $\displaystyle \int (x^{3/2} + x^{1/2})dx = \frac{2}{5}x^{5/2} + \frac{2}{3}x^{3/2} + C.$

5.2.15 $\displaystyle -\frac{7}{3t^3} + C.$

5.2.16 $\displaystyle \int (x + 4\sqrt{x} + 4)dx = \frac{x^2}{2} + \frac{8}{3}x^{3/2} + 4x + C.$

5.2.17 $\dfrac{x^4}{16} + \sin x + C.$

5.2.18 $\displaystyle\int (x^3 - 4x^2 + 4x)dx = \dfrac{x^4}{4} - \dfrac{4}{3}x^3 + 2x^2 + C.$

5.2.19 $-\dfrac{1}{x} + \tan x + 3x + C.$

5.2.20 $\dfrac{x^4}{4} + \csc x + 7x + C$

SECTION 5.3

5.3.1 Evaluate $\displaystyle\int 3x\sqrt{1-2x^2}dx$.

5.3.2 Evaluate $\displaystyle\int t^2(2-3t^3)^3dt$.

5.3.3 Evaluate $\displaystyle\int \frac{3x\,dx}{\sqrt[3]{3-7x^2}}$.

5.3.4 Evaluate $\displaystyle\int \sin 2x\cos 2x\,dx$.

5.3.5 Evaluate $\displaystyle\int \frac{dx}{\cos^2 2x}$.

5.3.6 Evaluate $\displaystyle\int (2+\sin 3t)^{1/2}\cos 3t\,dt$.

5.3.7 Evaluate $\displaystyle\int \csc 2t\cot 2t\,dt$.

5.3.8 Evaluate $\displaystyle\int \tan^3 5x\sec^2 5x\,dx$.

5.3.9 Evaluate $\displaystyle\int x^3\sqrt{5x^4-l8}dx$.

5.3.10 Evaluate $\displaystyle\int x\sqrt{x-5}dx$.

5.3.11 Evaluate $\displaystyle\int \frac{\sin x\,dx}{\cos^3 x}$.

5.3.12 Evaluate $\displaystyle\int [\tan(\tan\theta)]\sec^2\theta\,d\theta$.

5.3.13 Evaluate $\displaystyle\int \frac{1}{\sqrt{x}}\sin\sqrt{x}dx$.

5.3.14 Evaluate $\displaystyle\int (x^2+1)(x^3+3x)^{10}dx$.

5.3.15 Evaluate $\displaystyle\int \tan 3x\sec^2 3x\,dx$.

5.3.16 Evaluate $\displaystyle\int (x^3-x)(x^4-2x^2)^{15}dx$.

5.3.17 Evaluate $\displaystyle\int \frac{x^2\,dx}{\sqrt{x+1}}$.

5.3.18 Evaluate $\displaystyle\int \frac{4}{(x+4)^3}dx$.

5.3.19 Evaluate $\displaystyle\int \frac{x-2}{(x^2-4x+4)^2}dx$.

5.3.20 Evaluate $\displaystyle\int x\sec^2 x^2 dx$.

5.3.21 Evaluate $\displaystyle\int x\sqrt[n]{a+bx^2}\,dx$.

5.3.22 Evaluate $\displaystyle\int x^3\sin(x^4+2)dx$.

5.3.23 Evaluate $\displaystyle\int \frac{x^2\,dx}{\sqrt{1+x^3}}$.

5.3.24 Evaluate $\displaystyle\int x\sqrt[3]{x+1}\,dx$.

SOLUTIONS

SECTION 5.3

5.3.1 $u = 1 - 2x^2$, $du = -4x \, dx$, $x \, dx = \dfrac{-du}{4}$

$$-\frac{3}{4} \int u^{1/2} du = -\frac{1}{2} u^{3/2} + C = -\frac{1}{2}(1 - 3x^2)^{3/2} + C.$$

5.3.2 $u = 2 - 3t^3$, $du = -9t^2 dt$, $t^2 dt = -\dfrac{du}{9}$

$$-\frac{1}{9} \int u^3 du = -\frac{1}{36} u^4 + C = -\frac{1}{36}(2 - 3t^3)^4 + C.$$

5.3.3 $u = 3 - 7x^2$, $du = -14x \, dx$, $x \, dx = \dfrac{du}{-14}$

$$-\frac{3}{14} \int \frac{du}{u^{1/3}} = -\frac{9}{28} u^{2/3} + C = -\frac{9}{28}(3 - 7x^2)^{2/3} + C.$$

5.3.4 $u = \sin 2x$, $du = 2 \cos 2x \, dx$, $\cos 2x \, dx = \dfrac{du}{2}$

$$\frac{1}{2} \int u \, du = \frac{1}{4} u^2 + C = \frac{1}{4} \sin^2 2x + C.$$

5.3.5 $\displaystyle\int \frac{dx}{\cos^2 2x} = \int \sec^2 2x \, dx$, $u = 2x$, $du = 2dx$, $dx = \dfrac{du}{2}$

$$\frac{1}{2} \int \sec^2 u \, du = \frac{1}{2} \tan u + C = \frac{1}{2} \tan 2x = C.$$

5.3.6 $u = 2 + \sin 3t$, $du = 3 \cos 3t \, dt$, $\cos 3t \, dt = \dfrac{du}{3}$

$$\frac{1}{3} \int u^{1/2} du = \frac{2}{9} u^{3/2} + C = \frac{2}{9}(2 + \sin 3t)^{3/2} + C.$$

5.3.7 $u = 2t$, $du = 2dt$, $dt = \dfrac{du}{2}$

$$\frac{1}{2} \int \csc u \cot u \, du = -\frac{1}{2} \csc u + C = -\frac{1}{2} \csc 2t + C.$$

5.3.8 $u = \tan 5x$, $du = 5 \sec^2 5x \, dx$, $\sec^2 5x \, dx = \dfrac{du}{5}$

$$\frac{1}{5} \int u^3 du = \frac{1}{20} u^4 + C = \frac{1}{20} \tan^4 5x + C.$$

5.3.9 $u = 5x^4 - 18$, $du = 20x^3 dx$, $x^3 dx = \dfrac{du}{20}$

$$\frac{1}{20}\int u^{1/2}du = \frac{1}{30}u^{3/2} + C = \frac{1}{30}(5x^4 - 18)^{3/2} + C.$$

5.3.10 $u = x - 5$, $du = dx$, $x = u + 5$

$$\int (u+5)u^{1/2}du = \int (u^{3/2} + 5u^{1/2})du = \frac{2}{5}u^{5/2} + \frac{10}{3}u^{3/2} + C$$

$$= \frac{2}{5}(x-5)^{5/2} + \frac{10}{3}(x-5)^{3/2} + C.$$

5.3.11 $u = \cos x$, $du = -\sin x\, dx$, $\sin x\, dx = -du$

$$-\int u^{-3}du = \frac{1}{2u^2} + C = \frac{1}{2\cos^2 x} + C = \frac{1}{2}\sec^2 x + C.$$

5.3.12 $u = \tan\theta$, $du = \sec^2\theta d\theta$.

$$\int \tan u\, du = -\ln|\cos u| + C = -\ln|\cos(\tan\theta)| + C.$$

5.3.13 $u = \sqrt{x}$, $du = \dfrac{1}{2\sqrt{x}}dx$, $\dfrac{dx}{\sqrt{x}} = 2du$

$$2\int \sin u\, du = -2\cos u + C = -2\cos\sqrt{x} + C.$$

5.3.14 $u = x^3 + 3x$, $du = 3(x^2 + 1)dx$, $(x^2 + 1)dx = \dfrac{du}{3}$

$$\frac{1}{3}\int u^{10}du = \frac{1}{33}u^{11} + C = \frac{1}{33}(x^3 + 3x)^{11} + C.$$

5.3.15 $u = \tan 3x$, $du = 3\sec^2 3x\, dx$, $\dfrac{du}{3} = \sec^2 3x\, dx$

$$\frac{1}{3}\int u\, du = \frac{1}{3}\cdot\frac{u^2}{2} + C = \frac{1}{6}\tan^2 3x + C$$

5.3.16 $u = x^4 - 2x^2$, $du = 4(x^3 - x)dx$, $\dfrac{du}{4} = (x^3 - x)dx$

$$\frac{1}{4}\int u^{15}du = \frac{1}{64}u^{16} + C = \frac{1}{64}(x^4 - 2x^2)^{16} + C$$

5.3.17 $u = x+1,\ du = dx,\ x = u-1$

$$\int \frac{(u-1)^2}{u^{1/2}}\,du = \int \frac{(u^2-2u+1)}{u^{1/2}}\,du = \int (u^{3/2} - 2u^{1/2} + u^{-1/2})\,du$$

$$= \frac{2}{5}u^{5/2} - \frac{4}{3}u^{3/2} + 2u^{1/2} + C$$

$$= \frac{2}{5}(x+1)^{5/2} - \frac{4}{3}(x+1)^{3/2} + 2(x+1)^{1/2} + C.$$

5.3.18 $u = x+4,\ du = dx$

$$4\int u^{-3}\,du = -2u^{-2} + C = -\frac{2}{(x+4)^2} + C.$$

5.3.19 $u = x^2 - 4x + 4,\ du = 2(x-2)dx,\ (x-2)dx = \frac{du}{2}$

$$\frac{1}{2}\int \frac{du}{u^2} = -\frac{1}{2u} + C = -\frac{1}{2(x^2-4x+4)}C.$$

5.3.20 $u = x^2,\ du = 2x\,dx,\ x\,dx = \frac{du}{2}$

$$\frac{1}{2}\int \sec^2 u\,du = \frac{1}{2}\tan u + C = \frac{1}{2}\tan x^2 + C.$$

5.3.21 $u = a + bx^2,\ du = 2bx\,dx,\ \frac{du}{2b} = x\,dx$

$$\frac{1}{2b}\int u^{1/n}\,du = \frac{n}{2b(n+1)}u^{\frac{(n+1)}{n}} + C = \frac{n}{2b(n+1)}(a+bx^2)^{\frac{(n+1)}{n}} + C.$$

5.3.22 $u = x^4 + 2,\ du = 4x^3 dx,\ x^3 dx = \frac{du}{4}$

$$\frac{1}{4}\int \sin u\,du = -\frac{1}{4}\cos u + C = -\frac{1}{4}\cos(x^4+2) + C.$$

5.3.23 $u = 1 + x^3,\ du = 3x^2 dx,\ x^2 dx = du/3$

$$\frac{1}{3}\int u^{-1/2}\,du = \frac{2}{3}u^{1/2} + C = \frac{2}{3}\sqrt{1+x^3} + C.$$

5.3.24 $u = x+1,\ du = dx,\ x = u-1$

$$\int (u-1)u^{1/3}\,du = \int (u^{4/3} - u^{1/3})\,du = \frac{3}{7}u^{7/3} - \frac{3}{4}u^{\frac{4}{3}} + C = \frac{3}{7}(x+1)^{7/3} - \frac{3}{4}(x+1)^{\frac{4}{3}} + C$$

SECTION 5.4

5.4.1 Evaluate $\displaystyle\sum_{i=1}^{4}(i^2 + 2)$.

5.4.2 Evaluate $\displaystyle\sum_{j=1}^{4}(2j - 3)$.

5.4.3 Evaluate $\displaystyle\sum_{j=-2}^{2} 3j^2$.

5.4.4 Evaluate $\displaystyle\sum_{k=1}^{4}\frac{k}{k+1}$.

5.4.5 Evaluate $\displaystyle\sum_{n=1}^{4}\sin\frac{n\pi}{2}$.

5.4.6 Express $\dfrac{1}{2} + \dfrac{2}{3} + \dfrac{3}{4} + \dfrac{4}{5} + \dfrac{5}{6}$ in sigma notation.

5.4.7 Express $1 + \dfrac{2}{3} + \dfrac{3}{5} + \dfrac{4}{7} + \dfrac{5}{9}$ in sigma notation.

5.4.8 Evaluate $\displaystyle\sum_{k=1}^{10}(k+2)^2$ by first changing $f(k) = (k+2)^2$ to $f(k) = k^2$ and then, an appropriate change in the limits of summation.

5.4.9 Evaluate $\displaystyle\sum_{k=1}^{30}(k^2 + 2)$.

5.4.10 Evaluate $\displaystyle\sum_{k=1}^{10}(k+3)^3$ by first changing $f(k) = (k+3)^3$ to $f(k) = k^3$ and then by making an appropriate change in the limits of summation.

5.4.11 Evaluate $\displaystyle\sum_{k=1}^{30}(k^2 + 3k - 5)$.

5.4.12 Express $1 + 3 + 3^2 + 3^3 + 3^4 + 3^5$ in sigma notation with $j = 3$ as the lower limit.

5.4.13 Express $\displaystyle\sum_{k=5}^{10}\frac{1}{k}$ in sigma notation with $k = 1$ as the lower limit.

5.4.14 Express $\displaystyle\sum_{k=10}^{30}\frac{1}{k(k+1)}$ in sigma notation with $k = 4$ as the lower limit.

5.4.15 Evaluate $\displaystyle\sum_{k=1}^{n} \left(\frac{k}{n}\right)^2 \left(\frac{1}{n}\right)$.

5.4.16 Evaluate $\displaystyle\sum_{k=1}^{n} \left(\frac{k}{n}\right) \left(\frac{1}{n}\right)$.

5.4.17 Evaluate $\displaystyle\sum_{k=1}^{n} \left(1 + \frac{2k}{n}\right) \frac{1}{n}$.

5.4.18 Evaluate $\displaystyle\sum_{k=1}^{n} \left(\frac{k+2}{n}\right)^2 \frac{1}{n}$.

5.4.19 Evaluate $\displaystyle\sum_{k=2}^{20} k \left(1 - \frac{1}{k}\right)$.

SOLUTIONS

SECTION 5.4

5.4.1 $3 + 6 + 11 + 18 = 38.$

5.4.2 $-1 + 1 + 3 + 5 = 8.$

5.4.3 $12 + 3 + 0 + 3 + 12 = 30.$

5.4.4 $\dfrac{1}{2} + \dfrac{2}{3} + \dfrac{3}{4} + \dfrac{4}{5} = \dfrac{163}{60}.$

5.4.5 $1 + 0 - 1 + 0 = 0.$

5.4.6 $\displaystyle\sum_{k=1}^{5} \dfrac{k}{k+1}.$

5.4.7 $\displaystyle\sum_{k=1}^{5} \dfrac{k}{2k-1}.$

5.4.8 $\displaystyle\sum_{k=1}^{10}(k+2)^2 = \sum_{k=3}^{12} k^2 = \sum_{k=1}^{12} k^2 - \sum_{k=1}^{2} k^2 = \dfrac{12(13)(25)}{6} - \dfrac{2(3)(5)}{6} = 645$

5.4.9 $\displaystyle\sum_{k=1}^{30} k^2 + \sum_{k=1}^{30} 2 = \dfrac{1}{6}(30)(31)(61) + 2(30) = 9515.$

5.4.10 $\displaystyle\sum_{k=1}^{10}(k+3)^3 = \sum_{k=4}^{13} k^3 = \sum_{k=1}^{13} k^3 - \sum_{k=1}^{3} k^3 = \left[\dfrac{(13)(14)}{2}\right]^2 - \left[\dfrac{(3)(4)}{2}\right]^2 = 8245$

5.4.11 $\displaystyle\sum_{k=1}^{30} k^2 + 3\sum_{k=1}^{30} k - \sum_{k=1}^{30} 5 = \dfrac{1}{6}(30)(31)(61) + \dfrac{3}{2}(30)(31) - 5(30) = 10700.$

5.4.12 $\displaystyle\sum_{j=3}^{8} 3^{j-3}.$

5.4.13 $\displaystyle\sum_{k=1}^{6} \dfrac{1}{k+4}.$

5.4.14 $\displaystyle\sum_{k=4}^{24} \dfrac{1}{(k+6)(k+7)}.$

5.4.15 $\displaystyle\sum_{k=1}^{n} \dfrac{k^2}{n^3} = \dfrac{1}{n^3}\sum_{k=1}^{n} k^2 = \left(\dfrac{1}{n^3}\right) \cdot \dfrac{(n)(n+1)(2n+1)}{6} = \dfrac{(n+1)(2n+1)}{6n^2}.$

5.4.16 $\displaystyle\sum_{k=1}^{n} \dfrac{k}{n^2} = \dfrac{1}{n^2}\sum_{k=1}^{n} k = \left(\dfrac{1}{n^2}\right)\dfrac{n(n+1)}{2} = \dfrac{n+1}{2n}.$

5.4.17 $\displaystyle\sum_{k=1}^{n}\left(1+\frac{2k}{n}\right)\frac{1}{n} = \sum_{k=1}^{n}\left(\frac{1}{n}+\frac{2k}{n^2}\right) = \frac{1}{n}\sum_{k=1}^{n}1 + \frac{2}{n^2}\sum_{k=1}^{n}k$

$$= \frac{1}{n}(n) + \frac{2}{n^2}\frac{(n)(n+1)}{2} = \frac{2n+1}{n} = 2+\frac{1}{n}$$

5.4.18 $\displaystyle\sum_{k=1}^{n}\frac{(k+2)^2}{n^3} = \frac{1}{n^3}\sum_{k=1}^{n}(k+2)^2 = \frac{1}{n^3}\sum_{k=1}^{n}(k^2+4k+4)$

$$= \frac{1}{n^3}\left[\sum_{k=1}^{n}k^2 + 4\sum_{k=1}^{n}k + 4\sum_{k=1}^{n}1\right]$$

$$= \frac{1}{n^3}\left[\frac{n(n+1)(2n+1)}{6} + (4)\frac{n(n+1)}{2} + (4)(n)\right]$$

$$= \frac{2n^2+15n+37}{6n^2}.$$

5.4.19 $\displaystyle\sum_{k=2}^{20}(k-1) = \sum_{k=1}^{19}k = \frac{(19)(20)}{2} = 190.$

SECTION 5.5

5.5.1 Estimate the area under the curve $y = x^2$ by dividing the interval $[0, 2]$ into 4 subintervals of equal length and computing $\sum_{k=1}^{4} f(x_k^*)\Delta x$ with x_k^* as the left endpoint of each subinterval.

5.5.2 Estimate the area under the curve $y = x^2$ by dividing the interval $[0, 2]$ into 4 subintervals of equal length and computing $\sum_{k=1}^{4} f(x_k^*)\Delta x$ with x_k^* as the right endpoint of each subinterval.

5.5.3 Estimate the area under the curve $y = 1/x$ by dividing the interval $[1, 2]$ into 4 subintervals of equal length and computing $\sum_{k=1}^{4} f(x_k^*)\Delta x$ with x_k^* as the left endpoint of each subinterval.

5.5.4 Estimate the area under the curve $y = 1/x$ by dividing the interval $[1, 2]$ into 4 subintervals of equal length and computing $\sum_{k=1}^{4} f(x_k^*)\Delta x$ with x_k^* as the right endpoint of each subinterval.

5.5.5 Estimate the area under the curve $y = x^3 + 2$ by dividing the interval $[1, 4]$ into 3 subintervals of equal length and computing $\sum_{k=1}^{3} f(x_k^*)\Delta x$ with x_k^* as the left endpoint of each subinterval.

5.5.6 Estimate the area under the curve $y = x^3 + 2$ by dividing the interval $[1, 4]$ into 3 subintervals of equal length and computing $\sum_{k=1}^{3} f(x_k^*)\Delta x$ with x_k^* as the right endpoint of each subinterval.

5.5.7 Estimate the area under the curve $y = x^2 - x$ by dividing the interval $[3, 8]$ into 5 subintervals of equal length and computing $\sum_{k=1}^{5} f(x_k^*)\Delta x$ with x_k^* as the left endpoint of each subinterval.

5.5.8 Estimate the area under the curve $y = 1/x^2$ by dividing the interval $[1, 4]$ into 6 subintervals of equal length and computing $\sum_{k=1}^{6} f(x_k^*)\Delta x$ with x_k^* as the left endpoint of each subinterval.

5.5.9 Estimate the area under the curve $y = \sqrt{x}$ by dividing the interval $[0, 4]$ into 4 subintervals of equal length and computing $\sum_{k=1}^{4} f(x_k^*)\Delta x$ with x_k^* as the right endpoint of each subinterval.

5.5.10   Estimate the area under the line $y = 2x + 3$ by dividing the interval $[1, 9]$ into 4 subintervals of equal length and computing $\sum_{k=1}^{4} f(x_k^*)\Delta x$ with x_k^* as the left endpoint of each subinterval.

5.5.11   Estimate the area under the line $y = 2x + 3$ by dividing the interval $[1, 9]$ into 4 subintervals of equal length and computing $\sum_{k=1}^{4} f(x_k^*)\Delta x$ with x_k^* as the right endpoint of each subinterval.

5.5.12   Use $A = \lim\limits_{n \to +\infty} \sum_{k=1}^{n} f(x_k^*)\Delta x$ with x_k^* as the right endpoint of each subinterval to find the area under the line $y = 2x$ over the interval $[1, 3]$.

Hint: $\sum_{k=1}^{n} k = \dfrac{n(n+1)}{2}$.

5.5.13   Use $A = \lim\limits_{n \to +\infty} \sum_{k=1}^{n} f(x_k^*)\Delta x$ with x_k^* as the left endpoint of each subinterval to find the area under the line $y = 2x$ over the interval $[1, 3]$.

Hint: $\sum_{k=1}^{n} k = \dfrac{n(n+1)}{2}$.

5.5.14   Use $A = \lim\limits_{n \to +\infty} \sum_{k=1}^{n} f(x_k^*)\Delta x$ with x_k^* as the right endpoint of each subinterval to find the area under the line $y = 2x + 3$ over the interval $[1, 9]$.

Hint: $\sum_{k=1}^{n} k = \dfrac{n(n+1)}{2}$.

5.5.15   Use $A = \lim\limits_{n \to +\infty} \sum_{k=1}^{n} f(x_k^*)\Delta x$ with x_k^* as the left endpoint of each subinterval to find the area under the line $y = 3x + 4$ over the interval $[1, 2]$.

Hint: $\sum_{k=1}^{n} k = \dfrac{n(n+1)}{2}$.

5.5.16   Use $A = \lim\limits_{n \to +\infty} \sum_{k=1}^{n} f(x_k^*)\Delta x$ with x_k^* as the right endpoint of each subinterval to find the area under the curve $y = 2x^2$ over the interval $[0, 2]$.

Hint: $\sum_{k=1}^{n} k^2 = \dfrac{n(n+1)(2n+1)}{6}$.

5.5.17   Use $A = \lim\limits_{n\to+\infty} \sum\limits_{k=1}^{n} f(x_k^*)\Delta x$ with x_k^* as the left endpoint of each subinterval to find the area under the curve $y = 16 - x^2$ over the interval $[0, 4]$.

       Hint: $\sum\limits_{k=1}^{n} k^2 = \dfrac{n(n+1)(2n+1)}{6}$.

5.5.18   Use $A = \lim\limits_{n\to+\infty} \sum\limits_{k=1}^{n} f(x_k^*)\Delta x$ with x_k^* as the right endpoint of each subinterval to find the area under the curve $y = x^3 + 3$ over the interval $[0, 2]$.

       Hint: $\sum\limits_{k=1}^{n} k^3 = \left[\dfrac{n(n+1)}{2}\right]^2$.

5.5.19   Use $A = \lim\limits_{n\to+\infty} \sum\limits_{k=1}^{n} f(x_k^*)\Delta x$ with x_k^* as the right endpoint of each subinterval to find the area under the curve $y = x^2 - 1$ over the interval $[1, 2]$.

       Hint: $\sum\limits_{k=1}^{n} k = \dfrac{n(n+1)}{2}; \sum\limits_{k=1}^{n} k^2 = \dfrac{n(n+1)(2n+1)}{6}$.

SOLUTIONS

SECTION 5.5

5.5.1 $\Delta x = \dfrac{2-0}{4} = \dfrac{1}{2}$, $x_k^* = 0 + (k-1)\Delta x = \dfrac{k-1}{2}$

$$\sum_{k=1}^{4} f(x_k^*)\Delta x = \sum_{k=1}^{4} \left(\frac{k-1}{2}\right)^2 \left(\frac{1}{2}\right) = \frac{7}{4}.$$

5.5.2 $\Delta x = \dfrac{2-0}{4} = \dfrac{1}{2}$, $x_k^* = 0 + k\Delta x = \dfrac{k}{2}$

$$\sum_{k=1}^{4} f(x_k^*)\Delta x = \sum_{k=1}^{4} \left(\frac{k}{2}\right)^2 \left(\frac{1}{2}\right) = \frac{15}{4}.$$

5.5.3 $\Delta x = \dfrac{2-1}{4} = \dfrac{1}{4}$; $x_k^* = 1 + (k-1)\Delta x = \dfrac{3+k}{4}$

$$\sum_{k=1}^{4} f(x_k^*)\Delta x = \sum_{k=1}^{4} \left(\frac{1}{\frac{3+k}{4}}\right) \left(\frac{1}{4}\right) = \frac{319}{420}.$$

5.5.4 $\Delta x = \dfrac{2-1}{4} = \dfrac{1}{4}$; $x_k^* = 1 + k\Delta x = \dfrac{4+k}{4}$

$$\sum_{k=1}^{4} f(x_k^*)\Delta x = \sum_{k=1}^{4} \frac{1}{\left(\frac{4+k}{4}\right)} \left(\frac{1}{4}\right) = \frac{533}{840}.$$

5.5.5 $\Delta x = \dfrac{4-1}{3} = 1$; $x_k^* = 1 + (k-1)\Delta x = k$

$$\sum_{k=1}^{3} f(x_k^*)\Delta x = \sum_{k=1}^{3} (k^3 + 2)(1) = 42.$$

5.5.6 $\Delta x = \dfrac{4-1}{3} = 1$, $x_k^* = 1 + k\Delta x = 1 + k$

$$\sum_{k=1}^{3} f(x_k^*)\Delta x = \sum_{k=1}^{3} \left[(1+k)^3 + 2\right](1) = 105.$$

5.5.7 $\Delta x = \dfrac{8-3}{5} = 1, \ x_k^* = 3 + (k-1)\Delta x = 2 + k$

$$\sum_{k=1}^{5} f(x_k^*)\Delta x = \sum_{k=1}^{5} \left[(2+k)^2 - (2+k)\right](1) = 110.$$

5.5.8 $\Delta x = \dfrac{4-1}{6} = \dfrac{1}{2}, \ x_k^* = 1 + (k-1)\Delta x = \dfrac{1+k}{2}$

$$\sum_{k=1}^{6} f(x_k^*)\Delta x = \sum_{k=1}^{6} \frac{1}{\left(\dfrac{1+k}{2}\right)^2}\left(\frac{1}{2}\right) \approx 1.0236.$$

5.5.9 $\Delta x = \dfrac{4-0}{4} = 1, \ x_k^* = 0 + k\Delta x = k$

$$\sum_{k=1}^{4} f(x_k^*)\Delta x = \sum_{k=1}^{4} \sqrt{k}(1) \approx 6.1463.$$

5.5.10 $\Delta x = \dfrac{9-1}{4} = 2, \ x_k^* = 1 + (k-1)\Delta x = 2k - 1$

$$\sum_{k=1}^{4} f(x_k^*)\Delta x = \sum_{k=1}^{4} \left[2(2k-1)+3\right](2) = 88.$$

5.5.11 $\Delta x = \dfrac{9-1}{4} = 2, \ x_k^* = 1 + k\Delta x = 1 + 2k$

$$\sum_{k=1}^{4} f(x_k^*)\Delta x = \sum_{k=1}^{4} \left[2(1+2k)+3\right](2) = 120.$$

5.5.12 $\Delta x = \dfrac{3-1}{n} = \dfrac{2}{n}, \ x_k^* = 1 + k\Delta x = 1 + \dfrac{2k}{n}$

$$\sum_{k=1}^{n} f(x_k^*)\Delta x = \sum_{k=1}^{n} 2\left(1 + \frac{2k}{n}\right)\left(\frac{2}{n}\right) = \frac{4}{n}\sum_{k=1}^{n} 1 + \frac{8}{n^2}\sum_{k=1}^{n} k =$$

$$\frac{4}{n}(n) + \left(\frac{8}{n^2}\right)\frac{n(n+1)}{2} = 8 + \frac{4}{n}; \ A = \lim_{n \to +\infty}\left(8 + \frac{4}{n}\right) = 8.$$

5.5.13 $\Delta x = \dfrac{3-1}{n} = \dfrac{2}{n}, \ x_k^* = 1 + (k-1)\Delta x = 1 + \dfrac{2(k-1)}{n}$

$$\sum_{k=1}^{n} f(x_k^*)\Delta x = \sum_{k=1}^{n} 2\left[1 + \frac{2(k-1)}{n}\right]\left(\frac{2}{n}\right) = \frac{4}{n}\sum_{k=1}^{n} 1 + \frac{8}{n^2}\sum_{k=1}^{n}(k-1) =$$

$$\frac{4}{n}(n) + \left(\frac{8}{n^2}\right)\frac{(n-1)n}{2} = 8 - \frac{4}{n}; \ A = \lim_{n \to +\infty}\left(8 - \frac{4}{n}\right) = 8.$$

5.5.14 $\Delta x = \dfrac{9-1}{n} = \dfrac{8}{n},\ x_k^* = 1 + k\Delta x = 1 + \dfrac{8k}{n}$

$$\sum_{k=1}^{n} f(x_k^*)\Delta x = \sum_{k=1}^{n}\left[2\left(1+\frac{8k}{n}\right)+3\right]\left(\frac{8}{n}\right) = \frac{40}{n}\sum_{k=1}^{n}1 + \frac{128}{n^2}\sum_{k=1}^{n}k =$$

$$\frac{40}{n}(n) + \left(\frac{128}{n^2}\right)\frac{n(n+1)}{2} = 104 + \frac{64}{n};\ A = \lim_{n\to+\infty}\left(104 + \frac{64}{n}\right) = 104.$$

5.5.15 $\Delta x = \dfrac{2-1}{n} = \dfrac{1}{n},\ x_k^* = 1 + (k-1)\Delta x = 1 + \dfrac{(k-1)}{n}$

$$\sum_{k=1}^{n} f(x_k^*)\Delta x = \sum_{k=1}^{n}\left[3\left(1+\frac{k-1}{n}\right)+4\right]\left(\frac{1}{n}\right) = \frac{7}{n}\sum_{k=1}^{n}1 + \frac{3}{n^2}\sum_{k=1}^{n}(k-1) =$$

$$\frac{7}{n}(n) + \frac{3}{n^2}\frac{(n-1)n}{2} = \frac{17}{2} - \frac{3}{2n};\ A = \lim_{n\to+\infty}\left(\frac{17}{2} - \frac{3}{2n}\right) = \frac{17}{2}.$$

5.5.16 $\Delta x = \dfrac{2-0}{n} = \dfrac{2}{n},\ x_k^* = 0 + k\Delta x = \dfrac{2k}{n}$

$$\sum_{k=1}^{n} f(x_k^*)\Delta x = \sum_{k=1}^{n}2\left(\frac{2k}{n}\right)^2\left(\frac{2}{n}\right) = \frac{16}{n^3}\sum_{k=1}^{n}k^2$$

$$= \frac{16}{n^3}\frac{n(n+1)(2n+1)}{6},\ A = \lim_{n\to+\infty}\left[\frac{8}{3}\left(1+\frac{1}{n}\right)\left(2+\frac{1}{n}\right)\right] = \frac{16}{3}.$$

5.5.17 $\Delta x = \dfrac{4-0}{n} = \dfrac{4}{n},\ x_k^* = 0 + (k-1)\Delta x = \dfrac{4(k-1)}{n}$

$$\sum_{k=1}^{n} f(x_k^*)\Delta x = \sum_{k=1}^{n}\left\{16 - \left[\frac{4(k-1)}{n}\right]^2\right\}\left(\frac{4}{n}\right) = \frac{64}{n}\sum_{k=1}^{n}1 - \frac{64}{n^3}\sum_{k=1}^{n-1}(k-1)^2$$

$$= \frac{64}{n}(n) - \frac{64}{n^3}\sum_{k=1}^{n-1}k^2 = 64 - \frac{64}{n^3}\frac{(n-1)n(2n-1)}{6} = 64 - \frac{32}{3}\frac{(n-1)(2n-1)}{n^2}$$

$$A = \lim_{n\to+\infty}\left[64 - \frac{32}{3}\left(1-\frac{1}{n}\right)\left(2-\frac{1}{n}\right)\right] = 64 - \frac{64}{3} = \frac{128}{3}.$$

5.5.18 $\Delta x = \dfrac{2-0}{n} = \dfrac{2}{n},\ x_k^* = 0 + k\Delta x = \dfrac{2k}{n}$

$$\sum_{k=1}^{n} f(x_k^*)\Delta x = \sum\left[\left(\frac{2k}{n}\right)^3+3\right]\left(\frac{2}{n}\right) = \frac{16}{n^4}\sum_{k=1}^{n}k^3 + \frac{6}{n}\sum_{k=1}^{n}1 =$$

$$\frac{16}{n^4}\left[\frac{n(n+1)}{2}\right]^2 + \frac{6}{n}(n) = \frac{4(n+1)^2}{n^2} + 6$$

$$A = \lim_{n\to+\infty}\left[4\left(1+\frac{1}{n}\right)^2 + 6\right] = 10.$$

5.5.19 $\Delta x = \dfrac{2-1}{n} = \dfrac{1}{n}$, $x_k^* = 1 + k\Delta x = 1 + \dfrac{k}{n}$

$$\sum_{k=1}^{n} f(x_k^*)\Delta x = \sum_{k=1}^{n}\left[\left(1+\frac{k}{n}\right)^2 - 1\right]\left(\frac{1}{n}\right) = \sum_{k=1}^{n}\left[\frac{2k}{n^2} + \frac{k^2}{n^3}\right] =$$

$$\frac{2}{n^2}\sum_{k=1}^{n}k + \frac{1}{n^3}\sum_{k=1}^{n}k^2 = \frac{2}{n^2}\frac{n(n+1)}{2} + \frac{1}{n^3}\frac{n(n+1)(2n+1)}{6}$$

$$A = \lim_{n\to+\infty}\left[\left(1+\frac{1}{n}\right) + \frac{1}{6}\left(1+\frac{1}{n}\right)\left(2+\frac{1}{n}\right)\right] = 1 + \frac{2}{6} = \frac{4}{3}.$$

SECTION 5.6

5.6.1 Find the values of $\sum_{k=1}^{n} f(x_k^*)\Delta x_k$ and $\max \Delta x_k$ when $f(x) = 1/x$, $a = 1/2$, $b = 10$, $x_1 = 3/4$, $x_2 = 2$, $x_3 = 5$, $x_1^* = 1/2$, $x_2^* = 1$, $x_3^* = 2$, and $x_4^* = 5$, and $n = 4$.

5.6.2 Give a geometric interpretation for $\int_{-3}^{3} \sqrt{9 - x^2}\, dx$ and evaluate this definite integral.

5.6.3 Approximate the value $\int_{0}^{1} \dfrac{dx}{1 + x^2} = \dfrac{\pi}{4}$ by partitioning the interval $[0, 1]$ into 5 subintervals of equal width and choosing the midpoint of each subinterval to obtain the approximate Riemann sum.

5.6.4 Approximate the value of $\int_{1}^{3} \dfrac{dx}{x}$ by partitioning the interval $[1, 3]$ into 4 subintervals of equal width and choosing the midpoint of each subinterval to obtain the approximate Riemann sum.

5.6.5 Find $\int_{1}^{5} [2f(x) - g(x)]dx$ if $\int_{1}^{5} f(x)dx = 3$ and $\int_{1}^{5} g(x)dx = 10$

5.6.6 Let $f(x) = \begin{cases} 1 & x \geq 0 \\ x + 4 & x < 0 \end{cases}$. Sketch and give a geometric interpretation for $\int_{-4}^{2} f(x)dx$ and evaluate this definite integral.

5.6.7 Evaluate $\int_{0}^{5} \sqrt{25 - x^2}\, dx$ by interpreting the integral as an area.

5.6.8 Sketch, give a geometric interpretation for $\int_{-1}^{2} |x - 1|dx$ and evaluate this definite integral.

5.6.9 Calculate $\sum_{k=1}^{n} f(x_k^*)\Delta x_n$, if $f(x) = \dfrac{x^2}{2}$, $a = 1$, $b = 4$, $x_1 = 2$, $x_2 = 3$, $x_3 = 3.5$, $x_1^* = 2$, $x_2^* = 2.5$, $x_3^* = 3$ and $x_4^* = 4$.

5.6.10 Express $\displaystyle\lim_{\max \Delta x_k \to 0} \sum_{k=1}^{n} (2x_k^* - 3x_k^{*2})\Delta x_k$ as a definite integral with $a = 0$ and $b = 2/3$.

5.6.11 Calculate $\sum_{k=1}^{n} f(x_k^*)\Delta x$ if $f(x) = x^3$, $a = -3$, $b = 3$, $x_1 = -1$, $x_2 = 0$, $x_3 = 1$, $x_1^* = -2$, $x_2^* = 0$, $x_3^* = 0$, $x_4^* = 2$. Also, find $\max \Delta x_k$.

5.6.12   Calculate $\sum\limits_{k=1}^{n} f(x_k^*)\Delta x_k$ and $\max \Delta x_k$ when $f(x) = x^2 + 1$, $a = -1$, $b = 3$, $x_1 = 0$, $x_2 = 2$, $x_1^* = 0$ $x_2^* = 1/2$, and $x_3^* = 2$.

5.6.13   Calculate $\sum\limits_{k=1}^{n} f(x_k^*)\Delta x_k$ and $\max \Delta x_k$ when $f(x) = x^2 + x$, $a = 0$, $b = 3$, $x_1 = 1/2$, $x_2 = 2$, $x_1^* = 1/2$, $x_2^* = 1$, and $x_3^* = 2$.

5.6.14   Calculate $\sum\limits_{k=1}^{n} f(x_k^*)\Delta x_k$ and $\max \Delta x_k$ when $f(x) = 2x - 3$, $a = -1$, $b = 5$, $x_1 = 2$, $x_2 = 4$, $x_1^* = 0$, $x_2^* = 2$, and $x_3^* = 4$.

5.6.15   Prove that the function $f(x) = \begin{cases} \dfrac{1}{1-x} & 1 < x < 5 \\ 1 & x = 1 \end{cases}$ is not integrable on the interval $[0, 2]$.

5.6.16   Prove that the function $f(x) = \begin{cases} \cos\dfrac{1}{x} & x \neq 0 \\ 0 & x = 0 \end{cases}$ is integrable on the interval $[-1, 1]$.

SOLUTIONS

SECTION 5.6

5.6.1 $\displaystyle\sum_{k=1}^{n} f(x_k^*)\Delta x_k = (2)(1/4) + (1)(5/4) + (1/2)(3) + (1/5)(5) = \frac{17}{4}$

$\max \Delta x_k = 5.$

5.6.2 $\displaystyle\int_{-3}^{3} \sqrt{9-x^2}\,dx = \frac{1}{2}$ area of a circle of radius $3 = \frac{1}{2}\cdot \pi(3)^2 = \frac{9\pi}{2}.$

5.6.3 $\Delta x = \dfrac{1-0}{5} = \dfrac{1}{5}, \; x_k^* = 1/10, 3/10, 5/10, 7/10, 9/10, \; f(x) = \dfrac{1}{1+x^2}$

$\displaystyle\sum_{k=1}^{5} f(x_k^*)\Delta x = \left(\frac{100}{101} + \frac{100}{109} + \frac{100}{125} + \frac{100}{149} + \frac{100}{181}\right)\left(\frac{1}{5}\right) = 0.7862 \approx \frac{\pi}{4}.$

5.6.4 $f(x) = \dfrac{1}{x}, \; \Delta x = \dfrac{3-1}{4} = \dfrac{1}{2}, \; x_k^* = 5/4, 7/4, 9/4, 11/4$

$\displaystyle\sum_{k=1}^{4} f(x_k^*)\Delta x = \left(\frac{4}{5} + \frac{4}{7} + \frac{4}{9} + \frac{4}{11}\right)\left(\frac{1}{2}\right) = 1.089.$

5.6.5 $\displaystyle\int_{1}^{5} 2[f(x) - g(x)]dx = 2\int_{1}^{5} f(x)dx - \int_{1}^{5} g(x)dx = 2(3) - 10 = 4.$

5.6.6 $\displaystyle\int_{-4}^{2} f(x)dx = A_1 + A_2$

$\displaystyle\int_{-4}^{2} f(x)dx = \frac{1}{2}(4)(4) + 2(1) = 10.$

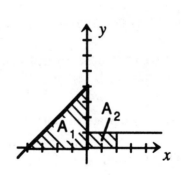

5.6.7 $\displaystyle\int_{0}^{5} \sqrt{25-x^2}\,dx = \frac{1}{4}$ area of a circle of radius $5 = \frac{1}{4}\pi(5)^2 = \frac{25\pi}{4}.$

5.6.8 $\displaystyle\int_{-1}^{2} |x-1|\,dx = A_1 + A_2$

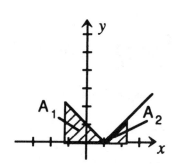

$$= \frac{1}{2}(2)(2) + \frac{1}{2}(1)(1) = 2\frac{1}{2}.$$

5.6.9 $\displaystyle\left(\frac{2^2}{2}\right)(1) + \left(\frac{(2\cdot 5)^2}{2}\right)(1) + \left(\frac{3^2}{2}\right)(0.5) + \left(\frac{4^2}{2}\right)(0.5) = 11.375.$

5.6.10 $\displaystyle\int_{0}^{2/3} (2x - 3x^2)\,dx.$

5.6.11 $(-2)^3(2) + (0)^3(1) + (0)^3(1) + (2)^3(2) = 0;\ \max \Delta x_k = 2.$

5.6.12 $[0^2 + 1](1) + \left[\left(\frac{1}{2}\right)^2 + 1\right](2) + [2^2 + 1](1) = \dfrac{17}{2};\ \max \Delta x_k = 2.$

5.6.13 $\left[\left(\frac{1}{2}\right)^2 + \left(\frac{1}{2}\right)\right]\left(\frac{1}{2}\right) + [(1)^2 + (1)]\left(\frac{3}{2}\right) + [(2)^2 + (2)](1) = \dfrac{75}{8};\ \max \Delta x_k = 3/2.$

5.6.14 $(2 \cdot 0 - 3)(3) + (2 \cdot 2 - 3)(2) + (2 \cdot 4 - 3)(1) = -2;\ \max \Delta x_k = 3.$

5.6.15 $f(x)$ is defined at all points on $[1,5]$ and f is not bounded on $[1,5]$ thus by Theorem 5.6.5(c) f is not integrable on $[1,5]$.

5.6.16 $f(x)$ is discontinuous at the point $x = 0$ because $\displaystyle\lim_{x\to 0}\cos\frac{1}{x}$ does not exist. f is continuous elsewhere. $-1 \le f(x) \le 1$ for x in $[-1,1]$ so f is bounded there. Thus by Theorem 5.6.5b f is integrable on $[-1,1]$.

SECTION 5.7

5.7.1 Evaluate $\int_{-1}^{1} (x^7 + 1)(3x^2 + 1)dx$.

5.7.2 Evaluate $\int_{1}^{3} \left(2x^2 + \dfrac{1}{2x^2} \right) dx$.

5.7.3 Evaluate $\int_{0}^{\pi/4} (x + 2\sec^2 x)dx$.

5.7.4 Evaluate $\int_{1}^{4} \left(1 + t\sqrt{t} - \dfrac{1}{t^2} \right) dt$.

5.7.5 Evaluate $\int_{0}^{1} (\sqrt[3]{t^2} + \sqrt{t})dt$.

5.7.6 Evaluate $\int_{0}^{\pi/3} \dfrac{1}{\cos^2 \phi} d\phi$.

5.7.7 Evaluate $\int_{\pi/4}^{\pi/2} \dfrac{1}{\sin^2 \theta} d\theta$.

5.7.8 Evaluate $\int_{\pi/6}^{\pi/3} (\cos \theta - \csc \theta \cot \theta)d\theta$.

5.7.9 Evaluate $\int_{0}^{\pi/4} \sec^2 \theta \, d\theta$.

5.7.10 Evaluate $\int_{\pi/3}^{\pi/2} (\theta - \csc \theta \cot \theta)d\theta$.

5.7.11 Evaluate $\int_{1}^{2} \left(\dfrac{1}{t^2} + \dfrac{1}{t^3} \right)\left(\dfrac{1}{t^2} - \dfrac{1}{t^3} \right) dt$.

5.7.12 Evaluate $\int_{-3}^{5} |x + 1|dx$.

5.7.13 Evaluate $\int_{-3}^{3} |2x + 1|dx$.

5.7.14 Evaluate $\int_{-1}^{3} f(x)dx$ if $f(x) = \begin{cases} x^2, & x > 2 \\ 1 - x, & x \leq 2 \end{cases}$

5.7.15 Evaluate $\int_{-2}^{2} f(x)dx$ if $f(x) = \begin{cases} x^3, & x > -1 \\ 1 - x^2, & x \leq -1 \end{cases}$

5.7.16 Find the area under the curve $y = x^2 + 2$ for $2 \leq x \leq 3$. Make a sketch of the region.

5.7.17 Find the area of the region between $y = 16 - x^2$ and the x-axis. Make a sketch of the region.

5.7.18 Find the area of the region between $y = x^2 - x - 6$ and the x-axis for $0 \leq x \leq 2$. Make a sketch of the region.

5.7.19 Find the average value of $f(x) = \sqrt{4x + 1}$ over the interval $0 \leq x \leq 2$.

5.7.20 Find the average value of $f(x) = x^2 \sec^2 x^3$ for $0 \leq x \leq \sqrt[3]{\pi/4}$.

5.7.21 (a) Find the average value of $f(x) = 3x + 1$ over $[0, 6]$.
(b) Find a point x^* in $[0, 6]$ such that $f(x^*) = f_{\text{ave}}$.
(c) Sketch the graph of $f(x) = 3x + 1$ over $[0, 6]$ and construct a rectangle over the interval whose area is the same as the area under the graph of f over the interval.

5.7.22 **(a)** Find the average value of $f(x) = (x+1)^2$ over $[-1, 2]$.

        (b) Find a point x^* in $[-1, 2]$ such that $f(x^*) = f_{ave}$.

        (c) Sketch the graph of $f(x) = (x+1)^2$ over $[-1, 2]$ and construct a rectangle over the interval whose area is the same as the area under the graph of f over the interval.

5.7.23 Find the average value of $f(x) = x \cos x^2$ for $0 \leq x \leq \sqrt{\dfrac{\pi}{2}}$.

5.7.24 Find the average value of $f(x) = x^3 \sqrt{3x^4 + 1}$ for $-1 \leq x \leq 2$.

5.7.25 Find the average value of $f(x) = x^3 + 1$ for $0 \leq x \leq 2$ and find all values of x^* described in the Mean Value Theorem for Integrals.

SOLUTIONS

SECTION 5.7

5.7.1 $\left[\dfrac{3x^{10}}{10} + \dfrac{x^8}{8} + \dfrac{3x^3}{3} + x\right]_{-1}^{l} = 4.$

5.7.2 $\displaystyle\int_1^3 \left(2x^2 + \dfrac{x^{-2}}{2}\right) dx = \left[\dfrac{2x^3}{3} - \dfrac{1}{2x}\right]_1^3 = \dfrac{53}{3}.$

5.7.3 $\left[\dfrac{x^2}{2} + \tan x\right]_0^{\pi/4} = \dfrac{\pi^2}{32} + 2.$

5.7.4 $\displaystyle\int_1^4 (1 + t^{3/2} - t^{-2})dt = \left[t + \dfrac{2t^{5/2}}{5} + \dfrac{1}{t}\right]_1^4 = \dfrac{293}{20}.$

5.7.5 $\displaystyle\int_0^1 (t^{2/3} + t^{1/2})dt = \left[\dfrac{3}{5}t^{5/3} + \dfrac{2}{3}t^{3/2}\right]_0^1 = \dfrac{19}{15}.$

5.7.6 $\displaystyle\int_0^{\pi/3} \sec^2\phi\,d\phi = \tan\phi\Big]_0^{\pi/3} = \sqrt{3}.$ **5.7.7** $\displaystyle\int_{\pi/4}^{\pi/2} \csc^2\theta\,d\theta = -\cot\theta\Big]_{\pi/4}^{\pi/2} = 1.$

5.7.8 $\left[-\sin\theta + \csc\theta\right]_{\pi/6}^{\pi/3} = \dfrac{7\sqrt{3} - 15}{6}.$ **5.7.9** $\tan\theta\Big]_0^{\pi/4} = 1.$

5.7.10 $\left[\dfrac{\theta^2}{2} + \csc\theta\right]_{\pi/3}^{\pi/2} = \dfrac{5\pi^2}{72} - \dfrac{2}{\sqrt{3}} + 1.$ **5.7.11** $\left[\dfrac{t^{-3}}{-3} - \dfrac{t^{-5}}{-5}\right]_1^2 = \dfrac{47}{480}.$

5.7.12 $\displaystyle\int_{-3}^{-1} -(x+1)dx + \int_{-1}^5 (x+1)dx = -\left(\dfrac{x^2}{2} + x\right)\Big]_{-3}^1 + \left[\dfrac{x^2}{2} + x\right]_{-1}^5 = 18$

5.7.13 $\displaystyle\int_{-3}^{-1/2} -(2x+1)dx + \int_{-1/2}^3 (2x+1)dx = -(x^2 + x)\Big]_{-3}^{-1/2} + \left[x^2 + x\right]_{-1/2}^3 = \dfrac{37}{2}$

5.7.14 $\displaystyle\int_{-1}^2 (1-x)dx + \int_2^3 x^2 dx = \left[x - \dfrac{x^2}{2}\right]_{-1}^2 + \dfrac{x^3}{3}\Big]_2^3 = \dfrac{47}{6}$

5.7.15 $\displaystyle\int_{-2}^{-1} (1-x^2)dx + \int_{-1}^2 x^3 dx = \left[x - \dfrac{x^3}{3}\right]_{-2}^{-1} + \dfrac{x^4}{4}\Big]_{-1}^2 = \dfrac{29}{12}$

5.7.16 $A = \displaystyle\int_2^3 (x^2 + 2)dx = \left[\dfrac{x^3}{3} + 2x\right]_2^3 = \dfrac{25}{3}.$

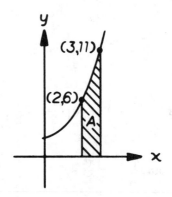

5.7.17 $A = \displaystyle\int_{-4}^4 (16 - x^2)dx = \left[16x - \dfrac{x^3}{3}\right]_{-4}^4 = \dfrac{256}{3}.$

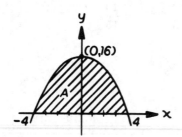

5.7.18 $A = \displaystyle\int_0^2 -(x^2 - x - 6)dx = \left[-\dfrac{x^3}{3} + \dfrac{x^2}{2} + 6x\right]_0^2 = \dfrac{34}{3}.$

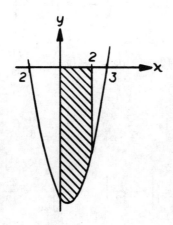

5.7.19 $\dfrac{1}{2-0}\displaystyle\int_0^2 \sqrt{4x+1}\,dx = \dfrac{1}{12}\left[(4x+1)^{3/2}\right]_0^2 = \dfrac{13}{6}.$

5.7.20 $\dfrac{1}{3\sqrt[3]{\dfrac{\pi}{4}}-0}\displaystyle\int_0^{\sqrt[3]{\frac{\pi}{4}}} x^2 \sec^2 x^3\,dx = \dfrac{1}{3\sqrt[3]{\dfrac{\pi}{4}}}\left[\tan x^3\right]_0^{\sqrt[3]{\frac{\pi}{4}}} = \dfrac{1}{3}\sqrt[3]{\dfrac{4}{\pi}}.$

5.7.21 **(a)** $\dfrac{1}{6-0}\displaystyle\int_0^6 (3x+1)\,dx = 10$ **(b)** $3x^* + 1 = 10,\ x^* = 3$

(c)

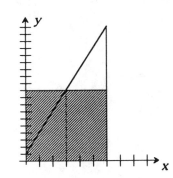

5.7.22 **(a)** $\dfrac{1}{2-(-1)}\displaystyle\int_{-1}^2 (x+1)^2\,dx = \dfrac{1}{9}(x+1)^3\Big|_{-1}^2 = 3$

(b) $(x^*+1)^2 = 3$ only $x^* = \sqrt{3}-1$ is in $[-1,2]$

(c)

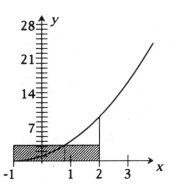

5.7.23 $\dfrac{1}{\sqrt{\dfrac{\pi}{2}}-0}\displaystyle\int_0^{\sqrt{\pi/2}} x\cos x^2\,dx = \dfrac{1}{2\sqrt{\dfrac{\pi}{2}}}\left[\sin x^2\right]_0^{\sqrt{\pi/2}} = \dfrac{1}{\sqrt{2\pi}}.$

5.7.24 $\dfrac{1}{2-(-1)}\displaystyle\int_{-1}^2 x^3\sqrt{3x^4+1}\,dx = \dfrac{1}{54}\left[(3x^4+1)^{3/2}\right]_{-1}^2 = \dfrac{335}{54}.$

5.7.25 $f_{av} = \dfrac{1}{2-0} \displaystyle\int_0^2 (x^3+1)dx = \dfrac{1}{2}\left[\dfrac{x^4}{4} + x\right]_0^2 = 3;$

$\displaystyle\int_0^2 (x^3+1)dx = f(x^*)(2-0) = (x^{*3}+1)(2)$

$6 = 2(x^{*3}+1)$

$x^* = \sqrt[3]{2}.$

SECTION 5.8

5.8.1 Evaluate $\displaystyle\int_{4/\pi}^{2/\pi} \frac{\sin(1/t)}{t^2}\,dt$.

5.8.2 Evaluate $\displaystyle\int_0^1 \frac{dx}{\sqrt{x+1}}$.

5.8.3 Evaluate $\displaystyle\int_1^2 (x^2+1)\sqrt{2x^3+6x}\,dx$.

5.8.4 Evaluate $\displaystyle\int_0^3 \sqrt{x^4+2x^2+1}\,dx$.

5.8.5 Evaluate $\displaystyle\int_0^1 x\sqrt{9x^2+16}\,dx$.

5.8.6 Evaluate $\displaystyle\int_0^1 \frac{x}{(1+x^2)^2}\,dx$.

5.8.7 Evaluate $\displaystyle\int_0^3 x\sqrt{9-x^2}\,dx$.

5.8.8 Evaluate $\displaystyle\int_0^4 \frac{x}{\sqrt{9+x^2}}\,dx$.

5.8.9 Evaluate $\displaystyle\int_0^{\pi/4} \tan^2 x \sec^2 x\,dx$.

5.8.10 Evaluate $\displaystyle\int_0^{\pi/9} \cos^2 3t \sin 3t\,dt$.

5.8.11 Evaluate $\displaystyle\int_1^2 x^2\sqrt{x-1}\,dx$.

5.8.12 Evaluate $\displaystyle\int_1^5 x\sqrt{2x-1}\,dx$.

5.8.13 Evaluate $\displaystyle\int_5^{10} \frac{x}{\sqrt{x-1}}\,dx$.

5.8.14 Evaluate $\displaystyle\int_0^{\sqrt{\pi/6}} \frac{t}{\cos^2 2t^2}\,dt$.

5.8.15 Evaluate $\displaystyle\int_{\sqrt{\pi/3}}^{\sqrt{\pi/2}} \frac{\theta}{\sin^2\left(\frac{\theta^2}{2}\right)}\,d\theta$.

5.8.16 Evaluate $\displaystyle\int_{\sqrt{\pi/6}}^{\sqrt{\pi/3}} x\csc x^2 \cot x^2\,dx$.

5.8.17 Evaluate $\displaystyle\int_0^1 \frac{x^2}{2}\sqrt{2x^3+7}\,dx$.

5.8.18 Evaluate $\displaystyle\int_0^{\pi/8} (2x+\sec 2x \tan 2x)\,dx$.

SOLUTIONS

SECTION 5.8

5.8.1 $u = \dfrac{1}{t}$, $du = \dfrac{1}{t^2}dt$, $\dfrac{dt}{t^2} = -du$

$$-\int_{\pi/4}^{\pi/2} \sin u \, du = \Big[\cos u\Big]_{\pi/4}^{\pi/2} = -\frac{\sqrt{2}}{2} \text{ or } \Big[\cos\left(\frac{1}{t}\right)\Big]_{4/\pi}^{2/\pi} = -\frac{\sqrt{2}}{2}.$$

5.8.2 $u = x + 1$, $du = dx$, $\displaystyle\int_1^2 u^{-1/2}du = \Big[2u^{1/2}\Big]_1^2 = 2\sqrt{2} - 2$ or

$$\Big[2\sqrt{x+1}\Big]_0^1 = 2\sqrt{2} - 2.$$

5.8.3 $u = 2x^3 + 6x$, $du = 6(x^2 + 1)dx$, $(x^2 + 1)dx = \dfrac{du}{6}$,

$$\frac{1}{6}\int_8^{28} u^{1/2}du = \frac{1}{9}u^{3/2}\Big]_8^{28} = \frac{56\sqrt{7} - 16\sqrt{2}}{9} \text{ or}$$

$$\frac{1}{9}\Big[(2x^3 + 6x)^{3/2}\Big]_1^2 = \frac{56\sqrt{7} - 16\sqrt{2}}{9}.$$

5.8.4 $\displaystyle\int_0^3 \sqrt{(x^2 + 1)^2}\,dx = \int_0^3 (x^2 + 1)dx = \Big[\frac{x^3}{3} + x\Big]_0^3 = 12.$

5.8.5 $u = 9x^2 + 16$, $du = 18x\,dx$, $x\,dx = \dfrac{du}{18}$,

$$\frac{1}{18}\int_{16}^{25} u^{1/2}du = \frac{1}{27}\Big[u^{3/2}\Big]_{16}^{25} = \frac{61}{27} \text{ or } \frac{1}{27}\Big[(9x^2 + 16)^{3/2}\Big]_0^1 = \frac{61}{27}.$$

5.8.6 $u = 1 + x^2$, $du = 2x\,dx$, $x\,dx = \dfrac{du}{2}$,

$$\frac{1}{2}\int_1^2 u^{-2}du = -\Big[\frac{1}{2u}\Big]_1^2 = \frac{1}{4} \text{ or } -\Big[\frac{1}{2(1 + x^2)}\Big]_0^1 = \frac{1}{4}.$$

5.8.7 $u = 9 - x^2$, $du = -2x\,dx$, $x\,dx = -\dfrac{du}{2}$,

$$\frac{1}{2}\int_0^9 u^{1/2}du = \frac{1}{3}\Big[u^{3/2}\Big]_0^9 = 9 \text{ or } -\frac{1}{3}\Big[(9 - x^2)^{3/2}\Big]_0^3 = 9.$$

5.8.8 $u = 9 + x^2,\ du = 2x\,dx,\ x\,dx = \dfrac{dy}{2}$

$$\frac{1}{2}\int_9^{25} u^{-1/2}\,du = \left[u^{1/2}\right]_9^{25} = 2 \text{ or } \left[\sqrt{9+x^2}\right]_0^4 = 2.$$

5.8.9 $u = \tan x,\ du = \sec^2 x\,dx,$

$$\int_0^1 u^2\,du = \left.\frac{u^3}{3}\right|_0^1 = \frac{1}{3} \text{ or } \left.\frac{\tan^3 x}{3}\right|_0^{\pi/4} = \frac{1}{3}.$$

5.8.10 $u = \cos 3t,\ du = -3\sin 3t\,dt,\ \sin 3t\,dt = -\dfrac{du}{3},$

$$\frac{1}{3}\int_{1/2}^1 u^2\,du = \frac{1}{9}\left[u^3\right]_{1/2}^1 = \frac{7}{72} \text{ or } -\frac{1}{9}\left[\cos^3 3t\right]_0^{\pi/9} = \frac{7}{72}.$$

5.8.11 $u = x - 1,\ du = dx,\ x = u + 1,$

$$\int_0^1 (u+1)^2 u^{1/2}\,du = \int_0^1 (u^{5/2} + 2u^{3/2} + u^{1/2})\,du = \left[\frac{2}{7}u^{7/2} + \frac{4}{5}u^{5/2} + \frac{2}{3}u^{3/2}\right]_0^1 = \frac{184}{105}$$

$$\text{or } \left[\frac{2}{7}(x-1)^{7/2} + \frac{4}{5}(x-1)^{5/2} + \frac{2}{3}(x-1)^{3/2}\right]_1^2 = \frac{184}{105}.$$

5.8.12 $u = 2x - 1,\ du = 2dx,\ dx = \dfrac{du}{2},\ x = \dfrac{u+1}{2},$

$$\frac{1}{2}\int_1^9 \left(\frac{u+1}{2}\right) u^{1/2}\,du = \frac{1}{4}\int_1^9 (u^{3/2} + u^{1/2})\,du = \frac{1}{4}\left(\frac{2}{5}u^{5/2} + \frac{2}{3}u^{3/2}\right)\Big]_1^9 = \frac{428}{15} \text{ or }$$

$$\frac{1}{4}\left[\frac{2}{5}(2x-1)^{5/2} + \frac{2}{3}(2x-1)^{3/2}\right]_1^5 = \frac{428}{15}.$$

5.8.13 $u = x - 1,\ du = dx,\ x = u + 1,$

$$\int_4^9 \frac{u+1}{u^{1/2}}\,du = \int_4^9 (u^{1/2} + u^{-1/2})\,du = \left[\frac{2}{3}u^{3/2} + 2u^{1/2}\right]_4^9 = \frac{44}{3} \text{ or }$$

$$\left[\frac{2}{3}(x-1)^{3/2} + 2(x-1)^{1/2}\right]_5^{10} = \frac{44}{3}.$$

5.8.14 $u = 2t^2,\ du = 4t\,dt,\ t\,dt = \dfrac{du}{4},$

$$\int_0^{\sqrt{\pi/6}} t\sec^2 2t^2\,dt = \frac{1}{4}\int_0^{\pi/3} \sec^2 u\,du = \frac{1}{4}\left[\tan u\right]_0^{\pi/3} = \frac{\sqrt{3}}{4} \text{ or }$$

$$\frac{1}{4}\left[\tan 2t^2\right]_0^{\sqrt{\pi/6}} = \frac{\sqrt{3}}{4}.$$

5.8.15 $u = \dfrac{\theta^2}{2}$, $du = \theta\, d\theta$

$$\int_{\sqrt{\pi/3}}^{\sqrt{\pi/2}} \theta \csc^2 \frac{\theta^2}{2}\, d\theta = \int_{\pi/6}^{\pi/4} \csc^2 u\, du = -\left[\cot u\right]_{\pi/6}^{\pi/4} = \sqrt{3} - 1 \text{ or}$$

$$-\left[\cot \frac{\theta^2}{2}\right]_{\sqrt{\pi/3}}^{\sqrt{\pi/2}} = \sqrt{3} - 1.$$

5.8.16 $u = x^2$, $du = 2x\, dx$, $x\, dx = \dfrac{du}{2}$,

$$-\frac{1}{2}\int_{\frac{\pi}{6}}^{\frac{\pi}{3}} -\csc u \cot u\, du = -\frac{1}{2}\csc u \Big|_{\frac{\pi}{6}}^{\frac{\pi}{3}} = -\frac{1}{2}\left[\frac{2}{\sqrt{3}} - 2\right] = -\frac{-3 - \sqrt{3}}{3} \text{ or}$$

$$-\frac{1}{2}\csc x^2 \Big|_{\sqrt{\frac{\pi}{6}}}^{\sqrt{\frac{\pi}{3}}} = \frac{3 - \sqrt{3}}{3}$$

5.8.17 $u = 2x^3 + 7$, $du = 6x^2\, dx$, $x^2\, dx = \dfrac{du}{6}$.

$$\frac{1}{12}\int_{7}^{9} u^{1/2}\, du = \frac{1}{18}\left[u^{3/2}\right]_{7}^{9} = \frac{27 - 7\sqrt{7}}{18}, \text{ or } \frac{1}{18}\left[(2x^3 + 7)^{3/2}\right]_{0}^{1} = \frac{27 - 7\sqrt{7}}{18}.$$

5.8.18 $u = 2x$, $dx = \dfrac{du}{2}$,

$$\frac{1}{2}\int_{0}^{\pi/4} (u + \sec u \tan u)\, du = \frac{1}{2}\left[\frac{u^2}{2} + \sec u\right]_{0}^{\pi/4} = \frac{\pi^2}{64} + \frac{\sqrt{2}}{2} - \frac{1}{2}$$

$$\text{or } \frac{1}{2}\left[2x^2 + \sec 2x\right]_{0}^{\pi/8} = \frac{\pi^2}{64} + \frac{\sqrt{2}}{2} - \frac{1}{2}.$$

SECTION 5.9

5.9.1 Given that $\displaystyle\int_{-2}^{4} f(x)dx = 4$, find

    (a) $\displaystyle\int_{-2}^{4} f(t)dt$
                                   (b) $\displaystyle\int_{-2}^{4} f(u)du$

5.9.2 Given that $\displaystyle\int_{-3}^{1} f(x)dx = -2$, find

    (a) $\displaystyle\int_{-3}^{1} f(t)dt$
                                   (b) $\displaystyle\int_{-3}^{1} f(w)dw$

5.9.3 Evaluate $\displaystyle\int_{3}^{x} t^3 dt$

5.9.4 Evaluate $\displaystyle\int_{-2\pi}^{x} \sin t\, dt$

5.9.5 Differentiate $\displaystyle\int_{1}^{x^3} \frac{\sin t}{t} dt.$
                               5.9.6 Find $F'(x)$ if $F(x) = \displaystyle\int_{0}^{x} \sin^2\theta\, d\theta.$

5.9.7 Find $\dfrac{d^2y}{dx^2}$ if $y = \displaystyle\int_{3}^{x} t^3 \cos^2 t\, dt.$

5.9.8 Find $F'(x)$ if $F(x) = \displaystyle\int_{1}^{x} (t^3 + 1)dt$. Check your work by first integrating and then differentiating.

5.9.9 Find $F'(x)$ if $F(x) = \displaystyle\int_{0}^{\sin x} \sqrt{1 - t^3}dt.$

5.9.10 If $F(x) = \displaystyle\int_{1}^{x} \frac{\sin 2t}{t} dt$, find $\displaystyle\lim_{x \to 0} F'(x).$

5.9.11 Find $F''(x)$ if $F(x) = \displaystyle\int_{0}^{x} \frac{1}{\sqrt{1 - 3t^2}} dt.$

5.9.12 Find $\dfrac{d}{dx}\left[\displaystyle\int_{0}^{x} (t+1)^{1/2}dt\right]$. Check your work by first integrating, then differentiating.

5.9.13 Express the antiderivative of $1/(1+x)$ on the interval $(0, +\infty)$ whose value at $x = 2$ is 0 as an integral, $F(x)$.

5.9.14 Express the antiderivative of $\dfrac{1}{x-2}$ on the interval $(3, +\infty)$ whose value at $x = 4$ is 0 as an integral, $F(x)$.

5.9.15 Express the antiderivative of $1/(4 + x^2)$ on the interval $(-\infty, +\infty)$ whose value at $x = -2$ is 0 as an integral, $F(x)$.

SOLUTIONS

SECTION 5.9

5.9.1 (a) 4   (b) 4

5.9.2 (a) -2 (b) -2

5.9.3 $\displaystyle\int_3^x t^3\,dt = \frac{t^4}{4}\bigg|_3^x = \frac{x^4}{4} - \frac{81}{4}$

5.9.4 $\displaystyle\int_{-2\pi}^x \sin t\,dt = -\cos t\,\bigg|_{-2\pi}^x = -\cos x - (-\cos(-2\pi))$

$$= -\cos x + 1 = 1 - \cos x$$

5.9.5 $\displaystyle\left(\frac{\sin x^3}{x^3}\right)\frac{d}{dx}[x^3] = \frac{\sin x^3}{x^3}(3x^2) = \frac{3\sin x^3}{x}.$

5.9.6 $\sin^2 x.$

5.9.7 $\displaystyle\frac{dy}{dx} = x^3\cos^2 x$

$\displaystyle\frac{d^2 y}{dx^2} = 3x^2\cos^2 x - 2x^3\sin x\cos x$

5.9.8 $\displaystyle F(x) = \left[\frac{t^4}{4} + t\right]_1^x = \frac{x^4}{4} + x - \frac{5}{4}; \text{ so } F'(x) = \frac{d}{dx}\left[\frac{x^4}{4} + x - \frac{5}{4}\right] = x^3 + 1.$

5.9.9 $\displaystyle\sqrt{1 - \sin^3 x}\,\frac{d}{dx}[\sin x] = \cos x\sqrt{1 - \sin^3 x}.$

5.9.10 $\displaystyle\lim_{x\to 0}\frac{\sin 2x}{x} = \lim_{x\to 0} 2\frac{\sin 2x}{2x} = 2\lim_{2x\to 0}\frac{\sin 2x}{2x} = 2(1) = 2;\ (2x \to 0 \text{ as } x \to 0).$

5.9.11 $\displaystyle F'(x) = \frac{1}{\sqrt{1 - 3x^2}} = (1 - 3x^2)^{-1/2};\ F''(x) = -\frac{1}{2}(1 - 3x^2)^{-3/2}(-6x) = 3x(1 - 3x^2)^{-3/2}.$

5.9.12 $\displaystyle\frac{d}{dx}\left[\int_0^x (t+1)^{1/2}dt\right] = \frac{d}{dx}\left[\frac{2}{3}(x+1)^{3/2} - \frac{2}{3}\right] = (x+1)^{1/2}.$

5.9.13 $\displaystyle F(x) = \int_2^x \frac{1}{1+t}dt.$ **5.9.14** $\displaystyle F(x) = \int_4^x \frac{1}{t-2}dt.$

5.9.15 $\displaystyle F(x) = \int_{-2}^x \frac{1}{4+t^2}dt.$

SUPPLEMENTARY EXERCISES, CHAPTER 5

In Exercises 1–10, evaluate the integrals and check your results by differentiation.

1. $\displaystyle\int \left[\frac{1}{x^3} + \frac{1}{\sqrt{x}} - 5\sin x\right] dx.$

2. $\displaystyle\int \frac{2t^4 - t + 2}{t^3} \, dt.$

3. $\displaystyle\int \frac{(\sqrt{5}+2)^8}{\sqrt{x}} \, dx.$

4. $\displaystyle\int x^3 \cos(2x^4 - 1) \, dx.$

5. $\displaystyle\int \frac{x\sin\sqrt{2x^2 - 5}}{\sqrt{2x^2 - 5}} \, dx.$

6. $\displaystyle\int \sqrt{\cos\theta}\,\sin(2\theta) \, d\theta.$

7. $\displaystyle\int \sqrt{x}(3 + \sqrt[3]{x^4}) \, dx.$

8. $\displaystyle\int \frac{x^{1/3}\,dx}{x^{8/3} + 2x^{4/3} + 1}.$

9. $\displaystyle\int \sec^2(\sin 5t)\cos 5t \, dt.$

10. $\displaystyle\int \frac{\cot^2 x}{\sin^2 x} \, dx.$

11. Evaluate $\displaystyle\int y(y^2 + 2)^2 \, dy$ two ways: (a) by multiplying out and integrating term by term; and (b) by using the substitution $u = y^2 + 2$. Show that your answers differ by a constant.

In Exercises 12–17, evaluate the definite integral by making the indicated substitution and changing the x-limits of integration to u-limits.

12. $\displaystyle\int_1^0 \sqrt[5]{1 - 2x}\, dx,\, u = 1 - 2x.$

13. $\displaystyle\int_0^{\pi/2} \sin^4 x \cos x \, dx,\, u = \sin x.$

14. $\displaystyle\int_0^{-3} \frac{x\,dx}{\sqrt{x^2 + 16}},\, u = x^2 + 16.$

15. $\displaystyle\int_2^5 \frac{x - 2}{\sqrt{x - 1}} \, dx,\, u = x - 1.$

16. $\displaystyle\int_{\pi/6}^{\pi/4} \frac{\sin 2x\,dx}{\sqrt{1 - \frac{3}{2}\cos 2x}},\, u = 1 - \frac{3}{2}\cos 2x.$

17. $\displaystyle\int_1^4 \frac{1}{\sqrt{x}}\cos\left(\frac{\pi\sqrt{x}}{2}\right) dx,\, u = \frac{\pi\sqrt{x}}{2}.$

In Exercises 18 and 19, evaluate $\int_{-2}^2 f(x)\, dx$.

18. $f(x) = \begin{cases} x^3 & \text{for } x \geq 0 \\ -x & \text{for } x < 0. \end{cases}$

19. $f(x) = |2x - 1|.$

In Exercises 20–22, solve for x.

20. $\displaystyle\int_1^x \frac{1}{\sqrt{t}}\, dt = 3.$

21. $\displaystyle\int_0^x \frac{1}{(3t + 1)^2}\, dt = \frac{1}{6}.$

22. $\displaystyle\int_2^x (4t - 1)\, dt = 9.$

23. (a) $\displaystyle\sum_{i=3}^{6} 5$ (b) $\displaystyle\sum_{i=n}^{n+3} 2$

(c) $\displaystyle\sum_{i=n}^{n+3} n$ (d) $\displaystyle\sum_{k=1}^{3} \left(\frac{k-1}{k+3}\right)$

(e) $\displaystyle\sum_{k=2}^{4} \frac{6}{k^2}$ (f) $\displaystyle\sum_{n=4}^{4} (2n+1)$

(g) $\displaystyle\sum_{k=0}^{4} \sin(k\pi/4)$ (h) $\displaystyle\sum_{k=1}^{4} \sin^k(\pi/4)$.

24. Express in sigma notation and evaluate:

(a) $3\cdot 1 + 4\cdot 2 + 5\cdot 3 + \cdots + 102\cdot 100$ (b) $200 + 198 + \cdots + 4 + 2$.

25. Express in sigma notation, first starting with $k = 1$, and then with $k = 2$. (Do not evaluate.)

(a) $\dfrac{1}{4} - \dfrac{4}{9} + \dfrac{9}{16} - \cdots - \dfrac{64}{81} + \dfrac{81}{100}$ (b) $\dfrac{\pi^2}{1} - \dfrac{\pi^3}{2} + \dfrac{\pi^4}{3} - \cdots + \dfrac{\pi^{12}}{11}$.

In Exercises 26–29, use the partition of $[a, b]$ into n subintervals of equal length, and find a closed form for the sum of the areas of (a) the inscribed rectangles and (b) the circumscribed rectangles. (c) Use your answer in either part (a) or part (b) to find the area under the curve $y = f(x)$ over the interval $[a, b]$. (Check your answer by integration.)

26. $f(x) = 6 - 2x; a = 1, b = 3$.   **27.** $f(x) = 16 - x^2; a = 0, b = 4$.

28. $f(x) = x^2 + 2; a = 1, b = 4$.   **29.** $f(x) = 6; a = -1, b = 1$.

30. Given that

$$\int_1^5 P(x)\,dx = -1, \quad \int_3^5 P(x)\,dx = 3,$$

and

$$\int_3^5 Q(x)\,dx = 4$$

evaluate the following:

(a) $\displaystyle\int_3^5 [2P(x) + Q(x)]\,dx$ (b) $\displaystyle\int_5^1 P(t)\,dt$

(c) $\displaystyle\int_{-3}^{-5} Q(-x)\,dx$ (d) $\displaystyle\int_3^1 P(x)\,dx$.

31. Suppose that f is continuous and $x^2 \leq f(x) \leq 6$ for all x in $[-1, 2]$. Find values of A and B such that

$$A \leq \int_{-1}^{2} f(x)\, dx \leq B$$

In Exercises 33–36, find the average value of $f(x)$ over the indicated interval and all values of x^* described in the Mean-Value Theorem for Integrals.

32. $f(x) = 3x^2;\ [-2, -1]$.

33. $f(x) = \dfrac{x}{\sqrt{x^2 + 9}};\ [0, 4]$.

34. $f(x) = 2 + |x|;\ [-3, 1]$.

35. $f(x) = \sin^2 x;\ [0, \pi]$.

[Hint: $\sin^2 x = \frac{1}{2}(1 - \cos 2x)$.]

SOLUTIONS

SUPPLEMENTARY EXERCISES, CHAPTER 5

1. $-x^{-2}/2 + 2\sqrt{x} + 5\cos x + C$

2. $\int (2t - 1/t^2 + 2/t^3)dt = t^2 + 1/t - 1/t^2 + C$

3. $u = \sqrt{x} + 2,\ 2\int u^8 du = \dfrac{2}{9}u^9 + C = \dfrac{2}{9}(\sqrt{x} + 2)^9 + C$

4. $u = 2x^4 - 1,\ \dfrac{1}{8}\int \cos u\, du = \dfrac{1}{8}\sin(2x^4 - 1) + C$

5. $u = \sqrt{2x^2 - 5},\ du = 2x/\sqrt{2x^2 - 5}dx,\ \dfrac{1}{2}\int \sin u\, du = -\dfrac{1}{2}\cos\sqrt{2x^2 - 5} + C$

6. $\int \sqrt{\cos\theta}(2\sin\theta\cos\theta)d\theta = 2\int \cos^{3/2}\theta\sin\theta\, d\theta = -\dfrac{4}{5}\cos^{5/2}\theta + C$

7. $\int (3x^{1/2} + x^{11/6})dx = 2x^{3/2} + \dfrac{6}{17}x^{17/6} + C$

8. $\int (x^{4/3} + 1)^{-2}x^{1/3}dx,\ u = x^{4/3} + 1,\ \dfrac{3}{4}\int u^{-2}du = (-3/4)/(x^{4/3} + 1) + C$

9. $u = \sin 5t,\ \dfrac{1}{5}\int \sec^2 u\, du = \dfrac{1}{5}\tan(\sin 5t) + C$

10. $\int \cot^2 x\csc^2 x\, dx,\ u = \cot x,\ -\int u^2 du = -\dfrac{1}{3}\cot^3 x + C$

11. **(a)** $\int (y^5 + 4y^3 + 4y)dy = \dfrac{1}{6}y^6 + y^4 + 2y^2 + C$

     (b) $\dfrac{1}{6}(y^2 + 2)^3 + C$

     [answer to (b)] − [answer to (a)]

$$= \dfrac{1}{6}(y^6 + 6y^4 + 12y^2 + 8) + C - \left(\dfrac{1}{6}y^6 + y^4 + 2y^2 + C\right) = 4/3$$

12. $-\dfrac{1}{2}\displaystyle\int_{-1}^{1} u^{1/5}du = -\dfrac{5}{12}u^{6/5}\Big]_{-1}^{1} = 0$

13. $\displaystyle\int_0^1 u^4 du = 1/5$

14. $\dfrac{1}{2}\displaystyle\int_{16}^{25} u^{-1/2}du = u^{1/2}\Big]_{16}^{25} = 1$

15. $u = x - 1,\ x = u + 1,\ \displaystyle\int_1^4 \dfrac{u-1}{\sqrt{u}}du = \int_1^4 (u^{1/2} - u^{-1/2})du = \dfrac{2}{3}u^{3/2} - 2u^{1/2}\Big]_1^4 = 8/3$

16. $\dfrac{1}{3}\displaystyle\int_{1/4}^1 u^{-1/2}du = \dfrac{2}{3}u^{1/2}\Big]_{1/4}^1 = 1/3$

17. $\dfrac{4}{\pi}\displaystyle\int_{\pi/2}^{\pi} \cos u\, du = \dfrac{4}{\pi}\sin u\Big]_{\pi/2}^{\pi} = -4/\pi$

18. $\displaystyle\int_{-2}^0 (-x)dx + \int_0^2 x^3 dx = -\dfrac{1}{2}x^2\Big]_{-2}^0 + \dfrac{1}{4}x^4\Big]_0^2 = 6$

19. $\int_{-2}^{1/2} -(2x-1)dx + \int_{1/2}^{2} (2x-1)dx = (-x^2+x)\Big]_{-2}^{1/2} + (x^2-x)\Big]_{1/2}^{2} = 17/2$

20. $\int_{1}^{x} \frac{1}{\sqrt{t}}dt = 2\sqrt{t}\Big]_{1}^{x} = 2(\sqrt{x}-1) = 3$, $\sqrt{x} = 5/2$, $x = 25/4$.

21. $\int_{0}^{x} \frac{1}{(3t+1)^2}dt = -\frac{1}{3(3t+1)}\Big]_{0}^{x} = -\frac{1}{3(3x+1)} + \frac{1}{3} = \frac{1}{6}$, $3x+1 = 2$, $x = 1/3$

22. $\int_{2}^{x} (4t-1)dt = (2t^2-t)\Big]_{2}^{x} = 2x^2 - x - 6 = 9$, $2x^2 - x - 15 = 0$,

$(2x+5)(x-3) = 0$, $x = -5/2$ and $x = 3$.

23. **(a)** $5+5+5+5 = 20$ **(b)** $2+2+2+2 = 8$

   (c) $n+n+n+n = 4n$ **(d)** $0+1/5+2/6 = 8/15$

   (e) $6/4+6/9+6/16 = 61/24$ **(f)** 9

   (g) $\sin(0) + \sin(\pi/4) + \sin(\pi/2) + \sin(3\pi/4) + \sin(\pi) = 0 + \sqrt{2}/2 + 1 + \sqrt{2}/2 + 0 = 1 + \sqrt{2}$

   (h) $\sqrt{2}/2 + (\sqrt{2}/2)^2 + (\sqrt{2}/2)^3 + (\sqrt{2}/2)^4 = 3\sqrt{2}/4 + 3/4$

24. **(a)** $\sum_{k=1}^{100}(k+2)k = \sum_{k=1}^{100}k^2 + 2\sum_{k=1}^{100}k = \frac{1}{6}(100)(101)(201) + 2\cdot\frac{1}{2}(100)(101) = 348,450$

   (b) $\sum_{k=1}^{100}(202-2k) = \sum_{k=1}^{100}202 - 2\sum_{k=1}^{100}k = (100)(202) - 2\cdot\frac{1}{2}(100)(101) = 10,100$

25. **(a)** $\sum_{k=1}^{9}(-1)^{k+1}\left(\frac{k}{k+1}\right)^2 = \sum_{k=2}^{10}(-1)^k\left(\frac{k-1}{k}\right)^2$

   (b) $\sum_{k=1}^{11}(-1)^{k+1}\frac{\pi^{k+1}}{k} = \sum_{k=2}^{12}(-1)^k\frac{\pi^k}{k-1}$

26. **(a)** $\Delta x = 2/n$, $c_k = 1 + 2k/n$

   $\sum_{k=1}^{n}f(c_k)\Delta x = \sum_{k=1}^{n}[6 - 2(1+2k/n)](2/n) = \frac{8}{n}\sum_{k=1}^{n}1 - \frac{8}{n^2}\sum_{k=1}^{n}k = 8 - 4\frac{n+1}{n}$

   (b) $d_k = 1 + 2(k-1)/n$

   $\sum_{k=1}^{n}f(d_k)\Delta x = \sum_{k=1}^{n}[6 - 2(1+2(k-1)/n)](2/n)$

   $= \frac{8}{n}\sum_{k=1}^{n}1 - \frac{8}{n^2}\sum_{k=1}^{n}(k-1) = 8 - 4\frac{n-1}{n}$

   (c) area $= \lim_{n\to+\infty}[8 - 4(1+1/n)] = 8 - 4 = 4$; $\int_{1}^{3}(6-2x)dx = 4$

27. (a) $\Delta x = 4/n$, $c_k = 4k/n$

$$\sum_{k=1}^{n} f(c_k)\Delta x = \sum_{k=1}^{n}(16 - 16k^2/n^2)(4/n)$$

$$= \frac{64}{n}\sum_{k=1}^{n}1 - \frac{64}{n^3}\sum_{k=1}^{n}k^2 = 64 - \frac{32}{3}\frac{(n+1)(2n+1)}{n^2}$$

(b) $d_k = 4(k-1)/n$

$$\sum_{k=1}^{n} f(d_k)\Delta x = \sum_{k=1}^{n}(16 - 16(k-1)^2/n^2)(4/n) = \frac{64}{n}\sum_{k=1}^{n}1 - \frac{64}{n^3}\sum_{k=1}^{n}(k-1)^2$$

$$= 64 - \frac{32}{3}\frac{(n-1)(2n-1)}{n^2}$$

(c) area $= \lim_{n \to +\infty}\left[64 - \frac{32}{3}\left(1 + \frac{1}{n}\right)\left(2 + \frac{1}{n}\right)\right] = 128/3;\ \int_{0}^{4}(16 - x^2)dx = 128/3$

28. (a) $\Delta x = 3/n$, $c_k = 1 + 3(k-1)/n$

$$\sum_{k=1}^{n} f(c_k)\Delta x = \sum_{k=1}^{n}[3 + 6(k-1)/n + 9(k-1)^2/n^2](3/n)$$

$$= \frac{9}{n}\sum_{k=1}^{n}1 + \frac{18}{n^2}\sum_{k=1}^{n}(k-1) + \frac{27}{n^3}\sum_{k=1}^{n}(k-1)^2$$

$$= 9 + 9\frac{n-1}{n} + \frac{9}{2}\frac{(n-1)(2n-1)}{n^2}$$

(b) $d_k = 1 + 3k/n$

$$\sum_{k=1}^{n} f(d_k)\Delta x = \sum_{k=1}^{n}(3 + 6k/n + 9k^2/n^2)(3/n) = \frac{9}{n}\sum_{k=1}^{n}1 + \frac{18}{n^2}\sum_{k=1}^{n}k + \frac{27}{n^3}\sum_{k=1}^{n}k^2$$

$$= 9 + 9\frac{n+1}{n} + \frac{9}{2}\frac{(n+1)(2n+1)}{n^2}$$

(c) $\lim_{n \to +\infty}[9 + 9(1 - 1/n) + (9/2)(1 - 1/n)(2 - 1/n)] = 27;\ \int_{1}^{4}(x^2 + 2)dx = 27$

29. (a) $\Delta x = 2/n$, because f is constant c_k can be chosen anywhere in the k-th subinterval so

$f(c_k) = 6$ and $\sum_{k=1}^{n} f(c_k)\Delta x = \sum_{k=1}^{n}(6)(2/n) = 12$

(b) same as for (a)

(c) area $= \lim_{n \to +\infty} 12 = 12;\ \int_{-1}^{1} 6\,dx = 12$

30. **(a)** $2\displaystyle\int_3^5 P(x)dx + \int_3^5 Q(x)dx = 2(3) + (4) = 10$

(b) $-\displaystyle\int_1^5 P(x)dx = -(-1) = 1$

(c) $-\displaystyle\int_3^5 Q(u)du = -\int_3^5 Q(x)dx = -4$

(d) $\displaystyle\int_3^5 P(x)dx + \int_5^1 P(x)dx = \int_3^5 P(x)dx - \int_1^5 P(x)dx = (3) - (-1) = 4$

31. If $x^2 \le f(x) \le 6$ then $\displaystyle\int_{-1}^2 x^2 dx \le \int_{-1}^2 f(x)dx \le \int_{-1}^2 6\,dx,\ 3 \le \int_{-1}^2 f(x)dx \le 18$

32. $f_{\text{ave}} = \displaystyle\int_{-2}^{-1} 3x^2 dx = 7;\ 3(x^*)^2 = 7,\ x^* = \pm\sqrt{7/3}$ but only $-\sqrt{7/3}$ is in $[-2, -1]$

33. $f_{\text{ave}} = \dfrac{1}{4}\displaystyle\int_0^4 x(x^2 + 9)^{-1/2}dx = \dfrac{1}{4}(x^2 + 9)^{1/2}\Big]_0^4 = 1/2;$

$\dfrac{x^*}{\sqrt{(x^*)^2 + 9}} = \dfrac{1}{2},\ 2x^* = \sqrt{(x^*)^2 + 9},\ 4(x^*)^2 = (x^*)^2 + 9,\ x^* = \pm\sqrt{3}$ but only $\sqrt{3}$ is in $[0, 4]$.

34. $f_{\text{ave}} = \dfrac{1}{4}\displaystyle\int_{-3}^1 (2 + |x|)dx = \dfrac{1}{4}\left[\int_{-3}^0 (2 - x)dx + \int_0^1 (2 + x)dx\right] = \dfrac{1}{4}[21/2 + 5/2] = 13/4;$

$2 + |x^*| = 13/4,\ |x^*| = 5/4,\ x^* = \pm 5/4$ but only $-5/4$ is in $[-3, 1]$

35. $f_{\text{ave}} = \dfrac{1}{\pi}\displaystyle\int_0^\pi \sin^2 x\,dx = \dfrac{1}{\pi}\int_0^\pi \dfrac{1}{2}(1 - \cos 2x)dx = \dfrac{1}{2\pi}\left(x - \dfrac{1}{2}\sin 2x\right)\Big]_0^\pi = 1/2;$

$\sin^2 x^* = 1/2,\ \sin x^* = \pm 1/\sqrt{2},\ x^* = \pi/4,\ 3\pi/4$ for x^* in $[0, \pi]$

CHAPTER 6
Applications of the Definite Integral

SECTION 6.1

6.1.1 Find the area of the region enclosed by $y = x + \dfrac{4}{x^2}$, the x axis, $x = 2$, and $x = 4$.

6.1.2 Find the area of the region enclosed by $y = 4x - x^2$ and $y = 3$.

6.1.3 Find the area of the region enclosed by $y = x^2 - 4x$ and $y = 16 - x^2$.

6.1.4 Find the area of the region enclosed by $x = y^2 - 4y$ and $x = y$.

6.1.5 Find the area of the region enclosed by $y = 3 - x^2$ and $y = -x + 1$ between $x = 0$ and $x = 2$.

6.1.6 Find the area of the region enclosed by $y = x^2 - 4x$ and $y = 2x - x^2$.

6.1.7 Find the area of the region enclosed by $x = y^2 - 4y + 2$ and $x = y - 2$.

6.1.8 Find the area of the region enclosed by $y = 2x - x^2$ and $y = -3$.

6.1.9 Find the area of the region enclosed by $x = 3y - y^2$ and $x + y = 3$.

6.1.10 Write a definite integral to represent the area of the shaded region in the diagram if one were to integrate with respect to x. <u>Do not evaluate</u>.

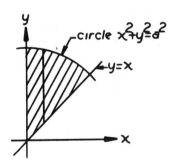

6.1.11 Find the area of the region enclosed by $2y = x^2$ and $y = x + 4$.

6.1.12 Find the area of the region enclosed by $y = x^2$ and $2x - y + 3 = 0$.

6.1.13 Find the area of the region enclosed by $x^2 = 8y$ and $x - 2y + 8 = 0$.

6.1.14 Find the area of the region enclosed by $x = y^2 - 5$ and $x = 3 - y^2$.

6.1.15   Find the area of the region enclosed by $y = x^2 - 4x + 4$ and $y = x$.

6.1.16   Find the area of the region enclosed by $y = x + 5$ and $y = x^2 - 1$.

6.1.17   Find the area of the region enclosed by $y = 2 - x^2$ and $y = -x$.

6.1.18   Find the area of the region enclosed by $y = x^3 + 1$, $x = -1$, $x = 2$, and the x axis.

SOLUTIONS

SECTION 6.1

6.1.1 $A = \int_2^4 \left(x + \frac{4}{x^2} \right) dx = \left[\frac{x^2}{2} - \frac{4}{x} \right]_2^4$

$= 7.$

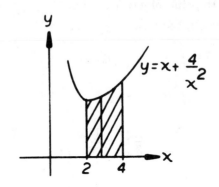

6.1.2 Equate $y = 4x - x^2$ and $y = 3$:

$$4x - x^2 = 3$$
$$x^2 - 4x + 3 = (x-1)(x-3) = 0$$
$$x = 1, 3$$

so the points of intersection
are $(1, 3)$ and $(3, 3)$.

$$A = \int_1^3 (4x - x^2 - 3)dx$$

$$= \left[2x^2 - \frac{x^3}{3} - 3x \right]_1^3 = \frac{4}{3}.$$

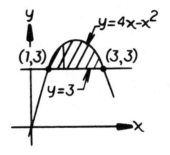

6.1.3 Equate $y = x^2 - 4x$ and $y = 16 - x^2$
to get $x^2 - 4x = 16 - x^2$,
$2x^2 - 4x - 16 = 2(x-4)(x+2) = 0$
so the points of intersection are
$(-2, 12)$ and $(4, 0)$, then,

$$A = \int_{-2}^4 [(16 - x^2) - (x^2 - 4x)]dx$$

$$= \int_{-2}^4 (16 + 4x - 2x^2)dx$$

$$= \left[16x + 2x^2 - \frac{2}{3}x^3 \right]_{-2}^4 = 72.$$

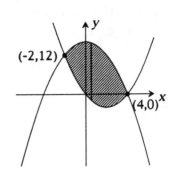

6.1.4   Equate $x = y^2 - 4y$ and

$y^2 - 4y = y$
$y^2 - 5y = y(y - 5) = 0$

so points of intersection are

$(0,0)$ and $(5,5)$.

$$A = \int_0^5 [y - (y^2 - 4y)]dy$$

$$= \int_0^5 (5y - y^2)dy$$

$$= \left[\frac{5}{2}y^2 - \frac{y^3}{3}\right]_0^5 = \frac{125}{6}.$$

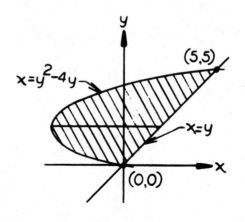

6.1.5 $$A = \int_0^2 [(3 - x^2) - (-x + 1)]dx$$

$$= \int_0^2 (2 + x - x^2)dx$$

$$= \left[2x + \frac{x^2}{2} - \frac{x^3}{3}\right]_0^2 = \frac{10}{3}.$$

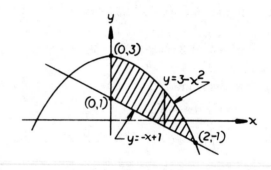

6.1.6   Equate $y = x^2 - 4x$ with $y = 2x - x^2$

to get $x^2 - 4x = 2x - x^2$,

$2x^2 - 6x = 2x(x - 3) = 0$,

so the points of intersection are

$(0,0)$ and $(3, -3)$, thus

$$A = \int_0^3 [(2x - x^2) - (x^2 - 4x)]dx$$

$$= \int_0^3 (6x - 2x^2)dx$$

$$= \left[3x^2 - \frac{2}{3}x^3\right]_0^3 = 9.$$

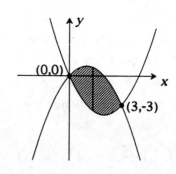

6.1.7 Equate $x = y^2 - 4y + 2$ and $x = y - 2$:

$$y^2 - 4y + 2 = y - 2$$
$$y^2 - 5y + 4 = (y - 1)(y - 4) = 0;$$

so the points of intersection are $(-1, 1)$ and $(2, 4)$.

$$A = \int_1^4 [(y - 2) - (y^2 - 4y + 2)]dy$$

$$= \int_1^4 (-y^2 + 5y - 4)dy$$

$$= \left[\frac{-y^3}{3} + \frac{5y^2}{2} - 4y \right]_1^4 = \frac{9}{2}.$$

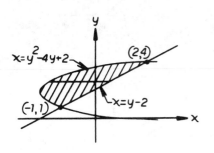

6.1.8 Equate $y = 2x - x^2$ and $y = -3$:

$$2x - x^2 = -3$$
$$x^2 - 2x - 3 = (x + 1)(x - 3) = 0;$$

so the points of intersection are $(-1, -3)$ and $(3, -3)$.

$$A = \int_{-1}^3 [(2x - x^2) - (-3)]dx$$

$$= \int_{-1}^3 (3 + 2x - x^2)dx$$

$$= \left[3x + x^2 - \frac{x^3}{3} \right]_{-1}^3 = \frac{32}{3}.$$

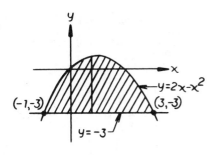

6.1.9 Equate $x = 3y - y^2$ and $x = 3 - y$:

$$3y - y^2 = 3 - y$$
$$y^2 - 4y + 3 = (y - 1)(y - 3) = 0;$$

so the points of intersection are $(0, 3)$ and $(2, 1)$.

$$A = \int_1^3 [(3y - y^2) - (3 - y)]dy$$

$$= \int_1^3 (-y^2 + 4y - 3)dy$$

$$= \left[\frac{-y^3}{3} + 2y^2 - 3y \right]_1^3 = \frac{4}{3}.$$

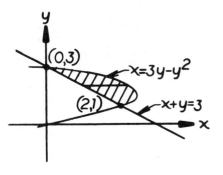

6.1.10 The line and the circle intersect at the point (a, a), thus,

$$A = \int_0^a (\sqrt{a^2 - x^2} - x)\,dx$$

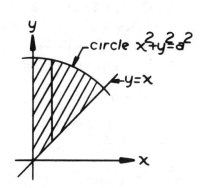

6.1.11 Equate $y = \dfrac{x^2}{2}$ and $y = x + 4$;

$$\frac{x^2}{2} = x + 4$$

$$x^2 - 2x - 8 = (x + 2)(x - 4) = 0;$$

so the points of intersection are $(-2, 2)$ and $(4, 8)$.

$$A = \int_{-2}^4 \left[(x + 4) - \frac{x^2}{2} \right] dx$$

$$= \int_{-2}^4 \left(4 + x - \frac{x^2}{2} \right) dx$$

$$= \left[4x + \frac{x^2}{2} - \frac{x^3}{6} \right]_{-2}^4 = 18.$$

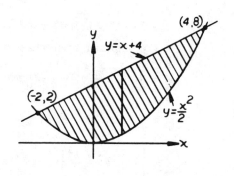

6.1.12 Equate $y = x^2$ and $y = 2x + 3$;

$$x^2 = 2x + 3$$

$$x^2 - 2x - 3 = (x + 1)(x - 3) = 0;$$

so the points of intersection are $(-1, 1)$ and $(3, 9)$.

$$A = \int_{-1}^3 [(2x + 3) - x^2]\,dx$$

$$= \int_{-1}^3 (3 + 2x - x^2)\,dx$$

$$= \left[3x + x^2 - \frac{x^3}{3} \right]_{-1}^3 = \frac{32}{3}.$$

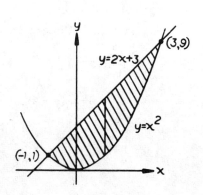

6.1.13 Equate $y = \dfrac{x^2}{8}$ and $y = \dfrac{x+8}{2}$:

$$\frac{x^2}{8} = \frac{x+8}{2}$$

$$x^2 - 4x - 32 = (x+4)(x-8) = 0;$$

so the points of intersection are $(-4, 2)$ and $(8, 8)$.

$$A = \int_{-4}^{8} \left[\left(\frac{x+8}{2} \right) - \frac{x^2}{8} \right] dx$$

$$= \int_{-4}^{8} \left(4 + \frac{x}{2} - \frac{x^2}{8} \right) dx$$

$$= \left[4x + \frac{x^2}{4} - \frac{x^3}{24} \right]_{-4}^{8} = 36.$$

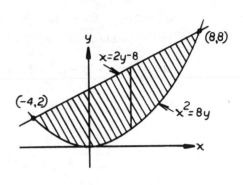

6.1.14 Equate $x = y^2 - 5$ with $x = 3 - y^2$ to get

$$y^2 - 5 = 3 - y^2,$$

$$2y^2 - 8 = 2(y-2)(y+2) = 0$$

so the intersection points are $(-1, -2)$ and $(-1, 2)$; then

$$A = \int_{-2}^{2} [(3 - y^2) - (y^2 - 5)] dy$$

$$= \int_{-2}^{2} (8 - 2y^2) dy = 8y - \frac{2}{3} y^3 \Big]_{-2}^{2}$$

$$= \frac{64}{3}$$

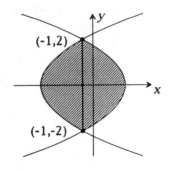

6.1.15 Equate $y = x^2 - 4x + 4$ and $y = x$;

$$x^2 - 4x + 4 = x$$
$$x^2 - 5x + 4 = (x-1)(x-4) = 0,$$

so the points of intersection are $(1, 1)$ and $(4, 4)$.

$$A = \int_{1}^{4} [x - (x^2 - 4x + 4)] dx$$

$$= \int_{1}^{4} (-x^2 + 5x - 4) dx$$

$$= \left[\frac{-x^3}{3} + \frac{5x^2}{2} - 4x \right]_{1}^{4} = \frac{9}{2}.$$

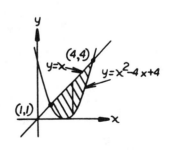

6.1.16   Equate $y = x + 5$ and $y = x^2 - 1$

$$x + 5 = x^2 - 1$$
$$x^2 - x - 6 = (x + 2)(x - 3) = 0;$$

so the points of intersection are $(-2, 3)$ and $(3, 8)$.

$$A = \int_{-2}^{3} [(x + 5) - (x^2 - 1)]dx$$

$$= \int_{-2}^{3} (6 + x - x^2)dx$$

$$= \left[6x + \frac{x^2}{2} - \frac{x^3}{3} \right]_{-2}^{3} = \frac{125}{6}.$$

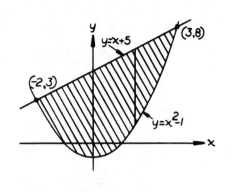

6.1.17   Equate $y = 2 - x^2$ and $y = -x$;

$$2 - x^2 = -x$$
$$x^2 - x - 2 = (x + 1)(x - 2) = 0,$$

so the points of intersection are $(-1, 1)$ and $(2, -2)$.

$$A = \int_{-1}^{2} [(2 - x^2) - (-x)]dx$$

$$= \int_{-1}^{2} (2 + x - x^2)dx$$

$$= \left[2x + \frac{x^2}{2} - \frac{x^3}{3} \right]_{-1}^{2} = \frac{9}{2}.$$

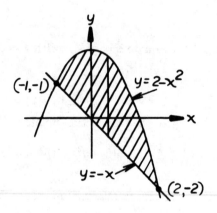

6.1.18 $A = \int_{-1}^{2} (x^3 + 1)dx = \left[\frac{x^4}{4} + x \right]_{-1}^{2} = \frac{27}{4}.$

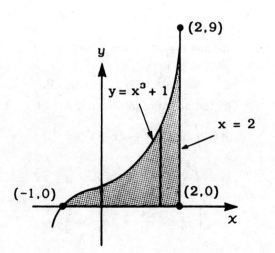

SECTION 6.2

6.2.1 Use the method of disks to find the volume of the solid that results when the area of the region enclosed by $y = x^2$, $x = 0$, and $y = 4$ is revolved about the y axis.

6.2.2 Find the volume of the solid that results when the area of the region enclosed by $x + y = 4$, $y = 0$, $x = 0$ is revolved about the x axis.

6.2.3 Use the method of disks to find the volume of the solid that results when the area of the region enclosed by $y^2 = x^3$, $x = 1$, and $y = 0$ is revolved about the x axis.

6.2.4 Use the method of washers to find the volume of the solid that results when the area of the region enclosed by $y = \sqrt{x}$, $y = 0$, and $x = 9$ is revolved about the y axis.

6.2.5 Use the method of washers to find the volume of the solid that results when the area of the region enclosed by $y^2 = 4x$, $x = 4$, and $y = 0$ is revolved about the y axis.

6.2.6 Use the method of washers to find the volume of the solid that results when the area enclosed by $y^2 = 4x$, $y = 2$, and $x = 4$ is revolved about the x axis.

6.2.7 Use the method of washers to find the volume of the solid that results when the area of the region enclosed by $y = 4 - x^2$ and $y = x + 2$ is revolved about the x axis.

6.2.8 Use the method of washers to find the volume of the solid that results when the area of the region enclosed by $y = x^2$, $y = 4$, and $x = 0$ is revolved about the x axis.

6.2.9 Use the method of washers to find the volume of the solid that results when the area of the region enclosed by $y^2 = x^3$, $x = 1$, and $y = 0$ is revolved about the y axis.

6.2.10 Use the method of disks to find the volume of the solid that results when the area of the region enclosed by $y = x^3$, $x = 2$, and $y = 0$ is revolved about the line $x = 2$.

6.2.11 Use the method of disks to find the volume of the solid that results when the area of the region enclosed by $y = x^3$, $x = 1$, and $y = -1$ is revolved about the line $y = -1$.

6.2.12 Use the method of disks to find the volume of the solid that results when the area of the region enclosed by $y = x^2$, $y = 0$, and $x = 2$ is revolved about $x = 2$.

6.2.13 Use the method of disks to find the volume of the solid that results when the area of the region enclosed by $y^2 = 2x$, $x = 2$, and $y = 1$ is revolved about the line $x = 2$.

6.2.14 Use the method of washers to find the volume of the solid that results when the area of the region enclosed by $y = 2x$, $x = 0$, and $y = 2$ is revolved about $x = 1$.

6.2.15 Use the method of washers to find the volume of the solid that results when the area of the region enclosed by $y^2 = 4x$ and $y = x$ is revolved about the line $x = 4$.

6.2.16 Use the method of washers to find the volume of the solid that results when the area of the region enclosed by $y = x^3/2$, $x = 2$, and $y = 0$ is revolved about the line $y = 4$.

6.2.17 The base of a solid is a circle of radius 2. All sections that are perpendicular to the diameter are squares. Find the volume of the solid.

6.2.18 The steeple of a church is constructed in the form of a pyramid 45 feet high. The cross sections are all squares and the base is a square of side 15 feet. Find the volume of the steeple.

SOLUTIONS

SECTION 6.2

6.2.1 $V = \pi \int_0^4 (\sqrt{y})^2 dy = \pi \int_0^4 y \, dy$

$\qquad = \pi \left. \dfrac{y^2}{2} \right]_0^4 = 8\pi.$

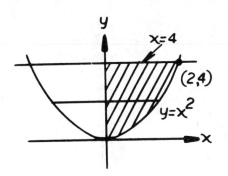

6.2.2 $V = \pi \int_0^4 (4 - x)^2 dx$

$\qquad = -\dfrac{\pi}{3} (4 - x)^3 \Big]_0^4 = \dfrac{64\pi}{3}.$

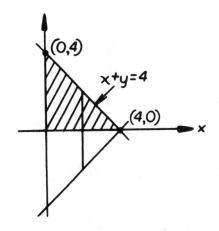

6.2.3 $V = \pi \int_0^1 x^3 dx = \pi \left. \dfrac{x^4}{4} \right]_0^1 = \dfrac{\pi}{4}.$

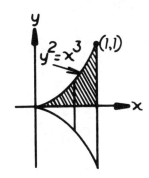

6.2.4 $V = \pi \displaystyle\int_0^3 [9^2 - (y^2)^2]dy$

$= \pi \displaystyle\int_0^3 (81 - y^4)dy$

$= \pi \left[81y - \dfrac{y^5}{5} \right]_0^3 = \dfrac{972\pi}{5}.$

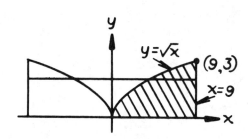

6.2.5 $V = \pi \displaystyle\int_0^4 \left[4^2 - \left(\dfrac{y^2}{4}\right)^2 \right] dy$

$= \pi \displaystyle\int_0^4 \left(16 - \dfrac{y^4}{16} \right) dy$

$= \pi \left[16y - \dfrac{y^5}{80} \right]_0^4 = \dfrac{256\pi}{5}.$

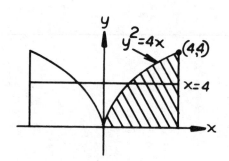

6.2.6 $V = \pi \displaystyle\int_1^4 \left[(\sqrt{4x})^2 - (2)^2 \right] dx$

$= \pi \displaystyle\int_1^4 (4x - 4)dx$

$= \pi \left[\dfrac{4x^2}{2} - 4x \right]_1^4 = 18\pi.$

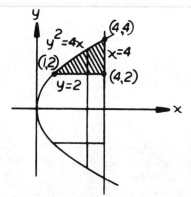

6.2.7   Equate $y = 4 - x^2$ and $y = x + 2$ to find point of intersection, thus,

$$4 - x^2 = x + 2$$
$$x^2 + x - 2 = (x + 2)(x - 1) = 0,$$

so the points of intersection are $(-2, 0)$ and $(1, 3)$.

$$V = \pi \int_{-2}^{1} [(4 - x^2)^2 - (x + 2)^2] dx$$

$$= \pi \int_{-2}^{1} (x^4 - 9x^2 - 4x + 12) dx$$

$$= \pi \left[\frac{x^5}{5} - \frac{9x^3}{3} - \frac{4x^2}{2} + 12x \right]_{-2}^{1} = \frac{108\pi}{5}.$$

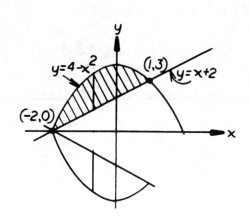

6.2.8 $V = \pi \int_{0}^{2} [(4)^2 - (x^2)^2] dx$

$$= \pi \int_{0}^{2} (16 - x^4) dx$$

$$= \pi \left[16x - \frac{x^5}{5} \right]_{0}^{2}$$

$$= \frac{128\pi}{5}.$$

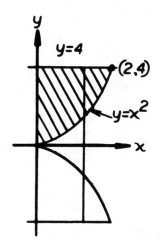

6.2.9 $V = \pi \int_{0}^{1} \left[(1)^2 - \left(y^{2/3} \right)^2 \right] dy$

$$= \pi \int_{0}^{1} (1 - y^{4/3}) dy$$

$$= \pi \left[y - \frac{3}{7} y^{7/3} \right]_{0}^{1} = \frac{4\pi}{7}.$$

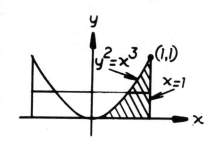

6.2.10 $V = \pi \displaystyle\int_0^8 \left(2 - y^{1/3}\right)^2 dy$

$= \pi \displaystyle\int_0^8 (4 - 4y^{1/3} + y^{2/3})dy$

$= \pi \left[4y - 4 \cdot \dfrac{3}{4}y^{4/3} + \dfrac{3}{5}y^{5/3}\right]_0^8$

$= \dfrac{16\pi}{5}.$

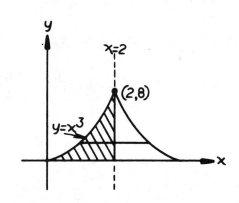

6.2.11 $V = \pi \displaystyle\int_{-1}^1 \left[x^3 - (-1)\right]^2 dx$

$= \pi \displaystyle\int_{-1}^1 (x^3 + 1)^2 dx$

$= \pi \displaystyle\int_{-1}^1 (x^6 + 2x^3 + 1)dx$

$V = \pi \left[\dfrac{x^7}{7} + \dfrac{2x^4}{4} + x\right]_{-1}^1 = \dfrac{16\pi}{7}.$

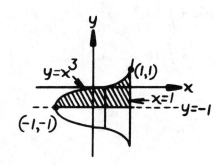

6.2.12 $V = \pi \displaystyle\int_0^4 (2 - \sqrt{y})^2 dy$

$= \pi \displaystyle\int_0^4 (4 - 4y^{1/2} + y)dy$

$= \pi \left[4y - 4 \cdot \dfrac{2}{3}y^{3/2} + \dfrac{y^2}{2}\right]_0^4 = \dfrac{8\pi}{3}.$

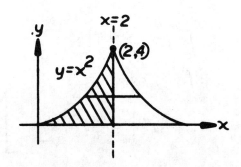

6.2.13 $\quad V = \pi \int_1^2 \left[\left(2 - \frac{y^2}{2} \right)^2 \right] dy$

$\qquad = \pi \int_1^2 \left(4 - 2y^2 + \frac{y^4}{4} \right) dy$

$\qquad = \pi \left[4y - \frac{2y^3}{3} + \frac{y^5}{20} \right]_1^2 = \frac{53\pi}{60}.$

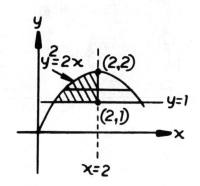

6.2.14 $\quad V = \pi \int_0^2 \left[1^2 - \left(\frac{y}{2} \right)^2 \right] dy$

$\qquad = \pi \int_0^2 \left(1 - \frac{y^2}{4} \right) dy$

$\qquad = \pi \left[y - \frac{y^3}{12} \right]_0^2 = \frac{4\pi}{3}.$

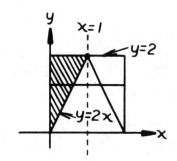

6.2.15 $\quad V = \pi \int_0^4 \left[\left(4 - \frac{y^2}{4} \right)^2 - (4 - y)^2 \right] dy$

$\qquad = \pi \int_0^4 \left(8y - 3y^2 + \frac{y^4}{16} \right) dy$

$\qquad = \pi \left[8\frac{y^2}{2} - 3\frac{y^3}{3} + \frac{1}{16}\frac{y^5}{5} \right]_0^4 = \frac{64\pi}{5}.$

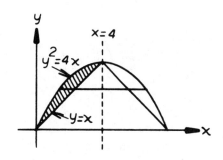

6.2.16 $V = \pi \int_0^2 \left[4^2 - \left(4 - \frac{x^3}{2} \right)^2 \right] dx$

$= \pi \int_0^2 \left(4x^3 - \frac{x^6}{4} \right) dx$

$= \pi \left[\frac{4x^4}{4} - \frac{x^7}{28} \right]_0^2 = \frac{80\pi}{7}.$

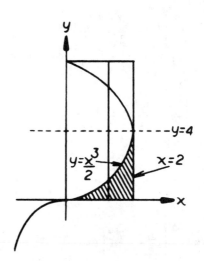

6.2.17   Let the circle $x^2 + y^2 = 4$ in the xy plane be the base of the solid. The area of each square cross section is $4y^2$ since each side is $2y$. Thus,

$A = 4y^2 = 4(4 - x^2)$

and

$V = 4 \int_{-2}^2 (4 - x^2) dx$

$= 4 \left[4x - \frac{x^3}{3} \right]_{-2}^2$

$= \frac{128}{3}.$

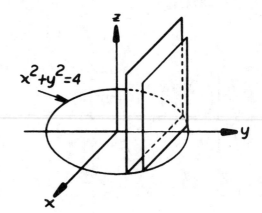

6.2.18   By similar triangles (see figure),

$\frac{45 - y}{45} = \frac{x/2}{15/2},$

$x = \frac{45 - y}{3},$

thus, the area of the cross section is

$A(y) = x^2 = \left(\frac{45 - y}{3} \right)^2$

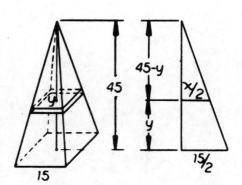

$$V = \int_0^{45} \left(\frac{45 - y}{3} \right)^2 dy$$

$$= -\frac{1}{27} \left[(45 - y)^3 \right]_0^{45}$$

$$= \frac{(45)^3}{27} = 3375 \text{ ft}^3.$$

SECTION 6.3

6.3.1   Use cylindrical shells to find the volume of the hemisphere that results when the region in the first quadrant enclosed by the circle $x^2 + y^2 = r^2$ is revolved about the x-axis.

6.3.2   Use cylindrical shells to find the volume of the solid that results when the area of the region enclosed by $x = 2y - y^2$ and $x = 0$ is revolved about the x-axis.

6.3.3   Use cylindrical shells to find the volume of the solid that results when the area of the region enclosed by $y = x^2$, $y = 4$, and $x = 0$ is revolved about the x-axis.

6.3.4   Use cylindrical shells to find the volume of the solid that results when the area of the region enclosed by $y^2 = x^3$, $x = 1$, and $y = 0$ is revolved about the x-axis.

6.3.5   Use cylindrical shells to find the volume of the solid that results when the region enclosed by $y = x^2$ and $x = y^2$ is revolved about the x-axis.

6.3.6   A storage tank is designed by rotating $y = -x^2 + 1$, $-1 \leq x \leq 1$, about the x-axis where x and y are measured in meters. Use cylindrical shells to determine how many cubic meters the tank will hold.

6.3.7   Use cylindrical shells to find the volume of the solid that results when the region enclosed by $xy = 3$, $x + y = 4$ is revolved about the y-axis.

6.3.8   Use cylindrical shells to find the volume of the solid that results when the region enclosed by $y = x^3 - 5x^2 + 6x$ over $[0, 2]$ is revolved about the y-axis.

6.3.9   Use cylindrical shells to find the volume of the solid that results when the region enclosed by $y^2 = 4x$, $x = 4$, and $y = 0$ is revolved about the y-axis.

6.3.10   Use cylindrical shells to find the volume of the solid that results when the area of the smaller region enclosed by $y^2 = 4x$, $y = 2$, and $x = 4$ is revolved about the y-axis.

6.3.11   Use cylindrical shells to find the volume of the solid that results when the area of the region enclosed by $y^2 = x^3$, $x = 1$, and $y = 0$ is revolved about the y-axis.

6.3.12   Use cylindrical shells to find the volume of the solid that results when the area of the region enclosed by $y = 2x + 3$, $x = 1$, $x = 4$, and $y = 0$ is revolved about the y-axis.

6.3.13   Use cylindrical shells to find the volume of the solid that results when the area of the region enclosed by $y = \sqrt{x+1}$, $x = 0$, $y = 0$, and $x = 3$ is revolved about the y-axis.

6.3.14   Use cylindrical shells to find the volume of the solid that results when the first quadrant region enclosed by $y = x^3$ and $y = x$ is revolved about the y-axis.

6.3.15   Use cylindrical shells to find the volume of the cone generated when the triangle with vertices $(0,0)$, $(r,0)$, $(0,h)$, where $r > 0$ and $h > 0$ is revolved about the y-axis.

6.3.16   Use cylindrical shells to find the volume of the solid that results when the region enclosed by $y^2 = 8x$ and $x = 2$ is revolved about the line $x = 4$.

6.3.17   Use cylindrical shells to find the volume of the solid that results when the area of the region enclosed by $y = 2x$, $x = 0$, and $y = 2$ is revolved about $x = 1$.

6.3.18   Use cylindrical shells to find the volume of the solid that results when the area of the region enclosed by $y^2 = 4x$ and $y = x$ is revolved about the line $x = 4$.

6.3.19   Use cylindrical shells to find the volume of the solid that results when the area of the region enclosed by $y = x^2$, $y = 0$, and $x = 2$ is revolved about $x = 2$.

6.3.20   Let a hemisphere of radius 5 be cut by a plane parallel to the base of the hemisphere thus forming a segment of height 2. Find its volume using cylindrical shells.

SOLUTIONS

SECTION 6.3

6.3.1 $\quad V = 2\pi \int_0^r y\sqrt{r^2 - y^2}\,dy$

$\qquad = \dfrac{2\pi}{3}\left[(r^2 - y^2)^{3/2}\right]_0^r = \dfrac{2\pi r^3}{3}.$

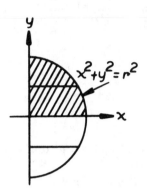

6.3.2 $\quad V = 2\pi \int_0^2 y(2y - y^2)\,dy$

$\qquad = 2\pi \int_0^2 (2y^2 - y^3)\,dy$

$\qquad = 2\pi \left[\dfrac{2y^3}{3} - \dfrac{y^4}{4}\right]_0^2 = \dfrac{8\pi}{3}.$

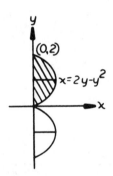

6.3.3 $\quad V = 2\pi \int_0^4 y(y^{1/2})\,dy$

$\qquad = 2\pi \int_0^4 y^{3/2}\,dy$

$\qquad = 2\pi \left[\dfrac{2}{5}y^{5/2}\right]_0^4 = \dfrac{128\pi}{5}.$

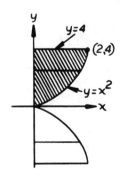

6.3.4 $V = 2\pi \displaystyle\int_0^1 y(1 - y^{2/3})dy$

$\qquad = 2\pi \displaystyle\int_0^1 (y - y^{5/3})dy$

$\qquad = 2\pi \left[\dfrac{y^2}{2} - \dfrac{3}{8}y^{8/3} \right]_0^1 = \dfrac{\pi}{4}.$

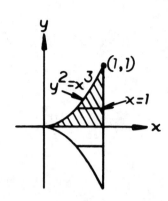

6.3.5 Equate $y = x^2$ and $x = y^2$ to find points of intersection.

$\qquad V = 2\pi \displaystyle\int_0^1 y(\sqrt{y} - y^2)dy$

$\qquad = 2\pi \displaystyle\int_0^1 (y^{3/2} - y^3)dy$

$\qquad = 2\pi \left[\dfrac{2}{5}y^{5/2} - \dfrac{y^4}{4} \right]_0^1 = \dfrac{3\pi}{10}.$

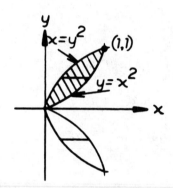

6.3.6 $V = 2\pi \displaystyle\int_0^1 y \left[\sqrt{1-y} - (-\sqrt{1-y}) \right] dy$

$\qquad = 4\pi \displaystyle\int_0^1 y\sqrt{1-y}\, dy$

$\qquad = -4\pi \displaystyle\int_1^0 (1-u)u^{1/2}\, du$

$\qquad$ (where $u = 1 - y$ and $y = 1 - u$)

$\qquad = -4\pi \displaystyle\int_1^0 (u^{1/2} - u^{3/2})du$

$\qquad = -4\pi \left[\dfrac{2}{3}u^{3/2} - \dfrac{2}{5}u^{5/2} \right]_1^0 = \dfrac{16\pi}{15}.$

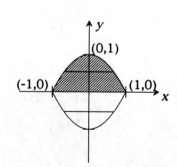

6.3.7 $V = 2\pi \int_1^3 x\left(4 - x - \dfrac{3}{x}\right) dx$

$$= 2\pi \int_1^3 (4x - x^2 - 3)dx$$

$$= 2\pi \left[2x^2 - \dfrac{x^3}{3} - 3x\right]_1^3 = \dfrac{8\pi}{3}.$$

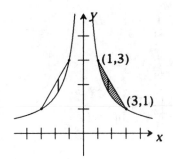

6.3.8 $V = 2\pi \int_0^2 x(x^3 - 5x^2 + 6x)dx$

$$= 2\pi \int_0^2 (x^4 - 5x^3 + 6x^2)dx$$

$$= 2\pi \left[\dfrac{x^5}{5} - \dfrac{5x^4}{4} + 2x^3\right]_0^2 = \dfrac{24\pi}{5}$$

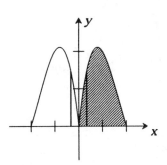

6.3.9 $V = 2\pi \int_0^4 x \cdot \sqrt{4x}\, dx$

$$= 4\pi \int_0^4 x^{3/2} dx$$

$$= 4\pi \left[\dfrac{2}{5}x^{5/2}\right]_0^4 = \dfrac{256\pi}{5}.$$

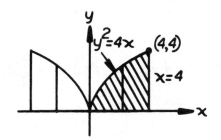

6.3.10 $V = 2\pi \int_1^4 x(\sqrt{4x} - 2)dx$

$$= 4\pi \int_1^4 (x^{3/2} - x)dx$$

$$= 4\pi \left[\dfrac{2}{5}x^{5/2} - \dfrac{x^2}{2}\right]_1^4 = \dfrac{98\pi}{5}.$$

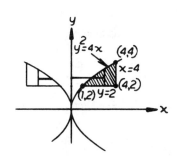

6.3.11

$$V = 2\pi \int_0^1 x \cdot x^{3/2} dx$$

$$= 2\pi \int_0^1 x^{5/2} dx$$

$$= 2\pi \left[\frac{2}{7} x^{7/2} \right]_0^1 = \frac{4\pi}{7}.$$

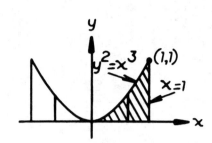

6.3.12

$$V = 2\pi \int_1^4 x(2x+3) dx$$

$$= 2\pi \int_1^4 (2x^2 + 3x) dx$$

$$= 2\pi \left[\frac{2x^3}{3} + \frac{3x^2}{2} \right]_1^4 = 129\pi.$$

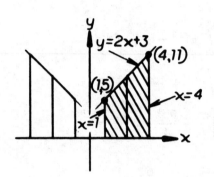

6.3.13

$$V = 2\pi \int_0^3 x\sqrt{x+1} dx$$

$$= 2\pi \int_1^4 (u-1)u^{1/2} du \text{ where } u = x-1)$$

$$= 2\pi \int_0^4 (u^{3/2} - u^{1/2}) du = 2\pi \left[\frac{2}{5} u^{5/2} - \frac{2}{3} u^{3/2} \right]_1^4$$

$$= \frac{232\pi}{15}.$$

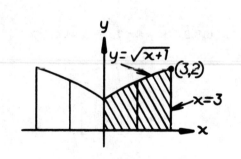

6.3.14 Equate $y = x^3$ and $y = x$ to find points of intersection.

$$V = 2\pi \int_0^1 x(x - x^3) dx$$

$$= 2\pi \int_0^1 (x^2 - x^4) dx$$

$$= 2\pi \left[\frac{x^3}{3} - \frac{x^5}{5} \right]_0^1 = \frac{4\pi}{15}.$$

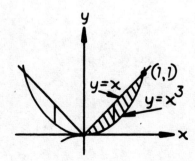

6.3.15 $y = \dfrac{h}{r}(r-x)$ is the equation of the line.

$$V = 2\pi \int_0^r x\left[\frac{h}{r}(r-x)\right]dx$$

$$= 2\pi \int_0^r \left[hx - \frac{h}{r}x^2\right]dx$$

$$= 2\pi \left[\frac{hx^2}{2} - \frac{hx^3}{3r}\right]_0^r = \frac{\pi r^2 h}{3}$$

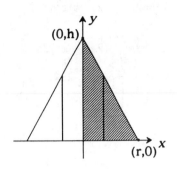

6.3.16 $V = 2\pi \int_0^2 (4-x)[\sqrt{8x} - (-\sqrt{8x})]dx$

$$= 8\sqrt{2}\pi \int_0^2 (4x^{1/2} - x^{3/2})dx$$

$$= 8\sqrt{2}\pi \left[4\cdot\frac{2}{3}x^{3/2} - \frac{2}{5}x^{5/2}\right]_0^2 = \frac{896\pi}{15}.$$

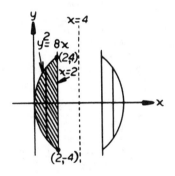

6.3.17 $V = 2\pi \int_0^1 (1-x)(2-2x)dy$

$$= 4\pi \int_0^1 (1-x)^2 dx$$

$$= -\frac{4\pi}{3}\left[(1-x)^3\right]_0^1 = \frac{4\pi}{3}.$$

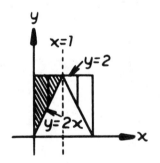

6.3.18 $V = 2\pi \int_0^4 (4-x)(\sqrt{4x} - x)dx$

$$= 2\pi \int_0^4 (8x^{1/2} - 4x - 2x^{3/2} + x^2)dx$$

$$= 2\pi \left[8\cdot\frac{2}{3}x^{3/2} - 4\cdot\frac{1}{2}x^2 - 2\cdot\frac{2}{5}x^{5/2} + \frac{x^3}{3}\right]_0^4$$

$$= \frac{64\pi}{5}.$$

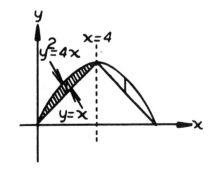

6.3.19 $V = 2\pi \displaystyle\int_0^2 (2-x)x^2\,dx$

$\qquad = 2\pi \displaystyle\int_0^2 (2x^2 - x^3)\,dx$

$\qquad = 2\pi \left[\dfrac{2x^3}{3} - \dfrac{x^4}{4}\right]_0^2 = \dfrac{8\pi}{3}.$

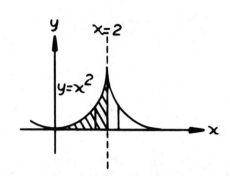

6.3.20 $V = 2\pi \displaystyle\int_0^4 x\left(\sqrt{25-x^2} - 3\right)dx$

$\qquad = 2\pi \displaystyle\int_0^4 \left(x\sqrt{25-x^2} - 3x\right)dx$

$\qquad = 2\pi \left[\displaystyle\int_0^4 x\sqrt{25-x^2}\,dx - 3\int_0^4 x\,dx\right]$

$\qquad = 2\pi \left[\dfrac{1}{-2}\displaystyle\int_{25}^9 u^{1/2}\,du - 3\int_0^4 x\,dx\right]$

(where $u = 25 - x^2$, $\dfrac{du}{-2} = x\,dx$)

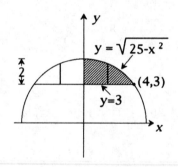

$\qquad = 2\pi \left[\left.\dfrac{1}{-2}\cdot\dfrac{2}{3}u^{3/2}\right|_{25}^9 - \left.\dfrac{3}{2}x^2\right|_0^4\right]$

$\qquad = \dfrac{52\pi}{3}.$

SECTION 6.4

6.4.1 Find the arc length of the curve $y = 2x^{3/2}$ from $x = 0$ to $x = \dfrac{8}{9}$.

6.4.2 Find the arc length of the curve $y = \dfrac{x^3}{6} + \dfrac{1}{2x}$ from $x = 1$ to $x = 3$.

6.4.3 Find the arc length of the curve $y = \dfrac{2}{3}(x+1)^{3/2}$ from $x = 1$ to $x = 2$.

6.4.4 Find the arc length of the curve $y = 2x^{3/2}$ from $x = 0$ to $x = 3$.

6.4.5 Find the arc length of the curve $y = \dfrac{x^3}{12} + \dfrac{1}{x}$ from $x = 1$ to $x = 2$.

6.4.6 Find the arc length of the curve $x = \dfrac{y^4}{8} + \dfrac{1}{4y^2}$ from $y = 1$ to $y = 4$.

6.4.7 Find the arc length of the curve $y = \dfrac{1}{3}(x^2+2)^{3/2}$ from $x = 0$ to $x = 3$.

6.4.8 Find the arc length of the curve $y^3 = x^2$ from $(0,0)$ to $(8,4)$.

6.4.9 Find the arc length of the curve $y = \dfrac{x^3}{3} + \dfrac{1}{4x}$ from $x = 1$ to $x = 3$.

6.4.10 Find the arc length of the curve $y = \dfrac{x^4}{4} + \dfrac{1}{8x^2}$ from $x = 1$ to $x = 2$.

6.4.11 Find the arc length of the curve $4y^3 = 9x^2$ from $(0,0)$ to $(2\sqrt{3},3)$.

6.4.12 Find the arc length of the curve $x = \dfrac{y^4}{4} + \dfrac{1}{8y^2}$ from $y = 1$ to $y = 2$.

6.4.13 Find the arc length of the curve $y = \dfrac{2}{3}x^{3/2} - \dfrac{1}{2}x^{1/2}$ from $x = 1$ to $x = 4$.

6.4.14 Find the arc length of the curve $x = \dfrac{3}{5}y^{5/3} - \dfrac{3}{4}y^{1/3}$ from $y = 1$ to $y = 8$.

6.4.15 Find the arc length of the curve $y = \dfrac{x^4}{8} + \dfrac{1}{4x^2}$ from $x = 1$ to $x = 2$.

6.4.16 Find the arc length of the curve $y = \dfrac{x^3}{24} + \dfrac{2}{x}$ from $x = 1$ to $x = 2$.

6.4.17 Find the arc length of the curve $x = \dfrac{y^3}{18} + \dfrac{3}{2y}$ from $y = 1$ to $y = 2$.

6.4.18 Find the arc length of the curve $y = \dfrac{x^4}{24} + \dfrac{3}{4x^2}$ from $x = 1$ to $x = 2$.

SOLUTIONS

SECTION 6.4

6.4.1 $f'(x) = 3x^{1/2}$, $1 + [f'(x)]^2 = 1 + 9x$

$$L = \int_0^{8/9} \sqrt{1 + 9x}\, dx = \frac{1}{9} \int_1^9 u^{1/2} du \text{ where } u = 1 + 9x$$

$$= \frac{1}{9} \left[\frac{2}{3} u^{3/2} \right]_1^9 = \frac{52}{27}.$$

6.4.2 $f'(x) = \dfrac{x^2}{2} - \dfrac{1}{2x^2}$, $[f'(x)]^2 = \dfrac{x^4}{4} - \dfrac{1}{2} + \dfrac{1}{4x^4}$,

$$1 + [f'(x)]^2 = 1 + \frac{x^4}{4} - \frac{1}{2} + \frac{1}{4x^4} = \frac{x^4}{4} + \frac{1}{2} + \frac{1}{4x^4}$$

$$L = \int_1^3 \sqrt{\frac{x^4}{4} + \frac{1}{2} + \frac{1}{4x^4}}\, dx = \int_1^3 \sqrt{\left(\frac{x^2}{2} + \frac{1}{2x^2} \right)^2}\, dx$$

$$= \int_1^3 \left(\frac{x^2}{2} + \frac{1}{2x^2} \right) dx = \left[\frac{x^3}{6} - \frac{1}{2x} \right]_1^3 = \frac{14}{3}.$$

6.4.3 $f'(x) = (x+1)^{1/2}$, $[f'(x)]^2 = x + 1$,

$1 + [f'(x)]^2 = 1 + x + 1 = x + 2$

$$L = \int_1^2 \sqrt{x + 2}\, dx = \int_3^4 \sqrt{u}\, du \text{ where } u = x + 2$$

$$= \frac{2}{3} u^{3/2} \Big]_3^4 = \frac{2}{3}(8 - 3\sqrt{3}).$$

6.4.4 $f'(x) = 3x^{1/2}$, $[f'(x)]^2 = 9x$, $1 + [f'(x)]^2 = 1 + 9x$

$$L = \int_0^3 \sqrt{1 + 9x}\, dx = \frac{1}{9} \int_1^{28} u^{1/2} du \text{ where } u = 1 + 9x$$

$$= \frac{1}{9} \left[\frac{2}{3} u^{3/2} \right]_1^{28} = \frac{2}{27}(56\sqrt{7} - 1).$$

6.4.5 $f'(x) = \dfrac{x^2}{4} - \dfrac{1}{x^2}$, $[f'(x)]^2 = \dfrac{x^4}{16} - \dfrac{1}{2} + \dfrac{1}{x^4}$,

$$1 + [f'(x)]^2 = 1 + \frac{x^4}{16} - \frac{1}{2} + \frac{1}{x^4} = \frac{x^4}{16} + \frac{1}{2} + \frac{1}{x^4}$$

$$L = \int_1^2 \sqrt{\frac{x^4}{16} + \frac{1}{2} + \frac{1}{x^4}}\, dx = \int_1^2 \sqrt{\left(\frac{x^2}{4} + \frac{1}{x^2} \right)^2}\, dx$$

$$= \int_1^2 \left(\frac{x^2}{4} + \frac{1}{x^2} \right) dx = \left[\frac{x^3}{12} - \frac{1}{x} \right]_1^2 = \frac{13}{12}.$$

6.4.6 $g'(y) = \dfrac{y^3}{2} - \dfrac{1}{2y^3},\ [g'(y)]^2 = \dfrac{y^6}{4} - \dfrac{1}{2} + \dfrac{1}{4y^6},$

$$1 + [g'(y)]^2 = 1 + \frac{y^6}{4} - \frac{1}{2} + \frac{1}{4y^6} = \frac{y^6}{4} + \frac{1}{2} + \frac{1}{4y^6}$$

$$L = \int_1^4 \sqrt{\frac{y^6}{4} + \frac{1}{2} + \frac{1}{4y^6}}\, dy = \int_1^4 \sqrt{\left(\frac{y^3}{2} + \frac{1}{2y^3}\right)^2}\, dy$$

$$\int_1^4 \left(\frac{y^3}{2} + \frac{1}{2y^3}\right) dy = \left[\frac{y^4}{8} - \frac{1}{4y^2}\right]_1^4 = \frac{2055}{64}.$$

6.4.7 $f'(x) = x(x^2 + 2)^{1/2},\ [f'(x)]^2 = x^2(x^2 + 2),$

$1 + [f'(x)]^2 = 1 + x^2(x^2 + 2) = x^4 + 2x^2 + 1$

$$L = \int_0^3 \sqrt{x^4 + 2x^2 + 1}\, dx = \int_0^3 \sqrt{(x^2 + 1)^2}\, dx$$

$$= \int_0^3 (x^2 + 1)dx = \left[\frac{x^3}{3} + x\right]_0^3 = 12.$$

6.4.8 $g(y) = y^{3/2},\ g'(y) = \dfrac{3}{2}y^{1/2},\ [g'(y)]^2 = \dfrac{9}{4}y,\ 1 + [g'(y)]^2 = 1 + \dfrac{9y}{4}$

$$L = \int_0^4 \sqrt{1 + \frac{9y}{4}}\, dy = \frac{4}{9}\int_1^{10} u^{1/2} du \text{ where } u = 1 + \frac{9y}{4}$$

$$= \frac{4}{9} \cdot \left[\frac{2}{3}u^{3/2}\right]_1^{10} = \frac{8}{27}(10\sqrt{10} - 1).$$

6.4.9 $f'(x) = x^2 - \dfrac{1}{4x^2},\ [f'(x)]^2 = x^4 - \dfrac{1}{2} + \dfrac{1}{16x^4},$

$$1 + [f'(x)]^2 = 1 + x^4 - \frac{1}{2} + \frac{1}{16x^4} = x^4 + \frac{1}{2} + \frac{1}{16x^4}$$

$$L = \int_1^3 \sqrt{x^4 + \frac{1}{2} + \frac{1}{16x^4}}\, dx = \int_1^3 \sqrt{\left(x^2 + \frac{1}{4x^2}\right)^2}\, dx$$

$$= \int_1^3 \left(x^2 + \frac{1}{4x^2}\right) dx = \left[\frac{x^3}{3} - \frac{1}{4x}\right]_1^3 = \frac{53}{6}.$$

6.4.10 $f'(x) = x^3 - \dfrac{1}{4x^3},\ [f'(x)]^2 = x^6 - \dfrac{1}{2} + \dfrac{1}{16x^6},$

$$1 + [f'(x)]^2 = 1 + x^6 - \frac{1}{2} + \frac{1}{16x^6} = x^6 + \frac{1}{2} + \frac{1}{16x^6}$$

$$L = \int_1^2 \sqrt{x^6 + \frac{1}{2} + \frac{1}{16x^6}}\, dx = \int_1^2 \sqrt{\left(x^3 + \frac{1}{4x^3}\right)^2}\, dx$$

$$= \int_1^2 \left(x^3 + \frac{1}{4x^3}\right) dx = \left[\frac{x^4}{4} - \frac{1}{8x^2}\right]_1^2 = \frac{123}{32}.$$

6.4.11   Let $g(y) = \dfrac{2}{3}y^{3/2}$ for y in $(0,3)$. $g'(y) = y^{1/2}$, $[g'(y)]^2 = y$, $1 + [g'(y)]^2 = 1 + y$

$$L = \int_0^3 \sqrt{1+y}\,dy = \dfrac{2}{3}\,(l+y)^{3/2}\Big]_0^3 = \dfrac{14}{3}.$$

6.4.12 $g'(y) = y^3 - \dfrac{1}{4y^3}$, $[g'(y)]^2 = y^6 - \dfrac{1}{2} + \dfrac{1}{16y^6}$,

$$1 + [g'(y)]^2 = 1 + y^6 - \dfrac{1}{2} + \dfrac{1}{16y^6} = y^6 + \dfrac{1}{2} + \dfrac{1}{16y^6}$$

$$L = \int_1^2 \sqrt{y^6 + \dfrac{1}{2} + \dfrac{1}{16y^6}}\,dy = \int_1^2 \sqrt{\left(y^3 + \dfrac{1}{4y^3}\right)^2}\,dy$$

$$= \int_1^2 \left(y^3 + \dfrac{1}{4y^3}\right)dy = \left[\dfrac{y^4}{4} - \dfrac{1}{8y^2}\right]_1^2 = \dfrac{123}{32}.$$

6.4.13 $f'(x) = x^{1/2} - \dfrac{1}{4x^{1/2}}$, $[f'(x)]^2 = x - \dfrac{1}{2} + \dfrac{1}{16x}$,

$$1 + [f'(x)]^2 = 1 + x - \dfrac{1}{2} + \dfrac{1}{16x} = x + \dfrac{1}{2} + \dfrac{1}{16x}$$

$$L = \int_1^4 \sqrt{x + \dfrac{1}{2} + \dfrac{1}{16x}}\,dx = \int_1^4 \sqrt{\left(x^{1/2} + \dfrac{1}{4x^{1/2}}\right)^2}\,dx$$

$$= \int_1^4 \left(x^{1/2} + \dfrac{1}{4x^{1/2}}\right)dx = \left[\dfrac{2}{3}x^{3/2} + \dfrac{1}{2}x^{1/2}\right]_1^4 = \dfrac{31}{6}.$$

6.4.14 $g'(y) = y^{2/3} - \dfrac{1}{4y^{2/3}}$, $[g'(y)]^2 = y^{4/3} - \dfrac{1}{2} + \dfrac{1}{16y^{4/3}}$,

$$1 + [g'(y)]^2 = 1 + y^{4/3} - \dfrac{1}{2} + \dfrac{1}{16y^{4/3}} = y^{4/3} + \dfrac{1}{2} + \dfrac{1}{16y^{4/3}},$$

$$L = \int_1^8 \sqrt{y^{4/3} + \dfrac{1}{2} + \dfrac{1}{16y^{4/3}}}\,dy = \int_1^8 \sqrt{\left(y^{2/3} + \dfrac{1}{4y^{2/3}}\right)^2}\,dy$$

$$= \int_1^8 \left(y^{2/3} + \dfrac{1}{4y^{2/3}}\right)dy = \left[\dfrac{3}{5}y^{5/3} + \dfrac{3}{4}y^{1/3}\right]_1^8 = \dfrac{387}{20}.$$

6.4.15 $f'(x) = \dfrac{x^3}{2} - \dfrac{1}{2x^3}$, $[f'(x)]^2 = \dfrac{x^6}{4} - \dfrac{1}{2} + \dfrac{1}{4x^6}$,

$$1 + [f'(x)]^2 = 1 + \dfrac{x^6}{4} - \dfrac{1}{2} + \dfrac{1}{4x^6} = \dfrac{x^6}{4} + \dfrac{1}{2} + \dfrac{1}{4x^6}$$

$$L = \int_1^2 \sqrt{\dfrac{x^6}{4} + \dfrac{1}{2} + \dfrac{1}{4x^6}}\,dx = \int_1^2 \sqrt{\left(\dfrac{x^3}{2} + \dfrac{1}{2x^3}\right)^2}\,dx$$

$$\int_1^2 \left(\dfrac{x^3}{2} + \dfrac{1}{2x^3}\right)dx = \left[\dfrac{x^4}{8} - \dfrac{1}{4x^2}\right]_1^2 = \dfrac{33}{16}.$$

6.4.16 $f'(x) = \dfrac{x^2}{8} - \dfrac{2}{x^2}$, $[f'(x)]^2 = \dfrac{x^4}{64} - \dfrac{1}{2} + \dfrac{4}{x^4}$.

$$1 + [f'(x)]^2 = 1 + \dfrac{x^4}{64} - \dfrac{1}{2} + \dfrac{4}{x^4} = \dfrac{x^4}{64} + \dfrac{1}{2} + \dfrac{4}{x^4}$$

$$L = \int_1^2 \sqrt{\dfrac{x^4}{64} + \dfrac{1}{2} + \dfrac{4}{x^4}}\, dx = \int_1^2 \sqrt{\left(\dfrac{x^2}{8} + \dfrac{2}{x^2}\right)^2}\, dx$$

$$= \int_1^2 \left(\dfrac{x^2}{8} + \dfrac{2}{x^2}\right) dx = \left[\dfrac{x^3}{24} - \dfrac{2}{x}\right]_1^2 = \dfrac{31}{24}.$$

6.4.17 $g'(y) = \dfrac{y^2}{6} - \dfrac{3}{2y^2}$, $[g'(y)]^2 = \dfrac{y^4}{36} - \dfrac{1}{2} + \dfrac{9}{4y^4}$,

$$1 + [g'(y)]^2 = 1 + \dfrac{y^4}{36} - \dfrac{1}{2} + \dfrac{9}{4y^4} = \dfrac{y^4}{36} + \dfrac{1}{2} + \dfrac{9}{4y^4}$$

$$L = \int_1^2 \sqrt{\dfrac{y^4}{36} + \dfrac{1}{2} + \dfrac{9}{4y^4}}\, dy = \int_1^2 \sqrt{\left(\dfrac{y^2}{6} + \dfrac{3}{2y^2}\right)^2}\, dy$$

$$= \int_1^2 \left(\dfrac{y^2}{6} + \dfrac{3}{2y^2}\right) dy = \left[\dfrac{y^3}{18} - \dfrac{3}{2y}\right]_1^2 = \dfrac{41}{36}.$$

6.4.18 $f'(x) = \dfrac{x^3}{6} - \dfrac{3}{2x^3}$, $[f'(x)]^2 = \dfrac{x^6}{36} - \dfrac{1}{2} + \dfrac{9}{4x^6}$,

$$1 + [f'(x)]^2 = 1 + \dfrac{x^6}{36} - \dfrac{1}{2} + \dfrac{9}{4x^6} = \dfrac{x^6}{36} + \dfrac{1}{2} + \dfrac{9}{4x^6}$$

$$L = \int_1^2 \sqrt{\dfrac{x^6}{36} + \dfrac{1}{2} + \dfrac{9}{4x^6}}\, dx - \int_1^2 \sqrt{\left(\dfrac{x^3}{6} + \dfrac{3}{2x^3}\right)^2}\, dx$$

$$= \int_1^2 \left(\dfrac{x^3}{6} + \dfrac{3}{2x^3}\right) dx = \left[\dfrac{x^4}{24} - \dfrac{3}{4x^2}\right]_1^2 = \dfrac{19}{16}.$$

SECTION 6.5

6.5.1 Find the area of the surface generated when $y = \sqrt{x}$ from $x = 1$ to $x = 6$ is revolved about the x-axis.

6.5.2 Find the area of the surface generated when $y = 8 - x$ from $x = 0$ to $x = 6$ is rotated around the x-axis.

6.5.3 Find the area of the surface generated when $x = \sqrt{4 - y}$ from $y = 0$ to $y = 3$ is rotated around the y-axis.

6.5.4 Find the area of the surface generated when $y^2 = 8x$ from $x = 1$ to $x = 2$ is rotated around the x-axis.

6.5.5 Find the area of the surface generated when $y = \dfrac{x^3}{6} + \dfrac{1}{2x}$ from $x = 1$ to $x = 3$ is rotated around the x-axis.

6.5.6 Find the area of the surface generated when $y^2 = 4x$ from $(1, 2)$ to $(4, 4)$ is rotated around the x-axis.

6.5.7 Find the area of the surface generated when $y = \sqrt{9 - x^2}$ from $x = 0$ to $x = 2$ is rotated around the x-axis.

6.5.8 Find the area of the surface generated when $y = \dfrac{x^4}{8} + \dfrac{1}{4x^2}$ from $x = 1$ to $x = 2$ is rotated around the x-axis.

6.5.9 Find the area of the surface generated when $y = x^3$ from $x = 0$ to $x = 1$ is rotated around the x-axis.

6.5.10 Find the area of the surface generated when $y = \dfrac{x^3}{3} + \dfrac{1}{4x}$ from $x = 1$ to $x = 2$ is rotated around the x-axis.

6.5.11 Find the area of the surface of a sphere of radius r which is generated by rotating a semicircle about a diameter.

6.5.12 Find the area of the surface generated when $x = \sqrt{16 - y^2}$ from $y = 0$ to $y = 3$ is rotated around the y-axis.

6.5.13 Find the area of the surface generated when $y = \dfrac{x^3}{12} + \dfrac{1}{x}$ from $x = 1$ to $x = 2$ is rotated around the x-axis.

6.5.14 Find the area of the surface generated when $x = \dfrac{y^4}{8} + \dfrac{1}{4y^2}$ from $y = 1$ to $y = 2$ is rotated around the y-axis.

6.5.15 Find the area of the surface generated when $y = \sqrt{9-x}$ from $x = 3$ to $x = 7$ is rotated around the x-axis.

6.5.16 Find the area of the surface generated when $x = \sqrt{16-y}$ from $y = 4$ to $y = 10$ is rotated around the y-axis.

6.5.17 Find the area of the surface generated when $y = \sqrt{25-x^2}$ from $x = 0$ to $x = 3$ is rotated around the x-axis.

6.5.18 Find the area of the surface generated when $x = \sqrt{36-y}$ from $y = 6$ to $y = 16$ is rotated around the y-axis.

SOLUTIONS

SECTION 6.5

6.5.1 $f'(x) = \dfrac{1}{2\sqrt{x}}$, $1 + [f'(x)]^2 = 1 + \dfrac{1}{4x}$

$$S = 2\pi \int_1^6 \sqrt{x}\sqrt{\frac{4x+1}{4x}}\, dx = \pi \int_1^6 \sqrt{4x+1}\, dx = \frac{\pi}{4}\int_5^{25} u^{1/2} du$$

$$= \frac{\pi}{4}\left[\frac{2}{3}u^{3/2}\right]_5^{25} = \frac{5\pi}{6}(25 - \sqrt{5}).$$

6.5.2 $f'(x) = -1$, $[f'(x)]^2 = 1$, $1 + [f'(x)]^2 = 1 + 1 = 2$

$$S = 2\pi \int_0^6 (8-x)\sqrt{2}\, dx = 2\sqrt{2}\,\pi\left[8x - \frac{x^2}{2}\right]_0^6 = 60\sqrt{2}\,\pi.$$

6.5.3 $g'(y) = \dfrac{1}{2\sqrt{4-y}}(-1)$, $[g'(y)]^2 = \dfrac{1}{4(4-y)}$,

$$1 + [g'(y)]^2 = 1 + \frac{1}{4(4-y)} = \frac{17-4y}{4(4-y)}$$

$$S = 2\pi\int_0^3 \sqrt{4-y}\sqrt{\frac{17-4y}{4(4-y)}}\, dy = \pi\int_0^3 \sqrt{17-4y}\, dy$$

$$= \frac{\pi}{4}\int_5^{17} u^{1/2} du \text{ where } u = 17 - 4y$$

$$= \frac{\pi}{4}\left[\frac{2}{3}u^{3/2}\right]_5^{17} = \frac{\pi}{6}(17\sqrt{17} - 5\sqrt{5}).$$

6.5.4 $f(x) = \sqrt{8x}$, $f'(x) = \sqrt{\dfrac{2}{x}}$, $[f'(x)]^2 = \dfrac{2}{x}$, $1 + [f'(x)] = 1 + \dfrac{2}{x}$

$$S = 2\pi\int_1^2 \sqrt{8x}\sqrt{1 + \frac{2}{x}}\, dx = 4\sqrt{2}\,\pi\int_1^2 \sqrt{x+2}\, dx$$

$$= \frac{8\sqrt{2}\,\pi}{3}\left[(x+2)^{3/2}\right]_1^2 = \frac{8\sqrt{2}\,\pi}{3}(8 - 3\sqrt{3}).$$

6.5.5 $f'(x) = \dfrac{x^2}{2} - \dfrac{1}{2x^2}$, $[f'(x)]^2 = \dfrac{x^4}{4} - \dfrac{1}{2} + \dfrac{1}{4x^4}$,

$$1 + [f'(x)]^2 = 1 + \frac{x^4}{4} - \frac{1}{2} + \frac{1}{4x^4} = \frac{x^4}{4} + \frac{1}{2} + \frac{1}{4x^4}$$

$$S = 2\pi \int_1^3 \left(\frac{x^3}{6} + \frac{1}{2x}\right) \sqrt{\frac{x^4}{4} + \frac{1}{2} + \frac{1}{4x^4}}\, dx = 2\pi \int_1^3 \left(\frac{x^3}{6} + \frac{1}{2x}\right) \sqrt{\left(\frac{x^2}{2} + \frac{1}{2x^2}\right)^2}\, dx$$

$$= 2\pi \int_1^3 \left(\frac{x^3}{6} + \frac{1}{2x}\right) \left(\frac{x^2}{2} + \frac{1}{2x^2}\right) dx = 2\pi \int_1^3 \left(\frac{x^5}{12} + \frac{x}{3} + \frac{1}{4x^3}\right) dx$$

$$= 2\pi \left[\frac{x^6}{72} + \frac{x^2}{6} - \frac{1}{8x^2}\right]_1^3 = \frac{208\pi}{9}.$$

6.5.6 $f(x) = 2\sqrt{x}$, $f'(x) = \dfrac{1}{\sqrt{x}}$, $[f'(x)]^2 = \dfrac{1}{x}$, $1 + [f'(x)]^2 = 1 + \dfrac{1}{x} = \dfrac{x+1}{x}$

$$S = 2\pi \int_1^4 2\sqrt{x}\sqrt{\frac{x+1}{x}}\, dx = 4\pi \int_1^4 \sqrt{x+1}\, dx = 4\pi \left[\frac{2}{3}(x+1)^{3/2}\right]_1^4$$

$$= \frac{8\pi}{3}(5\sqrt{5} - 2\sqrt{2}).$$

6.5.7 $f'(x) = \dfrac{-x}{\sqrt{9-x^2}}$, $[f'(x)]^2 = \dfrac{x^2}{9-x^2}$,

$$1 + [f'(x)]^2 = 1 + \frac{x^2}{9-x^2} = \frac{9}{9-x^2}$$

$$S = 2\pi \int_0^2 \sqrt{9-x^2}\sqrt{\frac{9}{9-x^2}}\, dx = 6\pi \int_0^2 dx = \left[6\pi x\right]_0^2 = 12\pi.$$

6.5.8 $f'(x) = \dfrac{x^3}{2} - \dfrac{1}{2x^3}$, $[f'(x)]^2 = \dfrac{x^6}{4} - \dfrac{1}{2} + \dfrac{1}{4x^6}$,

$$1 + [f'(x)]^2 = 1 + \frac{x^6}{4} - \frac{1}{2} + \frac{1}{4x^6} = \frac{x^6}{4} + \frac{1}{2} + \frac{1}{4x^6}$$

$$S = 2\pi \int_1^2 \left(\frac{x^4}{8} + \frac{1}{4x^2}\right) \sqrt{\frac{x^6}{4} + \frac{1}{2} + \frac{1}{4x^6}}\, dx = 2\pi \int_1^2 \left(\frac{x^4}{8} + \frac{1}{4x^2}\right) \sqrt{\left(\frac{x^3}{2} + \frac{1}{2x^3}\right)^2}\, dx$$

$$= 2\pi \int_1^2 \left(\frac{x^4}{8} + \frac{1}{4x^2}\right) \left(\frac{x^3}{2} + \frac{1}{2x^3}\right) dx = 2\pi \int_1^2 \left(\frac{x^7}{16} + \frac{3x}{16} + \frac{1}{8x^5}\right) dx$$

$$= 2\pi \left[\frac{x^8}{128} + \frac{3x^2}{32} - \frac{1}{32x^4}\right]_1^2 = \frac{1179\pi}{256}.$$

6.5.9 $f'(x) = 3x^2$, $[f'(x)]^2 = 9x^4$, $1 + [f'(x)]^2 = 1 + 9x^4$

$$S = 2\pi \int_0^1 x^3 \sqrt{1 + 9x^4}\, dx = \frac{\pi}{18} \int_1^{10} u^{1/2} du \text{ where } u = 1 + 9x^4$$

$$= \frac{\pi}{18} \left[\frac{3}{2}u^{3/2}\right]_1^{10} = \frac{\pi}{27}(10\sqrt{10} - 1).$$

6.5.10 $f'(x) = x^2 - \dfrac{1}{4x^2}$, $[f'(x)]^2 = x^4 - \dfrac{1}{2} + \dfrac{1}{16x^4}$,

$$1 + [f'(x)]^2 = 1 + x^4 - \frac{1}{2} + \frac{1}{16x^4} = x^4 + \frac{1}{2} + \frac{1}{16x^4}$$

$$S = 2\pi \int_1^2 \left(\frac{x^3}{3} + \frac{1}{4x}\right) \sqrt{x^4 + \frac{1}{2} + \frac{1}{16x^4}}\, dx = 2\pi \int_1^2 \left(\frac{x^3}{3} + \frac{1}{4x}\right) \sqrt{\left(x^2 + \frac{1}{4x^2}\right)^2}\, dx$$

$$= 2\pi \int_1^2 \left(\frac{x^3}{3} + \frac{1}{4x}\right)\left(x^2 + \frac{1}{4x^2}\right) dx = 2\pi \int_1^2 \left(\frac{x^5}{3} + \frac{x}{3} + \frac{1}{16x^3}\right) dx$$

$$= 2\pi \left[\frac{x^6}{18} + \frac{x^2}{6} - \frac{1}{32x^2}\right]_1^2 = \frac{515\pi}{64}.$$

6.5.11 Let $f(x) = \sqrt{r^2 - x^2}$, $f'(x) = \dfrac{-x}{\sqrt{r^2 - x^2}}$, $[f'(x)]^2 = \dfrac{x^2}{r^2 - x^2}$,

$$1 + [f'(x)]^2 = 1 + \frac{x^2}{r^2 - x^2}$$

$$S = 2\pi \int_{-r}^r \sqrt{r^2 - x^2}\sqrt{1 + \frac{x^2}{r^2 - x^2}}\, dx = 2\pi r \int_{-r}^r dx = \Big[2\pi r x\Big]_{-r}^r = 4\pi r^2.$$

6.5.12 $g'(y) = \dfrac{-y}{\sqrt{16 - y^2}}$, $[g'(y)]^2 = \dfrac{y^2}{16 - y^2}$,

$$1 + [g'(y)]^2 = 1 + \frac{y^2}{16 - y^2} = \frac{16}{16 - y^2}$$

$$S = 2\pi \int_0^3 \sqrt{16 - y^2}\sqrt{\frac{16}{16 - y^2}}\, dy = 8\pi \int_0^3 dy = \Big[8\pi y\Big]_0^3 = 24\pi.$$

6.5.13 $f'(x) = \dfrac{x^2}{4} - \dfrac{1}{x^2}$, $[f'(x)]^2 = \dfrac{x^4}{16} - \dfrac{1}{2} + \dfrac{1}{x^4}$.

$$1 + [f'(x)]^2 = 1 + \frac{x^4}{16} - \frac{1}{2} + \frac{1}{x^4} = \frac{x^4}{16} + \frac{1}{2} + \frac{1}{x^4}$$

$$S = 2\pi \int_1^2 \left(\frac{x^3}{12} + \frac{1}{x}\right) \sqrt{\frac{x^4}{16} + \frac{1}{2} + \frac{1}{x^4}}\, dx = 2\pi \int_1^2 \left(\frac{x^3}{12} + \frac{1}{x}\right) \sqrt{\left(\frac{x^2}{4} + \frac{1}{x^2}\right)^2}\, dx$$

$$= 2\pi \int_1^2 \left(\frac{x^3}{12} + \frac{1}{x}\right)\left(\frac{x^2}{4} + \frac{1}{x^2}\right) dx = 2\pi \int_1^2 \left(\frac{x^5}{48} + \frac{x}{3} + \frac{1}{x^3}\right) dx$$

$$= 2\pi \left[\frac{x^6}{288} + \frac{x^2}{6} - \frac{1}{2x^2}\right]_1^2 = \frac{35\pi}{16}.$$

6.5.14 $g'(y) = \dfrac{y^3}{2} - \dfrac{1}{2y^3}$, $[g'(y)]^2 = \dfrac{y^6}{4} - \dfrac{1}{2} + \dfrac{1}{4y^6}$,

$$1 + [g'(y)]^2 = 1 + \frac{y^6}{4} - \frac{1}{2} + \frac{1}{4y^6} = \frac{y^6}{4} + \frac{1}{2} + \frac{1}{4y^6}$$

$$S = 2\pi \int_1^2 \left(\frac{y^4}{8} + \frac{1}{4y^2} \right) \sqrt{\frac{y^6}{4} + \frac{1}{2} + \frac{1}{4y^6}}\, dy = 2\pi \int_1^2 \left(\frac{y^4}{8} + \frac{1}{4y^2} \right) \sqrt{\left(\frac{y^3}{2} + \frac{1}{2y^3} \right)^2}\, dy$$

$$= 2\pi \int_1^2 \left(\frac{y^4}{8} + \frac{1}{4y^2} \right) \left(\frac{y^3}{2} + \frac{1}{2y^3} \right) dy$$

$$S = 2\pi \int_1^2 \left(\frac{y^7}{16} + \frac{3y}{16} + \frac{1}{8y^5} \right) dy = 2\pi \left[\frac{y^8}{128} + \frac{3y^2}{32} - \frac{1}{32y^4} \right]_1^2 = \frac{1179\pi}{256}.$$

6.5.15 $f'(x) = \dfrac{-1}{2\sqrt{9-x}}$, $[f'(x)]^2 = \dfrac{1}{4(9-x)}$,

$$1 + [f'(x)]^2 = 1 + \frac{1}{4(9-x)} = \frac{37 - 4x}{4(9-x)}$$

$$S = 2\pi \int_3^7 \sqrt{9-x} \sqrt{\frac{37 - 4x}{4(9-x)}}\, dx = \pi \int_3^7 \sqrt{37 - 4x}\, dx$$

$$= \frac{\pi}{4} \int_9^{25} u^{1/2} du \text{ where } u = 37 - 4x,$$

$$= \frac{\pi}{4} \left[\frac{2}{3} u^{3/2} \right]_9^{25} = \frac{49\pi}{3}.$$

6.5.16 $g'(y) = \dfrac{-1}{2\sqrt{16-y}}$, $[g'(y)]^2 = \dfrac{1}{4(16-y)}$,

$$1 + [g'(y)]^2 = 1 + \frac{1}{4(16-y)} = \frac{65 - 4y}{4(16-y)}$$

$$S = 2\pi \int_4^{10} \sqrt{16-y} \sqrt{\frac{65 - 4y}{4(16-y)}}\, dy = \pi \int_4^{10} \sqrt{65 - 4y}\, dy$$

$$= \frac{\pi}{4} \int_{25}^{49} u^{1/2} du \text{ where } u = 65 - 4y$$

$$= \frac{\pi}{6} \left[u^{3/2} \right]_{25}^{49} = \frac{109\pi}{3}.$$

6.5.17 $f'(x) = \dfrac{-x}{\sqrt{25 - x^2}}$, $[f'(x)]^2 = \dfrac{x^2}{25 - x^2}$, $1 + [f'(x)]^2 = 1 + \dfrac{x^2}{25 - x^2} = \dfrac{25}{25 - x^2}$

$$S = 2\pi \int_0^3 \sqrt{25 - x^2} \sqrt{\frac{25}{25 - x^2}}\, dx = 10\pi \int_0^3 dx = 30\pi$$

6.5.18 $g'(y) = \dfrac{-1}{2\sqrt{36-y}}$, $[g'(y)]^2 = \dfrac{1}{4(36-y)}$, $1 + [g'(x)]^2 = 1 + \dfrac{1}{4(36-y)} = \dfrac{145-4y}{4(36-y)}$

$$S = 2\pi \int_6^{16} \sqrt{36-y}\sqrt{\dfrac{145-4y}{4(36-y)}}\, dy = \pi \int_6^{16} \sqrt{145-4y}\, dy$$

$$= \dfrac{\pi}{4}\int_{81}^{121} u^{\frac{1}{2}}\, du = \dfrac{\pi}{4}\left[\dfrac{2}{3}u^{3/2}\right]_{81}^{121} = \dfrac{301\pi}{3}$$

SECTION 6.6

6.6.1   A stone is thrown downward from the top of a 160 foot high cliff with an initial velocity of 48 feet per second. What is the speed of the stone upon impact with the ground?

6.6.2   A projectile is fired downward from a height of 128 feet and reaches the ground in 2 seconds. What was its initial velocity?

6.6.3   A projectile is launched vertically upward from the ground with an initial velocity of 80 feet per second.

    (a)   How long does it take the projectile to reach the ground?
    (b)   When is the projectile 64 feet above the ground?
    (c)   What is the velocity of the projectile when it is 64 feet above the ground?

6.6.4   A playful student drops your math book from a dormitory window and it hits the ground in 3 seconds. How high up is the window?

6.6.5   A particle is moving so that at any time, its acceleration is equal to $10t$ for $t \geq 0$. At the end of 3 seconds, the particle has moved 105 feet. What is its velocity at the end of 3 seconds?

6.6.6   A projectile fired upward from the ground is to reach 144 feet.

    (a)   What must be its initial velocity?
    (b)   What is the velocity of the projectile when it is 80 feet above the ground?

6.6.7   Find the position, velocity, speed, and acceleration at time $t = 1$ second of a particle if $v(t) = 2t - 4$; $s = 3$ when $t = 0$.

6.6.8   A ball is rolled across a level floor with an initial velocity of 28 feet per second. How far will the ball roll if the speed diminishes by 4 feet/sec^2 due to friction?

6.6.9   A particle, initially moving at 16 ft/sec. is slowing down at the rate of 0.8 ft/sec^2. How far will the particle travel before coming to rest?

6.6.10   A projectile is fired vertically upward from a point 20 feet above the ground with a velocity of 40 feet per second. Find the speed of the projectile when it is 36 feet above the ground.

6.6.11 A rapid transit trolley moves with a constant acceleration and covers the distance between two points 300 feet apart in 8 seconds. Its velocity as it passes the second point is 50 feet per second.

(a) What is its acceleration?

(b) What is the velocity of the trolley as it passes the first point?

6.6.12 A jet plane travels from rest to a velocity of 300 feet per second in a distance of 450 feet. What is its constant acceleration?

6.6.13 A particle is moving so that its velocity, $v(t) = t^2 - t - 2$ for $0 \leq t \leq 3$. Find the displacement and total distance travelled by the particle.

6.6.14 A particle is moving so that its velocity, $v(t) = 4 - t$ for $0 \leq t \leq 6$. Find the displacement and total distance travelled by the particle.

6.6.15 A particle is moving so that its velocity, $v(t) = 8 - 2t$ for $0 \leq t \leq 5$. Find the displacement and total distance travelled by the particle.

6.6.16 A particle is moving so that its velocity, $v(t) = t^2 - 3t + 2$ for $0 \leq t \leq 3$. Find the displacement and total distance travelled by the particle.

6.6.17 A particle is moving so that its velocity, $v(t) = t^2 - 4t + 3$ for $0 \leq t \leq 4$. Find the displacement and total distance travelled by the particle.

6.6.18 A particle is moving so that its velocity, $v(t) = t - 8/t^2$ for $1 \leq t \leq 3$. Find the displacement and total distance travelled by the particle.

6.6.19 The graph of a velocity function over the interval $[t_1, t_2]$ is as shown.

(a) Is the acceleration positive or is it negative?

(b) Is the acceleration increasing or is it decreasing?

(c) Is the displacement positive or is it negative?

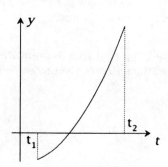

SOLUTIONS

SECTION 6.6

6.6.1 $s(t) = 0$ upon impact with the ground.

$s(t) = -16t^2 - 48t + 160 = -16(t^2 + 3t - 10) = -16(t+5)(t-2)$

$s(t) = 0$ when $t = 2$ sec.

$v(t) = -32t - 48$; $v(2) = -32(2) - 48 = -112$, the speed at impact is 112 ft/sec.

6.6.2 $s = 128$ when $t = 0$, so $s(t) = -16t^2 + v_0 t + 128 = 0$.

$s = 0$ when $t = 2$, $-16(2)^2 + v_0(2) + 128 = 0$

$v_0 = -32$ ft/sec.

6.6.3 **(a)** $s(t) = 0$ when the projectile hits the ground,

$s(t) = -16t^2 + 80t = -16t(t-5)$

$s(t) = 0$ when $t = 0$ and when $t = 5$ seconds.

(b) $-16t^2 + 80t = 64$

$-16(t^2 - 5t + 4) = -16(t-4)(t-1)$

The projectile is 64 feet above the ground when $t = 1$ and $t = 4$ seconds.

(c) $v(t) = -32t + 80$

$v(1) = -32(1) + 80 = 48$ ft/sec

$v(4) = -32(4) + 80 = -48$ ft/sec.

6.6.4 Let $h =$ height of the dormitory window, then, $s = h$ and $v = 0$ when $t = 0$, thus, $s(t) = -16t^2 + h$

$s(3) = 0$, so $-16(3)^2 + h = 0$ $h = 144$ feet.

6.6.5 $a(t) = 10t$

$v(t) = \displaystyle\int 10t\, dt = 5t^2 + C_1$, $v(0) = v_0$, so $C_1 = v_0$ and

$v(t) = 5t^2 + v_0$

$s(t) = \displaystyle\int (5t^2 + v_0)dt = \frac{5}{3}t^3 + v_0 t + C_2$

$s(0) = 0$, so, $C_2 = 0$

$s(3) = 105$ so $\dfrac{5}{3}(3)^3 + 3v_0 = 105$

$v_0 = 20$

thus $v(t) = 5t^2 + 20$

$v(3) = 5(3)^2 + 20 = 65$ ft/sec.

6.6.6   **(a)** $s(t) = -16t^2 + v_0 t$

$v(t) = -32t + v_0$

$v = 0$ when the projectile is at its maximum height, thus,

$-32t + v_0 = 0,$

$$t = \frac{v_0}{32}$$

$s(t) = 144$ feet when $t = \dfrac{v_0}{32}$ sec so $-16\left(\dfrac{v_0}{32}\right)^2 + v_0\left(\dfrac{v_0}{32}\right) = 144$

$v_0^2 = (64)(144)$

$v_0 = 96$ ft/sec (positive since fired upward),

     (b) $-16t^2 + 96t = 80$

$-16(t^2 - 6t + 5) = -16(t - 5)(t - 1) = 0$

The projectile is 80 feet above the ground when $t = 1$ and $t = 5$ seconds.

$v(t) = -32t + 96, \qquad v(1) = -32(1) + 96 = 64$ ft/sec,

$v(5) = -32(5) + 96 = -64$ ft/sec.

6.6.7 $s(t) = \displaystyle\int (2t - 4)dt = t^2 - 4t + C_1; \; s(0) = 3; \; c_1 = 3$

$s(t) = t^2 - 4t + 3, \; s(1) = (1)^2 - 4(1) + 3 = 0$

$v(1) = 2(1) - 4 = -2$

$|v(1)| = |-2| = 2$

$a(t) = \dfrac{dv}{dt} = 2, \; a(1) = 2.$

6.6.8 $v(t) = \displaystyle\int -4\,dt = -4t + C_1; \; v(0) = 28$ so $C_1 = 28, \; v(t) = -4t + 28$

$s(t) = \displaystyle\int (-4t + 28)dt = -2t^2 + 28t + C_2,$ if $s(0) = 0, \; C_2 = 0$ and

$s(t) = -2t^2 + 28t$

The ball comes to rest when $v = 0,$ so $-4t + 28 = 0, \; t = 7$ sec, thus

$s(7) = -2(7)^2 + 28(7) = 98.$ The ball rolls 98 feet before coming to rest.

6.6.9 $v(t) = \displaystyle\int -0.8\,dt = -0.8t + C_1, \; v(0) = 16$ so $C_1 = 16,$

$v(t) = -0.8t + 16$

$s(t) = \displaystyle\int v(t)dt = \int (-0.8t + 16)dt = -0.4t^2 + 16t + C_2,$ if $s(0) = 0$

$C_2 = 0, \; s(t) = -0.4t^2 + 16t$

The particle comes to rest when $v(t) = 0,$ so $-0.8t + 16 = 0, \; t = 20$ sec; thus,

$s(20) = -0.4(20)^2 + 16(20) = 160.$ The particle travels 160 feet before coming to rest.

6.6.10 $s = 20$ when $t = 0$, so, $s(t) = -16t^2 + 40t + 20$. When $s = 36$, $s(t) = -16t^2 + 40t + 20 = 36$
or $-8(2t-1)(t-2) = 0$; thus, the projectile is 36 feet above the ground when $t = 1/2$ second
and $t = 2$ second. $v(t) = -32t + 40$ so speed $= |-32(1/2) + 40| = |-32(2) + 40| = 24$ ft/sec.

6.6.11   Let a be the acceleration of the trolley, so that $v(t) = \int a\,dt$. $v(t) = at + c_1$. When $t = 0$,
$v(0) = v_0$ so that $c_1 = v_0$ and $v(t) = at + v_0$

$$s(t) = \int v(t)dt = \int (at + v_0)dt = \frac{at^2}{2} + v_0 t + C_2$$

Let $s(0) = 0$ so $c_2 = 0$ and $s(t) = \frac{at^2}{2} + v_0 t$

When $t = 8$ secs, $v = 50$ ft/sec and $s = 300$ ft so

$$\left.\begin{array}{r}\frac{a}{2}(8)^2 + v_0(8) = 300 \\[2mm] 8a + v_0 = 50\end{array}\right\} \text{ or } \left.\begin{array}{r}8a + 2v_0 = 75 \\[2mm] 8a + c_0 = 50\end{array}\right\} \text{ so that } a = \frac{25}{8} \text{ and } v_0 = 25$$

(a)   the acceleration of the trolley is $\dfrac{25}{8}$ ft/sec^2

(b)   the velocity of the trolley as it passes the first point is 25 ft/sec.

6.6.12   Let $a =$ acceleration of the jet plane so that $v(t) = \int a\,dt$, $v(t) = at + C_1$. When $t = 0$, $v = 0$
so $C_1 = 0$ and $v(t) = at$, $s(t) = \frac{at^2}{2} + C_2$. Let $s(0) = 0$ so $C_2 = 0$ and $s(t) = \frac{at^2}{2}$, thus, $v = at$
and $s = \frac{at^2}{2}$. When $s = 450$, $v = 300$, so $300 = at$, $t = \frac{300}{a}$ and $450 = \frac{a}{2}\left(\frac{300}{a}\right)^2$ or $a = 100$.
The acceleration of the jet plane is 100 ft/sec^2.

6.6.13   displacement $= \displaystyle\int_0^3 (t^2 - t - 2)dt = \left.\frac{t^3}{3} - \frac{t^2}{2} - 2t\right|_0^3 = -\frac{3}{2}$

distance $= \displaystyle\int_0^3 |t^2 - t - 2|dt = \int_0^2 -(t^2 - t - 2)dt + \int_2^3 (t^2 - t - 2)dt = \frac{31}{6}$.

6.6.14   displacement $= \displaystyle\int_0^6 (4 - t)dt = \left.4t - \frac{t^2}{2}\right|_0^6 = 6$

distance $= \displaystyle\int_0^6 |4 - t|dt = \int_0^4 (4 - t)dt + \int_4^6 -(4 - t)dt = 10$.

6.6.15   displacement $= \displaystyle\int_0^5 (8 - 2t)dt = \left.8t - \frac{2t^2}{2}\right|_0^5 = 15$

distance $= \displaystyle\int_0^5 |8 - 2t|dt = \int_0^4 (8 - 2t)dt + \int_4^5 -(8 - 2t)dt = 17$.

6.6.16 $\text{displacement} = \int_0^3 (t^2 - 3t + 2)dt = \dfrac{t^3}{3} - \dfrac{3t^2}{2} + 2t \Big]_0^3 = \dfrac{3}{2}$

$\text{distance} = \int_0^1 (t^2 - 3t + 2)dt + \int_1^2 -(t^2 - 3t + 2)dt + \int_2^3 (t^2 - 3t + 2)dt = \dfrac{11}{6}.$

6.6.17 $\text{displacement} = \int_0^4 (t^2 - 4t + 3)dt = \dfrac{t^3}{3} - \dfrac{4t^2}{2} + 3t \Big]_0^4 = \dfrac{4}{3}$

$\text{distance} = \int_0^4 |t^2 - 4t + 3|dt$

$= \int_0^1 (t^2 - 4t + 3)dt + \int_1^3 -(t^2 - 4t + 3)dt + \int_3^4 (t^2 - 4t + 3)dt = 4.$

6.6.18 $\text{displacement} = \int_1^3 \left(t - \dfrac{8}{t^2}\right) dt = \dfrac{t^2}{2} + \dfrac{8}{t} \Big]_1^3 = -\dfrac{4}{3}$

$\text{distance} = \int_1^3 \left|t - \dfrac{8}{t^2}\right| dt = \int_1^2 -\left(t - \dfrac{8}{t^2}\right) dt + \int_2^3 \left(t - \dfrac{8}{t^2}\right) dt = \dfrac{11}{3}.$

6.6.19 **(a)** acceleration is positive
      (b) acceleration is increasing
      (c) displacement is positive

SECTION 6.7

6.7.1 A spring exerts a force of 1N when stretched 5 cm. How much work, in Joules, is required to stretch the spring from a length of 10 cm to a length of 20 cm?

6.7.2 A spring whose natural length is 2.5 m exerts a force of 100 N when stretched 40 cm. How much work, in Joules, is required to stretch the spring from its natural length to 4 m?

6.7.3 A spring exerts a force of 1 ton when stretched 10 feet beyond its natural length. How much work is required to stretch the spring 8 feet beyond its natural length?

6.7.4 A spring whose natural length is 18 inches exerts a force of 10 pounds when stretched 16 inches. How much work is required to stretch the spring 4 inches beyond its natural length?

6.7.5 A dredger scoops a shovel full of mud weighing 2000 pounds from the bottom of a river at a constant rate. Water leaks out uniformly at such a rate that half the weight of the contents is lost when the scoop has been lifted 25 feet. How much work is done by the dredger in lifting the mud this distance?

6.7.6 A 60 foot length of steel chain weighing 10 pounds per foot is hanging from the top of a building. How much work is required to pull half of it to the top?

6.7.7 A 50 foot chain weighing 10 pounds per foot supports a steel beam weighing 1000 pounds. How much work is done in winding 40 feet of the chain onto a drum.

6.7.8 A bucket weighing 1000 pounds is to be lifted from the bottom of a shaft 20 feet deep. The weight of the cable used to hoist it is 10 pounds per foot. How much work is done lifting the bucket to the top of the shaft?

6.7.9 A cylindrical tank 8 feet in diameter and 10 feet high is filled with water weighing 62.4 lbs/ft^3. How much work is required to pump the water over the top of the tank?

6.7.10 A cylindrical tank is to be filled with gasoline weighing 50 lbs/ft^3. If the tank is 20 feet high and 10 feet in diameter, how much work is done by the pump in filling the tank through a hole in the bottom of the tank?

6.7.11 A cylindrical tank 5 feet in diameter and 10 feet high is filled with oil whose density is 48 lbs/ft^3. How much work is required to pump the oil over the top of the tank?

6.7.12 A conical tank has a diameter of 9 feet and is 12 feet deep. If the tank is filled with water of density 62.4 lbs/ft^3, how much work is required to pump the water over the top?

6.7.13 A conical tank has a diameter of 8 feet and is 10 feet deep. If the tank is filled to a depth of 6 feet with water of density 62.4 lbs/ft^3, how much work is required to pump the water over the top?

SOLUTIONS

SECTION 6.7

6.7.1 $F(x) = kx$; $F\left(\dfrac{1}{0.05}\right) = 1$, $k = 20$ N/m

$$W = \left[\int_{0.1}^{0.2} 20x\, dx = 10x^2\right]_{0.1}^{0.2} = 0.3j.$$

6.7.2 $F(x) = kx$; $k = \dfrac{F}{x} = \dfrac{100}{0.4} = 250$ N/m

$$W = \int_{2.5}^{4} 250x\, dx = 250\dfrac{x^2}{2}\Bigg]_{2.5}^{4} = 1218.75j.$$

6.7.3 $F(x) = kx$, $F(10) = 10k = 2000$, $k = 200$ lbs/ft

$$W = \int_{0}^{8} 200x\, dx = 200 \int_{0}^{8} x\, dx = 200 \left[\dfrac{x^2}{2}\right]_{0}^{8} = 6400 \text{ ft-lbs.}$$

6.7.4 $F(x) = kx$, $F\left(\dfrac{16}{12}\right) = \dfrac{16}{12}k = 10$, $k = \dfrac{15}{2}$ lbs/ft

$$W = \int_{0}^{4/12} \dfrac{15}{2}x\, dx = \dfrac{15}{2} \int_{0}^{1/3} x\, dx = \dfrac{15}{2} \left[\dfrac{x^2}{2}\right]_{0}^{1/3} = \dfrac{5}{12} \text{ ft-lbs.}$$

6.7.5   Weight $= 2000 - \dfrac{x}{25}(1000) = 40(50 - x)$

$$W = \int_{0}^{25} 40(50 - x)dx = 40 \int_{0}^{25} (50 - x)dx = 40 \left[50x - \dfrac{x^2}{2}\right]_{0}^{25} = 37,500 \text{ ft-lbs.}$$

6.7.6 $W = \displaystyle\int_{0}^{30} 10(60 - x)dx$

$$= 10 \int_{0}^{30} (60 - x)dx$$

$$= 10 \left[60x - \dfrac{x^2}{2}\right]_{0}^{30}$$

$$= 13,500 \text{ ft-lbs.}$$

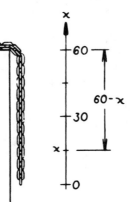

6.7.7 Total $wt = wt$ of beam $+ wt$ of chain
$$= 1000 + 10(50 - x) = 10(150 - x)$$

$$W = \int_0^{40} 10(150 - x)dx$$

$$= 10 \int_0^{40} (150 - x)dx$$

$$= 10 \left[150x - \frac{x^2}{2} \right]_0^{40} = 52,000 \text{ ft lbs.}$$

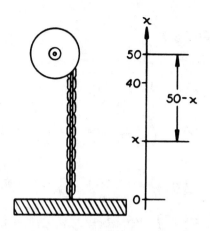

6.7.8 Total weight $=$ weight of bucket $+$ weight of cable
$$= 1000 + 10(20 - x) = 10(120 - x)$$

$$W = \int_0^{20} 10(120 - x)dx = 10 \int_0^{20} (120 - x)dx = 10 \left[120x - \frac{x^2}{2} \right]_0^{20} = 22,000 \text{ ft-lbs.}$$

6.7.9 $W = \int_0^{10} (10 - x)62.4(16\pi)dx$

$$= 998.4\pi \int_0^{10} (10 - x)dx$$

$$= 998.4\pi \left[10x - \frac{x^2}{2} \right]_0^{10}$$

$$= 49,920\pi \text{ ft-lbs.}$$

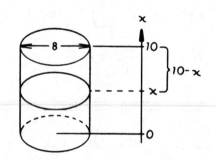

6.7.10 $W = \int_0^{20} x(50)(25\pi)dx$

$$= 1250\pi \int_0^{20} x\, dx$$

$$= 1250\pi \left[\frac{x^2}{2} \right]_0^{20} = 250,000\pi \text{ ft-lbs.}$$

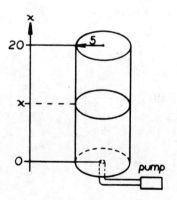

6.7.11 $W = \displaystyle\int_0^{10} (10-x)(48)\left(\dfrac{5}{2}\right)^2 \pi\, dx$

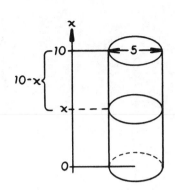

$\qquad\qquad = 300\pi \displaystyle\int_0^{10} (10-x)\, dx$

$\qquad\qquad = 300\pi \left[10x - \dfrac{x^2}{2}\right]_0^{10} = 15{,}000\pi \text{ ft-lbs.}$

6.7.12 By similar triangles, $\dfrac{r}{9/2} = \dfrac{x}{12}$, $r = \dfrac{3x}{8}$

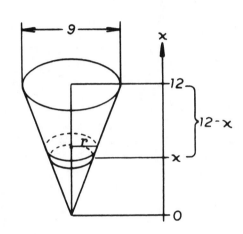

$\qquad\quad W = \displaystyle\int_0^{12} (12-x)62.4\pi \left(\dfrac{3x}{8}\right)^2 dx$

$\qquad\qquad = \dfrac{561.6\pi}{64} \displaystyle\int_0^{12} (12x^2 - x^3)\, dx$

$\qquad\qquad = \dfrac{561.6\pi}{64} \left[\dfrac{12x^3}{3} - \dfrac{x^4}{4}\right]_0^{12}$

$\qquad\qquad = 15163.2\pi \text{ ft-lbs.}$

6.7.13 By similar triangles, $\dfrac{r}{4} = \dfrac{x}{10}$, $r = \dfrac{2}{5}x$

$\qquad\quad W = \displaystyle\int_0^6 (10-x)62.4\pi \left(\dfrac{2x}{5}\right)^2 dx = \dfrac{249.6\pi}{25} \displaystyle\int_0^6 (10x^2 - x^3)\, dx = 3953.7\pi \text{ ft-lbs.}$

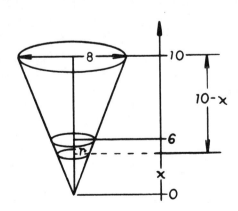

SECTION 6.8

6.8.1   A flat rectangular plate, 6 feet long and 3 feet wide is submerged in water (weight density 62.4 lbs/ft^3) with the 3 foot edge parallel to and 2 feet below the surface. Find the force against the surface of the plate.

6.8.2   A flat rectangular plate, 6 meters long and 3 meters wide is submerged in water (weight density 9810 N/m^3) with the 6 meter edge parallel to and 2 meters below the surface. Find the force against the surface of the plate.

6.8.3   A flat triangular plate whose dimensions are 5, 5, and 6 feet is submerged in water (weight density 62.4 lbs/ft^3) so that its longer side is at the surface and parallel to it. Find the force against the surface of the plate.

6.8.4   A flat triangular plate whose dimensions are 5, 5, and 6 feet is submerged in water (weight density 62.4 lbs/ft^3) so that its longer side is below the surface and parallel to it and its vertex is 2 feet below the water. Find the force against the surface of the plate.

6.8.5   A flat plate, shaped in the form of a semicircle 6 meters in diameter is submerged in water (weight density 9810 N/m^3) as shown. Find the force against the surface of the plate.

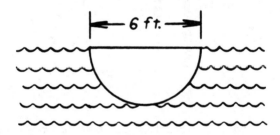

6.8.6   A horizontal cylindrical tank of diameter 8 feet is half full of a chemical (weight density 50 lbs/ft^3). Calculate the force against one end.

6.8.7   Liquid cement (weight density 250 lbs/ft^3) is poured into a form whose ends are 5 foot squares. Find the force on one end if the cement is 4 feet deep.

6.8.8 A trapazoidal gate in a dam is submerged in water (weight density 62.4 lbs/ft³) as shown in the figure. Find the force on the gate when the surface of the water is 2 feet above the top of the gate.

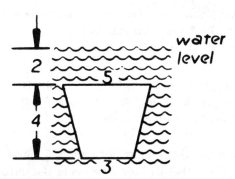

6.8.9 A trapazoidal gate in a dam is submerged in water (weight density 62.4 lbs/ft³) as shown in the figure. Find the force on the gate when the surface of the water is 2 feet above the top of the gate.

6.8.10 A flat plate in the form of a T is submerged in water (weight density 9810 N/m³) as shown in the figure. Find the force on the surface when the surface of the water is 2 meters above the cross piece.

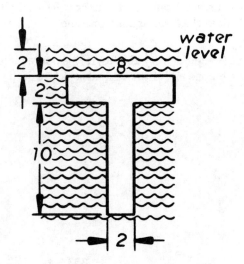

6.8.11 A swimming pool is 30 feet long and 15 feet wide. The bottom is flat but inclined as shown in the figure. The water is 10 feet deep on one end and 2 feet deep on the other. Find the force on one of the sides when the pool is filled with water (weight density 62.4 lbs/ft^3).

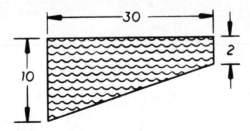

6.8.12 A swimming pool is 30 feet long and 15 feet wide. The bottom is flat but inclined as shown in the figure. The water is 10 feet deep on one end and 2 feet deep on the other end. Find the force on the bottom of the pool when it is filled with water (weight density 62.4 lbs/ft^3).

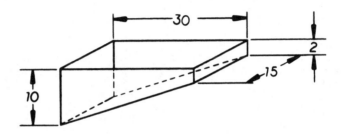

6.8.13 The face of the dam shown in the figure is an inclined rectangle. Find the fluid force on the face when the water (weight density 62.4 lbs/ft^3) is level with the top of the dam.

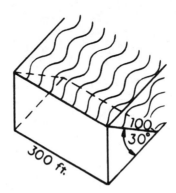

SOLUTIONS

SECTION 6.8

6.8.1 $F = \int_2^8 62.4x(3)dx$

$\qquad = 187.2 \int_2^8 x\, dx$

$\qquad = 187.2 \left(\dfrac{x^2}{2}\right)\Big]_2^8 = 5616$ lbs.

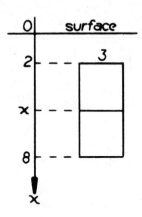

6.8.2 $F = \int_2^5 9810x(6)dx$

$\qquad = 58860 \int_2^5 x\, dx$

$\qquad = 58860 \left[\dfrac{x^2}{2}\right]_2^5 = 618030$ N.

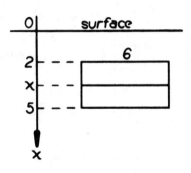

6.8.3 By similar triangles, $\dfrac{4-x}{4} = \dfrac{W(x)}{6}$,

$\qquad W(x) = 6 - \dfrac{3}{2}x$

$\qquad F = \int_0^4 (62.4)x\left(6 - \dfrac{3}{2}x\right) dx$

$\qquad = 62.4 \int_0^4 \left(6x - \dfrac{3x^2}{2}\right) dx$

$\qquad = 62.4 \left[\dfrac{6x^2}{2} - \dfrac{3}{2}\dfrac{x^3}{3}\right]_0^4 = 998.4$ lbs.

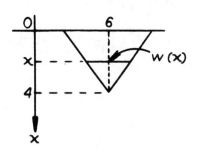

6.8.4 By similar triangles, $\dfrac{x}{4} = \dfrac{W(x)}{6}$,

$$W(x) = \frac{3x}{2}$$

$$F = \int_2^6 (62.4)x\left(\frac{3x}{2}\right)dx$$

$$= 93.6 \int_2^6 x^2 dx$$

$$= 93.6 \left[\frac{x^3}{3}\right]_2^6 = 6489.6 \text{ lbs.}$$

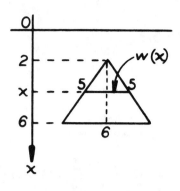

6.8.5 $F = \displaystyle\int_0^3 (9810)(x)(2\sqrt{9-x^2})dx$

$$= 19620 \int_0^3 x\sqrt{9-x^2}\,dx$$

$$= \frac{19620}{2} \int_0^9 u^{1/2}du \text{ where } u = 9 - x^2$$

$$= \frac{19620}{2}\left(\frac{2}{3}\right)\left[u^{3/2}\right]_0^9 = 176580 \text{ N.}$$

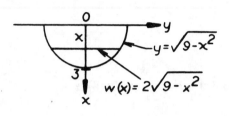

6.8.6 $F = \displaystyle\int_0^4 (50)(x)(2\sqrt{16-x^2})dx$

$$= 100 \int_0^4 x\sqrt{16-x^2}\,dx$$

$$= \frac{100}{2}\int_0^{16} u^{1/2}du \text{ where } u = 16 - x^2$$

$$= (50)\left(\frac{2}{3}\right)\left[u^{3/2}\right]_0^{16} = 2133.3 \text{ lbs.}$$

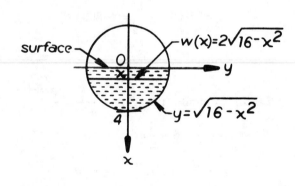

6.8.7 $F = \displaystyle\int_0^4 (250)(x)(5)dx$

$$= 1250 \int_0^4 x\,dx$$

$$= 1250\left[\frac{x^2}{2}\right]_0^4$$

$$= 10,000 \text{ lbs.}$$

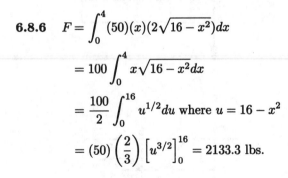

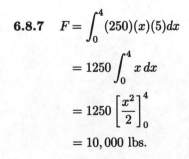

6.8.8 $W(x) = 3 + 2u(x)$; by similar triangles,

$$\frac{u(x)}{1} = \frac{6-x}{4} \text{ so } u(x) = \frac{1}{4}(6-x) \text{ and}$$

$$W(x) = 3 + 2\left[\frac{1}{4}(6-x)\right] = 6 - \frac{x}{2}$$

$$F = \int_2^6 62.4x\left(6 - \frac{x}{2}\right) dx$$

$$= 62.4 \int_2^6 \left(6x - \frac{x^2}{2}\right) dx$$

$$= 62.4 \left[\frac{6x^2}{2} - \frac{1}{2}\frac{x^3}{3}\right]_2^6$$

$$= 4492.8 \text{ lbs.}$$

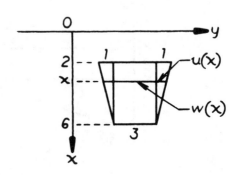

6.8.9 $W(x) = 3 + 2u(x)$, by similar triangles

$$\frac{u(x)}{1} = \frac{x}{4} \text{ so } u(x) = \frac{x}{4} \text{ and}$$

$$W(x) = 3 + 2 \cdot \frac{x}{4} = 3 + \frac{x}{2}$$

$$F = \int_0^4 62.4(2+x)\left(3 + \frac{x}{2}\right) dx$$

$$= 62.4 \int_0^4 \left(6 + 4x + \frac{x^2}{2}\right) dx$$

$$F = 62.4 \left[6x + 4\frac{x^2}{2} + \frac{1}{2}\frac{x^3}{3}\right]_0^4$$

$$= 4160 \text{ lbs.}$$

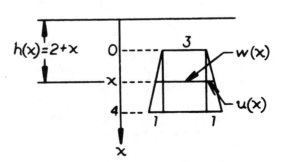

6.8.10 $F = \int_2^4 (9810)(x_1)(8)dx + \int_4^{14} (9810)x_2(2)dx$

$$= 2,246,680 \text{ N.}$$

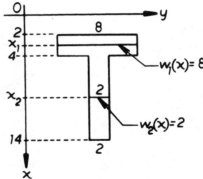

6.8.11   The equation of the line through $(2, 30)$ and $(10, 0)$ is

$$y = W_2(x) = \frac{15}{4}(10 - x)$$

$$F = \int_0^2 (62.4)(x_1)(30)dx_1 + \int_2^{10} (62.4)(x_2)\frac{15}{4}(10 - x_2)dx_2 = 38688 \text{ lbs}$$

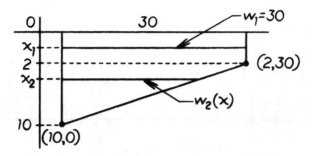

6.8.12   The length along the bottom inclined plane is

$$\sqrt{30^2 + 8^2} = 964 = 2\sqrt{241}$$

so $\dfrac{h(x) - 2}{8} = \dfrac{x}{2\sqrt{241}}$; $h(x) = \dfrac{4}{\sqrt{241}}x + 2$

$$F = \int_0^{2\sqrt{241}} 62.4\left(\frac{4x}{\sqrt{241}} + 2\right)(15)dx$$

$$= 936\int_0^{2\sqrt{241}} \left(\frac{4x}{\sqrt{241}} + 2\right)dx$$

$$= 936\frac{4x}{\sqrt{241}}\left[\frac{1}{2}x^2 + 2x\right]_0^{2\sqrt{241}}$$

$$= 11232\sqrt{241}$$
$$\approx 174367.53 \text{ lbs.}$$

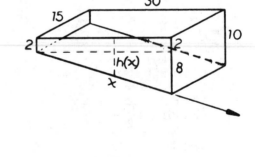

6.8.13 $h(x) = x\sin 30° = \dfrac{x}{2}$

$$F = \int_0^{100} 62.4\frac{x}{2}(300)dx$$

$$= 9360\int_0^{100} x\, dx$$

$$= 9360\left[\frac{x^2}{2}\right]_0^{100}$$

$$= 46,800,000 \text{ lbs.}$$

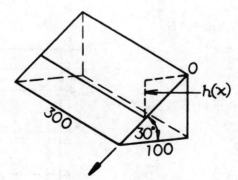

SUPPLEMENTARY EXERCISES, CHAPTER 6

In Exercises 1–3, set up, but do not evaluate, an integral or sum of integrals that gives the area of the region R. (Set up the integral with respect to x or y as directed.)

1. R is the region in the first quadrant enclosed by $y = x^2$, $y = 2 + x$, and $x = 0$.

    (a) Integrate with respect to x.   **(b)** Integrate with respect to y.

2. R is enclosed by $x = 4y - y^2$ and $y = \frac{1}{2}x$.

    (a) Integrate with respect to x.   **(b)** Integrate with respect to y.

3. R is enclosed by $x = 9$ and $x = y^2$.

    (a) Integrate with respect to x.   **(b)** Integrate with respect to y.

In Exercises 4–9, set up, but do not evaluate, an integral or sum of integrals that gives the stated volume. (Set up the integral with respect to x or y as directed.)

4. The volume generated by revolving the region in Exercise 1 about the x-axis.

    (a) Integrate with respect to x.   **(b)** Integrate with respect to y.

5. The volume generated by revolving the region in Exercise 1 about the y-axis.

    (a) Integrate with respect to x.   **(b)** Integrate with respect to y.

6. The volume generated by revolving the region in Exercise 2 about the x-axis.

    (a) Integrate with respect to x.   **(b)** Integrate with respect to y.

7. The volume generated by revolving the region in Exercise 2 about the y-axis.

    (a) Integrate with respect to x.   **(b)** Integrate with respect to y.

8. The volume generated by revolving the region in Exercise 3 about the x-axis.

    (a) Integrate with respect to x.   **(b)** Integrate with respect to y.

9. The volume generated by revolving the region in Exercise 3 about the y-axis.

    (a) Integrate with respect to x.   **(b)** Integrate with respect to y.

In Exercises 10 and 11, find (a) the area of the region described, and (b) the volume generated by revolving the region about the indicated line.

10. The region in the first quadrant enclosed by $y = \sin x$, $y = \cos x$, and $x = 0$; revolved about the x-axis. [*Hint:* $\cos^2 x - \sin^2 x = \cos 2x$.]

11. The region enclosed by the x-axis, the y-axis, and $x = \sqrt{4 - y}$; revolved about the y-axis.

12. Set up a sum of definite integrals that represents the total shaded area between the curves $y = f(x)$ and $y = g(x)$ below.

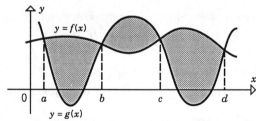

13. Find the *total* area bounded between $y = x^3$ and $y = x$ over the interval $[-1, 2]$. (See preceding exercise.)

In Exercises 14 and 15, find the volume generated by revolving the region described about the axis indicated.

14. The region bounded above by the curve $y = \cos x^2$, on the left by the y-axis, and below by the x-axis; revolved about the y-axis.

15. The region bounded by $y = \sqrt{x}$, $x = 4$, and $y = 0$; revolved about

    (a) the line $x = 4$ **(b)** the line $y = 2$.

16. An American football has the shape of the solid generated by revolving the region bounded between the x-axis and the parabola $y = 4R(x^2 - \frac{1}{2}L^2)/L^2$ about the x-axis. Find its volume.

In Exercises 17–20 find the arc length of the indicated curve.

17. $8y^2 = x^3$ between $(0, 0)$ and $(2, 1)$.   **18.** $y = \dfrac{1}{3}(x^2 + 2)^{3/2}, 0 \le x \le 3$.

19. $y = \dfrac{1}{10}x^5 + \dfrac{1}{6}x^{-3}, 1 \le x \le 2$.   **20.** $y = \dfrac{1}{3}x^3 + \dfrac{1}{4x}, 1 \le x \le 2$.

In Exercises 21 and 22, find the area of the surface generated by revolving the given curve about the indicated axis.

21. $y = x^3$ between $(1, 1)$ and $(2, 8)$; x-axis.

22. $y = \sqrt{2x - x^2}$ between $(\frac{1}{2}, \sqrt{3}/2$ and $(1, 1)$; x-axis.

23. Find the work done in stretching a spring from 8 in. to 10 in. if its natural length is 6 in., and a force of 2 lb is needed to hold it at a length of 10 in.

24. Find the spring constant if 180 in·lb of work are required to stretch a spring 3 in. from its natural length.

25. A 250-lb weight is suspended from a ledge by a uniform 40-ft cable weighing 30 lb. How much work is required to bring the weight up to the ledge?

26. A tank in the shape of a right-circular cone has a 6-ft diameter at the top and a height of 5 ft. It is filled with a liquid of weight density 64 lb/ft^3. How much work can be done by the liquid if it runs out of the bottom of the tank?

27. A vessel has the shape obtained by revolving about the y-axis the part of the parabola $y = 2(x^2-4)$ lying below the x-axis. If x and y are in feet, how much work is required to pump all the water in the full vessel to a point 4 ft above its top?

28. Two like magnetic poles repel each other with a force $F = k/x^2$ newtons, where k is a constant. Express the work needed to move them along a line from D meters apart to $D/3$ meters apart.

In Exercises 29 and 30, the flat surface shown is submerged vertically in a liquid of weight density ρ lb/ft^3. Find the fluid force against the surface.

29.   **30.**

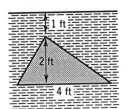

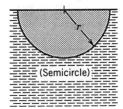

SOLUTIONS

SUPPLEMENTARY EXERCISES, CHAPTER 6

1. (a) $\displaystyle\int_0^2 (x+2-x^2)dx$

    (b) $\displaystyle\int_0^2 \sqrt{y}\,dy + \int_2^4 (\sqrt{y}-y+2)dy$

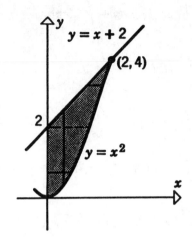

2. (a) solve $x = 4y - y^2$ for y:
    $y^2 - 4y + x = 0$,
    $$y = \frac{4 \pm \sqrt{16-4x}}{2} = 2 \pm \sqrt{4-x}$$
    so the lower boundary of the
    region is $y = 2 - \sqrt{4-x}$ because
    $y \le 2$, and the area is
    $$\int_0^4 (x/2 - 2 + \sqrt{4-x})dx$$

    (b) $\displaystyle\int_0^2 [(4y-y^2)-2y]dy = \int_0^2 (2y-y^2)dy$

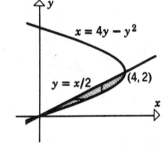

3. (a) $\displaystyle\int_0^9 [\sqrt{x}-(-\sqrt{x})]dx = \int_0^9 2\sqrt{x}\,dx$

    (b) $\displaystyle\int_{-3}^3 (9-y^2)dy$

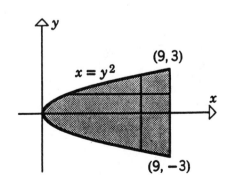

365

4. (a) $\displaystyle\int_0^2 \pi[(x+2)^2 - x^4]dx$

(b) $\displaystyle\int_0^2 2\pi y(\sqrt{y})dy + \int_2^4 2\pi y[\sqrt{y}-(y-2)]dy = \int_0^2 2\pi y^{3/2}dy + \int_2^4 2\pi y(\sqrt{y}-y+2)dy$

5. (a) $\displaystyle\int_0^2 2\pi x(x+2-x^2)dx$ **(b)** $\displaystyle\int_0^2 \pi y\, dy + \int_2^4 \pi[y-(y-2)^2]dy$

6. (a) $\displaystyle\int_0^4 \pi[x^2/4 - (2-\sqrt{4-x})^2]dx$

(b) $\displaystyle\int_0^2 2\pi y[(4y-y^2)-2y]dy = \int_0^2 2\pi y(2y-y^2)dy$

7. (a) $\displaystyle\int_0^4 2\pi x[x/2 - (2-\sqrt{4-x})]dx$ **(b)** $\displaystyle\int_0^2 \pi[(4y-y^2)^2 - 4y^2]dy$

8. (a) $\displaystyle\int_0^9 \pi x\, dx$ **(b)** $\displaystyle\int_0^3 2\pi y(9-y^2)dy$

9. (a) $\displaystyle\int_0^9 2\pi x(2\sqrt{x})dx = \int_0^9 4\pi x^{3/2}dx$ **(b)** $\displaystyle\int_{-3}^3 \pi(81-y^4)dy$

10. (a) $\displaystyle A = \int_0^{\pi/4}(\cos x - \sin x)dx$

$\qquad\quad = \sqrt{2}-1$

(b) $\displaystyle V = \int_0^{\pi/4}\pi(\cos^2 x - \sin^2 x)dx$

$\qquad\quad = \pi\int_0^{\pi/4}\cos 2x\, dx = \pi/2$

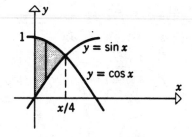

11. (a) $\displaystyle A = \int_0^4 \sqrt{4-y}\, dy = 16/3$

(b) $\displaystyle V = \int_0^4 \pi(4-y)dy = 8\pi$

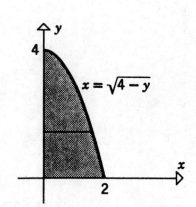

12. $\displaystyle\int_a^b [f(x) - g(x)]dx + \int_b^c [g(x) - f(x)]dx + \int_c^d [f(x) - g(x)]dx$

13. $\displaystyle A = \int_{-1}^0 (x^3 - x)dx + \int_0^1 (x - x^3)dx$

$\displaystyle \qquad + \int_1^2 (x^3 - x)dx$

$\displaystyle \qquad = 1/4 + 1/4 + 9/4 = 11/4$

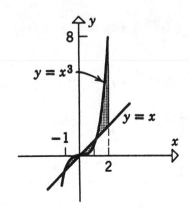

14. $\displaystyle V = \int_0^{\sqrt{\pi/2}} 2\pi x \cos(x^2)dx = \pi$

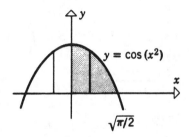

15. **(a)** $\displaystyle V = \int_0^4 2\pi(4 - x)\sqrt{x}\,dx$

$\displaystyle \qquad = 2\pi \int_0^4 (4x^{1/2} - x^{3/2})dx$

$\displaystyle \qquad = 256\pi/15$

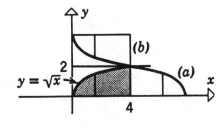

(b) $\displaystyle V = \int_0^4 \pi[4 - (2 - \sqrt{x})^2]dx$

$\displaystyle \qquad = \pi \int_0^4 (4x^{1/2} - x)dx = 40\pi/3$

16. $\displaystyle V = \int_{-L/2}^{L/2} \pi[4R(x^2 - L^2/4)/L^2]^2 dx$

$\displaystyle \qquad = \frac{2\pi R^2}{L^4} \int_0^{L/2} (16x^4 - 8L^2x^2 + L^4)dx$

$\displaystyle \qquad = 8\pi R^2 L/15$

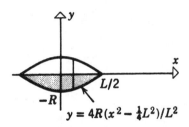

17. $y = \dfrac{x^{3/2}}{\sqrt{8}}$, $0 \le x \le 2$; $y' = \dfrac{3x^{1/2}}{2\sqrt{8}}$, $L = \displaystyle\int_0^2 \sqrt{1 + \dfrac{9}{32}x}\, dx = 61/27$

18. $y' = x(x^2 + 2)^{1/2}$, $1 + (y')^2 = 1 + x^2(x^2 + 2) = (x^2 + 1)^2$, $L = \displaystyle\int_0^3 (x^2 + 1)\,dx = 12$

19. $y' = \dfrac{1}{2}x^4 - \dfrac{1}{2}x^{-4}$, $1 + (y')^2 = 1 + \left(\dfrac{1}{4}x^{16} - \dfrac{1}{2} + \dfrac{1}{4}x^{-16}\right) = \left(\dfrac{1}{2}x^4 + \dfrac{1}{2}x^{-4}\right)^2$,

$L = \displaystyle\int_1^2 \left(\dfrac{1}{2}x^4 + \dfrac{1}{2}x^{-4}\right) dx = 779/240$

20. $y' = x^2 - \dfrac{1}{4}x^{-2}$, $1 + (y')^2 = 1 + \left(x^4 - \dfrac{1}{2} + \dfrac{1}{16}x^{-4}\right) = \left(x^2 + \dfrac{1}{4}x^{-2}\right)^2$,

$L = \displaystyle\int_1^2 \left(x^2 + \dfrac{1}{4}x^{-2}\right) dx = 59/24$

21. $y' = 3x^2$, $1 + (y')^2 = 1 + 9x^4$, $S = \displaystyle\int_1^2 2\pi x^3 \sqrt{1 + 9x^4}\, dx = \pi(145^{3/2} - 10^{3/2})/27$

22. $y' = (1 - x)/\sqrt{2x - x^2}$, $1 + (y')^2 = 1 + \dfrac{(1 - x)^2}{2x - x^2} = \dfrac{1}{2x - x^2}$,

$S = \displaystyle\int_{1/2}^1 2\pi \sqrt{2x - x^2}\,\dfrac{1}{\sqrt{2x - x^2}}\, dx = 2\pi \displaystyle\int_{1/2}^1 dx = \pi$

23. $F(x) = kx$, $F(4) = 4k = 2$, $k = 1/2$, $W = \displaystyle\int_2^4 \dfrac{1}{2}x\, dx = 3 \text{ in·lb}$

24. $W = \displaystyle\int_0^3 kx\, dx = 9k/2 = 180$, $k = 40\,\text{lb/in}$

25. $F(x) = 250 + \dfrac{3}{4}(40 - x) = 280 - \dfrac{3}{4}x$,

$W = \displaystyle\int_0^{40} \left(280 - \dfrac{3}{4}x\right) dx = 10{,}600 \text{ ft·lb}$

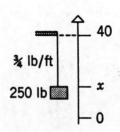

26. $r/3 = x/5,\ r = 3x/5$

$$W = \int_0^5 64x\pi(3x/5)^2 dx$$

$$= \frac{576}{25}\pi \int_0^5 x^3 dx$$

$$= 3600\pi \ \text{ft·lb}$$

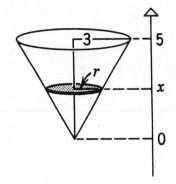

27. $A(y) = \pi x^2 = \pi(y/2 + 4),$

$$W = \int_{-8}^0 62.4(4 - y)[\pi(y/2 + 4)]dy$$

$$= 31.2\pi \int_{-8}^0 (32 - 4y - y^2)dy$$

$$= 6656\pi \ \text{ft·lb}$$

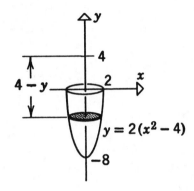

28. $x = D - y,\ F(y) = \dfrac{k}{(D - y)^2}$

$$W = \int_0^{2D/3} \frac{k}{(D - y)^2}dy$$

$$= k \int_0^{2D/3} (D - y)^{-2}dy$$

$$= 2k/D \ \text{J}$$

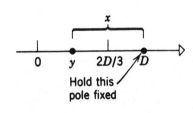

Hold this
pole fixed

29. By similar triangles

$$w(x)/4 = (x - 1)/2$$

$$w(x) = 2(x - 1)$$

$$F = \int_1^3 \rho x[2(x - 1)]dx$$

$$= 2\rho \int_1^3 (x^2 - x)dx = 28\rho/3 \ \text{lb}$$

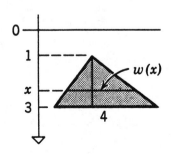

30. $[w(x)/2]^2 = r^2 - x^2$

$$w(x) = 2\sqrt{r^2 - x^2}$$

$$F = \int_0^r \rho x [2\sqrt{r^2 - x^2}]\,dx$$

$$= 2\rho \int_0^r x(r^2 - x^2)^{1/2}\,dx = 2\rho r^3/3 \,\text{lb}$$

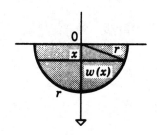

CHAPTER 7

Logarithmic and Exponential Functions

SECTION 7.1

7.1.1 Find the exact value for $\log_2 32$ without use of a calculator.

7.1.2 Find the exact value for $\log_{\sqrt{6}} 6$ without use of a calculator.

7.1.3 Find the exact value for $\log_2\left(\dfrac{1}{64}\right)$ without use of a calculator.

7.1.4 Solve for x if $5^x = 625$.

7.1.5 Solve for x if $6^x = 1/216$.

7.1.6 Find the domain of f if $f(x) = \log_{10}(4x - 3)$.

7.1.7 Find the domain of f if $f(x) = \log_5\left(x^2 - 4\right)$.

7.1.8 Show that, to any base, $2\log\sin\theta = \log(1 - \cos\theta) + \log(1 + \cos\theta)$, $0 < \theta < \pi$.

7.1.9 Show that $\log_a \dfrac{6}{5} - \log_a 300 + \log_a 125 = -\log_a 2$.

7.1.10 Show that $\log_a \dfrac{9}{32} + \log_a \dfrac{256}{3} + \log_a \dfrac{3}{8} + \log_a \dfrac{1}{3} = \log_a 3$.

7.1.11 Show that $\log_a 3\sqrt{x} - \log_a \dfrac{9}{\sqrt{x^3}} - \log_a \dfrac{1}{3} = 2\log_a x$.

7.1.12 Show that

$$\log_a \sqrt[3]{\frac{(x + 2)^3}{x^3 - 8}} = \log_a(x + 2) - \frac{1}{3}\log_a(x - 2) - \frac{1}{3}\log_a\left(x^2 + 2x + 4\right).$$

7.1.13 Solve for x if $3^x = 9^{2x-1}$.

7.1.14 Solve for x if $\log_a x + \log_a(x + 2) = 0$.

7.1.15 Solve for x if $\log_{10}(x + 1) - \log_{10}(x - 2) = 1$.

7.1.16   A radioactive isotope is transformed into another more stable isotope of a certain element by

$$A(t) = 0.0125e^{-t/500}$$

where t is the time in seconds and A is the amount present in mgms.

(a) How much of the isotope was originally present?
(b) When will half of the original amount be transformed?
(c) When will 0.005 mgms of the original isotope remain?

7.1.17   A bacterial population grows by an amount given by

$$N(t) = 135e^{t/125}$$

where N is the number of bacteria present and t is the time in minutes.

(a) How many bacteria were originally present?
(b) In how many minutes will the original number of bacteria double?
(c) In how many minutes will the original number of bacteria triple?
(d) When will there be 185 bacteria present?

SOLUTIONS

SECTION 7.1

7.1.1 $log_2 32 = log_2 \left(2^5\right) = 5.$ **7.1.2** $log_{\sqrt{6}} 6 = log_{\sqrt{6}} \left(\sqrt{6}\right)^2 = 2.$

7.1.3 $log_2 \dfrac{1}{64} = log_2 \left(\dfrac{1}{2^6}\right) = log_2 \left(2^{-6}\right) = -6.$

7.1.4 $5^x = 625 = (5)^4; \, x = 4$ **7.1.5** $6^x = 1/216 = (1/6)^3 = (6)^{-3}; \, x = -3$

7.1.6 $4x - 3 > 0, \, x > 3/4$ so the domain is $(3/4, +\infty)$.

7.1.7 $x^2 - 4 > 0, \, x < -2$ or $x > 2$ so the domain is $(-\infty, -2) \cup (2, +\infty)$.

7.1.8 $\log(1 - \cos\theta) + \log(1 + cos\theta) = \log(1 - \cos^2\theta)$
$$= \log \sin^2\theta = 2\log\sin\theta.$$

7.1.9 $\log_a \dfrac{6}{5} - \log_a 300 + \log_a 125 = \log_a 6 - \log_a 5 - \log_a(6)(25)(2) + \log_a 5^3$
$$= \log_a 6 - \log_a 5 - \log_a 6 - 2\log_a 5 - \log_a 2 + 3\log_a 5$$
$$= -\log_a 2.$$

7.1.10 $\log_a \dfrac{9}{32} + \log_a \dfrac{256}{3} + \log_a \dfrac{3}{8} + \log_a \dfrac{1}{3} = \log_a \dfrac{3^2}{2^5} + \log_a \dfrac{2^8}{3} + \log_a \dfrac{3}{2^3} + \log_a \dfrac{1}{3}$
$$= 2\log_a 3 - 5\log_a 2 + 8\log_a 2 - \log_a 3 + \log_a 3$$
$$-3\log_a 2 - \log_a 3$$
$$= \log_a 3.$$

7.1.11 $\log_a 3\sqrt{x} - \log_a \dfrac{9}{\sqrt{x^3}} - \log_a \dfrac{1}{3} = \log_a \left[\dfrac{3\sqrt{x}}{\dfrac{9}{\sqrt{x^3}}\dfrac{1}{3}}\right] = \log_a \sqrt{x^4} = \log_a x^2 = 2\log_a x.$

7.1.12 $\log_a \sqrt{\dfrac{(x+2)^3}{x^3 - 8}} = \dfrac{1}{3} \log_a \left[\dfrac{(x+2)^3}{x^3 - 8}\right]$
$$= \dfrac{1}{3}\left[\log_a(x+2)^3 - \log_a(x-2)\left(x^2 + 2x + 4\right)\right]$$
$$= \log_a(x+2) - \dfrac{1}{3}\log_a(x-2) - \dfrac{1}{3}\log_a\left(x^2 + 2x + 4\right).$$

7.1.13 $3^x = 9^{2x-1} = 3^{2(2x-1)}$ so $x = 2(2x-1)$; $x = 2/3$.

7.1.14 $\log_a x + \log_a(x+2) = \log_a x(x+2) = 0$ only if $x(x+2) = 1$, so, $x^2+2x-1 = 0$, $x = -\dfrac{2 \pm \sqrt{4+4}}{2}$,
$x = -1+\sqrt{2}$ or $x - 1 - \sqrt{2}$ so choose $x = \sqrt{2} - 1$.

7.1.15 $\log_{10}(x+1) - log_{10}(x-2) = 1$, $\log_{10}\dfrac{x+1}{x-2} = log_{10}10$, $\dfrac{x+1}{x-2} = 10$, $x = \dfrac{21}{9}$ or $\dfrac{7}{3}$.

7.1.16 **(a)** When $t = 0$, $A(0) = 0.0125$ so there was originally 0.0125 mgms present.
  (b) $0.00625 = 0.0125e^{-t/500}$,
  $$e^{-t/500} = \frac{1}{2}$$
  $t = 500 \ln 2$ sec or $t = 346.6$ sec
  (c) $0.005 = 0.0125e^{-t/500}$
  $t = -500 \ln 0.4$ sec ≈ 458.1 sec.

7.1.17 **(a)** When $t = 0$, $N(0) = 135$
  (b) $270 = 135e^{t/125}$
  $e^{t/125} = 2$
  $t = 125 \ln 2$ min ≈ 86.6 min
  (c) $405 = 135e^{t/125}$
  $e^{t/125} = 3$
  $t = 125 \ln 2$ min ≈ 137.3 min
  (d) $185 = 135e^{t/125}$
  $$t = 125 \ln \frac{185}{135} = 39.4 \text{ min}.$$

SECTION 7.2

7.2.1 Use implicit differentiation to find $\dfrac{dy}{dx}$ if $x \ln y = 1$.

7.2.2 Use implicit differentiation to find $\dfrac{dy}{dx}$ if $xy = \ln(x \tan y)$.

7.2.3 Find $f'(x)$ if $f(x) = \ln\left(2x\sqrt{2+x}\right)$.

7.2.4 Evaluate $\displaystyle\int \frac{4x-2}{x^2-x}\,dx$.

7.2.5 Evaluate $\displaystyle\int \frac{x^2}{3x^3-5}\,dx$.

7.2.6 Evaluate $\displaystyle\int \frac{x}{6x^2+1}\,dx$.

7.2.7 Evaluate $\displaystyle\int \frac{x}{1-3x^2}\,dx$.

7.2.8 Evaluate $\displaystyle\int \frac{\sin x + \cos x}{\sin x}\,dx$.

7.2.9 Evaluate $\displaystyle\int \cot 3x\,dx$.

7.2.10 Find $f'(x)$ if $f(x) = \ln(\tan x + \sec x)$.

7.2.11 Evaluate $\displaystyle\int \frac{\sin x}{2 - \cos x}\,dx$.

7.2.12 Find $f'(x)$ if $f(x) = x \ln \sin 2x + x^2$.

7.2.13 Use logarithmic differentiation to find $\dfrac{dy}{dx}$ if $y = \dfrac{x\sqrt{x^2+1}}{(x+1)^{2/3}}$.

7.2.14 Use logarithmic differentiation to find $\dfrac{dy}{dx}$ if $y = \sqrt[3]{\dfrac{(x^2+5)\cos^4 2x}{(x^3-8)^2}}$.

7.2.15 Find $f'(x)$ if $f(x) = \ln\left(3x\sqrt{3-x^2}\right)$.

7.2.16 Use logarithmic differentiation to find $\dfrac{dy}{dx}$ if $y = \sqrt[5]{\dfrac{\tan x}{(1+x^5)^3}}$.

7.2.17 Find $f'(x)$ if $f(x) = e^{x \sin x}$.

7.2.18 Find $f'(x)$ if $f(x) = e^{-2x} \sin 3x$.

7.2.19 Find $\dfrac{dy}{dx}$ if $y = (\sin x)^x$.

7.2.20 Find $f'(x)$ if $f(x) = \dfrac{e^{\ln 2x}}{2x}$.

7.2.21 Find $f'(x)$ if $f(x) = x^4 4^x$.

7.2.22 Find $\dfrac{dy}{dt}$ if $y = (\tan t)^t$.

7.2.23 Find $f'(x)$ if $f(x) = e^x + x^e$.

7.2.24 Find $f'(x)$ if $f(x) = (\sec x)^{\cos x}$.

7.2.25 Evaluate $\displaystyle\int \frac{e^\theta}{\sqrt{1-e^\theta}}\,d\theta$.

7.2.26 Evaluate $\displaystyle\int \frac{e^{2x}+e^{-2x}}{e^{2x}-e^{-2x}}\,dx$.

7.2.27   Evaluate $\displaystyle\int_0^1 \frac{e^{\sqrt{x}}}{\sqrt{x}}\,dx$.

7.2.28   Find $F'(x)$ if $F(x) = \displaystyle\int_1^x e^{-t^2}\,dt$.

7.2.29   Use implicit differentiation to find dy/dx if $\tan y = e^x + \ln x$.

7.2.30   Use implicit differentiation to find dy/dx if $e^{2x} = \sin(x + 3y)$.

7.2.31   Find the minimum value of $y = xe^{-2x}$

7.2.32   Find the area of the region enclosed by $y = e^{2x}$, $y = 4$, and $x = 1$.

7.2.33   Find the positive value of k such that the area under the graph of $y = e^{3x}$ over the interval $[0, k]$ is 3 square units.

SOLUTIONS

SECTION 7.2

7.2.1 $x \left(\dfrac{1}{y}\right)\dfrac{dy}{dx} + \ln y = 0;\ \dfrac{dy}{dx} = -\dfrac{y}{x}\ln y.$

7.2.2 $x\dfrac{dy}{dx} + y = \dfrac{1}{x\tan y}\left(x\sec^2 y\dfrac{dy}{dx} + \tan y\right)$

$$\left(x - \dfrac{\sec^2 y}{\tan y}\right)\dfrac{dy}{dx} = \dfrac{1}{x} - y$$

$$\dfrac{x\tan y - \sec^2 y}{\tan y}\dfrac{dy}{dx} = \dfrac{1 - xy}{x}$$

$$\dfrac{dy}{dx} = \dfrac{(1 - xy)\tan y}{x(x\tan y - \sec^2 y)}.$$

7.2.3 $f(x) = \ln 2 + \ln x + \dfrac{1}{2}\ln(2 + x),\ f'(x) = \dfrac{1}{x} + \left(\dfrac{1}{2}\right)\left(\dfrac{1}{2+x}\right) = \dfrac{1}{x} + \dfrac{1}{2(2+x)}$ or $\dfrac{4 + 3x}{2x(2+x)}.$

7.2.4 $u = x^2 - x,\ du = (2x - 1)dx,\ 2\displaystyle\int\dfrac{du}{u} = 2\ln|u| + C = 2\ln|x^2 - x| + C.$

7.2.5 $u = 3x^3 - 5,\ du = 9x^2 dx,\ x^2 dx = \dfrac{du}{9},\ \dfrac{1}{9}\displaystyle\int\dfrac{du}{u} = \dfrac{1}{9}\ln|u| + C = \dfrac{1}{9}\ln|3x^3 - 5| + C.$

7.2.6 $u = 6x^2 + 1,\ du = 12x dx,\ x dx = \dfrac{du}{12},\ \dfrac{1}{12}\displaystyle\int\dfrac{du}{u} = \dfrac{1}{12}\ln|u| + C = \dfrac{1}{12}\ln\left(6x^2 + 1\right) + C.$

7.2.7 $u = 1 - 3x^2,\ du = -6x dx,\ x dx = -\dfrac{du}{6},\ -\dfrac{1}{6}\displaystyle\int\dfrac{du}{u} = -\dfrac{1}{6}\ln|u| + C = -\dfrac{1}{6}\ln|1 - 3x^2| + C.$

7.2.8 $\displaystyle\int\dfrac{\sin x + \cos x}{\sin x}dx = \int\left(1 + \dfrac{\cos x}{\sin x}\right)dx = x + \ln|\sin x| + C.$

7.2.9 $\displaystyle\int\cot 3x dx = \int\dfrac{\cos 3x}{\sin 3x}dx;\ u = \sin 3x,\ du = 3\cos 3x dx,\ \cos 3x dx = \dfrac{du}{3},$

$$\dfrac{1}{3}\int\dfrac{du}{u} = \dfrac{1}{3}\ln|u| + C = \dfrac{1}{3}\ln|\sin 3x| + C.$$

7.2.10 $f'(x) = \dfrac{1}{\tan x + \sec x}\left(\sec^2 x + \sec x\tan x\right)$

$$= \dfrac{\sec x(\sec x + \tan x)}{\tan x + \sec x} = \sec x.$$

7.2.11 $u = 2 - \cos x$, $du = \sin x\, dx$, $\displaystyle\int \frac{du}{u} = \ln|u| + C = \ln|2 - \cos x| + C.$

7.2.12 $f'(x) = (x)\left(\dfrac{1}{\sin 2x}\right)(\cos 2x)(2) + \ln \sin 2x + 2x$

$\qquad\quad = 2x \cot 2x + \ln \sin 2x + 2x.$

7.2.13 $\ln|y| = \ln\left|\dfrac{x\sqrt{x^2+1}}{(x+1)^{2/3}}\right|$

$\qquad\quad \ln|y| = \ln|x| + \dfrac{1}{2}\ln(x^2+1) - \dfrac{2}{3}\ln|x+1|$

$\qquad\quad \dfrac{1}{y}\dfrac{dy}{dx} = \dfrac{1}{x} + \dfrac{x}{x^2+1} - \dfrac{2}{3(x+1)}$

$\qquad\quad \dfrac{dy}{dx} = y\left[\dfrac{1}{x} + \dfrac{x}{x^2+1} - \dfrac{2}{3(x+1)}\right] = \dfrac{x\sqrt{x^2+1}}{(x+1)^{2/3}}\left[\dfrac{1}{x} + \dfrac{x}{x^2+1} - \dfrac{2}{3(x+1)}\right].$

7.2.14 $\ln|y| = \ln\left|\sqrt[3]{\dfrac{(x^2+5)\cos^4 2x}{(x^3-8)^2}}\right|.$

$\qquad\quad \ln|y| = \dfrac{1}{3}\left[\ln(x^2+5) + 4\ln|\cos 2x| - 2\ln|x^3-8|\right]$

$\qquad\quad \dfrac{1}{y}\dfrac{dy}{dx} = \dfrac{1}{3}\left[\dfrac{2x}{x^2+5} + 4\left(\dfrac{1}{\cos 2x}\right)(-\sin 2x)(2) - 2\left(\dfrac{1}{x^3-8}\right)(3x^2)\right]$

$\qquad\qquad\quad = \dfrac{1}{3}\left(\dfrac{2x}{x^2+5} - 8\tan 2x - \dfrac{6x^2}{x^3-8}\right)$

$\qquad\quad \dfrac{dy}{dx} = \dfrac{1}{3}y\left(\dfrac{2x}{x^2+5} - 8\tan 2x - \dfrac{6x^2}{x^3-8}\right)$

$\qquad\qquad\quad = \dfrac{1}{3}\sqrt[3]{\dfrac{(x^2+5)\cos^4 2x}{(x^3-8)^2}}\left(\dfrac{2x}{x^2+5} - 8\tan 2x - \dfrac{6x^2}{x^3-8}\right).$

7.2.15 $f(x) = \ln 3x + \dfrac{1}{2}\ln(3-x^2)$

$\qquad\quad f'(x) = \dfrac{1}{x} - \dfrac{x}{3-x^2}$ or $\dfrac{3-2x^2}{x(3-x^2)}.$

7.2.16 $\ln|y| = \ln\left|\sqrt[5]{\dfrac{\tan x}{(1+x^5)^3}}\right| = \dfrac{1}{5}\left[\ln|\tan x| - 3\ln|(1+x^5)|\right]$

$\qquad\quad \dfrac{1}{y}\dfrac{dy}{dx} = \dfrac{1}{5}\left[\dfrac{1}{\tan x}(\sec^2 x) - \dfrac{3}{1+x^5}(5x^4)\right] = \dfrac{1}{5}\left(\dfrac{\sec^2 x}{\tan x} - \dfrac{15x^4}{1+x^5}\right)$

$\qquad\quad \dfrac{dy}{dx} = \dfrac{y}{5}\left(\dfrac{\sec^2 x}{\tan x} - \dfrac{15x^4}{1+x^5}\right) = \dfrac{1}{5}\sqrt[5]{\dfrac{\tan x}{(1+x^5)^3}}\left(\dfrac{\sec^2 x}{\tan x} - \dfrac{15x^4}{1+x^5}\right).$

7.2.17 $f'(x) = e^{x\sin x}\dfrac{d}{dx}[x\sin x] = e^{x\sin x}(x\cos x + \sin x).$

7.2.18 $f'(x) = e^{-2x}\dfrac{d}{dx}[\sin 3x] + \sin 3x\dfrac{d}{dx}\left[e^{-2x}\right]$

$$= e^{-2x}(\cos 3x)(3) + \sin 3x\left(e^{-2x}\right)(-2)$$

$$= e^{-2x}(3\cos 3x - 2\sin 3x).$$

7.2.19 $\ln|y| = x\ln|\sin x|$
$\qquad\dfrac{1}{y}\dfrac{dy}{dx} = x\left(\dfrac{1}{\sin x}\right)(\cos x) + \ln|\sin x|$

$\qquad\dfrac{dy}{dx} = (\sin x)^x(x\cot x + \ln|\sin x|).$

7.2.20 $f(x) = \dfrac{e^{\ln 2x}}{2x} = \dfrac{2x}{2x} = 1,\ f'(x) = 0.$

7.2.21 Let $y = x^4 4^x$, then $\ln y = 4\ln|x| + x\ln 4$
$\qquad\dfrac{1}{y}\dfrac{dy}{dx} = \dfrac{4}{x} + \ln 4$

$\qquad\dfrac{dy}{dx} = x^4 4^x\left(\dfrac{4}{x} + \ln 4\right).$

7.2.22 Let $y = (\tan t)^t$, then $\ln|y| = t\ln|\tan t|$

$\qquad\dfrac{1}{y}\dfrac{dy}{dt} = t\dfrac{d}{dt}[\ln|\tan t|] + \ln|\tan t|\dfrac{d}{dt}[t]$

$\qquad\dfrac{dy}{dt} = y\left[t\left(\dfrac{1}{\tan t}\right)\sec^2 t + \ln|\tan t|\right]$

$\qquad\dfrac{dy}{dt} = (\tan t)^t\left(\dfrac{t\sec^2 t}{\tan t} + \ln|\tan t|\right).$

7.2.23 $f'(x) = e^x + ex^{e-1}.$

7.2.24 Let $y = (\sec x)^{\cos x}$, then $\ln|y| = \cos x\ln|\sec x|$
$\qquad\dfrac{1}{y}\dfrac{dy}{dx} = \cos x\dfrac{d}{dx}[\ln|\sec x|] + \ln|\sec x|\dfrac{d}{dx}[\cos x]$

$\qquad\dfrac{dy}{dx} = y\left[\cos x\left(\dfrac{1}{\sec x}\right)(\sec x\tan x) + \ln|\sec x|(-\sin x)\right]$

$\qquad\dfrac{dy}{dx} = (\sec x)^{\cos x}\sin x(1 - \ln|\sec x|).$

7.2.25 $u = 1 - e^{\theta},\ du = -e^{\theta}d\theta,\ e^{\theta}d\theta = -du,\ -\displaystyle\int u^{-1/2}du = -2u^{1/2} + C = -2\sqrt{1 - e^{\theta}} + C.$

7.2.26 $u = e^{2x} - e^{-2x},\ du = 2\left(e^{2x} + e^{-2x}\right)dx,\ \left(e^{2x} + e^{-2x}\right)dx = \dfrac{du}{2},$

$\qquad\dfrac{1}{2}\displaystyle\int\dfrac{du}{u} = \dfrac{1}{2}\ln|u| + C = \dfrac{1}{2}\ln|e^{2x} - e^{-2x}| + C.$

7.2.27 $u = \sqrt{x}, \, du = \dfrac{dx}{2\sqrt{x}}, \, \dfrac{dx}{\sqrt{x}} = 2du$ **7.2.28** $F'(x) = e^{-x^2}.$

$2 \displaystyle\int_0^1 e^u \, du = 2 \, [c^u]_0^1 = 2(e - 1).$

7.2.29 $\sec^2 y \dfrac{dy}{dx} = e^x + \dfrac{1}{x},$ **7.2.30** $2e^{2x} = \cos(x + 3y)\left(1 + 3\dfrac{dy}{dx}\right)$

$\dfrac{dy}{dx} = \cos^2 x \left(e^x + \dfrac{1}{x}\right).$

$2e^{2x} = \cos(x + 3y) + 3\cos(x + 3y)\dfrac{dy}{dx}$

$\dfrac{dy}{dx} = \dfrac{2e^{2x} - \cos(x + 3y)}{3\cos(x + 3y)}$

$\dfrac{dy}{dx} = \dfrac{1}{3}\left[2e^{2x}\sec(x + 3y) - 1\right]$

7.2.31 $y = xe^{-2x}, \, y' = -2\left(x - \dfrac{1}{2}\right)e^{-2x}; \, y' = 0$ when $x = \dfrac{1}{2}$ at which there is a relative maximum
and hence the absolute maximum for $(-\infty, +\infty)$. The maximum value is $y = 0.18$.

7.2.32 Equate $y = e^{2x}$ and $y = 4$
$4 = e^{2x}, \, \ln 4 = \ln e^{2x}$
$\ln 4 = 2x, \, 2\ln 2 = 2x, \, x = \ln 2$
So the point of intersection is $(\ln 2, 4)$

$A = \displaystyle\int_{\ln 2}^1 (e^{2x} - 4)dx = \left[\dfrac{1}{2}e^{2x} - 4x\right]_{\ln 2}^1$

$= \dfrac{1}{2}e^2 - 6 + 4\ln 2 \approx 0.467.$

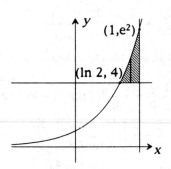

7.2.33 $A = \displaystyle\int_0^k e^{3x} \, dx = \dfrac{1}{3}e^{3x}\bigg|_0^k$

$= \dfrac{1}{3}[e^{3k} - 1]$ since $A = 3$

$A = \dfrac{1}{3}[e^{3k} - 1] = 3, \, e^{3k} = 10$

$\ln(e^{3k}) = \ln 10, \, 3k = \ln 10, \, k = \dfrac{\ln 10}{3}.$

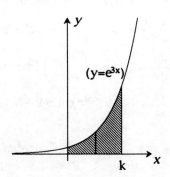

SECTION 7.3

7.3.1 Find the equation of the tangent to the graph of $y = xe^x$ at $(0,0)$.

7.3.2 Find the equation of the tangent to the graph of $y = e^{3x}$ at $(2, e^6)$.

7.3.3 Find the equation of the tangent to the graph of $y = \ln x^2$ at $(1,0)$.

7.3.4 Sketch the graph of $f(x) = \frac{1}{2}\left(e^x - e^{-x}\right)$. Determine all intervals where f is increasing and decreasing, locate and label all extrema and inflection points.

7.3.5 Sketch the graph of $f(x) = xe^{-x}$. Determine all intervals where f is increasing and decreasing, locate and label all extrema and inflection points.

7.3.6 Sketch the graph of $f(x) = e^{-x^2}$. Determine all intervals where f is increasing and decreasing, locate and label all extrema and inflection points. Find also $\lim\limits_{x \to +\infty} e^{-x^2}$ and $\lim\limits_{x \to -\infty} e^{-x^2}$.

7.3.7 Evaluate
(a) $\lim\limits_{x \to +\infty} 2e^{2x}$ (b) $\lim\limits_{x \to -\infty} 2e^{2x}$.

7.3.8 Evaluate
(a) $\lim\limits_{x \to +\infty} 2e^{-2x}$ (b) $\lim\limits_{x \to -\infty} 2e^{-2x}$.

7.3.9 Find the volume of the solid generated when the region in the first quadrant enclosed by $y = e^x$, $x = 2$, and $y = 0$ is revolved about the x-axis.

7.3.10 Find the area under $y = e^x$ for $0 \le x \le 1$.

7.3.11 Find the area enclosed by $y = e^x$, $y = e^{-x}$, and $x = 1$.

7.3.12 Find the average value of $f(x) = e^{2x}$ for $0 \le x \le 2$.

7.3.13 Find the area of the region enclosed by $y = xe^{x^2}$, $x = 0$, $x = 1$, and $y = 0$.

7.3.14 Find the length of the curve $y = \frac{1}{2}\left(e^x + e^{-x}\right)$ for $0 \le x \le 2$.

7.3.15 Find the area of the surface generated by revolving $y = \frac{1}{2}\left(e^x + e^{-x}\right)$ from $x = 0$ to $x = 2$ about the x-axis.

7.3.16 Find the volume of the solid generated when $y = \frac{1}{2}\left(e^x + e^{-x}\right)$, $x = 0$, $x = 2$, and $y = 0$ is revolved around the x-axis.

SOLUTIONS

SECTION 7.3

7.3.1 $f'(x) = xe^x + e^x$ and $m = f'(0) = 1$, so the equation of the tangent to the graph of $y = xe^x$ at $(0,0)$ is $y - 0 = 1(x - 0)$ or $y = x$.

7.3.2 $f'(x) = 3e^{3x}$ and $m = f'(2) = 3e^6$, so the equation of the tangent to the graph of $y = e^{3x}$ at $(2, e^6)$ is $y - e^6 = 3e^6(x - 2)$ or $y = e^6(3x - 5)$.

7.3.3 $f'(x) = \dfrac{2}{x}$ and $m = f'(1) = 2$, so the equation of the tangent to the graph of $y = \ln x^2$ at $(1, 0)$ is $y - 0 = 2(x - 1)$ or $y = 2x - 2$.

7.3.4 $f'(x) = \dfrac{1}{2}\left(e^x + e^{-x}\right)$

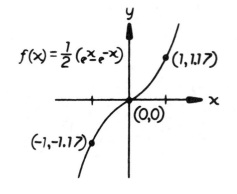

$$= \frac{e^{-x}}{2}\left(e^{2x} + 1\right) > 0,$$

so f is always increasing and never decreasing, thus, f is increasing for $(-\infty, \infty)$. There are no critical points and thus no extrema.

$$f''(x) = \frac{1}{2}\left(e^x - e^{-x}\right)$$

$$= \frac{e^{-x}}{2}\left(e^{2x} - 1\right) = 0 \text{ for } x = 0.$$

so $x = 0$, $y = 0$ is an inflection point.

7.3.5 $f'(x) = e^{-x} - xe^{-x} = e^{-x}(1 - x)$;
$f'(x) = 0$ when $x = 1$.
$f''(x) = e^{-x}(x - 2)$; $f''(x) = 0$
when $x = 2$.

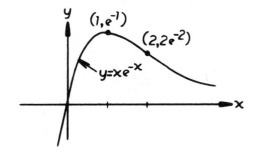

$x = 1$, $y = e^{-1}$ is a relative maximum and $x = 2$, $y = 2e^{-2}$
is an inflection point.

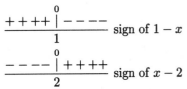

7.3.6 $f'(x) = e^{-x^2}(-2x) = -2xe^{-x^2}$,

$f'(x) = 0$ when $x = 0$.

$f''(x) = -2\left[xe^{-x^2}(-2x) + e^{-x^2}\right]$

$\qquad = -2e^{-x^2}\left(1 - 2x^2\right)$

$f''(x) = 0$ when $x = \pm\dfrac{1}{\sqrt{2}}$

Relative maximum at $(0,1)$ (2nd
derivative test)

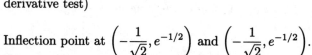

Inflection point at $\left(-\dfrac{1}{\sqrt{2}}, e^{-1/2}\right)$ and $\left(-\dfrac{1}{\sqrt{2}}, e^{-1/2}\right)$.

Also $\lim\limits_{x\to\pm\infty} f(x) = 0$.

7.3.7   **(a)** $+\infty$ **(b)** 0

7.3.8   **(a)** 0 **(b)** $+\infty$

7.3.9 $V = \pi\displaystyle\int_0^2 e^{2x}\,dx = \dfrac{\pi}{2}e^{2x}\Big]_0^2 = \dfrac{\pi}{2}\left(e^4 - 1\right)$.

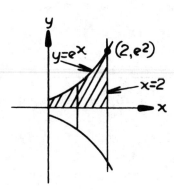

7.3.10 $A = \displaystyle\int_0^1 e^x\,dx = e^x\Big]_0^1 = e - 1$.

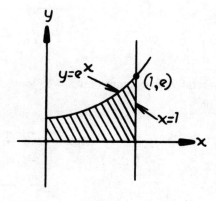

7.3.11 $A = \int_0^1 \left(e^x - e^{-x}\right) dx = \left[e^x + e^{-x}\right]_0^1$

$\qquad = e + \dfrac{1}{e} - 2.$

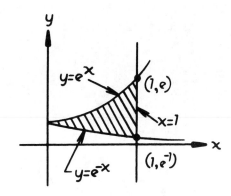

7.3.12 $f_{av} = \dfrac{1}{2-0} \int_0^2 e^{2x} dx = \dfrac{1}{4}\left[e^{2x}\right]_0^2 = \dfrac{1}{4}\left(e^4 - 1\right).$

7.3.13 $A = \int_0^1 xe^{x^2} dx = \dfrac{1}{2}\left[e^{x^2}\right]_0^1 = \dfrac{1}{2}(e-1).$

7.3.14 $f'(x) = \dfrac{1}{2}\left(e^x - e^{-x}\right),\ [f'(x)]^2 = \dfrac{1}{4}\left(e^{2x} - 2 + e^{-2x}\right),\ 1 + [f'(x)]^2 = \dfrac{1}{4}\left(e^{2x} + 2 + e^{-2x}\right);$

$\qquad L = \int_0^2 \sqrt{\dfrac{1}{4}\left(e^{2x} + 2 + e^{-2x}\right)}\, dx = \dfrac{1}{2}\int_0^2 \sqrt{\left(e^x + e^{-x}\right)^2}\, dx$

$\qquad = \dfrac{1}{2}\int_0^2 \left(e^x + e^{-x}\right) dx = \dfrac{1}{2}\left[e^x - e^{-x}\right]_0^2 = \dfrac{e^2 - e^{-2}}{2} \approx 3.6.$

7.3.15 $f'(x) = \dfrac{1}{2}\left(e^x - e^{-x}\right),\ [f'(x)]^2 = \dfrac{1}{4}\left(e^{2x} - 2 + e^{-2x}\right),\ 1 + [f'(x)]^2 = \dfrac{1}{4}\left(e^{2x} + 2 + e^{-2x}\right)$

$\qquad S = 2\pi \int_0^2 \dfrac{1}{2}\left(e^x + e^{-x}\right)\sqrt{\dfrac{1}{4}\left(e^{2x} + 2 + e^{-2x}\right)}\, dx$

$\qquad = \dfrac{\pi}{2}\int_0^2 \left(e^x + e^{-x}\right)\sqrt{\left(e^x + e^{-x}\right)^2}\, dx = \dfrac{\pi}{2}\int_0^2 \left(e^x + e^{-x}\right)\left(e^x + e^{-x}\right) dx$

$\qquad = \dfrac{\pi}{2}\int_0^2 \left(e^{2x} + 2 + e^{-2x}\right) dx = \dfrac{\pi}{2}\left[\dfrac{1}{2}e^{2x} + 2x - \dfrac{1}{2}e^{-2x}\right]_0^2$

$\qquad = \dfrac{\pi}{2}\left(\dfrac{1}{2}e^4 + 4 - \dfrac{1}{2}e^{-4}\right).$

7.3.16 $V = \pi \displaystyle\int_0^2 \left[\frac{1}{2} \left(e^x + e^{-x} \right) \right]^2 dx$

$= \dfrac{\pi}{4} \displaystyle\int_0^2 \left(e^{2x} + 2 + e^{-2x} \right) dx$

$= \dfrac{\pi}{4} \left[\frac{1}{2} e^{2x} + 2x - \frac{1}{2} e^{-2x} \right]_0^2$

$= \dfrac{\pi}{4} \left(\dfrac{e^4}{2} + 4 - \dfrac{1}{2e^4} \right) \approx 7.8\pi$

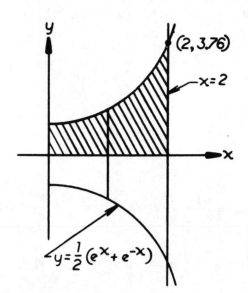

$(2, 3.76)$

$x = 2$

$y = \dfrac{1}{2}\left(e^x + e^{-x}\right)$

SECTION 7.4

7.4.1 Find $f^{-1}(x)$ if $f(x) = 4 + x^3$.

7.4.2 Determine whether or not $f(x) = (x-1)^2$ is a one to one function on $[2,4]$.

7.4.3 Determine whether or not $f(x) = 2x + 3$ is a one to one function and if so, find $f^{-1}(x)$.

7.4.4 Determine whether or not $g(x) = \sqrt{2x+1}$ is a one to one function and if so, find $g^{-1}(x)$ and specify its domain.

7.4.5 Show that $f(x) = x^2 + 4x + 9$ is not a one to one function. Modify the domain of f so that it will be a one to one function.

7.4.6 Show that $f(x) = \sqrt{4 - x^2}$ is not a one to one function. Modify the domain of f so that it will be a one to one function.

7.4.7 Find $f^{-1}(x)$ if $f(x) = \dfrac{1}{x^3 + 1}$ for $x \geq 0$ and specify the domain of f^{-1}.

7.4.8 Find $f^{-1}(-1)$ if $f(x) = -2x^5 + \dfrac{7}{8}$.

7.4.9 **(a)** Show that $f(x) = \dfrac{2x+3}{4x-2}$ is its own inverse.
  (b) What does the result in (a) tell you about the graph of f?

7.4.10 **(a)** Show that $g(x) = \dfrac{x-5}{2x-1}$ is its own inverse.
  (b) What does the result in (a) tell you about the graph of g?

7.4.11 Find $f^{-1}(x)$ if $f(x) = \sqrt[3]{2x+9}$.

7.4.12 Determine whether or not $f(x) = 2x^5 + x^3 + 7x - 5$ is a one to one function.

7.4.13 **(a)** Show that $f(x) = x^3 - 5x^2 + 6x + 1$ is not one to one on $(-\infty, +\infty)$.
  (b) Find the largest value of k such that f is one to one on the interval $(-k, k)$.

7.4.14 Find $g^{-1}(4)$ if $g(x) = 2x + 3$.

7.4.15 Find $f^{-1}(x)$ if $f(x) = 2\sqrt{x-1}$ and specify the domain of f^{-1}.

7.4.16 Find $f^{-1}(x)$ if $f(x) = \dfrac{\sqrt{x}}{3} + 4$ and specify the domain of f^{-1}.

7.4.17 Solve $xy + y^2 = 1$ for x, then use the relationship $\dfrac{dy}{dx} = \dfrac{1}{dx/dy}$ to find $\dfrac{dy}{dx}$. Now find $\dfrac{dy}{dx}$ by implicit differentiation and show that your answers to the first and second part are in agreement.

7.4.18 Solve $xy^2 + xy = 1$ for x, then use the relationship $\dfrac{dy}{dx} = \dfrac{1}{dx/dy}$ to find $\dfrac{dy}{dx}$. Now find $\dfrac{dy}{dx}$ by implicit differentiation and show that your answers to the first and second part are in agreement.

7.4.19 Let $f(x) = x^3 + 2x + 3$ and let $(3, 0)$ be a point on the graph of $y = f^{-1}(x)$. Find the slope of the tangent line to the graph of $y = f^{-1}(x)$ at the given point.

SOLUTIONS

SECTION 7.4

7.4.1 $y = f^{-1}(x)$, $x = f(y) = 4 + y^3$, $y^3 = x - 4$, $y = \sqrt[3]{x-4}$.

7.4.2 $f'(x) = 2(x-1) > 0$ for x in $[2,4]$, thus since f is an increasing function on $[2,4]$ it is a one to one function and does possess an inverse.

7.4.3 $f'(x) = 2 > 0$ so f is an increasing function on $(-\infty, +\infty)$ and is a one to one function. Let $y = f^{-1}(x)$, then $x = f(y) = 2y + 3$, $y = \dfrac{x-3}{2}$.

7.4.4 $g'(x) = \dfrac{1}{\sqrt{2x+1}} > 0$ thus, g is an increasing function on $(-1/2, +\infty)$ and is a one to one function. Let $y = g^{-1}(x)$, then $x = g(y) = \sqrt{2y+1}$, $2y + 1 = x^2$, $y = \dfrac{x^2+1}{2}$ for x in $[0, +\infty)$.

7.4.5 $f'(x) = 2x + 4$, sign of $2x + 4$, $\overset{\displaystyle 0}{\underset{-2}{---\,|+++}}$; thus, f is an increasing function on $(-2, +\infty)$ and a decreasing function on $(-\infty, -2)$ so f is not a one to one function on $(-\infty, +\infty)$, however, f is one to one on $(-\infty, -2]$ or $[-2, +\infty)$.

7.4.6 $f'(x) = \dfrac{-x}{\sqrt{4-x^2}}$, $\overset{\displaystyle 0}{\underset{-2 \quad 0 \quad 2}{+++\,|---}}$ sign of $(-x)$; thus, f is an increasing function on $(-2, 0)$ and a decreasing function on $(0, 2)$ so f is not a one to one function on $(-2, +2)$, however, f is a one to one function on $(-2, 0]$ or $[0, 2)$.

7.4.7 $y = f^{-1}(x)$, $x = f(y) = \dfrac{1}{y^3+1}$, $y^3 = \dfrac{1-x}{x}$, $y = \sqrt[3]{\dfrac{1-x}{x}}$, for $x \neq 0$.

7.4.8 $y = f^{-1}(x)$, $x = f(y) = -2y^5 + \dfrac{7}{8}$, $y = \sqrt[5]{\dfrac{7-8x}{16}}$, so $f^{-1}(-1) = \sqrt[5]{\dfrac{15}{16}}$.

7.4.9 **(a)** $f(f(x)) = \dfrac{2\left(\frac{2x+3}{4x-2}\right) + 3}{4\left(\frac{2x+3}{4x-2}\right) - 2} = \dfrac{4x + 6 + 12x - 6}{8x + 12 - 8x + 4} = x$ thus $f = f^{-1}$.

  (b) The graph of f is symmetric about the line $y = x$.

7.4.10 **(a)** $g(g(x)) = \dfrac{\left(\frac{x-5}{2x-1}\right) - 5}{2\left(\frac{x-5}{2x-1}\right) - 1} = \dfrac{x - 5 - 10x + 5}{2x - 10 - 2x + 1} = x$ thus $g = g^{-1}$

  (b) The graph of g is symmetric about the line $y = x$.

7.4.11 $y = f^{-1}(x)$, $x = f(y) = \sqrt[3]{2y+9}$, $x^3 = 2y + 9$, $y = \dfrac{x^3 - 9}{2}$.

7.4.12 $f'(x) = 10x^4 + 3x^2 + 7 > 0$ for $(-\infty, +\infty)$ then f is an increasing function and also a one to one function.

7.4.13 **(a)** $f(x) = x^3 - 5x^2 + 6x = x(x-2)(x-3)$ so $f(0) = f(2) = f(3) = 0$ thus f is not one to one.

(b) $f'(x) = 3x^2 - 10x + 6 = 0$ when $x = \dfrac{10 \pm \sqrt{100 - 72}}{6} = \dfrac{5 \pm \sqrt{7}}{3}$.

$f'(x) > 0$ if $x < \dfrac{5 - \sqrt{7}}{3}$ (f is increasing),

$f'(x) < 0$ if $\dfrac{5 - \sqrt{7}}{3} < x < \dfrac{5 + \sqrt{7}}{3}$ (f is decreasing),

so $f(x)$ takes on values less that $f\left(\dfrac{5 - \sqrt{7}}{3}\right)$ on both sides of $\dfrac{5 - \sqrt{7}}{3}$ thus $\dfrac{5 - \sqrt{7}}{3}$ is the largest value of k.

7.4.14 $y = g^{-1}(x)$, $x = g(y) = 2y + 3$, $y = \dfrac{x - 3}{2} = g^{-1}(x)$ so $g^{-1}(4) = 1/2$.

7.4.15 $y = f^{-1}(x)$, $x = f(y) = 2\sqrt{y - 1}$, $x^2 = 4(y - 1)$, $y = \dfrac{x^2 + 4}{4}$ for $x \geq 0$.

7.4.16 $y = f^{-1}(x)$, $x = f(y) = \dfrac{\sqrt{y}}{3} + 4$, $\sqrt{y} = 3(x - 4)$, $y = 9(x - 4)^2$ for $x \geq 4$.

7.4.17 $x = \dfrac{1 - y^2}{y}$, $\dfrac{dx}{dy} = -\dfrac{y^2 + 1}{y^2}$ so $\dfrac{dy}{dx} = -\dfrac{y^2}{y^2 + 1}$; $x\dfrac{dy}{dx} + y + 2y\dfrac{dy}{dx} = 0$, $\dfrac{dy}{dx} = -\dfrac{y}{x + 2y}$, but

$x = \dfrac{1 - y^2}{y}$ so $\dfrac{dy}{dx} = -\dfrac{y}{\dfrac{1 - y^2}{y} + 2y} = -\dfrac{y^2}{y^2 + 1}$.

7.4.18 $x = \dfrac{1}{y^2 + y}$, $\dfrac{dx}{dy} = -\dfrac{2y + 1}{(y^2 + y)^2}$ so $\dfrac{dy}{dx} = -\dfrac{(y^2 + y)^2}{2y + 1}$; $2xy\dfrac{dy}{dx} + y^2 + x\dfrac{dy}{dx} + y = 0$,

$\dfrac{dy}{dx} = -\dfrac{y^2 + y}{x(2y + 1)}$, but $x = \dfrac{1}{y^2 + y}$ so $\dfrac{dy}{dx} = -\dfrac{y^2 + y}{\dfrac{1}{y^2 + y}(2y + 1)} = -\dfrac{(y^2 + y)^2}{2y + 1}$.

7.4.19 Let $y = f^{-1}(x)$, then $f(y) = f(f^{-1}(x)) = x$. Thus $x = f(y) = y^3 + 2y + 3$, $\dfrac{dx}{dy} = 3y^2 + 2$.

$\dfrac{dy}{dx} = \dfrac{1}{dx/dy} = \dfrac{1}{3y^2 + 2} \cdot \dfrac{dy}{dx}\bigg|_{(3,0)} = \dfrac{1}{3(0) + 2} = \dfrac{1}{2}$. Thus the slope of the tangent line to

$f^{-1}(x)$ at $(3, 0)$ is $(f^{-1})'(3) = \dfrac{1}{2}$.

SECTION 7.5

7.5.1 Find the domain of $f(x) = \ln(3 - 4x)$.

7.5.2 Find the domain of $f(x) = \ln(9 - x^2)$.

7.5.3 Find the domain of $f(x) = \ln|1 + \ln x|$.

7.5.4 Simplify $e^{-\ln(x+2)}$.

7.5.5 Simplify $e^{\ln 2 + \ln x}$.   **7.5.6** Simplify $\ln(x^2 e^{-3x})$.

7.5.7 Let $f(x) = e^{-3x}$. Find the simplest exact value of $f(\ln \frac{1}{2})$.

SOLUTIONS

SECTION 7.5

7.5.1 $3 - 4x > 0, x < \frac{3}{4}$.

7.5.2 $9 - x^2 > 0$

$(3 - x)(3 + x) > 0$

$-3 < x < 3$, so the domain is $(-3, 3)$.

7.5.3 $x > 0$, so the domain is $(0, \infty)$.

7.5.4 $e^{-\ln(x+2)} = (x + 2)^{-1} = \dfrac{1}{x + 2}$

7.5.5 $e^{\ln 2x} = 2x$.

7.5.6 $2 \ln x - 3x$.

7.5.7 $f(\ln \frac{1}{2}) = e^{-3(\ln \frac{1}{2})} = e^{\ln 8} = 8$.

SECTION 7.6

7.6.1 Find $f'(x)$ if $f(x) = e^{2\ln\sinh 3x}$.

7.6.2 Establish that $\sinh(x - y) = \sinh x \cosh y - \cosh x \sinh y$.

7.6.3 Find $f'(x)$ if $f(x) = \cosh^2 3x$.

7.6.4 Find $f'(x)$ if $f(x) = x \sinh \sqrt{x}$.

7.6.5 Find $f'(x)$ if $f(x) = \text{sech}^2 x \tanh x$.

7.6.6 Evaluate $\displaystyle\int \frac{\sinh x}{1 + \cosh x} dx$.

7.6.7 Evaluate $\displaystyle\int \sqrt{\sinh 2x} \cosh 2x\, dx$.

7.6.8 Evaluate $\displaystyle\int x\, \text{sech}^2 x^2 dx$.

7.6.9 Find $\dfrac{dy}{dx}$ if $y = (\cosh 2x)^{3x}$.

7.6.10 Evaluate $\displaystyle\int \tanh^3 3x\, \text{sech}^2 3x\, dx$.

7.6.11 Find $f'(x)$ if $f(x) = e^{3x} \tanh 2x$.

7.6.12 Find $f'(x)$ if $f(x) = \cosh\left(e^{-2x}\right)$.

7.6.13 Evaluate $\int \sinh^5 \pi x \cosh \pi x\, dx$.

7.6.14 Evaluate without a calculator $\sinh(\ln 5)$.

7.6.15 Find $\dfrac{dy}{dx}$ if $y = 2^{\cosh 3x}$.

7.6.16 Show that $\sinh 2x = 2 \sinh x \cosh x$.

7.6.17 Find $f'(x)$ if $f(x) = \sinh 2x \cosh^2 2x$.

7.6.18 Evaluate without a calculator $\cosh(\ln 2)$.

7.6.19 If $\cosh x = \dfrac{5}{4}$, $\sinh x = ?$

SOLUTIONS

SECTION 7.6

7.6.1 $f(x) = e^{2\ln\sinh 3x} = e^{\ln(\sinh 3x)^2} = (\sinh 3x)^2$; $f'(x) = 2\sinh 3x(\cosh 3x)(3) = 6\sinh 3x \cosh 3x$
or $3\sinh 6x$.

7.6.2 $\sinh x \cosh y - \cosh x \sinh y = \left(\dfrac{e^x - e^{-x}}{2}\right)\left(\dfrac{e^y + e^{-y}}{2}\right) - \left(\dfrac{e^x + e^{-x}}{2}\right)\left(\dfrac{e^y - e^{-y}}{2}\right)$

$$= \frac{2e^{x-y} - 2e^{-x+y}}{4} = \frac{e^{(x-y)} - e^{-(x-y)}}{2} = \sinh(x - y).$$

7.6.3 $f'(x) = 2\cosh 3x(\sinh 3x)(3) = 6\sinh 3x \cosh 3x$ or $3\sinh 6x$.

7.6.4 $f'(x) = x\left(\cosh\sqrt{x}\right)\left(\dfrac{1}{2\sqrt{x}}\right) + \sinh\sqrt{x} = \dfrac{\sqrt{x}}{2}\cosh\sqrt{x} + \sinh\sqrt{x}$.

7.6.5 $f'(x) = \text{sech}^2 x \left(\text{sech}^2 x\right) + \tanh x(2\,\text{sech}\,x)(\text{sech}\,x \tanh x)$
$= \text{sech}^4 x + 2\,\text{sech}^2 x \tanh^2 x.$

7.6.6 $u = 1 + \cosh x$, $du = \sinh x\,dx$, $\displaystyle\int \frac{du}{u} = \ln|u| + C = \ln(1 + \cosh x) + C.$

7.6.7 $u = \sinh 2x$, $du = 2\cosh 2x\,dx$, $\cosh 2x\,dx = du/2$,
$$\frac{1}{2}\int u^{1/2}du = \frac{1}{3}u^{3/2} + C = \frac{1}{3}(\sinh 2x)^{3/2} + C.$$

7.6.8 $u = x^2$, $du = 2x\,dx$, $x\,dx = du/2$, $\dfrac{1}{2}\displaystyle\int \sec h^2 u\,du = \dfrac{1}{2}\tanh u + C = \dfrac{1}{2}\tanh x^2 + C.$

7.6.9 $\ln y = 3x\ln\cosh 2x$
$$\frac{dy}{dx} = (\cosh 2x)^{3x}\left[3x\left(\frac{1}{\cosh 2x}\right)(\sinh 2x)2 + \ln\cosh 2x(3)\right]$$
$$= (\cosh 2x)^{3x}(6x\tanh 2x + 3\ln\cosh 2x).$$

7.6.10 $u = \tanh 3x$, $du = 3\,\text{sech}^2 3x\,dx$, $\text{sech}^2 3x\,dx = \dfrac{du}{3}$
$$\frac{1}{3}\int u^3 du = \frac{1}{12}u^4 + C = \frac{1}{12}\tanh^4 3x + C.$$

7.6.11 $f'(x) = e^{3x}\left(\text{sech}^2 2x\right)(2) + \tanh 2x\left(e^{3x}\right)3$
$= 2e^{3x}\,\text{sech}^2 2x + 3e^{3x}\tanh 2x.$

7.6.12 $f'(x) = \left(\sinh e^{-2x}\right)\right)\left(e^{-2x}\right)(-2) = -2e^{-2x}\sinh\left(e^{-2x}\right).$

7.6.13 $u = \sinh \pi x,\ du = \pi\cosh \pi x dx,\ \cosh \pi x dx = du/\pi,\ \dfrac{1}{\pi}\displaystyle\int u^5 du = \dfrac{u^6}{6\pi} + C = \dfrac{1}{6\pi}\sinh^6 \pi x + C.$

7.6.14 $\sinh(\ln 5) = \dfrac{e^{\ln 5} - e^{-\ln 5}}{2} = \dfrac{5 - \dfrac{1}{5}}{2} = \dfrac{24}{10} = \dfrac{12}{5}.$

7.6.15 $\ln y = \cosh 3x \ln 2.\ \dfrac{dy}{dx} = 2^{\cosh 3x}(\ln 2)(\sinh 3x)(3) = 2^{\cosh 3x}(3\ln 2\sinh 3x).$

7.6.16 $2\sinh x \cosh x = 2\left(\dfrac{e^x - e^{-x}}{2}\right)\left(\dfrac{e^x + e^{-x}}{2}\right) = \dfrac{e^{2x} - e^{-2x}}{2} = \sinh 2x.$

7.6.17 $f'(x) = \sinh 2x(2\cosh 2x)(\sinh 2x)(2) + \cosh^2 2x(\cosh 2x)(2)$
$\qquad = 4\sinh^2 2x \cosh 2x + 2\cosh^3 2x.$

7.6.18 $\cosh(\ln 2) = \dfrac{e^{\ln 2} + e^{-\ln 2}}{2} = \dfrac{2 + \dfrac{1}{2}}{2} = \dfrac{5}{4}.$

7.6.19 $\dfrac{25}{16} - \sinh^2 x = 1,\ \sinh^2 x = \dfrac{9}{16},\ \sinh x = \pm\dfrac{3}{4}.$

SECTION 7.7

7.7.1 Solve the following differential equation:

$$\sin x\, dy - 2y \cos x\, dx = 0.$$

7.7.2 The population of a certain city increases at a rate proportional to the number of its inhabitants at any time. If the population of the city was originally 10,000 and it doubled in 15 years, in how many years will it triple?

7.7.3 Solve the following differential equation:

$$2x(y+1)dx + (x^2+1)dy = 0.$$

7.7.4 Solve the following differential equation:

$$\frac{dy}{dx} = \cos 2x.$$

7.7.5 Solve the following differential equation:

$$\frac{dy}{dx} = (x+3)^2.$$

7.7.6 Solve the following differential equation:

$$\frac{1}{x}\frac{dy}{dx} = 2y.$$

7.7.7 Solve the following differential equation:

$$\frac{dy}{dx} = \frac{1+x}{x^2 y^2}.$$

7.7.8 Solve the following differential equation:

$$\sin x \left(e^y + 1\right) dx = e^y(1 + \cos x)dy, \; y(0) = 0.$$

7.7.9 Find an equation of the curve in the xy-plane that passes through the point $(0,1)$ and whose tangent at (x,y) has the slope $= e^{2x} - y$.

7.7.10 Solve the following differential equation:

$$\left(1 + x^2\right) y' = -\left(xy + x^3 + x\right).$$

7.7.11 Solve the following differential equation:

$$\cos x \frac{dy}{dx} + y \sin x = 1.$$

7.7.12 Solve the following differential equation:

$$\frac{dy}{dx} + 5y = 20, \ y(0) = 2.$$

7.7.13 Solve the following differential equation:

$$\cos^2 x \frac{dy}{dx} + y = 1, \ y(0) = 4.$$

7.7.14 A certain radioactive substance has a half life of 1300 years. Assume an amount y_0 was initially present.

   (a) Find a formula for the amount of substance present at any time.

   (b) In how many years will only 1/10 of the original amount remain?

7.7.15 A tank initially contains 100 gal. of pure water. At time $t = 0$, a solution containing 4 lb. of dissolved salt per gal. flows into the tank at 3 gal./min. The well stirred mixture is pumped out of the tank at the same rate.

   (a) How much salt is present at the end of 30 min.?

   (b) How much salt is present after a very long time?

7.7.16 A tank initially contains 150 gal. of brine in which there is dissolved 30 lb. of salt. At $t = 0$, a brine solution containing 3 lb. of dissolved salt per gallon flows into the tank at 4 gal./min. The well stirred mixture flows out of the tank at the same rate. How much salt is in the tank at the end of 10 min.?

7.7.17 A particle moving along the x-axis encounters a resisting force that results in an acceleration of $a = \dfrac{dv}{dt} = -0.02v^2$. Given that $x = 0$ cm and $v = 35$ cm/sec at $t = 0$, find the velocity v and the position x as a function of t for $t \geq 0$.

SOLUTIONS

SECTION 7.7

7.7.1 $\dfrac{dy}{y} = 2\dfrac{\cos x}{\sin x}dx$

$\ln|y| = 2\ln|\sin x| + C_1 = \ln|C(\sin x)^2| \quad y = C\sin^2 x.$

7.7.2 From (27), $k = \dfrac{1}{T}\ln 2 = \dfrac{1}{15}\ln 2 = 0.0462$, so, the population at any time is

$P(t) = 10,000e^{0.0462t}$, thus, $30,000 = 10,000e^{0.0462t}$;

$$e^{.0462t} = 3$$
$$.0462t = \ln 3$$
$$t = \frac{\ln 3}{0.0462} = 23.8 \text{ years}$$

7.7.3 $\dfrac{dy}{y+1} = -\dfrac{2x}{x^2+1}dx$

$\ln|y+1| = -\ln(x^2+1) + C_1 = -\ln C(x^2+1)$

$y+1 = \dfrac{C}{x^2+1}$ and $y = \dfrac{C}{x^2+1} - 1.$

7.7.4 $dy = \cos 2x\, dx$

$y = \dfrac{1}{2}\sin 2x + C.$

7.7.5 $dy = (x+3)^2 dx$

$y = \dfrac{1}{3}(x+3)^3 + C.$

7.7.6 $\dfrac{dy}{y} = 2x\, dx$

$\ln|y| = x^2 + C_1$

$y = Ce^{x^2}.$

7.7.7 $y^2 dy = \left(\dfrac{1+x}{x^2}\right)dx = \left(\dfrac{1}{x^2} + \dfrac{1}{x}\right)dx$

$\dfrac{y^3}{3} = -\dfrac{1}{x} + \ln|x| + C_1$

or $y = \left(3\ln|x| - \dfrac{3}{x} + C\right)^{1/3}.$

7.7.8 $\dfrac{e^y}{e^y+1}dy = \dfrac{\sin x}{1+\cos x}dx$

$\ln|e^y+1| = -\ln|1+\cos x| + C_1 = \ln\left|\dfrac{C}{1+\cos x}\right| \quad e^y + 1 = \dfrac{C}{1+\cos x}$

$y(0) = 0$ so $1 + 1 = \dfrac{C}{1+1}, C = 4.$

$e^y = \dfrac{4}{1+\cos x} - 1$ or $y = \ln\left(\dfrac{3-\cos x}{1+\cos x}\right).$

401

7.7.9   Slope $= \dfrac{dy}{dx} = e^{2x} - y$, $\dfrac{dy}{dx} + y = e^{2x}$, $\mu = e^{\int dx} = e^x$. $\dfrac{d}{dx}[e^x y] = e^{3x}$,

$e^x y = \displaystyle\int e^{3x}\,dx = \dfrac{1}{3}e^{3x} + C$, $y = \dfrac{1}{3}e^{2x} + Ce^{-x}$ but $y = 1$ when $x = 0$ so

$1 = \dfrac{1}{3}(1) + C(1)$, $C = \dfrac{2}{3}$, $y = \dfrac{1}{3}e^{2x} + \dfrac{2}{3}e^{-x}$.

7.7.10 $\dfrac{dy}{dx} + \dfrac{x}{1+x^2}y = -x$

$\mu = e^{\int \frac{x}{1+x^2}\,dx} = e^{\frac{1}{2}\ln(1+x^2)} = \sqrt{1+x^2}$

$y\sqrt{1+x^2} = -\displaystyle\int x\sqrt{1+x^2}\,dx = -\dfrac{1}{2}\dfrac{(1+x^2)^{3/2}}{3/2} + C$

$y = -\dfrac{1}{3}\left(1+x^2\right) + C\left(1+x^2\right)^{-1/2}$.

7.7.11 $\dfrac{dy}{dx} + y\dfrac{\sin x}{\cos x} = \sec x$

$\mu = e^{\int \frac{\sin x}{\cos x}\,dx} = e^{-\ln\cos x} = \sec x$

$y\sec x = \displaystyle\int \sec^2 x\,dx + C = \tan x + C$

$y = \sin x + C\cos x$

7.7.12 $\mu = e^{5\int dx} = e^{5x}$

$ye^{5x} = \displaystyle\int 20e^{5x}\,dx + C = 4e^{5x} + C$

$y = 4 + Ce^{-5x}$; $y(0) = 2$, thus, $2 = 4 + C(1)$, $C = -2$

so, $y = 4 - 2e^{-5x}$.

7.7.13 $\dfrac{dy}{dx} + \left(\sec^2 x\right)y = \sec^2 x$ $\mu = e^{\int \sec^2 x\,dx} = e^{\tan x}$

$ye^{\tan x} = \displaystyle\int \left(\sec^2 x\right)e^{\tan x}\,dx + C = e^{\tan x} + C$

$y = 1 + Ce^{-\tan x}$; $y(0) = 4$, thus, $4 = 1 + C(1)$, $C = 3$,

so, $y = 1 + 3e^{-\tan x}$.

7.7.14   **(a)**   From (28), $k = -\dfrac{1}{T}\ln 2 = -\dfrac{1}{1300}\ln 2 = -0.0005332$ $y(t) = y_0 e^{-0.0005332t}$

(b) $\dfrac{y_0}{10} = y_0 e^{-0.0005332t}$; $-\ln 10 = -0.0005332t$, $t \approx 4319$ yrs.

7.7.15 **(a)** $\dfrac{dy}{dt}$ = $rate$ in-rate out, where y is the amount of salt present at time t,

$\dfrac{dy}{dt} = (4)(3) - \dfrac{y}{100}(3) = 12 - \dfrac{3y}{100}$, thus $\dfrac{dy}{dt} + \dfrac{3y}{100} = 12$ with $y(0) = 0$.

$\mu = e^{\int \frac{3}{100}dt} = e^{\frac{3t}{100}}$

$e^{\frac{3t}{100}}y = \int 12e^{\frac{3t}{100}}dt = 400e^{\frac{3t}{100}} + C$, $y(t) = 400 + Ce^{-\frac{3t}{100}}$ when $t = 0$,

$y = 0$, so $0 = 400 + C$, $C = -400$

$$y(t) = 400\left(1 - e^{-\frac{3t}{100}}\right)$$

$$y(30) = 400\left[1 - e^{-\frac{3}{100}(30)}\right] \approx 237.4 \text{ lb.}$$

(b) $\displaystyle\lim_{t\to+\infty} y(t) = \lim_{t\to+\infty} 400\left(1 - e^{-\frac{3t}{100}}\right) = 400 \text{ lb.}$

7.7.16 $\dfrac{dy}{dt}$ = rate in-rate out, where y is the amount of salt present at time t,

$\dfrac{dy}{dt} = (3)(4) - \dfrac{y}{150}(4) = 12 - \dfrac{2y}{75}$, thus, $\dfrac{dy}{dt} + \dfrac{2y}{75} = 12$ with $y(0) = 30$. $\mu = e^{\int \frac{2}{75}dt} = e^{\frac{2t}{75}}$;

$e^{\frac{2t}{75}}y = \int 12e^{\frac{2t}{75}}dt = 450e^{\frac{2t}{75}} + C$. $y(t) = 450 + Ce^{-\frac{2t}{75}}$; at $t = 0$, $y = 30$,

so, $30 = 450 + C$, $C = -420$. $y(t) = 450 - 420e^{-\frac{2t}{75}}$, when $t = 10$,

$y(10) = 450 - 420e^{-\frac{(2)(10)}{75}} \approx 128.3 \text{ lb.}$

7.7.17 $a = \dfrac{dv}{dt} = -0.02v^2$, $\dfrac{dv}{v^2} = -0.02dt$, $-v^{-1} = -0.02t + C$, $v = \dfrac{1}{0.02t - C}$.

Since $v = 35$ when $t = 0$, $35 = \dfrac{1}{0.02(0) - C}$, $C = -\dfrac{1}{35}$

$v = \dfrac{1}{0.02t - \frac{1}{35}} = \dfrac{35}{.7t + 1}$

$v = \dfrac{dx}{dt} = \dfrac{35}{.7t + 1}$, $dx = \dfrac{35}{.7t + 1}dt$

$\displaystyle\int dx = \int \dfrac{35}{.7t + 1}dt$, $x = 50\ln(.7t + 1) + C$. Since $x = 0$ when $t = 0$,

$0 = 50\ln(.7(0) + 1) + C$

$C = 0$, $x = 50\ln(.7t + 1)$.

SUPPLEMENTARY EXERCISES, CHAPTER 7

1. In each part determine where f and g are inverse functions.
   - (a) $f(x) = mx$ $\qquad$ $g(x) = 1/(mx)$
   - (b) $f(x) = 3/(x+1)$ $\quad$ $g(x) = (3-x)/x$
   - (c) $f(x) = x^3 - 8$ $\qquad$ $g(x) = \sqrt[3]{x} + 2$
   - (d) $f(x) = x^3 - 1$ $\qquad$ $g(x) = \sqrt[3]{x+1}$
   - (e) $f(x) = \sqrt{e^x}$ $\qquad$ $g(x) = 2\ln x.$

In Exercises 2–6, find $f^{-1}(x)$ if it exists.

2. $f(x) = 8x^3 - 1.$

3. $f(x) = x^2 - 2x + 1.$

4. $f(x) = x^2 - 2x + 1, x \geq 1.$

5. $f(x) = (e^x)^2 + 1.$

6. $f(x) = \exp(x^2) + 1.$

7. Let $f(x) = (ax + b)/(cx + d)$. What conditions on a, b, c, d guarantee that f^{-1} exists? Find $f^{-1}(x)$.

8. Show that $f(x) = (x+2)/(x-1)$ is its own inverse.

9. Find the largest open interval containing the origin on which f is one-to-one.
   - (a) $f(x) = |2x - 5|$
   - (b) $f(x) = x^2 + 4x$
   - (c) $f(x) = \cos(x - 2\pi/3).$

In Exercises 10–13, find $f^{-1}(x)$, and then use Formula (8) of Section 7.1 to obtain $(f^{-1})'(x)$. Check your work by differentiating $f^{-1}(x)$ directly.

10. $f(x) = x^3 - 8.$

11. $f(x) = 3/(x+1).$

12. $f(x) = mx + b(m \neq 0).$

13. $f(x) = \sqrt{e^x}.$

14. Prove that the line $y = x$ is the perpendicular bisector of the line segment joining (a, b) and (b, a).

15. If $r = \ln 2$ and $s = \ln 3$, express the following in terms of r and s:
   - (a) $\ln(1/12)$
   - (b) $\ln(9/\sqrt{8})$
   - (c) $\ln(\sqrt[4]{8/3}).$

16. Simplify:
   - (a) $e^{2 - \ln x}$
   - (b) $\exp(\ln x^2 - 2\ln y)$
   - (c) $\ln[x^3 \exp(-x^2)].$

17. Solve for x in terms of $\ln 3$ and $\ln 5$:
   - (a) $25^x = 3^{1-x}$
   - (b) $\sinh x = \frac{1}{4}\cosh x.$

18.   Express the following as a rational function of x: $3\ln(e^{2x}(e^x)^3) + 2\exp(\ln 1)$.

19.   If $\sinh x = -3/5$, find

    (a) $\cosh x$ **(b)** $\tanh x$ **(c)** $\sinh(2x)$.

20.   Express each of the following as a power of e:

    (a) 2^e **(b)** $(\sqrt{2})^{\pi}$.

In Exercises 21–42, find dy/dx. When appropriate, use implicit or logarithmic differentiation.

21. $y = 1/\sqrt{e^x}$.   **22.** $y = 1/e^{\sqrt{x}}$.

23. $y = x/\ln x$.   **24.** $y = e^x \ln(1/x)$.

25. $y = x/e^{\ln x}$.   **26.** $y = \ln\sqrt{x^2 + 2x}$.

27. $y = \ln(10^x/\sin x)$.   **28.** $y = \cos(e^{-2x})$

29. $y = e^{\tan x}e^{4\ln x}$.   **30.** $y = \ln\left|\dfrac{a+x}{a-x}\right|$.

31. $y = \ln|x + \sqrt{x^2 + a^2}|$.   **32.** $y = \ln|\tan 3x + \sec 3x|$.

33. $y = [\exp(x^2)]^3$.   **34.** $y = \ln(x^3/\sqrt{5 + \sin x})$.

35. $y = \sqrt{\ln(\sqrt{x})}$.   **36.** $y = e^{5x} + (5x)^e$.

37. $y = \pi^x x^{\pi}$.   **38** $y = 4(e^a)^3/\sqrt{\exp(5x)}$.

39. $y = \sinh[\tanh(5x)]$.   **40.** $x^4 + e^{xy} - y^2 = 20$.

41. $y = e^{3x}(1 + e^{-x})^2$.   **42.** $y = (\cosh x)^{x^3}$.

43.   Show that the function $y = e^{ax}\sin bx$ satisfies the equation $y'' - 2ay' + (a^2 + b^2)y = 0$ for any real constants a and b.

44.   Find dy if

    (a) $y = e^{-x}$ **(b)** $y = \ln(1 + x)$ **(c)** $y = 2^{x^2}$.

45.   If $y = 3^{2x}5^{7x}$, show that dy/dx is proportional to y.

46.   Use the chain rule and the Second Fundamental Theorem of Calculus to find the derivative.

    (a) $\dfrac{d}{dx}\left(\displaystyle\int_0^{\ln x} \dfrac{dt}{\sqrt{4 + e^t}}\right)$ **(b)** $\dfrac{d}{dx}\left(\displaystyle\int_1^{e^{5x}} \sqrt{\ln u + u}\,du\right)$.

    [*Hint*: See Exercise 44, Section 5.9.]

47. Suppose $F(-\pi) = 0$ and that $y = F(x)$ satisfies

$$dy/dx = (3\sin^2 2x + 2\cos^2 3x)^{1/2}$$

Express $F(x)$ as in integral.

48. If $y = Ce^{kt}$ (C, k constant) and $Y = \ln y$, show that the graph of Y versus t is a straight line.

49. Evaluate $\int e^{2x}(4 + e^{2x})dx$ two ways: (a) using the substitution $u = 4 + e^{2x}$, and (b) expanding the integrand. Verify that the antiderivatives in parts (a) and (b) differ by a constant.

In Exercises 50–65, evaluate the indicated integral.

50. $\displaystyle\int \frac{e^x}{1 + e^x}\,dx.$

51. $\displaystyle\int \frac{1 + e^x}{e^x}\,dx.$

52. $\displaystyle\int x^e\,dx.$

53. $\displaystyle\int \frac{x^2}{5 - 2x^3}\,dx.$

54. $\displaystyle\int \frac{4x^2 - 3x}{x^3}\,dx.$

55. $\displaystyle\int \frac{(\ln x^2)^2}{x}\,dx.$

56. $\displaystyle\int \frac{\sec x \tan x}{2\sec x - 1}\,dx.$

57. $\displaystyle\int \frac{e^{5x}}{3 + e^{5x}}\,dx.$

58. $\displaystyle\int (\cos 2x)\exp(\sin 2x)\,dx.$

59. $\displaystyle\int \tanh(3x + 1)\,dx.$

60. $\displaystyle\int \operatorname{sech}^2 x \tanh x\,dx.$

61. $\displaystyle\int e^{2x}(4 + e^{-3x})\,dx.$

62. $\displaystyle\int_e^{e^2} \frac{dx}{x \ln x}.$

63. $\displaystyle\int_0^1 \frac{dx}{\sqrt{e^x}}.$

64. $\displaystyle\int_0^{\pi/4} \frac{2^{\tan x}}{\cos^2 x}\,dx.$

65. $\displaystyle\int_1^4 \frac{dx}{\sqrt{xe^{\sqrt{x}}}}.$

66. Show that $\int e^{kx}\,dx = (e^{kx}/k) + C$ for any nonzero constant k.

67. Sketch $y = x^3 e^{-x}$, showing all relative extrema and inflection points.

68. Use the second derivative test to determine the nature of the critical points of $f(x) = x^2 e^{-x}$.

69. Show that if $y = f(x)$ satisfies $y' = e^y + 2y + x$, then every critical point of y is a relative minimum.

70. Sketch the curve $y = e^{-x/2}\sin 2x$ on $[\pi/2, 3\pi/2]$. Show all x-intercepts and points where $y = \pm e^{-x/2}$.

71. Let R be the region bounded by the curve $y = 4(2x - 1)^{-1/2}$, the x-axis, and the lines $x = 1$ and $x = 3$. Find the volume of the solid generated by revolving R about the x-axis.

72. Find the surface area generated when the curve $y = \cosh x, 0 \leq x \leq 1$, is revolved about the x-axis. [*Hint*: Equation (7b) of Section 7.6 will help to evaluate the integral.]

73. Suppose that a crystal dissolves at a rate proportional to the amount *un*dissolved, If 9 g are undissolved initially and 6 g remain undissolved after 1 min, how many grams remain undissolved after 3 min?

74. The population of the United States was 205 million in 1970. Assuming an annual growth rate of 1.8%, find (a) the population in the year 2000 and (b) the year in which the population will reach 1 billion.

SOLUTIONS

SUPPLEMENTARY EXERCISES CHAPTER 7

1. **(a)** $f(g(x)) = m\left(\dfrac{1}{mx}\right) = 1/x \neq x$; f and g are not inverses.

   (b) $f(g(x)) = \dfrac{3}{(3-x)/x+1} = x$, $g(f(x)) = \dfrac{3-3/(x+1)}{3/(x+1)} = x$; f and g are inverses.

   (c) $f(g(x)) = (x^{1/3}+2)^3 - 8 = x + 6x^{2/3} + 12x^{1/3} \neq x$; f and g are not inverses.

   (d) $f(g(x)) = x + 1 - 1 = x$, $g(f(x)) = \sqrt[3]{x^3-1+1} = x$; f and g are inverses.

   (e) $f(g(x)) = \sqrt{e^{2\ln x}} = \sqrt{x^2} = x$ where $x > 0$, $g(f(x)) = 2\ln\sqrt{e^x} = \ln e^x = x$; f and g are inverses.

2. $y = f^{-1}(x)$, $x = f(y) = 8y^3 - 1$, $y = \dfrac{1}{2}(x+1)^{1/3} = f^{-1}(x)$.

3. $f(0) = f(2)$; f is not one-to-one so $f^{-1}(x)$ does not exist.

4. $y = f^{-1}(x)$, $x = f(y) = y^2 - 2y + 1 = (y-1)^2$, $y - 1 = \sqrt{x}$, $y = 1 + \sqrt{x} = f^{-1}(x)$.

5. $y = f^{-1}(x)$, $x = f(y) = e^{2y} + 1$, $e^{2y} = x - 1$, $y = \dfrac{1}{2}\ln(x-1) = f^{-1}(x)$.

6. $f(-1) = f(1) = \exp(1) + 1$ so $f^{-1}(x)$ does not exist.

7. f^{-1} will exist if and only if f is one-to-one. Let x_1, x_2 be any two distinct points in the domain of $y = f(x) = (ax+b)/(cx+d)$.

   $y_1 = f(x_1) = (ax_1+b)/(cx_1+d)$, $y_2 = f(x_2) = (ax_2+b)/(cx_2+d)$,

   $$y_2 - y_1 = \frac{(ax_2+b)(cx_1+d) - (ax_1+b)(cx_2+d)}{(cx_2+d)(cx_1+d)}$$

   $$= \frac{ad(x_2-x_1) - bc(x_2-x_1)}{(cx_2+d)(cx_1+d)} = \frac{(ad-bc)(x_2-x_1)}{(cx_2+d)(cx_2+d)}$$

   f will be one-to-one if $y_1 \neq y_2$ (or equivalently $y_2 - y_1 \neq 0$) whenever $x_1 \neq x_2$, which occurs when $ad - bc \neq 0$. To find $f^{-1}(x)$ in this case, solve $y = (ax+b)/(cx+d)$ for x to get $x = (-dy+b)/(cy-a) = f^{-1}(y)$ so $f^{-1}(x) = (-dx+b)/(cx-a)$.

8. $f(f(x)) = \dfrac{\dfrac{x+2}{x-1}+2}{\dfrac{x+2}{x-1}-1} = x$

9. **(a)** $f(x) = \begin{cases} 2x - 5, & x \geq 5/2 \\ -2x + 5, & x < 5/2 \end{cases}$, $f'(x) = \begin{cases} 2, & x > 5/2 \\ -2, & x < 5/2 \end{cases}$

   and $f'(x)$ does not exist at $x = 5/2$. $f(x)$ is minimum when $x = 5/2$ and f is decreasing for $x < 5/2$ because $f'(x) < 0$, so f is one-to-one for x in the interval $(-\infty, 5/2)$

(b) $f'(x) = 2(x + 2)$, so f is decreasing for $x < -2$ and increasing for $x > -2$. f is one-to-one for x in $(-2, +\infty)$

(c) $f'(x) = -\sin(x - 2\pi/3)$, $f'(x) = 0$ when $x - 2\pi/3 = n\pi$, $x = 2\pi/3 + n\pi$ where n is an integer. $f'(-\pi/3) = f'(2\pi/3) = 0$ and $f'(x) > 0$ if $-\pi/3 < x < 2\pi/3$ so f is one-to-one for x in $(-\pi/3, 2\pi/3)$.

10. $y = f^{-1}(x)$, $x = f(y) = y^3 - 8$, $y = (x + 8)^{1/3} = f^{-1}(x)$; $f'(x) = 3x^2$,

     $f'(f^{-1}(x)) = 3[(x + 8)^{1/3}]^2 = 3(x + 8)^{2/3}$, $(f^{-1})'(x) = \dfrac{1}{3(x + 8)^{2/3}}$.

11. $y = f^{-1}(x)$, $x = f(y) = \dfrac{3}{y + 1}$, $y = \dfrac{3}{x} - 1 = f^{-1}(x)$; $f'(x) = -\dfrac{3}{(x + 1)^2}$,

     $f'(f^{-1}(x)) = -\dfrac{3}{(3/x)^2} = -\dfrac{x^2}{3}$, $(f^{-1})'(x) = -\dfrac{3}{x^2}$.

12. $y = f^{-1}(x)$, $x = f(y) = my + b$, $y = \dfrac{1}{m}(x - b) = f^{-1}(x)$; $f'(x) = m$, $f'(f^{-1}(x)) = m$,

     $(f^{-1})'(x) = \dfrac{1}{m}$.

13. $y = f^{-1}(x)$, $x = f(y) = e^{y/2}$, $y = 2\ln x = f^{-1}(x)$; $f'(x) = \dfrac{1}{2}e^{x/2}$, $f'(f^{-1}(x)) = \dfrac{1}{2}e^{\ln x} = \dfrac{x}{2}$,

     $(f^{-1})'(x) = \dfrac{2}{x}$.

14.   The midpoint of the line segment joining (a, b) and (b, a) is $\left(\dfrac{a + b}{2}, \dfrac{a + b}{2}\right)$ which lies on $y = x$. The slope of the line segment is $(a - b)/(b - a) = -1$ which is the negative reciprocal of the slope of $y = x$ so the lines are perpendicular.

15.   (a) $\ln(1/12) = -\ln 12 = -\ln(2^2 \cdot 3) = -(2\ln 2 + \ln 3) = -(2r + s)$

     (b) $\ln(9/\sqrt{8}) = \ln(3^2 \cdot 2^{-3/2}) = 2\ln 3 - \dfrac{3}{2}\ln 2 = 2s - 3r/2$

     (c) $\ln(\sqrt[4]{8/3}) = \dfrac{1}{4}\ln(2^3/3) = \dfrac{1}{4}(3\ln 2 - \ln 3) = (3r - s)/4$

16.   (a) $e^{2 - \ln x} = e^2/e^{\ln x} = e^2/x$

     (b) $\exp(\ln x^2 - 2\ln y) = \exp(\ln x^2)/\exp(\ln y^2) = x^2/y^2$

     (c) $\ln[x^3 \exp(-x^2)] = \ln x^3 + \ln[\exp(-x^2)] = 3\ln x - x^2$

17.   (a) $25^x = 3^{1-x}$, $(5^2)^x = 3^{1-x}$, $5^{2x} = 3^{1-x}$, $\ln 5^{2x} = \ln 3^{1-x}$,

        $2x\ln 5 = (1 - x)\ln 3$, $x = (\ln 3)/(2\ln 5 + \ln 3)$

     (b) $\sinh x = \dfrac{1}{4}\cosh x$, $\dfrac{1}{2}(e^x - e^{-x}) = \dfrac{1}{8}(e^x + e^{-x})$, $3e^x = 5e^{-x}$, $e^{2x} = 5/3$, $x = (\ln 5 - \ln 3)/2$

18. $3\ln(e^{2x}(e^x)^3) + 2\exp(\ln 1) = 3\ln e^{5x} + 2 = 15x + 2$

19.   **(a)** $\cosh x = (1 + \sinh^2 x)^{1/2} = (1 + 9/25)^{1/2} = \sqrt{34}/5$
     (b) $\tanh x = \sinh x/\cosh x = -3/\sqrt{34}$
     (c) $\sinh 2x = 2\sinh x \cosh x = -6\sqrt{34}/25$

20.   **(a)** $2^e = e^{e\ln 2}$ **(b)** $(\sqrt{2})^\pi = 2^{\pi/2} = e^{(\pi/2)\ln 2}$

21. $y = e^{-x/2}$, $dy/dx = -\dfrac{1}{2}e^{-x/2} = -1/(2\sqrt{e^x})$

22. $y = e^{-\sqrt{x}}$, $dy/dx = -e^{-\sqrt{x}}/(2\sqrt{x}) = -1/(2\sqrt{x}\,e^{\sqrt{x}})$

23. $dy/dx = (\ln x - 1)/(\ln x)^2$ **24.** $y = -e^x \ln x$, $dy/dx = -e^x(\ln x + 1/x)$

25. $y = x/x = 1$, $dy/dx = 0$

26. $y = \dfrac{1}{2}\ln(x^2 + 2x)$, $dy/dx = (x+1)/(x^2 + 2x)$

27. $y = x\ln 10 - \ln\sin x$, $dy/dx = \ln 10 - \cot x$

28. $dy/dx = 2e^{-2x}\sin(e^{-2x})$

29. $y = x^4 e^{\tan x}$, $dy/dx = x^3 e^{\tan x}(x\sec^2 x + 4)$

30. $y = \ln|a + x| - \ln|a - x|$, $dy/dx = 1/(a+x) + 1/(a-x) = 2a/(a^2 - x^2)$

31. $dy/dx = \dfrac{1 + x/\sqrt{x^2 + a^2}}{x + \sqrt{x^2 + a^2}} = 1/\sqrt{x^2 + a^2}$

32. $dy/dx = \dfrac{3\sec^2 3x + 3\sec 3x\tan 3x}{\tan 3x + \sec 3x} = 3\sec 3x$

33. $y = \exp(3x^2)$, $dy/dx = 6x\exp(3x^2)$

34. $y = 3\ln x - \dfrac{1}{2}\ln(5 + \sin x)$, $dy/dx = \dfrac{3}{x} - \dfrac{\cos x}{2(5 + \sin x)}$

35. $y = (\ln\sqrt{x})^{1/2}$, $dy/dx = \dfrac{1}{2}(\ln\sqrt{x})^{-1/2}\left(\dfrac{1}{2x}\right) = 1/(4x\sqrt{\ln\sqrt{x}}\,)$

36. $dy/dx = 5e^{5x} + 5e(5x)^{e-1}$

37. $dy/dx = \pi^x(\pi x^{\pi-1}) + x^\pi(\pi^x \ln \pi) = \pi^x x^{\pi-1}(\pi + x \ln \pi)$

38. $y = 4e^{3x}/e^{5x/2} = 4e^{x/2}, dy/dx = 2e^{x/2}$

39. $dy/dx = 5 \cosh[\tanh(5x)] \operatorname{sech}^2(5x)$

40. $4x^3 + e^{xy}(xy' + y) - 2yy' = 0, y' = (4x^3 + ye^{xy})/(2y - xe^{xy})$

41. $y = e^{3x}(1 + 2e^{-x} + e^{-2x}) = e^{3x} + 2e^{2x} + e^x, dy/dx = 3e^{3x} + 4e^{2x} + e^x.$

42. $\ln y = x^3 \ln \cosh x, y'/y = x^3 \tanh x + 3x^2 \ln \cosh x, y' = x^2(\cosh x)^{x^3}(x \tanh x + 3 \ln \cosh x)$

43. $y = e^{ax} \sin bx, y' = e^{ax}[b \cos bx + a \sin bx],$
$y'' = e^{ax}[2ab \cos bx + (a^2 - b^2) \sin bx]$ so $y'' - 2ay' + (a^2 + b^2)y = 0$

44. **(a)** $dy = -e^{-x}dx$ **(b)** $dy = \dfrac{1}{1+x}dx$ **(c)** $dy = 2x(\ln 2)2^{x^2} dx$

45. $y = 3^{2x}5^{7x} = e^{2x \ln 3}e^{7x \ln 5} = e^{(2\ln 3 + 7\ln 5)x}$ so $y = e^{kx}$ where $k = 2\ln 3 + 7\ln 5$ thus
$dy/dx = ke^{kx} = ky$

46. **(a)** $\dfrac{1}{\sqrt{4 + e^{\ln x}}}\left(\dfrac{1}{x}\right) = \dfrac{1}{x\sqrt{4 + x}}$ **(b)** $\sqrt{\ln e^{5x} + e^{5x}}(5e^{5x}) = 5e^{5x}\sqrt{5x + e^{5x}}$

47. $y = F(x) = \displaystyle\int_{-\pi}^x (3\sin^2 2t + 2\cos^2 3t)^{1/2}dt$

48. $Y = \ln y = \ln(Ce^{kt}) = \ln C + kt$ which is linear in t and Y so the graph is a straight line.

49. **(a)** $u = 4 + e^{2x}, du = 2e^{2x}dx, \dfrac{1}{2}\displaystyle\int u\,du = \dfrac{1}{4}(4 + e^{2x})^2 + C$

(b) $\displaystyle\int (4e^{2x} + e^{4x})dx = 2e^{2x} + \dfrac{1}{4}e^{4x} + C$

The results in (a) and (b) differ by a constant:

$$\left[\frac{1}{4}(4 + e^{2x})^2 + C\right] - \left[2e^{2x} + \frac{1}{4}e^{4x} + C\right] = \left(4 + 2e^{2x} + \frac{1}{4}e^{4x} + C\right) - \left[2e^{2x} + \frac{1}{4}e^{4x} + C\right] = 4$$

50. $u = 1 + e^x, \displaystyle\int \dfrac{1}{u}du = \ln(1 + e^x) + C$

51. $\displaystyle\int (e^{-x} + 1)dx = -e^{-x} + x + C$ **52.** $\dfrac{x^{e+1}}{e + 1} + C$

53. $u = 5 - 2x^3$, $-\dfrac{1}{6}\displaystyle\int \dfrac{1}{u}\,du = -\dfrac{1}{6}\ln|5 - 2x^3| + C$

54. $\displaystyle\int \left(\dfrac{4}{x} - \dfrac{3}{x^2} \right)dx = 4\ln|x| + 3/x + C$

55. $u = \ln x^2 = 2\ln|x|,\ \dfrac{1}{2}\displaystyle\int u^2\,du = \dfrac{1}{6}(\ln x^2)^3 + C$

56. $u = 2\sec x - 1,\ \dfrac{1}{2}\displaystyle\int \dfrac{1}{u}\,du = \dfrac{1}{2}\ln|2\sec x - 1| + C$

57. $\dfrac{1}{5}\ln(3 + e^{5x}) + C$

58. $u = \sin 2x,\ \dfrac{1}{2}\displaystyle\int \exp(u)\,du = \dfrac{1}{2}\exp(\sin 2x) + C$

59. $\displaystyle\int \dfrac{\sinh(3x + 1)}{\cosh(3x + 1)}\,dx = \dfrac{1}{3}\ln[\cosh(3x + 1)] + C$

60. $u = \tanh x,\ \displaystyle\int u\,du = \dfrac{1}{2}\tanh^2 x + C$

61. $\displaystyle\int (4e^{2x} + e^{-x})\,dx = 2e^{2x} - e^{-x} + C$ **62.** $u = \ln x,\ \displaystyle\int_1^2 \dfrac{1}{u}\,du = \ln 2$

63. $\displaystyle\int_0^1 e^{-x/2}\,dx = 2(1 - e^{-1/2})$ **64.** $u = \tan x,\ \displaystyle\int_0^1 2^u\,du = \dfrac{2^u}{\ln 2}\bigg]_0^1 = \dfrac{1}{\ln 2}$

65. $u = \sqrt{x},\ 2\displaystyle\int_1^2 e^{-u}\,du = 2(e^{-1} - e^{-2})$ **66.** $\dfrac{d}{dx}\left[\dfrac{e^{kx}}{k} + C \right] = e^{kx}$

67. $y = x^3 e^{-x}$
$y' = x^2(3 - x)e^{-x}$
$y' = 0$ when $x = 0, 3$
$y'' = x(x^2 - 6x + 6)e^{-x}$
$y'' = 0$ when $x = 0, 3 \pm \sqrt{3}$

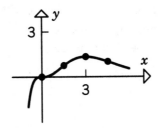

68. $f'(x) = x(2 - x)e^{-x}$; critical points $x = 0, 2$; $f''(x) = (x^2 - 4x + 2)e^{-x}$
$f''(0) > 0$, relative min at $x = 0$; $f''(2) < 0$, relative max at $x = 2$.

69. $y'' = e^y y' + 2y' + 1$. If x_0 is a critical point of y then when $x = x_0$, $y' = 0$ and
$y'' = e^{y_0}(0) + 2(0) + 1 = 1 > 0$ so a relative minimum occurs at x_0.

70. $-1 \leq \sin 2x \leq 1$ so $-e^{-x/2} \leq e^{-x/2} \sin 2x \leq e^{-x/2}$; $e^{-x/2} \sin 2x = e^{-x/2}$ when $\sin 2x = 1$,
$2x = \pi/2 + 2\pi n$ where n is an integer, $x = \pi/4 + \pi n$, so $x = \pi/4, 5\pi/4$ for x in $[-\pi/2, 3\pi/2]$.
Similarly $e^{-x/2} \sin 2x = -e^{-x/2}$ when $x = -\pi/4, 3\pi/4$. The x-intercepts occur when
$e^{-x/2} \sin 2x = 0$, $\sin 2x = 0$, $2x = n\pi$, $x = n\pi/2$, so $x = -\pi/2, 0, \pi/2, \pi, 3\pi/2$.

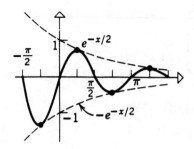

71. $V = \pi \displaystyle\int_1^3 16(2x - 1)^{-1}dx = 8\pi \ln(2x - 1)\Big]_1^3 = 8\pi \ln 5$

72. $y' = \sinh x$, $1 + (y')^2 = 1 + \sinh^2 x = \cosh^2 x$,
$$S = \int_0^1 2\pi \cosh x (\cosh x)dx = 2\pi \int_0^1 \cosh^2 x\, dx = \pi \int_0^1 (1 + \cosh 2x)dx$$
$$= \pi \left(x + \frac{1}{2}\sinh 2x\right)\Big]_0^1 = \pi(2 + \sinh 2)/2$$

73. Let $y =$ amount undissolved after t min, then $dy/dt = ky$ so $y = y_0 e^{kt} = 9e^{kt}$. But $y = 6$ when
$t = 1$ so $9e^k = 6$, $k = \ln(2/3)$. After 3 min $y = 9e^{3k} = 9e^{3\ln(2/3)} = 9(2/3)^3 = 8/3$ g.

74. Let $y =$ population (in millions) t years after 1970, then $y = 205e^{0.018t}$.
  (a) $t = 2000 - 1970 = 30$ for the year 2000 so $y = 205e^{(0.018)(30)} = 205e^{0.54} \approx 352$ million.
  (b) 1 billion $= 1000$ million, $205e^{0.018t} = 1000$ when $t = (1/0.018)\ln(1000/205) \approx 88$. The
population will reach one billion in the year $1970 + 88 = 2058$.

CHAPTER 8

Inverse Trigonometric and Hyperbolic Functions

SECTION 8.1

8.1.1   Find the exact value for $\cot\left[\sin^{-1}\left(-\frac{1}{4}\right)\right]$.

8.1.2   Find the exact value for $\tan\left[\sin^{-1}\left(-\frac{1}{4}\right)\right]$.

8.1.3   Find the exact value for $\tan\left(\sec^{-1}\frac{3}{2}\right)$.

8.1.4   Find the exact value for $\sin^{-1}\left(\cot\frac{\pi}{4}\right)$.

8.1.5   Find the exact value for $\sec\left(\sin^{-1}\frac{3}{4}\right)$.

8.1.6   Find the exact value for $\sin^{-1}\left[\sin\left(\frac{3\pi}{4}\right)\right]$.

8.1.7   Find the exact value for $\cos^{-1}\left[\cos\left(\frac{-\pi}{3}\right)\right]$.

8.1.8   If $\theta = \tan^{-1}(1/2)$, find:
     (a) $\cos\theta$ **(b)** $\csc\theta$

8.1.9   Find $\sin\theta$ if $\theta = \sec^{-1}\frac{17}{8}$.

8.1.10   Find the exact value for $\cos\left[\sin^{-1}\left(-\frac{3}{4}\right)\right]$.

8.1.11   Find the exact value for $\sin 2\left(\tan^{-1}1/3\right)$.

8.1.12   Simplify $\sin\left(\sec^{-1}x\right)$.

8.1.13 Find the exact value for $\sin\left[\tan^{-1}(-2/3)\right]$.

8.1.14 Find the exact value for $\cos\left[\sin^{-1}(-3/4)\right]$.

8.1.15 Evaluate $\tan^{-1}\left[\cot\left(\dfrac{\pi}{6}\right)\right]$.   **8.1.16** Simplify $\tan\left(\cos^{-1}2x\right)$.

8.1.17 Evaluate $\sin\left(\cos^{-1}\dfrac{2}{5}+\sin^{-1}\dfrac{2}{5}\right)$.   **8.1.18** Evaluate $\sin^{-1}\left(\cos\dfrac{3\pi}{4}\right)$.

SOLUTIONS

SECTION 8.1

8.1.1 Let $\theta = \sin^{-1}(-1/4)$ then,
$\sin\theta = -1/4$, and (see figure),

$$\cot\theta = \frac{\sqrt{15}}{-1} = -\sqrt{15}.$$

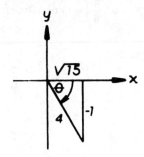

8.1.2 Let $\theta = \sin^{-1}(-1/4)$ then,
$\sin\theta = -1/4$, and (see figure),

$$\tan\theta = -\frac{1}{\sqrt{15}}.$$

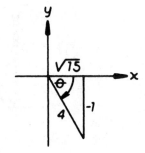

8.1.3 Let $\theta = \sec^{-1}\dfrac{3}{2}$ then,

$\sec\theta = \dfrac{3}{2}$; (see figure)

$$\tan\theta = \frac{\sqrt{5}}{2}.$$

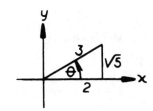

8.1.4 $\dfrac{\pi}{2}$

8.1.5 Let $\theta = \sin^{-1}\dfrac{3}{4}$, then,

$\sin\theta = \dfrac{3}{4}$ and $\sec\theta = \dfrac{4}{\sqrt{7}}$

(see figure).

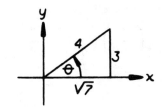

8.1.6 $\sin\dfrac{3\pi}{4} = \sin\left(\pi - \dfrac{\pi}{4}\right) = \sin\dfrac{\pi}{4}$, then, $\sin^{-1}\left(\sin\dfrac{3\pi}{4}\right) = \sin^{-1}\left(\sin\dfrac{\pi}{4}\right) = \dfrac{\pi}{4}$.

8.1.7 $\cos\left(\dfrac{-\pi}{3}\right) = \cos\left(\dfrac{\pi}{3}\right)$ so $\cos^{-1}\left[\cos(-\pi/3)\right] = \cos^{-1}\left[\cos\left(\dfrac{\pi}{3}\right)\right] = \dfrac{\pi}{3}$.

8.1.8   Refer to figure:

   (a) $\dfrac{2}{\sqrt{5}}$

   (b) $\sqrt{5}$

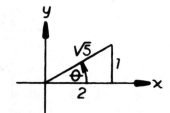

8.1.9 $\sin\theta = \dfrac{15}{17}$.

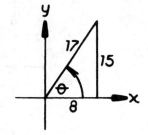

8.1.10   Let $\theta = \sin^{-1}(-3/4)$ then,

   $\sin\theta = -3/4$, and $\cos\theta = \dfrac{\sqrt{7}}{4}$

   (see figure).

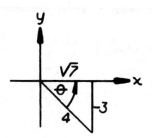

8.1.11   Let $\theta = \tan^{-1}1/3$, then

   $\sin 2\theta = 2\sin\theta\cos\theta$

   $= 2\left(\dfrac{1}{\sqrt{10}}\right)\left(\dfrac{3}{\sqrt{10}}\right) = \dfrac{3}{5}$.

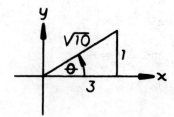

8.1.12   Let $\theta = \sec^{-1} x$, then
$\sec \theta = x$ and
$\sin \theta = \dfrac{\sqrt{x^2 - 1}}{x}$.
(see figure)

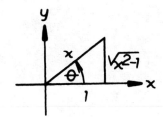

8.1.13   Let $\theta = \tan^{-1}(-2/3)$, then
$\tan \theta = -\dfrac{2}{3}$ and $\sin \theta = -\dfrac{2}{\sqrt{13}}$.
(see figure)

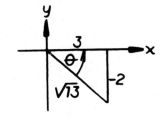

8.1.14   Let $\theta = \sin^{-1}\left(-\dfrac{3}{4}\right)$, then

$\sin \theta = -\dfrac{3}{4}$ and $\cos \theta = \dfrac{\sqrt{7}}{4}$.
(see figure)

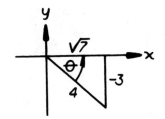

8.1.15 $\dfrac{\pi}{3}$

8.1.16   Let $\theta = \cos^{-1} 2x$, $\cos \theta = 2x$,
and $\tan \theta = \dfrac{\sqrt{1 - 4x^2}}{2x}$.
(see figure)

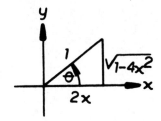

8.1.17   1   **8.1.18** $-\dfrac{\pi}{4}$

SECTION 8.2

8.2.1   Find $f'(x)$ if $f(x) = x \tan^{-1} 3x$.

8.2.2   Find $f'(x)$ if $f(x) = x \sin^{-1} 2x$.

8.2.3   Find $f'(x)$ if $f(x) = \sec(\tan^{-1} x)$.

8.2.4   Find $\dfrac{dy}{dx}$ if $y = \cos^{-1}(\cos x)$.

8.2.5   Find $f'(x)$ if $f(x) = e^{2x} \tan^{-1} 3x$.

8.2.6   Find $f'(x)$ if $f(x) = \tan^{-1}\left(\dfrac{x-1}{x+1}\right)$.

8.2.7   Find $f'(x)$ if $f(x) = \sin^{-1} x + \sqrt{1 - x^2}$.

8.2.8   Find $\dfrac{dy}{dx}$ if $y = \sin^{-1} \dfrac{1}{\sqrt{1 + x^2}}$.

8.2.9   Find $f'(x)$ if $f(x) = \ln(x^2 + 4) - x \tan^{-1} \dfrac{x}{2}$.

8.2.10   Find the equation of the tangent line to the graph $y = \sin^{-1} x$ at the point $(0, 0)$.

8.2.11   Evaluate $\displaystyle\int \dfrac{dx}{x\sqrt{x^2 - 9}}$.

8.2.12   Evaluate $\displaystyle\int \dfrac{x}{1 + x^4}\,dx$.

8.2.13   Evaluate $\displaystyle\int \dfrac{dx}{x(1 + (\ln x)^2)}$.

8.2.14   Find $f'(x)$ if $f(x) = \sin^{-1} e^x$.

8.2.15   Evaluate $\displaystyle\int_0^{\pi/2} \sqrt{\sin^{-1} x + \cos^{-1} x}\,dx$.

8.2.16   Evaluate $\displaystyle\int_0^1 \dfrac{e^x}{\sqrt{1 - e^{2x}}}\,dx$.

8.2.17   Find $f'(x)$ if $f(x) = \dfrac{1}{3} \tan^{-1} \dfrac{x^2}{4}$.

8.2.18   Evaluate $\displaystyle\int \dfrac{dx}{1 + 9x^2}$.

8.2.19   Find $f'(x)$ if $f(x) = x \sin^{-1} 2x$.

8.2.20   Find $f'(x)$ if $f(x) = x^2 \tan^{-1} 2x$.

8.2.21   Find $f'(x)$ if $f(x) = \sin^{-1}\left(\dfrac{x}{x-1}\right)$.

8.2.22   Evaluate $\displaystyle\int_{-1/2}^0 \dfrac{dx}{1 + 4x^2}$.

8.2.23   Evaluate $\displaystyle\int_0^{1/4} \dfrac{dt}{\sqrt{1 - 4t^2}}$.

SOLUTIONS

SECTION 8.2

8.2.1 $f'(x) = x\left(\dfrac{1}{1+9x^2}\right)(3) + \tan^{-1}3x(1) = \dfrac{3x}{1+9x^2} + \tan^{-1}3x.$

8.2.2 $f'(x) = (x)\left(\dfrac{1}{\sqrt{1-4x^2}}\right)(2) + (\sin^{-1}2x)(1) = \dfrac{2x}{\sqrt{1-4x^2}} + \sin^{-1}2x.$

8.2.3 Let $\theta = \tan^{-1}x$, then
$\tan\theta = x$ and
$\sec\theta = \sqrt{x^2+1} = f(x).$
$f'(x) = \dfrac{x}{\sqrt{x^2+1}}.$

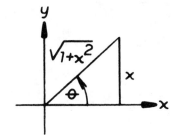

8.2.4 $\dfrac{dy}{dx} = 1.$

8.2.5 $f'(x) = e^{2x}\left(\dfrac{1}{1+9x^2}\right)(3) + 2e^{2x}\tan^{-1}3x = \dfrac{3e^{2x}}{1+9x^2} + 2e^{2x}\tan^{-1}3x.$

8.2.6 $f'(x) = \dfrac{1}{1+\left(\dfrac{x-1}{x+1}\right)^2}\left[\dfrac{(x+1)(1)-(x-1)(1)}{(x+1)^2}\right] = \dfrac{2}{2x^2+2} = \dfrac{1}{x^2+1}.$

8.2.7 $f'(x) = \dfrac{1}{\sqrt{1-x^2}} + \left(\dfrac{1}{2}\right)\left(\dfrac{1}{\sqrt{1-x^2}}\right)(-2x) = \dfrac{1-x}{\sqrt{1-x^2}} = \sqrt{\dfrac{1-x}{1+x}}.$

8.2.8 $\dfrac{dy}{dx} = \dfrac{1}{\sqrt{1-\dfrac{1}{1+x^2}}}\left(-\dfrac{1}{2}\right)(1+x^2)^{-3/2}(2x) = -\dfrac{x}{(1+x^2)^{3/2}\sqrt{\dfrac{x^2}{1+x^2}}} = -\dfrac{1}{1+x^2}.$

8.2.9 $f'(x) = \dfrac{1}{x^2+4}(2x) - (x)\dfrac{1}{1+\dfrac{x^2}{4}}\dfrac{1}{2} - \tan^{-1}\dfrac{x}{2}(1)$

$= \dfrac{2x}{x^2+4} - \dfrac{2x}{x^2+4} - \tan^{-1}\dfrac{x}{2} = \tan^{-1}\dfrac{x}{2}.$

8.2.10 $f'(x) = \dfrac{1}{\sqrt{1-x^2}}$ so $m = f'(0) = \dfrac{1}{\sqrt{1-0^2}} = 1$, then $y - 0 = 1(x-0)$ or $y = x$.

8.2.11 Using (12), $\dfrac{1}{3}\sec^{-1}\dfrac{x}{3} + C$.

8.2.12 $u = x^2$, $du = 2x\,dx$, $x\,dx = du/2$

$$\frac{1}{2}\int \frac{du}{1+u^2} = \frac{1}{2}\tan^{-1}u + C = \frac{1}{2}\tan^{-1}x^2 + C.$$

8.2.13 $u = \ln x$, $du = \dfrac{dx}{x}$;

$$\int \frac{du}{1+u^2} = \tan^{-1}u + C = \tan^{-1}(\ln x) + C.$$

8.2.14 $f'(x) = \dfrac{1}{\sqrt{1-(e^x)^2}}(e^x) = \dfrac{e^x}{\sqrt{1-e^{2x}}}$.

8.2.15 $\sin^{-1}x + \cos^{-1}x = \pi/2$, so $\displaystyle\int_0^{\pi/2}\sqrt{\frac{\pi}{2}}\,dx = \sqrt{\frac{\pi}{2}}\cdot\frac{\pi}{2} = \left(\frac{\pi}{2}\right)^{3/2}$.

8.2.16 $u = e^x$, $du = e^x\,dx$, $u = 1$ when $x = 0$ and $u = e$ when $x = 1$

$$\int_1^e \frac{du}{\sqrt{1-u^2}} = \sin^{-1}u\Big|_1^e = \sin^{-1}e - \sin^{-1}1 \text{ or } \sin^{-1}e - \frac{\pi}{2}.$$

8.2.17 $f'(x) = \left(\dfrac{1}{3}\right)\dfrac{1}{1+\dfrac{x^4}{16}}\dfrac{2x}{4} = \dfrac{8x}{3(16+x^4)}$.

8.2.18 $u = 3x$, $du = 3\,dx$, $dx = du/3$

$$\frac{1}{3}\int \frac{du}{1+u^2} = \frac{1}{3}\tan^{-1}u + C = \frac{1}{3}\tan^{-1}3x + C.$$

8.2.19 $f'(x) = (x)\left(\dfrac{1}{\sqrt{1-4x^2}}\right)(2) + \sin^{-1}2x = \dfrac{2x}{\sqrt{1-4x^2}} + \sin^{-1}2x$.

8.2.20 $f'(x) = (x^2)\left(\dfrac{1}{1+4x^2}\right)(2) + \left(\tan^{-1}2x\right)(2x) = \dfrac{2x^2}{1+4x^2} + 2x\tan^{-1}2x$.

8.2.21 $f'(x) = \dfrac{1}{\sqrt{1-\left(\dfrac{x}{x-1}\right)^2}}\left[\dfrac{(x-1)(1)-(x)(1)}{(x-1)^2}\right] = \dfrac{1}{(1-x)\sqrt{1-2x}}$.

8.2.22 $u = 2x$, $du = 2\,dx$, $u = -1$ when $x = -1/2$ and $u = 0$ when $x = 0$

$$\frac{1}{2} \int_{-1}^{0} \frac{du}{1+u^2} = \frac{1}{2} \tan^{-1} u \bigg]_{-1}^{0} = \frac{\pi}{8}.$$

8.2.23 $u = 2t$, $du = 2\,dt$, $u = 0$ when $t = 0$ and $u = 1/2$ when $t = 1/4$

$$\frac{1}{2} \int_{0}^{1/2} \frac{du}{\sqrt{1-u^2}} = \frac{1}{2} \sin^{-1} u \bigg]_{0}^{1/2} = \frac{\pi}{12}.$$

SECTION 8.3

8.3.1 Starting with the definition of $\cosh x$, derive the formula for $\cosh^{-1} x$ (in terms of logarithms). Be sure to carefully indicate the values of x for which your formula is valid.

8.3.2 Evaluate $\cosh(\cosh^{-1} 2)$.

8.3.3 Evaluate $\cosh(\sinh^{-1} 2)$.

8.3.4 Evaluate $\cosh(2\sinh^{-1} 2)$.

8.3.5 Evaluate $\cosh^{-1}[\cosh(-2)]$.

8.3.6 Simpify $\sinh(\cosh^{-1} x)$.

8.3.7 Find $f'(x)$ if $f(x) = \sinh^{-1}(2x - 1)$.

8.3.8 Find $f'(x)$ if $f(x) = \tanh^{-1}(3x + 2)$.

8.3.9 Find $f'(x)$ if $f(x) = x\sinh^{-1}\dfrac{2}{x}$.

8.3.10 Find $f'(x)$ if $f(x) = \sqrt{1 + x^2} + \sinh^{-1} x$.

8.3.11 Find $f'(x)$ if $f(x) = (\tanh^{-1} 3x)^2$.

8.3.12 Find $f'(x)$ if $f(x) = \ln\sqrt{x^2 - 1} + x\tanh^{-1} x$.

8.3.13 Find $f'(x)$ if $f(x) = \sinh^{-1}(\sin 2x)$.

8.3.14 Evaluate $\displaystyle\int \frac{dx}{\sqrt{1 + 2x^2}}$.

8.3.15 Evaluate $\displaystyle\int \frac{dx}{\sqrt{x^2 - 25}}$.

8.3.16 Evaluate $\displaystyle\int \frac{dx}{x\sqrt{4 + x^2}}$.

8.3.17 Evaluate $\displaystyle\int \frac{e^x}{\sqrt{e^{2x} + 1}}dx$.

8.3.18 Find $f'(x)$ if $f(x) = (\sinh^{-1} x)^x$.

SOLUTIONS

SECTION 8.3

8.3.1 Let $y = \cosh^{-1} x$, then $x = \cosh y = \dfrac{e^y + e^{-y}}{2}$, $e^y - 2x + e^{-y} = 0$ and so $e^{2y} - 2xe^y + 1 = 0$.

Solve for e^y, $e^y = \dfrac{2x \pm \sqrt{4x^2 - 4}}{2} = x \pm \sqrt{x^2 - 1}$. Since $y \geq 0$, $e^y \geq e^{-y}$ and

$x = \dfrac{e^y + e^{-y}}{2} \leq \dfrac{e^y + e^y}{2} = e^y$ or $e^y \geq x$ so choose $e^y = x + \sqrt{x^2 - 1}$ and

$y = \ln\left(x + \sqrt{x^2 - 1}\right)$, $\cosh^{-1} x = \ln(x + \sqrt{x^2 - 1})$ for $x \geq 1$.

8.3.2 2.

8.3.3 Let $u = \sinh^{-1} 2$, then $\sinh u = 2$, $\cosh^2 u = 1 + \sinh^2 u = 1 + 4 = 5$, $\cosh u = \sqrt{5}$.

8.3.4 Let $u = \sinh^{-1} 2$, $\sinh u = 2$; $\cosh 2u = 2\sinh^2 u + 1 = 2(2)^2 + 1 = 9$. Thus, $\cosh(2\sinh^{-1} 2) = 9$.

8.3.5 $\cosh(-2) = \cosh 2$ so $\cosh^{-1}[\cosh 2] = 2$.

8.3.6 Let $u = \cosh^{-1} x$, $x = \cosh u$, $\sinh^2 u = \cosh^2 u - 1 = x^2 - 1$ so $\sinh u = \sinh(\cosh^{-1} x) = \sqrt{x^2 - 1}$ since $\cosh^{-1} x \geq 0$.

8.3.7 $f'(x) = \dfrac{1}{\sqrt{1 + (2x-1)^2}}(2) = \dfrac{2}{\sqrt{4x^2 - 4x + 2}}$.

8.3.8 $f'(x) = \dfrac{1}{1 - (3x+2)^2}(3) = -\dfrac{3}{9x^2 + 12x + 3} = -\dfrac{1}{3x^2 + 4x + 1}$.

8.3.9 $f'(x) = (x)\dfrac{1}{\sqrt{1 + \left(\dfrac{2}{x}\right)^2}}\left(-\dfrac{2}{x^2}\right) + \sinh^{-1}\left(\dfrac{2}{x}\right)(1) = -\dfrac{2}{\sqrt{x^2 + 4}} + \sinh^{-1}\left(\dfrac{2}{x}\right)$.

8.3.10 $f'(x) = \left(\dfrac{1}{2}\right)\left(\dfrac{1}{\sqrt{1 + x^2}}\right)(2x) + \dfrac{1}{\sqrt{1 + x^2}} = \dfrac{x}{\sqrt{1 + x^2}} + \dfrac{1}{\sqrt{1 + x^2}} = \dfrac{x + 1}{\sqrt{1 + x^2}}$.

8.3.11 $f'(x) = 2\left(\tanh^{-1} 3x\right)\left(\dfrac{1}{1 - 9x^2}\right)(3) = \dfrac{6\tanh^{-1} 3x}{1 - 9x^2}$.

8.3.12 $f(x) = \dfrac{1}{2}\ln(x^2 - 1) + x\tanh^{-1}x;$

$$f'(x) = \left(\dfrac{1}{2}\right)\left(\dfrac{1}{x^2 - 1}\right)(2x) + (x)\left(\dfrac{1}{1 - x^2}\right)(1) + \tanh^{-1}x(1)$$

$$f'(x) = \dfrac{x}{x^2 - 1} + \dfrac{x}{1 - x^2} + \tanh^{-1}x = \dfrac{x}{x^2 - 1} - \dfrac{x}{x^2 - 1} + \tanh^{-1}x = \tanh^{-1}x.$$

8.3.13 $f'(x) = \left(\dfrac{1}{\sqrt{1 + \sin^2 2x}}\right)(\cos 2x)(2) = \dfrac{2\cos 2x}{\sqrt{1 + \sin^2 2x}}.$

8.3.14 $u = \sqrt{2}x,\ du = \sqrt{2}dx,\ \dfrac{1}{\sqrt{2}}\displaystyle\int\dfrac{du}{\sqrt{1 + u^2}} = \dfrac{1}{\sqrt{2}}\sinh^{-1}\sqrt{2}x + C.$

8.3.15 rewrite: $\displaystyle\int\dfrac{\dfrac{dx}{5}}{\sqrt{\dfrac{x^2}{25} - 1}} = \dfrac{1}{5}\int\dfrac{dx}{\sqrt{\dfrac{x^2}{25} - 1}};\ u = \dfrac{x}{5},\ du = \dfrac{dx}{5},\ dx = 5\,du$

$$= \dfrac{1}{5}\int\dfrac{5\,du}{\sqrt{u^2 - 1}} = \cosh^{-1}\dfrac{x}{5} + C.$$

8.3.16 rewrite: $\displaystyle\int\dfrac{\dfrac{dx}{2}}{x\sqrt{1 + \left(\dfrac{x}{2}\right)^2}} = \dfrac{1}{2}\int\dfrac{dx}{x\sqrt{1 + \left(\dfrac{x}{2}\right)^2}};$

let $u = \dfrac{x}{2},\ du = \dfrac{dx}{2},\ dx = 2\,du,\ \dfrac{1}{2}(2)\displaystyle\int\dfrac{du}{2u\sqrt{1 + u^2}} = \dfrac{-1}{2}\operatorname{csch}\left|\dfrac{x}{2}\right| + C.$

8.3.17 $u = e^x,\ du = e^x dx$

$$\int\dfrac{du}{\sqrt{u^2 + 1}} = \sinh^{-1}e^x + C.$$

8.3.18 Let $y = f(x)$ so $\ln|y| = x\ln|\sinh^{-1}x|$

$$\dfrac{1}{y}\dfrac{dy}{dx} = (x)\left(\dfrac{1}{\sinh^{-1}x}\right)\left(\dfrac{1}{\sqrt{1 + x^2}}\right) + \ln\sinh^{-1}x$$

$$\dfrac{dy}{dx} = \left(\sinh^{-1}x\right)^x\left(\dfrac{x\operatorname{csch}^{-1}x}{\sqrt{1 + x^2}} + \ln\sinh^{-1}x\right).$$

SUPPLEMENTARY EXERCISES, CHAPTER 8

In Exercises 1–3, find the exact value.

1. (a) $\cos^{-1}(-1/2)$ (b) $\cot^{-1}[\cot(3/4)]$

     (c) $\cos[\sin^{-1}(4/5)]$ (d) $\cos[\sin^{-1}(-4/5)]$.

2. (a) $\tan^{-1}(-1)$ (b) $\csc^{-1}(-2/\sqrt{3})$

     (c) $\cos^{-1}[\cos(-\pi/3)]$ (d) $\sin[-\sec^{-1}(2/\sqrt{3})]$.

3. (a) $\sin^{-1}(1/\sqrt{3})$ (b) $\sin^{-1}[\sin(5\pi/4)]$.

     (c) $\tan(\sec^{-1}5)$ (d) $\tan^{-1}[\cot(\pi/6)]$.

4. Use a double-angle formula to convert the given expression to an algebraic function of x.

     (a) $\sin(2\csc^{-1}x), |x| \geq 1$ (b) $\cos(2\sin^{-1}x), |x| \leq 1$ (c) $\sin(2\tan^{-1}x)$.

5. Simplify:

     (a) $\cos[\cos^{-1}(4/5) + \sin^{-1}(5/13)]$ (b) $\sin[\sin^{-1}(4/5) + \cos^{-1}(5/13)]$

     (c) $\tan[\tan^{-1}(1/3) + \tan^{-1}(2)]$.

6. If $u = \operatorname{csch}^{-1}(-5/12)$, find $\coth u$, $\sinh u$, $\cosh u$, and $\sinh(2u)$.

7. If $u = \tanh^{-1}(-3/5)$, find $\cosh u$, $\sinh u$, and $\cosh(2u)$.

In Exercises 8 and 9, sketch the graph of f.

8. (a) $f(x) = 3\sin^{-1}(x/2)$ (b) $f(x) = \cos^{-1}x - \pi/2$.

9. (a) $f(x) = 2\tan^{-1}(-3x)$ (b) $f(x) = \cos^{-1}x + \sin^{-1}x$.

In Exercises 10–23, find dy/dx, using implicit or logarithmic differentiation where convenient.

10. $y = \sin^{-1}(e^x) + 2\tan^{-1}(3x)$.   **11.** $y = \dfrac{1}{\sec^{-1}x^2}$

12. $y = x\sin^{-1}x + \sqrt{1 - x^2}$.   **13.** $y = \cosh^{-1}(\sec x)$.

14. $\tan^{-1} y = \sin^{-1} x$.

15. $y = x \tanh^{-1}(\ln x)$.

16. $y = \tan^{-1}\left(\dfrac{2x}{1 - x^2}\right)$.

17. $y = \sqrt{\sin^{-1} 3x}$.

18. $y = (\sin^{-1} 2x)^{-1}$.

19. $y = \exp(\sec^{-1} x)$.

20. $y = (\tan^{-1} x)/\ln x$.

21. $y = \pi^{\sin^{-1} x}$.

22. $y = (\sinh^{-1} x)^\pi$.

23. $y = \tanh^{-1}\left(\dfrac{1}{\coth x}\right)$.

24. Let $f(x) = \tan^{-1} x + \tan^{-1}(1/x)$.

    (a) By considering $f'(x)$, show that $f(x) = C_1$ on $(-\infty, 0)$ and $f(x) = C_2$ on $(0, +\infty)$, where C_1 and C_2 are constants.

    (b) Find C_1 and C_2 in part (a).

25. Show that $y = \tan^{-1} x$ satisfies $y'' = -2 \sin y \cos^3 y$.

In Exercises 26–35, evaluate the integral.

26. $\displaystyle\int \frac{dx}{\sqrt{9 - 4x^2}}$.

27. $\displaystyle\int \frac{e^x\, dx}{1 - e^{2x}}$.

28. $\displaystyle\int \frac{\cot x\, dx}{\sqrt{1 - \sin^2 x}}$.

29. $\displaystyle\int \frac{dx}{x\sqrt{(\ln x)^2 - 1}}$.

30. $\displaystyle\int \frac{dx}{e^x\sqrt{1 - e^{-2x}}}$.

31. $\displaystyle\int_{2\sqrt{3}/9}^{2/3} \frac{dx}{3x\sqrt{9x^2 - 1}}$.

32. $\displaystyle\int_0^{\sqrt{2}} \frac{x\, dx}{4 + x^4}$.

33. $\displaystyle\int \frac{dx}{x^{1/2} + x^{3/2}}$.

34. $\displaystyle\int \frac{x^2\, dx}{\sqrt{1 + x^6}}$.

35. $\displaystyle\int_{1/4}^{1/2} \frac{dx}{\sqrt{x}\sqrt{1 - x}}$.

36. The hypotenuse of a right triangle is growing at a rate of a cm/sec and one leg is decreasing at a rate of b cm/sec. How fast is the acute angle between the hypotenuse and the other leg changing at the instant when both legs are 1 cm?

37. For $f(x) = \tan^{-1}(2x) - \tan^{-1} x$ find the value(s) of x at which $f(x)$ assumes its maximum value on the interval $[0, +\infty)$.

38. Find the area between the curve $y = 1/(9 + x^2)$, the x-axis, and the lines $x = \pm\sqrt{3}$.

SOLUTIONS

SUPPLEMENTARY EXERCISES, CHAPTER 8

1. **(a)** $2\pi/3$ **(b)** $3/4$

    (c) $\cos[\sin^{-1}(4/5)] = 3/5$

    (d) $\cos[\sin^{-1}(-4/5)] = \cos[-\sin^{-1}(4/5)] = \cos[\sin^{-1}(4/5)] = 3/5$

2. **(a)** $-\pi/4$ **(b)** $-2\pi/3$

    (c) $\cos^{-1}[\cos(-\pi/3)] = \cos^{-1}[\cos(\pi/3)] = \pi/3$

    (d) $\sin[-\sec^{-1}(2/\sqrt{3})] = \sin(-\pi/6) = -1/2$

3. **(a)** $\pi/4$

    (b) $\sin^{-1}[\sin(5\pi/4)] = \sin^{-1}[\sin(-\pi/4)] = -\pi/4$

    (c) $\tan(\sec^{-1}5) = 2\sqrt{6}$

    (d) $\tan^{-1}[\cot(\pi/6)] = \tan^{-1}(\sqrt{3}) = \pi/3$

4. **(a)** let $\theta = \csc^{-1}x$, $\sin 2\theta = 2\sin\theta\cos\theta = 2(1/x)(\sqrt{x^2-1}/x) = 2\sqrt{x^2-1}/x^2$

    (b) let $\theta = \sin^{-1}x$, $\cos 2\theta = 1 - 2\sin^2\theta = 1 - 2x^2$

    (c) let $\theta = \tan^{-1}x$, $\sin 2\theta = 2\sin\theta\cos\theta = 2(x/\sqrt{1+x^2})(1/\sqrt{1+x^2}) = 2x/(1+x^2)$

5. **(a)** let $\alpha = \cos^{-1}(4/5)$, $\beta = \sin^{-1}(5/13)$

        $\cos(\alpha+\beta) = \cos\alpha\cos\beta - \sin\alpha\sin\beta = (4/5)(12/13) - (3/5)(5/13) = 33/65$

    (b) let $\alpha = \sin^{-1}(4/5)$, $\beta = \cos^{-1}(5/13)$

        $\sin(\alpha+\beta) = \sin\alpha\cos\beta + \cos\alpha\sin\beta = (4/5)(5/13) + (3/5)(12/13) = 56/65$

    (c) let $\alpha = \tan^{-1}(1/3)$, $\beta = \tan^{-1}(2)$; $\tan(\alpha+\beta) = \dfrac{\tan\alpha + \tan\beta}{1 - \tan\alpha\tan\beta} = \dfrac{1/3 + 2}{1 - (1/3)(2)} = 7$

6. $\operatorname{csch} u = -5/12, \sinh u = 1/\operatorname{csch} u = -12/5, \cosh^2 u = 1 + \sinh^2 u = 169/25, \cosh u = 13/5,$
$\coth u = \cosh u/\sinh u = -13/12, \sinh 2u = 2\sinh u\cosh u = -312/25.$

7. $\tanh u = -3/5, \operatorname{sech}^2 u = 1 - \tanh^2 u = 16/25, \operatorname{sech} u = 4/5, \cosh u = 5/4,$
$\sinh u = (\tanh u)(\cosh u) = -3/4, \cosh 2u = \cosh^2 u + \sinh^2 u = 34/16.$

8. **(a)**   **(b)**

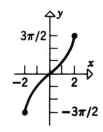

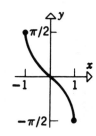

9. (a)

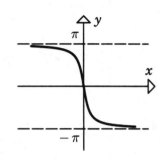

(b)

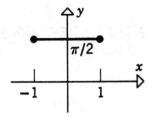

$$f(x) = \cos^{-1} x + \sin^{-1} x = \pi/2$$

10. $e^x/\sqrt{1 - e^{2x}} + 6/(1 + 9x^2)$

11. $-[\sec^{-1}(x^2)]^{-2} \dfrac{1}{x^2\sqrt{x^4 - 1}}(2x) = -\dfrac{2[\sec^{-1}(x^2)]^{-2}}{x\sqrt{x^4 - 1}}$

12. $x/\sqrt{1 - x^2} + \sin^{-1} x - x/\sqrt{1 - x^2} = \sin^{-1} x$

13. $(\sec x \tan x)/\sqrt{\sec^2 x - 1} = (\sec x \tan x)/|\tan x|$

14. $y'/(1 + y^2) = 1/\sqrt{1 - x^2}$, $y' = (1 + y^2)/\sqrt{1 - x^2}$

15. $1/[1 - (\ln x)^2] + \tanh^{-1}(\ln x)$

16. $\dfrac{1}{1 + 4x^2/(1 - x^2)^2} \dfrac{2(1 + x^2)}{(1 - x^2)^2} = \dfrac{2(1 + x^2)}{(1 - x^2)^2 + 4x^2} = \dfrac{2(1 + x^2)}{(1 + x^2)^2} = \dfrac{2}{1 + x^2}$

17. $3/[2\sqrt{\sin^{-1} 3x}\sqrt{1 - 9x^2}]$ **18.** $-2(\sin^{-1} 2x)^{-2}/\sqrt{1 - 4x^2}$

19. $\exp(\sec^{-1} x)/(x\sqrt{x^2 - 1})$

20. $\dfrac{(\ln x)/(1 + x^2) - (\tan^{-1} x)/x}{(\ln x)^2} = \dfrac{x \ln x - (1 + x^2)\tan^{-1} x}{x(1 + x^2)(\ln x)^2}$

21. $\pi^{\sin^{-1} x}(\ln \pi)/\sqrt{1 - x^2}$ **22.** $\pi(\sinh^{-1} x)^{\pi-1}/\sqrt{1 + x^2}$

23. $y = \tanh^{-1}(1/\coth x) = \tanh^{-1}(\tanh x)$ if $x \neq 0$, so $y = x$, $dy/dx = 1$ if $x \neq 0$.

24.   **(a)** $f'(x) = \dfrac{1}{1+x^2} + \dfrac{(-1/x^2)}{1+1/x^2} = 0$ for x in the intervals $(-\infty, 0)$ and $(0, +\infty)$ so $f(x) = C_1$ on $(-\infty, 0)$ and $f(x) = C_2$ on $(0, +\infty)$

    (b)   Let $x = -1$ then $C_1 = f(-1) = (-\pi/4) + (-\pi/4) = -\pi/2$;

      let $x = 1$ then $C_2 = f(1) = (\pi/4) + (\pi/4) = \pi/2$.

25. $y = \tan^{-1} x$, $y' = 1/(1+x^2)$, $y'' = -2x/(1+x^2)^2$; $\sin y = x/\sqrt{1+x^2}$, $\cos y = 1/\sqrt{1+x^2}$,

    $-2 \sin y \cos^3 y = -2 \dfrac{x}{(1+x^2)^{1/2}} \dfrac{1}{(1+x^2)^{3/2}} = -2x/(1+x^2)^2 = y''$

26. $u = 2x$, $\dfrac{1}{2} \displaystyle\int \dfrac{1}{\sqrt{9-u^2}} du = \dfrac{1}{2} \sin^{-1}(2x/3) + C$

27. $u = e^x$, $\displaystyle\int \dfrac{1}{1-u^2} du = \dfrac{1}{2} \ln \left| \dfrac{1+e^x}{1-e^x} \right| + C$

28. $u = \sin x$, $\displaystyle\int \dfrac{\cos x}{\sin x \sqrt{1 - \sin^2 x}} dx = \int \dfrac{1}{u\sqrt{1-u^2}} du = -\operatorname{sech}^{-1}|\sin x| + C$

29. $u = \ln x$, $\displaystyle\int \dfrac{1}{\sqrt{u^2-1}} du = \cosh^{-1}(\ln x) + C$, $x > e$

30. $u = e^{-x}$, $- \displaystyle\int \dfrac{1}{\sqrt{1-u^2}} du = -\sin^{-1}(e^{-x}) + C$

31. $u = 3x$, $\dfrac{1}{3} \displaystyle\int_{2/\sqrt{3}}^{2} \dfrac{1}{u\sqrt{u^2-1}} du = \dfrac{1}{3} \sec^{-1} u \Big]_{2/\sqrt{3}}^{2} = \dfrac{1}{3}(\pi/3 - \pi/6) = \pi/18$

32. $u = x^2$, $\dfrac{1}{2} \displaystyle\int_{0}^{2} \dfrac{1}{4+u^2} du = \dfrac{1}{4} \tan^{-1}(u/2) \Big]_{0}^{2} = \pi/16$

33. $\displaystyle\int \dfrac{1}{x^{1/2} + x^{3/2}} dx = \int \dfrac{1}{x^{1/2}(1+x)} dx$, $u = x^{1/2}$, $2 \displaystyle\int \dfrac{1}{1+u^2} du = 2 \tan^{-1} \sqrt{x} + C$

34. $u = x^3$, $\dfrac{1}{3} \displaystyle\int \dfrac{1}{\sqrt{1+u^2}} du = \dfrac{1}{3} \sinh^{-1}(x^3) + C$

35. $u = \sqrt{x}$, $2 \displaystyle\int_{1/2}^{1/\sqrt{2}} \dfrac{1}{\sqrt{1-u^2}} du = 2 \sin^{-1} u \Big]_{1/2}^{1/\sqrt{2}} = 2 \left(\dfrac{\pi}{4} - \dfrac{\pi}{6} \right) = \pi/6$

36. $dy/dt = a, dx/dt = -b, \theta = \sin^{-1}(x/y)$,

$$\frac{d\theta}{dt} = \frac{1}{\sqrt{1 - x^2/y^2}} \left[\frac{y(dx/dt) - x(dy/dt)}{y^2} \right]$$

$$= \frac{y(dx/dt) - x(dy/dt)}{y\sqrt{y^2 - x^2}}$$

When both legs are 1, $y = \sqrt{2}$ so

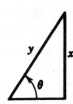

$$\left. \frac{d\theta}{dt} \right|_{x=1, y=\sqrt{2}} = \frac{\sqrt{2}(-b) - (1)(a)}{\sqrt{2}\sqrt{2-1}} = -(a + \sqrt{2}b)/\sqrt{2},$$

θ is decreasing at the rate of $(a + \sqrt{2}b)/\sqrt{2}$ radians/sec.

37. $f'(x) = \dfrac{2}{1 + 4x^2} - \dfrac{1}{1 + x^2} = \dfrac{1 - 2x^2}{(1 + 4x^2)(1 + x^2)}$, the critical point is $x = 1/\sqrt{2}$ for $x > 0$, $f'(x) > 0$ for $0 < x < 1/\sqrt{2}$ and $f'(x) < 0$ for $x > 1/\sqrt{2}$ so $f(x)$ assumes its maximum value at $x = 1/\sqrt{2}$

38. $A = \displaystyle\int_{-\sqrt{3}}^{\sqrt{3}} \frac{1}{9 + x^2} dx = \frac{1}{3} \tan^{-1}(x/3) \Big]_{-\sqrt{3}}^{\sqrt{3}} = \frac{1}{3}[(\pi/6) - (-\pi/6)] = \pi/9$

CHAPTER 9

Techniques of Integration

SECTION 9.1

9.1.1 **(a)** Use Endpaper Tables to evaluate $\int \dfrac{2x^2}{3+4x}\,dx$.

      (b) If you have access to a computer algebra system such as Mathematica, Maple, or Derive, use it to evaluate the integral.

      (c) Confirm that the results obtained in parts (a) and (b) are equivalent.

9.1.2 **(a)** Use Endpaper Tables to evaluate $\int \dfrac{1}{4x+3x^2}\,dx$.

      (b) If you have access to a computer algebra system such as Mathematica, Maple, or Derive, use it to evaluate the integral.

      (c) Confirm that the results obtained in parts (a) and (b) are equivalent.

9.1.3 Use Endpaper Tables to evaluate $\int \dfrac{7x}{\sqrt{9+3x}}\,dx$

9.1.4 Use Endpaper Tables to evaluate $\int \sqrt{x^2+9}\,dx$

9.1.5 Use Endpaper Tables to evaluate $\int \sqrt{9-2x^2}\,dx$

9.1.6 Use Endpaper Tables to evaluate $\int 4x^2\sqrt{16-x^2}\,dx$

9.1.7 Use Endpaper Tables to evaluate $\int \dfrac{\sqrt{50-2x^2}}{3x}\,dx$

9.1.8 Use Endpaper Tables to evaluate $\int \dfrac{5\,dx}{72+2x^2}$

9.1.9 Use Endpaper Tables to evaluate $\int \tan^3 5x\,dx$

9.1.10 Use Endpaper Tables to evaluate $\int \sin 5x \cos 3x\,dx$

9.1.11 Use Endpaper Tables to evaluate $\int x^3 \sin 4x\,dx$

9.1.12 Use Endpaper Tables to evaluate $\int x^2 e^{3x}\,dx$

9.1.13 Use Endpaper Tables to evaluate $\int \dfrac{1}{\sqrt{7 + 6x - x^2}}\,dx$

9.1.14 Use Endpaper Tables to evaluate $\int \sqrt{7 + 6x - x^2}\,dx$

9.1.15 Find the volume of the solid generated when the region enclosed by $y = \sqrt{2x - 8}$ for $4 \le x \le 6$ is revolved around the y-axis.

9.1.16 Find the arc length of the curve $y = \ln(x^4)$ for $3 \le x \le 8$.

SOLUTIONS

SECTION 9.1

9.1.1 $\dfrac{1}{64}\left(16x^2 - 24x - 27 + 18\ln|3+4x|\right) + C$

9.1.2 $\dfrac{1}{4}\ln\left|\dfrac{x}{4+3x}\right| + C.$ **9.1.3** $\dfrac{14}{9}(x-6)\sqrt{9+3x} + C.$

9.1.4 $\dfrac{x}{2}\sqrt{x^2+9} + \dfrac{9}{2}\ln\left|x + \sqrt{x^2+9}\right| + C.$

9.1.5 $\dfrac{1}{2}x\sqrt{9-2x^2} + \dfrac{9}{2\sqrt{2}}\sin^{-1}\dfrac{\sqrt{2}\,x}{3} + C.$

9.1.6 $x(x^2-8)\sqrt{16-x^2} + 128\sin^{-1}\dfrac{x}{4} + C.$

9.1.7 $\dfrac{\sqrt{2}}{3}\sqrt{25-x^2} - \dfrac{5}{3}\sqrt{2}\ln\left|\dfrac{5+\sqrt{25-x^2}}{x}\right| + C.$

9.1.8 $\dfrac{5}{12}\tan^{-1}\dfrac{x}{6} + C.$ **9.1.9** $\dfrac{1}{10}\tan^2 5x - \dfrac{1}{5}\ln|\sec 5x| + C.$

9.1.10 $-\dfrac{1}{16}\cos 8x - \dfrac{1}{4}\cos 2x + C.$

9.1.11 $2(3x - 8x^3)\cos 4x + \dfrac{3}{2}(8x^2 - 1)\sin 4x + C.$

9.1.12 $\dfrac{1}{3}e^{3x}(9x^2 - 6x + 2) + C.$ **9.1.13** $\sin^{-1}\left(\dfrac{x-3}{4}\right) + C.$

9.1.14 $\dfrac{1}{2}(x-3)\sqrt{7+6x-x^2} + 8\sin^{-1}\left(\dfrac{x-3}{4}\right) + C.$

9.1.15 $V = 2\pi\displaystyle\int_4^6 x\sqrt{2x-8}\,dx = \dfrac{\pi}{15}\left[(3x+8)(2x-8)^{3/2}\right]_4^6 = \dfrac{208\pi}{15} \approx 43.563$

9.1.16 $L = \displaystyle\int_3^8 \dfrac{\sqrt{x^2+16}}{x}\,dx = \left[\sqrt{x^2+16} - 4\ln\left|\dfrac{4+\sqrt{x^2+16}}{x}\right|\right]_3^8$

$= 4\sqrt{5} - 4\ln\left(\dfrac{1+\sqrt{5}}{2}\right) - 5 + 4\ln 3 \approx 6.414$

SECTION 9.2

9.2.1 Evaluate $\int \tan^{-1} 4x \, dx$.

9.2.2 Evaluate $\int_0^{\pi/6} x \cos 3x \, dx$.

9.2.3 Evaluate $\int \dfrac{\sin 2\pi x}{e^{2\pi x}} \, dx$.

9.2.4 Evaluate $\int \ln(1 + x^2) dx$.

9.2.5 Use integration by parts to evaluate $\int x\sqrt{1+x}\,dx$.

9.2.6 Evaluate $\int \sinh^{-1} 2x \, dx$.

9.2.7 Use integration by parts to evaluate $\int_0^4 x\sqrt{2x+1}\,dx$.

9.2.8 Evaluate $\int xe^{-2x} dx$.

9.2.9 Evaluate $\int x \sec^2 3x \, dx$.

9.2.10 Evaluate $\int e^{-x} \cos 2x \, dx$.

9.2.11 Evaluate $\int \ln^2 x \, dx$.

9.2.12 Evaluate $\int_{-1}^2 \ln(x+2) dx$.

9.2.13 Evaluate $\int x \cosh 2x \, dx$.

9.2.14 Evaluate $\int x \csc^2 2x \, dx$.

9.2.15 Evaluate $\int x \sinh 4x \, dx$.

9.2.16 Evaluate $\int \cosh^{-1} 4x \, dx$.

9.2.17 Evaluate $\int x \operatorname{sech}^2 2x \, dx$.

9.2.18 Evaluate $\int \dfrac{x^3}{\sqrt{x^2+1}} dx$.

9.2.19 $\int x \sin^{-1}\left(\dfrac{a}{x}\right) dx$.

SOLUTIONS

SECTION 9.2

9.2.1 $u = \tan^{-1} 4x$, $dv = dx$, $du = \dfrac{4}{1+16x^2}dx$, $v = x$

$$\int \tan^{-1} 4x\, dx = x\tan^{-1} 4x - \int (x)\left(\frac{4}{1+16x^2}\right)dx$$

$$= x\tan^{-1} 4x - \frac{1}{8}\ln(1+16x^2) + C$$

9.2.2 $u = x$, $dv = \cos 3x\, dx$, $du = dx$, $v = \dfrac{1}{3}\sin 3x$

$$\int_0^{\frac{\pi}{6}} x\cos 3x\, dx = \frac{1}{3}\Big[x\sin 3x\Big]_0^{\pi/6} - \frac{1}{3}\int_0^{\frac{\pi}{6}} \sin 3x\, dx$$

$$= \frac{\pi}{18} + \frac{1}{9}\Big[\cos 3x\Big]_0^{\pi/6} = \frac{\pi}{18} - \frac{1}{9}.$$

9.2.3 Rewrite as $\displaystyle\int e^{-2\pi x}\sin 2\pi x\, dx$ then $u = e^{-2\pi x}$, $dv = \sin 2\pi x\, dx$, $du = -2\pi e^{-2\pi x}dx$,

$v = \dfrac{-1}{2\pi}\cos 2\pi x$ $\displaystyle\int e^{-2\pi x}\sin 2\pi x\, dx = \frac{-1}{2\pi}e^{-2\pi x}\cos 2\pi x - \int e^{-2\pi x}\cos 2\pi x\, dx$

For $\displaystyle\int e^{-2\pi x}\cos 2\pi x\, dx$, let $u = e^{-2\pi x}$, $dv = \cos 2\pi x\, dx$ $du = -2\pi e^{-2\pi x}dx$,

$v = \dfrac{1}{2\pi}\sin 2\pi x$ so $\displaystyle\int e^{-2\pi x}\cos 2\pi x\, dx = \frac{1}{2\pi}e^{-2\pi x}\sin 2\pi x + \int e^{-2\pi x}\sin 2\pi x\, dx$

Thus $\displaystyle\int e^{-2\pi x}\sin 2\pi x\, dx = -\frac{1}{2\pi}e^{-2\pi x}\cos 2\pi x - \frac{1}{2\pi}e^{-2\pi x}\sin 2\pi x$

$+\displaystyle\int e^{-2\pi x}\sin 2\pi x\, dx; 2\int e^{-2\pi x}\sin 2\pi x\, dx = -\frac{1}{2\pi}e^{-2\pi x}\cos 2\pi x - \frac{1}{2\pi}e^{-2\pi x}\sin 2\pi x$

$\displaystyle\int e^{-2\pi x}\sin 2\pi x\, dx = -\frac{1}{2\pi}e^{-2\pi x}[\cos 2\pi x + \sin 2\pi x] + C.$

9.2.4 $u = \ln(1+x^2)$, $dv = dx$, $du = \dfrac{2x}{1+x^2}dx$, $v = x$

$$\int \ln(1+x^2)dx = x\ln(1+x^2) - 2\int \frac{x^2}{1+x^2}dx$$

$$= x\ln(1+x^2) - 2\int \left(1 - \frac{1}{1+x^2}\right)dx$$

$$= x\ln(1+x^2) - 2x + 2\tan^{-1}x + C.$$

9.2.5 $u = x, dv = \sqrt{1+x}\,dx, du = dx, v = \dfrac{2}{3}(1+x)^{3/2}$

$$\int x\sqrt{1+x}\,dx = \frac{2x}{3}(1+x)^{3/2} - \frac{2}{3}\int (1+x)^{3/2}dx$$

$$= \frac{2x}{3}(1+x)^{3/2} - \frac{4}{15}(1+x)^{5/2} + C.$$

9.2.6 $u = \sinh^{-1} 2x, dv = dx, du = \dfrac{2}{\sqrt{4x^2+1}}dx, v = x$

$$\int \sinh^{-1} 2x\,dx = x\sinh^{-1} 2x - 2\int \frac{x}{\sqrt{4x^2+1}}dx$$

$$= x\sinh^{-1} 2x - \frac{1}{2}\sqrt{4x^2+1} + C$$

9.2.7 $u = x, dv = \sqrt{2x+1}\,dx, du = dx, v = \dfrac{1}{3}(2x+1)^{3/2}$

$$\int_0^4 x\sqrt{2x+1}\,dx = \frac{1}{3}\left[x(2x+1)^{3/2}\right]_0^4 - \frac{1}{3}\int_0^4 (2x+1)^{3/2}dx$$

$$= 36 - \frac{1}{15}\left[(2x+1)^{5/2}\right]_0^4 = \frac{298}{15}.$$

9.2.8 $u = x, dv = e^{-2x}, du = dx, v = -\dfrac{1}{2}e^{-2x}$

$$\int xe^{-2x}dx = -\frac{x}{2}e^{-2x} + \frac{1}{2}\int e^{-2x}dx = -\frac{x}{2}e^{-2x} - \frac{1}{4}e^{-2x} + C.$$

9.2.9 $u = x, dv = \sec^2 3x\,dx, du = dx, v = \dfrac{1}{3}\tan 3x$

$$\int x\sec^2 3x\,dx = \frac{x}{3}\tan 3x - \frac{1}{3}\int \tan 3x\,dx = \frac{x}{3}\tan 3x + \frac{1}{9}\ln|\cos 3x| + C.$$

9.2.10 $u = e^{-x}, dv = \cos 2x\,dx, du = -e^{-x}. v = \dfrac{1}{2}\sin 2x$

$$\int e^{-x}\cos 2x\,dx = \frac{e^{-x}}{2}\sin 2x + \frac{1}{2}\int e^{-x}\sin 2x\,dx$$

For $\displaystyle\int e^{-x}\sin 2x\,dx$, let $u = e^{-x}, dv = \sin 2x\,dx, du = -e^{-x}dx,$

$v = -\dfrac{1}{2}\cos 2x$ so $\displaystyle\int e^{-x}\sin 2x\,dx = -\frac{e^{-x}}{2}\cos 2x - \frac{1}{2}\int e^{-x}\cos 2x\,dx.$ Thus

$$\int e^{-x} \cos 2x \, dx = \frac{e^{-x}}{2} \sin 2x + \frac{1}{2} \left[-\frac{e^{-x}}{2} \cos 2x - \frac{1}{2} \int e^{-x} \cos 2x \, dx \right] + C_1.$$

$$\frac{5}{4} \int e^{-x} \cos 2x \, dx = \frac{e^{-x}}{2} \sin 2x - \frac{e^{-x}}{4} \cos 2x + C_1$$

$$\int e^{-x} \cos 2x \, dx = \frac{e^{-x}}{5} (2 \sin 2x - \cos 2x) + C.$$

9.2.11 $u = \ln^2 x$, $dv = dx$, $du = \dfrac{2}{x} \ln x \, dx$, $v = x$

$$\int \ln^2 x \, dx = x \ln^2 x - 2 \int \ln x \, dx.$$

For $\displaystyle\int \ln x \, dx$, let $u = \ln x$, $dv = dx$, $du = \dfrac{dx}{x}$, $v = x$

so $\displaystyle\int \ln x \, dx = x \ln x - \int dx = x \ln x - x$; thus,

$$\int \ln^2 x \, dx = x \ln^2 x - 2(x \ln x - x) + C = x \ln^2 x - 2x \ln x + 2x + C.$$

9.2.12 $u = \ln(x+2)$, $dv = dx$, $du = \dfrac{dx}{x+2}$, $v = x$

$$\int_{-1}^{2} \ln(x+2) dx = x \ln(x+2) \Big]_{-1}^{2} - \int_{-1}^{2} \frac{x}{x+2} dx = 2 \ln 4 - \int_{-1}^{2} \left(1 - \frac{2}{x+2} \right) dx$$

$$= 2 \ln 4 - \left[x + 2 \ln(x+2) \right]_{-1}^{2} = 4 \ln 4 - 3.$$

9.2.13 $u = x$, $dv = \cosh 2x \, dx$, $du = dx$, $v = \dfrac{1}{2} \sinh 2x$

$$\int x \cosh 2x \, dx = \frac{x}{2} \sinh 2x - \frac{1}{2} \int \sinh 2x \, dx = \frac{x}{2} \sinh 2x - \frac{1}{4} \cosh 2x + C.$$

9.2.14 $u = x$, $dv = \csc^2 2x \, dx$, $du = dx$, $v = -\dfrac{1}{2} \cot 2x$

$$\int x \csc^2 2x \, dx = -\frac{x}{2} \cot 2x + \frac{1}{2} \int \cot 2x \, dx = -\frac{x}{2} \cot 2x + \frac{1}{4} \ln |\sin 2x| + C.$$

9.2.15 $u = x$, $dv = \sinh 4x \, dx$, $du = dx$, $v = \dfrac{1}{4} \cosh 4x$

$$\int x \sinh 4x \, dx = \frac{x}{4} \cosh 4x - \int \frac{1}{4} \cosh 4x \, dx$$

$$= \frac{x}{4} \cosh 4x - \frac{1}{16} \sinh 4x + C.$$

9.2.16 $u = \cosh^{-1} 4x$, $dv = dx$, $du = \dfrac{4\,dx}{\sqrt{16x^2 - 1}}$, $v = x$

$$\int \cosh^{-1} 4x\,dx = x\cosh^{-1} 4x - 4\int \frac{x}{\sqrt{16x^2 - 1}}dx = x\cosh^{-1} 4x - \frac{1}{4}\sqrt{16x^2 - 1} + C.$$

9.2.17 $u = x$, $dv = \operatorname{sech}^2 2x\,dx$, $du = dx$, $v = \dfrac{1}{2}\tanh 2x$

$$\int x\operatorname{sech}^2 2x\,dx = \frac{x}{2}\tanh 2x - \frac{1}{2}\int \tanh 2x\,dx = \frac{x}{2}\tanh 2x - \frac{1}{4}\ln\cosh 2x + C.$$

9.2.18 $u = x^2$, $dv = \dfrac{x}{\sqrt{x^2 + 1}}dx$, $du = 2x\,dx$, $v = \sqrt{x^2 + 1}$

$$\int \frac{x^3}{\sqrt{x^2 + 1}}\,dx = x^2\sqrt{x^2 + 1} - 2\int x\sqrt{x^2 + 1}\,dx = x^2(x^2 + 1)^{1/2} - \frac{2}{3}(x^2 + 1)^{3/2} + C.$$

9.2.19 $u = \sin^{-1}\left(\dfrac{a}{x}\right)$, $dv = x\,dx$, $du = \dfrac{x}{\sqrt{x^2 - a^2}}\left(\dfrac{-a}{x^2}\right)dx = \dfrac{-a\,dx}{x\sqrt{x^2 - a^2}}$, $v = \dfrac{x^2}{2}$

$$\int x\sin^{-1}\left(\frac{a}{x}\right)dx = \frac{x^2}{2}\sin^{-1}\left(\frac{a}{x}\right) + \frac{a}{2}\int \frac{x\,dx}{\sqrt{x^2 - a^2}}$$

$$= \frac{x^2}{2}\sin^{-1}\left(\frac{a}{x}\right) + \frac{a}{2}\sqrt{x^2 - a^2} + C.$$

SECTION 9.3

9.3.1 Evaluate $\int \cos^3 2x \sin^2 2x \, dx$.

9.3.2 Evaluate $\int \cos^2 3x \sin^2 3x \, dx$.

9.3.3 Evaluate $\int \sin^3 x \cos^5 x \, dx$.

9.3.4 Evaluate $\int \sin^2 \frac{x}{2} \cos \frac{x}{2} \, dx$.

9.3.5 Evaluate $\int \sin^4 \frac{\theta}{3} \cos^3 \frac{\theta}{3} \, d\theta$.

9.3.6 Evaluate $\int \sin^2 \frac{t}{2} \cos^5 \frac{t}{2} \, dt$.

9.3.7 Evaluate $\int \sin^3 3\theta \, d\theta$.

9.3.8 Evaluate $\int \frac{\sin x}{\cos^5 x} \, dx$.

9.3.9 Evaluate $\int_{\pi/4}^{\pi/3} \frac{dx}{\cos^2 x}$.

9.3.10 Evaluate $\int \sin^4 2x \, dx$.

9.3.11 Evaluate $\int \cos^4 2x \, dx$.

9.3.12 Evaluate $\int_{0}^{\pi/2} \sin^2 2\theta \cos^2 2\theta \, d\theta$.

9.3.13 Evaluate $\int x \sin^2 x^2 \cos^2 x^2 \, dx$.

9.3.14 Evaluate $\int_{\pi/3}^{2\pi/3} \sin^4 \theta \cot^3 \theta \, d\theta$.

9.3.15 Evaluate $\int \frac{\cot^2 \theta}{\csc^2 \theta} \, d\theta$.

9.3.16 Evaluate $\int \frac{\sin \theta \cos \theta}{\sin^2 \theta + 1} \, d\theta$.

9.3.17 Evaluate $\int \cosh^4 x \sinh^3 x \, dx$.

9.3.18 Evaluate $\int x \sin x \cos x \, dx$.

SOLUTIONS

SECTION 9.3

9.3.1 $\displaystyle\int \cos^3 2x \sin^2 2x \, dx = \int (1 - \sin^2 2x) \sin^2 2x \cos 2x \, dx$

$$= \int (\sin^2 2x - \sin^4 2x) \cos 2x \, dx = \frac{1}{6} \sin^3 2x - \frac{1}{10} \sin^5 2x + C.$$

9.3.2 $\displaystyle\int \cos^2 3x \sin^2 3x \, dx = \frac{1}{4} \int (1 + \cos 6x)(1 - \cos 6x) dx$

$$= \frac{1}{4} \int (1 - \cos^2 6x) dx = \frac{1}{4} \int \sin^2 6x \, dx$$

$$= \frac{1}{8} \int (1 - \cos 12x) dx = \frac{x}{8} - \frac{1}{96} \sin 12x + C$$

9.3.3 $\displaystyle\int \sin^3 x \cos^5 x \, dx = \int (1 - \cos^2 x) \cos^5 x \sin x \, dx$

$$= \int (\cos^5 x - \cos^7 x) \sin x \, dx = -\frac{1}{6} \cos^6 x + \frac{1}{8} \cos^8 x + C$$

9.3.4 $\displaystyle\int \sin^2 \frac{x}{2} \cos \frac{x}{2} dx = \frac{2}{3} \sin^3 \frac{x}{2} + C$

9.3.5 $\displaystyle\int \sin^4 \frac{\theta}{3} \cos^3 \frac{\theta}{3} d\theta = \int \sin^4 \frac{\theta}{3} \left(1 - \sin^2 \frac{\theta}{3}\right) \cos \frac{\theta}{3} d\theta$

$$= \int \left(\sin^4 \frac{\theta}{3} - \sin^6 \frac{\theta}{3}\right) \cos \frac{\theta}{3} d\theta = \frac{3}{5} \sin^5 \frac{\theta}{3} - \frac{3}{7} \sin^7 \frac{\theta}{3} + C$$

9.3.6 $\displaystyle\int \sin^2 \frac{t}{2} \cos^5 \frac{t}{2} dt = \int \sin^2 \frac{t}{2} \left(1 - \sin^2 \frac{t}{2}\right)^2 \cos \frac{t}{2} dt$

$$= \int \sin^2 \frac{t}{2} \left(1 - 2\sin^2 \frac{t}{2} + \sin^4 \frac{t}{2}\right) \cos \frac{t}{2} dt$$

$$= \int \left(\sin^2 \frac{t}{2} - 2\sin^4 \frac{t}{2} + \sin^6 \frac{t}{2}\right) \cos \frac{t}{2} dt$$

$$= \frac{2}{3} \sin^3 \frac{t}{2} - \frac{4}{5} \sin^5 \frac{t}{2} + \frac{2}{7} \sin^7 \frac{t}{2} + C$$

9.3.7 $\displaystyle\int \sin^3 3\theta \, d\theta = \int (1 - \cos^2 3\theta) \sin 3\theta \, d\theta = -\frac{1}{3} \cos 3\theta + \frac{1}{9} \cos^3 3\theta + C$

9.3.8 $\displaystyle\int \frac{\sin x}{\cos^5 x} dx = \frac{1}{4 \cos^4 x} + C \text{ or } \frac{1}{4} \sec^4 x + C$

449

9.3.9 $\displaystyle\int_{\pi/4}^{\pi/3} \frac{dx}{\cos^2 x} = \int_{\pi/4}^{\pi/3} \sec^2 x\, dx = \tan x\Big]_{\pi/4}^{\pi/3} = \tan\frac{\pi}{3} - \tan\frac{\pi}{4} = \sqrt{3} - 1$

9.3.10 $\displaystyle\int \sin^4 2x\, dx = \frac{1}{4}\int(1 - \cos 4x)^2 dx = \frac{1}{4}\int(1 - 2\cos 4x + \cos^2 4x)dx$

$\displaystyle = \frac{1}{4}\int\left[1 - 2\cos 4x + \frac{1}{2}(1 + \cos 8x)\right]dx = \frac{1}{4}\int\left(\frac{3}{2} - 2\cos 4x + \frac{1}{2}\cos 8x\right)dx$

$\displaystyle = \frac{3x}{8} - \frac{1}{8}\sin 4x + \frac{1}{64}\sin 8x + C$

9.3.11 $\displaystyle\int \cos^4 2x\, dx = \frac{1}{4}\int(1 + \cos 4x)^2 dx = \frac{1}{4}\int(1 + 2\cos 4x + \cos^2 4x)dx$

$\displaystyle = \frac{1}{4}\int\left[1 + 2\cos 4x + \frac{1}{2}(1 + \cos 8x)\right]dx = \frac{1}{4}\int\left(\frac{3}{2} + 2\cos 4x + \frac{1}{2}\cos 8x\right)dx$

$\displaystyle = \frac{3x}{8} + \frac{1}{8}\sin 4x + \frac{1}{64}\sin 8x + C$

9.3.12 $\displaystyle\int_0^{\pi/2} \sin^2 2\theta \cos^2 2\theta = \frac{1}{4}\int_0^{\pi/2}(1 - \cos 4\theta)(1 + \cos 4\theta)d\theta$

$\displaystyle = \frac{1}{4}\int_0^{\pi/2}(1 - \cos^2 4\theta)d\theta = \frac{1}{4}\int_0^{\pi/2}\sin^2 4\theta\, d\theta$

$\displaystyle = \frac{1}{8}\int_0^{\pi/2}(1 - \cos 8\theta)d\theta = \frac{1}{8}\left[\theta - \frac{1}{8}\sin 8\theta\right]_0^{\pi/2} = \frac{\pi}{16}$

9.3.13 $\displaystyle\int x\sin^2 x^2 \cos^2 x^2 dx = \frac{1}{4}\int x(1 - \cos 2x^2)(1 + \cos 2x^2)dx$

$\displaystyle = \frac{1}{4}\int x(1 - \cos^2 2x^2)dx = \frac{1}{4}\int x\sin^2 2x^2 dx$

$\displaystyle = \frac{1}{8}\int x(1 - \cos 4x^2)dx = \frac{x^2}{16} - \frac{1}{64}\sin 4x^2 + C$

9.3.14 $\displaystyle\int_{\pi/3}^{2\pi/3} \sin^4\theta \cot^3\theta\, d\theta = \int_{\pi/3}^{2\pi/3} \cos^3\theta \sin\theta\, d\theta = -\frac{1}{4}\cos^4\theta\Big]_{\pi/3}^{2\pi/3} = 0$

9.3.15 $\displaystyle\int \frac{\cot^2\theta}{\csc^2\theta}d\theta = \int \cos^2\theta\, d\theta = \frac{1}{2}\int(1 + \cos 2\theta)d\theta = \frac{\theta}{2} + \frac{1}{4}\sin 2\theta + C$

9.3.16 $u = \sin^2\theta + 1,\; du = 2\sin\theta\cos\theta\, d\theta$

$\displaystyle \frac{1}{2}\int \frac{du}{u} = \frac{1}{2}\ln(\sin^2\theta + 1) + C$

9.3.17 $\displaystyle\int \cosh^4 x \sinh^3 x\, dx = \int \cosh^4 x (\cosh^2 x - 1) \sinh x\, dx$

$$= \int (\cosh^6 x - \cosh^4 x) \sinh x\, dx = \frac{1}{7} \cosh^7 x - \frac{1}{5} \cosh^5 x + C$$

9.3.18 $u = x,\ dv = \sin x \cos x\, dx,\ du = dx,\ v = \dfrac{1}{2}\sin^2 x$

$$\int x \sin x \cos x\, dx = \frac{x}{2} \sin^2 x - \frac{1}{2} \int \sin^2 x\, dx$$

$$= \frac{x}{2} \sin^2 x - \frac{1}{4} \int (1 - \cos 2x)\, dx = \frac{x}{2} \sin^2 x - \frac{x}{4} + \frac{1}{8} \sin 2x + C$$

SECTION 9.4

9.4.1 Evaluate $\displaystyle\int_0^{\pi/4} \tan^2 x\, dx$.

9.4.2 Evaluate $\displaystyle\int \tanh^2 2x\, \text{sech}^2 2x\, dx$.

9.4.3 Evaluate $\displaystyle\int \tan^3 \frac{x}{2} \sec^4 \frac{x}{2}\, dx$.

9.4.4 Evaluate $\displaystyle\int \cot^2 2x\, dx$.

9.4.5 Evaluate $\displaystyle\int (\tan x + \sec x)^2\, dx$.

9.4.6 Evaluate $\displaystyle\int \cot^4 2\theta\, d\theta$.

9.4.7 Evaluate $\displaystyle\int \tan^5 t \sec^4 t\, dt$.

9.4.8 Evaluate $\displaystyle\int \text{sech}^3 3x \tanh 3x\, dx$.

9.4.9 Evaluate $\displaystyle\int (\tan^2 x - \sec^2 x)^4\, dx$.

9.4.10 Evaluate $\displaystyle\int \csc^3 4x \cot^3 4x\, dx$.

9.4.11 Evaluate $\displaystyle\int \tan^5 x\, dx$.

9.4.12 Evaluate $\displaystyle\int \tan^3 2x \sec^6 2x\, dx$.

9.4.13 Evaluate $\displaystyle\int \tan^3 3\theta\, d\theta$.

9.4.14 Evaluate $\displaystyle\int \sec^3 \frac{x}{2} \tan \frac{x}{2}\, dx$.

9.4.15 Evaluate $\displaystyle\int \sec^6 \frac{x}{3} \tan^2 \frac{x}{3}\, dx$.

9.4.16 Evaluate $\displaystyle\int \frac{1}{\cos^4 x}\, dx$.

9.4.17 Evaluate $\displaystyle\int \frac{1}{\sec 2x \tan 2x}\, dx$.

9.4.18 Evaluate $\displaystyle\int \frac{\sin x}{\cos^2 x}\, dx$.

SOLUTIONS

SECTION 9.4

9.4.1 $\displaystyle\int_0^{\pi/4} \tan^2 x\, dx = \int_0^{\pi/4} (\sec^2 x - 1)dx = \Big[\tan x - x \Big]_0^{\pi/4} = 1 - \frac{\pi}{4}.$

9.4.2 $\dfrac{1}{6}\tanh^3 2x + C.$

9.4.3 $\displaystyle\int \tan^3 \frac{x}{2}\sec^4 \frac{x}{2}dx = \int \tan^3 \frac{x}{2}\sec^2 \frac{x}{2}\sec^2 \frac{x}{2}dx = \int \tan^3 \frac{x}{2}\left(\tan^2 \frac{x}{2}+1\right)\sec^2 \frac{x}{2}dx$

$\displaystyle\qquad = \int \left(\tan^5 \frac{x}{2}+\tan^3 \frac{x}{2}\right)\sec^2 \frac{x}{2}dx = \frac{1}{3}\tan^6 \frac{x}{2}+\frac{1}{2}\tan^4 \frac{x}{2}+C$

9.4.4 $\displaystyle\int \cot^2 2x\, dx = \int (\csc^2 2x - 1)dx = -\frac{1}{2}\cot 2x - x + C$

9.4.5 $\displaystyle\int (\tan x + \sec x)^2 dx = \int (\tan^2 x + 2\sec x \tan x + \sec^2 x)dx$

$\displaystyle\qquad = \int (\sec^2 x - 1 + 2\sec x \tan x + \sec^2 x)dx$

$\displaystyle\qquad = \int (2\sec^2 x + 2\sec x \tan x - 1)dx$

$\displaystyle\qquad = 2\tan x + 2\sec x - x + C$

9.4.6 $\displaystyle\int \cot^4 2\theta\, d\theta = \int \cot^2 2\theta \cot^2 2\theta\, d\theta$

$\displaystyle\qquad = \int \cot^2 2\theta(\csc^2 2\theta - 1)d\theta = \int (\cot^2 2\theta \csc^2 2\theta - \cot^2 2\theta)d\theta$

$\displaystyle\qquad = \int (\cot^2 2\theta \csc^2 2\theta - \csc^2 2\theta + 1)d\theta = -\frac{1}{6}\cot^3 2\theta + \frac{1}{2}\cot 2\theta + \theta + C$

9.4.7 $\displaystyle\int \tan^5 t \sec^4 t\, dt = \int \tan^5 t \sec^2 t \sec^2 t\, dt = \int \tan^5 t(\tan^2 t + 1)\sec^2 t\, dt$

$\displaystyle\qquad = \int (\tan^7 t + \tan^5 t)\sec^2 t\, dt = \frac{1}{8}\tan^8 t + \frac{1}{6}\tan^6 t + C.$

9.4.8 $\displaystyle\int \operatorname{sech}^3 3x \tanh 3x\, dx = \int \operatorname{sech}^2 3x \operatorname{sech} 3x \tanh 3x\, dx = -\frac{1}{9}\operatorname{sech}^3 3x + C.$

9.4.9 $\displaystyle\int (\tan^2 x - \sec^2 x)^4 dx = \int (-1)^4 dx = x + C.$

9.4.10 $\displaystyle\int \csc^3 4x \cot^3 4x\,dx = \int \csc^2 4x(\csc^2 4x - 1)\csc 4x \cot 4x\,dx$

$$= \int (\csc^4 4x - \csc^2 4x)\csc 4x \cot 4x\,dx$$

$$= -\frac{1}{20}\csc^5 4x + \frac{1}{12}\csc^3 4x + C.$$

9.4.11 $\displaystyle\int \tan^5 x\,dx = \int \tan^3 x(\sec^2 x - 1)dx = \int \left[\tan^3 x \sec^2 x - \tan x(\sec^2 x - 1)\right]dx$

$$= \frac{1}{4}\tan^4 x - \frac{1}{2}\tan^2 x - \ln|\cos x| + C.$$

9.4.12 $\displaystyle\int \tan^3 2x \sec^6 2x\,dx = \int \tan^3 2x \sec^4 2x \sec^2 2x\,dx = \int \tan^3 2x(\tan^2 2x + 1)^2 \sec^2 2x\,dx$

$$= \int \tan^3 2x(\tan^4 2x + 2\tan^2 2x + 1)\sec^2 2x\,dx$$

$$= \int (\tan^7 2x + 2\tan^5 2x + \tan^3 2x)\sec^2 2x\,dx$$

$$= \frac{1}{16}\tan^8 2x + \frac{1}{6}\tan^6 2x + \frac{1}{8}\tan^4 2x + C.$$

9.4.13 $\displaystyle\int \tan^3 3\theta\,d\theta = \int \tan^2 3\theta \tan 3\theta\,d\theta = \int (\sec^2 3\theta - 1)\tan 3\theta\,d\theta$

$$= \int (\tan 3\theta \sec^2 3\theta - \tan 3\theta)d\theta = \frac{1}{6}\tan^2 3\theta + \frac{1}{3}\ln|\cos 3\theta| + C.$$

9.4.14 $\displaystyle\int \sec^3 \frac{x}{2}\tan \frac{x}{2}\,dx = \int \sec^2 \frac{x}{2}\sec \frac{x}{2}\tan \frac{x}{2}\,dx = \frac{2}{3}\sec^3 \frac{x}{2} + C.$

9.4.15 $\displaystyle\int \sec^6 \frac{x}{3}\tan^2 \frac{x}{3}\,dx = \int \sec^4 \frac{x}{3}\sec^2 \frac{x}{3}\tan^2 \frac{x}{3}\,dx = \int \left(\tan^2 \frac{x}{3} + 1\right)^2 \tan^2 \frac{x}{3}\sec^2 \frac{x}{3}\,dx$

$$= \int \left(\tan^6 \frac{x}{3} + 2\tan^4 \frac{x}{3} + \tan^2 \frac{x}{3}\right)\sec^2 \frac{x}{3}\,dx$$

$$= \frac{3}{7}\tan^7 \frac{x}{3} + \frac{6}{5}\tan^5 \frac{x}{3} + \tan^3 \frac{x}{3} + C.$$

9.4.16 $\displaystyle\int \frac{1}{\cos^4 x}\,dx = \int \sec^4 x\,dx = \int \sec^2 x \sec^2 x\,dx = \int (\tan^2 x + 1)\sec^2 x\,dx$

$$= \frac{1}{3}\tan^3 x + \tan x + C.$$

9.4.17 $\displaystyle\int \frac{1}{\sec 2x \tan 2x}\,dx = \int \frac{\cos^2 2x}{\sin 2x} = \int \frac{(1 - \sin^2 2x)}{\sin 2x}\,dx = \int (\csc 2x - \sin 2x)\,dx$

$$= \frac{1}{2}\ln|\csc 2x + \cot 2x| + \frac{1}{2}\cos 2x + C.$$

9.4.18 $\displaystyle\int \frac{\sin x}{\cos^2 x}\,dx = \int \sec x \tan x\,dx = \sec x + C.$

SECTION 9.5

9.5.1 Evaluate $\int \dfrac{x^3}{\sqrt{25 - 4x^2}}\,dx$.

9.5.2 Evaluate $\int \dfrac{1}{x^2\sqrt{4 - x^2}}\,dx$.

9.5.3 Evaluate $\int \dfrac{1}{(x^2 + 4)^{3/2}}\,dx$.

9.5.4 Evaluate $\int \dfrac{1}{x^2\sqrt{x^2 + 4}}\,dx$.

9.5.5 Evaluate $\int \dfrac{1}{(4x^2 - 9)^{3/2}}\,dx$.

9.5.6 Evaluate $\displaystyle\int_0^{\sqrt{2}} \dfrac{x^2}{\sqrt{4 - x^2}}\,dx$.

9.5.7 Evaluate $\displaystyle\int_5^{5\sqrt{3}} \dfrac{1}{x^2\sqrt{x^2 + 25}}\,dx$.

9.5.8 Evaluate $\int \dfrac{1}{x^2\sqrt{9 - x^2}}\,dx$.

9.5.9 Evaluate $\int \dfrac{1}{\sqrt{2 + 4x^2}}\,dx$.

9.5.10 Evaluate $\int \dfrac{1}{x\sqrt{x^2 - 4}}\,dx$.

9.5.11 Evaluate $\displaystyle\int_{2\sqrt{2}}^4 \dfrac{\sqrt{x^2 - 4}}{x}\,dx$.

9.5.12 Evaluate $\displaystyle\int_2^4 \dfrac{dx}{\sqrt{x^2 - 1}}\,dx$.

9.5.13 Evaluate $\int \dfrac{1}{(x^2 - 2x + 10)^{3/2}}\,dx$.

9.5.14 Evaluate $\displaystyle\int_{-1}^1 \dfrac{1}{\sqrt{x^2 + 2x + 2}}\,dx$.

9.5.15 Evaluate $\int \dfrac{1}{\sqrt{x^2 - 2x - 8}}\,dx$.

9.5.16 Evaluate $\int \dfrac{x}{\sqrt{x^2 - 2x - 8}}\,dx$.

9.5.17 Evaluate $\displaystyle\int_2^4 \dfrac{1}{x^2 - 4x + 8}\,dx$.

9.5.18 Evaluate $\int \dfrac{1}{(4x^2 - 24x + 27)^{3/2}}\,dx$.

SOLUTIONS

SECTION 9.5

9.5.1 $2x = 5\sin\theta$, $dx = \dfrac{5}{2}\cos\theta\, d\theta$

$$\frac{125}{16}\int \sin^3\theta\, d\theta = \frac{125}{16}\left(-\cos\theta + \frac{1}{3}\cos^3\theta\right) + C$$

$$= -\frac{25}{16}(25 - 4x^2)^{1/2} - \frac{1}{48}(25 - 4x^2)^{3/2} + C$$

9.5.2 $x = 2\sin\theta$, $dx = 2\cos\theta\, d\theta$

$$\frac{1}{4}\int \csc^2\theta\, d\theta = -\frac{1}{4}\cot\theta + C = -\frac{\sqrt{4 - x^2}}{4x} + C.$$

9.5.3 $x = 2\tan\theta$, $dx = 2\sec^2\theta\, d\theta$

$$\frac{1}{4}\int \cos\theta\, d\theta = \frac{1}{4}\sin\theta + C = \frac{x}{4\sqrt{x^2 + 4}} + C.$$

9.5.4 $x = 2\tan\theta$, $dx = 2\sec^2\theta\, d\theta$

$$\frac{1}{4}\int \cot\theta\csc\theta\, d\theta = -\frac{1}{4}\csc\theta + C = -\frac{\sqrt{4 + x^2}}{4x} + C.$$

9.5.5 $2x = 3\sec\theta$, $dx = \dfrac{3}{2}\sec\theta\tan\theta\, d\theta$

$$\frac{1}{18}\int \csc\theta\cot\theta\, d\theta = -\frac{1}{18}\csc\theta + C = -\frac{x}{9\sqrt{4x^2 - 9}} + C.$$

9.5.6 $x = 2\sin\theta$, $dx = 2\cos\theta\, d\theta$

$$4\int_0^{\pi/4} \sin^2\theta\, d\theta = 2\left(\theta - \sin\theta\cos\theta\right)\Big]_0^{\pi/4}$$

$$= 2\left(\frac{\pi}{4} - \frac{1}{2}\right) = \frac{\pi - 2}{2}$$

9.5.7 $x = 5\tan\theta$, $dx = 5\sec^2\theta\, d\theta$

$$\frac{1}{25}\int_{\pi/4}^{\pi/3} \cot\theta\csc\theta\, d\theta = -\frac{1}{25}\csc\theta\Big]_{\pi/4}^{\pi/3}$$

$$= -\frac{1}{25}\left[\frac{2}{\sqrt{3}} - \sqrt{2}\right]$$

9.5.8 $x = 3\sin\theta$, $dx = 3\cos\theta\, d\theta$

$$\frac{1}{9}\int \csc^2\theta\, d\theta = -\frac{1}{9}\cot\theta + C = -\frac{\sqrt{9 - x^2}}{9x} + C.$$

461

9.5.9 $2x = \sqrt{2}\tan\theta$, $dx = \dfrac{\sqrt{2}}{2}\sec^2\theta\,d\theta$

$$\frac{1}{2}\int \sec\theta\,d\theta = \frac{1}{2}\ln|\sec\theta + \tan\theta| + C = \frac{1}{2}\ln\left|\sqrt{2+4x^2} + 2x\right| + C.$$

9.5.10 $x = 2\sec\theta$, $dx = 2\sec\theta\tan\theta\,d\theta$

$$\frac{1}{2}\int d\theta = \frac{1}{2}\theta + C = \frac{1}{2}\sec^{-1}\frac{x}{2} + C.$$

9.5.11 $x = 2\sec\theta$, $dx = 2\sec\theta\tan\theta\,d\theta$

$$2\int_{\pi/4}^{\pi/3} \tan^2\theta\,d\theta = 2\left(\tan\theta - \theta\right)\Big]_{\pi/4}^{\pi/3} = 2\left[\left(\sqrt{3} - \frac{\pi}{3}\right) - \left(1 - \frac{\pi}{4}\right)\right]$$

$$= \frac{6\sqrt{3} - 12 + \pi}{6}$$

9.5.12 $x = \sec\theta$, $dx = \sec\theta\tan\theta\,d\theta$

$$\int_{\pi/3}^{\sec^{-1}4} \sec\theta\,d\theta = \ln|\sec\theta + \tan\theta|\Big]_{\pi/3}^{\sec^{-1}4}$$

$$= \ln(4 + \sqrt{15}) - \ln(2 + \sqrt{3}) = \ln\left(\frac{4 + \sqrt{15}}{2 + \sqrt{3}}\right).$$

9.5.13 $\displaystyle\int \frac{1}{[(x-1)^2 + 9]^{3/2}}dx$, let $u = x - 1$, $du = dx$;

$$\int \frac{du}{(u^2 + 9)^{3/2}}, u = 3\tan\theta, du = 3\sec^2\theta\,d\theta$$

$$\int \frac{3\sec^2\theta\,d\theta}{(9\tan^2\theta + 9)^{3/2}} = \frac{1}{9}\int \cos\theta\,d\theta = \frac{1}{9}\sin\theta + C.$$

$$\frac{x-1}{9(x^2 - 2x + 10)^{3/2}} + C.$$

9.5.14 $\displaystyle\int_{-1}^{1} \frac{dx}{\sqrt{(x+1)^2 + 1}}$, $u = x + 1$, $du = dx$, $\displaystyle\int_{0}^{2} \frac{du}{\sqrt{u^2 + 1}}$, $u = \tan\theta$, $du = \sec^2\theta$

$$\int_{0}^{\tan^{-1}2} \frac{\sec^2\theta\,d\theta}{\sqrt{\tan^2\theta + 1}} = \int_{0}^{\tan^{-1}2} \sec\theta\,d\theta = \left[\ln(\sec\theta + \tan\theta)\right]_{0}^{\tan^{-1}2} = \ln(\sqrt{5} + 2).$$

9.5.15 $\displaystyle\int \frac{1}{\sqrt{(x-1)^2 - 9}}dx$, $u = x - 1$, $du = dx$, $\displaystyle\int \frac{du}{\sqrt{u^2 - 9}}$, $u = 3\sec\theta$, $du = 3\sec\theta\tan\theta\,d\theta$

$$\int \frac{3\sec\theta\tan\theta\,d\theta}{\sqrt{9\sec^2\theta - 9}} = \int \sec\theta\,d\theta = \ln|\sec\theta + \tan\theta| + C = \ln\left|x - 1 + \sqrt{x^2 - 2x - 8}\right| + C.$$

9.5.16 $\displaystyle\int \frac{x\,dx}{\sqrt{(x-1)^2-9}}$, $u=x-1$, $du=dx$, $\displaystyle\int \frac{(u+1)du}{\sqrt{u^2-9}}$, $u=3\sec\theta$, $du=3\sec\theta\tan\theta\,d\theta$

$$\int \frac{(3\sec\theta+1)3\sec\theta\tan\theta\,d\theta}{\sqrt{9sec^2\theta-9}} = \int (3\sec^2\theta+\sec\theta)d\theta = 3\tan\theta + \ln|\sec\theta+\tan\theta| + C$$

$$= \sqrt{x^2-2x-8} + \ln\left|x-1+\sqrt{x^2-2x-8}\right| + C$$

9.5.17 $\displaystyle\int_2^4 \frac{dx}{(x-2)^2+4}$, $u=x-2$, $du=dx$, $\displaystyle\int_0^2 \frac{du}{u^2+4}$, $u=2\tan\theta$, $du=2\sec^2\theta\,d\theta$,

$$\int_0^{\pi/4} \frac{2\sec^2\theta\,d\theta}{4\tan^2\theta+4} = \frac{1}{2}\int_0^{\pi/4} d\theta = \frac{\pi}{8}.$$

9.5.18 $\displaystyle\int \frac{dx}{[4(x-3)^2-9]^{3/2}}$, $u=x-3$, $du=dx$, $\displaystyle\int \frac{du}{(4u^2-9)^{3/2}}$,

$2u=3\sec\theta$, $du=\dfrac{3}{2}\sec\theta\tan\theta\,d\theta$,

$$\int \frac{\frac{3}{2}\sec\theta\tan\theta\,d\theta}{(9\sec^2\theta-9)} = \frac{1}{18}\int \cot\theta\csc\theta\,d\theta$$

$$= -\frac{1}{18}\csc\theta + C = \frac{3-x}{9\sqrt{4x^2-24x+27}} + C.$$

SECTION 9.6

9.6.1 Evaluate $\int \dfrac{x^2 - 6}{x(x-1)^2}\,dx$.

9.6.2 Evaluate $\int \dfrac{x+3}{(x-1)(x^2 - 4x + 4)}\,dx$.

9.6.3 Evaluate $\int \dfrac{x+2}{x - x^3}\,dx$.

9.6.4 Evaluate $\int \dfrac{x+1}{x^2(x-1)}\,dx$.

9.6.5 Evaluate $\int \dfrac{x^2}{x^2 - 2x + 1}\,dx$.

9.6.6 Evaluate $\int \dfrac{x^4}{x^4 - 1}\,dx$.

9.6.7 Evaluate $\int \dfrac{4x^2 - 3x}{(x-2)(x^2 + 1)}\,dx$.

9.6.8 Evaluate $\int \dfrac{2x - 3}{x^3 - 3x^2 + 2x}\,dx$.

9.6.9 Evaluate $\int \dfrac{2x + 1}{x^3 + x^2 + 2x + 2}\,dx$.

9.6.10 Evaluate $\int \dfrac{\ln x}{(x+1)^2}\,dx$.

9.6.11 Evaluate $\int \dfrac{x + 4}{x^3 + 3x^2 - 10x}\,dx$.

9.6.12 Evaluate $\int \dfrac{x + 1}{x^2 + 2x - 3}\,dx$.

9.6.13 Evaluate $\int \dfrac{\cos \theta}{\sin^2 \theta + 4\sin \theta - 5}\,d\theta$.

9.6.14 Evaluate $\int \dfrac{4x}{x^3 - x^2 - x + 1}\,dx$.

9.6.15 Evaluate $\int \dfrac{x + 4}{x^3 + x}\,dx$.

9.6.16 Evaluate $\int \dfrac{x^2 + 3x - 1}{x^3 - 1}\,dx$.

9.6.17 Find the area of the region bounded by the curve $y = \dfrac{x - 4}{x^2 - 5x + 6}$, and the x-axis for $6 \le x \le 8$.

SOLUTIONS

SECTION 9.6

9.6.1 $\dfrac{x^2 - 6}{x(x-1)^2} = \dfrac{A}{x} + \dfrac{B}{x-1} + \dfrac{C}{(x-1)^2}$; $A = -6$, $B = 7$, $C = -5$

$$-6\int \frac{1}{x}dx + 7\int \frac{1}{x-1}dx - 5\int \frac{1}{(x-1)^2}dx = -6\ln|x| + 7\ln|x-1| + \frac{5}{x-1} + C.$$

9.6.2 $\dfrac{x+3}{(x-1)(x-2)^2} = \dfrac{A}{x-1} + \dfrac{B}{x-2} + \dfrac{C}{(x-2)^2}$; $A = 4$, $B = -4$, $C = 5$

$$4\int \frac{1}{x-1}dx - 4\int \frac{1}{x-2}dx + 5\int \frac{1}{(x-2)^2}dx = 4\ln|x-1| - 4\ln|x-2| - \frac{5}{x-2} + C$$

$$= 4\ln\left|\frac{x-1}{x-2}\right| - \frac{5}{(x-2)} + C.$$

9.6.3 $\dfrac{x+2}{x(1-x)(1+x)} = \dfrac{A}{x} + \dfrac{B}{1-x} + \dfrac{C}{1+x}$; $A = 2$, $B = 3/2$, $C = -1/2$

$$2\int \frac{1}{x}dx + \frac{3}{2}\int \frac{1}{1-x}dx - \frac{1}{2}\int \frac{1}{1+x}dx = 2\ln|x| - \frac{3}{2}\ln|1-x| - \frac{1}{2}\ln|1+x| + C.$$

9.6.4 $\dfrac{x+1}{x^2(x-1)} = \dfrac{A}{x} + \dfrac{B}{x^2} + \dfrac{C}{x-1}$; $A = -2$, $B = -1$, $C = 2$

$$-2\int \frac{1}{x}dx - \int \frac{1}{x^2}dx + 2\int \frac{1}{x-1}dx = 2\ln\left|\frac{x-1}{x}\right| + \frac{1}{x} + C.$$

9.6.5 $\dfrac{x^2}{x^2 - 2x + 1} = 1 + \dfrac{2x-1}{(x-1)^2}$

$\dfrac{2x-1}{(x-1)^2} = \dfrac{A}{x-1} + \dfrac{B}{(x-1)^2}$; $A = 2$, $B = 1$

$$\int dx + 2\int \frac{1}{x-1}dx + \int \frac{1}{(x-1)^2}dx = x + 2\ln|x-1| - \frac{1}{x-1} + C.$$

9.6.6 $\dfrac{x^4}{x^4 - 1} = 1 + \dfrac{1}{x^4 - 1} = 1 + \dfrac{1}{(x^2+1)(x+1)(x-1)}$

$\dfrac{1}{(x^2+1)(x+1)(x-1)} = \dfrac{Ax+B}{x^2+1} + \dfrac{C}{x+1} + \dfrac{D}{x-1}$; $A = 0$, $B = -\dfrac{1}{2}$, $C = -\dfrac{1}{4}$, $D = \dfrac{1}{4}$.

$$\int dx - \frac{1}{2}\int \frac{1}{x^2+1}dx - \frac{1}{4}\int \frac{1}{x+1}dx + \frac{1}{4}\int \frac{1}{x-1}dx = x - \frac{1}{2}\tan^{-1}x + \frac{1}{4}\ln\left|\frac{x-1}{x+1}\right| + C$$

9.6.7 $\dfrac{4x^2 - 3x}{(x-2)(x^2+1)} = \dfrac{A}{x-2} + \dfrac{Bx+C}{x^2+1}$; $A = 2$, $B = 2$, $C = 1$

$$2\int \frac{1}{x-2}dx + 2\int \frac{x}{x^2+1}dx + \int \frac{1}{x^2+1}dx = \ln|x-2| + \ln(x^2+1) + \tan^{-1}x + C.$$

9.6.8 $\dfrac{2x-3}{x(x-1)(x-2)} = \dfrac{A}{x} + \dfrac{B}{x-1} + \dfrac{C}{x-2}$; $A = -3/2$, $B = 1$, $C = 1/2$

$$-\frac{3}{2}\int \frac{1}{x}dx + \int \frac{1}{x-1}dx + \frac{1}{2}\int \frac{1}{x-2}dx = -\frac{3}{2}\ln|x| + \ln|x-1| + \frac{1}{2}\ln|x-2| + C.$$

9.6.9 $\dfrac{2x+1}{(x+1)(x^2+2)} = \dfrac{A}{x+1} + \dfrac{Bx+C}{x^2+2}$; $A = -\dfrac{1}{3}$, $B = \dfrac{1}{3}$, $C = \dfrac{5}{3}$

$$-\frac{1}{3}\int \frac{1}{x+1}dx + \frac{1}{3}\int \frac{x}{x^2+2}dx + \frac{5}{3}\int \frac{1}{x^2+2}dx = -\frac{1}{3}\ln$$

$$|x+1| + \frac{1}{6}\ln(x^2+2) + \frac{5\sqrt{2}}{6}\tan^{-1}\frac{x}{\sqrt{2}} + C.$$

9.6.10 $u = \ln x$, $dv = \dfrac{1}{(x+1)^2}dx$, $du = \dfrac{1}{x}dx$, $v = -\dfrac{1}{x+1}$

$$\int \frac{\ln x}{(x+1)^2}dx = -\frac{\ln x}{x+1} + \int \frac{1}{x(x+1)}dx; \quad \frac{1}{x(x+1)} = \frac{A}{x} + \frac{B}{x+1};$$

$$A = 1, B = -1 \int \frac{1}{x}dx - \int \frac{1}{x+1}dx = \ln|x| - \ln|x+1| + C = \ln\left|\frac{x}{x+1}\right| + C$$

so $\displaystyle\int \frac{\ln x}{(x+1)^2}dx = -\frac{\ln x}{x+1} + \ln\left|\frac{x}{x+1}\right| + C.$

9.6.11 $\dfrac{x+4}{x(x-2)(x+5)} = \dfrac{A}{x} + \dfrac{B}{x-2} + \dfrac{C}{x+5}$; $A = -\dfrac{2}{5}$, $B = \dfrac{3}{7}$, $C = -\dfrac{1}{35} - \dfrac{2}{5}$

$$\int \frac{1}{x}dx + \frac{3}{7}\int \frac{1}{x-2}dx - \frac{1}{35}\int \frac{1}{x+5}dx = -\frac{2}{5}\ln|x| + \frac{3}{7}\ln|x-2| - \frac{1}{35}\ln|x+5| + C.$$

9.6.12 $\displaystyle\int \frac{x+1}{x^2+2x-3}dx = \frac{1}{2}\ln|x^2+2x-3| + C$

9.6.13 $u = \sin\theta$, $du = \cos\theta\, d\theta$

$$\int \frac{1}{u^2+4u-5}\, du; \quad \frac{1}{(u-1)(u+5)} = \frac{A}{u-1} + \frac{B}{u+5}; \quad A = \frac{1}{6}, B = -\frac{1}{6}$$

$$\frac{1}{6}\int \frac{1}{u-1}du - \frac{1}{6}\int \frac{1}{u+5}du = \frac{1}{6}\ln|u-1| - \frac{1}{6}\ln|u+5| + C$$

$$= \frac{1}{6}\ln\left|\frac{u-1}{u+5}\right| + C = \frac{1}{6}\ln\left|\frac{\sin\theta-1}{\sin\theta+5}\right| + C.$$

9.6.14 $\dfrac{4x}{(x-1)^2(x+1)} = \dfrac{A}{x-1} + \dfrac{B}{(x-1)^2} + \dfrac{C}{x+1}$; $A=1$, $B=2$, $C=-1$

$$\int \frac{1}{x-1}dx + 2\int \frac{1}{(x-1)^2}dx - \int \frac{1}{x+1}dx = \ln|x-1| - \frac{2}{x-1} - \ln|x+1| + C.$$

9.6.15 $\dfrac{x+4}{x(x^2+1)} = \dfrac{A}{x} + \dfrac{Bx+C}{x^2+1}$; $A=4$, $B=-4$, $C=1$

$$4\int \frac{1}{x}dx - 4\int \frac{x}{x^2+1}dx + \int \frac{1}{x^2+1}dx = 4\ln|x| - 2\ln(x^2+1) + \tan^{-1}x + C.$$

9.6.16 $\dfrac{x^2+3x-1}{(x-1)(x^2+x+1)} = \dfrac{A}{x-1} + \dfrac{Bx+C}{x^2+x+1}$; $A=1$, $B=0$, $C=2$

$$\int \frac{1}{x-1}dx + 2\int \frac{1}{x^2+x+1}dx, \quad \int \frac{1}{x^2+x+1}dx = \int \frac{1}{(x+1/2)^2+3/4}dx, \quad u=x+1/2,$$

$$du = dx, \quad \int \frac{du}{u^2+\frac{3}{4}} = \frac{2}{\sqrt{3}}\tan^{-1}\left(\frac{2x+1}{\sqrt{3}}\right) + C$$

so $\displaystyle\int \frac{x^2+3x-1}{(x-1)(x^2+x+1)}dx = \ell n|x-1| + \frac{4}{\sqrt{3}}\tan^{-1}\left(\frac{2x+1}{\sqrt{3}}\right) + C.$

9.6.17 $A = \displaystyle\int_6^8 \frac{x-4}{x^2-5x+6}dx$ $\dfrac{x-4}{x^2-5x+6} = \dfrac{x-4}{(x-3)(x-2)} = \dfrac{A}{x-3} + \dfrac{B}{x-2}$; $A=-1$,

$$B = 2A = -\int_6^8 \frac{1}{x-3}dx + 2\int_6^8 \frac{1}{x-2}dx = \left[-\ln|x-3| + 2\ln|x-2| \right]_6^8$$

$$\ln\frac{108}{80} = \ln 1.35.$$

SECTION 9.7

9.7.1 Evaluate $\displaystyle\int \frac{x^{1/2}}{1+x^{3/4}}\,dx$

9.7.2 Evaluate $\displaystyle\int \frac{dx}{x-x^{5/3}}$

9.7.3 Evaluate $\displaystyle\int \frac{x^{1/2}}{x-x^{4/3}}\,dx$

9.7.4 Evaluate $\displaystyle\int \frac{x^{3/2}-x^{1/3}}{4x^{1/4}}\,dx$

9.7.5 Evaluate $\displaystyle\int \frac{e^{4x}}{(1+e^{2x})^{2/3}}\,dx$

9.7.6 Evaluate $\displaystyle\int \frac{e^{2x}}{(1+e^{x})^{1/3}}\,dx$

9.7.7 Evaluate $\displaystyle\int \frac{e^{2x}}{(1+e^{x})^{4/3}}\,dx$

9.7.8 Evaluate $\displaystyle\int \frac{dx}{1+\sin x+\cos x}$

9.7.9 Evaluate $\displaystyle\int \frac{dx}{3+2\cos x}$

9.7.10 Evaluate $\displaystyle\int \frac{dx}{2+3\cos x}$

SOLUTIONS

SECTION 9.7

9.7.1   Let $u = x^{1/4}$, $x = u^4$, $x^{1/2} = u^2$, $x^{3/4} = u^3$, $dx = 4u^3 du$

$$\int \frac{x^{1/2}}{1 + x^{3/4}} dx = \int \frac{4u^5}{1 + u^3} du = 4 \int \left(u^2 - \frac{u^2}{1 + u^3} \right) du = \frac{4}{3} u^3 - \frac{4}{3} \ln(1 + u^3) + C$$

thus $\displaystyle\int \frac{x^{1/2}}{1 + x^{3/4}} dx = \frac{4}{3} x^{3/4} - \frac{4}{3} \ln(1 + x^{3/4}) + C.$

9.7.2   Let $u = x^{1/3}$, $x = u^3$, $x^{5/3} = u^5$, $dx = 3u^2 du$

$$\int \frac{dx}{x - x^{5/3}} = \int \frac{3du}{u - u^3} = 3 \ln \left| \frac{u}{\sqrt{1 - u^2}} \right| + C$$

thus $\displaystyle\int \frac{dx}{x - x^{5/3}} = 3 \ln \left| \frac{x^{1/3}}{\sqrt{1 - x^{2/3}}} \right| + C.$

9.7.3   Let $u = x^{1/6}$, $x = u^6$, $x^{1/2} = u^3$, $x^{4/3} = u^8$, $dx = 6u^5 du$

$$\int \frac{x^{1/2}}{x - x^{4/3}} dx = \int \frac{6u^8}{u^6 - u^8} du = 6 \int \left(-1 + \frac{1}{1 - u^2} \right) du$$

$$= 6 \left[-u - \frac{1}{2} \ln |1 - u| + \frac{1}{2} \ln |1 + u| \right] + C, \text{ thus}$$

$$\int \frac{x^{1/2}}{x - x^{4/3}} dx = -6x^{1/6} + 3 \ln \left| \frac{1 + x^{1/6}}{1 - x^{1/6}} \right| + C$$

9.7.4   Let $u = x^{1/12}$, $x = u^{12}$, $x^{3/2} = u^{18}$, $x^{1/3} = u^4$, $x^{1/4} = u^3$, $dx = 12u^{11} du$

$$\int \frac{x^{3/2} - x^{1/3}}{4x^{1/4}} dx = 3 \int \frac{u^{29} - u^{15}}{u^3} du = \frac{1}{9} u^{27} - \frac{3}{13} u^{13} + C$$

thus $\displaystyle\int \frac{x^{3/2} - x^{1/3}}{4x^{1/4}} dx = \frac{1}{9} x^{9/4} - \frac{3}{13} x^{13/12} + C.$

9.7.5   Let $u = (1 + e^{2x})^{2/3}$, then $e^{2x} = u^{3/2} - 1$ and $x = \frac{1}{2} \ln(u^{3/2} - 1)$, thus,

$$dx = \frac{3}{4} \frac{u^{1/2}}{u^{3/2} - 1} du, \; e^{4x} = (u^{3/2} - 1)^2 \text{ and } \int \frac{e^{4x}}{(1 + e^{2x})^{2/3}} dx = \frac{3}{4} \int (u - u^{-1/2}) du$$

$$= \frac{3}{8} u^2 - \frac{3}{2} u^{1/2} + C, \text{ so } \int \frac{e^{4x}}{(1 + e^{2x})^{2/3}} dx = \frac{3}{8} (1 + e^{2x})^{4/3} - \frac{3}{2} (1 + e^{2x})^{1/3} + C$$

or $\dfrac{3}{8} (1 + e^{2x})^{1/3} (e^{2x} - 3) + C.$

9.7.6 Let $u = (1 + e^x)^{1/3}$, $u^3 = 1 + e^x$, $x = \ln(u^3 - 1)$, $dx = \dfrac{3u^2}{u^3 - 1}\,du$, thus,

$$\int \frac{e^{2x}}{(1 + e^x)^{1/3}}\,dx = 3\int (u^4 - u)\,du = \frac{3}{5}u^5 - \frac{3}{2}u^2 + C \text{ so}$$

$$\int \frac{e^{2x}}{(1 + e^x)^{1/3}}\,dx = \frac{3}{5}(1 + e^x)^{5/3} - \frac{3}{2}(1 + e^x)^{2/3} + C \text{ or } \frac{3}{10}(1 + e^x)^{2/3}(2e^x - 3) + C.$$

9.7.7 Let $u = (1 + e^x)^{4/3}$, $u^{3/4} = 1 + e^x$, $x = \ln(u^{3/4} - 1)$ and $dx = \dfrac{3}{4u^{1/4}(u^{3/4} - 1)}\,du$, thus,

$$\int \frac{e^{2x}}{(1 + e^x)^{4/3}}\,dx = \frac{3}{4}\int (u^{-1/2} - u^{-5/4})\,du = \frac{3}{2}u^{1/2} + 3u^{-1/4} + C$$

$$\int \frac{e^{2x}}{(1 + e^x)^{4/3}}\,dx = \frac{3}{2}(1 + e^x)^{2/3} + 3(1 + e^x)^{-1/3} + C \text{ or } \frac{3}{2}(1 + e^x)^{-1/3}(e^x + 3) + C$$

9.7.8 Let $\sin x = \dfrac{2u}{1 + u^2}$, $\cos x = \dfrac{1 - u^2}{1 + u^2}$, and $dx = \dfrac{2}{1 + u^2}\,du$, then

$$\int \frac{dx}{1 + \sin x + \cos x} = \int \frac{\dfrac{2}{1 + u^2}}{1 + \dfrac{2u}{1 + u^2} + \dfrac{1 - u^2}{1 + u^2}}\,du = \int \frac{du}{1 + u} = \ln|1 + u| + C \text{ so}$$

$$\int \frac{dx}{1 + \sin x + \cos x} = \ln\left|1 + \tan\frac{x}{2}\right| + C \text{ where } u = \tan\frac{x}{2}.$$

9.7.9 Let $\cos x = \dfrac{1 - u^2}{1 + u^2}$ and $dx = \dfrac{2}{1 + u^2}$, then $\displaystyle\int \frac{dx}{3 + 2\cos x} = \int \frac{\dfrac{2}{1 + u^2}}{3 + 2\left(\dfrac{1 - u^2}{1 + u^2}\right)}\,du =$

$$\int \frac{2}{5 + u^2}\,du. \text{ Let } u = \sqrt{5}\tan\theta, \ du = \sqrt{5}\sec^2\theta\,d\theta, \text{ thus } \int \frac{2}{5 + u^2}\,du =$$

$$\frac{2\sqrt{5}}{5}\int d\theta = \frac{2\sqrt{5}}{5}\theta + C = \frac{2\sqrt{5}}{5}\tan^{-1}\frac{u}{\sqrt{5}} \text{ so}$$

$$\int \frac{dx}{3 + 2\cos x} = \frac{2\sqrt{5}}{5}\tan^{-1}\left(\frac{1}{\sqrt{5}}\tan\frac{x}{2}\right) + C \text{ where } u = \tan\frac{x}{2}.$$

9.7.10 Let $\cos x = \dfrac{1-u^2}{1+u^2}$ and $dx = \dfrac{2}{1+u^2}\,du$, then

$$\int \frac{dx}{2+3\cos x} = \int \frac{\dfrac{2}{1+u^2}}{2+3\left(\dfrac{1-u^2}{1+u^2}\right)}\,du = \int \frac{2}{5-u^2}\,du$$

$$= \int \left(\frac{1/\sqrt{5}}{\sqrt{5}-u} + \frac{1/\sqrt{5}}{\sqrt{5}+u}\right)du = \frac{1}{\sqrt{5}}\ln\left(\frac{\sqrt{5}+u}{\sqrt{5}-u}\right) + C, \text{ so}$$

$$\int \frac{dx}{2+3\cos x} = \frac{1}{\sqrt{5}}\ln\left(\frac{\sqrt{5}+\tan\dfrac{x}{2}}{\sqrt{5}-\tan\dfrac{x}{2}}\right) + C \text{ where } u = \tan\frac{x}{2}.$$

SECTION 9.8

9.8.1 Use $n = 10$ subdivisions to approximate the value of $\displaystyle\int_0^8 \sqrt{x+1}\,dx$ by the midpoint approximation. Find the exact value of the integral and approximate the magnitude of the error. Express your answer to at least four decimal places.

9.8.2 Use $n = 10$ subdivisions to approximate the value of $\displaystyle\int_1^9 \frac{1}{\sqrt{x}}\,dx$ by the trapezoidal approximation. Find the exact value of the integral and approximate the magnitude of the error. Express your answers to at least four decimal places.

9.8.3 Use $n = 10$ subdivisions to approximate the value of $\displaystyle\int_{\pi/2}^{\pi} \sin x\,dx$ by Simpson's rule. Find the exact value of the integral and approximate the magnitude of the error. Express your answer to at least four decimal places.

9.8.4 Use $n = 10$ subdivisions to approximate the value of $\displaystyle\int_0^{1.5} \cos x\,dx$ by the midpoint approximation. Find the exact value of the integral and approximate the magnitude of the error. Express your answer to at least four decimal places.

9.8.5 Use $n = 10$ subdivisions to find the exact value of $\displaystyle\int_1^2 e^x\,dx$ by the trapezoidal approximation. Find the exact value of the integral and approximate the magnitude of the error. Express your answer to at least four decimal places.

9.8.6 Use $n = 10$ subdivisions to approximate the value of $\displaystyle\int_{-1}^1 \frac{1}{3x-4}\,dx$ by Simpson's rule. Find the exact value of the integral and approximate the magnitude of the error. Express your answers to at least four decimal places.

9.8.7 Use $n = 10$ subdivisions to approximate the value of $\displaystyle\int_0^2 \sqrt{2x+1}\,dx$ by Simpson's rule. Find the exact value of the integral and approximate the magnitude of the error. Express your answers to at least four decimal places.

9.8.8 Use $n = 10$ subdivisions to approximate the value of $\displaystyle\int_{\pi/4}^{\pi/2} \sin 2x\,dx$ by Simpson's rule. Find the exact value of the integral and approximate the magnitude of the error. Express your answers to at least four decimal places.

9.8.9 Use inequality (10) to find an upper bound on the magnitude of the error for the approximate value of $\displaystyle\int_0^2 \sqrt{2(x+1)}\,dx$ found by the midpoint approximation.

9.8.10 Use inequality (11) to find an upper bound on the magnitude of the error for the approximate value of $\int_1^9 \frac{1}{\sqrt{x}}\, dx$ found by the trapezoidal approximation.

9.8.11 Use inequality (12) to find an upper bound on the magnitude of the error for the approximate value of $\int_{\pi/2}^{\pi} \sin x\, dx$ found by Simpson's rule.

9.8.12 Use inequality (12) to find an upper bound on the magnitude of the error for the approximate value of $\int_{-1}^{1} \frac{1}{3x-4}\, dx$ found by Simpson's rule.

9.8.13 Use $n = 10$ subdivisions to approximate the value of the integral $\int_0^2 \sqrt{4+x^3}\, dx$ by

    (a) the midpoint approximation   **(b)** the Simpson's approximation

9.8.14 Use $n = 10$ subdivisions to approximate the value of the integral $\int_0^4 \sqrt{1+x^4}\, dx$ by

    (a) midpoint approximation   **(b)** Simpson's rule

9.8.15 Use $n = 10$ subdivisions to approximate the value of the integral $\int_0^2 \sqrt{x^3+3}\, dx$ by

    (a) trapezoidal approximation   **(b)** Simpson's rule

9.8.16 Use $n = 10$ subdivisions to approximate the value of the integral $\int_0^2 \frac{1}{1+x^2}\, dx$ by 1.10715

    (a) midpoint approximation   **(b)** Simpson's rule

9.8.17 Use $n = 10$ subdivisions to approximate the value of the integral $\int_0^{1/4} \frac{1}{\sqrt{1-3x^2}}\, dx$ by

    (a) trapezoidal approximation   **(b)** Simpson's rule

SOLUTIONS

SECTION 9.8

9.8.1 Exact Value $= 17.3333333$ $|E_M| \approx .008723707$
Midpoint Approximation $= 17.342057$

9.8.2 Exact Value $= 4$ $|E_T| = .02479$
Trapezoidal Approximation $= 4.02479$

9.8.3 Exact Value $= 1$ $|E_S| = .00251$
Simpson's Approximation $= .99749$

9.8.4 Exact Value $= .99749$ $|E_M| = .00094$
Midpoint Approximation $= .9984$

9.8.5 Exact Value $= 4.67077$ $|E_T| = .00389$
Trapezoidal Approximation $= 4.67467$

9.8.6 Exact Value $= -.64864$ $|E_S| = .0008$
Simpson's Approximation $= -.64871$

9.8.7 Exact Value $= 3.39345$ $|E_S| = 0$
Simpson's Approximation $= 3.39345$

9.8.8 Exact Value $= .5$ $|E_S| = 0$
Simpson's Approximation $= .5$

9.8.9 $|E_M| \leq \dfrac{2^3(1)}{2400} = .00333$

9.8.10 $|E_{\hat{T}}| \leq \dfrac{8^3(3/2)}{1200} = .64$

9.8.11 $|E_{\hat{T}}| \leq \dfrac{\left(\frac{\pi}{2}\right)^3 (1)}{1200} = .002056$

9.8.12 $|E_{\hat{S}}| \leq \dfrac{2^5 \left(\dfrac{24 \cdot 81}{2^5}\right)}{1800000} = .00108$

9.8.13 (a) 4.81827   (b) 4.82116

9.8.14 (a) 22.3912   (b) 22.444

9.8.15 (a) 3.24798   (b) 3.24131

9.8.16 (a) 1.10741   (b) 1.10715

9.8.17 (a) .25861   (b) .25856

SUPPLEMENTARY EXERCISES, CHAPTER 9

In Exercises 1–64, evaluate the integrals.

1. $\displaystyle\int x\cos 2x\, dx.$

2. $\displaystyle\int x\cos x^2\, dx.$

3. $\displaystyle\int \tan^3 x\sec x\, dx.$

4. $\displaystyle\int \sin^3 x\cos^2 x\, dx.$

5. $\displaystyle\int \tan^2 3t\sec^2 3t\, dt.$

6. $\displaystyle\int \cot 2x\csc^3 2x\, dx.$

7. $\displaystyle\int \frac{\sin^2 x\, dx}{1+\cos x}.$

8. $\displaystyle\int \frac{\sin 2x\, dx}{\cos x(1+\cos x)}.$

9. $\displaystyle\int x^2\cos^2 x\, dx.$

10. $\displaystyle\int \sin^2 2x\cos^2 2x\, dx.$

11. $\displaystyle\int \sec^5 x\sin x\, dx.$

12. $\displaystyle\int \tan^5 2x\, dx.$

13. $\displaystyle\int \sin^4 x\cos^2 x\, dx.$

14. $\displaystyle\int \frac{dx}{\sec^4 x}.$

15. $\displaystyle\int_0^{\pi/4} \sin 5x\sin 3x\, dx.$

16. $\displaystyle\int_{-\pi/10}^{0} \sin 2x\cos 3x\, dx.$

17. $\displaystyle\int_0^1 \sin^2 \pi x\, dx.$

18. $\displaystyle\int_0^{\pi/3} \sin^3 3x\, dx.$

19. $\displaystyle\int_0^{\sqrt{\pi/2}} x\sec^2(x^2)\, dx.$

20. $\displaystyle\int_0^{\pi/4} \frac{\sec^2 x\, dx}{\sqrt{1+3\tan x}}.$

21. $\displaystyle\int \frac{\sin(\cot^{-1} x)\, dx}{1+x^2}.$

22. $\displaystyle\int \frac{e^{\tan 3x}\, dx}{\cos^2 3x}.$

23. $\displaystyle\int e^x\sec(e^x)\, dx.$

24. $\displaystyle\int x\sec^2 3x\, dx.$

25. $\displaystyle\int \frac{e^{2x}}{\sqrt{e^{2x}+1}}\, dx.$

26. $\displaystyle\int \frac{e^{2x}}{e^{2x}+1}\, dx.$

27. $\displaystyle\int e^{3x}\sin 2x\, dx.$

28. $\displaystyle\int \ln(a^2+x^2)\, dx.$

29. $\displaystyle\int_1^2 \sin^{-1}(x/2)\, dx.$

30. $\displaystyle\int \frac{\cos 2\pi x}{e^{2\pi x}}\, dx.$

31. $\displaystyle\int \sin(3\ln x)\, dx.$

32. $\displaystyle\int x^3 e^{-x^2}\, dx.$

33. $\displaystyle\int \frac{x\, dx}{\sqrt{x^2-9}}.$

34. $\displaystyle\int_1^2 \frac{\sqrt{4x^2-1}}{x}\, dx.$

35. $\displaystyle\int_1^3 \frac{\sqrt{9-x^2}}{x}\, dx.$

36. $\displaystyle\int_0^{\pi/6} \frac{\cos 3x}{\sqrt{4-\sin^2 3x}}\, dx.$

37. $\displaystyle\int \frac{x^2\, dx}{\sqrt{2x+3}}.$

38. $\displaystyle\int \frac{1}{\sqrt{x}(x+9)}\, dx.$

39. $\displaystyle\int \frac{dt}{\sqrt{3-4t-4t^2}}.$

40. $\displaystyle\int \frac{dx}{\sqrt{6x-x^2}}.$

41. $\displaystyle\int \frac{dx}{x^2\sqrt{a^2-x^2}}.$

42. $\displaystyle\int \frac{x^3}{(x^2+4)^{1/3}}\, dx.$

43. $\displaystyle\int \sqrt{a^2 - x^2}\, dx.$

44. $\displaystyle\int x\sqrt{a^2 - x^2}\, dx.$

45. $\displaystyle\int \frac{x - 2}{\sqrt{4x - x^2}}\, dx.$

46. $\displaystyle\int_1^3 \frac{dx}{x^2 - 2x + 5}.$

47. $\displaystyle\int \frac{dx}{2x^2 + 3x + 1}.$

48. $\displaystyle\int \frac{dx}{(x^2 + 4)^2}.$

49. $\displaystyle\int \frac{x + 1}{x^3 + x^2 - 6x}\, dx.$

50. $\displaystyle\int \frac{x^3 + 1}{x - 2}\, dx.$

51. $\displaystyle\int \frac{x^2 - 1}{x^3 - 3x}\, dx.$

52. $\displaystyle\int \frac{x - 3}{x^3 - 1}\, dx.$

53. $\displaystyle\int \frac{2x^2 + 5}{x^4 - 1}\, dx.$

54. $\displaystyle\int \frac{x^4 - x^3 - x - 1}{x^3 - x^2}\, dx.$

55. $\displaystyle\int \frac{dx}{(x^2 + 4)(x - 3)}.$

56. $\displaystyle\int \frac{x\, dx}{(x + 1)^3}.$

57. $\displaystyle\int \frac{3x^2 + 12x + 2}{(x^2 + 4)^2}\, dx.$

58. $\displaystyle\int \frac{(4x + 2)\, dx}{x^4 + 2x^3 + x^2}.$

59. $\displaystyle\int \frac{x\, dx}{x^2 + 2x + 5}.$

60. $\displaystyle\int \frac{6x\, dx}{(x^2 + 9)^3}.$

61. $\displaystyle\int \frac{dx}{\sqrt{3 - 2x^2}}.$

62. $\displaystyle\int \frac{1 + t}{\sqrt{t}}\, dt.$

63. $\displaystyle\int \frac{\sqrt{t}\, dt}{1 + t}.$

64. $\displaystyle\int \frac{\sqrt{1 - x^2}}{x^2}\, dx.$

Exercises 65–70 relate to Section 9.7. Evaluate the integrals.

65. $\displaystyle\int_0^1 \frac{x^{2/3}}{1 + x^{1/3}}\, dx.$

66. $\displaystyle\int \frac{dx}{x^{1/2} + x^{1/4}}.$

67. $\displaystyle\int \frac{dx}{1 - \tan x}.$

68. $\displaystyle\int \frac{dx}{3\cos x + 5}.$

69. $\displaystyle\int \frac{dx}{\sin x - \tan x}.$

70. $\displaystyle\int_0^{\pi/3} \frac{dx}{5\sec x - 3}.$

71. Use partial fractions to show that

$$\int \frac{dx}{x^2 - a^2} = \frac{1}{2a} \ln \left| \frac{x - a}{x + a} \right| + C \quad (a \neq 0)$$

72. Find the arc length of (a) the parabola $y = x^2/2$ from $(0, 0)$ to $(2, 2)$ and (b) the curve $y = \ln(\sec x)$ from $(0, 0)$ to $(\pi/4, \frac{1}{2}\ln 2)$.

73. Let R be the region bounded by the curve $y = 1/(4 + x^2)$ and the lines $x = 0, y = 0$, and $x = 2$. Find (a) the area of R, (b) the volume of the solid obtained by revolving R about the x-axis, and (c) the volume of the solid obtained by revolving R about the y-axis.

74. Derive the following reduction formulas for $a \neq 0$:

(a) $\displaystyle \int x^n e^{ax}\, dx = \frac{x^n e^{ax}}{a} - \frac{n}{a} \int x^{n-1} e^{ax}\, dx.$

(b) $\displaystyle \int x^n \sin ax\, dx = \frac{-x^n \cos ax}{a} + \frac{n}{a} \int x^{n-1} \cos ax\, dx$

$\displaystyle \int x^n \cos ax\, dx = \frac{x^n \sin ax}{a} - \frac{n}{a} \int x^{n-1} \sin ax\, dx$

(c) $\displaystyle \int \sin^n ax \cos^m ax\, dx = -\frac{\sin^{n-1} ax \cos^{m+1} ax}{a(m+n)} + \frac{n-1}{m+n} \int \sin^{n-2} ax \cos^m ax\, dx$

$\displaystyle = \frac{\sin^{n+1} ax \cos^{m-1} ax}{a(m+n)} + \frac{m-1}{m+n} \int \sin^n ax \cos^{m-2} ax\, dx.$

75. Use Exercise 74 to evaluate the following integrals.

(a) $\displaystyle \int x^3 e^{2x}\, dx$

(b) $\displaystyle \int_0^{\pi/10} x^2 \sin 5x\, dx$

(c) $\displaystyle \int \sin^2 x \cos^4 x\, dx.$

76. Evaluate the following integrals assuming that $a \neq 0$.

(a) $\displaystyle \int x^n \ln ax\, dx\ (n \neq -1)$

(b) $\displaystyle \int \sec^n ax \tan ax\, dx\ (n \geq 1).$

77. Find $\int (\sin^3 \theta / \cos^5 \theta)\, d\theta$ two ways: (a) letting $u = \cos \theta$ and (b) expressing the integrand in terms of $\sec \theta$ and $\tan \theta$. Show that your answers differ by a constant.

In Exercises 78–81, approximate the integral using the given value of n and (a) the trapezoidal rule, (b) Simpson's rule. Use a calculator and express the answer to four decimal places.

78. $\displaystyle \int_0^1 \sqrt{x}\, dx, n = 4.$

79. $\displaystyle \int_{-4}^2 e^{-x}\, dx, n = 6.$

80. $\displaystyle \int_0^4 \sinh x\, dx, n = 4.$

81. $\displaystyle \int_4^{5.2} \ln x\, dx, n = 6.$

In Exercises 82 and 83, use Simpson's rule with $n = 10$ to approximate the given integral. Use a calculator and express the answer to five decimal places.

82. $\displaystyle \int_0^2 \cos(\sinh x)\, dx.$

83. $\displaystyle \int_1^2 \sin(\ln x)\, dx.$

84.   **(a)** Show that if $f(x)$ is continuous for $0 \leq x \leq 1$, then $\displaystyle\int_0^x x f(\sin x)\, dx = \frac{\pi}{2} \int_0^x f(\sin x)\, dx$

        [*Hint*: Let $x = \pi - u$.]

  (b) Use the result in part (a) to find $\displaystyle\int_0^\pi \frac{x \sin x}{2 - \sin^2 x}\, dx$

Evaluate the integral.

85. $\displaystyle\int \frac{1}{e^{ax} + 1}\, dx,\, a \neq 0.$

SOLUTIONS

SUPPLEMENTARY EXERCISES CHAPTER 9

1. $u = x$, $dv = \cos 2x\, dx$, $du = dx$, $v = \dfrac{1}{2}\sin 2x$; $\displaystyle\int x\cos 2x\, dx = \dfrac{1}{2}x\sin 2x + \dfrac{1}{4}\cos 2x + C$

2. $\dfrac{1}{2}\sin(x^2) + C$

3. $\displaystyle\int(\sec^2 x - 1)\sec x\tan x\, dx = \dfrac{1}{3}\sec^3 x - \sec x + C$

4. $\displaystyle\int(1 - \cos^2 x)\cos^2 x\sin x\, dx = -\dfrac{1}{3}\cos^3 x + \dfrac{1}{5}\cos^5 x + C$

5. $u = \tan 3t$, $\dfrac{1}{3}\displaystyle\int u^2\, du = \dfrac{1}{9}\tan^3 3t + C$ **6.** $\displaystyle\int\csc^2 2x(\csc 2x\cot 2x)dx = -\dfrac{1}{6}\csc^3 2x + C$

7. $\displaystyle\int\dfrac{1 - \cos^2 x}{1 + \cos x}dx = \int(1 - \cos x)dx = x - \sin x + C$

8. $\displaystyle\int\dfrac{2\sin x\cos x}{\cos x(1 + \cos x)}dx = \int\dfrac{2\sin x}{1 + \cos x}dx = -2\ln(1 + \cos x) + C$

9. $\displaystyle\int x^2\cos^2 x\, dx = \dfrac{1}{2}\int x^2(1 + \cos 2x)dx = \dfrac{1}{2}\int x^2\, dx + \dfrac{1}{2}\int x^2\cos 2x\, dx$,

use integration by parts twice to get

$\displaystyle\int x^2\cos 2x\, dx = \dfrac{1}{2}x^2\sin 2x + \dfrac{1}{2}x\cos 2x - \dfrac{1}{4}\sin 2x + C_1$

so $\displaystyle\int x^2\cos^2 x\, dx = \dfrac{1}{6}x^3 + \dfrac{1}{4}(x^2 - 1/2)\sin 2x + \dfrac{1}{4}x\cos 2x + C$

10. $\displaystyle\int\sin^2 2x\cos^2 2x\, dx = \dfrac{1}{4}\int(2\sin 2x\cos 2x)^2\, dx = \dfrac{1}{4}\int\sin^2 4x\, dx$

$= \dfrac{1}{8}\displaystyle\int(1 - \cos 8x)dx = \dfrac{1}{8}x - \dfrac{1}{64}\sin 8x + C$

11. $\displaystyle\int\cos^{-5}x\sin x\, dx = \dfrac{1}{4}\cos^{-4}x + C = \dfrac{1}{4}\sec^4 x + C$

12. Let $u = 2x$, $\dfrac{1}{2}\displaystyle\int\tan^5 u\, du = \dfrac{1}{8}\tan^4 2x - \dfrac{1}{4}\tan^2 2x - \dfrac{1}{2}\ln|\cos 2x| + C$

13. $\dfrac{1}{8}\displaystyle\int (1-\cos 2x)^2(1+\cos 2x)dx = \dfrac{1}{8}\int (1-\cos 2x)\sin^2 2x\,dx$

$\qquad = \dfrac{1}{8}\displaystyle\int \sin^2 2x\,dx - \dfrac{1}{8}\int \sin^2 2x \cos 2x\,dx$

$\qquad = \dfrac{1}{16}\displaystyle\int (1-\cos 4x)dx - \dfrac{1}{48}\sin^3 2x = \dfrac{1}{16}x - \dfrac{1}{64}\sin 4x - \dfrac{1}{48}\sin^3 2x + C$

14. $\displaystyle\int \cos^4 x\,dx = \dfrac{3}{8}x + \dfrac{1}{4}\sin 2x + \dfrac{1}{32}\sin 4x + C$ using (5) of 9.3.

15. $\displaystyle\int_0^{\pi/4} \sin 5x \sin 3x\,dx = \dfrac{1}{2}\int_0^{\pi/4}(\cos 2x - \cos 8x)dx = \dfrac{1}{4}\sin 2x - \dfrac{1}{16}\sin 8x \Big]_0^{\pi/4} = 1/4$

16. $\displaystyle\int_{-\pi/10}^0 \sin 2x \cos 3x\,dx = \dfrac{1}{2}\int_{-\pi/10}^0 (\sin 5x - \sin x)dx$

$\qquad\qquad\qquad = -\dfrac{1}{10}\cos 5x + \dfrac{1}{2}\cos x \Big]_{-\pi/10}^0 = \dfrac{2}{5} - \dfrac{1}{2}\cos(\pi/10)$

17. $\dfrac{1}{2}\displaystyle\int_0^1 (1-\cos 2\pi x)dx = \dfrac{1}{2}x - \dfrac{1}{4\pi}\sin 2\pi x \Big]_0^1 = 1/2$

18. $\displaystyle\int_0^{\pi/3}(1-\cos^2 3x)\sin 3x\,dx = -\dfrac{1}{3}\cos 3x + \dfrac{1}{9}\cos^3 3x \Big]_0^{\pi/3} = 4/9$

19. $\dfrac{1}{2}\tan(x^2) \Big]_0^{\sqrt{\pi/2}} = 1/2$

20. $u = 1 + 3\tan x,\ \dfrac{1}{3}\displaystyle\int_1^4 u^{-1/2}du = \dfrac{2}{3}u^{1/2}\Big]_1^4 = 2/3$

21. $u = \cot^{-1} x,\ -\displaystyle\int \sin u\,du = \cos(\cot^{-1} x) + C = x/\sqrt{1+x^2} + C$

22. $\displaystyle\int e^{\tan 3x} \sec^2 3x\,dx = \dfrac{1}{3}e^{\tan 3x} + C$ 　　　　**23.** $\ln|\sec(e^x) + \tan(e^x)| + C$

24. $u = x,\ dv = \sec^2 3x\,dx,\ du = dx,\ v = \dfrac{1}{3}\tan 3x,\ \displaystyle\int x \sec^2 3x\,dx = \dfrac{1}{3}x\tan 3x + \dfrac{1}{9}\ln|\cos 3x| + C$

25. $u = e^{2x} + 1,\ \dfrac{1}{2}\displaystyle\int u^{-1/2}du = \sqrt{e^{2x}+1} + C$

26. $u = e^{2x} + 1, \ \dfrac{1}{2} \displaystyle\int \dfrac{1}{u} du = \dfrac{1}{2} \ln(e^{2x} + 1) + C$

27. Use integration by parts with $u = e^{3x}$, $dv = \sin 2x \, dx$ to get

$$\int e^{3x} \sin 2x \, dx = -\dfrac{1}{2} e^{3x} \cos 2x + \dfrac{3}{2} \int e^{3x} \cos 2x \, dx \text{ and again with}$$

$u = e^{3x}$, $dv = \cos 2x \, dx$ to get $\displaystyle\int e^{3x} \cos 2x \, dx = \dfrac{1}{2} e^{3x} \sin 2x - \dfrac{3}{2} \int e^{3x} \sin 2x \, dx$ so, with

$$I = \int e^{3x} \sin 2x \, dx, \ I = -\dfrac{1}{2} e^{3x} \cos 2x + \dfrac{3}{4} e^{3x} \sin 2x - \dfrac{9}{4} I, \ I = \dfrac{1}{13} e^{3x} (3 \sin 2x - 2 \cos 2x) + C$$

28. $u = \ln(a^2 + x^2), \ dv = dx, \ du = \dfrac{2x}{a^2 + x^2} dx, \ v = x$

$$\int \ln(a^2 + x^2) dx = x \ln(a^2 + x^2) - 2 \int \dfrac{x^2}{a^2 + x^2} dx$$

but $\displaystyle\int \dfrac{x^2}{a^2 + x^2} dx = \int \left(1 - \dfrac{a^2}{a^2 + x^2}\right) dx = x - a \tan^{-1}(x/a) + C_1$

so $\displaystyle\int \ln(a^2 + x^2) dx = x \ln(a^2 + x^2) - 2x + 2a \tan^{-1}(x/a) + C$

29. $u = \sin^{-1}(x/2), \ dv = dx, \ du = 1/\sqrt{4 - x^2} dx, \ v = x$

$$\int_1^2 \sin^{-1}(x/2) dx = x \sin^{-1}(x/2) \Big|_1^2 - \int_1^2 x(4 - x^2)^{-1/2} dx$$

$$= (2)(\pi/2) - (1)(\pi/6) + (4 - x^2)^{1/2} \Big|_1^2 = 5\pi/6 - \sqrt{3}$$

30. Rewrite as $\displaystyle\int e^{-2\pi x} \cos 2\pi x \, dx$ then $u = e^{-2\pi x}$, $dv = \cos 2\pi x \, dx$,

$du = -2\pi e^{-2\pi x} dx, \ v = \dfrac{1}{2\pi} \sin 2\pi x$

$$\int e^{-2\pi x} \cos 2\pi x \, dx = \dfrac{1}{2\pi} e^{-2\pi x} \sin 2\pi x + \int e^{-2\pi x} \sin 2\pi x \, dx.$$

For $\displaystyle\int e^{-2\pi x} \sin 2\pi x \, dx$ use $u = e^{-2\pi x}$, $dv = \sin 2\pi x \, dx$ to get

$$\int e^{-2\pi x} \sin 2\pi x \, dx = -\dfrac{1}{2\pi} e^{-2\pi x} \cos 2\pi x - \int e^{-2\pi x} \cos 2\pi x \, dx \text{ so}$$

$$\int e^{-2\pi x} \cos 2\pi x \, dx = \dfrac{1}{2\pi} e^{-2\pi x} \sin 2\pi x - \dfrac{1}{2\pi} e^{-2\pi x} \cos 2\pi x - \int e^{-2\pi x} \cos 2\pi x \, dx,$$

$$\int e^{-2\pi x} \cos 2\pi x \, dx = \dfrac{1}{4\pi} e^{-2\pi x} (\sin 2\pi x - \cos 2\pi x) + C$$

31. $u = \sin(3\ln x)$, $dv = dx$, $du = \dfrac{3}{x}\cos(3\ln x)dx$, $v = x$

$$\int \sin(3\ln x)dx = x\sin(3\ln x) - 3\int \cos(3\ln x)dx. \text{ Use } u = \cos(3\ln x),\ dv = dx \text{ to get}$$

$$\int \cos(3\ln x)dx = x\cos(3\ln x) + 3\int \sin(3\ln x)dx \text{ so}$$

$$\int \sin(3\ln x)dx = x\sin(3\ln x) - 3x\cos(3\ln x) - 9\int \sin(3\ln x)dx,$$

$$\int \sin(3\ln x)dx = \frac{1}{10}x[\sin(3\ln x) - 3\cos(3\ln x)] + C$$

32. $u = x^2$, $dv = xe^{-x^2}dx$, $du = 2x\,dx$, $v = -\dfrac{1}{2}e^{-x^2}$

$$\int x^3 e^{-x^2}\,dx = -\frac{1}{2}x^2 e^{-x^2} + \int xe^{-x^2}\,dx = -\frac{1}{2}x^2 e^{-x^2} - \frac{1}{2}e^{-x^2} + C$$

33. $\displaystyle\int x(x^2 - 9)^{-1/2}dx = \sqrt{x^2 - 9} + C$

34. $x = \dfrac{1}{2}\sec\theta$, $dx = \dfrac{1}{2}\sec\theta\tan\theta\,d\theta$

$$\int_{\pi/3}^{\sec^{-1}4} \tan^2\theta\,d\theta = \ \tan\theta - \theta\Big]_{\pi/3}^{\sec^{-1}4} = \tan(\sec^{-1}4) - \sec^{-1}4 - \sqrt{3} + \pi/3$$

$$= \sqrt{15} - \sec^{-1}4 - \sqrt{3} + \pi/3$$

35. $x = 3\sin\theta$, $dx = 3\cos\theta\,d\theta$

$$3\int_{\sin^{-1}(1/3)}^{\pi/2} \frac{\cos^2\theta}{\sin\theta}\,d\theta = 3\int_{\sin^{-1}(1/3)}^{\pi/2} \frac{1 - \sin^2\theta}{\sin\theta}\,d\theta = 3\int_{\sin^{-1}(1/3)}^{\pi/2} (\csc\theta - \sin\theta)d\theta$$

$$= \ -3\ln|\csc\theta + \cot\theta| + 3\cos\theta\Big]_{\sin^{-1}(1/3)}^{\pi/2}$$

$$= -3\ln(1) + 3\ln|3 + \sqrt{8}| - 3(\sqrt{8}/3) = 3\ln(3 + \sqrt{8}) - \sqrt{8}$$

36. $u = \sin 3x$, $du = 3\cos 3x\,dx$

$$\frac{1}{3}\int_0^1 \frac{1}{\sqrt{4 - u^2}}\,du = \frac{1}{3}\sin^{-1}\frac{u}{2}\Big]_0^1 = \frac{1}{3}(\pi/6) = \pi/18$$

37. $u = \sqrt{2x+3}$, $x = (u^2 - 3)/2$, $dx = u\,du$

$$\frac{1}{4}\int (u^2 - 3)^2 du = \frac{1}{4}\int (u^4 - 6u^2 + 9)du$$

$$= \frac{1}{4}\left(\frac{1}{5}u^5 - 2u^3 + 9u\right) + C = \frac{1}{20}u(u^4 - 10u^2 + 45) + C$$

$$= \frac{1}{20}\sqrt{2x+3}(4x^2 + 12x + 9 - 20x - 30 + 45) + C$$

$$= \frac{1}{5}(x^2 - 2x + 6)\sqrt{2x+3} + C$$

38. $u = \sqrt{x}$, $du = \frac{1}{2\sqrt{x}}dx$; $2\int \frac{1}{u^2 + 9}du = \frac{2}{3}\tan^{-1}\frac{\sqrt{x}}{3} + C$

39. $\frac{1}{2}\int \frac{1}{\sqrt{1-(t+1/2)^2}}dt = \frac{1}{2}\sin^{-1}(t+1/2) + C$

40. $\int \frac{1}{\sqrt{9-(x-3)^2}}dx = \sin^{-1}\frac{x-3}{3} + C$

41. $x = a\sin\theta$, $dx = a\cos\theta\,d\theta$; $\frac{1}{a^2}\int \csc^2\theta\,d\theta = -\frac{1}{a^2}\cot\theta + C = -\frac{\sqrt{a^2 - x^2}}{a^2 x} + C$

42. $u = (x^2 + 4)^{1/3}$, $x^2 = u^3 - 4$, $2x\,dx = 3u^2 du$, $x\,dx = \frac{3}{2}u^2 du$

$$\frac{3}{2}\int (u^3 - 4)u\,du = \frac{3}{2}\int (u^4 - 4u)du = \frac{3}{2}\left(\frac{1}{5}u^5 - 2u^2\right) + C$$

$$= \frac{3}{10}u^2(u^3 - 10) + C = \frac{3}{10}(x^2 + 4)^{2/3}(x^2 - 6) + C$$

43. $x = a\sin\theta$, $dx = a\cos\theta\,d\theta$

$$a^2\int \cos^2\theta\,d\theta = \frac{1}{2}a^2\theta + \frac{1}{4}a^2\sin 2\theta + C = \frac{1}{2}a^2\sin^{-1}(x/a) + \frac{1}{2}x\sqrt{a^2 - x^2} + C$$

44. $-\frac{1}{3}(a^2 - x^2)^{3/2} + C$

45. $\int \frac{x-2}{\sqrt{4-(x-2)^2}}dx = \int \frac{u}{\sqrt{4-u^2}}du$ $(u = x - 2) = -\sqrt{4 - u^2} + C = -\sqrt{4x - x^2} + C$

46. $\int_1^3 \frac{1}{(x-1)^2 + 4}dx = \frac{1}{2}\tan^{-1}\frac{x-1}{2}\Big]_1^3 = \pi/8$

47. $2x^2 + 3x + 1 = (2x+1)(x+1)$, $\dfrac{1}{(2x+1)(x+1)} = \dfrac{2}{2x+1} - \dfrac{1}{x+1}$

$$\int \frac{dx}{(2x+1)(x+1)} = \ln\left|\frac{2x+1}{x+1}\right| + C$$

48. $x = 2\tan\theta$, $dx = 2\sec^2\theta\, d\theta$

$$\frac{1}{8}\int \cos^2\theta\, d\theta = \frac{1}{16}\theta + \frac{1}{32}\sin 2\theta + C = \frac{1}{16}\tan^{-1}(x/2) + \frac{x}{8(x^2+4)} + C$$

49. $\dfrac{x+1}{x(x+3)(x-2)} = \dfrac{-1/6}{x} + \dfrac{-2/15}{x+3} + \dfrac{3/10}{x-2}$

$$\int \frac{x+1}{x^3+x^2-6x}dx = -\frac{1}{6}\ln|x| - \frac{2}{15}\ln|x+3| + \frac{3}{10}\ln|x-2| + C$$

50. $\displaystyle\int \frac{x^3+1}{x-2}dx = \int\left(x^2 + 2x + 4 + \frac{9}{x-2}\right)dx = \frac{1}{3}x^3 + x^2 + 4x + 9\ln|x-2| + C$

51. $u = x^3 - 3x$, $\dfrac{1}{3}\displaystyle\int \frac{1}{u}du = \frac{1}{3}\ln|x^3 - 3x| + C$

52. $x^3 - 1 = (x-1)(x^2+x+1)$,

$$\frac{x-3}{(x-1)(x^2+x+1)} = \frac{-2/3}{x-1} + \frac{(2/3)x + (7/3)}{x^2+x+1}$$

$$\frac{1}{3}\int \frac{2x+7}{x^2+x+1}dx = \frac{1}{3}\int \frac{2x+7}{(x+1/2)^2 + 3/4}dx = \frac{1}{3}\int \frac{2u+6}{u^2+3/4}du \quad (u = x+1/2)$$

$$= \frac{1}{3}\ln(u^2 + 3/4) + \frac{4}{\sqrt{3}}\tan^{-1}(2u/\sqrt{3}) + C_1$$

so $\displaystyle\int \frac{x-3}{x^3-1}dx = -\frac{2}{3}\ln|x-1| + \frac{1}{3}\ln(x^2+x+1) + \frac{4}{\sqrt{3}}\tan^{-1}\frac{2x+1}{\sqrt{3}} + C$

53. $x^4 - 1 = (x+1)(x-1)(x^2+1)$

$$\int \frac{2x^2+5}{x^4-1}dx = \int\left[\frac{-7/4}{x+1} + \frac{7/4}{x-1} + \frac{-3/2}{x^2+1}\right]dx = \frac{7}{4}\ln\left|\frac{x-1}{x+1}\right| - \frac{3}{2}\tan^{-1}x + C$$

54. $\displaystyle\int \frac{x^4 - x^3 - x - 1}{x^3 - x^2}dx = \int\left[x - \frac{x+1}{x^2(x-1)}\right]dx = \int\left[x - \left(\frac{-2}{x} + \frac{-1}{x^2} + \frac{2}{x-1}\right)\right]dx$

$$= \frac{1}{2}x^2 + 2\ln\left|\frac{x}{x-1}\right| - \frac{1}{x} + C$$

55. $\displaystyle\int \frac{dx}{(x^2+4)(x-3)} = \int \left[\frac{(-1/13)x - (3/13)}{x^2+4} + \frac{1/13}{x-3}\right]dx$

$$= -\frac{1}{26}\ln(x^2+4) - \frac{3}{26}\tan^{-1}\frac{x}{2} + \frac{1}{13}\ln|x-3| + C$$

56. $u = x+1$, $\displaystyle\int \frac{u-1}{u^3}du = \int (u^{-2} - u^{-3})du = -u^{-1} + \frac{1}{2}u^{-2} + C = -\frac{1}{x+1} + \frac{1}{2(x+1)^2} + C$

57. $\displaystyle\frac{3x^2 + 12x + 2}{(x^2+4)^2} = \frac{3}{x^2+4} + \frac{12x - 10}{(x^2+4)^2} = \frac{3}{x^2+4} + \frac{12x}{(x^2+4)^2} - \frac{10}{(x^2+4)^2}$

$\displaystyle\int \frac{1}{(x^2+4)^2} = \frac{1}{8}\int \cos^2\theta\, d\theta \quad (x = 2\tan\theta)$

$$= \frac{1}{16}\theta + \frac{1}{32}\sin 2\theta + C_1 = \frac{1}{16}\tan^{-1}(x/2) + \frac{x}{8(x^2+4)} + C_1$$

so $\displaystyle\int \frac{3x^2 + 12x + 2}{(x^2+4)^2}dx = \frac{3}{2}\tan^{-1}\frac{x}{2} - \frac{6}{x^2+4} - \frac{5}{8}\tan^{-1}\frac{x}{2} - \frac{5x}{4(x^2+4)} + C$

$$= \frac{7}{8}\tan^{-1}\frac{x}{2} - \frac{5x + 24}{4(x^2+4)} + C$$

58. $x^4 + 2x^3 + x^2 = x^2(x+1)^2$,

$\displaystyle\frac{4x+2}{x^2(x+1)^2} = \frac{0}{x} + \frac{2}{x^2} + \frac{0}{x+1} + \frac{-2}{(x+1)^2} = 2/x^2 - 2/(x+1)^2$

$\displaystyle\int \frac{4x+2}{x^4 + 2x^3 + x^2}dx = -\frac{2}{x} + \frac{2}{x+1} + C = -\frac{2}{x(x+1)} + C$

59. $\displaystyle\int \frac{x}{(x+1)^2 + 4}dx = \int \frac{u-1}{u^2+4}du \quad (u = x+1)$

$$= \frac{1}{2}\ln(u^2+4) - \frac{1}{2}\tan^{-1}\frac{u}{2} + C = \frac{1}{2}\ln(x^2 + 2x + 5) - \frac{1}{2}\tan^{-1}\frac{x+1}{2} + C$$

60. $\displaystyle -\frac{3}{2(x^2+9)^2} + C$

61. $u = \sqrt{2}x$, $\displaystyle\frac{1}{\sqrt{2}}\int \frac{1}{\sqrt{3-u^2}}du = \frac{1}{\sqrt{2}}\sin^{-1}\sqrt{2/3}\,x + C$

62. $\displaystyle\int (t^{-1/2} + t^{1/2})dt = 2t^{1/2} + \frac{2}{3}t^{3/2} + C$

63. $u = \sqrt{t}$, $t = u^2$, $dt = 2u\,du$; $\displaystyle\int \frac{2u^2}{u^2+1}du = 2\int \left[1 - \frac{1}{u^2+1}\right]du = 2\sqrt{t} - 2\tan^{-1}\sqrt{t} + C$

64. $x = \sin\theta$, $dx = \cos\theta\, d\theta$; $\displaystyle\int \cot^2\theta\, d\theta = -\cot\theta - \theta + C = -\sqrt{1-x^2}/x - \sin^{-1}x + C$

65. $u = 1 + x^{1/3}$, $x = (u-1)^3$, $dx = 3(u-1)^2 du$

$$3\int_1^2 \frac{(u-1)^4}{u}\, du = 3\int_1^2 \left[u^3 - 4u^2 + 6u - 4 + \frac{1}{u}\right] du$$

$$= 3\left[\frac{1}{4}u^4 - \frac{4}{3}u^3 + 3u^2 - 4u + \ln u\right]_1^2 = -7/4 + 3\ln 2$$

66. $u = x^{1/4}$, $x = u^4$, $dx = 4u^3 du$

$$4\int \frac{u^2}{u+1}\, du = 4\int \left(u - 1 + \frac{1}{u+1}\right) du = 2x^{1/2} - 4x^{1/4} + 4\ln(x^{1/4} + 1) + C$$

67. $u = \tan(x/2)$, $\tan x = \sin x / \cos x = 2u/(1 - u^2)$

$$\int \frac{dx}{1 - \tan x} = \int \frac{2u^2 - 2}{(u^2 + 1)(u^2 + 2u - 1)}\, du$$

$$\frac{2u^2 - 2}{(u^2 + 1)(u^2 + 2u - 1)} = \frac{u + 1}{u^2 + 1} + \frac{-u - 1}{u^2 + 2u - 1}$$

$$\int \frac{dx}{1 - \tan x} = \frac{1}{2}\ln(u^2 + 1) + \tan^{-1}u - \frac{1}{2}\ln|u^2 + 2u - 1| + C$$

$$= \tan^{-1}u - \frac{1}{2}\ln\left|\frac{u^2 + 2u - 1}{u^2 + 1}\right| + C = \tan^{-1}u - \frac{1}{2}\ln\left|\frac{2u}{1 + u^2} - \frac{1 - u^2}{1 + u^2}\right| + C$$

$$= \frac{x}{2} - \frac{1}{2}\ln|\sin x - \cos x| + C$$

68. $u = \tan(x/2)$, $\displaystyle\int \frac{dx}{3\cos x + 5} = \int \frac{1}{u^2 + 4}\, du = \frac{1}{2}\tan^{-1}\left[\frac{1}{2}\tan(x/2)\right] + C$

69. $u = \tan(x/2)$, $\tan x = 2u/(1 - u^2)$

$$\int \frac{dx}{\sin x - \tan x} = \frac{1}{2}\int \frac{u^2 - 1}{u^3}\, du = \frac{1}{2}\int (1/u - u^{-3})\, du = \frac{1}{2}\ln\left|\tan\frac{x}{2}\right| + \frac{1}{4}\cot^2\frac{x}{2} + C$$

70. $u = \tan(x/2)$, $\sec x = 1/\cos x = (1 + u^2)/(1 - u^2)$

$$\int_0^{\pi/3} \frac{dx}{5\sec x - 3} = \int_0^{1/\sqrt{3}} \frac{1 - u^2}{(u^2 + 1)(4u^2 + 1)}\, du = \int_0^{1/\sqrt{3}} \left(\frac{-2/3}{u^2 + 1} + \frac{5/3}{4u^2 + 1}\right) du$$

$$= \left[-\frac{2}{3}\tan^{-1}u + \frac{5}{6}\tan^{-1}2u\right]_0^{1/\sqrt{3}} = -\pi/9 + \frac{5}{6}\tan^{-1}(2/\sqrt{3})$$

71. $\displaystyle\int \frac{dx}{x^2 - a^2} = \int \left[\frac{1/(2a)}{x-a} + \frac{-1/(2a)}{x+a}\right] dx$

$$= \frac{1}{2a} \ln|x-a| - \frac{1}{2a} \ln|x+a| + C = \frac{1}{2a} \ln\left|\frac{x-a}{x+a}\right| + C$$

72. **(a)** $\displaystyle L = \int_0^2 \sqrt{1+x^2}\,dx = \int_0^{\tan^{-1}2} \sec^3\theta\,d\theta, \quad x = \sec\theta$

$$= \frac{1}{2} \sec\theta\tan\theta + \frac{1}{2}\ln|\sec\theta + \tan\theta|\Big]_0^{\tan^{-1}2} = \sqrt{5} + \frac{1}{2}\ln(\sqrt{5}+2)$$

(b) $\displaystyle L = \int_0^{\pi/4} \sqrt{1+\tan^2 x}\,dx = \int_0^{\pi/4} \sec x\,dx = \ln|\sec x + \tan x|\Big]_0^{\pi/4} = \ln(\sqrt{2}+1)$

73. **(a)** $\displaystyle A = \int_0^2 \frac{1}{4+x^2}\,dx = \frac{1}{2}\tan^{-1}\frac{x}{2}\Big]_0^2 = \pi/8$

(b) $\displaystyle V = \pi \int_0^2 \frac{1}{(4+x^2)^2}\,dx = \frac{\pi}{8}\int_0^{\pi/4} \cos^2\theta\,d\theta \quad (x = 2\tan\theta)$

$$= \frac{\pi}{16}\left(\theta + \frac{1}{2}\sin 2\theta\right)\Big]_0^{\pi/4} = \pi(\pi+2)/64$$

(c) $\displaystyle V = 2\pi \int_0^2 \frac{x}{4+x^2}\,dx = \pi\ln(4+x^2)\Big]_0^2 = \pi\ln 2$

74. **(a)** $u = x^n,\ dv = e^{ax}dx,\ du = nx^{n-1}dx,\ v = \dfrac{1}{a}e^{ax}$

$$\int x^n e^{ax}\,dx = \frac{1}{a}x^n e^{ax} - \frac{n}{a}\int x^{n-1} e^{ax}\,dx$$

(b) $u = x^n,\ dv = \sin ax\,dx,\ du = nx^{n-1}dx,\ v = -\dfrac{1}{a}\cos ax$

$$\int x^n \sin ax\,dx = -\frac{1}{a}x^n\cos ax + \frac{n}{a}\int x^{n-1}\cos ax\,dx.$$

The second formula is obtained in a similar way.

(c) $u = \sin^{n-1} ax,\ dv = \sin ax\cos^m ax\,dx$

$$du = a(n-1)\sin^{n-2} ax\cos ax\,dx,\ v = -\frac{\cos^{m+1} ax}{a(m+1)}$$

$$\int \sin^n ax\cos^m ax\,dx = -\frac{\sin^{n-1} ax\cos^{m+1} ax}{a(m+1)} + \frac{n-1}{m+1}\int \sin^{n-2} ax\cos^{m+2} ax\,dx$$

but $\displaystyle\int \sin^{n-2} ax \cos^{m+2} ax\, dx = \int \sin^{n-2} ax(1 - \sin^2 ax) \cos^m ax\, dx$

$$= \int \sin^{n-2} ax \cos^m ax\, dx - \int \sin^n ax \cos^m ax\, dx \text{ so}$$

$$\frac{m+n}{m+1} \int \sin^n ax \cos^m ax\, dx = -\frac{\sin^{n-1} ax \cos^{m+1} ax}{a(m+1)} + \frac{n-1}{m+1} \int \sin^{n-2} ax \cos^m ax\, dx$$

and $\displaystyle\int \sin^n ax \cos^m ax\, dx = -\frac{\sin^{n-1} ax \cos^{m+1} ax}{a(m+n)} + \frac{n-1}{m+n} \int \sin^{n-2} ax \cos^m ax\, dx.$

Similarly, take $u = \cos^{m-1} ax$, $dv = \sin^n ax \cos ax\, dx$ to get the second equality.

75. **(a)** $\displaystyle\int x^3 e^{2x}\, dx = \frac{1}{2}x^3 e^{2x} - \frac{3}{2}\int x^2 e^{2x}\, dx = \frac{1}{2}x^3 e^{2x} - \frac{3}{2}\left[\frac{1}{2}x^2 e^{2x} - \int x e^{2x}\, dx\right]$

$$= \frac{1}{2}x^3 e^{2x} - \frac{3}{4}x^2 e^{2x} + \frac{3}{2}\left[\frac{1}{2}x e^{2x} - \frac{1}{2}\int e^{2x}\, dx\right]$$

$$= \frac{1}{2}x^3 e^{2x} - \frac{3}{4}x^2 e^{2x} + \frac{3}{4}x e^{2x} - \frac{3}{8}e^{2x} + C$$

(b) $\displaystyle\int_0^{\pi/10} x^2 \sin 5x\, dx = -\frac{1}{5}x^2 \cos 5x\bigg]_0^{\pi/10} + \frac{2}{5}\int_0^{\pi/10} x \cos 5x\, dx$

$$= 0 + \frac{2}{5}\left[\frac{1}{5}x \sin 5x\right]_0^{\pi/10} - \frac{2}{25}\int_0^{\pi/10} \sin 5x\, dx$$

$$= \frac{2}{25}(\pi/10) + \frac{2}{125}\cos 5x\bigg]_0^{\pi/10} = \pi/125 + \frac{2}{125}(0 - 1) = (\pi - 2)/125$$

(c) $\displaystyle\int \sin^2 x \cos^4 x\, dx = \frac{1}{6}\sin^3 x \cos^3 x + \frac{1}{2}\int \sin^2 x \cos^2 x\, dx$

$$= \frac{1}{6}\sin^3 x \cos^3 x + \frac{1}{2}\left[\frac{1}{4}\sin^3 x \cos x + \frac{1}{4}\int \sin^2 x\, dx\right]$$

$$= \frac{1}{6}\sin^3 x \cos^3 x + \frac{1}{8}\sin^3 x \cos x + \frac{1}{8}\left[-\frac{1}{2}\sin x \cos x + \frac{1}{2}\int dx\right]$$

$$= \frac{1}{6}\sin^3 x \cos^3 x + \frac{1}{8}\sin^3 x \cos x - \frac{1}{16}\sin x \cos x + \frac{x}{16} + C$$

76. **(a)** $u = \ln ax$, $dv = x^n dx$, $du = \dfrac{1}{x}dx$, $v = \dfrac{x^{n+1}}{n+1}$

$$\int x^n \ln ax\, dx = \frac{1}{n+1}x^{n+1}\ln ax - \frac{1}{n+1}\int x^n dx$$

$$= \frac{1}{n+1}x^{n+1}\ln ax - \frac{1}{(n+1)^2}x^{n+1} + C$$

(b) $\displaystyle\int \sec^{n-1} ax(\sec ax \tan ax)dx = \frac{1}{an}\sec^n ax + C$

77. **(a)** $\displaystyle \int \frac{1 - \cos^2 \theta}{\cos^5 \theta} \sin \theta \, d\theta = \int (\cos^{-5} \theta - \cos^{-3} \theta) \sin \theta \, d\theta$

$$= \frac{1}{4} \cos^{-4} \theta - \frac{1}{2} \cos^{-2} \theta + C = \frac{1}{4} \sec^4 \theta - \frac{1}{2} \sec^2 \theta + C$$

(b) $\displaystyle \int \tan^3 \theta \sec^2 \theta \, d\theta = \frac{1}{4} \tan^4 \theta + C$ but $\displaystyle \frac{1}{4} \tan^4 \theta = \frac{1}{4} (\sec^2 \theta - 1)^2 = \frac{1}{4} (\sec^4 \theta - 2 \sec^2 \theta + 1)$

so the answers to (a) and (b) differ by $1/4$.

78. **(a)** 0.6433   **(b)** 0.6565   **79.** **(a)** 58.9275   **(b)** 54.7328

80. **(a)** 28.4649   **(b)** 26.4386   **81.** **(a)** 1.8277   **(b)** 1.8278

82. 0.35593   **83.** 0.36972

84. **(a)** $\displaystyle \int_0^{\pi} x f(\sin x) \, dx = - \int_{\pi}^0 (\pi - u) f(\sin(\pi - u)) \, du = \int_0^{\pi} (\pi - u) f(\sin u) \, du$

$$= \pi \int_0^{\pi} f(\sin u) \, du - \int_0^{\pi} u f(\sin u) \, du = \pi \int_0^{\pi} f(\sin x) \, dx - \int_0^{\pi} x f(\sin x) \, dx,$$

$$2 \int_0^{\pi} x f(\sin x) \, dx = \pi \int_0^{\pi} f(\sin x) \, dx, \quad \int_0^{\pi} x f(\sin x) \, dx = \frac{\pi}{2} \int_0^{\pi} f(\sin x) \, dx.$$

(b) $\displaystyle \int_0^{\pi} \frac{x \sin x}{2 - \sin^2 x} \, dx = \frac{\pi}{2} \int_0^{\pi} \frac{\sin x}{2 - \sin^2 x} \, dx = \frac{\pi}{2} \int_0^{\pi} \frac{\sin x}{1 + \cos^2 x} \, dx, \text{ let } u = \cos x,$

$$= -\frac{\pi}{2} \int_1^{-1} \frac{1}{1 + u^2} \, du = \frac{\pi}{2} \tan^{-1} u \Big]_{-1}^{1} = \frac{1}{4} \pi^2$$

85. $\displaystyle \int \frac{1}{e^{ax} + 1} \, dx = \int \frac{e^{-ax}}{1 + e^{-ax}} \, dx = -\frac{1}{a} \ln(1 + e^{-ax}) + C$

CHAPTER 10

Improper Integrals; L'Hôpital's Rule

SECTION 10.1

10.1.1 Evaluate $\int_0^1 \dfrac{x}{\sqrt{1-x^2}}\,dx$.

10.1.2 Evaluate $\int_0^2 \dfrac{1}{x^2}\,dx$.

10.1.3 Evaluate $\int_0^{\pi/4} \dfrac{\sec^2 x}{\sqrt{\tan x}}\,dx$.

10.1.4 Evaluate $\int_{-2}^0 \dfrac{1}{x+2}\,dx$.

10.1.5 Evaluate $\int_1^\infty \dfrac{dx}{x^3}$.

10.1.6 Evaluate $\int_1^4 \dfrac{1}{(x-1)^3}\,dx$.

10.1.7 Evaluate $\int_{-\infty}^\infty \dfrac{e^x}{1+e^{2x}}\,dx$.

10.1.8 Evaluate $\int_1^\infty \dfrac{1}{\sqrt{x}}\,dx$.

10.1.9 Evaluate $\int_1^4 \dfrac{1}{\sqrt[3]{x-3}}\,dx$.

10.1.10 Evaluate $\int_1^2 \dfrac{1}{x\ln x}\,dx$.

10.1.11 Evaluate $\int_0^3 \dfrac{x}{(x^2-1)^{2/3}}\,dx$.

10.1.12 Evaluate $\int_3^4 \dfrac{1}{(x-4)^3}\,dx$.

10.1.13 Evaluate $\int_0^\infty \dfrac{1}{x^{1/3}}\,dx$.

10.1.14 Evaluate $\int_0^8 \dfrac{1}{x^{1/3}}\,dx$.

10.1.15 Evaluate $\int_2^\infty \dfrac{1}{(x-1)^3}\,dx$.

10.1.16 Evaluate $\int_0^\infty xe^{-x^2}\,dx$.

10.1.17 Evaluate $\int_{-\infty}^1 e^{(x-e^x)}\,dx$.

10.1.18 Evaluate $\int_0^1 \dfrac{1}{\sqrt{1-x^2}}\,dx$.

10.1.19 $\int_1^3 \dfrac{3\,dx}{x^2-3x}$

SOLUTIONS

SECTION 10.1

10.1.1 $\quad -\lim\limits_{\ell\to1^-}\sqrt{1-x^2}\Big]_0^\ell = -\left(\lim\limits_{\ell\to1^-}\sqrt{1-\ell^2}-\sqrt{1}\right)=1.$

10.1.2 $\quad \int_0^2 \dfrac{dx}{x^2} = \lim\limits_{\ell\to0^+}-\dfrac{1}{x}\Big]_\ell^2 = -\left(\dfrac{1}{2}-\lim\limits_{\ell\to0^+}\dfrac{1}{\ell}\right)=\infty,$ thus $\int_0^2\dfrac{dx}{x^2}$ is divergent.

10.1.3 $\quad \lim\limits_{\ell\to0^+}2\sqrt{\tan x}\Big]_\ell^{\pi/4}=2\left(\sqrt{\tan\dfrac{\pi}{4}}-\lim\limits_{\ell\to0^+}\sqrt{\tan\ell}\right)=2\left(\sqrt{1}-0\right)=2.$

10.1.4 $\quad \lim\limits_{\ell\to-2^+}\ln(x+2)\Big]_\ell^0=\ln2-\lim\limits_{\ell\to-2^+}\ln(\ell+2)=+\infty,$ divergent.

10.1.5 $\quad \lim\limits_{\ell\to+\infty}-\dfrac{1}{2x^2}\Big]_1^\ell=\dfrac{1}{2}\left(\lim\limits_{n\to+\infty}\dfrac{1}{\ell^2}-\dfrac{1}{1}\right)=\dfrac{1}{2}.$

10.1.6 $\quad \int_1^4\dfrac{1}{(x-1)^3}dx=\lim\limits_{\ell\to1^+}-\dfrac{1}{2(x-1)^2}\Big]_\ell^4=-\dfrac{1}{2}\left[\dfrac{1}{(4-1)^2}-\lim\limits_{\ell\to1^+}\dfrac{1}{(\ell-1)^2}\right]=+\infty,$

$\qquad$ thus $\int_0^4\dfrac{1}{(x-1)^3}dx$ is divergent.

10.1.7 $\quad \int_{-\infty}^0\dfrac{e^x}{1+e^{2x}}dx=\tan^{-1}e^0-\lim\limits_{\ell\to-\infty}\tan^{-1}e^\ell=\dfrac{\pi}{4}-0=\dfrac{\pi}{4}$

$\qquad$ similarly, $\int_0^\infty\dfrac{e^x}{1+e^{2x}}dx=\lim\limits_{\ell\to+\infty}\tan^{-1}e^\ell-\tan^{-1}e^0=\dfrac{\pi}{2}-\dfrac{\pi}{4}=\dfrac{\pi}{4}$

$\qquad$ so, $\int_{-\infty}^\infty\dfrac{e^x}{1+e^{2x}}dx=\dfrac{\pi}{4}+\dfrac{\pi}{4}=\dfrac{\pi}{2}.$

10.1.8 $\quad \lim\limits_{\ell\to+\infty}2\sqrt{x}\Big]_1^\ell=2\left(\lim\limits_{n\to+\infty}\sqrt{\ell}-\sqrt{1}\right)=+\infty,$ divergent.

10.1.9 $\displaystyle\int_3^4 \frac{1}{\sqrt[3]{x-3}}dx = \lim_{\ell\to 3^+} \frac{3}{2}(x-3)^{2/3}\Big]_{\ell}^4 = \frac{3}{2}\left[(4-3)^{2/3} - \lim_{\ell\to 3^+}(\ell-3)^{2/3}\right] = \frac{3}{2},$

similarly, $\displaystyle\int_1^3 \frac{1}{3\sqrt[3]{x-3}}dx = \lim_{\ell\to 3^-} \frac{3}{2}(x-3)^{2/3}\Big]_1^{\ell}$

$$= \frac{3}{2}\left[\lim_{\ell\to 3^-}(\ell-3)^{2/3} - (1-3)^{2/3}\right] = -\frac{3}{2}\sqrt[3]{4}$$

so, $\displaystyle\int_1^4 \frac{1}{\sqrt[3]{x-3}}dx = \frac{3}{2} - \frac{3}{2}\sqrt[3]{4} = \frac{3}{2}\left(1-\sqrt[3]{4}\right).$

10.1.10 $\displaystyle\lim_{\ell\to 1^+} \ln(\ln x)\Big]_{\ell}^2 = \ln(\ln 2) - \lim_{\ell\to 1^+}\ln(\ln\ell) = +\infty,$ divergent.

10.1.11 $\displaystyle\int_0^1 \frac{x}{(x^2-1)^{2/3}}dx = \lim_{\ell\to 1^-} \frac{3}{2}\left(x^2-1\right)^{1/3}\Big]_0^1$

$$= \frac{3}{2}\left[\lim_{\ell\to 1^-}\left(\ell^2-1\right)^{1/3} - (-1)^{1/3}\right] = \frac{3}{2},\ \text{similarly,}$$

$$\int_1^3 \frac{x}{(x^2-1)^{2/3}}dx = \lim_{\ell\to 1^+} \frac{3}{2}\left(x^2-1\right)^{1/3}\Big]_{\ell}^3$$

$$= \frac{3}{2}\left[(9-1)^{1/3} - \lim_{\ell\to 1^+}\left(\ell^2-1\right)^{1/3}\right] = 3,$$

thus, $\displaystyle\int_0^3 \frac{x}{(x^2-1)^{2/3}}dx = \frac{3}{2} + 3 = \frac{9}{2}.$

10.1.12 $\displaystyle\lim_{\ell\to 4^-} -\frac{1}{2(x-4)^2}\Big]_3^{\ell} = -\frac{1}{2}\left[\lim_{\ell\to 4^-}\frac{1}{(\ell-4)^2} - \frac{1}{(-1)^2}\right] = -\infty,$ divergent.

10.1.13 $\displaystyle\int_1^\infty \frac{1}{x^{1/3}}dx = \lim_{\ell\to +\infty} \frac{3}{2}x^{2/3}\Big]_1^{\ell} = \frac{3}{2}\left[\lim_{\ell\to +\infty}\ell^{2/3} - (1)^{2/3}\right] = +\infty,$ thus

$\displaystyle\int_0^\infty \frac{1}{x^{1/3}}dx$ is divergent.

10.1.14 $\displaystyle\lim_{\ell\to 0^-} \frac{3}{2}x^{2/3}\Big]_{\ell}^8 = \frac{3}{2}\left[(8)^{2/3} - \lim_{\ell\to 0^+}\ell^{2/3}\right] = \frac{3}{2}(4) = 6.$

10.1.15 $\displaystyle\lim_{\ell\to +\infty} \frac{-1}{2(x-1)^2}\Big]_2^{\ell} = -\frac{1}{2}\left[\lim_{\ell\to +\infty}\frac{1}{(\ell-1)^2} - \frac{1}{(2-1)}\right] = \frac{1}{2}.$

10.1.16 $\quad \lim\limits_{\ell\to+\infty} \left.-\dfrac{1}{2e^{x^2}}\right]_0^\ell = -\dfrac{1}{2}\left(\lim\limits_{\ell\to+\infty} -\dfrac{1}{e^{\ell^2}} - \dfrac{1}{e^0}\right) = \dfrac{1}{2}.$

10.1.17 $\quad \lim\limits_{\ell\to-\infty^+} \left.-\dfrac{1}{e^{e^x}}\right]_\ell^1 = -\left(\dfrac{1}{e^e} \lim\limits_{\ell\to-\infty^+} -\dfrac{1}{e^{e^e}}\right) = 1 - \dfrac{1}{e^e}.$

10.1.18 $\quad \lim\limits_{\ell\to1^-} \left.\sin^{-1} x\right]_0^\ell = \lim\limits_{\ell\to1^-} \sin^{-1} x - \sin^{-1} 0 = \dfrac{\pi}{2}.$

10.1.19 $\quad \displaystyle\int_1^3 \dfrac{3\,dx}{x^2 - 3x} = \text{(by partial fractions)}$

$\displaystyle\int_1^3 \left(\dfrac{1}{x-3} - \dfrac{1}{x}\right)dx = \lim\limits_{\ell\to3^-} \left[\ln|x-3| - \ln|x|\right]_{11}^\ell = -\infty, \text{ diverges.}$

SECTION 10.2

10.2.1 Evaluate $\lim\limits_{x \to e} \left[\dfrac{\ln(\ln x)}{\ln x - 1} \right]$.

10.2.2 Evaluate $\lim\limits_{x \to 4} \dfrac{x^2 - 16}{x^2 + x - 20}$.

10.2.3 Evaluate $\lim\limits_{x \to +\infty} \dfrac{x^{-3/2}}{\sin \dfrac{1}{x}}$.

10.2.4 Evaluate $\lim\limits_{x \to 1} \dfrac{\ln x}{x - 1}$.

10.2.5 Evaluate $\lim\limits_{x \to 0} \dfrac{x}{1 + \sin x}$.

10.2.6 Evaluate $\lim\limits_{x \to 1} \dfrac{x^2 + 2x - 3}{x^2 + 3x - 4}$.

10.2.7 Evaluate $\lim\limits_{x \to 0} \dfrac{\sin x - x}{\tan x - x}$.

10.2.8 Evaluate $\lim\limits_{x \to 0} \dfrac{\sinh x}{\cosh x - 1}$.

10.2.9 Evaluate $\lim\limits_{x \to 0} \dfrac{\tan x}{x}$.

10.2.10 Evaluate $\lim\limits_{x \to 0} \dfrac{xe^x}{1 - e^x}$.

10.2.11 Evaluate $\lim\limits_{x \to 0} \dfrac{x - \sin x}{2 + 2x + x^2 - 2e^x}$.

10.2.12 Evaluate $\lim\limits_{x \to 0} \dfrac{5^x - 3^x}{x}$.

10.2.13 Evaluate $\lim\limits_{x \to a} \dfrac{\dfrac{1}{x} - \dfrac{1}{a}}{x - a}$.

10.2.14 Evaluate $\lim\limits_{x \to 0} \left[\dfrac{x - \ln(1 + x)}{x^2} \right]$.

10.2.15 Evaluate $\lim\limits_{x \to \frac{\pi}{4}} \dfrac{1 - \tan x}{\cos 2x}$.

10.2.16 Evaluate $\lim\limits_{x \to 0} \dfrac{x - \tan x}{1 - \cos x}$.

10.2.17 Evaluate $\lim\limits_{x \to 0} \dfrac{e^{2x} - 1}{x^2 - \sin x}$.

10.2.18 Evaluate $\lim\limits_{x \to 0} \dfrac{xe^{3x}}{1 - e^{3x}}$.

SOLUTIONS

SECTION 10.2

10.2.1 $\displaystyle\lim_{x\to e}\frac{\frac{1}{x\ln x}}{\frac{1}{x}}=\lim_{x\to e}\frac{1}{\ln x}=1.$

10.2.2 $\displaystyle\lim_{x\to 4}\frac{2x}{2x+1}=\frac{8}{9}.$

10.2.3 $\displaystyle\frac{3}{2}\lim_{x\to+\infty}\frac{1}{x^{1/2}\cos\frac{1}{x}}=0.$

10.2.4 $\displaystyle\lim_{x\to 1}\frac{1/x}{1}=1.$

10.2.5 $0.$

10.2.6 $\displaystyle\lim_{x\to 1}\frac{2x+2}{2x+3}=\frac{4}{5}.$

10.2.7 $\displaystyle\lim_{x\to 0}\frac{\cos x-1}{\sec^2 x-1}=\lim_{x\to 0}-\frac{\sin x}{2\sec^2 x\tan x}=-\frac{1}{2}\lim_{x\to 0}\cos^3 x=-\frac{1}{2}.$

10.2.8 $\displaystyle\lim_{x\to 0}\frac{\cosh x}{\sinh x}=+\infty.$

10.2.9 $\displaystyle\lim_{x\to 0}\frac{\sec^2 x}{1}=1.$

10.2.10 $\displaystyle\lim_{x\to 0}\frac{xe^x+e^x}{-e^x}=-1.$

10.2.11 $\displaystyle\lim_{x\to 0}\frac{1-\cos x}{2+2x-2e^x}=\lim_{x\to 0}\frac{\sin x}{2-2e^x}=\lim_{x\to 0}\frac{\cos x}{-2e^x}=-\frac{1}{2}.$

10.2.12 $\displaystyle\lim_{x\to 0}\frac{5^x\ln 5-x^3\ln 3}{1}=\ln\frac{5}{3}.$

10.2.13 $\displaystyle\lim_{x\to a}\frac{-\frac{1}{x^2}}{1}=-\frac{1}{a^2}.$

10.2.14 $\displaystyle\lim_{x\to 0}\frac{1-\frac{1}{1+x}}{2x}=\lim_{x\to 0}\frac{\frac{1}{(1+x)^2}}{2}=\frac{1}{2}.$

10.2.15 $\displaystyle\lim_{x\to\frac{\pi}{4}}\frac{-\sec^2 x}{-2\sin 2x}=1.$

10.2.16 $\displaystyle\lim_{x\to 0}\frac{1-\sec^2 x}{\sin x}=\lim_{x\to 0}\frac{-2\sec^2 x\tan x}{\cos x}=0.$

10.2.17 $\displaystyle\lim_{x\to 0}\frac{2e^{2x}}{2x-\cos x}=-2.$

10.2.18 $\displaystyle\lim_{x\to 0}\frac{e^{3x}+3xe^{3x}}{-3e^{3x}}=-\frac{1}{3}.$

SECTION 10.3

10.3.1 Evaluate $\displaystyle\lim_{x\to+\infty}\frac{x^3-2x+1}{4x^3+2}$.

10.3.2 Evaluate $\displaystyle\lim_{x\to+\infty}\frac{e^x}{x^3}$.

10.3.3 Evaluate $\displaystyle\lim_{x\to\frac{\pi}{2}^-}(\sec x-\tan x)$.

10.3.4 Evaluate $\displaystyle\lim_{x\to0}\left(\csc x-\frac{1}{x}\right)$.

10.3.5 Evaluate $\displaystyle\lim_{x\to0}\frac{1}{x^2}-\frac{1}{x}$.

10.3.6 Evaluate $\displaystyle\lim_{x\to1^-}(x-1)\tan\frac{\pi x}{2}$.

10.3.7 Evaluate $\displaystyle\lim_{x\to0^+}\sin x\ln x$.

10.3.8 Evaluate $\displaystyle\lim_{x\to0^+}x\ln x$.

10.3.9 Evaluate $\displaystyle\lim_{x\to0^+}x\ln\sin x$.

10.3.10 Evaluate $\displaystyle\lim_{x\to+\infty}(x+e^x)^{2/x}$.

10.3.11 Evaluate $\displaystyle\lim_{x\to\frac{\pi}{2}^+}(\tan x)^{\cos x}$.

10.3.12 Evaluate $\displaystyle\lim_{x\to+\infty}\left(1+\frac{1}{x^2}\right)^{x^2}$.

10.3.13 Evaluate $\displaystyle\lim_{x\to0^+}(\sin x)^x$.

10.3.14 Evaluate $\displaystyle\lim_{x\to0}(\cos\ 3x)^{1/x}$.

10.3.15 Evaluate $\displaystyle\lim_{x\to0}(\sin 2x+1)^{1/x}$.

10.3.16 Evaluate $\displaystyle\lim_{x\to+\infty}\left(2e^x+x^2\right)^{3/x}$.

10.3.17 Evaluate $\displaystyle\lim_{x\to0}(\cosh x)^{4/x}$.

10.3.18 Evaluate $\displaystyle\lim_{x\to0}(e^x+3x)^{1/x}$.

SOLUTIONS

SECTION 10.3

10.3.1 $\displaystyle\lim_{x\to+\infty}\frac{3x^2-2}{12x^2}=\lim_{x\to+\infty}\frac{6x}{24x}=\frac{1}{4}.$

10.3.2 $\displaystyle\lim_{x\to+\infty}\frac{e^x}{3x^2}=\lim_{x\to+\infty}\frac{e^x}{6x}=\lim_{x\to+\infty}\frac{e^x}{6}=+\infty.$

10.3.3 $\displaystyle\lim_{x\to\frac{\pi}{2}^-}\frac{1-\sin x}{\cos x}=\lim_{x\to\frac{\pi}{2}^-}\frac{-\cos x}{-\sin x}=0.$

10.3.4 $\displaystyle\lim_{x\to0}\frac{x-\sin x}{x\sin x}=\lim_{x\to0}\frac{1-\cos x}{x\cos x+\sin x}=\lim_{x\to0}\frac{\sin x}{2\cos x-x\sin x}=0.$

10.3.5 $\displaystyle\lim_{x\to0}\frac{1-x}{x^2}=+\infty.$

10.3.6 $\displaystyle\lim_{x\to1^-}\frac{x-1}{\cot\frac{\pi}{2}x}=\lim_{x\to1^-}-\frac{2}{\pi}\sin^2\frac{\pi}{2}x=-\frac{2}{\pi}.$

10.3.7 $\displaystyle\lim_{x\to0^+}\frac{\ln x}{\csc x}=\lim_{x\to0^+}\frac{\frac{1}{x}}{-\csc x\cot x}=\lim_{x\to0^+}\left(\frac{\sin x}{x}\right)(-\tan x)=0.$

10.3.8 $\displaystyle\lim_{x\to0^+}\frac{\ln x}{\dfrac{1}{x}}=\lim_{x\to0^+}\frac{\frac{1}{x}}{\dfrac{1}{x^2}}=0.$

10.3.9 $\displaystyle\lim_{x\to0^+}\frac{\ln\sin x}{\dfrac{1}{x}}=\left(\frac{x}{\sin x}\right)(-x\cos x)=0.$

10.3.10 Let $y=(x+e^x)^{2/x}$,

$$\lim_{x\to+\infty}\ln y=\lim_{x\to+\infty}\frac{2\ln(x+e^x)}{x}=\lim_{x\to+\infty}\frac{2(1+e^x)}{x+e^x}$$

$$=\lim_{x\to+\infty}\frac{2e^x}{1+e^x}=\lim_{x\to+\infty}\frac{2e^x}{e^x}=2;\ \lim_{y\to+\infty}y=e^2.$$

10.3.11 Let $y=(\tan x)^{\cos x}$,

$$\lim_{x\to\frac{\pi}{2}^+}\ln y=\lim_{x\to\frac{\pi}{2}^+}\cos x\ln\tan x=\lim_{x\to\frac{\pi}{2}^+}\frac{\ln\tan x}{\sec x}$$

$$=\lim_{x\to\frac{\pi}{2}^+}\frac{\sec x}{\tan^2 x}=\lim_{x\to\frac{\pi}{2}^+}\frac{\cos x}{\sin^2 x}=0;\ \lim_{x\to\frac{\pi}{2}^+}y=e^0=1.$$

10.3.12 Let $y = \left(1 + \dfrac{1}{x^2}\right)^{x^2}$,

$$\lim_{x \to +\infty} \ln y = \lim_{x \to +\infty} x^2 \ln\left(1 + \frac{1}{x^2}\right) = \lim_{x \to +\infty} \frac{\ln\left(1 + \dfrac{1}{x^2}\right)}{\dfrac{1}{x^2}}$$

$$= \lim_{x \to +\infty} \frac{x^2}{x^2 + 1} = 1; \ \lim_{x \to +\infty} y = e.$$

10.3.13 Let $y = (\sin x)^x$,

$$\lim_{x \to 0^+} \ln y = \lim_{x \to 0^+} x \ln \sin x = \lim_{x \to 0^+} \frac{\ln \sin x}{\dfrac{1}{x}}$$

$$= \lim_{x \to 0^+} -\left(\frac{x}{\sin x}\right)(x \cos x) = 0; \ \lim_{x \to 0} y = e^0 = 1.$$

10.3.14 Let $y = (\cos 3x)^{1/x}$,

$$\lim_{x \to 0} \ln y = \lim_{x \to 0} \frac{\ln \cos 3x}{x} = \lim_{x \to 0} \frac{-3 \tan 3x}{1} = 0; \ \lim_{x \to 0} y = e^0 = 1.$$

10.3.15 Let $y = (\sin 2x + 1)^{1/x}$,

$$\lim_{x \to 0} \ln y = \lim_{x \to 0} \frac{\ln(\sin 2x + 1)}{x} = \lim_{x \to 0} \frac{2 \cos 2x}{\sin 2x + 1} = 2; \ \lim_{x \to 0} y = e^2.$$

10.3.16 Let $y = (2e^x + x^2)^{3/x}$,

$$\lim_{x \to +\infty} \ln y = \lim_{x \to +\infty} \frac{3 \ln(2e^x + x^2)}{x} = \lim_{x \to +\infty} \frac{3(2e^x + 2x)}{2e^x + x^2}$$

$$= \lim_{x \to +\infty} \frac{3(2e^x + 2)}{2e^x + 2x} = 3; \ \lim_{x \to +\infty} y = e^3.$$

10.3.17 Let $y = (\cosh x)^{4/x}$,

$$\lim_{x \to 0} \ln y = \lim_{x \to 0} \frac{4 \ln \cosh x}{x} = \lim_{x \to 0} \frac{4 \tanh x}{1} = 0; \ \lim_{x \to 0} y = e^0 = 1.$$

10.3.18 Let $y = (e^x + 3x)^{1/x}$,

$$\lim_{x \to 0} \ln y = \lim_{x \to 0} \frac{\ln(e^x + 3x)}{x} = \lim_{x \to 0} \frac{e^x + 3}{e^x + 3x} = \frac{4}{1} = 4; \ \lim_{x \to 0} y = e^4.$$

SUPPLEMENTARY EXERCISES, CHAPTER 10

In Exercises 1–13, evaluate the integrals that converge.

1. $\displaystyle\int_{-\infty}^{+\infty} \frac{dx}{x^2 + 4}$.

2. $\displaystyle\int_{4}^{6} \frac{dx}{4 - x}$.

3. $\displaystyle\int_{-1}^{1} \frac{dx}{\sqrt{x^2}}$.

4. $\displaystyle\int_{0}^{\pi/2} \sec^2 x\, dx$.

5. $\displaystyle\int_{-\infty}^{+\infty} x e^{-x^2}\, dx$.

6. $\displaystyle\int_{-\infty}^{0} x e^x\, dx$.

7. $\displaystyle\int_{0}^{\pi/2} \cot x\, dx$.

8. $\displaystyle\int_{0}^{+\infty} \frac{dx}{x^5}$.

9. $\displaystyle\int_{e}^{+\infty} \frac{dx}{x(\ln x)^2}$.

10. $\displaystyle\int_{0}^{1} \sqrt{x}\, \ln x\, dx$.

11. $\displaystyle\int_{0}^{+\infty} \frac{dx}{x^2 + 2x + 2}$.

12. $\displaystyle\int_{0}^{4} \frac{e^{-\sqrt{x}}}{\sqrt{x}}\, dx$.

13. For what values of n does $\displaystyle\int_{0}^{1} x^n \ln x\, dx$ converge? What is its value?

In Exercises 14–27, find the limit.

14. $\displaystyle\lim_{x \to 0} \frac{x e^{3x} - x}{1 - \cos 2x}$.

15. $\displaystyle\lim_{x \to 0^+} x^2 e^{1/x}$.

16. $\displaystyle\lim_{x \to +\infty} (\sqrt{x^2 + x} - x)$.

17. $\displaystyle\lim_{x \to 0^-} x^2 e^{1/x}$.

18. $\displaystyle\lim_{x \to 0^-} (1 - x)^{2/x}$.

19. $\displaystyle\lim_{\theta \to 0} \left(\frac{\csc \theta}{\theta} - \frac{1}{\theta^2} \right)$.

20. $\displaystyle\lim_{x \to 0} \frac{x - \tan^{-1} x}{x^4}$.

21. $\displaystyle\lim_{x \to 2} \frac{x - 1 - e^{x-2}}{1 - \cos 2\pi x}$.

22. $\displaystyle\lim_{x \to 0} \frac{9^x - 3^x}{x}$.

23. $\displaystyle\lim_{x \to 0} \frac{\displaystyle\int_{0}^{x} \sin t^2\, dt}{\sin x^2}$.

24. $\displaystyle\lim_{x \to +\infty} x^{1/x}$.

25. $\displaystyle\lim_{x \to +\infty} \frac{(\ln x)^3}{x}$.

26. $\displaystyle\lim_{x \to +\infty} \left(\frac{x}{x - 3} \right)^x$.

27. $\displaystyle\lim_{x \to 0^+} (1 + x)^{\ln x}$.

28. Find the area of the region in the first quadrant between the curves $y = e^{-x}$ and $y = e^{-2x}$.

In Exercises 29 and 30, find (a) the area of the region R, and (b) the volume obtained by revolving R about the x-axis.

29. The region R bounded between the x-axis and the curve $y = x^{-2/3}, x \geq 8$.

30. The region R bounded between the x-axis and the curve $y = x^{-1/3}, 0 \leq x \leq 1$.

31. Find the total arc length of the graph of the function $f(x) = \sqrt{x - x^2} - \sin^{-1}\sqrt{x}$. [*Hint*: First find the domain of f.]

SOLUTIONS

SUPPLEMENTARY EXERCISES CHAPTER 10

1. $\displaystyle\int_0^{+\infty}\frac{dx}{x^2+4}=\lim_{\ell\to+\infty}\frac{1}{2}\tan^{-1}(x/2)\Big]_0^{\ell}=\lim_{\ell\to+\infty}\frac{1}{2}\tan^{-1}(\ell/2)=\pi/4,$

$\displaystyle\int_{\ell}^0\frac{dx}{x^2+4}=\lim_{\ell\to-\infty}\frac{1}{2}\tan^{-1}(x/2)\Big]_{\ell}^0=\pi/4$ so $\displaystyle\int_{-\infty}^{+\infty}\frac{dx}{x^2+4}=\pi/2$

2. $\displaystyle\lim_{\ell\to4^+}-\ln|4-x|\,\Big]_{\ell}^6=\lim_{\ell\to4^+}(-\ln2+\ln|4-\ell|)=+\infty,$ diverges

3. $\displaystyle\int_0^1 x^{-2/3}dx=\lim_{\ell\to0^+}3x^{1/3}\Big]_{\ell}^1=3,\ \int_{-1}^0 x^{-2/3}dx=\lim_{\ell\to0^-}3x^{1/3}\Big]_{-1}^{\ell}=3,\ \int_{-1}^1 x^{-2/3}dx=6$

4. $\displaystyle\lim_{\ell\to\pi/2^-}\tan x\,\Big]_0^{\ell}=+\infty,$ diverges

5. $\displaystyle\lim_{\ell\to+\infty}-\frac{1}{2}e^{-x^2}\Big]_0^{\ell}=\lim_{\ell\to+\infty}\frac{1}{2}(-e^{-\ell^2}+1)=1/2,$

$\displaystyle\lim_{\ell\to-\infty}-\frac{1}{2}e^{-x^2}\Big]_{\ell}^0=\lim_{\ell\to-\infty}\frac{1}{2}(-1+e^{-\ell^2})=-1/2,$ so $\displaystyle\int_{-\infty}^{+\infty}xe^{-x^2}dx=1/2-1/2=0$

6. $\displaystyle\lim_{\ell\to-\infty}(xe^x-e^x)\Big]_{\ell}^0=\lim_{\ell\to-\infty}(-1-\ell e^{\ell}+e^{\ell})=-1$ because

$\displaystyle\lim_{\ell\to-\infty}\ell e^{\ell}=\lim_{\ell\to-\infty}\frac{\ell}{e^{-\ell}}=\lim_{\ell\to-\infty}\frac{1}{-e^{-\ell}}=0$ and $\displaystyle\lim_{\ell\to-\infty}e^{\ell}=0$

7. $\displaystyle\lim_{\ell\to0^+}\ln|\sin x|\,\Big]_{\ell}^{\pi/2}=\lim_{\ell\to0^+}-\ln|\sin\ell|=+\infty,$ diverges

8. $\displaystyle\int_0^{+\infty}x^{-5}dx=\int_0^1 x^{-5}dx+\int_1^{+\infty}x^{-5}dx,\ \int_0^1 x^{-5}dx=\lim_{\ell\to0^+}-1/(4x^4)\Big]_{\ell}^1=+\infty$ so

$\displaystyle\int_0^{+\infty}x^{-5}dx$ is divergent.

9. $\lim\limits_{\ell\to+\infty} -1/\ln x \Big]_e^\ell = \lim\limits_{\ell\to+\infty} (-1/\ln\ell + 1) = 1$

10. $\lim\limits_{\ell\to 0^+} \left(\dfrac{2}{3}x^{3/2}\ln x - \dfrac{4}{9}x^{3/2}\right)\Big]_\ell^1 = \lim\limits_{\ell\to 0^+}\left(-\dfrac{4}{9} - \dfrac{2}{3}\ell^{3/2}\ln\ell + \dfrac{4}{9}\ell^{3/2}\right) = -4/9$ because

$\lim\limits_{\ell\to 0^+} \ell^{3/2}\ln\ell = \lim\limits_{\ell\to 0^+} \dfrac{\ln\ell}{\ell^{-3/2}} = \lim\limits_{\ell\to 0^+} \dfrac{1/\ell}{(-3/2)\ell^{-5/2}} = \lim\limits_{\ell\to 0^+} -\dfrac{2}{3}\ell^{3/2} = 0$ and $\lim\limits_{\ell\to 0^+}\ell^{3/2} = 0$.

11. $\lim\limits_{\ell\to+\infty}\tan^{-1}(x+1)\Big]_e^\ell = \lim\limits_{\ell\to+\infty}[\tan^{-1}(\ell+1) - \tan^{-1}(1)] = \pi/2 - \pi/4 = \pi/4$.

12. $\lim\limits_{\ell\to 0^+} -2e^{-\sqrt{x}}\Big]_\ell^4 = \lim\limits_{\ell\to 0^+} 2(-e^{-2} + e^{-\sqrt{\ell}}) = 2(1 - e^{-2})$

13. If $n = -1$ then $\displaystyle\int_0^1 \dfrac{\ln x}{x}dx = \lim\limits_{\ell\to 0^+}\dfrac{1}{2}(\ln x)^2\Big]_\ell^1 = -\infty$ so the integral diverges. If $n \neq -1$ then

$\displaystyle\int_0^1 x^n\ln x\, dx = \lim\limits_{\ell\to 0^+}\left[\dfrac{x^{n+1}}{n+1}\ln x - \dfrac{x^{n+1}}{(n+1)^2}\right]_\ell^1 = \lim\limits_{\ell\to 0^+}\left[-\dfrac{1}{(n+1)^2} - \dfrac{\ell^{n+1}}{n+1}\ln\ell + \dfrac{\ell^{n+1}}{(n+1)^2}\right]$

If $n < -1$ then $n + 1 < 0$, $\lim\limits_{\ell\to 0^+}\ell^{n+1}\ln\ell = -\infty$ and $\lim\limits_{\ell\to 0^+}\ell^{n+1} = +\infty$ so the integral diverges.

If $n > -1$ then $\lim\limits_{\ell\to 0^+}\ell^{n+1}\ln\ell = \lim\limits_{\ell\to 0^+}\dfrac{\ln\ell}{\ell^{-(n+1)}} = \lim\limits_{\ell\to 0^+} -\dfrac{\ell^{n+1}}{n+1} = 0$ and $\lim\limits_{\ell\to 0^+}\ell^{n+1} = 0$ so the integral converges to $-1/(n+1)^2$.

14. $\lim\limits_{x\to 0}\dfrac{3xe^{3x} + e^{3x} - 1}{2\sin 2x} = \lim\limits_{x\to 0}\dfrac{9xe^{3x} + 6e^{3x}}{4\cos 2x} = 3/2$

15. $\lim\limits_{x\to 0^+}\dfrac{e^{1/x}}{1/x^2} = \lim\limits_{x\to 0^+}\dfrac{(-1/x^2)e^{1/x}}{-2/x^3} = \lim\limits_{x\to 0^+}\dfrac{e^{1/x}}{2/x} = \lim\limits_{x\to 0^+}\dfrac{(-1/x^2)e^{1/x}}{-2/x^2} = \lim\limits_{x\to 0^+}(1/2)e^{1/x} = +\infty$

16. $\lim\limits_{x\to+\infty}\dfrac{(x^2+x) - x^2}{\sqrt{x^2+x}+x} = \lim\limits_{x\to+\infty}\dfrac{x}{\sqrt{x^2+x}+x} = \lim\limits_{x\to+\infty}\dfrac{1}{\sqrt{1+1/x}+1} = 1/2$

17. $\lim\limits_{x\to 0^-} x^2 e^{1/x} = (0)(0) = 0$

18. $y = (1-x)^{2/x}$, $\lim\limits_{x\to 0^-}\ln y = \lim\limits_{x\to 0^-}\dfrac{2\ln(1-x)}{x} = \lim\limits_{x\to 0^-}\dfrac{-2}{1-x} = -2$, $\lim\limits_{x\to 0^-} y = e^{-2}$

19. $\displaystyle\lim_{\theta\to 0}\left(\frac{1}{\theta\sin\theta}-\frac{1}{\theta^2}\right)=\lim_{\theta\to 0}\frac{\theta-\sin\theta}{\theta^2\sin\theta}=\lim_{\theta\to 0}\frac{1-\cos\theta}{\theta^2\cos\theta+2\theta\sin\theta}$

$$=\lim_{\theta\to 0}\frac{\sin\theta}{-\theta^2\sin\theta+4\theta\cos\theta+2\sin\theta}$$

$$=\lim_{\theta\to 0}\frac{\cos\theta}{-\theta^2\cos\theta-6\theta\sin\theta+6\cos\theta}=1/6$$

20. $\displaystyle\lim_{x\to 0}\frac{1-1/(1+x^2)}{4x^3}=\lim_{x\to 0}\frac{1}{4x(1+x^2)},$ which does not exist

21. $\displaystyle\lim_{x\to 2}\frac{1-e^{x-2}}{2\pi\sin 2\pi x}=\lim_{x\to 2}\frac{-e^{x-2}}{4\pi^2\cos 2\pi x}=-1/(4\pi^2)$

22. $\displaystyle\lim_{x\to 0}\frac{9^x\ln 9-3^x\ln 3}{1}=\ln 9-\ln 3=\ln 3$

23. $\displaystyle\lim_{x\to 0}\frac{\sin(x^2)}{2x\cos(x^2)}=\lim_{x\to 0}\frac{2x\cos(x^2)}{-4x^2\sin(x^2)+2\cos(x^2)}=0$

24. $y=x^{1/x},\ \displaystyle\lim_{x\to+\infty}\ln y=\lim_{x\to+\infty}\frac{\ln x}{x}=\lim_{x\to+\infty}\frac{1}{x}=0,\ \lim_{x\to+\infty}y=e^0=1$

25. $\displaystyle\lim_{x\to+\infty}\frac{3(\ln x)^2}{x}=\lim_{x\to+\infty}\frac{6\ln x}{x}=\lim_{x\to+\infty}\frac{6}{x}=0$

26. $y=\left(\dfrac{x}{x-3}\right)^x,\ \displaystyle\lim_{x\to+\infty}\ln y=\lim_{x\to+\infty}\frac{\ln\dfrac{x}{x-3}}{1/x}=\lim_{x\to+\infty}\frac{3x}{x-3}=3,\ \lim_{x\to+\infty}y=e^3$

27. $y=(1+x)^{\ln x},\ \displaystyle\lim_{x\to 0^+}\ln y=\lim_{x\to 0^+}\ln x\ln(1+x)=\lim_{x\to 0^+}\frac{\ln(1+x)}{1/\ln x}$

$$=\lim_{x\to 0^+}\frac{1/(1+x)}{-1/[x(\ln x)^2]}=\lim_{x\to 0^+}\frac{-x(\ln x)^2}{1+x},$$

but $\displaystyle\lim_{x\to 0^+}x(\ln x)^2=\lim_{x\to 0^+}\frac{(\ln x)^2}{1/x}=\lim_{x\to 0^+}\frac{(2\ln x)/x}{-1/x^2}$

$$=\lim_{x\to 0^+}\frac{2\ln x}{-1/x}=\lim_{x\to 0^+}\frac{2/x}{-1/x^2}=\lim_{x\to 0^+}(-2x)=0$$

so $\displaystyle\lim_{x\to 0^+}\frac{-x(\ln)^2}{1+x}=\frac{0}{1}=0$ and $\displaystyle\lim_{x\to 0^+}y=e^0=1$

28. $A = \displaystyle\int_0^{+\infty} (e^{-x} - e^{-2x})dx = \lim_{\ell \to +\infty} \left. \left(-e^{-x} + \frac{1}{2}e^{-2x}\right)\right]_0^{\ell}$

$$= \lim_{\ell \to +\infty} \left(-e^{-\ell} + \frac{1}{2}e^{-2\ell} + \frac{1}{2}\right) = 1/2$$

29. **(a)** $A = \displaystyle\int_8^{+\infty} x^{-2/3}dx = \lim_{\ell \to +\infty} \left. 3x^{1/3}\right]_8^{\ell} = +\infty$

(b) $V = \displaystyle\int_8^{+\infty} \pi x^{-4/3}dx = \lim_{\ell \to +\infty} \left. -3\pi x^{-1/3}\right]_8^{\ell} = 3\pi/2$

30. **(a)** $A = \displaystyle\int_0^1 x^{-1/3}dx = \lim_{\ell \to 0^+} \left. \frac{3}{2}x^{2/3}\right]_{\ell}^1 = 3/2$

(b) $V = \displaystyle\int_0^1 \pi x^{-2/3}dx = \lim_{\ell \to 0^+} \left. 3\pi x^{1/3}\right]_{\ell}^1 = 3\pi$

31. $f(x) = \sqrt{x - x^2} - \sin^{-1}\sqrt{x} = \sqrt{x(1-x)} - \sin^{-1}\sqrt{x}$, the domain of f is $0 \leq x \leq 1$,

$f'(x) = \frac{1}{2}(x - x^2)^{-1/2}(1 - 2x) - \frac{1}{\sqrt{1-x}}\left(\frac{1}{2}x^{-1/2}\right) = -\frac{x}{\sqrt{x - x^2}}$,

$1 + [f'(x)]^2 = 1 + \frac{x^2}{x - x^2} = \frac{1}{1-x}$,

$L = \displaystyle\int_0^1 \frac{1}{\sqrt{1-x}}dx = \lim_{\ell \to 1^-} \left. -2\sqrt{1-x}\right]_0^{\ell} = \lim_{\ell \to 1^-} 2(-\sqrt{1-\ell} + 1) = 2$

CHAPTER 11

Infinite Series

SECTION 11.1

11.1.1 Express the following sequence in bracket notation:

$$1, \frac{4}{3}, \frac{6}{4}, \frac{8}{5}, \frac{10}{6}, \dots$$

Determine if the sequence converges, and if so, find its limit.

11.1.2 Express the following sequence in bracket notation:

$$1, 2/3, 3/5, 4/7, \dots$$

Determine if the sequence converges, and if so, find its limit.

11.1.3 Express the following sequence in bracket notation:

$$1/2, -3/4, 7/8, -15/16, \dots$$

Determine if the sequence converges, and if so, find its limit.

11.1.4 Write the first five terms of the sequence given by

$$\left\{ (-1)^{n+1} \frac{n}{n+2} \right\}_{n=1}^{+\infty}$$

Determine if the sequence converges, and if so, find its limit.

11.1.5 Does the sequence given by

$$\left\{ (1+n)^{\frac{1}{n}} \right\}_{n=1}^{+\infty}$$

converge or diverge? If it converges, what is its limit?

11.1.6 List the first five terms of the sequence given by

$$\left\{ \frac{n}{n+2} \right\}_{n=1}^{+\infty}$$

Determine if the sequence converges, and if so, find its limit.

11.1.7 List the first five terms of the sequence given by

$$\left\{ 1 + (-1)^n \right\}_{n=1}^{+\infty}$$

Determine if the sequence converges, and if so, find its limit.

11.1.8 Does the sequence given by

$$\left\{ \frac{n^3 + 6n^2 + 11n + 6}{2n^3 + 3n^2 + 1} \right\}_{n=1}^{+\infty}$$

converge or diverge? If it converges, what is its limit?

11.1.9 List the first five terms of the sequence given by

$$\left\{ \frac{1}{n} \sin \frac{\pi}{n} \right\}_{n=1}^{+\infty}$$

Determine if the sequence converges, and if so, find its limit.

11.1.10 Determine if the sequence given by

$$\left\{ \frac{\ln n}{n} \right\}_{n=1}^{+\infty}$$

converges or diverges? If it converges, find its limit.

11.1.11 Determine if the sequence given by

$$\left\{ \frac{\sqrt{n}}{\ln n} \right\}_{n=2}^{+\infty}$$

converges or diverges? If it converges, find its limit.

11.1.12 Determine if the sequence given by

$$\left\{ \frac{1 - n^2}{2 + 3n^2} \right\}_{n=1}^{+\infty}$$

converges or diverges? If it converges, find its limit.

11.1.13 Determine if the sequence given by

$$\left\{ n \sin \frac{1}{n} \right\}_{n=1}^{+\infty}$$

converges or diverges? If it converges, find its limit.

11.1.14 Express the following sequence in bracket notation:

$$0, -\frac{1}{3^2}, \frac{2}{3^3}, -\frac{3}{3^4}, \ldots$$

Determine if the sequence converges, and if so, find its limit.

11.1.15   Does the sequence given by

$$\left\{ \frac{2n}{\sqrt{n^2 - 1}} \right\}_{n=2}^{+\infty}$$

converge or diverge? If it converges, find its limit.

11.1.16   Does the sequence given by

$$\left\{ \frac{n}{2n + 1} \right\}_{n=1}^{+\infty}$$

converge or diverge? If it converges, find its limit.

11.1.17   Express the following sequence in bracket notation:

$$1/2, 4/3, 9/4, 16/5, 25/6, \ldots$$

Determine if the sequence converges. If it converges, find its limit.

11.1.18   Determine if the sequence given by

$$\left\{ \frac{\sin n}{n} \right\}_{n=1}^{+\infty}$$

converges or diverges? If it converges, find its limit.

SOLUTIONS

SECTION 11.1

11.1.1 $\left\{\dfrac{2n}{n+1}\right\}_{n=1}^{+\infty}$, $\displaystyle\lim_{n\to+\infty}\dfrac{2n}{n+1}=2$, converges.

11.1.2 $\left\{\dfrac{n}{2n-1}\right\}_{n=1}^{+\infty}$, $\displaystyle\lim_{n\to+\infty}\dfrac{n}{2n-1}=\dfrac{1}{2}$, converges.

11.1.3 $\left\{(-1)^{n+1}\dfrac{2^n-1}{2^n}\right\}_{n=1}^{+\infty}$, diverges because odd numbered terms approach $+1$ and even number terms approach -1.

11.1.4 $1/3,-2/4,3/5,-4/6,5/7,\ldots$; diverges because the odd numbered terms approach $+1$, and the even numbered terms approach -1.

11.1.5 Let $y=(1+x)^{1/x}$, $\displaystyle\lim_{x\to+\infty}\ln y=\lim_{x\to+\infty}\dfrac{\ln(1+x)}{x}=\lim_{x\to+\infty}\dfrac{\dfrac{1}{1+x}}{1}=0$,

$\displaystyle\lim_{n\to+\infty}(1+n)^{1/n}=\lim_{x\to+\infty}y=e^0=1$, converges.

11.1.6 $1/3,2/4,3/5,4/6,5/7$; $\displaystyle\lim_{n\to+\infty}\dfrac{n}{n+2}=1$, converges.

11.1.7 $0,2,0,2,0,\ldots$; diverges.

11.1.8 $\displaystyle\lim_{n\to+\infty}\dfrac{n^3+6n^2+1ln+6}{2n^3+3n^2+1}=\dfrac{1}{2}$, converges.

11.1.9 $\dfrac{1}{1}\sin\pi,\dfrac{1}{2}\sin\dfrac{\pi}{2},\dfrac{1}{3}\sin\dfrac{\pi}{3},\dfrac{1}{4}\sin\dfrac{\pi}{4},\dfrac{1}{5}\sin\dfrac{\pi}{5}$; $\displaystyle\lim_{n\to+\infty}\dfrac{1}{n}\sin\dfrac{\pi}{n}=0$, converges.

11.1.10 Let $y=\dfrac{\ln x}{x}$, $\displaystyle\lim_{x\to+\infty}\dfrac{\ln x}{x}=\lim_{x\to+\infty}\dfrac{1/x}{1}=0$, so $\displaystyle\lim_{n\to+\infty}\dfrac{\ln x}{x}=\lim_{x\to+\infty}y=0$, converges.

11.1.11 Let $y=\dfrac{\sqrt{x}}{\ln x}$, $\displaystyle\lim_{x\to+\infty}\dfrac{\sqrt{x}}{\ln x}=\lim_{x\to+\infty}\dfrac{\dfrac{1}{2\sqrt{x}}}{\dfrac{1}{x}}=\lim_{x\to+\infty}\dfrac{\sqrt{x}}{2}=\infty$, so $\displaystyle\lim_{n\to+\infty}\dfrac{\sqrt{n}}{\ln n}=+\infty$, diverges.

11.1.12 $\displaystyle\lim_{n\to+\infty}\dfrac{1-n^2}{2+3n^2}=-\dfrac{1}{3}$, converges.

11.1.13 $\displaystyle\lim_{n\to+\infty} n\sin\left(\frac{1}{n}\right) = \lim_{n\to+\infty} \frac{\sin(1/n)}{\frac{1}{n}} = \lim_{n\to+\infty} \frac{(-1/n^2)\cos(1/n)}{-1/n^2} = 1$, converges.

11.1.14 $\left\{(-1)^{n+1}\dfrac{n-1}{3^n}\right\}_{n=1}^{+\infty}$, $\displaystyle\lim_{n\to+\infty}(-1)^{n+1}\frac{n-1}{3^n} = 0$, converges.

11.1.15 $\displaystyle\lim_{n\to+\infty} \frac{2n}{\sqrt{n^2-1}} = \lim_{n\to+\infty} \frac{2}{\sqrt{1-\dfrac{1}{n^2}}} = 2$, converges.

11.1.16 $\displaystyle\lim_{n\to+\infty} \frac{n}{2n+1} = \lim_{n\to+\infty} \frac{1}{2+\dfrac{1}{n}} = \frac{1}{2}$, converges.

11.1.17 $\left\{\dfrac{n^2}{n+1}\right\}_{n=1}^{+\infty}$, $\displaystyle\lim_{n\to+\infty}\frac{n^2}{n+1} = \lim_{n\to+\infty}\frac{2n}{1} = +\infty$, diverges.

11.1.18 $\displaystyle\lim_{n\to+\infty} \frac{\sin n}{n} = 0$, converges.

SECTION 11.2

11.2.1 Use $a_{n+1} - a_n$ to show that the sequence given by $\left\{ \dfrac{2n}{n+1} \right\}_{n=1}^{+\infty}$ is strictly monotone and classify it as increasing or decreasing.

11.2.2 Use $a_{n+1} - a_n$ to show that the sequence given by $\left\{ \dfrac{2n}{2n-1} \right\}_{n=1}^{+\infty}$ is strictly monotone and classify it as increasing or decreasing.

11.2.3 Use $a_{n+1} - a_n$ to show that the sequence given by $\left\{ \dfrac{2n-5}{3n+2} \right\}_{n=1}^{+\infty}$ is strictly monotone and classify it as increasing or decreasing.

11.2.4 Use $a_{n+1} - a_n$ to show that the sequence given by $\left\{ 1 - \dfrac{2}{n} \right\}_{n=1}^{+\infty}$ is strictly monotone and classify it as increasing or decreasing.

11.2.5 Use $a_{n+1} - a_n$ to show that the sequence given by $\left\{ n - 3^n \right\}_{n=1}^{+\infty}$ is strictly monotone and classify it as increasing or decreasing.

11.2.6 Use a_{n+1}/a_n to show that the sequence given by $\left\{ \dfrac{3^n}{e^n} \right\}_{n=1}^{+\infty}$ is strictly monotone and classify it as increasing or decreasing.

11.2.7 Use a_{n+1}/a_n to show that the sequence given by $\left\{ \dfrac{(n+1)^2}{4^n} \right\}_{n=1}^{+\infty}$ is strictly monotone and classify it as increasing or decreasing.

11.2.8 Use a_{n+1}/a_n to show that the sequence given by $\left\{ \dfrac{2^n}{4^n+1} \right\}_{n=1}^{+\infty}$ is strictly monotone and classify it as increasing or decreasing.

11.2.9 Use any method to show that the sequence given by $\left\{ \dfrac{3^n}{(n+1)!} \right\}_{n=1}^{+\infty}$ is eventually increasing or eventually decreasing.

11.2.10 Use differentiation to show that the sequence given by $\left\{ \dfrac{2n^2}{n^2+1} \right\}_{n=1}^{+\infty}$ is strictly monotone and classify it as increasing or decreasing.

11.2.11   Use differentiation to show that the sequence given by $\left\{\dfrac{e^n}{\sqrt{n}}\right\}_{n=1}^{+\infty}$ is strictly monotone and classify it as increasing or decreasing.

11.2.12   Use differentiation to show that the sequence given by $\left\{\dfrac{n+2}{e^n}\right\}_{n=1}^{\infty}$ is strictly monotone and classify it as increasing or decreasing.

11.2.13   Determine whether the sequence given by $\left\{\dfrac{(n!)^2}{(2n)!}\right\}_{n=1}^{+\infty}$ is monotone. If so, classify it as increasing or decreasing.

11.2.14   Determine whether the sequence given by $\left\{\left(\dfrac{9}{10}\right)^n\right\}_{n=1}^{+\infty}$ is monotone. If so, classify it as increasing or decreasing.

11.2.15   Determine whether the sequence given by $\left\{\ln\left(\dfrac{2n}{n+1}\right)\right\}_{n=1}^{+\infty}$ is strictly monotone and classify it as increasing or decreasing.

11.2.16   Use any method to show that the sequence given by $\left\{3n^2 - 16n\right\}_{n=1}^{+\infty}$ is eventually increasing or eventually decreasing.

11.2.17   Use any method to show that the sequence given by $\left\{\dfrac{n!}{4^n}\right\}_{n=1}^{+\infty}$ is eventually increasing or eventually decreasing.

11.2.18   Use any method to show that the sequence given by $\left\{n + \dfrac{e}{n}\right\}_{n=1}^{+\infty}$ is eventually increasing or eventually decreasing.

SOLUTIONS

SECTION 11.2

11.2.1 $\quad a_{n+1} - a_n = \dfrac{2n+2}{n+2} - \dfrac{2n}{n+1} = \dfrac{2}{n^2+3n+2} > 0$ for $n \geq 1$, increasing.

11.2.2 $\quad a_{n+1} - a_n = \dfrac{2n+2}{2n+1} - \dfrac{2n}{2n-1} = \dfrac{-2}{4n^2-1} < 0$ for $n \geq 1$, decreasing.

11.2.3 $\quad a_{n+1} - a_n = \dfrac{2n-3}{3n+5} - \dfrac{2n-5}{3n+2} = \dfrac{19}{9n^2+21n+10} > 0$ for $n \geq 1$, increasing.

11.2.4 $\quad a_{n+1} - a_n = \left(1 - \dfrac{2}{n+1}\right) - \left(1 - \dfrac{2}{n}\right) = \dfrac{2}{n^2+n} > 0$ for $n \geq 1$, increasing.

11.2.5 $\quad a_{n+1} - a_n = \left((n+1) - 3^{n+1}\right) - (n - 3^n) = 1 - 2(3^n) < 0$ for $n \geq 1$, decreasing.

11.2.6 $\quad \dfrac{a_{n+1}}{a_n} = \dfrac{\frac{3^{n+1}}{e^{n+1}}}{\frac{3^n}{e^n}} = \dfrac{3^{n+1}}{e^{n+1}} \cdot \dfrac{e^n}{3^n} = \dfrac{3}{e} > 1$, for $n \geq 1$, increasing.

11.2.7 $\quad \dfrac{a_{n+1}}{a_n} = \dfrac{\frac{(n+2)^2}{4^{n+1}}}{\frac{(n+1)^2}{4^n}} = \dfrac{(n+2)^2}{4^{n+1}} \cdot \dfrac{4^n}{(n+1)^2} = \dfrac{n^2+4n+4}{4n^2+8n+4} < 1$, for $n \geq 1$, decreasing.

11.2.8 $\quad \dfrac{a_{n+1}}{a_n} = \dfrac{\frac{2^{n+1}}{4^{n+1}+1}}{\frac{2^n}{4^n+1}} = \dfrac{2^{n+1}}{2^n} \cdot \dfrac{4^n+1}{4^{n+1}+1} = \dfrac{2(4^n+1)}{4^{n+1}+1} < 1$ for $n \geq 1$, decreasing.

11.2.9 $\quad \dfrac{a_{n+1}}{a_n} = \dfrac{\frac{3^{n+1}}{(n+2)!}}{\frac{3^n}{(n+1)!}} = \dfrac{3^{n+1}}{3^n} \cdot \dfrac{(n+1)!}{(n+2)!} = \dfrac{3}{n+2} < 1$, $n \geq 2$, so the sequence is eventually decreasing.

11.2.10 $\quad$ Let $f(x) = \dfrac{2x^2}{x^2+1}$, then $f'(x) = \dfrac{4x}{(x^2+1)^2} > 0$ for $x \geq 1$, increasing.

11.2.11 $\quad$ Let $f(x) = \dfrac{e^x}{\sqrt{x}}$, $f'(x) = \dfrac{2xe^x - e^x}{2x^{3/2}} > 0$ for $x \geq 1$, increasing.

11.2.12 $\quad$ Let $f(x) = \dfrac{x+2}{e^x}$, then $f'(x) = \dfrac{-(x+1)}{e^x} < 0$ for $x \geq 1$, decreasing.

11.2.13 $\dfrac{a_{n+1}}{a_n} = \dfrac{\dfrac{[(n+1)!]^2}{(2n+2)!}}{\dfrac{(n!)^2}{(2n)!}} = \dfrac{[(n+1)n!]^2}{(2n+2)(2n+1)(2n)!} \cdot \dfrac{(2n)!}{(n!)^2} = \dfrac{n+1}{4n+2} < 1$

for $n \geq 1$, decreasing.

11.2.14 $\dfrac{a_{n+1}}{a_n} = \dfrac{\left(\dfrac{9}{10}\right)^{n+1}}{\left(\dfrac{9}{10}\right)^n} = \dfrac{9}{10} < 1$, so, decreasing.

11.2.15 Let $f(x) = \ln\left(\dfrac{2x}{x+1}\right) = \ln 2 + \ln x - \ln(x+1)$, then

$f'(x) = \dfrac{1}{x} - \dfrac{1}{x+1} = \dfrac{x+1}{x(x+1)} > 0$ for $x \geq 1$, increasing.

11.2.16 Let $f(x) = 3x^2 - 16x$, then $f'(x) = 6x - 16 > 0$ for $x \geq 3$, so the sequence is eventually increasing.

11.2.17 $\dfrac{a_{n+1}}{a_n} = \dfrac{\dfrac{(n+1)!}{4^{n+1}}}{\dfrac{n!}{4^n}} = \dfrac{4^n}{4^{n+1}} \cdot \dfrac{(n+1)!}{n!} = \dfrac{n+1}{4} > 1$, for $n \geq 4$, so the sequence is eventually increasing.

11.2.18 Let $f(x) = x + \dfrac{e}{x}$, then $f'(x) = 1 - \dfrac{e}{x^2} > 0$ for $x \geq 2$, so the sequence is eventually increasing.

SECTION 11.3

11.3.1 Determine whether $\displaystyle\sum_{k=1}^{\infty} \frac{1}{2k(k+1)}$ converges or diverges. If it converges, find its sum.

11.3.2 Express $0.315315315315\ldots$ as the quotient of two integers.

11.3.3 Determine whether $\displaystyle\sum_{k=2}^{\infty} \frac{(-1)^k}{5^k}$ converges or diverges. If it converges, find its sum.

11.3.4 Determine whether $\displaystyle\sum_{k=0}^{\infty} \frac{3^{k+2}}{4^{k+1}}$ converges or diverges. If it converges, find its sum.

11.3.5 Determine whether $\displaystyle\sum_{k=0}^{\infty} \frac{3}{10^k}$ converges or diverges. If it converges, find its sum.

11.3.6 Determine whether $\displaystyle\sum_{k=1}^{\infty} \frac{3}{e^k}$ converges or diverges. If it converges, find its sum.

11.3.7 Express $0.342342342342\ldots$ as the quotient of two integers.

11.3.8 Determine whether $\displaystyle\sum_{k=1}^{\infty} \frac{1}{k^2 + 5k + 6}$ converges or diverges. If it converges, find its sum.

11.3.9 Determine whether $\displaystyle\sum_{k=1}^{\infty} \frac{2^k}{5}$ converges or diverges. If it converges, find its sum.

11.3.10 Determine whether the series given by $\displaystyle\sum_{k=0}^{\infty} u_k = 1 - \frac{2}{5} + \frac{4}{25} - \frac{8}{125} + \cdots$ converges or diverges. If it converges, find its sum.

11.3.11 Determine whether the series given by $\displaystyle\sum_{k=1}^{\infty} \frac{(-1)^{k+1}}{4^k}$ converges or diverges. If it converges, find its sum.

11.3.12 Express $0.21212121\ldots$ as the quotient of two integers.

11.3.13 Determine whether $\displaystyle\sum_{k=3}^{\infty} \left(\frac{1}{2}\right)^k$ converges or diverges. If it converges, find its sum.

11.3.14   Determine whether $\displaystyle\sum_{k=1}^{\infty} \frac{2}{(2k-1)(2k+1)}$ converges or diverges. If it converges, find its sum.

11.3.15   Determine whether $\displaystyle\sum_{k=1}^{\infty} \left(-\frac{2}{7}\right)^{k+1}$ converges or diverges. If it converges, find its sum.

11.3.16   Determine whether $\displaystyle\sum_{k=1}^{\infty} 4^{k-1}$ converges or diverges. If it converges, find its sum.

11.3.17   Determine whether $\displaystyle\sum_{k=1}^{\infty} \left(-\frac{2}{3}\right)^{k+1}$ converges or diverges. If it converges, find its sum.

11.3.18   Determine whether $\displaystyle\sum_{k=1}^{\infty} \frac{1}{(k+1)(k+2)}$ converges or diverges. If it converges, find its sum.

SOLUTIONS

SECTION 11.3

11.3.1 $Sn = \sum_{k=1}^{n} \left(\dfrac{1}{2k} - \dfrac{1}{2(k+1)} \right) = \dfrac{1}{2} - \dfrac{1}{2(n+1)}, \ \lim\limits_{n \to +\infty} Sn = \dfrac{1}{2}.$

11.3.2 $0.315315315\ldots = 0.315 + 0.000315 + 0.000000315\ldots = \dfrac{0.315}{1 - 0.001} = \dfrac{315}{999} = \dfrac{35}{111}.$

11.3.3 geometric series, $a = 1/5^2$, $r = -1/5$, sum $= \dfrac{\dfrac{1}{5^2}}{1 + \dfrac{1}{5}} = \dfrac{1}{30}.$

11.3.4 geometric series, $a = 9/4$, $r = 3/4$, sum $= \dfrac{\dfrac{9}{4}}{1 - \dfrac{3}{4}} = \dfrac{9}{1} = 9.$

11.3.5 geometric series, $a = 3$, $r = 1/10$, sum $= \dfrac{3}{1 - \dfrac{1}{10}} = \dfrac{30}{9} = \dfrac{10}{3}.$

11.3.6 geometric series, $a = 3/e$, $r = 1/e$, sum $= \dfrac{\dfrac{3}{e}}{1 - \dfrac{1}{e}} = \dfrac{3}{e - 1}.$

11.3.7 $0.342342342\ldots = 0.342 + 0.000342 + 0.000000342\ldots = \dfrac{0.342}{1 - 0.001} = \dfrac{38}{111}.$

11.3.8 $Sn = \sum_{k=1}^{n} \dfrac{1}{(k+2)(k+3)} = \sum_{k=1}^{n} \left(\dfrac{1}{k+2} - \dfrac{1}{k+3} \right) = \dfrac{1}{3} - \dfrac{1}{n+3}, \ \lim\limits_{n \to +\infty} Sn = \dfrac{1}{3}.$

11.3.9 geometric series, $r = 2 > 1$, diverges.

11.3.10 geometric series, $\sum_{k=0}^{\infty}(-1)^k \left(\dfrac{2}{5} \right)^k$, $a = 1$, $r = -2/5$, sum $= \dfrac{1}{1 + \dfrac{2}{5}} = \dfrac{5}{7}.$

11.3.11 geometric series, $a = 1/4$, $r = -1/4$, sum $= \dfrac{1/4}{1 + 1/4} = \dfrac{1}{5}.$

11.3.12 $0.21212121\ldots = 0.21 + 0.0021 + 0.000021\ldots = \dfrac{0.21}{1 - 0.01} = \dfrac{21}{99} = \dfrac{7}{33}.$

11.3.13 geometric series, $a = 1/8$, $r = 1/2$, sum $= \dfrac{\dfrac{1}{8}}{1 - \dfrac{1}{2}} = \dfrac{1}{4}$.

11.3.14 $Sn = \displaystyle\sum_{k=1}^{n} \left(\dfrac{1}{2k-1} - \dfrac{1}{2k+1} \right) = 1 - \dfrac{1}{2n+1}$, $\displaystyle\lim_{n \to +\infty} Sn = 1$.

11.3.15 geometric series, $a = \dfrac{4}{49}$, $r = \left(-\dfrac{2}{7} \right)$, sum $= \dfrac{\dfrac{4}{49}}{1 + \dfrac{2}{7}} = \dfrac{4}{63}$

11.3.16 geometric series, $r = 4 > 1$, diverges.

11.3.17 geometric series, $a = 4/9$, $r = -2/3$, sum $= \dfrac{\dfrac{4}{9}}{1 + \dfrac{2}{3}} = \dfrac{4}{15}$.

11.3.18 $Sn = \displaystyle\sum_{k=1}^{n} \left(\dfrac{1}{k+1} - \dfrac{1}{k+2} \right) = \dfrac{1}{2} - \dfrac{1}{n+2}$, $\displaystyle\lim_{n \to +\infty} Sn = \dfrac{1}{2}$.

SECTION 11.4

11.4.1 Determine whether $\sum_{k=1}^{\infty} \dfrac{k}{3k+2}$ converges or diverges. Justify your answer by citing a relevant test.

11.4.2 Determine whether $\sum_{k=1}^{\infty} \dfrac{1}{k\sqrt{k}}$ converges or diverges. Justify your answer by citing a relevant test.

11.4.3 Determine whether $\sum_{k=1}^{\infty} \dfrac{1}{3k+4}$ converges or diverges. Justify your answer by citing a relevant test.

11.4.4 Determine whether $\sum_{k=1}^{\infty} \dfrac{k^2}{2k+1}$ converges or diverges. Justify your answer by citing a relevant test.

11.4.5 Determine whether $\sum_{k=1}^{\infty} \dfrac{k}{\sqrt{2k^2+1}}$ converges or diverges. Justify your answer by citing a relevant test.

11.4.6 Determine whether $\sum_{k=1}^{\infty} \dfrac{3}{5^k}$ converges or diverges. Justify your answer by citing a relevant test.

11.4.7 Determine whether $\sum_{k=1}^{\infty} \dfrac{\tan^{-1} k}{1+k^2}$ converges or diverges. Justify your answer by citing a relevant test.

11.4.8 Determine whether $\sum_{k=2}^{\infty} \dfrac{1}{k(\ln k)^3}$ converges or diverges. Justify your answer by citing a relevant test.

11.4.9 Determine whether $\sum_{k=1}^{\infty} \dfrac{1}{(2k+3)^3}$ converges or diverges. Justify your answer by citing a relevant test.

11.4.10 Determine whether $\sum_{k=1}^{\infty} \dfrac{k+1}{k(k+2)}$ converges or diverges. Justify your answer by citing a relevant test.

11.4.11 Find the sum of $\sum_{k=0}^{\infty} \left(\dfrac{5}{10^k} - \dfrac{6}{100^k} \right)$.

11.4.12 Determine whether $\sum_{k=1}^{\infty} \dfrac{1}{\sqrt{(k+1)^3}}$ converges or diverges. Justify your answer by citing a relevant test.

11.4.13 Determine whether $\sum_{k=1}^{\infty} \dfrac{2k}{1+k^4}$ converges or diverges. Justify your answer by citing a relevant test.

11.4.14 Determine whether $\sum_{k=1}^{\infty} \dfrac{1}{k\sqrt{k^2-1}}$ converges or diverges. Justify your answer by citing a relevant test.

11.4.15 Determine whether $\sum_{k=1}^{\infty} \dfrac{k}{e^k}$ converges or diverges. Justify your answer by citing a relevant test.

11.4.16 Determine whether $\sum_{k=1}^{\infty} e^{-k} \sin k$ converges or diverges. Justify your answer by citing a relevant test.

11.4.17 Determine whether $\sum_{k=1}^{\infty} \dfrac{1}{\cosh^2 k}$ converges or diverges. Justify your answer by citing a relevant test.

11.4.18 Determine whether $\sum_{k=1}^{\infty} \dfrac{\ln k}{k}$ converges or diverges. Justify your answer by citing a relevant test.

11.4.19 Which of the following statements about series is true?

    (a) If $\lim\limits_{k \to +\infty} u_k = 0$, then $\sum u_k$ converges.

    (b) If $\lim\limits_{k \to +\infty} u_k \neq 0$, then $\sum u_k$ diverges.

    (c) If $\sum u_k$ diverges, then $\lim\limits_{k \to +\infty} u_k \neq 0$.

    (d) $\sum u_k$ converges if and only if $\lim\limits_{k \to +\infty} u_k = 0$.

    (e) None of the preceding.

SOLUTIONS

SECTION 11.4

11.4.1 $\lim\limits_{k \to +\infty} \dfrac{k}{3k+2} = \dfrac{1}{3}$, series diverges since $\lim\limits_{k \to +\infty} u_k \neq 0$.

11.4.2 $\sum\limits_{k=1}^{\infty} \dfrac{1}{k^{3/2}}$ converges since the p series with $p = 3/2 > 1$ converges.

11.4.3 $\displaystyle\int_1^{\infty} \dfrac{1}{3x+4}\,dx = \lim\limits_{\ell \to +\infty} \dfrac{1}{3} \ln(3x+4) \Big]_1^{\ell} = +\infty$, series diverges by integral test.

11.4.4 $\lim\limits_{k \to +\infty} \dfrac{k^2}{2k+1} = +\infty$, series diverges since $\lim\limits_{k \to +\infty} u_k \neq 0$.

11.4.5 $\lim\limits_{k \to +\infty} \dfrac{k}{\sqrt{2k^2+1}} = \lim\limits_{k \to +\infty} \dfrac{1}{\sqrt{2+1/k^2}} = \dfrac{1}{\sqrt{2}}$, series diverges since $\lim\limits_{k \to +\infty} u_k \neq 0$.

11.4.6 geometric series, converges, since the geometric series with $r = \dfrac{1}{e} < 1$ converges.

11.4.7 $\displaystyle\int_1^{\infty} \dfrac{\tan^{-1} x}{1+x^2}\,dx = \lim\limits_{\ell \to +\infty} \dfrac{(\tan^{-1} x)^2}{2} \Big]_1^{\ell} = \dfrac{3\pi^2}{32}$, series converges by integral test.

11.4.8 $\displaystyle\int_2^{\infty} \dfrac{1}{x(\ln x)^3}\,dx = \lim\limits_{\ell \to +\infty} -\dfrac{1}{2(\ln x)^2} \Big]_2^{\ell} = \dfrac{1}{2(\ln 2)^2}$, converges by integral test.

11.4.9 $\displaystyle\int_1^{\infty} \dfrac{1}{(2x+3)^3}\,dx = \lim\limits_{\ell \to +\infty} -\dfrac{1}{4(2x+3)^2} \Big]_1^{\ell} = \dfrac{1}{100}$, converges by integral test.

11.4.10 $\displaystyle\int_1^{\infty} \dfrac{x+1}{x(x+2)}\,dx = \lim\limits_{\ell \to +\infty} \dfrac{1}{2} \ln(x^2+2x) \Big]_1^{\ell} = +\infty$, diverges by integral test.

11.4.11 $\sum\limits_{k=0}^{\infty} \dfrac{5}{10^k} = \dfrac{5/1}{1 - \dfrac{1}{10}} = \dfrac{50}{9}$; $\sum\limits_{k=0}^{\infty} \dfrac{6}{100^k} = \dfrac{6/1}{1 - \dfrac{1}{100}} = \dfrac{600}{99}$

$\sum\limits_{k=0}^{\infty} \left(\dfrac{5}{10^k} - \dfrac{6}{100^k} \right) = -\dfrac{50}{99}$.

11.4.12 $\sum\limits_{k=1}^{\infty} \dfrac{1}{\sqrt{(k+1)^3}} = \sum\limits_{k=2}^{\infty} \dfrac{1}{k^{3/2}}$ converges since the p series with $p = 3/2 > 1$ converges.

11.4.13 $\displaystyle\int_1^\infty \frac{2x}{1+x^4}\,dx = \lim_{\ell\to+\infty} \left. \tan^{-1} x^2 \right]_1^\ell = \frac{\pi}{4}$, converges by integral test.

11.4.14 $\displaystyle\int_1^\infty \frac{1}{x\sqrt{x^2-1}}\,dx = \lim_{\ell\to+\infty} \left. \sec^{-1} x \right]_1^\ell = \frac{\pi}{2}$, converges by integral test.

11.4.15 $\displaystyle\int_1^\infty \frac{x}{e^x}\,dx = \lim_{\ell\to+\infty} \left. -e^{-x}(x+1) \right]_1^\ell = \frac{2}{e}$, series converges by integral test.

11.4.16 $\displaystyle\int_1^\infty e^{-\pi x} \sin \pi x\,dx = \lim_{\ell\to\infty} \left. -\frac{1}{2\pi}e^{-\pi x}\left[\cos \pi x + \sin \pi x \right] \right]_1^\ell = -\frac{1}{2\pi e^\pi}$, series converges by integral test.

11.4.17 $\displaystyle\int_1^\infty \frac{1}{\cosh^2 x}\,dx = \lim_{\ell\to+\infty} \left. \tanh x \right]_1^\ell = 1 - \tanh 1$, series converges by integral test.

11.4.18 $\displaystyle\int_1^\infty \frac{\ln x}{x}\,dx = \lim_{\ell\to+\infty} \left. \frac{(\ln x)^2}{2} \right]_1^\ell = +\infty$, series diverges by integral test.

11.4.19 (b)

SECTION 11.5

11.5.1 Determine whether $\sum\limits_{k=1}^{\infty} \dfrac{k^2}{e^k}$ converges or diverges. Justify your answer by citing a relevant test.

11.5.2 Determine whether $\sum\limits_{k=1}^{\infty} \dfrac{k}{2^k}$ converges or diverges. Justify your answer by citing a relevant test.

11.5.3 Determine whether $\sum\limits_{k=1}^{\infty} \dfrac{k!}{10^{4k}}$ converges or diverges. Justify your answer by citing a relevant test.

11.5.4 Determine whether $\sum\limits_{k=1}^{\infty} \dfrac{1}{2+3^{-k}}$ converges or diverges. Justify your answer by citing a relevant test.

11.5.5 Determine whether $\sum\limits_{k=1}^{\infty} \left(\dfrac{k}{2k+100} \right)^k$ converges or diverges. Justify your answer by citing a relevant test.

11.5.6 Determine whether $\sum\limits_{k=1}^{\infty} \left(\dfrac{3k}{2k+1} \right)^k$ converges or diverges. Justify your answer by citing a relevant test.

11.5.7 Determine whether $\sum\limits_{k=0}^{\infty} \dfrac{2^{k-1}}{3^k(k+1)}$ converges or diverges. Justify your answer by citing a relevant test.

11.5.8 Determine whether $\sum\limits_{k=1}^{\infty} \dfrac{k!}{2^k}$ converges or diverges. Justify your answer by citing a relevant test.

11.5.9 Determine whether $\sum\limits_{k=0}^{\infty} \dfrac{k^k}{5^{k+1}}$ converges or diverges. Justify your answer by citing a relevant test.

11.5.10 Determine whether $\sum\limits_{k=1}^{\infty} \dfrac{1}{2k+9}$ converges or diverges. Justify your answer by citing a relevant test.

11.5.11 Determine whether $\sum_{k=1}^{\infty} \dfrac{e^k}{k!}$ converges or diverges. Justify your answer by citing a relevant test.

11.5.12 Determine whether $\sum_{k=1}^{\infty} \dfrac{10^k}{k!}$ converges or diverges. Justify your answer by citing a relevant test.

11.5.13 Determine whether $\sum_{k=1}^{\infty} \dfrac{k^3}{3^k}$ converges or diverges. Justify your answer by citing a relevant test.

11.5.14 Determine whether $\sum_{k=1}^{\infty} \dfrac{k!}{k^2}$ converges or diverges. Justify your answer by citing a relevant test.

11.5.15 Determine whether $\sum_{k=1}^{\infty} \left(\dfrac{\ln k}{k} \right)^k$ converges or diverges. Justify your answer by citing a relevant test.

11.5.16 Determine whether $\sum_{k=1}^{\infty} \dfrac{k^k}{k!}$ converges or diverges. Justify your answer by citing a relevant test.

11.5.17 Determine whether $\sum_{k=1}^{\infty} \dfrac{3^{2k}}{(2k)!}$ converges or diverges. Justify your answer by citing a relevant test.

SOLUTIONS

SECTION 11.5

11.5.1 $\rho = \lim\limits_{k \to +\infty} \dfrac{\dfrac{(k+1)^2}{e^{k+1}}}{\dfrac{k^2}{e^k}} = \lim\limits_{k \to +\infty} \dfrac{e^k}{e^{k+1}} \cdot \dfrac{(k+1)^2}{k^2} = \dfrac{1}{e} \lim\limits_{k \to +\infty} \left(\dfrac{k+1}{k}\right)^2 = \dfrac{1}{e} < 1$ so the series

converges by ratio test.

11.5.2 $\rho = \lim\limits_{k \to +\infty} \dfrac{\dfrac{k+1}{2^{k+1}}}{\dfrac{k}{2^k}} = \lim\limits_{k \to +\infty} \dfrac{2^k}{2^{k+1}} \cdot \dfrac{k+1}{k} = \dfrac{1}{2} \lim\limits_{k \to +\infty} \left(\dfrac{k+1}{k}\right) = \dfrac{1}{2} < 1$ so series converges by

ratio test.

11.5.3 $\rho = \lim\limits_{k \to +\infty} \dfrac{\dfrac{(k+1)!}{10^{4(k+1)}}}{\dfrac{k!}{10^{4k}}} = \lim\limits_{k \to +\infty} \dfrac{10^{4k}}{10^{4k+4}} \cdot \dfrac{(k+1)!}{k!} = \dfrac{1}{10^4} \lim\limits_{k \to +\infty} (k+1) = +\infty$, so series diverges

by ratio test.

11.5.4 $\lim\limits_{k \to +\infty} \dfrac{1}{2 + 3^{-k}} = \dfrac{1}{2}$, so series diverges since $\lim\limits_{k \to +\infty} u_k \neq 0$, by the divergence test.

11.5.5 $\rho = \lim\limits_{k \to +\infty} \dfrac{k}{2k + 100} = \dfrac{1}{2} < 1$ so series converges by root test.

11.5.6 $\rho = \lim\limits_{k \to +\infty} \dfrac{3k}{2k + 1} = \dfrac{3}{2} > 1$ so series diverges by root test.

11.5.7 $\lim\limits_{k \to +\infty} \dfrac{\dfrac{2^k}{3^{k+1}(k+2)}}{\dfrac{2^{k-1}}{3^k(k+1)}} = \lim\limits_{k \to +\infty} \dfrac{2^k}{3^{k+1}(k+2)} \cdot \dfrac{3^k(k+1)}{2^{k-1}} = \lim\limits_{k \to +\infty} \dfrac{2}{3} \dfrac{k+1}{k+2} = \dfrac{2}{3}$, converges by

ratio test.

11.5.8 $\rho = \lim\limits_{k \to +\infty} \dfrac{\dfrac{(k+1)!}{2^{k+1}}}{\dfrac{k!}{2^k}} = \lim\limits_{k \to +\infty} \dfrac{2^k}{2^{k+1}} \cdot \dfrac{(k+1)!}{k!} = \dfrac{1}{2} \lim\limits_{k \to +\infty} (k+1) = +\infty$, so series diverges by

ratio test.

11.5.9 $\sum = \lim\limits_{k \to +\infty} \dfrac{k}{5} = +\infty$, diverges by root test.

11.5.10 $\displaystyle\int_1^\infty \dfrac{dx}{2x + 9} = \left[\lim\limits_{\ell \to +\infty} \dfrac{1}{2} \ln(2x + 9)\right]_1^\ell = +\infty$, so series diverges by integral test.

11.5.11 $\rho = \lim\limits_{k\to+\infty} \dfrac{\frac{e^{k+1}}{(k+1)!}}{\frac{e^k}{k!}} = \lim\limits_{k\to+\infty} \dfrac{e^{k+1}}{e^k} \cdot \dfrac{k!}{(k+1)!} = e \lim\limits_{k\to+\infty} \dfrac{1}{k+1} = 0 < 1$ so series converges by ratio test.

11.5.12 $\rho = \lim\limits_{k\to+\infty} \dfrac{\frac{10^{k+1}}{(k+1)!}}{\frac{10^k}{k!}} = \lim\limits_{k\to+\infty} \dfrac{10^{k+1}}{10^k} \cdot \dfrac{k!}{(k+1)!} = 10 \lim\limits_{k\to+\infty} \dfrac{1}{k+1} = 0 < 1$ so series converges by ratio test.

11.5.13 $\rho = \lim\limits_{k\to+\infty} \dfrac{\frac{(k+1)^3}{3^{k+1}}}{\frac{k^3}{3^k}} = \lim\limits_{k\to+\infty} \dfrac{3^k}{3^{k+1}} \cdot \dfrac{k^3}{(k+1)^3}$

11.5.13 $\rho = \lim\limits_{k\to+\infty} \dfrac{\frac{(k+1)^3}{3^{k+1}}}{\frac{k^3}{3^k}} = \lim\limits_{k\to+\infty} \dfrac{3^k}{3^{k+1}} \cdot \dfrac{(k+1)^3}{k^3} = \dfrac{1}{3} \lim\limits_{k\to+\infty} \left(\dfrac{k}{k+1}\right)^3 = \dfrac{1}{3} < 1$ so series converges by ratio test.

11.5.14 $\rho = \lim\limits_{k\to+\infty} \dfrac{\frac{(k+1)!}{(k+1)^2}}{\frac{k!}{k^2}} = \lim\limits_{k\to+\infty} \dfrac{k^2}{(k+1)^2} \cdot \dfrac{(k+1)!}{k!} = \lim\limits_{k\to+\infty} \dfrac{k^2}{k+1} = +\infty$, so series diverges by ratio test.

11.5.15 $\rho = \lim\limits_{k\to+\infty} \dfrac{\ln k}{k} = \lim\limits_{k\to+\infty} \dfrac{1}{k} = 0 < 1$ so series converges by root test.

11.5.16 $\rho = \lim\limits_{k\to+\infty} \dfrac{\frac{(k+1)^{k+1}}{(k+1)!}}{\frac{k^k}{k!}} = \lim\limits_{k\to+\infty} \dfrac{k!}{(k+1)!} \cdot \dfrac{(k+1)^k}{k^k} = \lim\limits_{k\to+\infty} \dfrac{(k+1)^k}{k^k}$

$= \lim\limits_{k\to+\infty} \left(1 + \dfrac{1}{k}\right)^k = e > 1$ so series diverges by ratio test.

11.5.17 $\rho = \lim\limits_{k\to+\infty} \dfrac{\frac{3^{2(k+1)}}{[2(k+1)]!}}{\frac{3^{2k}}{(2k)!}} = \lim\limits_{k\to+\infty} \dfrac{3^{2k+2}}{3^{2k}} \cdot \dfrac{(2k)!}{(2k+2)!}$

$= 3^2 \lim\limits_{k\to+\infty} \dfrac{1}{(2k+2)(2k+1)} = 0 < 1$, so series converges by ratio test.

SECTION 11.6

11.6.1 Determine whether $\displaystyle\sum_{k=1}^{\infty} \frac{\sqrt{k}}{k^2 + 1}$ converges or diverges. Justify your answer by citing a relevant test.

11.6.2 Determine whether $\displaystyle\sum_{k=1}^{\infty} \frac{k^2}{(2k^2 + 1)^2}$ converges or diverges. Justify your answer by citing a relevant test.

11.6.3 Determine whether $\displaystyle\sum_{k=1}^{\infty} \frac{1}{(k+3)(k+4)}$ converges or diverges. Justify your answer by citing a relevant test.

11.6.4 Determine whether $\displaystyle\sum_{k=1}^{\infty} \frac{1}{(k+3)(k-4)}$ converges or diverges. Justify your answer by citing a relevant test.

11.6.5 Determine whether $\displaystyle\sum_{k=1}^{\infty} \frac{1}{3k + 2}$ converges or diverges. Justify your answer by citing a relevant test.

11.6.6 Determine whether $\displaystyle\sum_{k=1}^{\infty} \frac{1}{3^k + 2}$ converges or diverges. Justify your answer by citing a relevant test.

11.6.7 Determine whether $\displaystyle\sum_{k=1}^{\infty} \frac{\ln k}{k}$ converges or diverges. Justify your answer by citing a relevant test.

11.6.8 Determine whether $\displaystyle\sum_{k=1}^{\infty} \frac{k^2}{(k+2)(k+4)}$ converges or diverges. Justify your answer by citing a relevant test.

11.6.9 Determine whether $\displaystyle\sum_{k=1}^{\infty} \frac{k+1}{k^3 + 1}$ converges or diverges. Justify your answer by citing a relevant test.

11.6.10 Determine whether $\displaystyle\sum_{k=1}^{\infty} \frac{1}{1 + \sqrt{k}}$ converges or diverges. Justify your answer by citing a relevant test.

11.6.11 Determine whether $\displaystyle\sum_{k=1}^{\infty} \frac{3 + |\cos k|}{k^4}$ converges or diverges. Justify your answer by citing a relevant test.

11.6.12 Determine whether $\displaystyle\sum_{k=1}^{\infty} \frac{1}{3^k - 2}$ converges or diverges. Justify your answer by citing a relevant test.

11.6.13 Determine whether $\displaystyle\sum_{k=1}^{\infty} \frac{3^k + k}{k! + 3}$ converges or diverges. Justify your answer by citing a relevant test.

11.6.14 Determine whether $\displaystyle\sum_{k=1}^{\infty} \frac{1}{(2 + k)^{3/5}}$ converges or diverges. Justify your answer by citing a relevant test.

11.6.15 Determine whether $\displaystyle\sum_{k=2}^{\infty} \frac{1}{k - \ln k}$ converges or diverges. Justify your answer by citing a relevant test.

11.6.16 Determine whether $\displaystyle\sum_{k=1}^{\infty} \frac{1}{3k^{3/2} + 1}$ converges or diverges. Justify your answer by citing a relevant test.

11.6.17 Determine whether $\displaystyle\sum_{k=1}^{\infty} \frac{k^2 + 3}{k(k + 1)(k + 2)}$ converges or diverges. Justify your answer by citing a relevant test.

11.6.18 Which of the following statements about $\displaystyle\sum_{k=2}^{\infty} \frac{1}{k \ln k}$ is true:

   (a) converges because $\displaystyle\lim_{k \to +\infty} \frac{1}{k \ln k} = 0$.
   (b) converges because $\displaystyle\frac{1}{k \ln k} < \frac{1}{k}$.
   (c) converges by ratio test.
   (d) diverges by ratio test.
   (e) diverges by integral test.

SOLUTIONS

SECTION 11.6

11.6.1 $\dfrac{\sqrt{k}}{k^2+1} < \dfrac{1}{k^{3/2}}$, $\displaystyle\sum_{k=1}^{\infty} \dfrac{1}{k^{3/2}}$ converges (p series, $p > 1$), $\displaystyle\sum_{k=1}^{\infty} \dfrac{\sqrt{k}}{k^2+1}$ converges by the comparison test.

11.6.2 $\dfrac{k^2}{(2k^2+1)^2} < \dfrac{1}{4k^2}$, $\dfrac{1}{4}\displaystyle\sum_{k=1}^{\infty} \dfrac{1}{k^2}$, converges ($p$ series, $p > 1$), $\displaystyle\sum_{k=1}^{\infty} \dfrac{k^2}{(2k^2+1)^2}$ converges by the comparison test.

11.6.3 $\dfrac{1}{(k+3)(k+4)} < \dfrac{1}{k^2}$, $\displaystyle\sum_{k=1}^{\infty} \dfrac{1}{k^2}$, converges ($p$ series, $p > 1$), $\displaystyle\sum_{k=1}^{\infty} \dfrac{1}{(k+3)(k+4)}$ converges by the comparison test.

11.6.4 Limit comparison test, compare with the convergent p series $\displaystyle\sum_{k=1}^{\infty} \dfrac{1}{k^2}$,

$\rho = \displaystyle\lim_{k\to+\infty} \dfrac{k^2}{k^2-k-12} = 1$, series converges.

11.6.5 Limit comparison test, compare with the divergent harmonic series $\dfrac{1}{3}\displaystyle\sum_{k=1}^{\infty} \dfrac{1}{k}$,

$\rho = \displaystyle\lim_{k\to+\infty} \dfrac{3k}{3k+2} = 1$, series diverges.

11.6.6 $\dfrac{1}{3^k+2} < \dfrac{1}{3^k}$, $\displaystyle\sum_{k=1}^{\infty} \dfrac{1}{3^k}$ is a convergent geometric series, $\displaystyle\sum_{k=1}^{\infty} \dfrac{1}{3^k+2}$ converges by comparison.

11.6.7 $\dfrac{\ln k}{k} > \dfrac{1}{k}$ for $k > 3$, $\displaystyle\sum \dfrac{1}{k}$ divergent harmonic series, $\displaystyle\sum_{k=3}^{\infty} \dfrac{\ln k}{k}$ diverges by comparison.

11.6.8 $\displaystyle\lim_{k\to+\infty} \dfrac{k^2}{(k+2)(k+4)} = 1$, series diverges since $\displaystyle\lim_{k\to+\infty} u_k \neq 0$.

11.6.9 Limit comparison test, compare with the convergent p series $\displaystyle\sum_{k=1}^{\infty} \dfrac{1}{k^2}$,

$\rho = \displaystyle\lim_{k\to\infty} \dfrac{k^2(k+1)}{k^3+1} = 1$, series converges.

11.6.10 Limit comparison test, compare with the divergent p series $\displaystyle\sum_{k=1}^{\infty} \frac{1}{\sqrt{k}}$,

$\rho = \displaystyle\lim_{k\to+\infty} \frac{\sqrt{k}}{1+\sqrt{k}} = 1$, series diverges.

11.6.11 $\dfrac{3+|\cos k|}{k^4} \le \dfrac{4}{k^4}$, $4\displaystyle\sum_{k=1}^{\infty}\frac{1}{k^4}$ converges (p series, $p>1$), $\displaystyle\sum_{k=1}^{\infty}\frac{3+|\cos k|}{k^4}$ converges by the

comparison test.

11.6.12 Limit comparison test, compare with the convergent geometric series $\displaystyle\sum_{k=1}^{\infty}\frac{1}{3^k}$,

$\rho = \displaystyle\lim_{k\to+\infty}\frac{3^k}{3^k-2} = 1$, series converges.

11.6.13 $\dfrac{3^k+k}{k!+3} < \dfrac{3^k}{2k!}$, $\dfrac{1}{2}\displaystyle\sum_{k=1}^{\infty}\frac{3^k}{k!}$ converges (ratio test), $\displaystyle\sum_{k=1}^{\infty}\frac{3^k+k}{k!+3}$ converges by the comparison test.

11.6.14 Limit comparison test, compare with the divergent p series $\displaystyle\sum_{k=1}^{\infty}\frac{1}{k^{3/5}}$.

$\rho = \displaystyle\lim_{k\to+\infty}\frac{k^{3/5}}{(2+k)^{3/5}} = 1$, series diverges.

11.6.15 $\dfrac{1}{k-\ln k} > \dfrac{1}{k}$, $\displaystyle\sum_{k=2}^{\infty}\frac{1}{k}$ diverges (harmonic series), $\displaystyle\sum_{k=2}^{\infty}\frac{1}{k-\ln k}$ diverges by the comparison test.

11.6.16 $\dfrac{1}{3k^{3/2}+1} < \dfrac{1}{3k^{3/2}}$, $\dfrac{1}{3}\displaystyle\sum_{k=1}^{\infty}\frac{1}{k^{3/2}}$ converges (p series, $p>1$), $\displaystyle\sum_{k=1}^{\infty}\frac{1}{3k^{3/2}+1}$

converges by the comparison test.

11.6.17 Limit comparison test, compare with the divergent harmonic series

$\displaystyle\sum_{k=1}^{\infty}\frac{1}{k}$, $\rho = \displaystyle\lim_{k\to+\infty}\frac{k(k^2+3)}{k(k+1)(k+2)} = 1$, series diverges.

11.6.18 (e).

SECTION 11.7

11.7.1 Determine whether $\sum_{k=1}^{\infty} \dfrac{(-1)^k}{k + \sqrt{k}}$ converges absolutely, converges conditionally, or diverges. Justify your answer by citing a relevant test.

11.7.2 Determine whether $\sum_{k=1}^{\infty} \dfrac{(-1)^{k+1}(k+2)}{k(k+1)}$ converges absolutely, converges conditionally, or diverges. Justify your answer by citing a relevant test.

11.7.3 Determine whether $\sum_{k=1}^{\infty} \dfrac{(-1)^{k+1}2^k}{3^k + 1}$ converges absolutely, converges conditionally, or diverges. Justify your answer by citing a relevant test.

11.7.4 Determine whether $\sum_{k=2}^{\infty} \dfrac{(-1)^k \ln k}{k}$ converges absolutely, converges conditionally, or diverges. Justify your answer by citing a relevant test.

11.7.5 Determine whether $\sum_{k=1}^{\infty} \dfrac{(-1)^{k+1}2^k}{k^2}$ converges absolutely, converges conditionally, or diverges. Justify your answer by citing a relevant test.

11.7.6 Determine whether $\sum_{k=1}^{\infty} \dfrac{(-1)^k}{\sqrt{k}}$ converges absolutely, converges conditionally, or diverges. Justify your answer by citing a relevant test.

11.7.7 Determine whether $\sum_{k=1}^{\infty} \dfrac{(-1)^{k+1}}{3^k}$ converges absolutely, converges conditionally, or diverges. Justify your answer by citing a relevant test.

11.7.8 Determine whether $\sum_{k=1}^{\infty} \dfrac{(-1)^k k^2}{(2k^2 + 1)^2}$ converges absolutely, converges conditionally, or diverges. Justify your answer by citing a relevant test.

11.7.9 Determine whether $\sum_{k=1}^{\infty} \dfrac{(-1)^k k^3}{3^k}$ converges absolutely, converges conditionally, or diverges. Justify your answer by citing a relevant test.

11.7.10 Determine whether $\sum_{k=1}^{\infty} \dfrac{(-1)^k 2}{e^k}$ converges absolutely, converges conditionally, or diverges. Justify your answer by citing a relevant test.

11.7.11   Determine whether $\displaystyle\sum_{k=1}^{\infty} \frac{(-1)^{k+1}}{3k+4}$ converges absolutely, converges conditionally, or diverges. Justify your answer by citing a relevant test.

11.7.12   Determine whether $\displaystyle\sum_{k=1}^{\infty} \frac{(-1)^{k+1}k^2}{2k+1}$ converges absolutely, converges conditionally, or diverges. Justify your answer by citing a relevant test.

11.7.13   Determine whether $\displaystyle\sum_{k=1}^{\infty} \frac{(-1)^k k!}{(2k+3)!}$ converges absolutely, converges conditionally, or diverges. Justify your answer by citing a relevant test.

11.7.14   Determine whether $\displaystyle\sum_{k=1}^{\infty} \frac{(-1)^k k^2}{e^k}$ converges absolutely, converges conditionally, or diverges. Justify your answer by citing a relevant test.

11.7.15   The series $\displaystyle\sum_{k=1}^{\infty} (-1)^{k+1} \frac{k}{3^k}$ satisfies the conditions of the alternating series test. For $n = 7$ use Theorem 11.7.2 to find an upper bound on the magnitude of the error that results if the sum of the series is approximated by s_7.

11.7.16   The series $\displaystyle\sum_{k=1}^{\infty} (-1)^{k+1} \frac{2^k}{k!}$ satisfies the conditions of the alternating series test. For $n = 5$ use Theorem 11.7.2 to find an upper bound on the magnitude of the error that results if the sum of the series is approximated by s_5.

11.7.17   The series $\displaystyle\sum (-1)^{k+1} \frac{k}{3^k}$ satisfies the conditions of the alternating series test. Use Theorem 11.7.2 to find a value of n for which the nth partial sum is ensured to approximate the sum of the series such that the $|\text{error}| < 0.001$.

11.7.18   Use Theorem 11.7.2 to find an upper bound on the magnitude of the error that results if s_{10} is used to approximate the sum of the geometric series, $1 - \dfrac{3}{4} + \dfrac{9}{16} - \dfrac{27}{64} + \cdots$. Compute s_{10} rounded to four decimal places and compare this value with the exact sum of the series.

SOLUTIONS

SECTION 11.7

11.7.1 Converges conditionally, $\sum_{k=1}^{\infty} \frac{(-1)^k}{k + \sqrt{k}}$ converges by alternating series test but $\sum_{k=1}^{\infty} \frac{1}{k + \sqrt{k}}$

diverges by limit comparison test with $\sum_{k=1}^{\infty} \frac{1}{k}$.

11.7.2 Converges conditionally, $\sum_{k=1}^{\infty} \frac{(-1)^k (k + 2)}{k(k + 1)}$ converges by alternating series test but

$\sum_{k=1}^{\infty} \frac{k + 2}{k(k + 1)}$ diverges by limit comparison test with $\sum_{k=1}^{\infty} \frac{1}{k}$.

11.7.3 Converges absolutely, $\sum_{k=1}^{\infty} \frac{2^k}{3^{k+1}}$ converges by comparison with the geometric series

$\sum_{k=1}^{\infty} \left(\frac{2}{3} \right)^k$.

11.7.4 Converges conditionally, $\sum_{k=1}^{\infty} \frac{(-1)^k \ln k}{k}$ converges by alternating series test but

$\sum_{k=1}^{\infty} \frac{\ln k}{k}$ diverges by the integral test.

11.7.5 $\rho = 2 \lim_{k \to +\infty} \left(\frac{k}{k + 1} \right)^2 = 2 > 1$, diverges by ratio test for absolute convergence.

11.7.6 Converges conditionally, $\sum_{k=1}^{\infty} \frac{(-1)^k}{\sqrt{k}}$ converges by alternating series test but $\sum_{k=1}^{\infty} \frac{1}{\sqrt{k}}$ is a divergent p series.

11.7.7 Converges absolutely, $\sum_{k=1}^{\infty} \frac{(-1)^{k+1}}{3^k}$ is a geometric series with $|r| = 1/3 < 1$.

11.7.8 Converges absolutely, $\sum_{k=1}^{\infty} \frac{k^2}{(2k^2 + 1)^2}$ converges by comparison with the p series $\frac{1}{4} \sum_{k=1}^{\infty} \frac{1}{k^2}$.

11.7.9 $\rho = \frac{1}{3} \lim_{k \to +\infty} \left(\frac{k + 1}{k} \right)^3 = \frac{1}{3} < 1$, so series converges absolutely by ratio test for absolute convergence.

11.7.10 Converges absolutely, $2\sum\limits_{k=1}^{\infty}\dfrac{(-1)^k}{e^k}$ is a geometric series with $|r|=\dfrac{1}{e}<1$.

11.7.11 Converges conditionally, $\sum\limits_{k=1}^{\infty}\dfrac{(-1)^{k+1}}{3k+4}$ converges by alternating series test but $\sum\limits_{k=1}^{\infty}\dfrac{1}{3k+4}$ diverges by integral test.

11.7.12 Diverges since $\lim\limits_{k\to+\infty}\dfrac{k^2}{2k+1}=+\infty$.

11.7.13 $\rho=\lim\limits_{k\to+\infty}\dfrac{k+1}{(2k+5)(2k+4)}=0<1$ so series converges absolutely by ratio test for absolute convergence.

11.7.14 $\rho=\dfrac{1}{e}\lim\limits_{k\to+\infty}\left(\dfrac{k+1}{k}\right)^2=\dfrac{1}{e}<1$ so series converges absolutely by ratio test for absolute convergence.

11.7.15 $|\text{error}|\le a_8=\dfrac{8}{3^8}<.00121$.

11.7.16 $|\text{error}|\le a_6=\dfrac{2^6}{6!}<.0889$

11.7.17 $|\text{error}|<0.001$ if $a_{n+1}<0.001$, $\dfrac{n+1}{3^{n+1}}<0.001$, $\dfrac{3^{n+1}}{n+1}>1000$. But $\dfrac{3^8}{8}=820.125$, $\dfrac{3^9}{9}=2187$ so $\dfrac{3^{n+1}}{n+1}>1000$ if $n+1\ge 9$, $n\ge 8$; $n=8$.

11.7.18 Write $1-\dfrac{3}{4}+\dfrac{9}{16}-\dfrac{27}{64}+\cdots=\sum\limits_{n=1}^{\infty}\left(-\dfrac{3}{4}\right)^{n-1}$ $|\text{error}|\le a_{11}=\left(-\dfrac{3}{4}\right)^{10}<.0563$

$$S_{10}=\dfrac{1-1\left(-\dfrac{3}{4}\right)^{10}}{1-\left(-\dfrac{3}{4}\right)}=.5392,\quad S=\dfrac{1}{1-\left(-\dfrac{3}{4}\right)}=.5714,$$

$S-S_{10}=.0322<|\text{error}|\le a_{11}<.0563.$

SECTION 11.8

11.8.1 Find the interval of convergence for $\displaystyle\sum_{k=1}^{\infty} \frac{k}{2^k}(x-1)^k$.

11.8.2 Find the interval of convergence for $\displaystyle\sum_{k=1}^{\infty} \frac{(-1)^k 2^k}{k^2}(x-2)^k$.

11.8.3 Find the interval of convergence for $\displaystyle\sum_{k=1}^{\infty} \frac{(-1)^k (x-2)^k}{k+1}$.

11.8.4 Find the interval of convergence for $\displaystyle\sum_{k=1}^{\infty} (-1)^k \frac{x^k}{k}$.

11.8.5 Find the interval of convergence for $\displaystyle\sum_{k=1}^{\infty} \frac{2^k x^k}{\sqrt{k}}$.

11.8.6 Find the interval of convergence for $\displaystyle\sum_{k=1}^{\infty} \frac{2^k x^k}{3^k}$.

11.8.7 Find the interval of convergence for $\displaystyle\sum_{k=1}^{\infty} \frac{(x-1)^k}{2k+1}$.

11.8.8 Find the interval of convergence for $\displaystyle\sum_{k=1}^{\infty} \frac{(2x+3)^k}{\sqrt{2k+3}}$.

11.8.9 Find the interval of convergence for $\displaystyle\sum_{k=1}^{\infty} \frac{(x-3)^k}{(k+1)!}$.

11.8.10 Find the interval of convergence for $\displaystyle\sum_{k=1}^{\infty} \frac{k!}{(2k)!} x^k$.

11.8.11 Find the interval of convergence for $\displaystyle\sum_{k=1}^{\infty} k3^k (x-2)^k$.

11.8.12 Find the interval of convergence for $\displaystyle\sum_{k=1}^{\infty} \frac{(-1)^k x^k}{e^k}$.

11.8.13 Find the interval of convergence for $\displaystyle\sum_{k=0}^{\infty} \left(\frac{x-1}{3}\right)^k$.

11.8.14 Find the interval of convergence for $\displaystyle\sum_{k=1}^{\infty} \frac{k}{4^{2k-1}}(x-2)^k$.

11.8.15 Find the interval of convergence for $\displaystyle\sum_{k=1}^{\infty} \frac{2^k(x-3)^k}{k^2}$. For which values of x is the convergence absolute?

11.8.16 Find the interval of convergence for $\displaystyle\sum_{k=1}^{\infty} \frac{(-1)^k(x-1)^k}{3^k \sqrt[3]{k}}$.

11.8.17 Find the interval of convergence for $\displaystyle\sum_{k=1}^{\infty} \frac{k^2+k}{x^k}$.

11.8.18 Find the interval of convergence for $\displaystyle\sum_{k=1}^{\infty} \frac{(x+1)^k}{3^k k^2}$.

11.8.19 Find the interval of convergence for $\displaystyle\sum_{k=0}^{\infty} \frac{k!x^k}{2^k}$.

SOLUTIONS

SECTION 11.8

11.8.1 $\rho = \dfrac{|x-1|}{2} \lim\limits_{k \to +\infty} \left(\dfrac{k+1}{k}\right) = \dfrac{|x-1|}{2}$; converges if $|x-1| < 2$ or $-1 < x < 3$, diverges if $|x-1| > 2$; if $x = -1$, $\sum\limits_{k=1}^{\infty}(-1)^k k$ diverges, if $x = 3$, $\sum\limits_{k=1}^{\infty} k$ diverges so the interval of convergence is $(-1, 3)$.

11.8.2 $\rho = 2|x-2| \lim\limits_{k \to +\infty} \left(\dfrac{k}{k+1}\right)^2 = 2|x-2|$; converges if $|x-2| < 1/2$ or $\dfrac{3}{2} < x < \dfrac{5}{2}$, diverges if $|x-2| > 1/2$; if $x = 3/2$, $\sum\limits_{k=1}^{\infty} \dfrac{1}{k^2}$ converges, if $x = \dfrac{5}{2}$, $\sum\limits_{k=1}^{\infty} \dfrac{(-1)^k}{k^2}$ converges so the interval of convergence is $\left[\dfrac{3}{2}, \dfrac{5}{2}\right]$.

11.8.3 $\rho = |x-2| \lim\limits_{k \to +\infty} \left(\dfrac{k+1}{k+2}\right) = |x-2|$; converges if $|x-2| < 1$ or $1 < x < 3$, diverges if $|x-2| > 1$; if $x = 1$, $\sum\limits_{k=1}^{\infty} \dfrac{1}{k+1}$ diverges, if $x = 3$, $\sum\limits_{k=1}^{\infty} \dfrac{(-1)^k}{k+1}$ converges so the interval of convergence is $(1, 3]$.

11.8.4 $\rho = |x| \lim\limits_{k \to +\infty} \left(\dfrac{k}{k+1}\right) = |x|$; converges if $|x| < 1$, diverges if $|x| > 1$, if $x = -1$, $\sum\limits_{k=1}^{\infty} \dfrac{1}{k}$ diverges, if $x = 1$, $\sum\limits_{k=1}^{\infty} \dfrac{(-1)^k}{k}$ converges so the interval of convergence is $(-1, 1]$.

11.8.5 $\rho = 2|x| \lim\limits_{k \to +\infty} \sqrt{\dfrac{k}{k+1}} = 2|x|$; converges if $|x| < 1/2$ or $-1/2 < x < 1/2$, diverges if $|x| > 1/2$, if $x = -1/2$, $\sum\limits_{k=1}^{\infty} \dfrac{(-1)^k}{\sqrt{k}}$ converges, if $x = \dfrac{1}{2}$, $\sum\limits_{k=1}^{\infty} \dfrac{1}{\sqrt{k}}$ diverges so the interval of convergence is $[-1/2, 1/2)$.

11.8.6 $\rho = \lim\limits_{k \to +\infty} \left|\dfrac{2}{3}x\right| = \dfrac{2}{3}|x|$; converges if $|x| < 3/2$ or $-3/2 < x < 3/2$, diverges if $|x| > 3/2$, if $x = -3/2$, $\sum\limits_{k=1}^{\infty}(-1)^k$ diverges, if $x = 3/2$, $\sum\limits_{k=1}^{\infty}(1)^k$ diverges so the interval of convergence is $(-3/2, 3/2)$.

11.8.7 $\rho = |x-1| \lim_{k \to +\infty} \left(\dfrac{2k+1}{2k+3} \right) = |x-1|$; converges if $|x-1| < 1$ or $0 < x < 2$, diverges if $|x-1| > 1$, if $x = 0$, $\displaystyle\sum_{k=1}^{\infty} \dfrac{(-1)^k}{2k+1}$ converges, if $x = 2$, $\displaystyle\sum_{k=1}^{\infty} \dfrac{1}{2k+1}$ diverges so the interval of convergence is $[0, 2)$.

11.8.8 $\displaystyle\sum = |2x+3| \lim_{k \to +\infty} \sqrt{\dfrac{2k+3}{2k+5}} = |2x+3|$; converges if $|2x+3| < 1$ or $-2 < x < -1$, diverges if $|2x+3| > 1$; if $x = -2$, $\displaystyle\sum_{k=1}^{\infty} \dfrac{(-1)^k}{\sqrt{2k+3}}$ converges, if $x = -1$, $\displaystyle\sum_{k=1}^{\infty} \dfrac{1}{\sqrt{2k+3}}$ diverges, so the interval of convergence is $[-2, -1)$.

11.8.9 $\rho = |x-3| \lim_{k \to +\infty} \left(\dfrac{1}{k+2} \right) = 0$ so the interval of convergence is $(-\infty, +\infty)$.

11.8.10 $\rho = \dfrac{|x|}{2} \lim_{k \to +\infty} \left(\dfrac{1}{2k+1} \right) = 0$ so the interval of convergence is $(-\infty, +\infty)$.

11.8.11 $\rho = 3|x+2| \lim_{k \to +\infty} \left(\dfrac{k+1}{k} \right) = 3|x-2|$; converges if $|x-2| < 1/3$ or $\dfrac{5}{3} < x < \dfrac{7}{3}$, diverges if $|x-2| > 1/3$, if $x = \dfrac{5}{3}$, $\displaystyle\sum_{k=0}^{\infty} (-1)^k k$ diverges, if $x = \dfrac{7}{3}$, $\displaystyle\sum_{k=0}^{\infty} k$ diverges so the interval of convergence is $\left(\dfrac{5}{3}, \dfrac{7}{3} \right)$.

11.8.12 $\rho = \dfrac{|x|}{e}$, converges if $|x| < e$, diverges if $|x| > e$, if $x = -e$, $\displaystyle\sum_{k=1}^{\infty} 1$ diverges, if $x = e$, $\displaystyle\sum_{k=1}^{\infty} (-1)^k$ diverges so the interval of convergence is $(-e, e)$.

11.8.13 $\rho = \dfrac{|x-1|}{3}$, converges if $|x-1| < 3$, or $-2 < x < 4$, diverges if $|x-1| > 3$, if $x = -2$, $\displaystyle\sum_{k=0}^{\infty} (-1)^k$ diverges, if $x = 4$, $\displaystyle\sum_{k=0}^{\infty} 1$ diverges so the interval of convergence is $(-2, 4)$.

11.8.14 $\rho = \dfrac{|x-2|}{4^2} \lim_{k \to +\infty} \left(\dfrac{k+1}{k} \right) = \dfrac{|x-2|}{4^2}$; converges if $|x-2| < 16$ or $-14 < x < 18$; diverges if $|x-2| > 16$; if $x = -14$, $\displaystyle\sum_{k=1}^{\infty} (-1)^k 4k$ diverges; if $x = 18$, $\displaystyle\sum_{k=1}^{\infty} 4k$ diverges, so the interval of convergence is $(-14, 18)$.

11.8.15 $\rho = 2|x - 3| \lim\limits_{k \to +\infty} \left(\dfrac{k}{k+1}\right)^2 = 2|x - 3|$; converges if $|x - 3| < 1/2$ or $\dfrac{5}{2} < x < \dfrac{7}{2}$, diverges

if $|x - 3| > 1/2$, if $x = \dfrac{5}{2}$, $\sum\limits_{k=1}^{\infty} \dfrac{(-1)^k}{k^2}$ converges absolutely, if $x = \dfrac{7}{2}$, $\sum\limits_{k=1}^{\infty} \dfrac{1}{k^2}$ converges so the

interval of absolute convergence is $\left[\dfrac{5}{2}, \dfrac{7}{2}\right]$.

11.8.16 $\rho = \dfrac{|x - 1|}{3} \lim\limits_{k \to +\infty} \sqrt[3]{\dfrac{k}{k+1}} = \dfrac{|x - 1|}{3}$; converges if $|x - 1| < 3$ or $-2 < x < 4$, diverges if

$|x - 1| > 3$, if $x = -2$, $\sum\limits_{k=1}^{\infty} \dfrac{1}{\sqrt[3]{k}}$ diverges, if $x = 4$, $\sum\limits_{k=1}^{\infty} \dfrac{(-1)^k}{\sqrt[3]{k}}$ converges so the interval of

convergence is $(-2, 4]$.

11.8.17 $\rho = \dfrac{1}{|x|} \lim\limits_{k \to +\infty} \left(\dfrac{k+2}{k}\right) = \dfrac{1}{|x|}$, converges if $|x| > 1$, diverges if $|x| < 1$,

if $x = -1 \sum\limits_{k=1}^{\infty} (-1)^k(k^2 + k)$ diverges, if $x = 1$, $\sum\limits_{k=1}^{\infty} (k^2 + k)$ diverges so the interval of conver-

gence is $(-\infty, -1) \cup (1, +\infty)$.

11.8.18 $\rho = \dfrac{|x + 1|}{3} \lim\limits_{k \to +\infty} \left(\dfrac{k}{k+1}\right)^2 = \dfrac{|x + 1|}{3}$; converges if $|x + 1| < 3$ or $-4 < x < 2$, diverges if

$|x + 1| > 3$, if $x = -4$, $\sum\limits_{k=1}^{\infty} \dfrac{(-1)^k}{k^2}$ converges, if $x = 2$, $\sum\limits_{k=1}^{\infty} \dfrac{1}{k^2}$ converges so the interval of

convergence is $[-4, 2]$.

11.8.19 $\rho = \dfrac{|x|}{2} \lim\limits_{k \to +\infty} (k + 1) = +\infty$, the series converges only at $x = 0$.

SECTION 11.9

11.9.1   Find the fourth Taylor polynomial about $x = 2$ for $\ln x$.

11.9.2   Find the Taylor series for $f(x) = e^x$ in powers of $(x - 3)$. Express your answer in sigma notation.

11.9.3   Find the fifth Maclaurin polynomial for $f(x) = \sin x$.

11.9.4   Find the fifth Maclaurin polynomial for $f(x) = \cos x$.

11.9.5   Find the Maclaurin series for $f(x) = \ln(1 + x)$. Express your answer in sigma notation.

11.9.6   Find the fifth Maclaurin polynomial for $f(x) = \sinh x$.

11.9.7   Find the third Maclaurin polynomial for $f(x) = \sin^{-1} x$.

11.9.8   Find the Taylor series for $f(x) = \ln(x - 1)$ about $a = 2$. Express your answer in sigma notation.

11.9.9   Find the Taylor series for $f(x) = \dfrac{1}{x}$ about $a = 3$. Express your answer in sigma notation.

11.9.10   Find the fourth Maclaurin polynomial for $f(x) = \sqrt{1 + x}$.

11.9.11   Find the fifth Maclaurin polynomial for $f(x) = e^{x^2}$.

11.9.12   Find the third Taylor polynomial for $f(x) = \cos x$ about $x = \pi/3$.

11.9.13   Find the fourth Maclaurin polynomial for $f(x) = \sqrt[3]{1 + x}$.

11.9.14   Find the third Maclaurin polynomial for $f(x) = \sin^{-1} 3x$.

11.9.15   Find the third Taylor polynomial for $f(x) = \tan x$ about $x = \pi/3$.

11.9.16   Find the fourth Maclaurin polynomial for $f(x) = (1 + x)^{-2}$.

11.9.17   Find the third Taylor polynomial for $f(x) = e^x \sin \pi x$ about $x = 1$.

11.9.18   Find the fourth Taylor polynomial for $f(x) = \left(\dfrac{1}{2 + x} \right)$ about $x = 1$.

SOLUTIONS

SECTION 11.9

11.9.1 $\ln 2 + \frac{1}{2}(x-2) - \frac{1}{8}(x-2)^2 + \frac{1}{24}(x-2)^3 - \frac{1}{64}(x-2)^4$.

11.9.2 $f^{(k)}(x) = e^x$, $f^{(k)}(3) = e^3$; $\displaystyle\sum_{k=0}^{\infty} \frac{e^3}{k!}(x-3)^k$.

11.9.3 $x - \frac{x^3}{3!} + \frac{x^5}{5!}$

11.9.4 $1 - \frac{x^2}{2!} + \frac{x^4}{4!}$.

11.9.5 $\displaystyle\sum_{k=1}^{\infty} (-1)^{k+1} \frac{x^k}{k}$.

11.9.6 $x + \frac{x^3}{3!} + \frac{x^5}{5!}$.

11.9.7 $x + \frac{x^3}{3!}$.

11.9.8 $\displaystyle\sum_{k=1}^{\infty} \frac{(-1)^{k+1}}{k}(x-2)^k$.

11.9.9 $\displaystyle\sum_{k=0}^{\infty} \frac{(-1)^k (x-3)^k}{3^{k+1}}$.

11.9.10 $1 + \frac{x}{2} - \frac{x^2}{8} + \frac{x^3}{16} - \frac{5x^4}{128}$.

11.9.11 $1 + x^2 + \frac{x^4}{2!} + \frac{x^6}{3!}$.

11.9.12 $\frac{1}{2} - \frac{\sqrt{3}}{2}\left(x - \frac{\pi}{3}\right) - \frac{1}{4}\left(x - \frac{\pi}{3}\right)^2 + \frac{\sqrt{3}}{12}\left(x - \frac{\pi}{3}\right)^3$.

11.9.13 $1 + \frac{x}{3} - \frac{x^2}{9} + \frac{5x^3}{81} - \frac{10x^4}{243}$.

11.9.14 $3x + \frac{9}{2}x^3$.

11.9.15 $\sqrt{3} + 4\left(x - \frac{\pi}{3}\right) + 4\sqrt{3}\left(x - \frac{\pi}{3}\right)^2 + \frac{40}{3}\left(x - \frac{\pi}{3}\right)^3$.

11.9.16 $1 - 2x + 3x^2 - 4x^3 + 5x^4$.

11.9.17 $e\left[-\pi(x-1) - \pi(x-1)^2 - \frac{\pi(-3+\pi^2)}{3!}(x-1)^3\right]$.

11.9.18 $\frac{1}{3} - \frac{1}{9}(x-1) + \frac{1}{27}(x-1)^2 - \frac{1}{81}(x-1)^3 + \frac{1}{243}(x-1)^4$.

SECTION 11.10

11.10.1   Find the Lagrange form of the remainder term $R_n(x)$ for the function $f(x) = \sin x$, $a = 0$, $n = 4$.

11.10.2   Find the Lagrange form of the remainder term $R_n(x)$ for the function $f(x) = e^x$, $a = 1$, $n = 4$.

11.10.3   Find the Lagrange form of the remainder term $R_n(x)$ for the function $f(x) = \ln(1 + x)$, $a = 0$, $n = 4$.

11.10.4   Find the Maclaurin series for $f(x) = \dfrac{1}{1+x}$ by division. Indicate the interval on which the expansion is valid.

11.10.5   Given that $\cosh x = 1 + \dfrac{x^2}{2!} + \dfrac{x^4}{4!} + \cdots$, find the Maclaurin series for $\operatorname{sech} x$.

11.10.6   Find the Lagrange form of the remainder term, $R_n(x)$ for the function $f(x) = \sqrt{1 + x}$ with $a = 0$ and $n = 4$.

11.10.7   Find the Lagrange form of the remainder term, $R_n(x)$ for the function $f(x) = \cosh x$ with $a = 0$ and $n = 5$.

11.10.8   Find the Lagrange form of the remainder term, $R_n(x)$ for the function $f(x) = \ln \cos x$ with $a = 0$ and $n = 3$.

11.10.9   Prove that the Taylor series for $\sin x$ about $x = \pi/6$ converges to $\sin x$ for all x.

11.10.10   Prove that the Taylor series for $\cos x$ about $x = \dfrac{\pi}{3}$ converges to $\cos x$ for all x.

11.10.11   Given that the Maclaurin series for $e^x = \displaystyle\sum_{k=0}^{\infty} \dfrac{x^k}{k!}$ on $(-\infty, \infty)$, derive a Maclaurin series for $f(x) = xe^{-x}$. Indicate the interval of validity for the new series.

11.10.12   Given that the Maclaurin series for $\cos x = \displaystyle\sum_{k=0}^{\infty} (-1)^k \dfrac{x^{2k}}{(2k)!}$, valid for $(-\infty, +\infty)$, derive a Maclaurin series for $f(x) = x \cos \sqrt{x}$. Indicate the interval of validity for the new series.

11.10.13   Given that the Maclaurin series for $\dfrac{1}{1-x} = \displaystyle\sum_{k=0}^{\infty} x^k$, valid on $(-1, 1)$, derive a Maclaurin series for $\dfrac{x^2}{1 + 2x}$. Indicate the interval of validity for the new series.

11.10.14   Given that the Maclaurin series for $\cosh x = \sum\limits_{k=0}^{\infty} \dfrac{x^{2k}}{(2k)!}$ valid on $(-\infty, \infty)$, derive a Maclaurin series for $\cosh(x^2)$. Indicate the interval of validity for the new series.

11.10.15   Given that the Maclaurin series for $\tan^{-1} x = \sum\limits_{k=0}^{\infty} (-1)^k \dfrac{x^{2k+1}}{2k+1}$ valid on $[-1, 1]$, derive a Maclaurin series for $\tan^{-1} 2x$. Indicate the interval of validity for the new series.

11.10.16   Given that the Maclaurin series for $(1 + x)^m = 1 + \sum\limits_{k=1}^{\infty} \dfrac{m(m-1)\cdots(m-k+1)}{k!} x^k$ on $(-1, 1)$, find the first four nonzero terms in the Maclaurin series for the function $\sqrt[3]{1+k}$ and give the radius of convergence.

11.10.17   Given that the Maclaurin series for $(1 + x)^m = 1 + \sum\limits_{k=1}^{\infty} \dfrac{m(m-1)\cdots(m-k+1)}{k!} x^k$ on $(-1, 1)$, find the first four nonzero terms in the Maclaurin series for the function $\dfrac{1}{\sqrt{4 + x^2}}$.

11.10.18   Given that the Maclaurin series for $\ln(1 + x) = \sum\limits_{k=0}^{\infty} (-1)^k \dfrac{x^{k+1}}{k+1} = x - \dfrac{x^2}{2} + \dfrac{x^3}{3} - \dfrac{x^4}{4} + \cdots$ on $(-1, 1)$, derive a Maclaurin series for $f(x) = \ln(1 - x^2)$. Indicate the interval of validity for the new series.

SOLUTIONS

SECTION 11.10

11.10.1 $f^{(5)}(x) = \cos x$, $R_4(x) = \dfrac{\cos c}{5!}x^5$.

11.10.2 $f^{(5)}(x) = e^x$, $R_4(x) = \dfrac{e^c}{5!}(x-1)^5$.

11.10.3 $f^{(5)}(x) = \dfrac{4!}{(x+1)^5}$, $R_5(x) = \dfrac{1}{5(c+1)^5}x^5$.

11.10.4

$$
\begin{array}{r}
1-x+x^2-x^3 \;\cdots\cdots \\
1+x\,\overline{\smash{\big)}\,1+x} \\
-x \\
\underline{-x-x^2} \\
x^2 \\
x^2+x^3 \\
\underline{-x^3} \\
-x^3-\;x^4 \\
\hline
\end{array}
\qquad \text{Valid on } (-1,1).
$$

11.10.5 $\operatorname{sech} x = \dfrac{1}{1+\dfrac{x^2}{2!}+\dfrac{x^4}{4!}+\cdots} = 1-\dfrac{x^2}{2}+\dfrac{5}{24}x^4-\cdots$.

11.10.6 $f^{(5)}(x) = \dfrac{105}{32(1+x)^{9/2}}$, $R_4(x) = \dfrac{7}{256(c+1)^{9/2}}x^5$.

11.10.7 $f^{(6)}(x) = \cosh x$, $R_5(x) = \dfrac{\cosh c}{6!}x^6$.

11.10.8 $f^{(4)}(x) = -2\sec^4 x - 4\sec^2 x \tan^2 x$,

$R_3(x) = \dfrac{-2\sec^4 c - 4\sec^2 c\tan^2 c}{4!}x^4$.

11.10.9 $f(x) = \sin x$, $f^{(n+1)}(x) = \pm\sin x$ or $\pm\cos x$, but, $\left|f^{(n+1)}(x)\right| \le 1$, thus,

$$0 \le |R_n(x)| = \frac{|f^{(n+1)}(c)|}{(n+1)!}\left|x-\frac{\pi}{6}\right|^{n+1} \le \frac{1}{(n+1)!}\left|x-\frac{\pi}{6}\right|^{n+1}, \quad \lim_{n\to+\infty}\frac{|x-\pi/6|^{n+1}}{(n+1)!} = 0,$$

by the pinching theorem, $\displaystyle\lim_{n\to+\infty}|R_n(x)| = 0$ and thus, $\displaystyle\lim_{n\to a}R_n(x) = 0$ for all x.

11.10.10 $f(x) = \cos x$, $f^{(n+1)}(x) = \pm \cos x$ or $\pm \sin x$, but $\left| f^{(n+1)}(x) \right| \le 1$, thus,

$$0 \le |R_n(x)| = \frac{|f^{(n+1)}(c)|}{(n+1)!} \left| x - \frac{\pi}{3} \right|^{n+1} \le \frac{1}{(n+1)!} \left| x - \frac{\pi}{3} \right|^{n+1}, \quad \lim_{n \to +\infty} \frac{|x - \pi/3|^{n+1}}{(n+1)!} = 0,$$

by the pinching theorem, $\lim\limits_{n \to +\infty} |R_n(x)| = 0$ and thus, $\lim\limits_{n \to +\infty} R_n(x) = 0$ for all x.

11.10.12 $\cos\sqrt{x} = \sum\limits_{k=0}^{\infty} (-1)^k \dfrac{x^k}{(2k)!}$, $\; x\cos\sqrt{x} = \sum\limits_{k=0}^{\infty} (-1)^k \dfrac{x^{k+1}}{(2k)!}$, valid on $[0, \infty)$.

11.10.13 $\dfrac{1}{1+2x} = \sum\limits_{k=0}^{\infty} (-2x)^k = \sum\limits_{k=0}^{\infty} (-1)^k 2^k x^k;$

$$\frac{x^2}{1+2x} = x^2 \sum_{k=0}^{\infty} (-1)^k 2^k x^k = \sum_{k=0}^{\infty} (-1)^k 2^k x^{k+2}, \text{ valid for } (-1/2, 1/2).$$

11.10.14 $\cosh(x^2) = \sum\limits_{k=0}^{\infty} \dfrac{(x^2)^{2k}}{(2k)!} = \sum\limits_{k=0}^{\infty} \dfrac{x^{4k}}{(2k)!}$, valid on $(-\infty, +\infty)$.

11.10.15 $\tan^{-1} 2x = \sum\limits_{k=0}^{\infty} (-1)^k \dfrac{(2x)^{2k+1}}{2k+1} = \sum\limits_{k=0}^{\infty} (-1)^k \dfrac{2^{2k+1} x^{2k+1}}{2k+1}$, valid on $[-1/2, 1/2]$.

11.10.16 $\sqrt[3]{1+x} = 1 + \dfrac{1}{3}x + \left(\dfrac{1}{3}\right)\left(-\dfrac{2}{3}\right)x^2 + \left(\dfrac{1}{3}\right)\left(-\dfrac{2}{3}\right)\left(-\dfrac{5}{3}\right)x^3$

$$= 1 + \frac{1}{3}x - \frac{2}{9}x^2 + \frac{10}{27}x^3 + \cdots, R = 1.$$

11.10.17 $\dfrac{1}{\sqrt{4+x^2}}$ rewrite as $\dfrac{1}{2}\left(1 + \left(\dfrac{x}{2}\right)^2\right)^{-1/2}$

$$\frac{1}{2}\left(1 + \left(\frac{x}{2}\right)^2\right)^{-1/2} = \frac{1}{2}\left[1 + \left(-\frac{1}{2}\right)\left(\frac{x}{2}\right)^2 + \frac{\left(-\frac{1}{2}\right)\left(-\frac{3}{2}\right)\left(\frac{x}{2}\right)^4}{2!}\right.$$

$$\left. + \frac{\left(-\frac{1}{2}\right)\left(-\frac{3}{2}\right)\left(-\frac{5}{2}\right)\left(\frac{x}{2}\right)^6}{3!} \right.$$

$$= \frac{1}{2}\left[1 - \frac{x^2}{8} + \frac{3x^4}{128} - \frac{15x^6}{3072} + \cdots\right].$$

11.10.18 $\ln(1+x) = \sum_{k=0}^{\infty} (-1)^k \frac{x^{k+1}}{k+1},$

$\ln(1-x^2) = \ln(1+(-x^2)) = \sum_{k=0}^{\infty} (-1)^k \frac{(-x^2)^{k+1}}{k+1} = \sum_{k=0}^{\infty} (-1)^k (-1)^{k+1} \frac{x^{2(k+1)}}{k+1}$

$= \sum_{k=0}^{\infty} (-1)^{2k+1} \frac{x^{2(k+1)}}{k+1} = -\sum_{k=0}^{\infty} \frac{x^{2(k+1)}}{k+1},$ valid if $-1 < (-x^2) < 1$, $-1 < x < 1$.

SECTION 11.11

11.11.1   Use an appropriate series to approximate the cos 1 to two decimal place accuracy.

11.11.2   Use an appropriate series to approximate cos 40° to 3 decimal place accuracy.

11.11.3   Use an appropriate series to approximate the sin(0.1) to four decimal place accuracy.

11.11.4   Use series 16 to approximate ln 1.2 to 3 decimal place accuracy.

11.11.5   Use an appropriate series to approximate the cos 10° to four decimal place accuracy.

11.11.6   Use an appropriate series to approximate the sin 61° to four decimal place accuracy.

11.11.7   Use a Maclaurin series to approximate $\tan^{-1}(0.2)$ to three decimal place accuracy. Use the fact that the resulting series is an alternating series.

11.11.8   Use $x = -1/2$ in the Maclaurin series for e^x to approximate $\dfrac{1}{\sqrt{e}}$ to four decimal place accuracy.

11.11.9   Use series 16 to approximate ln 1.4 to 3 decimal place accuracy.

11.11.10   Use an appropriate series to approximate the sin 37° to four decimal place accuracy.

11.11.11   Use a Maclaurin series to approximate the sinh 0.1 to four decimal place accuracy.

11.11.12   Use a Maclaurin series to approximate the cosh 0.2 to four decimal place accuracy.

11.11.13   Use series 16 to approximate ln 1.6 to 3 three decimal place accuracy.

11.11.14   Use an appropriate series to approximate $\tan^{-1} 0.9$ to 3 decimal place accuracy.

11.11.15   Use series 16 to approximate ln 1.8 to 3 decimal place accuracy.

11.11.16   Use an appropriate series to approximate cos 1.5 to four decimal place accuracy.

SOLUTIONS

SECTION 11.11

11.11.1 Use a Taylor series expansion about $\pi/3$,

$$\cos x = \frac{1}{2} - \frac{\sqrt{3}}{2}(x - \pi/3) - \frac{1}{4}\left(x - \frac{\pi}{3}\right)^2 + \frac{\sqrt{3}}{12}(x - \pi/3)^3 + \cdots$$

$$|Rn(1)| \le \frac{|1 - \pi/3|^{n+1}}{(n+1)!} < 0.5 \times 10^{-2} \text{ if } n = 1$$

$$\cos 1 = \frac{1}{2} - \frac{\sqrt{3}}{2}\left(1 - \frac{\pi}{3}\right) \approx 0.54.$$

11.11.2 Use a Taylor series expansion about $\pi/4$.

$$\cos x = \frac{\sqrt{2}}{2} - \frac{\sqrt{2}}{2}\frac{(x - \pi/4)}{1!} - \frac{\sqrt{2}}{2}\frac{(x - \pi/4)^2}{2!} + \frac{\sqrt{2}}{2}\frac{(x - \pi/4)^3}{3!} + \cdots, 40° = \frac{2\pi}{9} \text{ radian,}$$

$$\left|R_n\left(\frac{2\pi}{9}\right)\right| \le \frac{\left|\frac{2\pi}{9} - \frac{\pi}{4}\right|^{n+1}}{(n+1)!} = \frac{\left|\frac{-\pi}{36}\right|^{n+1}}{(n+1)!} < 0.5 \times 10^{-3} \text{ if } n = 2;$$

$$\cos\left(\frac{2\pi}{9}\right) \approx \frac{\sqrt{2}}{2} - \frac{\sqrt{2}}{2.1!}\left(\frac{-\pi}{36}\right) - \frac{\sqrt{2}}{2.2!}\left(\frac{-\pi}{36}\right)^2 = 0.766$$

11.11.3 Use a Maclaurin series, $\sin x = x - \frac{x^3}{3!} + \frac{x^5}{5!} + \cdots$

$$|Rn(0.1)| \le \frac{(0.1)^{n+1}}{(n+1)!} < 0.5 \times 10^{-4} \text{ if } n = 3$$

$$\sin(0.1) \approx 0.1 - \frac{(0.1)^3}{3!} \approx 0.0998.$$

11.11.4 Let $x = 1/11$ in series 16 to get $\ln 1.2 \approx 0.182$.

11.11.5 Use a Maclaurin series, $\cos x = 1 - \frac{x^2}{2!} + \frac{x^4}{4!} + \cdots$

$$10° = \frac{\pi}{18} \text{ radians, } |Rn(x)| \le \frac{(\pi/18)^{n+1}}{(n+1)!} < 0.5 \times 10^{-4} \text{ if } n = 3$$

$$\cos 10° \approx 1 - \frac{(\pi/18)^2}{2!} \approx 0.9848.$$

11.11.6 Use a Taylor series expansion about $\pi/3$,

$$\sin x = \frac{\sqrt{3}}{2} + \frac{1}{2}(x - \pi/3) - \frac{\sqrt{3}}{4}(x - \pi/3)^2 - \cdots$$

$$61° = \frac{61\pi}{180}, \left| Rn\left(\frac{61\pi}{180}\right)\right| < \frac{(\pi/180)^{n+1}}{(n+1)!} < 0.5 \times 10^{-4} \text{ if } n = 2$$

$$\sin 61° \approx \frac{\sqrt{3}}{2} + \frac{1}{2}\left(\frac{\pi}{180}\right) - \frac{\sqrt{3}}{4}\left(\frac{\pi}{180}\right)^2 \approx 0.8746.$$

11.11.7 $\tan^{-1} x = x - \dfrac{x^3}{3} + \dfrac{x^5}{5} - \cdots \dfrac{(-1)^{k+1}x^{2k+1}}{2k+1},$

$$a_n = \frac{(0.2)^{2n+1}}{2n+1} \text{ for } n = 0, 1, 2, \ldots$$

$$|\text{error}| < a_{n+1} = \frac{(0.2)^{2n+3}}{2n+3} < 0.5 \times 10^{-3} \text{ if } n = 1$$

$$\tan^{-1}(0.2) \approx 0.2 - \frac{(0.2)^3}{3} \approx 0.197.$$

11.11.8 The series is alternating, $a_n = \dfrac{1}{2^n n!}$ for $n = 0, 1, 2, \ldots$

$$|\text{error}| < a_{n+1} = \frac{1}{2^{n+1}(n+1)!} < 0.5 \times 10^{-4} \text{ if } n = 5,$$

$$\frac{1}{\sqrt{e}} \approx 1 - \frac{1}{2} + \frac{1}{8} - \frac{1}{48} + \frac{1}{384} - \frac{1}{3840} = 0.6065.$$

11.11.9 Let $x = 1/6$ in series 16 to get $\ln 1.4 \approx 0.336$.

11.11.10 Use a Taylor series expansion about $\pi/6$,

$$\sin x = \frac{1}{2} + \frac{\sqrt{3}}{2}(x - \pi/6) - \frac{1}{4}(x - \pi/6)^2 - \frac{\sqrt{3}}{12}(x - \pi/6)^3 + \cdots$$

$$\left| Rn\left(\frac{7\pi}{180}\right)\right| < \frac{\left|\dfrac{37\pi}{180} - \dfrac{\pi}{6}\right|^{n+1}}{(n+1)!} \leq 0.5 \times 10^{-4} \text{ if } n = 3$$

$$\sin\frac{37\pi}{180} \approx \frac{1}{2} + \frac{\sqrt{3}}{2}\left(\frac{7\pi}{180}\right) - \frac{1}{4}\left(\frac{7\pi}{180}\right)^2 - \frac{\sqrt{3}}{12}\left(\frac{7\pi}{180}\right)^3 \approx 0.6018.$$

11.11.11 $\sinh x = x + \dfrac{x^3}{3!} + \dfrac{x^5}{5!} + \cdots |Rn(0.1)| \leq \dfrac{(0.1)^{n+1}}{(n+1)!} < 0.5 \times 10^{-4} \text{ if } n = 3,$

$$\sinh 0.1 \approx 0.1 + \frac{(0.1)^3}{3!} \approx 0.1002.$$

11.11.12 $\cosh x = 1 + \dfrac{x^2}{2!} + \dfrac{x^4}{4!} + \cdots |Rn(0.2)| \leq \dfrac{(0.2)^{n+1}}{(n+1)!} < 0.5 \times 10^{-4}$ if $n = 4$

$\cosh 0.2 \approx 1 + \dfrac{(0.2)^2}{2!} + \dfrac{(0.2)^4}{4!} \approx 1.0201.$

11.11.13   Let $x = 3/13$ in series 16 to get $\ln 1.6 \approx 0.470.$

11.11.14   Use a Taylor series expansion about 1.

$\tan^{-1} x = \dfrac{\pi}{4} + \dfrac{1}{2}(x-1) - \dfrac{1}{4}(x-1)^2 + \dfrac{1}{12}(x-1)^3 + \cdots$

$|R_n(0.9)| \leq \dfrac{|0.9 - 1|^{n+1}}{(n+1)!} = \dfrac{|-0.1|^{n+1}}{(n-1)!} < 0.5 \times 10^{-3}$ if $n = 3,$

$\tan^{-1}(0.9) \approx \dfrac{\pi}{4} + \dfrac{1}{2}(-0.1) - \dfrac{1}{4}(-0.1)^2 + \dfrac{1}{12}(-0.1)^3 = 0.7328.$

11.11.15   Let $x = 2/7$ in series 16 to get $\ln 1.8 \approx 0.587$

11.11.16   Use a Taylor series expansion about $\pi/2$, $\cos x = -\left(x - \dfrac{\pi}{2}\right) + \dfrac{1}{3!}\left(x - \dfrac{\pi}{2}\right)^3 - \cdots$

$|Rn(1.5)| \leq \dfrac{\left|1.5 - \dfrac{\pi}{2}\right|^{n+1}}{(n+1)!} < 0.5 \times 10^{-4}$ if $n = 3$

$\cos 1.5 \approx \left(1.5 - \dfrac{\pi}{2}\right) + \dfrac{1}{3!}\left(1.5 - \dfrac{\pi}{2}\right)^3 \approx -0.0786.$

SECTION 11.12

11.12.1 Find the first four nonzero terms of the Maclaurin series for $e^x \sin x$.

11.12.2 Find the first four nonzero terms of the Maclaurin series for $e^{-x} \cos x$.

11.12.3 Find the first four nonzero terms of the Maclaurin series for $\dfrac{e^x}{1-x}$.

11.12.4 Find the first four nonzero terms of the Maclaurin series for $\dfrac{\cos x}{\sqrt{1+x}}$.

11.12.5 Find the first four nonzero terms of the Maclaurin series for $\dfrac{\sin x}{1+x}$.

11.12.6 Find the first four nonzero terms of the Maclaurin series for $\coth x$.

11.12.7 Obtain the series for $\sec^2 x$ by first obtaining a series for the $\tan x$ and then by differentiating this series.

11.12.8 Use a series to approximate $\displaystyle\int_0^1 \cos x^2 \, dx$ to four decimal place accuracy.

11.12.9 Use a series to approximate $\displaystyle\int_0^1 \frac{\sin x}{x} \, dx$ to four decimal place accuracy.

11.12.10 Use a series to approximate $\displaystyle\int_0^{1/2} \cos x^3 \, dx$ to four decimal place accuracy.

11.12.11 Use a series to approximate $\displaystyle\int_0^1 \sin x^3 \, dx$ to four decimal place accuracy.

11.12.12 Use a series to show $\displaystyle\lim_{x \to 0} \frac{\sin x}{x} = 1$.

11.12.13 Use a series to show $\displaystyle\lim_{x \to 0} \frac{\tan x}{x} = 1$ by first obtaining a series for the $\tan x$.

11.12.14 Use a series to show $\displaystyle\lim_{x \to 0} \frac{e^x - 1}{\sin x} = 1$.

11.12.15 Use a series to show $\displaystyle\lim_{x \to 0} \frac{\cos x - 1}{x} = 0$.

SOLUTIONS

SECTION 11.12

11.12.1 $\left(1 + x + \dfrac{x^2}{2!} + \dfrac{x^3}{3!} + \dfrac{x^4}{4!} + \cdots\right)\left(x - \dfrac{x^3}{3!} + \dfrac{x^5}{5!} - \dfrac{x^7}{7!} + \cdots\right) = x + x^2 + \dfrac{x^3}{3} - \dfrac{3x^5}{40} \cdots$

11.12.2 $e^{-x}\cos x = \left(1 - x + \dfrac{x^2}{2!} - \dfrac{x^3}{3!} + \dfrac{x^4}{4!} - \cdots\right)\left(1 - \dfrac{x^2}{2!} + \dfrac{x^4}{4!} - \dfrac{x^6}{6!} + \cdots\right)$

$$= 1 - x + \dfrac{x^3}{3} - \dfrac{5x^4}{24} \cdots$$

11.12.3 $\left(\dfrac{1}{1-x}\right)e^x = \left(1 + x + x^2 + x^3 + \cdots\right)\left(1 + x + \dfrac{x^2}{2!} + \dfrac{x^3}{3!} + \cdots\right)$

$$= 1 + 2x + \dfrac{5}{2}x^2 + \dfrac{8}{3}x^3 + \cdots$$

11.12.4 Substitute $m = -1/2$ into the binomial series to get

$$\dfrac{1}{\sqrt{1+x}} = 1 - \dfrac{1}{2}x + \dfrac{3}{8}x^2 - \dfrac{5}{16}x^3 + \cdots, \text{ thus}$$

$$\dfrac{\cos x}{\sqrt{1+x}} = \dfrac{1}{\sqrt{1+x}}\cos x$$

$$= \left(1 - \dfrac{1}{2}x + \dfrac{3}{8}x^2 - \dfrac{5}{16}x^3 + \cdots\right)\left(1 - \dfrac{x^2}{2!} + \dfrac{x^4}{4!} - \dfrac{x^6}{6!} + \cdots\right)$$

$$= 1 - \dfrac{1}{2}x - \dfrac{1}{8}x^2 - \dfrac{1}{16}x^3 \cdots$$

11.12.5 $\left(\dfrac{1}{1+x}\right)\sin x = \left(1 - x + x^2 - x^3 + \cdots\right)\left(x - \dfrac{x^3}{3!} + \dfrac{x^5}{5!} - \dfrac{x^7}{7!} + \cdots\right)$

$$= x - x^2 + \dfrac{5x^3}{6} - \dfrac{5x^4}{6} + \cdots$$

11.12.6 $\coth x = \dfrac{\cosh x}{\sinh x} = \dfrac{1 + \dfrac{x^2}{2!} + \dfrac{x^4}{4!} + \dfrac{x^6}{6!} + \cdots}{x + \dfrac{x^3}{3!} + \dfrac{x^5}{5!} + \dfrac{x^7}{7!} + \cdots} = \dfrac{1}{x} + \dfrac{x}{3} - \dfrac{x^3}{45} + \dfrac{2x^5}{945} \cdots$

11.12.7 $\tan x = \dfrac{\sin x}{\cos x} = \dfrac{x - \dfrac{x^3}{3!} + \dfrac{x^5}{5!} - \dfrac{x^7}{7!} + \cdots}{1 - \dfrac{x^2}{2!} + \dfrac{x^4}{4!} - \dfrac{x^6}{6!} + \cdots} = x + \dfrac{x^3}{3} + \dfrac{2x^5}{15} + \dfrac{17x^7}{315} + \cdots$

$$\sec^2 x = \dfrac{d}{dx}[\tan x] = \dfrac{d}{dx}\left[x + \dfrac{x^3}{3} + \dfrac{2x^5}{15} + \dfrac{17x^7}{315} + \cdots\right] = 1 + x^2 + \dfrac{2x^4}{3} + \dfrac{17x^6}{2205} + \cdots$$

11.12.8 $\cos x^2 = 1 - \dfrac{x^4}{2!} + \dfrac{x^8}{4!} - \dfrac{x^{12}}{6!} + \dfrac{x^{16}}{8!} - \cdots$

$\displaystyle\int_0^1 \left(1 - \frac{x^4}{2!} + \frac{x^8}{4!} + \frac{x^{12}}{6!} + \frac{x^{16}}{8!} - \cdots\right) dx = x - \frac{x^5}{5\cdot 2!} + \frac{x^9}{9\cdot 4!} - \frac{x^{13}}{13\cdot 6!} + \frac{x^{17}}{17\cdot 8!} - \cdots \Big]_0^1$

$\qquad\qquad\qquad = 1 - \dfrac{1}{5\cdot 2!} + \dfrac{1}{9\cdot 4!} - \dfrac{1}{13\cdot 6!} + \dfrac{1}{17\cdot 8!} - \cdots$

$\dfrac{1}{17\cdot 8!} < 0.5 \times 10^{-4}$, so use the first four terms, to get

$\displaystyle\int_0^1 \cos x^2 dx \approx 1 - \frac{1}{5\cdot 2!} + \frac{1}{9\cdot 4!} - \frac{1}{13\cdot 6!} \approx 0.9045.$

11.12.9 $\displaystyle\int_0^1 \frac{\sin x}{x} dx = \int_0^1 \frac{1}{x}\left(x - \frac{x^3}{3!} + \frac{x^5}{5!} - \frac{x^7}{7!} + \frac{x^9}{9!} - \cdots\right) dx$

$\qquad = \displaystyle\int_0^1 \left(1 - \frac{x^2}{3!} + \frac{x^4}{5!} - \frac{x^6}{7!} + \frac{x^8}{9!} - \cdots\right) dx$

$\qquad = x - \dfrac{x^3}{3\cdot 3!} + \dfrac{x^5}{5\cdot 5!} - \dfrac{x^7}{7\cdot 7!} + \dfrac{x^9}{9\cdot 9!} - \cdots \Big]_0^1$

$\qquad = 1 - \dfrac{1}{3\cdot 3!} + \dfrac{1}{5\cdot 5!} - \dfrac{1}{7\cdot 7!} + \dfrac{1}{9\cdot 9!} - \cdots$

but, $\dfrac{1}{7\cdot 7!} < 0.5 \times 10^{-4}$, so use the first three terms to get,

$\displaystyle\int_0^1 \frac{\sin x}{x} dx \approx 1 - \frac{1}{3\cdot 3!} + \frac{1}{5\cdot 5!} \approx 0.9461.$

11.12.10 $\displaystyle\int_0^{1/2} \cos x^3 dx = \int_0^{1/2}\left(1 - \frac{x^6}{2!} + \frac{x^{12}}{4!} - \frac{x^{18}}{6!} + \frac{x^{24}}{8!} - \cdots\right) dx$

$\qquad = x - \dfrac{x^7}{7\cdot 2!} + \dfrac{x^{13}}{13\cdot 4!} - \dfrac{x^{19}}{19\cdot 6!} + \dfrac{x^{25}}{25\cdot 8!} \cdots \Big]_0^{1/2}$

$\qquad = \dfrac{1}{2} - \dfrac{1}{2^7\cdot 7\cdot 2!} + \dfrac{1}{2^{13}\cdot 13\cdot 4!} - \dfrac{1}{2^{19}\cdot 19\cdot 6!} + \dfrac{1}{2^{25}\cdot 25\cdot 8!} - \cdots$

but, $\dfrac{1}{2^{13}\cdot 13\cdot 4!} < 0.5 \times 10^{-4}$, so, use the first two terms to get

$\displaystyle\int_0^{1/2} \cos x^3 dx \approx \frac{1}{2} - \frac{1}{2^7\cdot 7\cdot 2!} \approx 0.4994.$

11.12.11 $\displaystyle\int_0^1 \sin x^3 dx = \left(x^3 - \frac{x^9}{3!} + \frac{x^{15}}{5!} - \frac{x^2}{7!} + \frac{x^{27}}{9!} - \cdots\right) dx$

$\qquad = \dfrac{x^4}{4} - \dfrac{x^{40}}{10\cdot 3!} + \dfrac{x^{16}}{16\cdot 5!} - \dfrac{x^{22}}{22\cdot 7!} + \cdots \Big]_0^1$

but $\dfrac{1}{22 \cdot 7!} < .5 \times 10^{-4}$ so use the first three terms to get

$$\int_0^1 \sin x^3 \, dx = \frac{1}{4} - \frac{1}{60} + \frac{1}{16 \cdot 5!} \approx .2338541$$

11.12.12 $\quad \dfrac{\sin x}{x} = \dfrac{1}{x}\left(x - \dfrac{x^3}{3!} + \dfrac{x^5}{5!} - \dfrac{x^7}{7!} + \cdots\right) = 1 - \dfrac{x^2}{3!} + \dfrac{x^4}{5!} - \dfrac{x^6}{7!} + \cdots$

$\qquad \displaystyle\lim_{x\to 0} \frac{\sin x}{x} = \lim_{x\to 0}\left(1 - \frac{x^2}{3!} + \frac{x^4}{5!} - \frac{x^6}{7!} + \cdots\right) = 1.$

11.12.13 $\quad \tan x = \dfrac{\sin x}{\cos x} = \dfrac{x - \dfrac{x^3}{3!} + \dfrac{x^5}{5!} - \dfrac{x^7}{7!} + \cdots}{1 - \dfrac{x^2}{2!} + \dfrac{x^4}{4!} - \dfrac{x^6}{6!} - \cdots} = x + \dfrac{x^3}{3} + \dfrac{2x^5}{15} + \dfrac{17x^7}{315} + \cdots$

$\qquad \dfrac{\tan x}{x} = \dfrac{1}{x}\left(x + \dfrac{x^3}{3} + \dfrac{2x^5}{15} + \dfrac{17x^7}{315} + \cdots\right) = 1 + \dfrac{x^2}{3} + \dfrac{2x^4}{15} + \dfrac{17x^6}{315} + \cdots$

$\qquad \displaystyle\lim_{x\to 0} \frac{\tan x}{x} = \lim_{x\to 0}\left(1 + \frac{x^2}{3} + \frac{2x^4}{15} + \frac{17x^6}{315} + \cdots\right) = 1.$

11.12.14 $\quad \displaystyle\lim_{x\to 0} \frac{e^x - 1}{\sin x} = \lim_{x\to 0} \frac{1 + x + \dfrac{x^2}{2!} + \dfrac{x^3}{3!} + \dfrac{x^4}{4!} + \cdots - 1}{x - \dfrac{x^3}{3!} + \dfrac{x^5}{5!} - \dfrac{x^7}{7!} + \cdots}$

$\qquad\qquad = \displaystyle\lim_{x\to 0} \frac{x + \dfrac{x^2}{2!} + \dfrac{x^3}{3!} + \dfrac{x^4}{4!} + \cdots}{x - \dfrac{x^3}{3!} + \dfrac{x^5}{5!} - \dfrac{x^7}{7!} + \cdots}$

$\qquad\qquad = \displaystyle\lim_{x\to 0} \frac{1 + x + \dfrac{x^2}{2!} + \dfrac{x^3}{3!} + \cdots}{1 - \dfrac{x^2}{2!} + \dfrac{x^4}{4!} - \dfrac{x^6}{6!} + \cdots} = 1.$

11.12.15 $\quad \dfrac{\cos x - 1}{x} = \dfrac{1}{x}\left(1 - \dfrac{x^2}{2!} + \dfrac{x^4}{4!} - \dfrac{x^6}{6!} + \cdots - 1\right)$

$\qquad\qquad = -\dfrac{x}{2!} + \dfrac{x^3}{4!} - \dfrac{x^5}{6!} + \cdots \displaystyle\lim_{x\to 0} \frac{\cos x - 1}{x}$

$\qquad \displaystyle\lim_{x\to 0} \frac{\cos x - 1}{x} = \lim_{x\to 0}\left(-\frac{x^2}{2!} + \frac{x^3}{4!} - \frac{x^5}{6!} + \cdots\right) = 0$

SUPPLEMENTARY EXERCISES, CHAPTER 11

In Exercises 1–6, find $L = \lim\limits_{n\to+\infty}$ if it exists.

1. $a_n = (-1)^n/e^n$.

2. $a_n = e^{1/n}$.

3. $a_n = \dfrac{1}{\sqrt{n}} - \dfrac{1}{\sqrt{n+1}}$.

4. $a_n = \sin(\pi n)$.

5. $a_n = \sin\left(\dfrac{(2n-1)\pi}{2}\right)$.

6. $a_n = \dfrac{n+1}{n(n+2)}$.

7. Which of the sequences $\{a_n\}_{n=1}^{+\infty}$ in Exercises 1–6 are (a) decreasing, (b) nondecreasing, and (c) alternating?

8. Suppose $f(x)$ satisfies

$$f'(x) > 0 \quad \text{and} \quad f(x) \le 1 - e^{-x}$$

for all $x \ge 1$. What can you conclude about the convergence of $\{a_n\}$ if $a_n = f(n), n = 1, 2, \ldots$?

9. Use your knowledge of geometric series and p-series to determine all values of q for which the following series converge.

(a) $\displaystyle\sum_{k=0}^{\infty} \pi^k/q^{2k}$

(b) $\displaystyle\sum_{k=1}^{\infty} (1/k^q)^3$

(c) $\displaystyle\sum_{k=2}^{\infty} 1/(\ln q^k)$

(d) $\displaystyle\sum_{k=2}^{\infty} 1/(\ln q)^k$.

10. (a) Use a suitable test to find all values of q for which $\displaystyle\sum_{k=2}^{\infty} 1/[k\,(\ln k)^q]$ converges.

(b) Why can't you use the integral test for the series $\displaystyle\sum_{k=1}^{\infty} (2 + \cos k)/k^2$? Test for convergence using a test that does apply.

11. Express $1.3636\ldots$ as (a) an infinite series in sigma notation, and (b) a ratio of integers.

12. In parts (a)–(d), use the comparison test to determine whether the series converges.

(a) $\displaystyle\sum_{k=1}^{\infty} \dfrac{2k-1}{3k^2-k}$

(b) $\displaystyle\sum_{k=1}^{\infty} \dfrac{2k+1}{3k^2+k}$

(c) $\displaystyle\sum_{k=1}^{\infty} \dfrac{2k-1}{3k^3-k^2}$

(d) $\displaystyle\sum_{k=1}^{\infty} \dfrac{2k+1}{3k^3+k^2}$.

13. Find the sum of the series (if it converges).

(a) $\displaystyle\sum_{k=1}^{\infty} \dfrac{2^k + 3^k}{6^{k+1}}$

(b) $\displaystyle\sum_{k=2}^{\infty} \ln\left(1 + \dfrac{1}{k}\right)$

(c) $\displaystyle\sum_{k=1}^{\infty} [k^{-1/2} - (k+1)^{-1/2}]$.

In Excrciscs 14–21, determine whether the series converges or diverges. You may use the following limits without proof:

$$\lim_{k \to +\infty} (1 + 1/k)^k = e, \quad \lim_{k \to +\infty} \sqrt[k]{k} = 1, \quad \lim_{k \to +\infty} \sqrt[k]{a} = 1$$

14. $\displaystyle\sum_{k=0}^{\infty} e^{-k}.$

15. $\displaystyle\sum_{k=1}^{\infty} ke^{-k^2}.$

16. $\displaystyle\sum_{k=1}^{\infty} \frac{k}{k^2 + 2k + 7}.$

17. $\displaystyle\sum_{k=1}^{\infty} \frac{\sqrt{k}}{k^2 + 7}.$

18. $\displaystyle\sum_{k=1}^{\infty} \left(\frac{k}{k+1}\right)^k.$

19. $\displaystyle\sum_{k=0}^{\infty} \frac{3^k k!}{(2k)!}.$

20. $\displaystyle\sum_{k=0}^{\infty} \frac{k^6 3^k}{(k+1)!}.$

21. $\displaystyle\sum_{k=1}^{\infty} \left(\frac{5k}{2k+1}\right)^{3k}.$

In Exercises 22–25, determine whether the given series is absolutely convergent, conditionally convergent, or divergent.

22. $\displaystyle\sum_{k=1}^{\infty} (-1)^k / e^{1/k}.$

23. $\displaystyle\sum_{k=0}^{\infty} (-2)^k / (3^k + 1).$

24. $\displaystyle\sum_{k=0}^{\infty} (-1)^k / (2k + 1).$

25. $\displaystyle\sum_{k=0}^{\infty} (-1)^k 3^k / 2^{k+1}.$

26. Find a value of n to ensure that the nth partial sum approximates the sum of the series to the stated accuracy.

(a) $\displaystyle\sum_{k=1}^{\infty} \frac{(-1)^k}{k^2 + 1};$ $|\text{error}| < 0.0001$

(b) $\displaystyle\sum_{k=1}^{\infty} \frac{(-1)^k}{5^k + 1};$ $|\text{error}| < 0.00005.$

In Exercises 27–30, determine the radius of convergence and the interval of convergence of the given power series.

27. $\displaystyle\sum_{k=1}^{\infty} \frac{(x-1)^k}{k\sqrt{k}}.$

28. $\displaystyle\sum_{k=1}^{\infty} \frac{k^2(x+2)^k}{(k+1)!}.$

29. $\displaystyle\sum_{k=1}^{\infty} \frac{k!(x-1)^k}{5^k}.$

30. $\displaystyle\sum_{k=1}^{\infty} \frac{(2k)!x^k}{(2k+1)!}.$

In Exercises 31–33, find

(a) the nth Taylor polynomial for f about $x = a$ (for the stated values of n and a);

(b) Lagrange's form of $R_n(x)$ (for the stated values of n and a);

(c) an upper bound on the absolute value of the error if $f(x)$ is approximated over the given interval by the Taylor polynomial obtained in part (a).

31. $f(x) = \ln(x - 1); a = 2; n = 3; [\frac{3}{2}, 2].$ **32.** $f(x) = e^{x/2}; a = 0; n = 4; [-1, 0].$

33. $f(x) = \sqrt{x}; a = 1; n = 2; [\frac{4}{9}, 1].$

34. **(a)** Use the identity $a - x = a(1 - x/a)$ to find the Maclaurin series for $1/(a - x)$ from the geometric series. What is its radius of convergence?

   (b) Find the Maclaurin series and radius of convergence of $1/(3 + x)$.

   (c) Find the Maclaurin series and radius of convergence of $2x/(4 + x^2)$.

   (d) Use partial fractions to find the Maclaurin series and radius of convergence of

$$\frac{1}{(1 - x)(2 - x)}$$

35. Use the known Maclaurin series for $\ln(1+x)$ to find the Maclaurin series and radius of convergence of $\ln(a + x)$ for $a > 0$.

36. Use the identity $x = a + (x - a)$ and the known Maclaurin series for e^x, $\sin x$, $\cos x$, and $1/(1-x)$ to find the Taylor series about $x = a$ for (a) e^x, (b) $\sin x$, and (c) $1/x$.

37. Use the series of Example 8 in Section 11.10 to find the Maclaurin series and radius of convergence of $1/\sqrt{9 + x}$.

In Exercises 38–43, use any method to find the first three nonzero terms of the Maclaurin series.

38. $e^{\tan x}$.   **39.** $\sec x$.   **40.** $(\sin x)/(e^x - x)$.

41. $\sqrt{\cos x}$.   **42.** $e^x \ln(1 - x)$.   **43.** $\ln(1 + \sin x)$.

44. Find a power series for $\dfrac{1 - \cos 3x}{x^2}$ and use it to evaluate $\displaystyle\lim_{x \to 0} \dfrac{1 - \cos 3x}{x^2}$.

45. Find a power series for $\dfrac{\ln(1 - 2x)}{x}$ and use it to evaluate $\displaystyle\lim_{x \to 0} \dfrac{\ln(1 - 2x)}{x}$.

46. How many decimal places of accuracy can be guaranteed if we approximate $\cos x$ by $1 - x^2/2$ for $-0.1 < x < 0.1$?

47. For what values of x can $\sin x$ be replaced by $x - x^3/6 + x^5/120$ with an ensured accuracy of 6×10^{-4}?

In Exercises 48–51, approximate the indicated quantity to three decimal-place accuracy.

48. $\cos(10°)$.

49. $\displaystyle\int_0^1 \frac{(1 - e^{-t/2})}{t}\, dt.$

50. $\displaystyle\int_0^1 \frac{\sin x}{\sqrt{x}}\, dx.$

51. Show that $y = \displaystyle\sum_{n=0}^{\infty} k^n x^n/n!$ satisfies $y' - ky = 0$ for any fixed k.

SOLUTIONS

SUPPLEMENTARY EXERCISES CHAPTER 11

1. $L = 0$

2. $L = e^0 = 1$

3. $L = 0 - 0 = 0$

4. $\sin \pi n = 0$ for all n so $L = 0$

5. $\sin[(2n-1)\pi/2]$ is alternately 1 and -1 so the limit does not exist

6. $L = 0$

7. $a_n = (-1)^n/e^n$ is alternating; $a_n = e^{1/n}$ is decreasing;

$a_n = \dfrac{1}{\sqrt{n}} - \dfrac{1}{\sqrt{n+1}} = \dfrac{\sqrt{n+1} - \sqrt{n}}{\sqrt{n}\sqrt{n+1}} = \dfrac{1}{\sqrt{n}\sqrt{n+1}\left(\sqrt{n+1} + \sqrt{n}\right)}$ is decreasing;

$a_n = \sin \pi n = 0$ is nondecreasing; $a_n = \sin[(2n-1)\pi/2]$ is alternating;

$a_n = \dfrac{n+1}{n(n+2)}$, let $f(x) = \dfrac{x+1}{x^2+2x}$ then $f'(x) = -\dfrac{x^2+2x+2}{(x^2+2x)^2} < 0$ if $x \geq 1$ so a_n is decreasing.

8. a_n is increasing because $f'(x) > 0$, $a_n \leq 1 - e^{-x} < 1$ so $\{a_n\}$ converges by Theorem 11.2.2a.

9. **(a)** $\displaystyle\sum_{k=0}^{\infty} \left(\pi/q^2\right)^k$ is a geometric series which converges for $\pi/q^2 < 1$, $q^2 > \pi$, $|q| > \sqrt{\pi}$

(b) $\displaystyle\sum_{k=1}^{\infty} 1/k^{3q}$ is a p-series with $p = 3q$, converges for $3q > 1$, $q > 1/3$

(c) $\displaystyle\sum_{k=2}^{\infty} 1/(k \ln q) = \sum_{k=2}^{\infty}(1/\ln q)(1/k)$ diverges for all q because $\displaystyle\sum_{k=2}^{\infty} 1/k$ diverges

(d) $\displaystyle\sum_{k=2}^{\infty}(1/\ln q)^k$ is a geometric series which converges for $|1/\ln q| < 1$, $|\ln q| > 1$,

$q > e$ or $0 < q < e^{-1}$

10. **(a)** If $q = 1$, $\displaystyle\int_{2}^{+\infty} \dfrac{1}{x \ln x}\, dx = \lim_{\ell \to +\infty} \ln(\ln x)\Big]_{2}^{\ell} = +\infty$, the series diverges. If $q \neq 1$,

$\displaystyle\int_{2}^{+\infty} \dfrac{1}{x}(\ln x)^{-q}\, dx = \lim_{\ell \to +\infty} \dfrac{(\ln x)^{1-q}}{1-q}\Big]_{2}^{\ell} = \begin{cases} +\infty & q < 1 \\ \dfrac{l}{(q-1)(\ln 2)^{q-1}}, & q > 1 \end{cases}$

so the series converges for $q > 1$.

(b) $(2 + \cos x)/x^2$ is not a decreasing function. The series converges because

$(2 + \cos k)/k^2 \leq 3/k^2$ and $\displaystyle\sum_{k=1}^{\infty} 3/k^2$ converges.

11. **(a)** $1.3636 \cdots = 1 + \displaystyle\sum_{k=1}^{\infty} 36(0.01)^k$

(b) $1.3636 \cdots = 1 + \dfrac{0.36}{1 - 0.01} = 1 + 36/99 = 1 + 4/11 = 15/11$

12. **(a)** $\dfrac{2k - 1}{3k^2 - k} \geq \dfrac{2k - k}{3k^2} = \dfrac{1}{3k}, \displaystyle\sum_{k=1}^{\infty} 1/(3k)$ diverges so $\displaystyle\sum_{k=1}^{\infty} \dfrac{2k - 1}{3k^2 - k}$ diverges.

(b) $\dfrac{2k + 1}{3k^2 + k} > \dfrac{2k}{3k^2 + k^2} = \dfrac{1}{2k}, \displaystyle\sum_{k=1}^{\infty} 1/(2k)$ diverges so $\displaystyle\sum_{k=1}^{\infty} \dfrac{2k + 1}{3k^2 + k}$ diverges.

(c) $\dfrac{2k - 1}{3k^3 - k^2} < \dfrac{2k}{3k^3 - k^3} = 1/k^2, \displaystyle\sum_{k=1}^{\infty} 1/k^2$ converges so $\displaystyle\sum_{k=1}^{\infty} \dfrac{2k - 1}{3k^3 - k^2}$ converges.

(d) $\dfrac{2k + 1}{3k^3 + k^2} < \dfrac{2k + k}{3k^3} = 1/k^2, \displaystyle\sum_{k=1}^{\infty} 1/k^2$ converges so $\displaystyle\sum_{k=1}^{\infty} \dfrac{2k + 1}{3k^3 + k^2}$ converges.

13. **(a)** $\dfrac{1}{6}\left[\displaystyle\sum_{k=1}^{\infty}\left(\dfrac{1}{3}\right)^k + \displaystyle\sum_{k=1}^{\infty}\left(\dfrac{1}{2}\right)^k \right] = \dfrac{1}{6}\left[\dfrac{1/3}{1 - 1/3} + \dfrac{1/2}{1 - 1/2} \right] = 1/4$

(b) $\displaystyle\sum_{k=2}^{\infty} \ln\dfrac{k + 1}{k} = \displaystyle\sum_{k=2}^{\infty}[\ln(k + 1) - \ln k],$

$s_n = [\ln 3 - \ln 2] + [\ln 4 - \ln 3] + \cdots + [\ln(n + 2) - \ln(n + 1)]$
$= \ln(n + 2) - \ln 2, \displaystyle\lim_{n \to +\infty} s_n = +\infty,$ diverges

(c) $s_n = [1^{-1/2} - 2^{-1/2}] + [2^{-1/2} - 3^{-1/2}] + \cdots + [n^{-1/2} - (n + 1)^{-1/2}]$

$= 1 - (n + 1)^{-1/2}, \displaystyle\lim_{n \to +\infty} s_n = 1$

14. converges (geometric series, $a = 1, r = e^{-1}$)

15. converges (integral test, $\displaystyle\int_{1}^{\infty} xe^{-x^2}\, dx$ converges)

16. diverges (limit comparison test with $\Sigma 1/k, \rho = 1$)

17. converges (comparison test, $\dfrac{\sqrt{k}}{k^2 + 7} < \dfrac{\sqrt{k}}{k^2} = \dfrac{1}{k^{3/2}}$)

18. diverges $\left(\displaystyle\lim_{k \to +\infty}\left(\dfrac{k}{k + 1}\right)^k = \displaystyle\lim_{k \to +\infty}\dfrac{1}{(1 + 1/k)^k} = 1/e \neq 0 \right)$

19. converges (ratio test, $\rho = 0$)   **20.** converges (ratio test, $\rho = 0$)

21. diverges (root test, $\rho = (5/2)^3 > 1$)   **22.** diverges $\left(\displaystyle\lim_{k \to +\infty} |u_k| = 1 \neq 0 \right)$

23. absolutely convergent (comparison test, $2^k/(3^k+1) < 2^k/3^k = (2/3)^k$, $\sum(2/3)^k$ is a convergent geometric series)

24. conditionally convergent (the series converges by the alternating series test but $\sum 1/(2k+1)$ diverges)

25. diverges $\left(\lim_{k\to+\infty} |u_k| = \lim_{k\to+\infty} \frac{1}{2}(3/2)^k = +\infty\right)$

26. (a) $1/[(n+1)^2+1] \le 0.0001$, $(n+1)^2 \ge 9999$, $n+1 \ge 100$, $n \ge 99$; take $n = 99$

(b) $1/(5^{n+1}+1) \le 0.00005$, $5^{n+1}+1 \ge 20{,}000$, $5^{n+1} \ge 19{,}999$, $(n+1)\ln 5 \ge \ln 19{,}999$,

$n \ge \dfrac{\ln 19{,}999}{\ln 5} - 1 \approx 5.15$; take $n = 6$

27. $\rho = \lim_{k\to+\infty} \dfrac{k^{3/2}|x-1|}{(k+1)^{3/2}} = |x-1|$, converges if $|x-1| < 1$, diverges if $|x-1| > 1$.

If $x = 0$, $\displaystyle\sum_{k=1}^{\infty} \dfrac{(-1)^k}{k^{3/2}}$ converges; if $x = 2$, $\displaystyle\sum_{k=1}^{\infty} \dfrac{1}{k^{3/2}}$ converges. $R = 1$, interval of

convergence $[0,2]$.

28. $\rho = \lim_{k\to+\infty} \dfrac{(k+1)^2|x-2|}{k^2(k+1)} = 0$, $R = +\infty$, interval of convergence $(-\infty, +\infty)$.

29. $\rho = \lim_{k\to+\infty} \dfrac{1}{5}(k+1)|x-1| = +\infty$, $R = 0$, converges only for $x = 1$.

30. $\rho = \lim_{k\to+\infty} \dfrac{2k+1}{2k+3}|x| = |x|$, converges if $|x| < 1$, diverges if $|x| > 1$. If $x = -1$,

$\displaystyle\sum_{k=1}^{\infty} \dfrac{(-1)^k}{2k+1}$ converges; if $x = 1$, $\displaystyle\sum_{k=1}^{\infty} 1/(2k+1)$ diverges. $R = 1$, interval of convergence $[-1,1)$.

31. (a) $(x-2) - \dfrac{1}{2}(x-2)^2 + \dfrac{1}{3}(x-2)^3$

(b) $R_3(x) = -\dfrac{(x-2)^4}{4(c-1)^4}$, c between 2 and x

(c) $|R_3(x)| = \dfrac{|x-2|^4}{4|c-1|^4} < \dfrac{|3/2-2|^4}{4|3/2-1|^4} = 1/4$

32. (a) $1 + x/2 + x^2/8 + x^3/48 + x^4/384$ (b) $R_4(x) = \dfrac{e^{c/2}x^5}{2^5 5!}$, c between 0 and x

(c) $|R_4(x)| = \dfrac{e^{c/2}|x|^5}{2^5 5!} < \dfrac{1}{2^5 5!} < 0.000261$

33. **(a)** $1 + \dfrac{1}{2}(x-1) - \dfrac{1}{8}(x-1)^2$ **(b)** $R_2(x) = \dfrac{(x-1)^3}{16c^{5/2}}$, c between 1 and x

  (c) $|R_2(x)| = \dfrac{|x-1|^3}{16c^{5/2}} < \dfrac{|4/9 - 1|^3}{16(4/9)^{5/2}} = \dfrac{(5/9)^3}{16(2/3)^5} < 0.0814$

34. **(a)** $\dfrac{1}{a-x} = \dfrac{1}{a}\left[\dfrac{1}{1-x/a}\right] = \dfrac{1}{a}\displaystyle\sum_{k=0}^{\infty}(x/a)^k = \sum_{k=0}^{\infty}\dfrac{x^k}{a^{k+1}}$, converges if $|x/a| < 1$,

  $|x| < |a|$ so $R = |a|$

  (b) $\dfrac{1}{3+x} = \dfrac{1}{3}\left[\dfrac{1}{1+x/3}\right] = \dfrac{1}{3}\displaystyle\sum_{k=0}^{\infty}(-1)^k(x/3)^k = \sum_{k=0}^{\infty}(-1)^k\dfrac{x^k}{3^{k+1}}$, $R = 3$

  (c) $\dfrac{2x}{4+x^2} = \dfrac{x}{2}\left[\dfrac{1}{1+x^2/4}\right] = \dfrac{x}{2}\displaystyle\sum_{k=0}^{\infty}(-1)^k(x^2/4)^k = \sum_{k=0}^{\infty}(-1)^k(x/2)^{2k+1}$,

  converges if $x^2/4 < 1$, $x^2 < 4$, $|x| < 2$ so $R = 2$

  (d) $\dfrac{1}{(1-x)(2-x)} = \dfrac{1}{1-x} - \dfrac{1}{2-x} = \dfrac{1}{1-x} - \dfrac{1}{2}\left[\dfrac{1}{1-x/2}\right]$

  $= \displaystyle\sum_{k=0}^{\infty}x^k - \dfrac{1}{2}\sum_{k=0}^{\infty}(x/2)^k = \sum_{k=0}^{\infty}(1 - 2^{-k-1})x^k$,

  the series for $1/(1-x)$ converges if $|x| < 1$ and that for $1/(2-x)$ if $|x| < 2$ so both will converge if $|x| < 1$ thus $R = 1$

35. $\ln(a+x) = \ln a(1 + x/a) = \ln a + \ln(1 + x/a) = \ln a + \displaystyle\sum_{k=0}^{\infty}(-1)^k\dfrac{(x/a)^{k+1}}{k+1}$,

  converges if $|x/a| < 1$, $|x| < |a| = a$ so $R = a$

36. **(a)** $e^x = e^{a+(x-a)} = e^a e^{x-a} = e^a \displaystyle\sum_{k=0}^{\infty}\dfrac{(x-a)^k}{k!} = \sum_{k=0}^{\infty}\dfrac{e^a(x-a)^k}{k!}$

  (b) $\sin x = \sin[a + (x-a)] = \sin a \cos(x-a) + \cos a \sin(x-a)$

  $= (\sin a)\displaystyle\sum_{k=0}^{\infty}(-1)^k\dfrac{(x-a)^{2k}}{(2k)!} + (\cos a)\sum_{k=0}^{\infty}(-1)^k\dfrac{(x-a)^{2k+1}}{(2k+1)!}$

  (c) $\dfrac{1}{x} = \dfrac{1}{a+(x-a)} = \dfrac{1}{a}\left[\dfrac{1}{1+(x-a)/a}\right]$

  $= \dfrac{1}{a}\displaystyle\sum_{k=0}^{\infty}(-1)^k\dfrac{(x-a)^k}{a^k} = \sum_{k=0}^{\infty}(-1)^k\dfrac{(x-a)^k}{a^{k+1}}$, $a \neq 0$

37. $1/(9+x)^{1/2} = \dfrac{1}{3}(1 + x/9)^{-1/2} = \dfrac{1}{3}\left[1 + \displaystyle\sum_{k=1}^{\infty}(-1)^k\dfrac{1 \cdot 3 \cdot 5 \cdots (2k-1)}{2^k k!}(x/9)^k\right]$,

  converges if $|x/9| < 1$, $|x| < 9$ so $R = 9$

38. $f(x) = e^{\tan x}$, $f'(x) = e^{\tan x} \sec^2 x$, $f''(x) = e^{\tan x} \left(2 \sec^2 x \tan x + \sec^4 x\right)$, $f(0) = 1$, $f'(0) = 1$, $f''(0) = 1$ so the Maclaurin series is $1 + x + x^2/2 + \cdots$

39. $\sec x = 1/\cos x = 1/(1 - x^2/2! + x^4/4! - \cdots) = 1 + x^2/2 + 5x^4/24 + \cdots$

40. $\dfrac{\sin x}{e^x + x} = \dfrac{x - x^3/3! + x^5/5! - \cdots}{(1 + x + x^2/2! + \cdots) + x} = \dfrac{x - x^3/3! + x^5/5! - \cdots}{1 + 2x + x^2/2! + \cdots} = x - 2x^2 + \dfrac{10}{3}x^3 + \cdots$

41. $[\cos x]^{1/2} = \left[1 - x^2/2! + x^4/4! - \cdots\right]^{1/2} = \left[1 + \left(-x^2/2! + x^4/4! - \cdots\right)\right]^{1/2}$

$\qquad = 1 + \dfrac{1}{2}\left(-x^2/2! + x^4/4! - \cdots\right) - \dfrac{1}{8}\left(-x^2/2! + x^4/4! - \cdots\right)^2 + \cdots$

$\qquad = 1 - x^2/4 - x^4/96 + \cdots$

42. $e^x \ln(1 - x) = \left(1 + x + x^2/2! + \cdots\right)\left(-x - x^2/2 - x^3/3 - \cdots\right) = -x - 3x^2/2 - 4x^3/3 + \cdots$

43. $f(x) = \ln(1 + \sin x)$, $f'(x) = \dfrac{\cos x}{1 + \sin x}$, $f''(x) = -\dfrac{1}{1 + \sin x}$, $f'''(x) = \dfrac{\cos x}{(1 + \sin x)^2}$;

$\qquad f(0) = 0$, $f'(0) = 1$, $f''(0) = -1$, $f'''(0) = 1$; $\ln(1 + \sin x) = x - \dfrac{1}{2}x^2 + \dfrac{1}{6}x^3 + \cdots$

44. $\dfrac{1 - \cos 3x}{x^2} = \dfrac{1}{x^2}\left[1 - \left(1 - \dfrac{9x^2}{2!} + \dfrac{81x^4}{4!} - \cdots\right)\right] = \dfrac{9}{2!} - \dfrac{81x^2}{4!} + \cdots$; $\displaystyle\lim_{x \to 0} \dfrac{1 - \cos 3x}{x^2} = \dfrac{9}{2}$

45. $\dfrac{\ln(1 - 2x)}{x} = \dfrac{1}{x}\left(-2x - 2x^2 - \dfrac{8}{3}x^3 - \cdots\right) = -2 - 2x - \dfrac{8}{3}x^2 - \cdots$, $\displaystyle\lim_{x \to 0} \dfrac{\ln(1 - 2x)}{x} = -2$

46. $\cos x = 1 - x^2/2 + (0)x^3 + R_3(x)$, $|R_3(x)| \leq \dfrac{|x|^4}{4!} < \dfrac{(0.1)^4}{4!} < 0.5 \times 10^{-5}$, so 5 decimal place accuracy is guaranteed.

47. $\sin x = x - x^3/3! + x^5/5! + (0)x^6 + R_6(x)$, $|R_6(x)| \leq \dfrac{|x|^7}{7!} < 6 \times 10^{-4}$ if $|x|^7 < 3.024$, $|x| < (3.024)^{1/7} \approx 1.17$

48. $\cos x = 1 - x^2/2! + x^4/4! - \cdots$, $|R_n(x)| \leq \dfrac{|x|^{n+1}}{(n + 1)!}$, $10° = \pi/18$ radians,

$\qquad |R_n(\pi/18)| \leq \dfrac{(\pi/18)^{n+1}}{(n + 1)!} < 0.5 \times 10^{-3}$ if $n = 3$, $\cos 10° \approx 1 - (\pi/18)^2/2 \approx 0.985$

49. $\displaystyle\int_0^1 \frac{1-e^{-t/2}}{t}\,dt = \int_0^1 \frac{\left[1 - \left(1 - \dfrac{t}{2} + \dfrac{t^2}{8} - \dfrac{t^3}{48} + \dfrac{t^4}{384} - \dfrac{t^5}{3840} + \cdots\right)\right]}{t}\,dt$

$$= \int_0^1 \left(\frac{1}{2} - \frac{t}{8} + \frac{t^2}{48} - \frac{t^3}{384} + \frac{t^4}{3840} - \cdots\right)dt$$

$$= \frac{t}{2} - \frac{t^2}{16} + \frac{t^3}{144} - \frac{t^4}{1436} + \frac{t^5}{19200} - \cdots\Bigg]_0^1$$

$$= 1/2 - 1/16 + 1/144 - 1/1436 + 1/19200 - \cdots,$$

but $1/19200 < 0.5 \times 10^{-3}$ so $\displaystyle\int_0^1 \frac{1-e^{-t/2}}{t}\,dt \approx 1/2 - 1/16 + 1/144 - 1/1436 \approx 0.444$

50. $\displaystyle\int_0^1 \frac{\sin x}{\sqrt{x}}\,dx = \int_0^1 x^{-1/2}(x - x^3/3! + x^5/5! - x^7/7! + \cdots)dx$

$$= \int_0^1 \left(x^{1/2} - \frac{1}{3!}x^{5/2} + \frac{1}{5!}x^{9/2} - \frac{1}{7!}x^{13/2} + \cdots\right)dx$$

$$= \frac{2}{3}x^{3/2} - \frac{2}{7\cdot 3!}x^{7/2} + \frac{2}{11\cdot 5!}x^{11/2} - \frac{2}{15\cdot 7!}x^{15/2} + \cdots\Bigg]_0^1$$

$$= \frac{2}{3} - \frac{2}{7\cdot 3!} + \frac{2}{11\cdot 5!} - \frac{2}{15\cdot 7!} + \cdots,$$

but $2/(15\cdot 7!) < 0.5 \times 10^{-3}$ so $\displaystyle\int_0^1 \frac{\sin x}{\sqrt{x}}\,dx \approx \frac{2}{3} - \frac{2}{7\cdot 3!} + \frac{2}{11\cdot 5!} \approx 0.621$

51. $\displaystyle y' = \sum_{n=1}^{\infty} \frac{k^n x^{n-1}}{(n-1)!} = \sum_{n=0}^{\infty} \frac{k^{n+1} x^k}{n!} = k\sum_{n=0}^{\infty} \frac{k^n x^k}{n!} = ky$, so $y' - ky = 0$

CHAPTER 12

Topics in Analytic Geometry

SECTION 12.2

12.2.1 Find the equation of the parabola with vertex at the origin and directrix $x = 5/2$.

12.2.2 Find the equation of the parabola with focus at (6,-2) and directrix $x = 2$. Sketch.

12.2.3 State the definition of a parabola. Use your definition to derive the equation of the parabola whose focus is at $F(3,0)$ and directrix $x = 1$.

12.2.4 Find the equation of the curve consisting of all points in the plane equidistant from the y−axis and the point (1,0). Identify the curve.

12.2.5 Find the equation of the curve consisting of all points in the plane equidistant from the line $y = 2$ and the point $(1,1)$. Identify the curve.

12.2.6 Find the equation of the parabola whose vertex is at the origin and axis along the x−axis if it goes through the point $(1,4)$. Sketch.

12.2.7 Sketch the parabola $y^2 - 4y - 7x + 11 = 0$ showing the focus, vertex, and directrix.

12.2.8 Sketch the parabola $y^2 + 6y + 6x = 0$ showing the focus, vertex and directrix.

12.2.9 Sketch the parabola $x^2 - 4x - 2y - 8 = 0$ showing the focus, vertex and directrix.

12.2.10 Sketch the parabola $2x^2 - 10x + 5y = 0$ showing the focus, vertex and directrix.

12.2.11 Sketch the parabola $3y^2 = 8x - 16$ showing the focus, vertex and directrix.

12.2.12 Find the equation for the parabola whose directrix is $x = -2$ and vertex at $(1,3)$. Where is the focus located?

12.2.13 Find the equation for the parabola whose directrix is $y = 3$ and vertex at $(-2,2)$. Where is the focus located?

12.2.14 Find the equation for the parabola whose directrix is $y = 0$ and focus at $(3,1)$. Where is the vertex located?

12.2.15 Find the equation for the parabola whose directrix is $x = 5$ and focus at $(-1,0)$. Where is the vertex located?

12.2.16   Find the equation for the parabola whose vertex is at $(-5/2, 1)$ and focus at $(0, 1)$. What is the equation for the directrix?

12.2.17   Find the equation for the parabola whose vertex is at $(2, 1)$ and passing through $(5, -2)$ if its axis of symmetry is parallel to the x-axis.

12.2.18   Find the equation of the parabola whose vertex is at the origin and passing through $(2,3)$ if its axis of symmetry is parallel to the y-axis.

12.2.19   Find the equation for the parabola whose vertex is at $(2, -3/2)$ and focus at $(2, 1)$. What is the equation for the directrix? Sketch.

SOLUTIONS

SECTION 12.2

12.2.1 $y^2 = 4px$, $p = -\dfrac{5}{2}$, $y^2 = -10x$

12.2.2   The vertex is half way between the focus and directrix so the vertex is at $(4, -2)$ and $p = 2$ thus, $(y + 2)^2 = 8(x - 4)$.

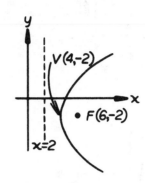

12.2.3   Use definition 12.2.1

$$PF = PQ$$

$$\sqrt{(x-3)^2 + y^2} = \sqrt{(x-1)^2}$$

$$x^2 - 6x + 9 + y^2 = x^2 - 2x + 1$$

$$y^2 = 4(x - 2)$$

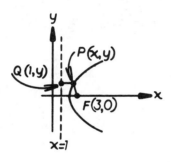

12.2.4   The curve is a parabola. The vertex is halfway between the focus and directrix so that the vertex is at $(1/2, 0)$ and $p = 1/2$, thus $y^2 = 2(x - 1/2)$.

12.2.5   The curve is a parabola. The vertex is halfway between the focus and the directrix so that the vertex is at $(1, 3/2)$ and $p = \dfrac{1}{2}$, thus $(x - 1)^2 = -2(y - 3/2)$.

12.2.6 $y^2 = 4px$
$4^2 = 4p(1)$
$p = 4$
$y^2 = 16x$

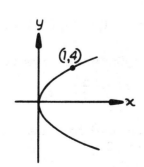

12.2.7

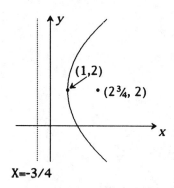

12.2.8

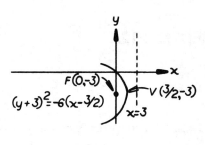

12.2.9

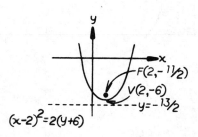

12.2.10

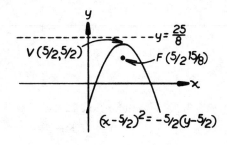

12.2.11

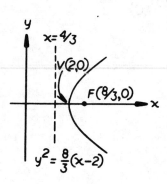

12.2.12 $p = 3$ so equation of parabola is $(y-3)^2 = 12(x-1)$; $F(4,3)$.

12.2.13 $p = -1$ so equation of parabola is $(x+2)^2 = -4(y-2)$; $F(-2,1)$.

12.2.14 The vertex is halfway between the focus and directrix so the vertex is at $(3, 1/2)$ and $p = 1/2$.
The equation of the parabola is $(x-3)^2 = 2\left(y - \frac{1}{2}\right)$.

12.2.15 The vertex is halfway between the focus and directrix so the vertex is at $(2,0)$ and $p = 3$.
The equation of the parabola is $y^2 = -12(x-2)$.

12.2.16 $p = 5/2$ so the equation of the parabola is $(y-1)^2 = 10\left(x + \dfrac{5}{2}\right)$. The directrix is $x = -5$.

12.2.17 $(y-1)^2 = 4p(x-2)$; $(-2-1)^2 = 4p(5-2)$; $p = \dfrac{3}{4}$, $(y-1)^2 = 3(x-2)$.

12.2.18 $x^2 = 4py$, $(2)^2 = 4p(3)$, $p = \dfrac{1}{3}$, $x^2 = \dfrac{4}{3}y$.

12.2.19 $p = \dfrac{5}{2}$ so the equation of the parabola is $(x-2)^2 = 10\left(y + \dfrac{3}{2}\right)$. The directrix is $y = -4$.

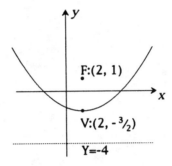

SECTION 12.3

12.3.1 Sketch the graph of $9x^2 + 16y^2 - 36x + 96y + 36 = 0$. Find the foci, the ends of the major and minor axes, and the eccentricity.

12.3.2 Sketch the graph of $9x^2 + 5y^2 + 36x - 30y + 36 = 0$. Find the foci, the ends of the major and minor axes, and the eccentricity.

12.3.3 Sketch the graph of $4x^2 + 9y^2 - 24x - 36y + 36 = 0$. Find the foci, the ends of the major and minor axes, and the eccentricity.

12.3.4 Sketch the graph of $x^2 + 2y^2 + 4x - 8y + 10 = 0$. Find the foci, the ends of the major and minor axes, and the eccentricity.

12.3.5 Find the equation of the ellipse whose center is at the origin and its major axis is on $y = 0$ if it passes through $(4, 3)$ and $(6, 2)$. Where are the foci located?

12.3.6 Find the equation of the ellipse whose major axis is 8 and foci at $(\pm 2, 0)$.

12.3.7 Find the equation of the ellipse whose minor axis is 10 and foci at $(0, \pm 3)$.

12.3.8 Find the equation of the ellipse whose major axis is 18 and foci at $\left(0, \pm 3\sqrt{6}\right)$.

12.3.9 Find the equation of the ellipse whose minor axis is $2\sqrt{3}$ and foci at $\left(0, \pm\sqrt{3}\right)$.

12.3.10 Find the volume of the solid that is generated when

$$\frac{x^2}{4} + \frac{y^2}{9} = 1$$

is revolved around the x axis.

12.3.11 Find the volume of the solid that is generated when

$$\frac{x^2}{4} + \frac{y^2}{9} = 1$$

is revolved around the y axis.

12.3.12 Sketch the graph of $4x^2 + 9y^2 - 16x + 18y - 11 = 0$. Label the foci and ends of the major and minor axes.

12.3.13 Find the equation of the ellipse whose major axis is 10 and foci at $(0, 2)$ and $(8, 2)$.

12.3.14 Find the equation of the ellipse whose minor axis is 6 and foci at $(1, -1)$ and $(7, -1)$.

12.3.15 Find the equation of the ellipse whose major axis is 16 and foci at $(-1, 1)$ and $(-1, 5)$.

12.3.16 Find the equation of the ellipse whose minor axis is 10 and foci at, $\left(5, 3 - \sqrt{3}\right)$, $\left(5, 3 + \sqrt{3}\right)$.

12.3.17 Find the equation of the parabola with vertex at the origin which passes through the ends of the minor axis of the ellipse $x^2 - 10x + 25y^2 = 0$.

12.3.18 Find the equation of the parabola with vertex at the origin which passes through the ends of the minor axis of the ellipse $y^2 - 10y + 25x^2 = 0$.

SOLUTIONS

SECTION 12.3

12.3.1

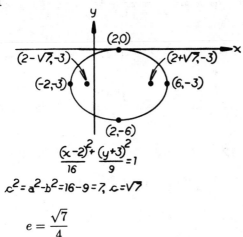

$$\frac{(x-2)^2}{16} + \frac{(y+3)^2}{9} = 1$$

$$c^2 = a^2 - b^2 = 16 - 9 = 7, \ c = \sqrt{7}$$

$$e = \frac{\sqrt{7}}{4}$$

12.3.2

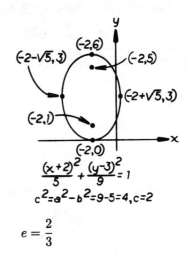

$$\frac{(x+2)^2}{5} + \frac{(y-3)^2}{9} = 1$$

$$c^2 = a^2 - b^2 = 9 - 5 = 4, \ c = 2$$

$$e = \frac{2}{3}$$

12.3.3

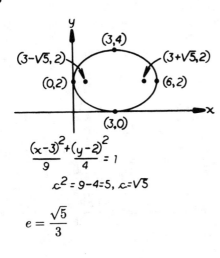

$$\frac{(x-3)^2}{9} + \frac{(y-2)^2}{4} = 1$$

$$c^2 = 9 - 4 = 5, \ c = \sqrt{5}$$

$$e = \frac{\sqrt{5}}{3}$$

12.3.4

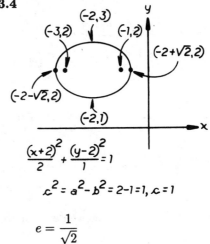

$$\frac{(x+2)^2}{2} + \frac{(y-2)^2}{1} = 1$$

$$c^2 = a^2 - b^2 = 2 - 1 = 1, \ c = 1$$

$$e = \frac{1}{\sqrt{2}}$$

12.3.5 $\dfrac{x^2}{a^2} + \dfrac{y^2}{b^2} = 1$; solve $\begin{cases} \dfrac{16}{a^2} + \dfrac{9}{b^2} = 1 \\ \dfrac{36}{a^2} + \dfrac{4}{b^2} = 1 \end{cases}$ to get $a^2 = 52$, $b^2 = 13$ so the equation of the ellipse is

$\dfrac{x^2}{52} + \dfrac{y^2}{13} = 1$. $c^2 = 52 - 13 = 39$, $c = \sqrt{39}$; the foci are at $\left(\pm\sqrt{39}, 0\right)$.

12.3.6 $a = 8/2 = 4$, $c = 2$, $b^2 = a^2 - c^2 = 16 - 4 = 12$; $\dfrac{x^2}{16} + \dfrac{y^2}{12} = 1$.

12.3.7 $b = 10/2 = 5$, $c = 3$, $a^2 = b^2 + c^2 = 25 + 9 = 34$; $\dfrac{x^2}{25} + \dfrac{y^2}{34} = 1$.

12.3.8 $a = 18/2 = 9$, $c = 3\sqrt{6}$, $b^2 = a^2 - c^2 = 81 - 54 = 27$; $\dfrac{x^2}{27} + \dfrac{y^2}{81} = 1$.

12.3.9 $b = 2\sqrt{3}/2 = \sqrt{3}$, $c = \sqrt{3}$, $a^2 = b^2 + c^2 = 3 + 3 = 6$; $\dfrac{x^2}{6} + \dfrac{y^2}{3} = 1$.

12.3.10 $V = \pi \displaystyle\int_{-2}^{2} \dfrac{9}{4}(4 - x^2)dx = 24\pi$ **12.3.11** $V = \pi \displaystyle\int_{-3}^{+3} \dfrac{4}{9}(9 - y^2)\,dy = 16\pi$

12.3.12

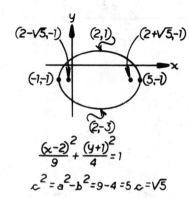

$$\frac{(x-2)^2}{9} + \frac{(y+1)^2}{4} = 1$$

$$c^2 = a^2 - b^2 = 9 - 4 = 5 \;\; c = \sqrt{5}$$

12.3.13 $a = 10/2 = 5$, the center is

at $(4, 2)$ so $c = 4$,

$b^2 = 25 - 16 = 9$

$$\frac{(x-4)^2}{25} + \frac{(y-2)^2}{9} = 1$$

12.3.14 $b = 6/2 = 3$, the center is at $(4, -1)$ so $c = 3$, $a^2 = 9 + 9 = 18$, $\dfrac{(x-4)^2}{18} + \dfrac{(y+1)^2}{9} = 1$.

12.3.15 $a = 16/2 = 8$, the center is at $(-1, 3)$ so $c = 2$, $b^2 = 64 - 4 = 60$, $\dfrac{(x+1)^2}{60} + \dfrac{(y-3)^2}{64} = 1$.

12.3.16 $b = 10/2 = 5$, the center is at $(5, 3)$ so $c = \sqrt{3}$, $a^2 = 25 + 3 = 28$, $\dfrac{(x-5)^2}{25} + \dfrac{(y-3)^2}{28} = 1$.

12.3.17   Place in standard form to get $\dfrac{(x-5)^2}{25} + y^2 = 1$, so $b = 1$ and the ends of the minor axis are
$(5, -1)$ and $(5, 1)$, thus, $y^2 = 4px$, $(-1)^2 = 4p(5)$, $p = 1/20$, $y^2 = \dfrac{1}{5}x$.

12.3.18   Place in standard form to get $x^2 + \dfrac{(y-5)^2}{25} = 1$, so $b = 1$ and the ends of the minor axis are
$(-1, 5)$ and $(1, 5)$, thus, $x^2 = 4py$, $(-1)^2 = 4p(5)$, $p = 1/20$, $x^2 = \dfrac{1}{5}y$.

SECTION 12.4

12.4.1 Sketch the hyperbola $9x^2 - 16y^2 = 144$. Find the coordinates of the vertices and foci, and find the equation of the asymptotes.

12.4.2 Sketch the hyperbola $16x^2 - 9y^2 - 160x - 72y + 112 = 0$. Find the coordinates of the vertices and foci, and find the equation of the asymptotes.

12.4.3 Sketch the hyperbola $\dfrac{(x-3)^2}{9} - \dfrac{(y+4)^2}{16} = 1$. Find the coordinates of the vertices and foci, and find the equation of the asymptotes.

12.4.4 Find the equation of the hyperbola whose vertices are 4 units apart and whose foci are at $(\pm 3, 0)$.

12.4.5 Find the equation of the hyperbola whose vertices are 8 units apart and whose foci are at $(\pm 5, 0)$.

12.4.6 Find the equation of the hyperbola whose vertices are 2 units apart and whose foci are at $\left(0, \pm 2\sqrt{5}\right)$.

12.4.7 Find the equation of the hyperbola whose asymptotes are $\pm \dfrac{3}{4}x$ and whose foci are at $(\pm 10, 0)$.

12.4.8 Sketch the hyperbola $4y^2 - 9x^2 - 36x - 8y - 68 = 0$. Find the eccentricity, coordinates of vertices and foci, and find the equation of the asymptotes.

12.4.9 Sketch the hyperbola $9x^2 - 4y^2 + 36x + 24y + 36 = 0$. Find the eccentricity, coordinates of vertices and foci, and find the equation of the asymptotes.

12.4.10 Sketch the hyperbola $3x^2 - y^2 - 12x - 6y = 0$. Find the eccentricity, coordinates of the vertices and foci, and find the equation of the asymptotes.

12.4.11 Sketch the hyperbola $4x^2 - y^2 + 24x + 4y + 28 = 0$. Find the eccentricity, coordinates of the vertices and foci, and find the equation of the asymptotes.

12.4.12 Find the equation of the ellipse whose major and minor axes are coincident with the focal and conjugate axes of the hyperbola $4x^2 - 25y^2 - 8x - 100y - 196 = 0$. Where are the foci of the ellipse located?

12.4.13 Find the equation of the ellipse whose major and minor axes are coincident with the focal and conjugate axes of the hyperbola $9x^2 - 4y^2 + 36x + 24y + 36 = 0$. Where are the foci of the ellipse located?

12.4.14   Find the equation of the hyperbola whose vertices are 10 units apart and whose foci are at $(1, -16)$ and $(1, 10)$.

12.4.15   Find the equation of the hyperbola whose center is at $(-3, -1)$, a vertex at $(1, -1)$, and a focus at $(2, -1)$.

12.4.16   Find the equation of the hyperbola whose center is at $(2, 2)$, a vertex at $(2, 10)$, and a focus at $(2, 11)$.

12.4.17   Find the equation of the hyperbola whose vertices are at $(7, -1)$ and $(-5, -1)$ if a focus is located at $(9, -1)$.

12.4.18   Find the equation of the hyperbola whose center is at $(-4, 6)$, a vertex at $(-4, 9)$ and a focus at $(-4, 11)$.

SOLUTIONS

SECTION 12.4

12.4.1

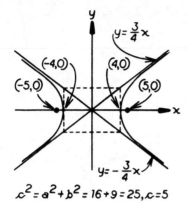

$c^2 = a^2 + b^2 = 16 + 9 = 25, c = 5$

12.4.2

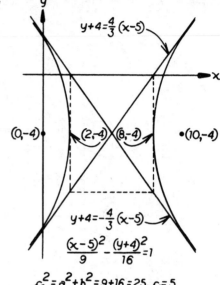

$$\frac{(x-5)^2}{9} - \frac{(y+4)^2}{16} = 1$$

$c^2 = a^2 + b^2 = 9 + 16 = 25, c = 5$

12.4.3

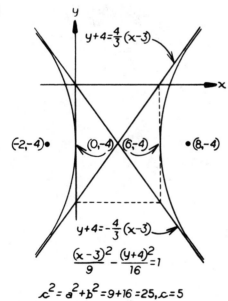

$$\frac{(x-3)^2}{9} - \frac{(y+4)^2}{16} = 1$$

$c^2 = a^2 + b^2 = 9 + 16 = 25, c = 5$

12.4.4 $a = 4/2 = 2$ and $c = 3$, thus $b^2 = c^2 - a^2 = 9 - 4 = 5$ and $\dfrac{x^2}{4} - \dfrac{y^2}{5} = 1$.

12.4.5 $a = \dfrac{8}{2} = 4$, $c = 5$, thus $b^2 = c^2 - a^2 = 25 - 16 = 9$ and $\dfrac{x^2}{16} - \dfrac{y^2}{9} = 1$.

12.4.6 $a = 2/2 = 1$, $c = 2\sqrt{5}$, thus, $b^2 = c^2 - a^2 = 20 - 1 = 19$ and $\dfrac{y^2}{1} - \dfrac{x^2}{19} = 1$.

12.4.7 $c = 10$, $\dfrac{b}{a} = \dfrac{3}{4}$, $a^2 + b^2 = 100$, so $a^2 + \dfrac{9}{16}a^2 = 100$, $a^2 = 64$, $b^2 = 36$

$$\frac{x^2}{64} - \frac{y^2}{36} = 1$$

12.4.8

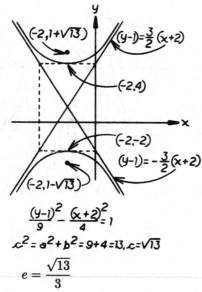

$$\frac{(y-1)^2}{9} - \frac{(x+2)^2}{4} = 1$$
$$c^2 = a^2 + b^2 = 9 + 4 = 13, c = \sqrt{13}$$
$$e = \frac{\sqrt{13}}{3}$$

12.4.9

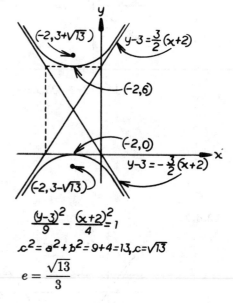

$$\frac{(y-3)^2}{9} - \frac{(x+2)^2}{4} = 1$$
$$c^2 = a^2 + b^2 = 9 + 4 = 13, c = \sqrt{13}$$
$$e = \frac{\sqrt{13}}{3}$$

12.4.10

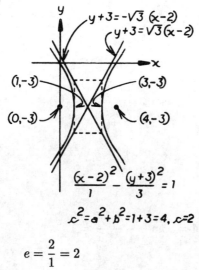

$$\frac{(x-2)^2}{1} - \frac{(y+3)^2}{3} = 1$$
$$c^2 = a^2 + b^2 = 1 + 3 = 4, c = 2$$
$$e = \frac{2}{1} = 2$$

12.4.11

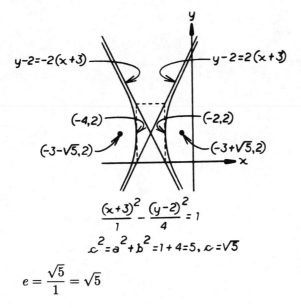

$$y-2=-2(x+3) \qquad y-2=2(x+3)$$

$(-4,2) \qquad (-2,2)$

$(-3-\sqrt{5},2) \qquad (-3+\sqrt{5},2)$

$$\frac{(x+3)^2}{1} - \frac{(y-2)^2}{4} = 1$$

$$c^2 = a^2 + b^2 = 1 + 4 = 5, \; c = \sqrt{5}$$

$$e = \frac{\sqrt{5}}{1} = \sqrt{5}$$

12.4.12 The equation of the hyperbola is $\dfrac{(x-1)^2}{25} - \dfrac{(y+2)^2}{4} = 1$ and thus the equation of the ellipse is $\dfrac{(x-1)^2}{25} + \dfrac{(y+2)^2}{4} = 1$; $c^2 = a^2 - b^2 = 25 - 4 = 21$, $c = \sqrt{21}$, the foci are at $\left(1 \pm \sqrt{21}, -2\right)$.

12.4.13 The equation of the hyperbola is $\dfrac{(y-3)^2}{9} - \dfrac{(x+2)^2}{4} = 1$ and thus the equation of the ellipse is $\dfrac{(x+2)^2}{4} + \dfrac{(y-3)^2}{9} = 1$; $c^2 = a^2 - b^2 = 9 - 4 = 5$, $c = \sqrt{5}$, the foci are at $\left(-2, 3 \pm \sqrt{5}\right)$.

12.4.14 The center of the hyperbola is halfway between the foci at $(1, -3)$, $a = 10/2 = 5$, $c = 13$, thus, $b^2 = c^2 - a^2 = 169 - 25 = 144$, $b = 12$, so $\dfrac{(y+3)^2}{25} - \dfrac{(x-1)^2}{144} = 1$.

12.4.15 $a = 4$, $c = 5$, so $b^2 = c^2 - a^2 = 25 - 16 = 9$, $b = 3$ and $\dfrac{(x+3)^2}{16} - \dfrac{(y+1)^2}{9} = 1$.

12.4.16 $a = 8$, $c = 9$, so $b^2 = c^2 - a^2 = 81 - 64 = 17$ and $\dfrac{(y-2)^2}{64} - \dfrac{(x-2)^2}{17} = 1$.

12.4.17 The center of the hyperbola is halfway between the vertices at $(1, -1)$; $a = 6$ and $c = 8$, so $b^2 = c^2 - a^2 = 64 - 36 = 28$ and $\dfrac{(x-1)^2}{36} - \dfrac{(y+1)^2}{28} = 1$.

12.4.18 $a = 3$, $c = 5$, so $b^2 = c^2 - a^2 = 25 - 9 = 16$ and $\dfrac{(y-6)^2}{9} - \dfrac{(x+4)^2}{16} = 1$.

SECTION 12.5

12.5.1 Rotate the coordinate axes to remove the xy term. Name and sketch the conic given by $13x^2 - 10xy + 13y^2 = 72$.

12.5.2 Rotate the coordinate axes to remove the xy term. Name and sketch the conic given by $x^2 + 2xy + y^2 - 2x + 2y = 0$.

12.5.3 Rotate the coordinate axes to remove the xy term. Name and sketch the conic given by $x^2 + 5xy + y^2 = 6$.

12.5.4 Rotate the coordinate axes to remove the xy term. Name and sketch the conic given by $x^2 - 3xy + y^2 = 5$.

12.5.5 Rotate the coordinate axes to remove the xy term. Name and sketch the conic given by $5x^2 - 6xy + 5y^2 = 8$.

12.5.6 Rotate the coordinate axes to remove the xy term. Name and sketch the conic given by $2x^2 + 4\sqrt{3}xy - 2y^2 = 4$.

12.5.7 Rotate the coordinate axes to remove the xy term. Name and sketch the conic given by $x^2 + 2xy + y^2 + 8x - 8y = 0$.

12.5.8 Rotate the coordinate axes to remove the xy term. Name and sketch the conic given by $16y^2 - 24xy + 9x^2 - 60y - 80x = 0$.

12.5.9 Rotate the coordinate axes to remove the xy term. Name and sketch the conic given by $180x^2 + 120xy + 145y^2 - 900 = 0$.

12.5.10 Rotate the coordinate axes to remove the xy term. Name and sketch the conic given by $145x^2 - 120xy + 180y^2 - 900 = 0$.

12.5.11 Rotate the coordinate axes to remove the xy term. Name and sketch the conic given by $3x^2 + 4xy = 16$.

12.5.12 Rotate the coordinate axes to remove the xy term. Name and sketch the conic given by $16x^2 - 24xy + 9y^2 - 240y - 180x = 0$.

12.5.13 Rotate the coordinate axes to remove the xy term. Name and sketch the conic given by $9x^2 + 24xy + 16y^2 + 80x - 60y = 0$.

12.5.14 Rotate the coordinate axes to remove the xy term. Name and sketch the conic given by $7x^2 - 6\sqrt{3}xy + 13y^2 - 64 = 0$.

SOLUTIONS

SECTION 12.5

12.5.1 $\cot 2\theta = \dfrac{13-13}{-10} = 0,\ 2\theta = 90°,\ \theta = 45°$

$$x = \frac{x'}{\sqrt{2}} - \frac{y'}{\sqrt{2}} = \frac{x'-y'}{\sqrt{2}};\ y = \frac{x'}{\sqrt{2}} + \frac{y'}{\sqrt{2}} = \frac{x'+y'}{\sqrt{2}}$$

$$13\left(\frac{x'-y'}{\sqrt{2}}\right)^2 - 10\left(\frac{x'-y'}{\sqrt{2}}\right)\left(\frac{x'+y'}{\sqrt{2}}\right) + 13\left(\frac{x'+y'}{\sqrt{2}}\right)^2 = 72$$

$$\frac{x'^2}{9} + \frac{y'^2}{4} = 1,\ \text{ellipse.}$$

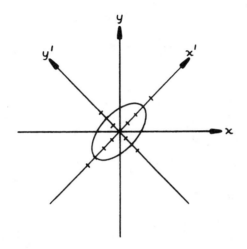

12.5.2 $\cot 2\theta = \dfrac{1-1}{2} = 0,\ 2\theta = 90°,\ \theta = 45°$

$$x = \frac{x'}{\sqrt{2}} - \frac{y'}{\sqrt{2}} = \frac{x'-y'}{\sqrt{2}};$$

$$y = \frac{x'}{\sqrt{2}} + \frac{y'}{\sqrt{2}} = \frac{x'+y'}{\sqrt{2}}$$

$$\left(\frac{x'-y'}{\sqrt{2}}\right)^2 + 2\left(\frac{x'-y'}{\sqrt{2}}\right)\left(\frac{x'+y'}{\sqrt{2}}\right) + \left(\frac{x'+y'}{\sqrt{2}}\right)^2 - 2\left(\frac{x'-y'}{\sqrt{2}}\right)$$
$$+ 2\left(\frac{x'+y'}{\sqrt{2}}\right) = 0$$

$x'^2 = -\sqrt{2}y'$, parabola.

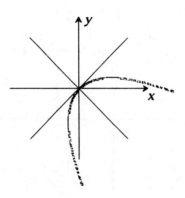

12.5.3 $\cot 2\theta = \dfrac{1-1}{-5} = 0$, $2\theta = 90°$,

$\theta = 45°$

$x = \dfrac{x'}{\sqrt{2}} - \dfrac{y'}{\sqrt{2}} = \dfrac{x'-y'}{\sqrt{2}}$;

$y = \dfrac{x'}{\sqrt{2}} + \dfrac{y'}{\sqrt{2}} = \dfrac{x'+y'}{\sqrt{2}}$

$\left(\dfrac{x'-y'}{\sqrt{2}}\right)^2 + 5\left(\dfrac{x'-y'}{\sqrt{2}}\right)\left(\dfrac{x'+y'}{\sqrt{2}}\right) + \left(\dfrac{x'+y'}{\sqrt{2}}\right)^2 = 6$

$\dfrac{x'^2}{12/7} - \dfrac{y'^2}{4} = 1$, hyperbola.

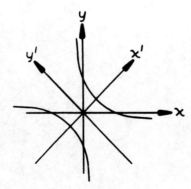

12.5.4 $\cot 2\theta = \dfrac{1-1}{-3} = 0,\ 2\theta = 90°,$

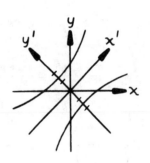

$\theta = 45°$

$x = \dfrac{x'}{\sqrt{2}} - \dfrac{y'}{\sqrt{2}} = \dfrac{x'-y'}{\sqrt{2}};$

$y = \dfrac{x'}{\sqrt{2}} + \dfrac{y'}{\sqrt{2}} = \dfrac{x'+y'}{\sqrt{2}}$

$\left(\dfrac{x'-y'}{\sqrt{2}}\right)^2 - 3\left(\dfrac{x'-y'}{\sqrt{2}}\right)\left(\dfrac{x'+y'}{\sqrt{2}}\right) + \left(\dfrac{x'+y'}{\sqrt{2}}\right)^2 = 5$

$\dfrac{y'^2}{2} - \dfrac{x'^2}{10} = 1,$ hyperbola.

12.5.5 $\cot 2\theta = \dfrac{5-5}{-6} = 0,\ 2\theta = 90°,$

$\theta = 45°$

$x = \dfrac{x'}{\sqrt{2}} - \dfrac{y'}{\sqrt{2}} = \dfrac{x'-y'}{\sqrt{2}};$

$y = \dfrac{x'}{\sqrt{2}} + \dfrac{y'}{\sqrt{2}} = \dfrac{x'+y'}{\sqrt{2}}$

$5\left(\dfrac{x'-y'}{\sqrt{2}}\right)^2 - 6\left(\dfrac{x'-y'}{\sqrt{2}}\right)\left(\dfrac{x'+y'}{\sqrt{2}}\right) + 5\left(\dfrac{x'+y'}{\sqrt{2}}\right)^2 = 8$

$\dfrac{x'^2}{4} + \dfrac{y'^2}{1} = 1,$ ellipse.

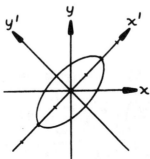

12.5.6 $\cot 2\theta = \dfrac{2+2}{4\sqrt{3}} = \dfrac{1}{\sqrt{3}},\ 2\theta = 60°,$

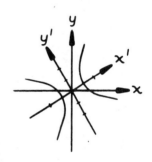

$\theta = 30°$

$x = \dfrac{\sqrt{3}}{2}x' - \dfrac{1}{2}y' = \dfrac{\sqrt{3}x' - y'}{2};$

$y = \dfrac{1}{2}x' + \dfrac{\sqrt{3}}{2}y' = \dfrac{x' + \sqrt{3}y'}{2}$

$2\dfrac{\sqrt{3}x' - y'}{2}^2 + 4\sqrt{3}\dfrac{\sqrt{3}x' - y'}{2}\dfrac{x' + \sqrt{3}y'}{2} - 2\dfrac{x' + \sqrt{3}y'}{2} = 4$

$\dfrac{x'^2}{1} - \dfrac{y'^2}{1} = 1,$ hyperbola.

12.5.7 $\cot 2\theta = \dfrac{1-1}{2} = 0,\ 2\theta = 90°,$

$\theta = 45°$

$x = \dfrac{x'}{\sqrt{2}} - \dfrac{y'}{\sqrt{2}} = \dfrac{x'-y'}{\sqrt{2}};$

$y = \dfrac{x'}{\sqrt{2}} + \dfrac{y'}{\sqrt{2}} = \dfrac{x'+y'}{\sqrt{2}}$

$\left(\dfrac{x'-y'}{\sqrt{2}}\right)^2 + 2\left(\dfrac{x'-y'}{\sqrt{2}}\right)\left(\dfrac{x'+y'}{\sqrt{2}}\right) + \left(\dfrac{x'+y'}{\sqrt{2}}\right)^2$

$+8\left(\dfrac{x'-y'}{\sqrt{2}}\right) - 8\left(\dfrac{x'+y'}{\sqrt{2}}\right) = 0$

$x'^2 = 4\sqrt{2}y'$, parabola.

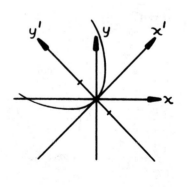

12.5.8 $\cot 2\theta = \dfrac{9-16}{-24} = \dfrac{7}{24}$

$\cos 2\theta = \dfrac{7}{25}$

$\cos\theta = \sqrt{\dfrac{1+7/25}{2}} = \dfrac{4}{5}$

$\sin\theta = \sqrt{\dfrac{1-7/25}{2}} = \dfrac{3}{5}$

$x = \dfrac{4x'}{5} - \dfrac{3y'}{5} = \dfrac{4x'-3y'}{5}$

$y = \dfrac{3x'}{5} + \dfrac{4y'}{5} = \dfrac{3x'+4y'}{5}$

$16\left(\dfrac{3x'+4y'}{5}\right)^2 - 24\left(\dfrac{4x'-3y'}{5}\right)\left(\dfrac{3x'+4y'}{5}\right) + 9\left(\dfrac{4x'-3y'}{5}\right)^2$

$-60\left(\dfrac{3x'+4y'}{5}\right) - 80\left(\dfrac{4x'-3y'}{5}\right) = 0$

$y'^2 = 4x'$, parabola.

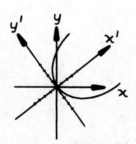

12.5.9 $\cot 2\theta = \dfrac{180 - 145}{120} = \dfrac{7}{24}$,

$\cos 2\theta = \dfrac{7}{25}$

$\cos \theta = \sqrt{\dfrac{1 + 7/25}{2}} = \dfrac{4}{5}$

$\sin \theta = \sqrt{\dfrac{1 - 7/25}{2}} = \dfrac{3}{5}$

$x = \dfrac{4}{5}x' - \dfrac{3}{5}y' = \dfrac{4x' - 3y'}{5}$

$y = \dfrac{3}{5}x' + \dfrac{4}{5}y' = \dfrac{3x' + 4y'}{5}$

$180\left(\dfrac{4x' - 3y'}{5}\right)^2 + 120\left(\dfrac{4x' - 3y'}{5}\right)\left(\dfrac{3x' + 4y'}{5}\right)$

$+145\left(\dfrac{3x' + 4y'}{5}\right)^2 - 900 = 0$

$\dfrac{y'^2}{9} + \dfrac{x'^2}{4} = 1$, ellipse.

12.5.10 $\cot 2\theta = \dfrac{145 - 180}{-120} = \dfrac{7}{24}$,

$\cos 2\theta = \dfrac{7}{25}$

$\cos \theta = \sqrt{\dfrac{1 + 7/25}{2}} = \dfrac{4}{5}$

$\sin \theta = \sqrt{\dfrac{1 - 7/25}{2}} = \dfrac{3}{5}$

$x = \dfrac{4x'}{5} - \dfrac{3y'}{5} = \dfrac{4x' - 3y'}{5}$

$y = \dfrac{3x'}{5} + \dfrac{4y'}{5} = \dfrac{3x' + 4y'}{5}$

$145\left(\dfrac{4x' + 3y'}{5}\right)^2 - 120\left(\dfrac{4x' - 3y'}{5}\right)\left(\dfrac{3x' + 4y'}{5}\right)$

$+180\left(\dfrac{3x' + 4y'}{5}\right)^2 - 900 = 0$

$\dfrac{x'^2}{9} + \dfrac{y'^2}{4} = 1$, ellipse.

12.5.11 $\cot 2\theta = \dfrac{3-0}{4} = \dfrac{3}{4},$

$\cos 2\theta = \dfrac{3}{5}$

$\cos\theta = \sqrt{\dfrac{1+\frac{3}{5}}{2}} = \dfrac{2}{\sqrt{5}}$

$\sin\theta = \sqrt{\dfrac{1-\frac{3}{5}}{2}} = \dfrac{1}{\sqrt{5}}$

$x = \dfrac{2}{\sqrt{5}}x' - \dfrac{1}{\sqrt{5}}y' = \dfrac{2x'-y'}{\sqrt{5}}$

$y = \dfrac{1}{\sqrt{5}}x' + \dfrac{2}{\sqrt{5}}y' = \dfrac{x'+2y'}{\sqrt{5}}$

$3\left(\dfrac{x'+2y'}{\sqrt{5}}\right)^2 + 4\left(\dfrac{2x'-y'}{\sqrt{5}}\right)\left(\dfrac{x'+2y'}{\sqrt{5}}\right) = 16$

$\dfrac{x'^2}{4} - \dfrac{y'^2}{16} = 1$, hyperbola.

12.5.12 $\cot 2\theta = \dfrac{16-9}{-24} = -\dfrac{7}{24},$

$\cos 2\theta = \dfrac{-7}{25}$

$\cos\theta = \sqrt{\dfrac{1+(-7/25)}{2}} = \dfrac{3}{5}$

$\sin\theta = \sqrt{\dfrac{1-(-7/25)}{2}} = \dfrac{4}{5}$

$x = \dfrac{3}{5}x' - \dfrac{4}{5}y' = \dfrac{3x'-4y'}{5}$

$y = \dfrac{4x'}{5} + \dfrac{3y'}{5} = \dfrac{4x'+3y'}{5}$

$16\left(\dfrac{3x'-4y'}{5}\right)^2 - 24\left(\dfrac{3x'-4y'}{5}\right)\left(\dfrac{4x'+3y'}{5}\right) + 9\left(\dfrac{4x'+3y'}{5}\right)^2$

$-240\left(\dfrac{4x'+3y'}{5}\right) - 180\left(\dfrac{3x'-4y'}{5}\right) = 0$

$y'^2 = 12x$, parabola.

12.5.13 $\cot 2\theta = \dfrac{9-16}{24} = -\dfrac{7}{24}$,

$\cos 2\theta = \dfrac{-7}{25}$

$\cos\theta = \sqrt{\dfrac{1+(-7/25)}{2}} = \dfrac{3}{5}$

$\sin\theta = \sqrt{\dfrac{1-(-7/25)}{2}} = \dfrac{4}{5}$

$x = \dfrac{3x'}{5} - \dfrac{4y'}{5} = \dfrac{3x'-4y'}{5}$

$y = \dfrac{4x'}{5} + \dfrac{3y'}{5} = \dfrac{4x'+3y'}{5}$

$9\left(\dfrac{3x'-4y'}{5}\right)^2 + 24\left(\dfrac{3x'-4y'}{5}\right)\left(\dfrac{4x'+3y'}{5}\right) + 16\left(\dfrac{4x'+3y'}{5}\right)^2$

$+80\left(\dfrac{3x'-4y'}{5}\right) - 60\left(\dfrac{4x'+3y'}{5}\right) = 0$

$x'^2 = 4y'$, parabola.

12.5.14 $\cot 2\theta = \dfrac{7-13}{-6\sqrt{3}} = \dfrac{1}{\sqrt{3}}$, $2\theta = 60°$, $\theta = 30°$

$x = \dfrac{\sqrt{3}}{2}x' - \dfrac{1}{2}y' = \dfrac{\sqrt{3}x'-y'}{2}$

$y = \dfrac{1}{2}x' + \dfrac{\sqrt{3}}{2}y' = \dfrac{x'+\sqrt{3}y'}{2}$

$7\left(\dfrac{\sqrt{3}x'-y'}{2}\right)^2 - 6\sqrt{3}\left(\dfrac{\sqrt{3}x'-y'}{2}\right)\left(\dfrac{x'+\sqrt{3}y'}{2}\right) + 13\left(\dfrac{x'+\sqrt{3}y'}{2}\right)^2 - 64 = 0$

$\dfrac{x'^2}{16} + \dfrac{y'^2}{4} = 1$, ellipse.

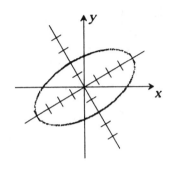

SUPPLEMENTARY EXERCISES, CHAPTER 12

In Exercises 1–8, identify the curve as a parabola, ellipse, or hyperbola, and give the following information:

Parabola: The coordinates of the vertex and focus; the equation of the directrix.

Ellipse: The coordinates of the center and foci; the lengths of the major and minor axes.

Hyperbola: The coordinates of the center, foci, and vertices, the equations of the asymptotes.

1. $y^2 + 12x - 6y + 33 = 0$.

2. $x^2 - 4y^2 = -1$.

3. $9x^2 + 4y^2 + 36x - 8y + 4 = 0$.

4. $6x + 8y - x^2 - 4y^2 = 12$.

5. $x^2 - 9y^2 - 4x + 18y - 14 = 0$.

6. $4y = x^2 + 2x - 7$.

7. $3x + 2y^2 - 4y - 7 = 0$.

8. $4x^2 = y^2 - 4y$.

In Exercises 9–16, find an equation for the curve described.

9. The parabola with vertex at $(1, 3)$ and directrix $x = -3$.

10. The ellipse with major axis of length 12 and foci at $(2, 7)$ and $(2, -1)$.

11. The hyperbola with foci $(0, \pm 5)$ and vertices 6 units apart.

12. The parabola with axis $y = 1$, vertex $(2, 1)$, and passing through $(3, -1)$.

13. The ellipse with foci $(\pm 3, 0)$ and such that the distances from the foci to $P(x, y)$ on the ellipse add up to 10 units.

14. The hyperbola with vertices $(0, \pm 2)$ and asymptotes $y = \pm 3x$.

15. The curve C with the property that the distance between the point $(3, 4)$ and any point $P(x, y)$ on C is equal to the distance between P and the line $y = 2$.

16. The hyperbola with vertices $(-3, 2)$ and $(1, 2)$ and perpendicular asymptotes.

In Exercises 17–20, sketch the curve whose equation is given in the stated exercise.

17. Exercise 1.   18. Exercise 2.   19. Exercise 3.   20. Exercise 6.

In Exercises 21–26, find the rotation angle θ needed to remove the xy-term; then name the conic and give its equation in $x'y'$-coordinates after the xy-term is removed.

21. $3x^2 - 2xy + 3y^2 = 4.$

22. $7x^2 - 8xy + y^2 = 9.$

23. $11x^2 + 10\sqrt{3}\,xy + y^2 = 4.$

24. $x^2 + 4xy + 4y^2 - 2\sqrt{5}\,x + \sqrt{5}\,y = 0.$

25. $16x^2 - 24xy + 9y^2 - 60x - 80y + 100 = 0.$

26. $73x^2 - 72xy + 52y^2 - 100 = 0.$

In Exercises 27–29, sketch the graph of the equation in the stated exercise, showing the xy- and $x'y'$-axes.

27. Exercise 21.

28. Exercise 23.

29. Exercise 24.

30. For the curve $xy + x - y = 1$,

(a) find the rotation angle necessary to remove the xy-term

(b) find the equation in $x'y'$-coordinates obtained by rotating the xy-coordinate axes through an angle of $\theta = \tan^{-1}(3/4)$ radians and verify the identities $A' + C' = A + C$ and $B'^2 - 4A'C' = B^2 - 4AC$.

SOLUTIONS

SUPPLEMENTARY EXERCISES CHAPTER 12

1. parabola, $(y-3)^2 = -12(x+2)$, $p = 3$; vertex $(-2,3)$, focus $(-5,3)$, directrix $x = 1$.

2. hyperbola, $y^2/(1/4) - x^2/1 = 1$, $a = 1/2$, $b = 1$, $c = \sqrt{5}/2$; center $(0,0)$, foci $(0, \pm\sqrt{5}/2)$, vertices $(0, \pm 1/2)$, asymptotes $y = \pm x/2$.

3. ellipse, $(x+2)^2/4 + (y-1)^2/9 = 1$, $a = 3$, $b = 2$, $c = \sqrt{5}$; center $(-2,1)$, foci $(-2, 1 \pm \sqrt{5})$, major axis 6, minor axis 4.

4. ellipse, $(x-3)^2/1 + (y-1)^2/(1/4) = 1$, $a = 1$, $b = 1/2$, $c = \sqrt{3}/2$; center $(3,1)$, foci $(3 \pm \sqrt{3}/2, 1)$, major axis 2, minor axis 1.

5. hyperbola, $(x-2)^2/9 - (y-1)^2/1 = 1$, $a = 3$, $b = 1$, $c = \sqrt{10}$, center $(2,1)$, foci $(2 \pm \sqrt{10}, 1)$, vertices $(-1,1)$ and $(5,1)$, asymptotes $y - 1 = \pm(x-2)/3$.

6. parabola, $(x+1)^2 = 4(y+2)$, $p = 1$; vertex $(-1,-2)$, focus $(-1,-1)$, directrix $y = -3$.

7. parabola, $(y-1)^2 = (-3/2)(x-3)$, $p = 3/8$; vertex $(3,1)$, focus $(21/8, 1)$, directrix $x = 27/8$.

8. hyperbola, $(y-2)^2/4 - x^2/1 = 1$, $a = 2$, $b = 1$, $c = \sqrt{5}$; center $(0,2)$, foci $(0, 2 \pm \sqrt{5})$, vertices $(0,0)$ and $(0,4)$, asymptotes $y - 2 = \pm 2x$.

9. $p = 4$; $(y-3)^2 = 16(x-1)$.

10. center $(2,3)$, $c = 4$, $a = 12/2 = 6$, $b^2 = 20$; $(x-2)^2/20 + (y-3)^2/36 = 1$.

11. center $(0,0)$, $c = 5$, $a = 6/2 = 3$, $b^2 = 16$, $y^2/9 - x^2/16 = 1$.

12. $(y-1)^2 = a(x-2)$, $(-2)^2 = a(1)$, $a = 4$; $(y-1)^2 = 4(x-2)$.

13. center $(0,0)$, $c = 3$, $a = 10/2 = 5$, $b^2 = 16$; $x^2/25 + y^2/16 = 1$.

14. center $(0,0)$, $a = 2$, $a/b = 3$ so $b = 2/3$; $y^2/4 - x^2/(4/9) = 1$.

15. The curve is a parabola with focus at $(3,4)$ and directrix $y = 2$ so the vertex is at $(3,3)$ and $p = 1$; $(x-3)^2 = 4(y-3)$.

16. center $(-1,2)$, $a = 2$, asymptotes $y = \pm(b/a)x$ where $(b/a)(-b/a) = -1$ because the asymptotes are perpendicular so $-b^2/a^2 = -1$, $b^2 = a^2 = 4$; $(x+1)^2/4 - (y-2)^2/4 = 1$.

17.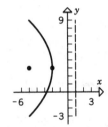

18.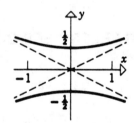

19.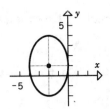

20.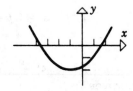

21. $\cot 2\theta = (3-3)/(-2) = 0$, $\theta = 45°$; use $x = (\sqrt{2}/2)(x'-y')$, $y = (\sqrt{2}/2)(x'+y')$ to get
$x'^2/2 + y'^2/1 = 1$; ellipse.

22. $\cot 2\theta = (7-1)/(-8) = -3/4$ so $\cos 2\theta = -3/5$, $\sin\theta = \sqrt{(1+3/5)/2} = 2/\sqrt{5}$,
$\cos\theta = \sqrt{(1-3/5)/2} = 1/\sqrt{5}$, $\theta = \tan^{-1} 2$; use $x = (1/\sqrt{5})(x'-2y')$, $y = (1/\sqrt{5})(2x'+y')$
to get $y'^2/1 - x'^2/9 = 1$; hyperbola.

23. $\cot 2\theta = (11\quad 1)/(10\sqrt{3}) = 1/\sqrt{3}$, $\theta - 30°$; use $x = (1/2)(\sqrt{3}x'-y')$, $y = (1/2)(x'+\sqrt{3}y')$ to
get $x'^2/(1/4) - y'^2/1 = 1$; hyperbola.

24. $\cot 2\theta = (1-4)/4 = -3/4$, $\sin\theta = 2/\sqrt{5}$, $\cos\theta = 1/\sqrt{5}$, $\theta = \tan^{-1} 2$; use $x = (1/\sqrt{5})(x'-2y')$,
$y = (1/\sqrt{5})(2x'+y')$ to get $y' = -x'^2$; parabola.

25. $\cot 2\theta = (16-9)/(-24) = -7/24$, $\cos 2\theta = -7/25$, $\sin\theta = \sqrt{(1+7/25)/2} = 4/5$,
$\cos\theta = \sqrt{(1-7/25)/2} = 3/5$, $\theta = \tan^{-1}(4/3)$; use $x = (1/5)(3x'-4y')$, $y = (1/5)(4x'+3y')$
to get $y'^2 = 4(x'-1)$; parabola.

26. $\cot 2\theta = (73-52)/(-72) = -7/24$, $\sin\theta = 4/5$, $\cos\theta = 3/5$, $\theta = \tan^{-1}(4/3)$; use
$x = (1/5)(3x'-4y')$, $y = (1/5)(4x'+3y')$ to get $x'^2/4 + y'^2/1 = 1$; ellipse

27.

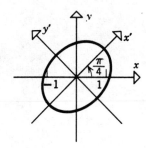

28.

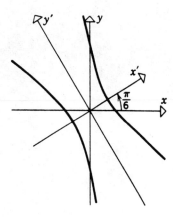

29.

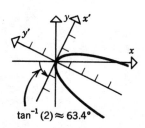

$\tan^{-1}(2) \approx 63.4°$

30.   **(a)** $\cot 2\theta = 0$, $\theta = 45°$

    (b) $\theta = \tan^{-1}(3/4)$, $\sin\theta = 3/5$, $\cos\theta = 4/5$; use $x = (1/5)(4x' - 3y')$, $y = (1/5)(3x' + 4y')$ to

    get $\dfrac{12}{25}x'^2 + \dfrac{7}{25}x'y' - \dfrac{12}{25}y'^2 + \dfrac{1}{5}x' - \dfrac{7}{5}y' = 1$; $A = C = 0$, $B = 1$, $A' = 12/25$, $B' = 7/25$,

    $C' = -12/25$ so $A' + C' = 0 = A + C$, $B'^2 - 4A'C' = 1 = B^2 - 4AC$.

CHAPTER 13

Polar Coordinates and Parametric Equations

SECTION 13.1

13.1.1   Identify the curve given by $r = 4\sin\theta - 6\cos\theta$ by transforming to rectangular coordinates.

13.1.2   Identify the curve given by $r = 3\cos\theta - \sin\theta$ by transforming to rectangular coordinates.

13.1.3   Identify the curve given by $r = \dfrac{1}{\cos\theta - 1}$ by transforming to rectangular coordinates.

13.1.4   Identify the curve given by $r = \dfrac{10}{2 + \cos\theta}$ by transforming to rectangular coordinates.

13.1.5   Identify the curve given by $r = \dfrac{1}{1 - \cos\theta}$ by transforming to rectangular coordinates.

13.1.6   Find the rectangular coordinates of the point whose polar coordinates are $(4, 2\pi/3)$.

13.1.7   Find three other pairs of polar coordinates for the point $(4, 2\pi/3)$ for $-2\pi < \theta < 2\pi$.

13.1.8   Find three other pairs of polar coordinates for the point $\left(-4, \dfrac{\pi}{6}\right)$ for $-2\pi < \theta < 2\pi$.

13.1.9   Express the equation $x^2 + y^2 - x + 3y - 3 = 0$ in polar form.

13.1.10   Express the equation $(x + y)^2 = x - y$ in polar form.

13.1.11   Express the equation $x^2 + y^2 - 6x = 0$ in polar form.

13.1.12   Express the polar equation $r = 4\sec\theta\tan\theta$ in rectangular form and identify.

13.1.13   Express the polar equation $r = 4\csc\theta\cot\theta$ in rectangular form and identify.

13.1.14   Express the equation $y^3 + x^2 y = x$ in polar form.

13.1.15   Express the $y^6 + x^2 y^4 = x^4$ in polar form.

13.1.16   Express $\left(x^2 + y^2\right)^2 = 4xy$ in polar form.

13.1.17   Express $x\left(x^2 + y^2\right) = 2\left(3x^2 - y^2\right)$ in polar form.

13.1.18   Express $\left(x^2 + y^2\right)^{3/2} = x^2 - y^2 - 2xy$ in polar form.

SOLUTIONS

SECTION 13.1

13.1.1 $r^2 = 4r\sin\theta - 6r\cos\theta$; $x^2 + y^2 = 4y - 6x$; $(x+3)^2 + (y-2)^2 = 13$, circle.

13.1.2 $r^2 = 3r\cos\theta - r\sin\theta$; $x^2 + y^2 = 3x - y$; $(x-3/2)^2 + (y+1/2)^2 = \dfrac{5}{2}$, circle.

13.1.3 $r\cos\theta - r = 1$; $r = r\cos\theta - 1 = x - 1$; $r^2 = (x-1)^2$; $x^2 + y^2 = x^2 - 2x + 1$;
$y^2 = -2(x - 1/2)$; parabola.

13.1.4 $2r + r\cos\theta = 10$; $r = \dfrac{10 - r\cos\theta}{2} = 5 - \dfrac{x}{2}$; $r^2 = \left(5 - \dfrac{x}{2}\right)^2$;
$x^2 + y^2 = 25 - 5x + \dfrac{x^2}{4}$; $\dfrac{(x + 10/3)^2}{400/9} + \dfrac{y^2}{400/12} = 1$, ellipse.

13.1.5 $r - r\cos\theta = 1$; $r = r\cos\theta + 1 = x + 1$; $r^2 = (x+1)^2$; $x^2 + y^2 = x^2 + 2x + 1$;
$y^2 = 2(x + 1/2)$; parabola.

13.1.6 $\left(-2, 2\sqrt{3}\right)$. $(-4, 5\pi/3)$, $\left(-4, -\dfrac{\pi}{3}\right)$, $\left(4, -\dfrac{4\pi}{3}\right)$.

13.1.8 $(4, 7\pi/6)$, $(4, -5\pi/6)$, $\left(-4, -\dfrac{11\pi}{6}\right)$.

13.1.9 $r^2 - r\cos\theta + 3r\sin\theta - 3 = 0$; $r^2 + r(3\sin\theta - \cos\theta) - 3 = 0$.

13.1.10 $(r\cos\theta + r\sin\theta)^2 = r\cos\theta - r\sin\theta$
$r = \dfrac{\cos\theta - \sin\theta}{1 + 2\sin\theta\cos\theta}$.

13.1.11 $r^2 - 6r\cos\theta = 0$; $r = 6\cos\theta$.

13.1.12 $r = 4\sec\theta\tan\theta = 4\left(\dfrac{1}{\cos\theta}\right)\left(\dfrac{\sin\theta}{\cos\theta}\right)$, $r\cos^2\theta = 4\sin\theta$, $r^2\cos^2\theta = 4r\sin\theta$,
thus, $x^2 = 4y$

13.1.13 $r = 4\csc\theta\cot\theta = 4\left(\dfrac{1}{\sin\theta}\right)\left(\dfrac{\cos\theta}{\sin\theta}\right)$, $r\sin^2\theta = 4\cos\theta$, $r^2\sin^2\theta = 4r\cos\theta$,
thus, $y^2 = 4x$

13.1.14 $r^3\sin^3\theta + \left(r^2\cos^2\theta\right)(r\sin\theta) = r\cos\theta$; $r^2\sin\theta\left(\sin^2\theta + \cos^2\theta\right) = \cos\theta$; $r^2 = \cot\theta$.

13.1.15 $r^6 \sin^6 \theta + \left(r^2 \cos^2 \theta\right)\left(r^4 \sin^4 \theta\right) = r^4 \cos^4 \theta;\ r^2 \sin^4 \theta \left(\sin^2 \theta + \cos^2 \theta\right) = \cos^4 \theta;\ r = \cot^2 \theta.$

13.1.16 $\left(r^2\right)^2 = 4\left(r \cos \theta\right)\left(r \sin \theta\right);\ r^4 = 4r^2 \cos \theta \sin \theta;\ r^2 = 2 \sin 2\theta.$

13.1.17 $r \cos \theta \left(r^2\right) = 2\left(3r^2 \cos^2 \theta - r^2 \sin^2 \theta\right);\ r \cos \theta = 2\left(3 \cos^2 \theta - \sin^2 \theta\right);$
$r = 2(3 \cos \theta - \sin \theta \tan \theta).$

13.1.18 $\left(r^2\right)^{3/2} = r^2 \cos^2 \theta - r^2 \sin^2 \theta - 2(r \cos \theta)(r \sin \theta);\ r = \cos 2\theta - \sin 2\theta.$

SECTION 13.2

13.2.1 Sketch and identify the graph of the polar curve $r^2 = 9 \sin 2\theta$.

13.2.2 Sketch and identify the graph of the polar curve $r = 1 + \cos \theta$.

13.2.3 Sketch and identify the graph of the polar curve $r = -2 \cos \theta$.

13.2.4 Sketch the graph of the polar curve $r = \sin 3\theta$.

13.2.5 Sketch and identify the graph of the polar curve $r = 4 + 4 \cos \theta$.

13.2.6 Sketch and identify the graph of the polar curve $r = \sqrt{3}$.

13.2.7 Sketch and identify the graph of the polar curve $r^2 = 4 \cos 2\theta$.

13.2.8 Sketch and identify the graph of the polar curve $r = 2 - 4 \sin \theta$.

13.2.9 Sketch the graph of the polar curve $r = -\cos 3\theta$.

13.2.10 Sketch and identify the graph of the polar curve $r = 2 \sin 2\theta$.

13.2.11 Sketch and identify the graph of $r = 2 + 4 \sin \theta$.

13.2.12 Sketch and identify the graph of $r = 4 + 2 \sin \theta$.

13.2.13 Sketch and identify the graph of $r = 3 \sin \theta$.

13.2.14 Sketch and identify the graph of $r = 1 - 2 \cos \theta$.

13.2.15 Sketch and identify the graph of $r = 2 + 4 \cos \theta$.

13.2.16 Sketch and identify the graph of $r = 3 + 2 \cos \theta$.

13.2.17 Sketch and identify the graph of $r = 4(1 - \cos \theta)$.

13.2.18 Sketch and identify the graph of $r = 4(1 - \sin \theta)$.

SOLUTIONS

SECTION 13.2

13.2.1

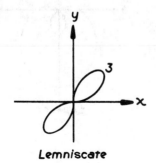

Lemniscate

13.2.2

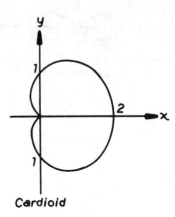

Cardioid

13.2.3

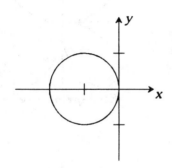

13.2.4

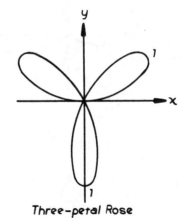

Three-petal Rose

13.2.5

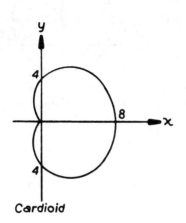

Cardioid

13.2.6

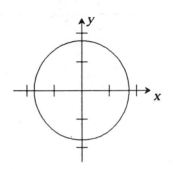

13.2.7

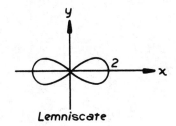

Lemniscate

13.2.8

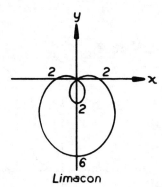

Limacon

13.2.9

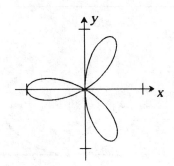

13.2.10

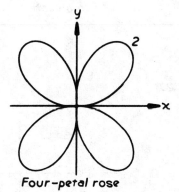

Four-petal rose

13.2.11

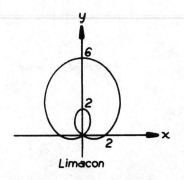

Limacon

13.2.12

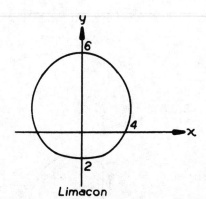

Limacon

13.2.13

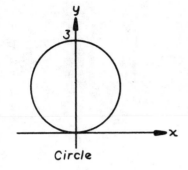

Circle

13.2.14

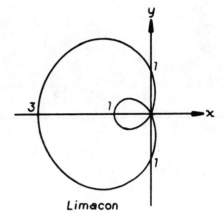

Limacon

13.2.15

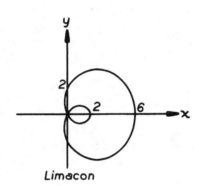

Limacon

13.2.16

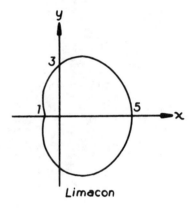

Limacon

13.2.17

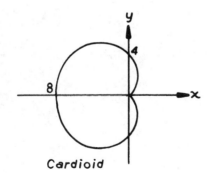

Cardioid

13.2.18

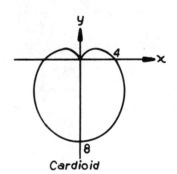

Cardioid

SECTION 13.3

13.3.1   Find the area of the region enclosed by $r = 4\sin 3\theta$.

13.3.2   Find the area of the region enclosed by $r = 2 + \sin\theta$.

13.3.3   Find the area of the region inside $r = 5\sin\theta$ and outside $r = 2 + \sin\theta$.

13.3.4   Find the area of the region that is common to $r = a(1 + \sin\theta)$ and $r = a(1 - \sin\theta)$.

13.3.5   Find the area of the region that is common to $r = 3\cos\theta$ and $r = 1 + \cos\theta$.

13.3.6   Find the area of the region that is common to $r = 1 + \sin\theta$ and $r = 1$.

13.3.7   Find the area of the region that is inside $r = 1$ and outside $r = 1 - \cos\theta$.

13.3.8   Find the area of the region enclosed by $r = 2 + \cos\theta$.

13.3.9   Find the area of the region that is inside $r = 2\cos\theta$ and outside $r = \sin\theta$.

13.3.10   Find the area of the region enclosed by $r = 1 - \sin\theta$.

13.3.11   Find the area of the region that is common to $r = 3a\cos\theta$ and $r = a(1 + \cos\theta)$.

13.3.12   Find the area of the region that is inside $r = 2$ and outside $r = 1 + \cos\theta$.

13.3.13   Find the area of the region enclosed by $r = 2\cos 3\theta$.

13.3.14   Find the area of the region that is inside $r = 3(1 + \sin\theta)$, and outside $r = 3\sin\theta$.

13.3.15   Find the area of the region that is common to $r = a\cos 3\theta$ and $r = a/2$.

13.3.16   Find the area of the region enclosed by $r^2 = \cos 2\theta$.

13.3.17   Find the area of the region enclosed by the inner loop of $r = 1 - 2\sin\theta$.

13.3.18   Find the area of the region enclosed by $r = \cos 2\theta$.

13.3.19   Find the area of the region enclosed by $r = \theta$ from $\theta = 0$ to $\theta = \dfrac{3\pi}{2}$.

SOLUTIONS

SECTION 13.3

13.3.1 $A = 6 \displaystyle\int_0^{\pi/6} \frac{1}{2} \left(16 \sin^2 3\theta\right) d\theta = 4\pi.$

13.3.2 $A = 2 \displaystyle\int_0^{\pi} \frac{1}{2} \left(2 + \sin\theta\right)^2 d\theta = \frac{9\pi}{2} + 8.$

13.3.3 $A = 2 \displaystyle\int_{\frac{\pi}{6}}^{\frac{\pi}{2}} \frac{1}{2} [(5\sin\theta)^2 - (2 + \sin\theta)^2] d\theta = \frac{8\pi}{3} + \sqrt{3}.$

13.3.4 $A = 2 \displaystyle\int_{-\pi/2}^{0} \frac{1}{2} (a(1 + \sin\theta))^2 d\theta + 2 \int_0^{\frac{\pi}{2}} (a(1 - \sin\theta))^2 d\theta = \frac{a^2}{2}[3\pi - 8].$

13.3.5 $A = 2 \displaystyle\int_0^{\pi/3} \frac{1}{2} (1 + \cos\theta)^2 d\theta + 2 \int_{\pi/3}^{\pi/2} \frac{1}{2} (3\cos\theta)^2 d\theta = \frac{5\pi}{4}.$

13.3.6 $A = 2 \displaystyle\int_{-\frac{\pi}{2}}^{0} \frac{1}{2} (1 + \sin\theta)^2 d\theta + 2 \int_0^{\pi/2} \frac{1}{2} (1)^2 d\theta = \frac{5\pi}{4} - 2.$

13.3.7 $A = 2 \displaystyle\int_0^{\pi/2} \frac{1}{2} \left[(1)^2 - (1 - \cos\theta)^2\right] d\theta = 2 - \frac{\pi}{4}.$

13.3.8 $A = \displaystyle\int_0^{2\pi} \frac{1}{2} (2 + \cos\theta)^2 d\theta = \frac{9\pi}{4}.$

13.3.9 $A = \displaystyle\int_{-\frac{\pi}{2}}^{0} \frac{1}{2} (2\cos\theta)^2 d\theta + \int_0^{\tan^{-1} 2} \frac{1}{2} \left[(2\cos\theta)^2 - (\sin\theta)^2\right] d\theta$

$\qquad = \dfrac{\pi}{2} + \dfrac{3}{4} \tan^{-1} 2 + \dfrac{1}{2}.$

13.3.10 $A = \displaystyle\int_0^{2\pi} \frac{1}{2} (1 - \sin\theta)^2 d\theta = \frac{3\pi}{2}.$

13.3.11 $A = 2 \displaystyle\int_0^{\pi/3} \frac{1}{2} [a(1 + \cos\theta)]^2 d\theta + 2 \int_{\pi/3}^{\pi/2} \frac{1}{2} (3a\cos\theta)^2 d\theta = \frac{5\pi a^2}{4}.$

13.3.12 $A = 2 \displaystyle\int_0^{\pi} \frac{1}{2} \left[2^2 - (1 + \cos\theta)^2 d\theta = \frac{5\pi}{2}.\right.$

13.3.13 $A = 6 \int_0^{\pi/6} \frac{1}{2}(2\cos 3\theta)^2 d\theta = \pi.$

13.3.14 $A = \int_0^{2\pi} \frac{1}{2}([3(1+\sin\theta)]^2 d\theta - \int_0^{\pi} \frac{1}{2}(3\sin\theta)^2 d\theta = \frac{45\pi}{4}.$

13.3.15 $A = 6 \int_0^{\pi/9} \frac{1}{2}\left[(a\cos 3\theta)^2 - \frac{a^2}{2}\right] d\theta + 6 \int_{\pi/9}^{\pi/6} \frac{1}{2}(a\cos 3\theta)^2 d\theta = \frac{\pi}{6}.$

13.3.16 $A = 4 \int_0^{\pi/4} \frac{1}{2}(\cos 2\theta) d\theta = 1.$

13.3.17 $A = \int_{\pi/6}^{5\pi/6} \frac{1}{2}(1 - 2\sin\theta)^2 d\theta = \pi - \frac{3\sqrt{3}}{2}.$

13.3.18 $A = 8 \int_0^{\pi/4} \frac{1}{2}(\cos 2\theta)^2 d\theta = 4 \int_0^{\pi/4} \left(\frac{1+\cos 4\theta}{2}\right)^2 d\theta = \pi/2.$

13.3.19 $A = \int_0^{3\pi/2} \frac{1}{2}\theta^2 d\theta = \frac{9\pi^3}{16}.$

SECTION 13.4

13.4.1   Sketch and identify the curve

$$x = 2 + \sin\theta$$
$$,0 \le \theta \le 2\pi$$
$$y = 3 - \cos\theta$$

by eliminating the parameter θ, and label the direction of increasing θ.

13.4.2   Find the arc length of the curve

$$x = 2 + \sin\theta$$
$$,0 \le \theta \le \frac{\pi}{2}$$
$$y = 3 - \cos\theta.$$

13.4.3   Find the arc length of the curve

$$x = e^t \sin t$$
$$,0 \le \theta \le \pi$$
$$y = e^t \cos t.$$

13.4.4   Find dy/dx at the point where $t = 1$ without eliminating t for

$$x = e^t - 1$$
$$y = 3 + e^{2t}.$$

13.4.5   Find $\dfrac{d^2y}{dx^2}$ at the point where $t = 1$ without eliminating t for

$$x = e^t - 1$$
$$y = 3 + e^{2t}.$$

13.4.6   Sketch and identify the curve

$$x = e^t - 1$$
$$y = 3 + e^{2t}$$

by elimnating the parameter t, and label the direction of increasing t.

13.4.7   Sketch the curve

$$x = 2\cos t$$
$$,0 \le t \le \pi/4$$
$$y = 3\sin t$$

by eliminating the parameter t and label the direction of increasing t.

13.4.8   Sketch the curve

$$x = 3 + \cosh t$$
$$y = 2 + \sinh t$$

by eliminating the parameter t, and label the direction of increasing t.

13.4.9   Find dy/dx at the point where $t = \pi/4$ without eliminating t for

$$x = 5 - 2\cos t$$
$$y = 3 + \sin t.$$

13.4.10   Find d^2y/dx^2 at the point where $t = \dfrac{3\pi}{4}$ without eliminating t for

$$x = 5 - 2\cos t$$
$$y = 3 + \sin t.$$

13.4.11   Sketch and identify the curve

$$x = 3 + \cos t$$
$$,0 \le t \le 2\pi$$
$$y = 3 - 2\sin t$$

by eliminating the parameter t and label the direction of increasing t.

13.4.12   Sketch and identify the curve

$$x = 3\sec t + 4 \qquad \text{for} \qquad -\pi/2 < t < \pi/2$$
$$y = 2\tan t - 3$$

by eliminating the parameter t, and label the direction of increasing t.

13.4.13   Sketch and identify the curve

$$x = \cos 2\theta$$
$$, 0 \le \theta \le 2\pi$$
$$y = \sin \theta$$

by eliminating the parameter θ and label the direction of increasing θ.

13.4.14   Find d^2y/dx^2 at the point where $\theta = \dfrac{\pi}{4}$ without eliminating θ for

$$x = \cos 2\theta$$
$$y = \sin \theta.$$

13.4.15   Sketch the curve

$$x = -1 + 3\cos\theta$$
$$, -\pi \le \theta \le 0$$
$$y = \sin\theta$$

by eliminating the parameter θ and label the direction of increasing θ.

13.4.16   Find $\dfrac{d^2y}{dx^2}$ at the point where $\theta = \dfrac{\pi}{6}$ without eliminating θ for

$$y = \sin\theta$$
$$x = -1 + 3\cos\theta.$$

13.4.17   Find the arc length of the curve

$$x = t^2 + 2$$
$$, \text{from } (2, -3) \text{ to } (3, -2)$$
$$y = t^3 - 3.$$

13.4.18   Find the arc length of the curve

$$x = 3(t-1)^2$$
$$, 0 \le t \le 1$$
$$y = 8t^{3/2}.$$

SOLUTIONS

SECTION 13.4

13.4.1 $\sin\theta = 2 - x$,
$\cos\theta = 3 - y$,
$(x-2)^2 + (y-3)^2 = 1$,
circle.

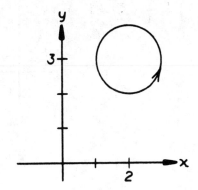

13.4.2 $\left(\dfrac{dx}{d\theta}\right)^2 + \left(\dfrac{dy}{d\theta}\right)^2 = (\cos\theta)^2 + (\sin\theta)^2 = 1;\ L = \displaystyle\int_0^{\pi/2} d\theta = \dfrac{\pi}{2}$.

13.4.3 $\left(\dfrac{dx}{dt}\right)^2 + \left(\dfrac{dy}{dt}\right)^2 = \left[e^t(\cos t + \sin t)\right]^2 + \left[e^t(\cos t - \sin t)\right]^2 = 2e^{2t}$;

$L = \displaystyle\int_0^{\pi} \sqrt{2e^{2t}}\,dt = \int_0^{\pi} \sqrt{2}\,e^t\,dt = \sqrt{2}\,(e^{\pi} - 1)$.

13.4.4 $\dfrac{dy}{dx} = \dfrac{2e^{2t}}{e^t} = 2e^t,\ \left.\dfrac{dy}{dx}\right|_{t=1} = 2e$.

13.4.5 $\dfrac{dy}{dx} = \dfrac{2e^{2t}}{e^t} = 2e^t,\ \dfrac{d^2y}{dx^2} = \dfrac{2e^t}{e^t} = 2,\ \left.\dfrac{d^2y}{dx^2}\right|_{t=1} = 2$.

13.4.6 $e^t = x + 1$,
$e^{2t} = y - 3$,
$(x+1)^2 = (y-3)$,
parabola.

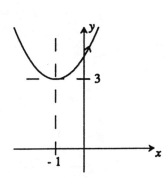

13.4.7 $\cos t = \dfrac{x}{2}$, $\sin t = \dfrac{y}{3}$,

$$\left(\dfrac{x}{2}\right)^2 + \left(\dfrac{y}{3}\right)^2 = \dfrac{x^2}{4} + \dfrac{y^2}{9} = 1.$$

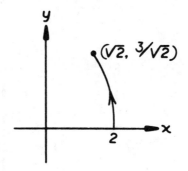

13.4.8 $\cosh t = x - 3$,

$\sinh t = y - 2$,

$(x-3)^2 - (y-2)^2 = 1.$

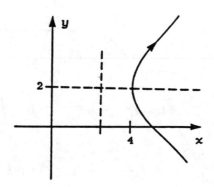

13.4.9 $\dfrac{dy}{dx} = \dfrac{\cos t}{2 \sin t} = \dfrac{1}{2} \cot t$, $\dfrac{dy}{dx}\bigg]_{t=\pi/4} = \dfrac{1}{2}.$

13.4.10 $\dfrac{dy}{dx} = \dfrac{\cos t}{2 \sin t} = \dfrac{1}{2} \cot t$, $\dfrac{d^2y}{dx^2} = \dfrac{-\dfrac{1}{2}\csc^2 t}{2 \sin t} = -\dfrac{1}{4}\csc^3 t$, $\dfrac{d^2y}{dx^2}\bigg]_{t=\pi/4} = -\dfrac{\sqrt{2}}{2}.$

13.4.11 $\cos t = x - 3$, $\sin t = \dfrac{3-y}{2}$

$(x-3)^2 + \dfrac{(y-3)^2}{4} = 1$

ellipse.

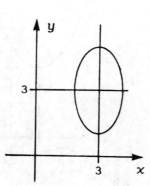

13.4.12 $\tan t = \dfrac{y+3}{2}$

$\sec t = \dfrac{x-4}{3}$

$\left(\dfrac{y+3}{2}\right)^2 + 1 = \left(\dfrac{x-4}{3}\right)^2$ or $\dfrac{(x-4)^2}{9} - \dfrac{(y+3)^2}{4} = 1$, hyperbola

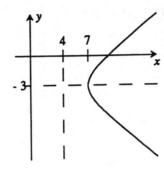

13.4.13 $\cos 2\theta = 1 - 2\sin^2 \theta$,

$x = 1 - 2y^2$, parabola.

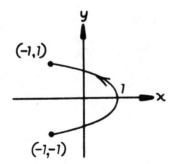

13.4.14 $\dfrac{dy}{dx} = \dfrac{\cos\theta}{-2\sin 2\theta} = \dfrac{\cos\theta}{-4\sin\theta\cos\theta} = -\dfrac{1}{4}\csc\theta,\; \dfrac{d^2y}{dx^2} = \dfrac{1/4\,\csc\theta\cot\theta}{-2\sin 2\theta} = -\dfrac{1}{16}\csc^3\theta$

$\dfrac{d^2y}{dx^2}\bigg]_{\theta=\pi/4} = -\dfrac{\sqrt{2}}{8}.$

13.4.15 $\cos\theta = \dfrac{x+1}{3}$,

$\sin\theta = y$,

$\left(\dfrac{x+1}{3}\right)^2 + y^2 = 1$, or

$\dfrac{(x+1)^2}{9} + \dfrac{y^2}{1} = 1.$

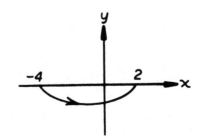

13.4.16 $\dfrac{dy}{dx} = \dfrac{\cos\theta}{-3\sin\theta} = -\dfrac{1}{3}\cot\theta,\ \dfrac{d^2y}{dx^2} = \dfrac{1/3\csc^2\theta}{-3\sin\theta} = -\dfrac{1}{9}\csc^3\theta,\ \dfrac{d^2y}{dx^2}\bigg]_{\theta=\pi/6} = -\dfrac{8}{9}.$

13.4.17 $\left(\dfrac{dx}{dt}\right)^2 + \left(\dfrac{dy}{dt}\right)^2 = (2t)^2 + \left(3t^2\right)^2 = t^2\left(4 + 9t^2\right);\ L = \displaystyle\int_0^1 t\sqrt{4+9t^2}\,dt = \dfrac{1}{27}\left[13\sqrt{13} - 8\right].$

13.4.18 $\left(\dfrac{dx}{dt}\right)^2 + \left(\dfrac{dy}{dt}\right)^2 = 36(t-1)^2 + 144t$

$L = \displaystyle\int_0^1 \sqrt{36t^2 + 72t + 36}\,dt = 9.$

SECTION 13.5

13.5.1   Find the slope of the tangent to the curve $r = 2\sin\theta$ at the point $\theta = \pi/3$

13.5.2   Find the slope of the tangent to the curve $r = \dfrac{4}{\theta}$ at the point $\theta = 4$

13.5.3   Find the slope of the tangent to the curve $r = \cos 5\theta$ at the point $\theta = \pi/5$

13.5.4   Find the slope of the tangent to the curve $r = 3 - 2\sin\theta$ at the point $\theta = \pi$

13.5.5   Find the arclength of the curve $r = e^{4\theta}$ from $\theta = 0$ to $\theta = 4$

13.5.6   Find the arclength of the curve $r = a$ from $\theta = 0$ to $\theta = \pi$

13.5.7   Find the arclength of the curve $r = \sin^2\left(\dfrac{\theta}{2}\right)$ from $\theta = 0$ to $\theta = \dfrac{\pi}{2}$

13.5.8   Find the arclength of the curve $r = 3a\cos\theta$ from $\theta = 0$ to $\theta = \pi$

13.5.9   Find all the points on the limaçon $r = a(1 + \sin\theta)$ where the tangent is horizontal

13.5.10   Find all points on the circle $r = 2\cos\theta$ where the tangent is (a) horizontal, (b) vertical

13.5.11   Find all points on the circle $r = 4\sin\theta$ where the tangent is (a) horizontal, (b) vertical

SOLUTIONS

SECTION 13.5

13.5.1 $r = 2\sin\theta$, $\theta = \pi/3$, $\dfrac{dr}{d\theta} = 2\cos\theta$, so

$$\frac{dy}{dx} = \frac{2\sin\theta(\cos\theta) + \sin\theta(2\cos\theta)}{-2\sin\theta(\sin\theta) + \cos\theta(2\cos\theta)} = \tan 2\theta, \text{ at } \theta = \pi/3,$$

the slope of the tangent line is $m = \dfrac{dy}{dx}\bigg|_{\theta=\pi/3} = \tan\dfrac{2\pi}{3} = -\sqrt{3}$.

13.5.2 $r = \dfrac{4}{\theta}$, $\theta = 4$, $\dfrac{dr}{d\theta} = -\dfrac{4}{\theta^2}$, so

$$\frac{dy}{dx} = \frac{\dfrac{4}{\theta}(\cos\theta) + \sin\theta\left(-\dfrac{4}{\theta^2}\right)}{-\dfrac{4}{\theta}(\sin\theta) + \cos\theta\left(-\dfrac{4}{\theta^2}\right)} = \frac{\tan\theta - \theta}{\theta\tan\theta + 1}. \text{ At } \theta = 4,$$

the slope of the tangent line is $m = \dfrac{dy}{dx}\bigg|_{\theta=4} = \dfrac{\tan 4 - 4}{4\tan 4 + 1} \approx -0.5047$

13.5.3 $r = \cos 5\theta$, $\theta = \pi/5$, $\dfrac{dr}{d\theta} = -5\sin 5\theta$, so

$$\frac{dy}{dx} = \frac{\cos 5\theta(\cos\theta) + \sin\theta(-5\sin 5\theta)}{-\cos 5\theta(\sin\theta) + \cos\theta(-5\sin 5\theta)}; \text{ at } \theta = \pi/5,$$

the slope of the tangent line is $m = \dfrac{dy}{dx}\bigg|_{\theta=\pi/5} = -\cot\dfrac{\pi}{5} \approx -1.3764$

13.5.4 $r = 3 - 2\sin\theta$, $\theta = \pi$, $\dfrac{dr}{d\theta} = -2\cos\theta$, so

$$\frac{dy}{dx} = \frac{(3 - 2\sin\theta)\cos\theta + \sin\theta(-2\cos\theta)}{-(3 - 2\sin\theta)\sin\theta + \cos\theta(-2\cos\theta)}. \text{ At } \theta = \pi,$$

the slope of the tangent line is $m = \dfrac{dy}{dx}\bigg|_{\theta=\pi} = \dfrac{3}{2}$.

13.5.5 $r^2 + (dr/d\theta)^2 = e^{8\theta} + 16e^{8\theta} = 17e^{8\theta}$

$$L = \int_0^4 \sqrt{17}e^{4\theta}d\theta = \frac{\sqrt{17}}{4}e^{4\theta}\bigg|_0^4 = \frac{\sqrt{17}}{4}\left(e^{16} - 1\right)$$

13.5.6 $r^2 + \left(\dfrac{dr}{d\theta}\right)^2 = a^2$

$$L = \int_0^\pi a\,d\theta = a\theta\bigg|_0^\pi = a\pi$$

13.5.7 $r^2 + (dr/d\theta)^2 = \sin^4\left(\dfrac{\theta}{2}\right) + \sin^2\left(\dfrac{\theta}{2}\right)\cos^2\left(\dfrac{\theta}{2}\right) = \sin^2\left(\dfrac{\theta}{2}\right)$

$L = \displaystyle\int_0^{\frac{\pi}{2}} \sin\left(\dfrac{\theta}{2}\right) d\theta = -2\cos\left(\dfrac{\theta}{2}\right)\Big|_0^{\frac{\pi}{2}} = -2\left(\dfrac{\sqrt{2}}{2} - 1\right) = 2 - \sqrt{2}$

13.5.8 $r^2 + (dr/d\theta)^2 = 9a^2\cos^2\theta + 9a^2\sin^2\theta = 9a^2$

$L = \displaystyle\int_0^{\pi} 3a\,d\theta = 3a\theta\Big|_0^{\pi} = 3a\pi$

13.5.9 $\dfrac{dx}{d\theta} = -a(1+\sin\theta)\sin\theta + (a\cos\theta)\cos\theta = a(\cos 2\theta - 1)$

$\dfrac{dy}{d\theta} = a(1+\sin\theta)\cos\theta + (a\cos\theta)\sin\theta = a\cos\theta(1+2\sin\theta).$

There is a horizontal tangent where $\dfrac{dy}{d\theta} = 0$ and $\dfrac{dx}{d\theta} \neq 0$ or when $\theta = \dfrac{\pi}{2}, \dfrac{3\pi}{2}, \dfrac{7\pi}{6},$ and $\dfrac{11\pi}{6}$.

13.5.10 $\dfrac{dx}{d\theta} = -2\cos\theta(\sin\theta) + (-2\sin\theta)\cos\theta = -2\sin 2\theta;$

$\dfrac{dy}{d\theta} = 2\cos\theta(\cos\theta) + (-2\sin\theta)\sin\theta = 2\cos 2\theta.$

(a) There is a horizontal tangent where $\dfrac{dy}{d\theta} = 0$ and $\dfrac{dx}{d\theta} \neq 0$ or when $\theta = \pi/4, 3\pi/4$.

(b) There is a vertical tangent where $\dfrac{dx}{d\theta} \neq$ and $\dfrac{dy}{d\theta} \neq 0$ or when $\theta = 0, \pi/2$.

13.5.11 $\dfrac{dx}{d\theta} = -4\sin\theta(\sin\theta) + 4\cos\theta(\cos\theta) > 4\cos 2\theta; \quad \dfrac{dy}{d\theta} = 4\sin\theta(\cos\theta) + 4\cos\theta(\sin\theta) = 4\sin 2\theta.$

(a) There is a horizontal tangent where $\dfrac{dy}{d\theta} = 0$ and $\dfrac{dx}{d\theta} \neq 0$ or when $\theta = 0, \pi/2$.

(b) There is a vertical tangent where $\dfrac{dx}{d\theta} = 0$ and $\dfrac{dy}{d\theta} \neq 0$ or when $\theta = \dfrac{\pi}{4}, \dfrac{3\pi}{4}$.

SUPPLEMENTARY EXERCISES, CHAPTER 13

1. Find the rectangular coordinates of the points with the given polar coordinates.

   (a) $(-2, 4\pi/3)$

   (b) $(2, -\pi/2)$

   (c) $(0, -\pi)$

   (d) $(-\sqrt{2}, -\pi/4)$

   (e) $(3, \pi)$

   (f) $(1, \tan^{-1}(-\frac{4}{3}))$.

2. In parts (a)–(c), points are given in rectangular coordinates. Express them in polar coordinates in three ways:

   (i) With $r \geq 0$ and $0 \leq \theta < 2\pi$

   (ii) With $r \geq 0$ and $-\pi < \theta \leq \pi$

   (iii) With $r \leq 0$ and $0 \leq \theta < 2\pi$.

   (a) $(-\sqrt{3}, -1)$

   (b) $(-3, 0)$

   (c) $(1, -1)$.

3. Sketch the region in polar coordinates determined by the given inequalities.

   (a) $1 \leq r \leq 2$, $\cos\theta \leq 0$

   (b) $-1 \leq r \leq 1$, $\pi/4 \leq \theta \leq \pi/2$.

In Exercises 4–11, identify the curve by transforming to rectangular coordinates.

4. $r = 2/(1 - \cos\theta)$.

5. $r^2 \sin(2\theta) = 1$.

6. $r = \pi/2$.

7. $r = -4\csc\theta$.

8. $r = 6/(3 - \sin\theta)$.

9. $\theta = \pi/3$.

10. $r = 2\sin\theta + 3\cos\theta$.

11. $r = 0$.

In Exercises 12–15, express the given equation in polar coordinates.

12. $x^2 + y^2 = kx$.

13. $x = -3$.

14. $y^2 = 4x$.

15. $y = 3x$.

In Exercises 16–23, sketch the curve in polar coordinates.

16. $r = -4\sin 3\theta$.

17. $r = -1 - 2\cos\theta$.

18. $r = 5\cos\theta$.

19. $r = 4 - \sin\theta$.

20. $r = 3(\cos\theta - 1)$.

21. $r = \theta/\pi$ $(\theta \geq 0)$.

22. $r = \sqrt{2}\,\cos(\theta/2)$.

23. $r = e^{-\theta/\pi}$ $(\theta \geq 0)$.

In Exercises 24–26, sketch the curves in the same polar coordinate system, and find all points of intersection.

24. $r = 3\cos\theta$, $r = 1 + \cos\theta$.   **25.** $r = a\cos(2\theta)$, $r = a/2$ $(a > 0)$.

26. $r = 2\sin\theta$, $r = 2 + 2\cos\theta$.

In Exercises 27–29, set up, but *do not evaluate*, definite integrals for the stated area and arc length.

27. **(a)** The area inside both the circle and cardioid in Exercise 24.

(b) The arc length of that part of the cardioid outside the circle in Exercise 24.

28. **(a)** The area inside the rose and outside the circle in Exercise 25.

(b) The arc length of that part of the rose lying inside the circle in Exercise 25.

29. **(a)** The area inside the circle and outside the cardioid in Exercise 26.

(b) The arc length of that portion of the circle lying inside the cardioid in Exercise 26.

In Exercises 30–33, find the area of the region described.

30. One petal of the rose $r = a\sin 3\theta$.

31. The region outside the circle $r = a$ and inside the lemniscate $r^2 = 2a^2\cos 2\theta$.

32. The region in part (a) of Exercise 28.   **33.** The region in part (a) of Exercise 29.

In Exercises 34–37:

(a) Sketch the curve and indicate the direction of increasing parameter.

(b) Use the parametric equations to find $dy/dx, d^2y/dx^2$, and the equation of the tangent line at the point on the curve corresponding to the parameter value t_0 (or θ_0).

34. $x = 3 - t^2$, $y = 2 + t$, $0 \le t \le 3$; $t_0 = 1$.

35. $x = 1 + 3\cos\theta$, $y = -1 + 2\sin\theta$, $0 \le \theta \le \pi$; $\theta_0 = \pi/2$.

36. $x = 2\tan\theta$, $y = \sec\theta$, $-\pi/2 < \theta < \pi/2$; $\theta_0 = \pi/3$.

37. $x = 1/t$, $y = \ln t$, $1 \le t \le e$; $t_0 = 2$.

In Exercises 38–43, find the arc length of the curve described.

38. $x = 2t^3$, $y = 3t^2$, $-4 \le t \le 4$.

39. $x = \ln \cos 2t$, $y = 2t$, $0 \le t \le \pi/6$.

40. $x = 3\cos t - 1$, $y = 3\sin t + 4$, $0 \le t \le \pi$.

41. $x = 3t^2$, $y = t^3 - 3t$, $0 \le t \le 1$.

42. $x = 1 - \cos t$, $y = t - \sin t$, $-\pi \le t \le \pi$.

43. $r = e^\theta$, $0 \le \theta \le 2\pi$.

In Exercises 44–46, $x(t)$ and $y(t)$ describe the motion of a particle. Find the coordinates of the particle when the instantaneous direction of motion is (a) horizontal and (b) vertical.

44. $x = -2t^2$, $y = t^3 - 3t + 5$.

45. $x = 1 - 2\sin t$, $y = t + 2\cos t$, $0 \le t \le \pi$.

46. $x = \ln t$, $y = t^2 - 4t$, $t > 0$.

47. At what instant does the trajectory described in Exercise 46 have a point of inflection?

48. Find a set of parametric equations for

(a) the line $y = 2x + 3$

(b) the ellipse $4(x - 2)^2 + y^2 = 4$.

SOLUTIONS

SUPPLEMENTARY EXERCISES CHAPTER 13

1. **(a)** $(1, \sqrt{3})$ **(b)** $(0, -2)$ **(c)** $(0,0)$
   (d) $(-1, 1)$ **(e)** $(-3, 0)$ **(f)** $(3/5, -4/5)$

2. **(a)** **(i)** $(2, 7\pi/6)$ **(ii)** $(2, -5\pi/6)$ **(iii)** $(-2, \pi/6)$
   (b) **(i)** $(3, \pi)$ **(ii)** $(3, \pi)$ **(iii)** $(-3, 0)$
   (c) **(i)** $(\sqrt{2}, 7\pi/4)$ **(ii)** $(\sqrt{2}, -\pi/4)$ **(iii)** $(-\sqrt{2}, 3\pi/4)$

3. **(a)** 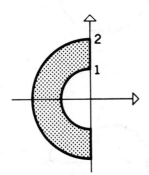   **(b)**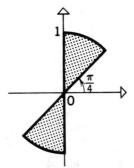

4. $r = 2/(1 - \cos\theta)$, $r - r\cos\theta = 2$, $r - x = 2$, $r = x + 2$, $r^2 = (x+2)^2$, $x^2 + y^2 = x^2 + 4x + 4$, $y^2 = 4x + 4$; parabola.

5. $r^2 \sin 2\theta = 1$, $r^2(2\sin\theta\cos\theta) = 1$, $2(r\sin\theta)(r\cos\theta) = 1$, $2yx = 1$; hyperbola.

6. $r = \pi/2$, $r^2 = \pi^2/4$, $x^2 + y^2 = \pi^2/4$; circle.

7. $r = -4\csc\theta$, $r = -4/\sin\theta$, $r\sin\theta = -4$, $y = -4$, line.

8. $r = 6/(3 - \sin\theta)$, $3r - r\sin\theta = 6$, $3r - y = 6$, $3r = y+6$, $9r^2 = (y+6)^2$, $9(x^2+y^2) = y^2 + 12y + 36$, $9x^2 + 8y^2 - 12y = 36$; ellipse.

9. $\theta = \pi/3$, $\tan\theta = \sqrt{3}$, $y = \sqrt{3}x$; line.

10. $r = 2\sin\theta + 3\cos\theta$, $r^2 = 2r\sin\theta + 3r\cos\theta$, $x^2 + y^2 = 2y + 3x$; circle.

11. $r = 0$, $x = 0$ and $y = 0$; point.   12. $x^2 + y^2 = kx$, $r^2 = kr\cos\theta$, $r = k\cos\theta$

13. $x = -3$, $r\cos\theta = -3$

14. $y^2 = 4x$, $(r\sin\theta)^2 = 4r\cos\theta$, $r\sin^2\theta = 4\cos\theta$, $r = 4\csc\theta\cot\theta$

15. $y = 3x$, $\tan\theta = 3$, $\theta = \tan^{-1}3$

16.

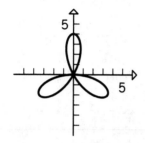

17.

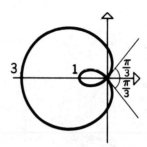

18.

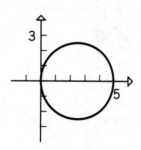

19.

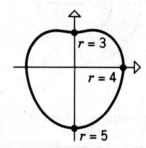

20.

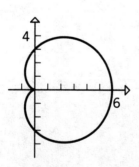

21.

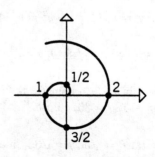

22.

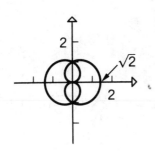

23.

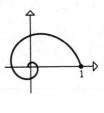

24. $3\cos\theta = 1 + \cos\theta$,
$\cos\theta = 1/2$, $\theta = \pm\pi/3$.
The curves intersect at
$(3/2, \pi/3)$, $(3/2, -\pi/3)$,
and also at the origin
(see sketch).

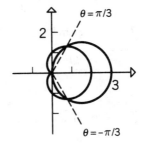

25. $a\cos 2\theta = a/2$, $\cos 2\theta = 1/2$;
one solution is $2\theta = \pi/3$, $\theta = \pi/6$
and from the symmetry of the graphs
the others are $\theta = -\pi/6$, $\pm\pi/3$, $\pm 2\pi/3$,
$\pm 5\pi/6$. The points of intersection are
$(a/2, \pm\pi/6)$, $(a/2, \pm\pi/3)$, $(a/2, \pm 2\pi/3)$,
$(a/2, \pm 5\pi/6)$.

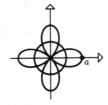

26. By inspection of the graphs, the
curves intersect at $(2, \pi/2)$ and
the origin.

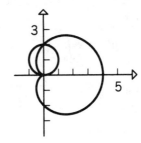

27. **(a)** $A = 2\left[\dfrac{1}{2}\displaystyle\int_0^{\pi/3}(1+\cos\theta)^2 d\theta + \dfrac{1}{2}\int_{\pi/3}^{\pi/2}(3\cos\theta)^2 d\theta\right]$

$\qquad = \displaystyle\int_0^{\pi/3}(1+\cos\theta)^2 d\theta + \int_{\pi/3}^{\pi/2} 9\cos^2\theta\, d\theta$

(b) $r^2 + (dr/d\theta)^2 = (1+\cos\theta)^2 + (-\sin\theta)^2 = 2(1+\cos\theta)$, $L = 2\displaystyle\int_{\pi/3}^{\pi}\sqrt{2(1+\cos\theta)}\,d\theta$

28. **(a)** $A = 8\displaystyle\int_0^{\pi/6}\dfrac{1}{2}[(a\cos2\theta)^2 - (a/2)^2]d\theta = 4a^2\int_0^{\pi/6}(\cos^2 2\theta - 1/4)d\theta$

(b) $r^2 + (dr/d\theta)^2 = (a\cos2\theta)^2 + (-2a\sin2\theta)^2 = a^2(\cos^2 2\theta + 4\sin^2 2\theta)$,

$\qquad L = 8\displaystyle\int_{\pi/6}^{\pi/4} a\sqrt{\cos^2 2\theta + 4\sin^2 2\theta}\,d\theta$

29. **(a)** $A = \displaystyle\int_{\pi/2}^{\pi}\dfrac{1}{2}[(2\sin\theta)^2 - (2+2\cos\theta)^2]d\theta = 2\int_{\pi/2}^{\pi}[\sin^2\theta - (1+\cos\theta)^2]d\theta$

(b) $r^2 + (dr/d\theta)^2 = (2\sin\theta)^2 + (2\cos\theta)^2 = 4$, $L = \displaystyle\int_0^{\pi/2} 2 d\theta$

30. $A = \displaystyle\int_0^{\pi/3}\dfrac{1}{2}(a\sin3\theta)^2 d\theta = \dfrac{a^2}{2}\int_0^{\pi/3}\sin^2 3\theta\, d\theta = \pi a^2/12$

31. $A - 4\displaystyle\int_0^{\pi/6}\dfrac{1}{2}(2a^2\cos2\theta - a^2)d\theta = 2a^2\int_0^{\pi/6}(2\cos2\theta - 1)d\theta = a^2(\sqrt{3} - \pi/3)$

32. $A = 4a^2\displaystyle\int_0^{\pi/6}(\cos^2 2\theta - 1/4)d\theta = a^2(\pi/6 + \sqrt{3}/4)$

33. $A = 2\displaystyle\int_{\pi/2}^{\pi}[\sin^2\theta - (1+\cos\theta)^2]d\theta = 2\int_{\pi/2}^{\pi}(-1 - 2\cos\theta - \cos2\theta)d\theta = 4 - \pi$

34. **(a)** Eliminate the parameter to get
$\qquad x = 3 - (y-2)^2$, $(y-2)^2 = -(x-3)$
$\qquad$ for $-6 \le x \le 3$ and $2 \le y \le 5$.

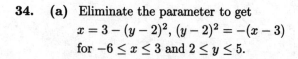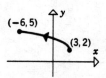

(b) $dy/dx = \dfrac{1}{-2t} = -\dfrac{1}{2}t^{-1}$, $d^2y/dx^2 = \dfrac{(1/2)t^{-2}}{-2t} = -\dfrac{1}{4}t^{-3}$; at $t_0 = 1$, $dy/dx = -1/2$ and
$d^2y/dx^2 = -1/4$, $x = 2$, $y = 3$ so the tangent line is $y - 3 = (-1/2)(x-2)$, $y = -x/2 + 4$.

35. **(a)** eliminate the parameter to get
$(x-1)^2/9 + (y+1)^2/4 = 1$ for
$-2 \le x \le 4$ and $-1 \le y \le 1$.

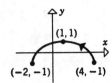

(b) $dy/dx = \dfrac{2\cos\theta}{-3\sin\theta} = -\dfrac{2}{3}\cot\theta$, $d^2y/dx^2 = \dfrac{(2/3)\csc^2\theta}{-3\sin\theta} = -\dfrac{2}{9}\csc^3\theta$; at $\theta_0 = \pi/2$,
$dy/dx = 0$ and $d^2y/dx^2 = -2/9$, $x=1$, $y=1$ so the tangent line is $y=1$.

36. **(a)** Eliminate the parameter to get
$y^2 - x^2/4 = 1$ for $-\infty < x < +\infty$
and $y \ge 1$.

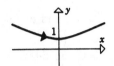

(b) $dy/dx = \dfrac{\sec\theta\tan\theta}{2\sec^2\theta} = \dfrac{1}{2}\sin\theta$, $d^2y/dx^2 = \dfrac{(1/2)\cos\theta}{2\sec^2\theta} = \dfrac{1}{4}\cos^3\theta$; at $\theta_0 = \pi/3$,
$dy/dx = \sqrt{3}/4$ and $d^2y/dx^2 = 1/32$, $x = 2\sqrt{3}$, $y=2$ so the tangent line is
$y - 2 = (\sqrt{3}/4)(x - 2\sqrt{3})$, $y = \sqrt{3}x/4 + 1/2$.

37. **(a)** Eliminate the parameter to get
$y = \ln(1/x) = -\ln x$ for
$1 \le x \le e^{-1}$ and $0 \le y \le 1$.

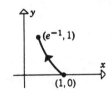

(b) $dy/dx = \dfrac{1/t}{-1/t^2} = -t$, $d^2y/dx^2 = \dfrac{-1}{-1/t^2} = t^2$; at $t_0 = 2$, $dy/dx = -2$ and $d^2y/dx^2 = 4$,
$x = 1/2$, $y = \ln 2$ so the tangent line is $y - \ln 2 = -2(x - 1/2)$, $y = -2x + 1 + \ln 2$.

38. $(dx/dt)^2 + (dy/dt)^2 = (6t^2)^2 + (6t)^2 = 36t^2(t^2 + 1)$,
$$L = \int_{-4}^{4} 6|t|\sqrt{t^2+1}\, dt = 12\int_{0}^{4} t(t^2+1)^{1/2}dt = 4(17^{3/2} - 1)$$

39. $(dx/dt)^2 + (dy/dt)^2 = (-2\tan 2t)^2 + 2^2 = 4\sec^2 2t$, $L = \displaystyle\int_{0}^{\pi/6} 2\sec 2t\, dt = \ln(2 + \sqrt{3})$

40. $(dx/dt)^2 + (dy/dt)^2 = (-3\sin t)^2 + (3\cos t)^2 = 9$, $L = \displaystyle\int_{0}^{\pi} 3dt = 3\pi$

41. $(dx/dt)^2 + (dy/dt)^2 = (6t)^2 + (3t^2 - 3)^2 = (3t^2 + 3)^2$, $L = \int_0^1 (3t^2 + 3)dt = 4$

42. $(dx/dt)^2 + (dy/dt)^2 = (\sin t)^2 + (1 - \cos t)^2 = 4\sin^2(t/2)$,

$L = \int_{-\pi}^{\pi} 2|\sin(t/2)|dt = 4\int_0^\pi \sin(t/2)dt = 8$

43. $r^2 + (dr/d\theta)^2 = (e^\theta)^2 + (e^\theta)^2 = 2e^{2\theta}$, $L = \int_0^{2\pi} \sqrt{2}\, e^\theta d\theta = \sqrt{2}(e^{2\pi} - 1)$

44. $dx/dt = -4t$, $dy/dt = 3t^2 - 3$

(a) horizontal when $dy/dt = 0$ and $dx/dt \neq 0$; $3t^2 - 3 = 0$, $t^2 = 1$, $t = \pm 1$ so (x, y) is $(-2, 3)$ or $(-2, 7)$.

(b) vertical when $dx/dt = 0$ and $dy/dt \neq 0$; $-4t = 0$, $t = 0$ so (x, y) is $(0, 5)$.

45. $dx/dt = -2\cos t$, $dy/dt = 1 - 2\sin t$

(a) horizontal when $dy/dt = 0$ and $dx/dt \neq 0$; $1 - 2\sin t = 0$, $\sin t = 1/2$, $t = \pi/6, 5\pi/6$ so (x, y) is $(0, \pi/6 + \sqrt{3})$ or $(0, 5\pi/6 - \sqrt{3})$.

(b) vertical when $dx/dt = 0$ and $dy/dt \neq 0$; $-2\cos t = 0$, $\cos t = 0$, $t = \pi/2$ so (x, y) is $(-1, \pi/2)$.

46. $dx/dt = 1/t$, $dy/dt = 2t - 4$

(a) horizontal when $dy/dt = 0$ and $dx/dt \neq 0$; $2t - 4 = 0$, $t = 2$ so (x, y) is $(\ln 2, -4)$.

(b) vertical when $dx/dt = 0$ and $dy/dt \neq 0$; $1/t = 0$ has no solution so the instantaneous direction of motion is never vertical.

47. $dy/dx = (2t - 4)/(1/t) = 2t^2 - 4t$, $d^2y/dx^2 = (4t - 4)/(1/t) = 4t(t - 1)$. For $t > 0$, $d^2y/dx^2 = 0$ when $t = 1$ and d^2y/dx^2 changes sign there so an inflection point occurs at $t = 1$.

48. (a) Let $x = t$, then $y = 2t + 3$.

(b) $(x - 2)^2 + (y/2)^2 = 1$, let $x - 2 = \cos t$ and $y/2 = \sin t$ to get $x = 2 + \cos t$, $y = 2\sin t$

CHAPTER 14

Three Dimensional Space; Vectors

SECTION 14.1

14.1.1 Describe the surface whose equation is given by $x^2 + y^2 + z^2 - 8y = 0$.

14.1.2 Find the distance between $P(2, 7, 8)$ and $Q(3, 9, 7)$ and the midpoint of a line segment joining P and Q.

14.1.3 Find the distance between $P(-3, -2, 4)$ and $Q(9, 7, 2)$ and the midpoint of a line segment joining P and Q.

14.1.4 Find the standard equation of the sphere with a diameter whose endpoints are $(1, 2, -3)$ and $(1, -4, 5)$.

14.1.5 Find the standard equation of the sphere with a diameter whose endpoints are $(4, 6, 12)$ and $(-2, 2, 10)$.

14.1.6 Find the equation for the sphere with center $(2, -3, 5)$ tangent to the x, y plane.

14.1.7 Show that $(4, 6, 12)$, $(2, 7, 6)$, and $(-2, 5, 7)$ are vertices of a right triangle.

14.1.8 Find the perimeter of the triangle whose vertices are $(6, 1, 5)$, $(0, 3, 2)$, and $(6, 1, -7)$

14.1.9 Show that $(5, 1, 5)$, $(4, 3, 2)$, and $(-3, -2, 1)$ are vertices of a right triangle.

14.1.10 Show that $(3, 7, -2)$, $(-1, 8, 3)$, and $(-3, 4, -2)$ are vertices of an isosceles triangle.

14.1.11 Show that $(4, 2, 4)$, $(10, 2, -2)$, and $(2, 0, -4)$ are vertices of an equilateral triangle.

14.1.12 Find the equation of the sphere whose center is located at $(2, 1, 3)$ and has a radius of 4.

14.1.13 Find the equation of the sphere whose center is located at $(-4, 0, 6)$ and passes through $(2, 2, 3)$.

14.1.14 Find the equation of the sphere whose center is located at $(5, 1, -4)$ and passes through $(3, -5, -1)$.

14.1.15 Describe the surface whose equation is given by $x^2 + y^2 + z^2 - 4x - 6y - 8z = 2$.

14.1.16   Describe the surface whose equation is given by $x^2 + y^2 + z^2 - 4x + 12y + 6z = 0$.

14.1.17   Sketch the surface whose equation is given by $x^2 + y^2 = 9$.

14.1.18   Sketch the surface whose equation is given by $y = 4x^2$.

SOLUTIONS

SECTION 14.1

14.1.1 $x^2 + (y-4)^2 + z^2 = 16$; sphere $C(0,4,0)$, $r = 4$.

14.1.2 $d = \sqrt{(3-2)^2 + (9-7)^2 + (7-8)^2} = \sqrt{1+4+1} = \sqrt{6}$; midpoint $\left(\dfrac{5}{2}, 8, \dfrac{15}{2}\right)$.

14.1.3 $d = \sqrt{(9+3)^2 + (7+2)^2 + (2-4)^2} = \sqrt{144+81+4} = \sqrt{229}$; midpoint $(3, 5/2, 3)$.

14.1.4 $r = \dfrac{1}{2}\sqrt{(1-1)^2 + (2+4)^2 + (-3-5)^2} = \dfrac{1}{2}\sqrt{100} = 5$,

center $(1,-1,1)$, $(x-1)^2 + (y+1)^2 + (z-1)^2 = 25$

14.1.5 $r = \dfrac{1}{2}\sqrt{(4+2)^2 + (6-2)^2 + (12-10)^2} = \dfrac{1}{2}\sqrt{56} = \sqrt{14}$

center $\left(\dfrac{4-2}{2}, \dfrac{6+2}{2}, \dfrac{12+10}{2}\right) = (1, 4, 11)$

$(x-1)^2 + (y-4)^2 + (z-11)^2 = 14$

14.1.6 $(x-2)^2 + (y+3)^2 + (z-5)^2 = r^2$

$r^2 = 5^2 = 25$

$(x-2)^2 + (y+3)^2 + (z-5)^2 = 25$

14.1.7 The sides have length $\sqrt{41}$, $\sqrt{62}$, and $\sqrt{21}$. It is a right triangle because the sides satisfy the Pythagorean Theorem, $\left(\sqrt{62}\right)^2 = \left(\sqrt{41}\right)^2 + \left(\sqrt{21}\right)^2$.

14.1.8 The sides have lengths 7, 12, and 11 so the perimeter is 30.

14.1.9 The sides have lengths $\sqrt{14}$, $\sqrt{89}$, and $\sqrt{75}$. It is a right triangle because the sides satisfy the Pythagorean Theorem, $\left(\sqrt{89}\right)^2 = \left(\sqrt{14}\right)^2 + \left(\sqrt{75}\right)^2$.

14.1.10 The sides have lengths $\sqrt{42}$, $\sqrt{45}$, and $\sqrt{45}$. Since two sides are equal, the triangle is isosceles.

14.1.11 The sides have lengths $\sqrt{72}$, $\sqrt{72}$, and $\sqrt{72}$. Since all three sides are equal, the triangle is equilateral.

14.1.12 $(x-2)^2 + (y-1)^2 + (z-3)^2 = 16$.

14.1.13 $r = \sqrt{(2+4)^2 + (2-0)^2 + (3-6)^2} = 7$; $(x+4)^2 + y^2 + (z-6)^2 = 49$.

14.1.14 $r = \sqrt{(3-5)^2 + (-5-1)^2 + (-1+4)^2} = 7$; $(x-5)^2 + (y-1)^2 + (z+4)^2 = 49$.

14.1.15 $(x-2)^2 + (y-3)^2 + (z-4)^2 = 31$; sphere $C(2,3,4)$, $r = \sqrt{31}$.

14.1.16 $(x-2)^2 + (y+6)^2 + (z+3)^2 = 49$; sphere $C(2,-6,-3)$, $r = 7$.

14.1.17   **14.1.18**

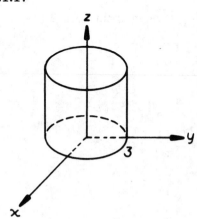   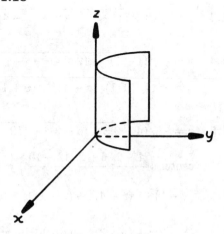

SECTION 14.2

14.2.1 Find the norm of $\mathbf{A} + \mathbf{B}$ if $\mathbf{A} = \langle 1, 2 \rangle$ and $\mathbf{B} = \langle -1, 0 \rangle$.

14.2.2 Find the components of the vector $\overrightarrow{P_1 P_2}$ if $P_1(1, 2)$ and $P_2(3, -4)$.

14.2.3 Find the norm of $\overrightarrow{P_1 P_2}$ if $P_1(4, -3)$ and $P_2(0, 5)$.

14.2.4 Find the norm of $2\mathbf{A} + \mathbf{C}$ if $\mathbf{A} = \langle 2, -1, 3 \rangle$ and $\mathbf{C} = \langle -2, 1, 0 \rangle$.

14.2.5 Express the vector from $P_1(-2, 3, 5)$ to $P_2(3, 5, -2)$ in the form $a\mathbf{i} + b\mathbf{j} + c\mathbf{k}$.

14.2.6 Find a vector with the same direction as $\mathbf{v} = 2\mathbf{i} - \mathbf{j} + 2\mathbf{k}$ but with twice the length.

14.2.7 Find a unit vector in the direction from $P_1(3, 0, -5)$ to $P_2(-1, 2, 3)$. Express your answer in $\mathbf{i}, \mathbf{j}, \mathbf{k}$ form.

14.2.8 Find a unit vector in the direction from $P_1(2, 9, 1)$ to $P_2(1, 7, 8)$. Express your answer in component form.

14.2.9 Find a unit vector in the direction of $\mathbf{u} + \mathbf{v}$ if $\mathbf{u} = 4\mathbf{i} + \mathbf{j} + 3\mathbf{k}$ and $\mathbf{v} = 2\mathbf{i} - 2\mathbf{j} + \mathbf{k}$.

14.2.10 Find the norm of $\mathbf{u} = 3\mathbf{i} - \mathbf{j} + \mathbf{k}$ and find a unit vector in a direction opposite to that of $\mathbf{u}$.

14.2.11 Find a unit vector in a direction $\mathbf{v} - \mathbf{u}$ if $\mathbf{v} = \langle 3, 4, 2 \rangle$ and $\mathbf{u} = \langle 5, -12, 1 \rangle$.

14.2.12 Find two unit vectors in 2-space parallel to the line $2x + y = 3$.

14.2.13 Use vectors to determine whether $P_1(1, 4, 2)$, $P_2(4, -3, 5)$ and $P_3(-5, -10, -8)$ are collinear.

14.2.14 Use vectors to determine whether $P_1(3, 1, 3)$, $P_2(1, 5, -1)$, and $P_3(4, -1, 5)$ are collinear.

14.2.15 Find the terminal point of $\mathbf{v} = 2\mathbf{i} - 5\mathbf{j}$ if the initial point is $(1, -2)$.

14.2.16 Find the initial point of $\mathbf{v} = \langle -1, 3 \rangle$ if the terminal point is $(1, 1)$.

14.2.17 Find $\mathbf{u}$ and $\mathbf{v}$ if $3\mathbf{u} + 2\mathbf{v} = \langle 9, -4 \rangle$ and $\mathbf{u} - 3\mathbf{v} = \langle -8, -5 \rangle$.

14.2.18 Find $\mathbf{u}$ and $\mathbf{v}$ if $2\mathbf{u} + 3\mathbf{v} = \langle -5, 13 \rangle$ and $\mathbf{u} + \mathbf{v} = \langle -1, 6 \rangle$.

SOLUTIONS

SECTION 14.2

14.2.1 $\|\mathbf{A} + \mathbf{B}\| = \|\langle 0, 2\rangle\| = 2.$ **14.2.2** $\langle 3 - 1, -4 - 2\rangle = \langle 2, -6\rangle.$

14.2.3 $\|\langle 0 - 4, 5 + 3\rangle\| = \|\langle -4, 8\rangle\| = 4\sqrt{5}.$

14.2.4 $\|2\langle 2, -1, 3\rangle + \langle -2, 1, 0\rangle\| = \|\langle 2, -1, 6\rangle\| = \sqrt{41}.$

14.2.5 $(3 + 2)\mathbf{i} + (5 - 3)\mathbf{j} + (-2 - 5)\mathbf{k} = 5\mathbf{i} + 2\mathbf{j} - 7\mathbf{k}.$

14.2.6 The required vector is $2\mathbf{v} = 4\mathbf{i} - 2\mathbf{j} + 4\mathbf{k}.$

14.2.7 $\overrightarrow{P_1P_2} = (-1 - 3)\mathbf{i} + (2 - 0)\mathbf{j} + (3 + 5)\mathbf{k} = -4\mathbf{i} + 2\mathbf{j} + 8\mathbf{k}, \, \| \overrightarrow{P_1P_2} \| = 2\sqrt{21}$ so the unit vector is $-\dfrac{2}{\sqrt{21}}\mathbf{i} + \dfrac{1}{\sqrt{21}}\mathbf{j} + \dfrac{4}{\sqrt{21}}\mathbf{k}.$

14.2.8 $\overrightarrow{P_1P_2} = \langle 1 - 2, \ 7 - 9, \ 8 - 1\rangle = \langle -1, -2, 7\rangle; \, \| \overrightarrow{P_1P_2} \| = 3\sqrt{6}$ so the unit vector is $\left\langle -\dfrac{1}{3\sqrt{6}}, \dfrac{-2}{3\sqrt{6}}, \dfrac{7}{3\sqrt{6}} \right\rangle.$

14.2.9 $\mathbf{u} + \mathbf{v} = (4 + 2)\mathbf{i} + (1 - 2)\mathbf{j} + (3 + 1)\mathbf{k} = 6\mathbf{i} - \mathbf{j} + 4\mathbf{k}, \|\mathbf{u} + \mathbf{v}\| = \sqrt{53}$ so the unit vector is $\dfrac{6}{\sqrt{53}}\mathbf{i} - \dfrac{1}{\sqrt{53}}\mathbf{j} + \dfrac{4}{\sqrt{53}}\mathbf{k}.$

14.2.10 $\|\mathbf{u}\| = \sqrt{11}$ so the required unit vector is $-\dfrac{\mathbf{u}}{\|\mathbf{u}\|} = -\dfrac{3}{\sqrt{11}}\mathbf{i} + \dfrac{1}{\sqrt{11}}\mathbf{j} - \dfrac{1}{\sqrt{11}}\mathbf{k}.$

14.2.11 $\mathbf{v} - \mathbf{u} = \langle 3 - 5, 4 + 12, 2 - 1\rangle = \langle -2, 16, 1\rangle$ and $\|\mathbf{v} - \mathbf{u}\| = 3\sqrt{29}$ so the required unit vector is $\left\langle -\dfrac{2}{3\sqrt{29}}, \dfrac{16}{3\sqrt{29}}, \dfrac{1}{3\sqrt{29}} \right\rangle.$

14.2.12 Choose two points on the line, for example $P_1(0, 3)$ and $P_2(1, 1)$ then $\overrightarrow{P_1P_2} = \langle 1, -2\rangle$ is parallel to the line, $\|\langle 1, -2\rangle\| = \sqrt{5}$ so $\langle 1/\sqrt{5}, -2/\sqrt{5}\rangle$ and $\langle -1/\sqrt{5}, 2\sqrt{5}\rangle$ are unit vectors parallel to the line.

14.2.13 The points are collinear if $\overrightarrow{P_1P_2}$ is parallel to $\overrightarrow{P_2P_3}$; $\overrightarrow{P_1P_2} = \langle 4 - 1, -3 - 4, 5 - 2\rangle = \langle -3, -7, 3\rangle,$ $\overrightarrow{P_2P_3} = \langle -5 - 4, \ -10 + 3, \ -8 - 5\rangle = \langle -9, -7, -13\rangle, \, \overrightarrow{P_1P_2}$ is not parallel to $\overrightarrow{P_2P_3}$ so the points are not collinear.

14.2.14   The points are collinear if $\overrightarrow{P_1P_2}$ is parallel to

$\overrightarrow{P_2P_3}$; $\overrightarrow{P_1P_2} = \langle 1-3,\ 5-1,\ -1-3\rangle = \langle -2, 4, -4\rangle$,

$\overrightarrow{P_2P_3} = \langle 4-1,\ -1-5,\ 5+1\rangle = \langle 3, -6, 6\rangle$, so $\overrightarrow{P_1P_2}$ is parallel to $\overrightarrow{P_2P_3}$ and the points are collinear.

14.2.15   Let $P(x, y)$ be the terminal point, then $x - 1 = 2$, $x = 3$; $y + 5 = -2$, $y = -7$ so the terminal point is $(3, -7)$.

14.2.16   Let $P(x, y)$ be the initial point, then $1 - x = -1$, $x = 2$; $1 - y = 3$, $y = -2$, so the initial point is $(2, -2)$.

14.2.17   Solve the system

$$\left.\begin{array}{l} 3u + 2v = \langle -9, -4\rangle \\[2mm] u - 3v = \langle -8, -5\rangle \end{array}\right\} \text{ to get } u = \langle 1, -2\rangle \text{ and } v = \langle 3, 1\rangle.$$

14.2.18   Solve the system

$$\left.\begin{array}{l} 2u + 3v = \langle -5, 13\rangle \\[2mm] u + v = \langle -1, 6\rangle \end{array}\right\} \text{ to get } u = \langle 2, 5\rangle \text{ and } v = \langle -3, 1\rangle.$$

SECTION 14.3

14.3.1   Find the direction cosines of the vector $\overrightarrow{P_1P_2}$ if $P_1(2,3,3)$ and $P_2(3,1,8)$.

14.3.2   Find $u \cdot v$ if $u = 3i + 3j - k$ and $v = i + 4j + k$.

14.3.3   Find the vector component of $\mathbf{u} = \mathbf{i} + 2\mathbf{j} + \mathbf{k}$ along $\mathbf{a} = 2\mathbf{i} + 2\mathbf{j} - \mathbf{k}$.

14.3.4   Let $\mathbf{u} = \langle 1,0,2 \rangle$, $\mathbf{v} = \langle 2,-1,3 \rangle$, and $\mathbf{w} = \langle -2,1,0 \rangle$, find:

   (a) $\mathbf{u} - \mathbf{v}$;
   (b) a unit vector in the direction of $\mathbf{w}$;
   (c) the vector component of $\mathbf{v}$ along $\mathbf{u}$.

14.3.5   Find $\mathbf{u} \cdot \mathbf{v}$ and the vector component of $\mathbf{u}$ along $\mathbf{a}$ if $\mathbf{u} = \langle 1,2,1 \rangle$ and $\mathbf{a} = \langle 1,-2,2 \rangle$.

14.3.6   Let $\mathbf{u} = \langle 3,-2,1 \rangle$, $\mathbf{a} = \langle 0,2,-1 \rangle$, and $\mathbf{w} = \langle 0,1,2 \rangle$, find:

   (a) $\mathbf{w} \cdot \mathbf{u}$;
   (b) a unit vector in the direction of the vector component of $\mathbf{u}$ along $\mathbf{a}$.

14.3.7   Find the cosine of the angle between $u = 3i + 3j - k$ and $v = -6i + 4j - k$.

14.3.8   Show that $\mathbf{u} = 2\mathbf{i} - \mathbf{j}$ and $\mathbf{v} = 3\mathbf{i} + 6\mathbf{j}$ are orthogonal.

14.3.9   Let $\mathbf{u} = \langle 1,-2,4 \rangle$ and $\mathbf{a} = \langle 0,2,3 \rangle$, find:

   (a) the vector component of $\mathbf{u}$ along $\mathbf{a}$;
   (b) the vector component of $\mathbf{u}$ orthogonal to $\mathbf{a}$;
   (c) the length of the component in part (b).

14.3.10   Find the vector component of $\mathbf{u} = \langle -1,4,2 \rangle$ orthogonal to $\mathbf{a} = \langle 2,-2,-1 \rangle$.

14.3.11   Find the vector component of $\mathbf{u} = \langle 2,-1,3 \rangle$ along $\mathbf{a} = \langle 3,0,4 \rangle$.

14.3.12   Find the cosine of the angle between $u = 2i - 2j - k$ and $v = 2i + 2j + k$.

14.3.13   Find a vector whose norm is 4 and whose direction angles are $\alpha = 30°$, $\beta = 120°$, and $\gamma = 135°$.

14.3.14 Find the cosine of the angle between $\mathbf{u} = 2\mathbf{i}+3\mathbf{j}+7\mathbf{k}$ and $\mathbf{v} = 2\mathbf{j} + \mathbf{k}$.

14.3.15 Find a vector whose norm is 6 and whose direction angles are $\alpha = 30°$, $\beta = 60°$, $\gamma = 120°$.

14.3.16 Find the vector component of $\mathbf{u} = \langle 3, -4, 4 \rangle$ orthogonal to $\mathbf{a} = \langle 2, 2, 1 \rangle$.

14.3.17 Find the direction cosines of the vector that is parallel to $(1, 4, 4)$ and $(3, 5, 4)$.

SOLUTIONS

SECTION 14.3

14.3.1 $\overrightarrow{P_1P_2} = \langle 1, -2, 5 \rangle$, $\| \overrightarrow{P_1P_2} \| = \sqrt{30}$, so $\cos\alpha = \dfrac{1}{\sqrt{30}}$, $\cos\beta = \dfrac{-2}{\sqrt{30}}$, $\cos\alpha = \dfrac{5}{\sqrt{30}}$.

14.3.2 $(3)(1) + (3)(4) + (-1)(1) = 14.$ **14.3.3** $\dfrac{10}{9}\mathbf{i} + \dfrac{10}{9}\mathbf{j} - \dfrac{5}{9}\mathbf{k}.$

14.3.4 (a) $\langle -1, 1, -1 \rangle$ (b) $\left\langle \dfrac{-2}{\sqrt{5}}, \dfrac{1}{\sqrt{5}}, 0 \right\rangle$ (c) $\left\langle \dfrac{8}{5}, 0, \dfrac{16}{5} \right\rangle.$

14.3.5 $\mathbf{u} \cdot \mathbf{a} = (1)(1) + (2)(-2) + (1)(2) = -1$, $\left\langle -\dfrac{1}{9}, \dfrac{2}{9}, \dfrac{-2}{9} \right\rangle$ is the required vector.

14.3.6 (a) $(0)(3) + (1)(-2) + (2)(1) = 0.$

    (b) $\text{Proj}_{\mathbf{a}}\mathbf{u} = \langle 0, -2, 1 \rangle$; $\|\text{Proj}_{\mathbf{a}}\mathbf{u}\| = \sqrt{5}$, so $\left\langle 0, \dfrac{-2}{\sqrt{5}}, \dfrac{1}{\sqrt{5}} \right\rangle$ is the required vector.

14.3.7 $\cos\theta = \dfrac{u \cdot v}{\|u\|\,\|v\|} = \dfrac{7}{\sqrt{19}\sqrt{53}}.$

14.3.8 $\mathbf{u}$ and $\mathbf{v}$ are orthogonal if $\mathbf{u} \cdot \mathbf{v} = 0$ and if $\mathbf{u} \neq \mathbf{0}$ and $\mathbf{v} \neq \mathbf{0}$, $\mathbf{u} \cdot \mathbf{v} = (2)(3) + (-1)(6) = 0$, as the vectors are orthogonal.

14.3.9 (a) $\left\langle 0, \dfrac{16}{13}, \dfrac{24}{13} \right\rangle$

    (b) $\langle 1, -2, 4 \rangle - \langle 0, \dfrac{16}{13}, \dfrac{24}{13} \rangle = \left\langle 1, -\dfrac{42}{13}, \dfrac{28}{13} \right\rangle$

    (c) $\left\| \left\langle 1, -\dfrac{42}{13}, \dfrac{28}{13} \right\rangle \right\| = \sqrt{\dfrac{209}{13}}.$

14.3.10 $\text{Proj}_{\mathbf{a}}\mathbf{u} = \left\langle -\dfrac{8}{3}, \dfrac{8}{3}, \dfrac{4}{3} \right\rangle$ so the required vector is $\langle -1, 4, 2 \rangle - \left\langle -\dfrac{8}{3}, \dfrac{8}{3}, \dfrac{4}{3} \right\rangle = \left\langle \dfrac{5}{3}, \dfrac{4}{3}, \dfrac{2}{3} \right\rangle.$

14.3.11 $\left\langle \dfrac{54}{25}, 0, \dfrac{72}{25} \right\rangle.$ **14.3.12** $\cos\theta = \dfrac{u \cdot v}{\|u\|\,\|v\|} = \dfrac{-1}{9}.$

14.3.13 $\mathbf{v} = 4\langle \cos 30°, \cos 120°, \cos 135° \rangle = 4\left\langle \dfrac{\sqrt{3}}{2}, -\dfrac{1}{2}, -\dfrac{1}{\sqrt{2}} \right\rangle$

$$= \left\langle 2\sqrt{3}, -2, -2\sqrt{2} \right\rangle.$$

14.3.14 $\cos\theta = \dfrac{\mathbf{u}\cdot\mathbf{v}}{\|\mathbf{u}\|\,\|\mathbf{v}\|} = \dfrac{13}{\sqrt{62}\sqrt{5}}.$

14.3.15 $\mathbf{v} = 6\left\langle\cos 30°, \cos 60°, \cos 120°\right\rangle = 6\left\langle\dfrac{\sqrt{3}}{2}, \dfrac{1}{2}, -\dfrac{1}{2}\right\rangle = \left\langle 3\sqrt{3}, 3, -3\right\rangle.$

14.3.16 $\text{Proj}_{\mathbf{a}}\mathbf{u} = \left\langle\dfrac{4}{9}, \dfrac{4}{9}, \dfrac{1}{9}\right\rangle$, so, the required vector is $\langle 3, -4, 4\rangle - \left\langle\dfrac{4}{9}, \dfrac{4}{9}, \dfrac{1}{9}\right\rangle = \left\langle\dfrac{23}{9}, -\dfrac{40}{9}, \dfrac{35}{9}\right\rangle.$

14.3.17 Let P_1 be the first point and P_2 the second point so $\overrightarrow{P_1P_2} = \langle 2, 1, 0\rangle, \|\overrightarrow{P_1P_2}\| = \sqrt{5}$ so $\cos\alpha = \dfrac{2}{\sqrt{5}}, \cos\beta = \dfrac{1}{\sqrt{5}}, \cos\gamma = 0$ are the direction cosines.

SECTION 14.4

14.4.1 Let $\mathbf{u} = \langle 1, 2, -1 \rangle$, $\mathbf{v} = \langle 2, -1, 3 \rangle$, and $\mathbf{w} = \left\langle 0, \frac{1}{2}, -3 \right\rangle$. Evaluate (a) $(\mathbf{u} \cdot \mathbf{v}) \times \mathbf{w}$ and (b) $(\mathbf{u} \cdot \mathbf{v})\mathbf{w}$ if the expressions make sense.

14.4.2 Let $\mathbf{u} = \langle 1, 2, -1 \rangle$, $\mathbf{v} = \langle 1, 1, 1 \rangle$, and $\mathbf{w} = \langle 1, 2, 2 \rangle$. Evaluate (a) $\|\mathbf{u} \cdot \mathbf{v}\|$ and (b) $\|\mathbf{u} \times \mathbf{w}\|$ if the expressions make sense.

14.4.3 Let $\mathbf{a} = 3\mathbf{i}-4\mathbf{j} - \mathbf{k}$, $\mathbf{b} = \mathbf{i}-2\mathbf{j}+2\mathbf{k}$, and $\mathbf{c} = 2\mathbf{i}-6\mathbf{j}$; find $\mathbf{a} \times (\mathbf{c} \times \mathbf{b})$.

14.4.4 Let $\mathbf{a} = \langle 3, -4, 0 \rangle$, $\mathbf{b} = \langle 1, -2, 2 \rangle$, and $\mathbf{c} = \langle 1, -1, 0 \rangle$; find $\mathbf{c} \cdot (\mathbf{a} \times \mathbf{b})$.

14.4.5 Find all unit vectors parallel to the $y\,z$-plane that are perpendicular to the vector $2i - 3j - k$.

14.4.6 Let $\mathbf{u} = 3\mathbf{i}+2\mathbf{k}$, $\mathbf{v} = 2\mathbf{i}+2\mathbf{j} - \mathbf{k}$, and $\mathbf{w} = \mathbf{i}-2\mathbf{j}+3\mathbf{k}$; find $\mathbf{u} \times (\mathbf{v} \times \mathbf{w})$.

14.4.7 Find unit vectors that are orthogonal to both $\mathbf{a} = \langle 1, 2, 3 \rangle$ and $\mathbf{b} = \langle -1, 0, 1 \rangle$.

14.4.8 Find unit vectors that are orthogonal to both $\mathbf{a} = \langle 2, -2, -1 \rangle$ and $\mathbf{b} = \langle 1, 1, 1 \rangle$.

14.4.9 Find the sine of the angle between $\mathbf{a} = \langle 1, 1, 1 \rangle$ and $\mathbf{b} = \langle 2, -1, 3 \rangle$.

14.4.10 Determine whether $\mathbf{u} = \langle -1, 0, 2 \rangle$, $\mathbf{v} = \langle 0, 1, 1 \rangle$ and $\mathbf{w} = \langle -2, 1, -1 \rangle$ lie in the same plane.

14.4.11 Find the area of the parallelogram determined by $\mathbf{a} = 3\mathbf{i}-4\mathbf{j} - \mathbf{k}$ and $\mathbf{b} = \mathbf{i}-2\mathbf{j}+2\mathbf{k}$.

14.4.12 Find a vector perpendicular to the plane determined by $P_1(1, 0, 2)$, $P_2(3, 1, 1)$, and $P_3(5, 1, 3)$.

14.4.13 Find the area of the triangle whose vertices are $P(1, -2, 3)$, $Q(2, 4, 1)$, and $R(2, 0, 1)$.

14.4.14 Let $\mathbf{a} = \langle 2, 0, 1 \rangle$, $\mathbf{b} = \langle 3, 2, 5 \rangle$, and $\mathbf{c} = \langle -1, 0, 2 \rangle$; find the volume of the parallelpiped determined by $\mathbf{a}$, $\mathbf{b}$, and $\mathbf{c}$.

14.4.15 Determine whether $\mathbf{u} = \langle -1, 0, 0 \rangle$, $\mathbf{v} = \langle 0, 1, 1 \rangle$ and $\mathbf{w} = \langle -1, 1, 1 \rangle$ lie in the same plane.

14.4.16 Find the volume of the parallelpiped whose vertices are $A(0, 0, 0)$, $B(1, -1, 1)$, $C(2, 1, -2)$, and $D(-1, 2, -1)$.

14.4.17 Find the volume of the parallelpiped whose edges are determined by
$\mathbf{a} = \langle 1, 0, 2 \rangle$, $\mathbf{b} = \langle 4, 6, 2 \rangle$, and $\mathbf{c} = \langle 3, 3, -6 \rangle$.

SOLUTIONS

SECTION 14.4

14.4.1 (a) $(\mathbf{u} \cdot \mathbf{v}) \times \mathbf{w}$ does not make sense

(b) $\mathbf{u} \cdot \mathbf{v} = \langle 1, 2, -1 \rangle \cdot \langle 2, -1, 3 \rangle = -3$, so
$(\mathbf{u} \cdot \mathbf{v})\mathbf{w} = -3\langle 0, 1/2, -3 \rangle = \langle 0, -3/2, 9 \rangle$.

14.4.2 (a) $\|\mathbf{u} \cdot \mathbf{v}\|$ does not make sense

(b) $\|\mathbf{u} \times \mathbf{w}\| = \|\langle 1, 2, -1 \rangle \times \langle 1, 2, 2 \rangle\| = \|\langle 6, -3, 0 \rangle\| = 3\sqrt{5}$.

14.4.3 $\mathbf{c} \times \mathbf{b} = -12\mathbf{i} - 4\mathbf{j} + 2\mathbf{k}$; $\mathbf{a} \times (\mathbf{c} \times \mathbf{b}) = -12\mathbf{i} + 6\mathbf{j} - 60\mathbf{k}$.

14.4.4 -2.

14.4.5 A vector parallel to the y-z plane must be perpendicular to i; $i \times (2i - 3j - k) = j - 3k$,
$\|j - 3k\| = \sqrt{10}$, the unit vectors are $\pm \dfrac{(j - 3k)}{\sqrt{10}}$

14.4.6 $\mathbf{v} \times \mathbf{w} = 4\mathbf{i} - 7\mathbf{j} - 6\mathbf{k}$, so $\mathbf{u} \times (\mathbf{v} \times \mathbf{w}) = 14\mathbf{i} + 26\mathbf{j} - 21\mathbf{k}$.

14.4.7 $\pm \dfrac{\langle 1, 2, 3 \rangle \times \langle -1, 0, 1 \rangle}{\|\langle 1, 2, 3 \rangle \times \langle -1, 0, 1 \rangle\|} = \pm \dfrac{\langle 2, -4, 2 \rangle}{\|\langle 2, -4, 2 \rangle\|} = \pm \left\langle \dfrac{1}{\sqrt{6}}, -\dfrac{2}{\sqrt{6}}, \dfrac{1}{\sqrt{6}} \right\rangle$.

14.4.8 $\pm \dfrac{\langle 2, -2, 1 \rangle \times \langle 1, 1, 1 \rangle}{\|\langle 2, -2, 1 \rangle \times \langle 1, 1, 1 \rangle\|} = \pm \dfrac{\langle -1, -3, 4 \rangle}{\|\langle -1, -3, 4 \rangle\|} = \pm \left\langle \dfrac{-1}{\sqrt{26}}, -\dfrac{3}{\sqrt{26}}, \dfrac{4}{\sqrt{26}} \right\rangle$.

14.4.9 $\sin\theta = \dfrac{\|a \times b\|}{\|a\| \|b\|} = \dfrac{\|\langle 4, 1, 3 \rangle\|}{\|\langle 1, 1, 1 \rangle\| \|\langle 2, -1, 3 \rangle\|} = \dfrac{\sqrt{26}}{\sqrt{3}\sqrt{14}}$.

14.4.10 $\mathbf{u} \cdot (\mathbf{v} \times \mathbf{w}) = 6 \neq 0$, no.

14.4.11 area $= \|\mathbf{a} \times \mathbf{b}\| = \| -10\mathbf{i} - 7\mathbf{j} - 2\mathbf{k}\| = 3\sqrt{17}$.

14.4.12 $\overrightarrow{P_1 P_2} \times \overrightarrow{P_1 P_3} = \langle 2, 1, -1 \rangle \times \langle 4, 1, 1 \rangle = \langle 2, -6, -2 \rangle$ or any nonzero scalar multiple.

14.4.13 $A = \dfrac{1}{2} \| \overrightarrow{PQ} \times \overrightarrow{PR} \| = \dfrac{1}{2}\|\langle 1, 6, -2 \rangle \times \langle 1, 2, -2 \rangle\| = \dfrac{1}{2}\|\langle -8, 0, -4 \rangle\| = 2\sqrt{5}$.

14.4.14 $V = |\mathbf{a} \cdot (\mathbf{b} \times \mathbf{c})| = 10$.   **14.4.15** $u \cdot (\mathbf{v} \times \mathbf{w}) = 0$, yes.

14.4.16 $\overrightarrow{AB} = \langle 1, -1, 1 \rangle$, $\overrightarrow{AC} = \langle 2, 1, -2 \rangle$, and $\overrightarrow{AD} = \langle -1, 2, -1 \rangle$,

so $V = \left| \overrightarrow{AB} \cdot \left(\overrightarrow{AC} \times \overrightarrow{AD} \right) \right| = 0$.

14.4.17 $V = |\mathbf{a} \cdot (\mathbf{b} \times \mathbf{c})| = 54$.

SECTION 14.5

14.5.1 Find any two points which lie on the line $x - 5 = 2t$, $y = -5t$, $z = -t$.

14.5.2 Find a unit vector that is parallel to $x - 4 = t$, $y + 2 = 2t$, $z - 5 = -2t$.

14.5.3 Find the parametric and vector equations of the line which passes through $P_1(4, 0, 5)$ and $P_2(2, 3, 1)$.

14.5.4 Find the parametric and vector equations of the line which passes through $P_1(3, 3, 1)$ and $P_2(4, 0, 2)$; also find two other points which lie on the line.

14.5.5 Find the parametric and vector equations of the line which passes through $P_1(3, 0, -5)$ and $P_2(-1, 2, 3)$.

14.5.6 Find the parametric and vector equations of the line which passes through $P_1(1, 4, 6)$ and $P_2(2, -1, 3)$; also find two other points on the line.

14.5.7 Do the lines that pass through $(2, 3, 3)$ and $(3, 1, 8)$ and through $(-3, -1, 0)$ and $(-1, 2, 1)$ intersect? If so, find their point of intersection.

14.5.8 Find the point of intersection of the lines

$$
\begin{array}{ccc}
x = 3 - t & & x = 8 + 2t \\
y = 5 + 3t & \text{and} & y = -6 - 4t \\
z = -1 - 4t & & z = 5 + t
\end{array}
$$

14.5.9 Find the point where the line which passes through $(1, 4, 2)$ and is parallel to $\langle 3, 2, -2 \rangle$ pierces the xy-plane.

14.5.10 Find the point where the line which passes through $(3, 5, -1)$ and is parallel to $\langle 1, -1, 1 \rangle$ pierces the xz-plane.

14.5.11 Show that the line determined by $(3, 1, 0)$ and $(1, 4, -3)$ is perpendicular to $x = 3t$, $y = 3 + 8t$, $z = -7 + 6t$.

14.5.12 Find the vector equation of the line which passes through $(2, -4, 5)$ and is perpendicular to the pair of lines which pass through $(2, -4, 5)$, $(5, 3, 0)$ and $(4, -3, 1)$, $(3, -4, 1)$.

14.5.13 Find the cosine of the angle between the lines $x = 2 + t$, $y = 3 + t$, $z = -1 + 2t$ and $x = 2 + 2t$, $y = 3 - t$, $z = -1 + 3t$.

14.5.14   Find the cosine of the angle between the line $x = 2t$, $y = 3t$, $z = t$ and the y–axis.

14.5.15   Find the point of intersection of $x = 2t$, $y = t$, $z = 3t$ and the plane $2x + y - z = 7$.

14.5.16   Find the point of intersection of $x = 2+t$, $y = 3-2t$, $z = -4t$ and the plane $2x - 3y + 4z = 10$.

14.5.17   A vector whose direction cosines are $1/2$, $\sqrt{3}/2$, $-1/2$ is parallel to the line which passes through $(2, 5, 2)$. Find the vector equation of the line.

SOLUTIONS

SECTION 14.5

14.5.1   If $t = 0$, then $(5,0,0)$ is one point and if $t = 1$, $(7,-5,-1)$ is another.

14.5.2   Vectors parallel to the line are $\pm\langle 1, 2, -2\rangle$ whose norms are $\|\langle 1, 2, -2\rangle\| = 3$, thus the required unit vector is $\langle 1/3, 2/3, -2/3\rangle$ or $\langle -1/3, -2/3, 2/3\rangle$.

14.5.3 $\overrightarrow{P_1P_2} = \langle -2, 3, -4\rangle$ so the parametric equation is $x = 4 - 2t$, $y = 3t$, $z = 5 - 4t$ and the vector equation is $x = \langle 4, 0, 5\rangle + t\langle -2, 3, -4\rangle$.

14.5.4 $\overrightarrow{P_1P_2} = \langle 1, -3, 1\rangle$ so the parametric equation is $x = 3 + t$, $y = 3 - 3t$, $z = 1 + t$ and the vector equation is $x = \langle 3, 3, 1\rangle + t\langle 1, -3, 1\rangle$; if $t = -1$, then $(2, 6, 0)$ is one point and if $t = 2$, $(5, -3, 3)$ is another.

14.5.5 $\overrightarrow{P_1P_2} = \langle -4, 2, 8\rangle$ or using $\langle -2, 1, 4\rangle$ for convenience then the parametric equation is
$x = 3 - 2t$, $y = t$, $z = -5 + 4t$ and the vector equation is $x = \langle 3, 0, -5\rangle + t\langle -2, 1, 4\rangle$.

14.5.6 $\overrightarrow{P_1P_2} = \langle 1, -5, -3\rangle$ so the parametric equation is $x = 1 + t$, $y = 4 - 5t$, $z = 6 - 3t$ and the vector equation is $x = \langle 1, 4, 6\rangle + t\langle 1, -5, -3\rangle$; if $t = -1$, then $(0, 9, 9)$ is one point and if $t = 2$, $(3, -6, 0)$ is another.

14.5.7   The equation of the line through $(2, 3, 3)$ and $(3, 1, 8)$ is $x = 2 + t$, $y = 3 - 2t$, $z = 3 + 5t$; the equation of the line through $(-3, -1, 0)$ and $(-1, 2, 1)$ is $x = -3 + 2t$, $y = -1 + 3t$,

$$z = t; \text{ solve the equations } \left.\begin{array}{c} 2 + t_1 = -3 + 2t_2 \\ 3 - 2t_1 = -1 + 3t_2 \\ 3 + 5t_1 = t_2 \end{array}\right\} \text{ for } t_1 \text{ and } t_2; \text{ solution of the first two}$$

equations yields $t_1 = -1$ and $t_2 = 2$ which do not satisfy the third equation so the lines do not intersect.

14.5.8   Solve the equations $\left.\begin{array}{c} t_1 + 2t_2 = -5 \\ 3t_1 + 4t_2 = -11 \\ 4t_1 + t_2 = -6 \end{array}\right\}$ for t_1 and t_2; solution of the first two equations

yields $t_1 = -1$ and $t_2 = -2$ which also satisfies the third equation, so the point of intersection is $(4, 2, 3)$.

14.5.9   The equation of the line is $x = 1 + 3t$, $y = 4 + 2t$, $z = 2 - 2t$. On the xy−plane, $z = 0$ so $2 - 2t = 0$, $t = 1$ and the point is $(4, 6, 0)$.

14.5.10 The equation of the line is $x = 3 + t$, $y = 5 - t$, $z = -1 + t$. On the xz–plane, $y = 0$ so $5 - t = 0$, $t = 5$ and the point is $(8, 0, 4)$.

14.5.11 A vector parallel to the line through $(3, 1, 0)$ and $(1, 4, -3)$ is $\langle -2, 3, -3 \rangle$; a vector parallel to $x = 3t$, $y = 3 + 8t$, $z = -7 + 6t$ is $\langle 3, 8, 6 \rangle$ thus $\langle -2, 3, -3 \rangle \cdot \langle 3, 8, 6 \rangle = 0$, so the lines are perpendicular.

14.5.12 A vector parallel to the line through $(2, -4, 5)$ and $(5, 3, 0)$ is $\langle 3, 7, -5 \rangle$; a vector parallel to the line through $(4, -3, 1)$ and $(3, -4, 1)$ is $\langle -1, -1, 0 \rangle$; a vector perpendicular to $\langle 3, 7, -5 \rangle$ and $\langle -1, -1, 0 \rangle$ is $\langle 3, 7, -5 \rangle \times \langle -1, -1, 0 \rangle = \langle -5, 5, 4 \rangle$, so the vector equation of the desired line is $x = \langle 2, -4, 5 \rangle + t \langle -5, 5, 4 \rangle$.

14.5.13 $\langle 1, 1, 2 \rangle$ is parallel to $x = 2 + t$, $y = 3 + t$, $z = t$ and $\langle 2, -1, 3 \rangle$ is parallel to $x = 2 + 2t$, $y = 3 - t$, $z = -1 + 3t$ so $\cos \theta = \dfrac{\langle 1, 1, 2 \rangle \cdot \langle 2, -1, 3 \rangle}{\|\langle 1, 1, 2 \rangle\| \, \|\langle 2, -1, 3 \rangle\|} = \dfrac{7}{\sqrt{6}\sqrt{14}}$.

14.5.14 $\langle 2, 3, 1 \rangle$ is parallel to $x = 2t$, $y = 3t$, $z = t$ and $\langle 0, 1, 0 \rangle$ is parallel to the y–axis so

$$\cos \theta = \frac{\langle 2, 3, 1 \rangle \cdot \langle 0, 1, 0 \rangle}{\|\langle 2, 3, 1 \rangle\| \, \|\langle 0, 1, 0 \rangle\|} = \frac{3}{\sqrt{14}\sqrt{1}}.$$

14.5.15 Substitute $x = 2t$, $y = t$, $z = 3t$ into $2x + y - z = 7$ to get $2(2t) + t - (3t) = 7$, $t = 7/2$ so the point is $(7, 7/2, 21/2)$.

14.5.16 Substitute $x = 2 + t$, $y = 3 - 2t$, $z = -4t$ into $2x - 3y + 4z = 10$ to get $t = -\dfrac{15}{8}$ so the point is $\left(\dfrac{1}{8}, \dfrac{27}{4}, \dfrac{15}{2} \right)$.

14.5.17 $x = \langle 2, 5, 2 \rangle + t \left\langle \dfrac{1}{2}, \dfrac{\sqrt{3}}{2}, -\dfrac{1}{2} \right\rangle$ or $x = \langle 2, 5, 2 \rangle + t \langle 1, \sqrt{3}, -1 \rangle$.

SECTION 14.6

14.6.1   Find the equation of the plane through $P(2,1,3)$, $Q(3,3,5)$, and $R(1,3,6)$.

14.6.2   Show that the line $x = \langle 0,1,1 \rangle + t\langle 2,4,-1 \rangle$ is parallel to the plane $2x - 3y - 8z = 0$.

14.6.3   Show that the line $x = 1 + 2t$, $y = -1 + 3t$, $z = 2 + 4t$ is parallel to the plane $x - 2y + z = 5$.

14.6.4   Find the equation of the plane through $P(1,1,1)$, $Q(2,4,3)$, and $R(-1,-2,-1)$.

14.6.5   Find the equation of the plane through $(1,2,-3)$ and perpendicular to $x = 1 + 2t$, $y = 2 + t$, $z = -3 - 5t$.

14.6.6   Find the equation of the plane that contains the point $(2,1,5)$ and the line $x = -1 + 3t$, $y = -2t$, $z = 2 + 4t$.

14.6.7   Find the equation of the plane through $(3,-2,-1)$ and parallel to $2x + y + 6z + 8 = 0$.

14.6.8   Find the direction cosines of a vector perpendicular to $3x - 2y + z - 7 = 0$.

14.6.9   Find the vector equation of the line through $(1,1,1)$ that is parallel to the line of intersection of the planes $3x - 4y + 2z - 2 = 0$ and $4x - 3y - z - 5 = 0$.

14.6.10   Find the parametric equations of the line through $(2,0,-3)$ that is parallel to the line of intersection of the planes $x + 2y + 3z + 4 = 0$ and $2x - y - z - 5 = 0$.

14.6.11   Find the vector equation of the line of intersection of the planes $x + y + z - 4 = 0$ and $2x - y + z - 2 = 0$.

14.6.12   Show that the equation of the plane which intercepts the coordinate axes at $(a,0,0)$, $(0,b,0)$, and $(0,0,c)$ can be written as

$$\frac{x}{a} + \frac{y}{b} + \frac{z}{c} = 1.$$

14.6.13   Find the equation of the plane through $(3,0,1)$ and perpendicular to the line $x = 2t$, $y = 1 - t$, $z = 4 - 3t$.

14.6.14   Find the equation of the plane that contains the point $(-2,1,1)$ and the line $x = \langle 2,1,1 \rangle + t\langle -1,4,4 \rangle$.

14.6.15   Find the parametric equations of the line of intersection of the planes $3x - 2y + z = 0$ and $8x + 2y + z - 11 = 0$.

14.6.16   Find the equation of the plane that contains $P_1(1, 1, 1)$ and $P_2(-1, 2, 1)$ and is parallel to the line of intersection of the planes $2x + y - z - 4 = 0$ and $3x - y + z - 2 = 0$.

14.6.17   Find the equation of the plane that contains $P_1(3, 1, 2)$ and $P_2(-1, 2, -1)$ and is parallel to the line of intersection of the planes $2x - y - z - 2 = 0$ and $3x + 2y - 2z - 4 = 0$.

SOLUTIONS

SECTION 14.6

14.6.1 $\overrightarrow{PQ} = \langle 1, 2, 2 \rangle$, $\overrightarrow{PR} = \langle -1, 2, 3 \rangle$ so $\langle 1, 2, 2 \rangle \times \langle -1, 2, 3 \rangle = \langle 2, -5, 4 \rangle$ is normal to the plane whose equation is $2(x - 2) - 5(y - 1) + 4(z - 3) = 0$ or $2x - 5y + 4z - 11 = 0$.

14.6.2 $\langle 2, 4, -1 \rangle$ is parallel to the line and $\langle 2, -3, -8 \rangle$ is normal to the plane, thus, $\langle 2, 4, -1 \rangle \cdot \langle 2, -3, -8 \rangle = 0$, so the line is parallel to the plane.

14.6.3 $\langle 2, 3, 4 \rangle$ is parallel to the line and $\langle 1, -2, 1 \rangle$ is normal to the plane, thus, $\langle 2, 3, 4 \rangle \cdot \langle 1, -2, 1 \rangle = 0$ so the line is parallel to the plane.

14.6.4 $\overrightarrow{PQ} = \langle 1, 3, 2 \rangle$, $\overrightarrow{PR} = \langle -2, -3, -2 \rangle$ so $\langle 1, 3, 2 \rangle \times \langle -2, -3, -2 \rangle = \langle 0, -2, 3 \rangle$ is normal to the plane whose equation is $0(x - 1) - 2(y - 1) + 3(z - 1) = 0$ or $2y - 3z + 1 = 0$.

14.6.5 $\langle 2, 1, -5 \rangle$ is parallel to the line and therefore perpendicular to the plane whose equation is $2(x - 1) + 1(y - 2) - 5(z + 3) = 0$ or $2x + y - 5z - 19 = 0$.

14.6.6 Find two other points on the plane by setting $t = 0$ and $t = 1$ to get $P_1(-1, 0, 2)$ and $P_2(2, -2, 6)$. Let $P_0(2, 1, 5)$ be the given point then $\overrightarrow{P_0P_1} = \langle -3, -1, -3 \rangle$ and $\overrightarrow{P_0P_2} = \langle 0, -3, 1 \rangle$, thus, $\overrightarrow{P_0P_1} \times \overrightarrow{P_0P_2} = \langle -10, 3, 9 \rangle$ is normal to the plane whose equation is $-10(x - 2) + 3(y - 1) + 9(z - 5) = 0$ or $10x - 3y - 9z - 28 = 0$.

14.6.7 Since the two planes are parallel, $\langle 2, 1, 6 \rangle$ is normal to both planes so the equation of the desired plane is $2(x - 3) + 1(y + 2) + 6(z + 1) = 0$ or $2x + y + 6z + 2 = 0$.

14.6.8 $\langle 3, -2, 1 \rangle$ or any scalar multiple is perpendicular to $3x - 2y + z - 7 = 0$. $\|\langle 3, -2, 1 \rangle\| = \sqrt{14}$ so the direction cosines of the normal vector are: $\cos \alpha = \dfrac{3}{\sqrt{14}}$, $\cos \beta = \dfrac{-2}{\sqrt{14}}$, $\cos \gamma = \dfrac{1}{\sqrt{14}}$.

14.6.9 $\langle 3, -4, 2 \rangle$ and $\langle 4, -3, -1 \rangle$ are respectively normal to the given planes. $\langle 3, -4, 2 \rangle \times \langle 4, -3, -1 \rangle = \langle 10, 11, 7 \rangle$ or any scalar multiple is parallel to the line of intersection of the given planes and is thus parallel to the line whose vector equation is $x = \langle 1, 1, 1 \rangle + t\langle 10, 11, 7 \rangle$.

14.6.10 $\langle 1, 2, 3 \rangle$ and $\langle 2, -1, -1 \rangle$ are respectively normal to the given planes. $\langle 1, 2, 3 \rangle \times \langle 2, -1, -1 \rangle = \langle 1, 7, -5 \rangle$ or any scalar multiple is parallel to the line of intersection of the given planes and is thus parallel to the line whose parametric equations are $x = 2 + t$, $y = 7t$, $z = -3 - 5t$.

14.6.11 $\langle 1, 1, 1 \rangle$ and $\langle 2, -1, 1 \rangle$ are respectively normal to the given planes.

$\langle 1, 1, 1 \rangle \times \langle 2, -1, , 1 \rangle = \langle 2, 1, -3 \rangle$ or any scalar multiple is parallel to the line of intersection of the given planes. Find a point on the line by setting $z = 0$ in both equations and solve
$$\left. \begin{array}{l} x + y = 4 \\ 2x - y = 2 \end{array} \right\}$$
to get $x = 2$, $y = 2$, $z = 0$, thus, the vector equation of the line of intersection of the two planes is $x = \langle 2, 2, 0 \rangle + t \langle 2, 1, -3 \rangle$.

14.6.12 Let $P_1(a, 0, 0)$, $P_2(0, b, 0)$, and $P_3(0, 0, c)$ be the given intercepts, then $\overrightarrow{P_1 P_2} = \langle -a, b, 0 \rangle$ and $\overrightarrow{P_1 P_3} = \langle -a, 0, c \rangle$, thus, $\langle -a, b, 0 \rangle \times \langle -a, 0, c \rangle = \langle bc, ac, ab \rangle$ is normal to the plane whose equation is $bc(x - a) + ac(y - 0) + ab(z - 0) = 0$ or $bcx + acy + abz = abc$ which can be written as $\dfrac{x}{a} + \dfrac{y}{b} + \dfrac{z}{c} = 1$.

14.6.13 $\langle 2, -1, -3 \rangle$ is parallel to the line and hence normal to the plane whose equation is
$$2(x - 3) - 1(y - 0) - 3(z - 1) = 0 \text{ or } 2x - y - 3z - 3 = 0.$$

14.6.14 Find two other points on the plane by setting $t = 0$ and $t = 1$ to get $P_1(2, 1, 1)$ and $P_2(1, 5, 5)$. Let $P_0(-2, 1, 1)$ be the given point, then $\overrightarrow{P_0 P_1} = \langle 4, 0, 0 \rangle$ and

$\overrightarrow{P_0 P_2} = \langle 3, 4, 4 \rangle$, thus, $\overrightarrow{P_0 P_1} \times \overrightarrow{P_0 P_2} = \langle 0, -16, 16 \rangle$ or any scalar multiple such as $\langle 0, 1, -1 \rangle$ is normal to the plane whose equation is $0(x + 2) + 1(y - 1) - 1(z - 1) = 0$ or $y - z = 0$.

14.6.15 $\langle 3, -2, 1 \rangle$ and $\langle 8, 2, 1 \rangle$ are respectively normal to the given planes.

$\langle 3, -2, 1 \rangle \times \langle 8, 2, 1 \rangle = \langle -4, 5, 22 \rangle$ or any scalar multiple is parallel to the line of intersection of the given planes. Find a point on the line by setting $z = 0$ in both equations and solve
$$\left. \begin{array}{l} 3x - 2y = 0 \\ 8x + 2y = 11 \end{array} \right\}$$
to get $x = 1$, $y = 3/2$, $z = 0$, thus, the parametric equations of the line of

intersection of the two planes are $x = 1 - 4t$, $y = \dfrac{3}{2} + 5t$, $z = 22t$.

14.6.16 $\langle 2, 1, -1 \rangle$ and $\langle 3, -1, 1 \rangle$ are respectively normal to the given planes.

$\langle 2, 1, -1 \rangle \times \langle 3, -1, 1 \rangle = \langle 0, -5, -5 \rangle$ or any scalar multiple such as $\langle 0, 1, 1 \rangle$ is parallel to the line of intersection of the two planes. Thus $\overrightarrow{P_1 P_2} = \langle -2, 1, 0 \rangle$ and $\langle 0, 1, 1 \rangle$ lie on the required plane whose normal is

$\langle -2, 1, 0 \rangle \times \langle 0, 1, 1 \rangle = \langle 1, 2, -2 \rangle$ and whose equation is
$$1(x - 1) + 2(y - 1) - 2(z - 1) = 0 \text{ or } x + 2y - 2z - 1 = 0.$$

14.6.17 $\langle 2, -1, -1 \rangle$ and $\langle 3, 2, -2 \rangle$ are respectively normal to the given planes.

$\langle 2, -1, -1 \rangle \times \langle 3, 2, -2 \rangle = \langle 4, 1, 7 \rangle$ or any scalar multiple is parallel to the line of intersection of the two planes. Thus $\overrightarrow{P_1 P_2} = \langle -4, 1, 3 \rangle$ and $\langle 4, 1, 7 \rangle$ lie on the required plane whose normal is $\langle -4, 1, -3 \rangle \times \langle 4, 1, 7 \rangle = \langle 10, 16, -8 \rangle$ and whose equation is $10(x - 3) + 16(y - 1) - 8(z - 2) = 0$ or $5x + 8y - 4z - 15 = 0$.

SECTION 14.7

14.7.1   Name and sketch $2x^2 + y^2 - 4z = 0$.

14.7.2   Name and sketch $x^2 - y^2 + z^2 + 2y = 1$.

14.7.3   Describe the surface given by $9x^2 + 4y^2 - 54x - 16y - 36z + 277 = 0$.

14.7.4   Describe and sketch the surface given by $6x^2 + 4y^2 - 3z^2 + 36x - 16y + 24z + 10 = 0$.

14.7.5   Name and sketch $z^2 = x^2 + y^2$.

14.7.6   Describe and sketch $x^2 + y^2 + z - 5 = 0$.

14.7.7   Describe and sketch $x^2 + 4y^2 + z^2 - 8y = 0$.

14.7.8   Describe and sketch $x^2 + y^2 - z^2 - 2x + 4y - 2z = 0$.

14.7.9   Name and sketch $z = 4x^2 + y^2$.

14.7.10   Name and sketch $\dfrac{x^2}{4} - \dfrac{y^2}{9} + \dfrac{z^2}{16} = 1$.

14.7.11   Describe the surface given by $2y^2 - 3x^2 + 4y + 30x - 6z - 85 = 0$.

14.7.12   Describe the surface given by $5x^2 + 4y^2 + 20z^2 - 20x + 32y + 40z + 56 = 0$.

14.7.13   Describe the surface given by $2y^2 + 5z^2 - 12y - 20z - 10x + 48 = 0$.

14.7.14   Describe the surface given by $6x^2 + 4y^2 - 2z^2 - 6x - 4y + z = 0$.

14.7.15   Describe the surface given by $3x^2 - 2y^2 - z^2 - 6x + 8y - 2z + 6 = 0$.

14.7.16   Identify and sketch $9x^2 + 4z^2 - 36 = 0$.

SOLUTIONS

SECTION 14.7

14.7.1 $z = \dfrac{x^2}{2} + \dfrac{y^2}{4}$

Elliptic paraboloid.

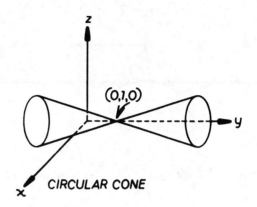

14.7.2 $x^2 + z^2 = (y-1)^2$

Circular cone.

14.7.3 Elliptic paraboloid, $\dfrac{(x-3)^2}{4} + \dfrac{(y-2)^2}{9} = z - 5$, $C(3, 2, 5)$.

14.7.4 $\dfrac{(x+3)^2}{2} + \dfrac{(y-2)^2}{3} - \dfrac{(z-4)^2}{4} = 1$

Hyperboloid of one sheet.

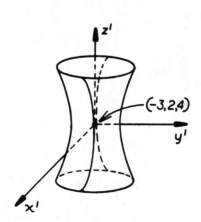

14.7.5   Circular cone.

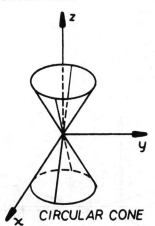

CIRCULAR CONE

14.7.6 $z - 5 = -(x^2 + y^2)$
Circular paraboloid, $C(0, 0, 5)$.

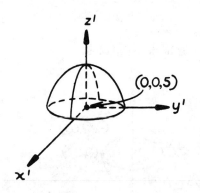

14.7.7 $\dfrac{x^2}{4} + (y - 1)^2 + \dfrac{z^2}{4} = 1$ Ellipsoid $C(0, 1, 0)$.

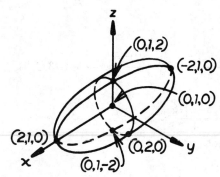

14.7.8 $\dfrac{(x - 1)^2}{4} + \dfrac{(y + 2)^2}{4} - \dfrac{(z + 1)^2}{4} = 1$
Hyperboloid of one sheet.

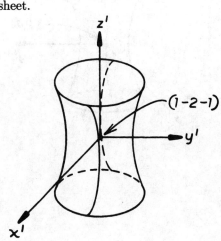

14.7.9 $z = \dfrac{x^2}{1/4} + y^2$
Elliptic paraboloid.

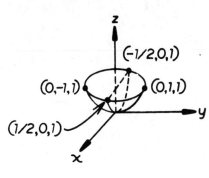

14.7.10   Hyperboloid of one sheet.

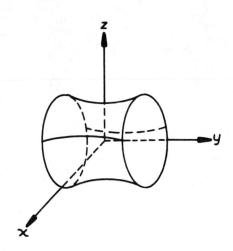

14.7.11 $z + 2 = \dfrac{(y+1)^2}{3} - \dfrac{(x-5)^2}{2}$, hyperbolic paraboloid, $C(5, -1, -2)$.

14.7.12 $\dfrac{(x-2)^2}{48/5} + \dfrac{(y+4)^2}{12} + \dfrac{(z+1)^2}{12/5} = 1$, ellipsoid, $C(2, -4, -1)$.

14.7.13 $x - 1 = \dfrac{(y-3)^2}{5} + \dfrac{(z-2)^2}{2}$, elliptic paraboloid, $C(1, 3, 2)$.

14.7.14 $\dfrac{(x-1/2)^2}{19/48} + \dfrac{(y-1/2)^2}{19/32} - \dfrac{(z-1/4)^2}{19/16} = 1$, hyperboloid of one sheet, $C(1/2, 1/2, 1/4)$.

14.7.15 $\dfrac{(y-2)^2}{6} + \dfrac{(z+1)^2}{12} - \dfrac{(x-1)^2}{4} = 1$, hyperboloid of one sheet, $C(1, 2, -1)$.

14.7.16 $\dfrac{x^2}{4} + \dfrac{z^2}{9} = 1$
Cylinder
$C(0, 0, 0)$.

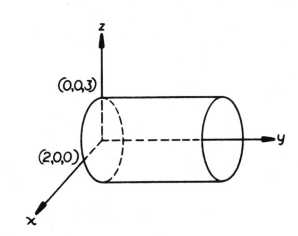

SECTION 14.8

14.8.1 Convert $(3, 3\pi/4, 2\pi/3)$ from spherical coordinates to cylindrical coordinates.

14.8.2 Convert $(3, 3\pi/4, 2\pi/3)$ from spherical coordinates to rectangular coordinates.

14.8.3 Convert $(2, 2\pi/3, \pi/2)$ from spherical coordinates to cylindrical coordinates.

14.8.4 Convert $(2, 2\pi/3, \pi/2)$ from spherical coordinates to rectangular coordinates.

14.8.5 Convert $\left(4, \dfrac{\pi}{6}, 5\right)$ in cylindrical coordinates to rectangular coordinates.

14.8.6 Convert $(4, \pi/6, 5)$ in cylindrical coordinates to spherical coordinates.

14.8.7 Convert $(3, \pi/6, 2\pi/3)$ in spherical coordinates to rectangular coordinates.

14.8.8 Convert $\left(3, \dfrac{\pi}{6}, \dfrac{2\pi}{3}\right)$ in spherical coordinates to cylindrical coordinates.

14.8.9 Convert $\left(-\dfrac{3}{2}, \dfrac{\sqrt{3}}{2}, 0\right)$ in rectangular coordinates to cylindrical coordinates.

14.8.10 Convert $\left(-\dfrac{3}{2}, \dfrac{\sqrt{3}}{2}, 0\right)$ in rectangular coordinates to spherical coordinates.

14.8.11 Convert $\left(-\sqrt{3}, 1, 2\sqrt{3}\right)$ in rectangular coordinates to cylindrical coordinates.

14.8.12 Convert $\left(-\sqrt{3}, 1, 2\sqrt{3}\right)$ in rectangular coordinates to spherical coordinates.

14.8.13 Transform $r^2 \cos 2\theta = z^2$ in cylindrical coordinates to rectangular coordinates. Name the resulting surface.

14.8.14 Transform $\rho = 2 \csc \phi$ in spherical coordinates to rectangular coordinates. Name the resulting surface.

14.8.15 Transform $r^2 + z^2 = 1$ in cylindrical coordinates to rectangular coordinates. Name the resulting surface.

14.8.16   Transform $z = \dfrac{1}{4}(x^2 + y^2)$ from rectangular coordinates to cylindrical coordinates.

14.8.17   Transform $z = \dfrac{1}{4}\left(x^2 + y^2\right)$ from rectangular coordinates to spherical coordinates.

14.8.18   Transform $\dfrac{x^2}{6} + \dfrac{y^2}{6} + \dfrac{z^2}{3} = 1$ from rectangular coordinates to cylindrical coordinates.

SOLUTIONS

SECTION 14.8

14.8.1 $\left(\dfrac{3\sqrt{3}}{2}, \dfrac{3\pi}{4}, -\dfrac{3}{2}\right)$.

14.8.2 $\left(-\dfrac{3\sqrt{3}}{2\sqrt{2}}, \dfrac{3\sqrt{3}}{2\sqrt{2}}, -\dfrac{3}{2}\right)$.

14.8.3 $(2, 2\pi/3, 0)$.

14.8.4 $(-1, \sqrt{3}, 0)$.

14.8.5 $\left(2\sqrt{3}, 2, 5\right)$.

14.8.6 $\left(\sqrt{41}, \dfrac{\pi}{6}, \tan^{-1}\dfrac{4}{5}\right)$.

14.8.7 $\left(\dfrac{9}{4}, \dfrac{3\sqrt{3}}{4}, -\dfrac{3}{2}\right)$.

14.8.8 $\left(\dfrac{3\sqrt{3}}{2}, \dfrac{\pi}{6}, -\dfrac{3}{2}\right)$.

14.8.9 $\left(\sqrt{3}, \dfrac{5\pi}{6}, 0\right)$.

14.8.10 $\left(\sqrt{3}, \dfrac{5\pi}{6}, \dfrac{\pi}{2}\right)$.

14.8.11 $\left(2, \dfrac{5\pi}{6}, 2\sqrt{3}\right)$.

14.8.12 $\left(4, \dfrac{5\pi}{6}, \dfrac{\pi}{6}\right)$.

14.8.13 $x^2 - y^2 = z^2$, hyperbolic paraboloid.

14.8.14 $x^2 + y^2 = 4$, cylinder.

14.8.15 $x^2 + y^2 + z^2 = 1$, sphere.

14.8.16 $r^2 = 4z$.

14.8.17 $\rho = 4\cot\phi\csc\phi$.

14.8.18 $r^2 + 2z^2 = 6$.

SUPPLEMENTARY EXERCISES, CHAPTER 14

1. Find $\overrightarrow{P_1P_2}$ and $\| \overrightarrow{P_1P_2} \|$.

    (a) $P_1(2,3)$, $P_2(5,-1)$ (b) $P_1(2,-1)$, $P_2(1,3)$.

In Exercises 2–7, find all vectors satisfying the given conditions.

2. A vector of length 1 in 2-space that is perpendicular to the line $x + y = -1$.

3. The vector oppositely directed to $3\mathbf{i} - 4\mathbf{j}$, and having the same length.

4. The vector obtained by rotating $\mathbf{i}$ counterclockwise through an angle θ in 2-space.

5. The vector with initial point $(1,2)$ and a terminal point that is $3/5$ of the way from $(1,2)$ to $(5,5)$.

6. A vector of length 2 that is parallel to the tangent to the curve $y = x^2$ at $(-1,1)$.

7. The vector of length 12 in 2-space that makes an angle of $120°$ with the x-axis.

8. Solve for c_1 and c_2 given that $c_1\langle -2,5 \rangle + 3c_2\langle 1,3 \rangle = \langle -6,-51 \rangle$.

9. Solve for $\mathbf{u}$ if $3\mathbf{u} - (\mathbf{i} + \mathbf{j}) = \mathbf{i} + \mathbf{u}$.

10. Solve for $\mathbf{u}$ and $\mathbf{v}$ if $3\mathbf{u} - 4\mathbf{v} = 3\mathbf{v} - 2\mathbf{u} = \langle 1,2 \rangle$.

11. Two forces $\mathbf{F}_1 = 2\mathbf{i} - \mathbf{j}$ and $\mathbf{F}_2 = -3\mathbf{i} - 4\mathbf{j}$ are applied at a point. What force $\mathbf{F}_3$ must be applied at the point to cancel the effect of $\mathbf{F}_1$ and $\mathbf{F}_2$?

12. Given the points $P(3,4)$, $Q(1,1)$, and $R(5,2)$, use vector methods to find the coordinates of the fourth vertex of the parallelogram whose adjacent sides are $\overrightarrow{PQ}$ and $\overrightarrow{QR}$.

In Exercises 13 and 14, find

(a) $\|\mathbf{a}\|$ (b) $\mathbf{a} \cdot \mathbf{b}$ (c) $\mathbf{a} \times \mathbf{b}$ (d) $\mathbf{b} \times \mathbf{a}$

(e) the area of the triangle with sides $\mathbf{a}$ and $\mathbf{b}$ (f) $3\mathbf{a} - 2\mathbf{b}$.

13. $\mathbf{a} = \langle 1,2,-1 \rangle$, $\mathbf{b} = \langle 2,-1,3 \rangle$.   14. $\mathbf{a} = \langle 1,-2,2 \rangle$, $\mathbf{b} = \langle 3,4,-5 \rangle$.

In Exercises 15 and 16, find

(a) $\|\text{proj}_{\mathbf{b}}\mathbf{a}\|$

(b) $\|\text{proj}_{\mathbf{a}}\mathbf{b}\|$

(c) the angle between $\mathbf{a}$ and $\mathbf{b}$

(d) the direction cosines of $\mathbf{a}$.

15. $\mathbf{a} = 3\mathbf{i} - 4\mathbf{j}$, $\mathbf{b} = 2\mathbf{i} + 2\mathbf{j} - \mathbf{k}$.

16. $\mathbf{a} = -\mathbf{j}$, $\mathbf{b} = \mathbf{i} + \mathbf{j}$.

17. Verify the identity $\mathbf{a} \times (\mathbf{b} \times \mathbf{c}) = (\mathbf{a} \cdot \mathbf{c})\mathbf{b} - (\mathbf{a} \cdot \mathbf{b})\mathbf{c}$ for $\mathbf{a} = \mathbf{i} + \mathbf{j}$, $\mathbf{b} = 2\mathbf{i} - \mathbf{k}$, $\mathbf{c} = \mathbf{j} - \mathbf{k}$.

18. Find the vector with length 5 and direction angles $\alpha = 60°$, $\beta = 120°$, $\gamma = 135°$.

19. Find the vector with length 3 and direction cosines $-1/\sqrt{2}$, 0, and $1/\sqrt{2}$.

20. For the points $P(6, 5, 7)$ and $Q(7, 3, 9)$, find

(a) the midpoint of the line segment PQ

(b) the length and direction cosines of $\overrightarrow{PQ}$.

21. If $\mathbf{u} = \mathbf{i} + 2\mathbf{j} - 3\mathbf{k}$ and $\mathbf{v} = \mathbf{i} + \mathbf{j} + 2\mathbf{k}$, find

(a) the vector component of $\mathbf{u}$ along $\mathbf{v}$

(b) the vector component of $\mathbf{u}$ orthogonal to $\mathbf{v}$.

22. Find the vector component of $\mathbf{i}$ along $3\mathbf{i} - 2\mathbf{j} + \mathbf{k}$.

23. A diagonal of a box makes angles of 50° and 70° with two of its edges. Find, to the nearest degree, the angle that it makes with the third edge.

24. Consider the points $O(0, 0, 0)$, $A(0, a, a)$, and $B(-3, 4, 2)$. Find all nonzero values of a that make $\overrightarrow{OA}$ orthogonal to $\overrightarrow{AB}$.

25. Under what conditions are $\mathbf{u} + \mathbf{v}$ and $\mathbf{u} - \mathbf{v}$ orthogonal?

26. If $M(3, -1, 5)$ is the midpoint of the line segment PQ and if the coordinates of P are $(1, 2, 3)$, find the coordinates of Q.

27. Find all possible vectors of length 1 orthogonal to both $\mathbf{a} = \langle 3, -2, 1 \rangle$ and $\mathbf{b} = \langle -2, 1, -3 \rangle$.

28. Find the distance from the point $P(2, 3, 4)$ to the plane containing the points $A(0, 0, 1)$, $B(1, 0, 0)$, and $C(0, 2, 0)$.

In Exercises 29–32, find an equation for the plane that satisfies the given conditions.

29. The plane through $A(1, 2, 3)$ and $B(2, 4, 2)$ that is parallel to $\mathbf{v} = \langle -3, -1, -2 \rangle$.

30. The plane through $P_0(-1, 2, 3)$ that is perpendicular to the planes $2x - 3y + 5 = 0$ and $3x - y - 4z + 6 = 0$.

31. The plane that passes through $P(1, 1, 1)$, $Q(2, 3, 0)$, and $R(2, 1, 2)$.

32. The plane with intercepts $x = 2$, $y = -3$, $z = 10$.

33. Let L be the line through $P(1, 2, 8)$ that is parallel to $\mathbf{v} = \langle 3, -1, -4 \rangle$.

(a) For what values of k and l will the point $Q(k, 3, l)$ be on L?

(b) If L' has parametric equations $x = -8 - 3t$, $y = 5 + t$, $z = 0$, show that L' intersects L and find the point of intersection.

(c) Find the point at which L intersects the plane through $R(-4, 0, 3)$ having a normal vector $\langle 3, -2, 6 \rangle$.

34. Consider the lines L_1 and L_2 with symmetric equations

$$L_1 : \frac{x - 1}{2} = \frac{y + \frac{3}{2}}{1} = \frac{z + 1}{2}$$

$$L_2 : \frac{4 - x}{1} = \frac{3 - y}{2} = \frac{4 + z}{2}$$

(see Exercise 34, Section 14.5).

(a) Are L_1 and L_2 parallel? Perpendicular?

(b) Find parametric equations for L_1 and L_2.

(c) Do L_1 and L_2 intersect? If so, where?

35. Find parametric equations for the line through P_1 and P_2.

(a) $P_1(1, -1, 2)$, $P_2(3, 2, -1)$ (b) $P_1(1, -3, 4)$, $P_2(1, 2, -3)$.

36. For points $A(1, -1, 2)$, $B(2, -3, 0)$, $C(-1, -2, 0)$, and $D(2, 1, -1)$, find

(a) $\overrightarrow{AB} \times \overrightarrow{AC}$ (b) the area of triangle ABC

(c) the volume of the parallelepiped determined by the vectors $\overrightarrow{AB}$, $\overrightarrow{AC}$, $\overrightarrow{AD}$

(d) the distance from D to the plane containing A, B, and C.

37. **(a)** Find parametric equations for the intersection of the planes $2x + y - z = 3$ and $x + 2y + z = 3$.

   (b) Find the acute angle between the two planes.

In Exercises 38–40, describe the region satisfying the given conditions.

38. **(a)** $x^2 + 9y^2 + 4z^2 > 36$ **(b)** $x^2 + y^2 + z^2 - 6x + 2y - 6 < 0$.

39. **(a)** $z > 4x^2 + 9y^2$ **(b)** $x^2 + 4y^2 + z^2 = 0$.

40. **(a)** $y^2 + 4z^2 = 4,\ 0 \le x \le 2$ **(b)** $9x^2 + 4y^2 + 36x - 8y = -60$.

In Exercises 41–45, identify the quadric surface whose equation is given.

41. $100x^2 + 225y^2 - 36z^2 = 0$.   **42.** $x^2 - z^2 + y = 0$.

43. $400x^2 + 25y^2 + 16z^2 = 400$.   **44.** $4x^2 - y^2 + 4z^2 = 4$.

45. $-16x^2 - 100y^2 + 25z^2 = 400$.

46. Identify the surface by completing the squares.

   (a) $x^2 + 4y^2 - z^2 - 6x + 8y + 4z = 0$ **(b)** $x^2 + y^2 + z^2 + 6x - 4y + 12z = 0$.

47. Find the work done by a constant force $\mathbf{F} = 3\mathbf{i} - 4\mathbf{j} + \mathbf{k}$ (pounds) acting on a particle that moves along the line segment from $P(5, 7, 0)$ to $Q(6, 6, 6)$ (units in feet).

48. Two forces $\mathbf{F}_1 = \mathbf{i} - 3\mathbf{j} + \mathbf{k}$ and $\mathbf{F}_2 = \mathbf{i} + 2\mathbf{j} + 2\mathbf{k}$ (pounds) act on a particle as it moves in a straight line from $P(-1, -2, 3)$ to $Q(0, 2, 0)$ (units in feet). How much work is done?

49. Convert $(\sqrt{2}, \pi/4, 1)$ from cylindrical coordinates to

   (a) rectangular coordinates   **(b)** spherical coordinates.

50. Convert from rectangular coordinates to (i) cylindrical coordinates, (ii) spherical coordinates.

   (a) $(2, 2, 2\sqrt{6})$ **(b)** $(1, \sqrt{3}, 0)$.

51. Express the equation in terms of rectangular coordinates.

   (a) $z = r^2 \cos 2\theta$ **(b)** $\rho^2 \sin \phi \cos \phi \cos \theta = 1$.

52. Sketch the set of points defined by the given conditions.

(a) $0 \leq \theta \leq \pi/2$, $0 \leq r \leq \cos\theta$, $0 \leq z \leq 2$ (cylindrical coordinates)

(b) $0 \leq \theta \leq \pi/2$, $0 \leq \phi \leq \pi/4$, $0 \leq \rho \leq 2\sec\phi$ (spherical coordinates)

(c) $r = 2\sin\theta$, $0 \leq z \leq 2$ (cylindrical coordinates)

(d) $\rho = 2\cos\phi$ (spherical coordinates).

SOLUTIONS

SUPPLEMENTARY EXERCISES CHAPTER 14

1. **(a)** $\overrightarrow{P_1P_2}= (5-2)\mathbf{i}+(-1-3)\mathbf{j} = 3\mathbf{i}-4\mathbf{j},\ \|\overrightarrow{P_1P_2}\| = 5$

   (b) $\overrightarrow{P_1P_2}= (1-2)\mathbf{i}+(3+1)\mathbf{j} = -\mathbf{i}+4\mathbf{j},\ \|\overrightarrow{P_1P_2}\| = \sqrt{17}$

2. The slope of the line $x+y = -1$ is -1 so the slope of a line perpendicular to it is 1 thus $\mathbf{i}+\mathbf{j}$ is a vector perpendicular to the given line, $\|\mathbf{i}+\mathbf{j}\| = \sqrt{2}$ so $(\mathbf{i}+\mathbf{j})/\sqrt{2}$ is a vector of length 1 that is perpendicular to the given line, another such vector is $-(\mathbf{i}+\mathbf{j})/\sqrt{2}$.

3. $-(3\mathbf{i}-4\mathbf{j}) = -3\mathbf{i}+4\mathbf{j}$

4. Let $\mathbf{v}$ be the desired vector, then $\|\mathbf{v}\| = \|\mathbf{i}\| = 1$ and $\phi = 0+\theta = \theta$ so $\mathbf{v} = \cos\theta\mathbf{i}+\sin\theta\mathbf{j}$.

5. $4\mathbf{i}+3\mathbf{j}$ is the vector from (1,2) to (5,5) so the desired vector is $(3/5)(4\mathbf{i}+3\mathbf{j})$

6. $dy/dx = 2x$, the slope of the tangent at $(-1,1)$ is $2(-1) = -2$ so the vector $\mathbf{i}-2\mathbf{j}$ is parallel to the tangent, $\|\mathbf{i}-2\mathbf{j}\| = \sqrt{5}$ so $2(\mathbf{i}-2\mathbf{j})/\sqrt{5}$ is a vector of length 2 that is parallel to the tangent, another such vector is $-2(\mathbf{i}-2\mathbf{j})/\sqrt{5}$.

7. $12\cos 120°\mathbf{i}+12\sin 120°\mathbf{j} = -6\mathbf{i}+6\sqrt{3}\mathbf{j}$

8. $c_1\langle-2,5\rangle+3c_2\langle1,3\rangle = \langle-2c_1,5c_1\rangle+\langle3c_2,9c_2\rangle = \langle-2c_1+3c_2, 5c_1+9c_2\rangle$ so $-2c_1+3c_2 = -6$ and $5c_1+9c_2 = -51$, solve to get $c_1 = -3,\ c_2 = -4$.

9. $3\mathbf{u}-(\mathbf{i}+\mathbf{j}) = \mathbf{i}+\mathbf{u},\ 2\mathbf{u} = 2\mathbf{i}+\mathbf{j},\ \mathbf{u} = \mathbf{i}+(1/2)\mathbf{j}$

10. If $3\mathbf{u}-4\mathbf{v} = 3\mathbf{v}-2\mathbf{u}$ then $\mathbf{v} = (5/7)\mathbf{u},\ 3\mathbf{u}-4\mathbf{v} = (3-20/7)\mathbf{u} = (1/7)\mathbf{u} = \langle1,2\rangle,\ \mathbf{u} = \langle7,14\rangle,$ $\mathbf{v} = (5/7)\langle7,14\rangle = \langle5,10\rangle$.

11. The effect of $\mathbf{F}_1$ and $\mathbf{F}_2$ is the same as the effect of $\mathbf{F}_1+\mathbf{F}_2 = -\mathbf{i}-5\mathbf{j}$ acting at the point, to cancel the effect a force $\mathbf{F}_3 = -(\mathbf{F}_1+\mathbf{F}_2) = \mathbf{i}+5\mathbf{j}$ must be applied at the point.

12. Let $S(x,y)$ be the fourth vertex, then $\overrightarrow{PS}=\overrightarrow{QR},\ \langle x-3,\ y-4\rangle = \langle4,1\rangle$, so $x-3 = 4$ and $y-4 = 1$, $x = 7$ and $y = 5$.

13. **(a)** $\sqrt{6}$ **(b)** -3 **(c)** $\langle5,-5,-5\rangle$

    (d) $\langle-5,5,5\rangle$ **(e)** $5\sqrt{3}/2$ **(f)** $\langle-1,8,-9\rangle$

14. **(a)** 3   **(b)** -15 **(c)** $\langle2,11,10\rangle$

    (d) $\langle-2,-11,-10\rangle$ **(e)** $15/2$ **(f)** $\langle-3,-14,16\rangle$

15. (a) $2/3$ (b) $2/5$ (c) $\cos^{-1}(-2/15)$ (d) $3/5, -4/5, 0$

16. (a) $1/\sqrt{2}$ (b) 1 (c) $3\pi/4$ (d) $0, -1, 0$

17. Both sides reduce to $2\mathbf{i} - 2\mathbf{j} + \mathbf{k}$

18. $\mathbf{v} = 5\langle \cos 60°, \cos 120°, \cos 135° \rangle = \langle 5/2, -5/2, -5/\sqrt{2} \rangle$

19. $\langle -3/\sqrt{2}, 0, 3/\sqrt{2} \rangle$

20. (a) Let $M(m_1, m_2, m_3)$ be the midpoint, then $\overrightarrow{PM} = (1/2)\,\overrightarrow{PQ}$,
$\langle m_1 - 6, m_2 - 5, m_3 - 7 \rangle = \langle 1/2, -1, 1 \rangle$, equate corresponding components to get
$m_1 = 13/2$, $m_2 = 4$, $m_3 = 8$ so the midpoint is $(13/2, 4, 8)$.

    (b) $\overrightarrow{PQ} = \langle 1, -2, 2 \rangle$, $\|\overrightarrow{PQ}\| = 3$, $\cos\alpha = 1/3$, $\cos\beta = -2/3$, $\cos\gamma = 2/3$

21. (a) $\text{proj}_{\mathbf{v}}\mathbf{u} = \dfrac{\mathbf{u}\cdot\mathbf{v}}{\|\mathbf{v}\|^2}\mathbf{v} = \dfrac{(-3)}{6}(\mathbf{i}+\mathbf{j}+2\mathbf{k}) = -\dfrac{1}{2}\mathbf{i} - \dfrac{1}{2}\mathbf{j} - \mathbf{k}$

    (b) $\mathbf{u} - \text{proj}_{\mathbf{v}}\mathbf{u} = \dfrac{3}{2}\mathbf{i} + \dfrac{5}{2}\mathbf{j} - 2\mathbf{k}$

22. With $\mathbf{u} = \mathbf{i}$ and $\mathbf{a} = 3\mathbf{i} - 2\mathbf{j} + \mathbf{k}$, $\text{proj}_{\mathbf{a}}\mathbf{u} = \dfrac{\mathbf{u}\cdot\mathbf{a}}{\|\mathbf{a}\|^2}\mathbf{a} = \dfrac{3}{14}(3\mathbf{i} - 2\mathbf{j} + \mathbf{k})$.

23. $\cos^2\alpha + \cos^2\beta + \cos^2\gamma = 1$, let $\alpha = 50°$, $\beta = 70°$,
$\cos^2\gamma = 1 - \cos^2(50°) - \cos^2(70°) \approx 0.46985$, $\gamma \approx 62°$.

24. $\overrightarrow{OA} \cdot \overrightarrow{AB} = 0$, $\langle 0, a, a \rangle \cdot \langle -3, 4 - a, 2 - a \rangle = 0$, $6a - 2a^2 = 0$, $a = 0$ or 3.

25. (a) $(\mathbf{u}+\mathbf{v})\cdot(\mathbf{u}-\mathbf{v}) = 0$, $\mathbf{u}\cdot\mathbf{u} - \mathbf{u}\cdot\mathbf{v} + \mathbf{v}\cdot\mathbf{u} - \mathbf{v}\cdot\mathbf{v} = 0$, $\|\mathbf{u}\|^2 - \|\mathbf{v}\|^2 = 0$, $\|\mathbf{u}\| = \|\mathbf{v}\|$

    (b) $(\mathbf{a}\cdot\mathbf{b})^2 + \|\mathbf{a}\times\mathbf{b}\|^2 = (\|\mathbf{a}\|\,\|\mathbf{b}\|\cos\theta)^2 + (\|\mathbf{a}\|\,\|\mathbf{b}\|\sin\theta)^2$
$$= \|\mathbf{a}\|^2\|\mathbf{b}\|^2(\cos^2\theta + \sin^2\theta) = \|\mathbf{a}\|^2\|\mathbf{b}\|^2$$

26. $\overrightarrow{PQ} = 2\overrightarrow{PM}$, $\langle q_1 - 1, q_2 - 2, q_3 - 3 \rangle = \langle 4, -6, 4 \rangle$, $q_1 = 5$, $q_2 = -4$, $q_3 = 7$ so Q has coordinates $(5, -4, 7)$.

27. $\mathbf{a} \times \mathbf{b} = \langle 5, 7, -1 \rangle$ is orthogonal to both $\mathbf{a}$ and $\mathbf{b}$, $\|\mathbf{a} \times \mathbf{b}\| = 5\sqrt{3}$ so
$\pm\langle 1/\sqrt{3}, 7/(5\sqrt{3}), -1/(5\sqrt{3}) \rangle$ are unit vectors orthogonal to both $\mathbf{a}$ and $\mathbf{b}$.

28. $2x + y + 2z = 2$ is an equation of the plane containing A, B, and C so
$D = |2(2) + (3) + 2(4) - 2|/\sqrt{4 + 1 + 4} = 13/3$.

29. The plane contains $\overrightarrow{AB}$ and is parallel to $\mathbf{v}$ thus $\mathbf{v} \times \overrightarrow{AB}$ is normal to the plane,

$\mathbf{v} \times \overrightarrow{AB} = \langle 5, -5, -5 \rangle$ so $\langle 1, -1, -1 \rangle$ is also a normal to the plane whose equation is

$x - y - z = -4$.

30. $\mathbf{n}_1 = \langle 2, -3, 0 \rangle$ and $\mathbf{n}_2 = \langle 3, -1, -4 \rangle$ are normals to the given planes so $\mathbf{n}_1 \times \mathbf{n}_2 = \langle 12, 8, 7 \rangle$ is normal to the desired plane whose equation is $12x + 8y + 7z = 25$.

31. $\overrightarrow{PQ} \times \overrightarrow{PR} = \langle 1, 2, -1 \rangle \times \langle 1, 0, 1 \rangle = \langle 2, -2, -2 \rangle$ is normal to the plane and hence so is $\langle 1, -1, -1 \rangle$, an equation of the plane is $x - y - z = -1$.

32. The intercepts correspond to the points $A(2, 0, 0)$, $B(0, -3, 0)$ and $C(0, 0, 10)$;

$\overrightarrow{AB} \times \overrightarrow{AC} = \langle -30, 20, -6 \rangle$ is normal to the plane and hence so is $\langle 15, -10, 3 \rangle$, an equation of the plane is $15x - 10y + 3z = 30$.

33. **(a)** Parametric equations of L are $x = 1 + 3t$, $y = 2 - t$, $z = 8 - 4t$. If Q is on L then for some t_0, $k = 1 + 3t_0$, $3 = 2 - t_0$, $\ell = 8 - 4t_0$. The second of these equations yields $t_0 = -1$ so $k = -2$, $\ell = 12$.

(b) Use parametric equations in part (a) for L, solve the system $1 + 3t_1 = -8 - 3t_2$, $2 - t_1 = 5 + t_2$, $8 - 4t_1 = 0$ to get $t_1 = 2$, $t_2 = -5$ so L' intersects L at $(7,0,0)$.

(c) An equation of the plane is $3x - 2y + 6z = 6$, use the parametric equations in part (a) to get $3(1 + 3t) - 2(2 - t) + 6(8 - 4t) = 6$, $t = 41/13$ so L intersects the plane at $(136/13, -15/13, -60/13)$.

34. **(a)** $\mathbf{v}_1 = \langle 2, 1, 2 \rangle$ and $\mathbf{v}_2 = \langle -1, -2, 2 \rangle$ are parallel, respectively, to L_1 and L_2. $\mathbf{v}_1 \cdot \mathbf{v}_2 = 0$ so the lines are perpendicular in the sense that $\mathbf{v}_1$ and $\mathbf{v}_2$ are perpendicular.

(b) $(1, -3/2, -1)$ and $(4, 3, -4)$ are points on L_1 and L_2, respectively, so parametric equations are $L_1 : x = 1 + 2t, y = -3/2 + t, z = -1 + 2t$; $L_2 : x = 4 - t, y = 3 - 2t, z = -4 + 2t$

(c) Solve the system $1 + 2t_1 = 4 - t_2$, $-3/2 + t_1 = 3 - 2t_2$, $-1 + 2t_1 = -4 + 2t_2$ to get $t_1 = 1/2$, $t_2 = 2$ so the lines intersect at $(2, -1, 0)$.

35. **(a)** $\overrightarrow{P_1 P_2} = \langle 2, 3, -3 \rangle$, use P_1 to get $x = 1 + 2t$, $y = -1 + 3t$, $z = 2 - 3t$.

(b) $\overrightarrow{P_1 P_2} = \langle 0, 5, -7 \rangle$, use P_1 to get $x = 1$, $y = -3 + 5t$, $z = 4 - 7t$.

36. **(a)** $\overrightarrow{AB} \times \overrightarrow{AC} = \langle 1, -2, -2 \rangle \times \langle -2, -1, -2 \rangle = \langle 2, 6, -5 \rangle$

(b) area $= \| \overrightarrow{AB} \times \overrightarrow{AC} \| / 2 = \sqrt{65}/2$

(c) volume $= | \overrightarrow{AD} \cdot (\overrightarrow{AB} \times \overrightarrow{AC}) | = | \langle 1, 2, -3 \rangle \cdot \langle 2, 6, -5 \rangle | = 29$

(d) $\overrightarrow{AB} \times \overrightarrow{AC}$ is normal to the plane so $2x + 6y - 5z + 14 = 0$ is an equation of the plane. The distance from D to the plane is $|2(2) + 6(1) - 5(-1) + 14| / \sqrt{4 + 36 + 25} = 29/\sqrt{65}$.

37. (a) $\mathbf{n}_1 = \langle 2, 1, -1 \rangle$ and $\mathbf{n}_2 = \langle 1, 2, 1 \rangle$ are normals to the planes so $\mathbf{n}_1 \times \mathbf{n}_2 = \langle 3, -3, 3 \rangle$ is parallel
to the line of intersection and hence so is $\langle 1, -1, 1 \rangle$. To find a point on the line of intersection,
let $x = 0$ in the equations of the planes to get $y - z = 3$ and $2y + z = 3$ which yield $y = 2$,
$z = -1$ so $(0, 2, -1)$ is on the line whose equations are $x = t$, $y = 2 - t$, $z = -1 + t$.

(b) $\mathbf{n}_1 \cdot \mathbf{n}_2 = 3 > 0$ so $\cos \theta = \dfrac{\mathbf{n}_1 \cdot \mathbf{n}_2}{\|\mathbf{n}_1\| \, \|\mathbf{n}_2\|} = \dfrac{3}{\sqrt{6}\sqrt{6}} = 1/2$, $\theta = 60°$.

38. (a) the region outside the ellipsoid $x^2/36 + y^2/4 + z^2/9 = 1$

(b) Complete the square to get $(x-3)^2 + (y+1)^2 + z^2 < 16$ which is the region inside the sphere
of radius 4 centered at $(3, -1, 0)$.

39. (a) the region above the elliptic paraboloid $z = 4x^2 + 9y^2$

(b) the point $(0,0,0)$

40. (a) The portion of the elliptic cylinder $y^2/4 + z^2 = 1$ that extends from $x = 0$ to $x = 2$.

(b) Complete the square to get $9(x+2)^2 + 4(y-1)^2 = -20$ which has no real solutions.

41. $z^2 = \dfrac{x^2}{(36/100)} + \dfrac{y^2}{(36/225)}$, elliptic cone

42. $y = z^2 - x^2$, hyperbolic paraboloid   **43.** $x^2 + y^2/16 + z^2/25 = 1$, ellipsoid

44. $x^2 - y^2/4 + z^2 = 1$, hyperboloid of one sheet

45. $x^2/25 + y^2/4 - z^2/16 = -1$, hyperboloid of two sheets

46. (a) $(x-3)^2 + 4(y+1)^2 - (z-2)^2 = 9$, hyperboloid of one sheet centered at $(3, -1, 2)$

(b) $(x+3)^2 + (y-2)^2 + (z+6)^2 = 49$, the sphere of radius 7 centered at $(-3, 2, -6)$

47. $W = \mathbf{F} \cdot \overrightarrow{PQ} = (3\mathbf{i} - 4\mathbf{j} + \mathbf{k}) \cdot (\mathbf{i} - \mathbf{j} + 6\mathbf{k}) = 13$ ft · lb

48. $W = (\mathbf{F}_1 + \mathbf{F}_2) \cdot \overrightarrow{PQ} = (2\mathbf{i} - \mathbf{j} + 3\mathbf{k}) \cdot (\mathbf{i} + 4\mathbf{j} - 3\mathbf{k}) = -11$ ft · lb

49. (a) $(1,1,1)$ (b) $(\sqrt{3}, \pi/4, \tan^{-1} \sqrt{2})$

50. (a) (i) $(2\sqrt{2}, \pi/4, 2\sqrt{6})$ (ii) $(4\sqrt{2}, \pi/4, \pi/6)$ (b) (i) $(2, \pi/3, 0)$ (ii) $(2, \pi/3, \pi/2)$

51. (a) $z = r^2(\cos^2 \theta - \sin^2 \theta)$, $z = x^2 - y^2$ (b) $(\rho \sin \phi \cos \theta)(\rho \cos \phi) = 1$, $xz = 1$

52. **(a)**

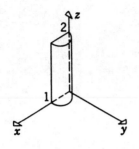

(b)

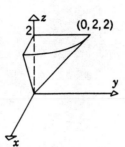

(c)

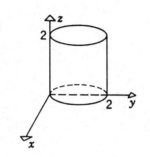

(d)

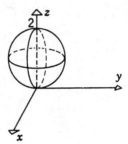

CHAPTER 15

Vector-Valued Functions

SECTION 15.1

15.1.1 Find the domain and $\mathbf{r}(0)$ for $r(t) = e^t \mathbf{i} - t e^t \mathbf{j}$.

15.1.2 Find the domain for $r(t) = \langle \sin t, \ln t, \tan^{-1} 2t \rangle$.

15.1.3 Find the domain for $r(t) = \ln \sqrt{1 + t} \mathbf{i} + \sqrt{4 + t^2} \mathbf{j} + t \mathbf{k}$.

15.1.4 Express $x = \cos^{-1} t$, $y = \sin 2t$, $z = t^2$ as a single vector equation.

15.1.5 Express $x = \sin 2t$, $y = \dfrac{1}{t}$, $z = t^2$ as a single vector equation.

15.1.6 Describe the graph of $\mathbf{r}(t) = (2 - 3t)\mathbf{i} + (1 + t)\mathbf{j} + (1 - t)\mathbf{k}$.

15.1.7 Describe the graph of $\mathbf{r}(t) = t^2 \mathbf{i} + t \mathbf{j}$.

15.1.8 Describe the graph of $\mathbf{r}(t) = \cos 3t \mathbf{i} + \sin 3t \mathbf{j} - 3 \mathbf{k}$.

15.1.9 Describe the graph of $\mathbf{r}(t) = 2 \cos t \mathbf{i} + \sin t \mathbf{j} + \mathbf{k}$.

15.1.10 Sketch the graph of $\mathbf{r}(t) = 4 \cos t \mathbf{i} + 3 \sin t \mathbf{j}$; $0 \le t \le \pi$ and show the direction of increasing t.

15.1.11 Sketch the graph of $\mathbf{r}(t) = (3 + 5 \cosh 2t)\mathbf{i} + (2 + 3 \sin h2t)\mathbf{j}$ and show the direction of increasing t.

15.1.12 Sketch the graph of $\mathbf{r}(t) = 2 \sin t \mathbf{i} + 3 \cos t \mathbf{j}$, $0 \le t \le \dfrac{\pi}{2}$ and show the direction of increasing t.

15.1.13 Sketch the graph of $\mathbf{r}(t) = (2 + \sin t)\mathbf{i} + (3 - \cos t)\mathbf{j}$, $0 \le t \le \pi$ and show the direction of increasing t.

15.1.14 Sketch the graph of $\mathbf{r}(t) = \sec t \mathbf{i} + \tan t \mathbf{j}$, $-\dfrac{\pi}{2} < 0 < \dfrac{\pi}{2}$, and show the direction of increasing t.

15.1.15 Sketch the graph of $\mathbf{r}(t) = \langle \sqrt{t + 1}, t \rangle$ and show the direction of increasing t.

15.1.16 Describe the graph of $\mathbf{r} = \cos t\,\mathbf{i} + \sin t\,\mathbf{j} + t\,\mathbf{k}$.

15.1.17 Describe the graph of $\mathbf{r} = \langle 3\cos 2t,\ 2\sin 2t,\ t\rangle$.

15.1.18 Describe the graph of $\mathbf{r} = \langle 2\cos 2t,\ 3\sin 2t,\ 2\rangle$.

SOLUTIONS

SECTION 15.1

15.1.1 The domain is $(-\infty, \infty)$, $\mathbf{r}(0) = \mathbf{i}$.

15.1.2 $(0, +\infty)$.   **15.1.3** $(-1, +\infty)$.

15.1.4 $\mathbf{r}(t) = \cos^{-1} t\mathbf{i} + \sin 2t\mathbf{j} + t^2\mathbf{k}$.   **15.1.5** $\mathbf{r}(t) = \sin 2t\mathbf{i} + \dfrac{1}{t}\mathbf{j} + t^2\mathbf{k}$.

15.1.6 The line in 3-space whose parameter equation is $x = 2 - 3t$, $y = 1 + t$, $z = 1 - t$ and which passes through the point $(2, 1, 1)$ and is parallel to $\langle -3, 1, -1 \rangle$.

15.1.7 The corresponding parametric equations are $x = t^2$, $y = t$ which describe the parabola $x = y^2$.

15.1.8 The corresponding parametric equations are $x = \cos 3t$, $y = \sin 3t$, $z = -3$. Eliminating t in the first two equations yields $x^2 + y^2 = 1$ so the graph is a circle of radius 1 in the plane $z = -3$.

15.1.9 The corresponding parametric equations are $x = 2\cos t$, $y = \sin t$, $z = 1$. Eliminating t in the first two equations yields $\dfrac{x^2}{4} + y^2 = 1$ so the graph is an ellipse in the plane $z = 1$.

15.1.10   **15.1.11**

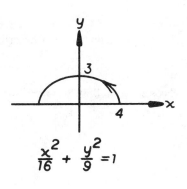

$$\frac{x^2}{16} + \frac{y^2}{9} = 1$$

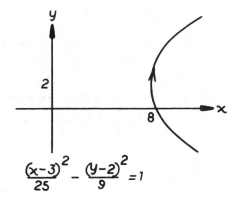

$$\frac{(x-3)^2}{25} - \frac{(y-2)^2}{9} = 1$$

15.1.12

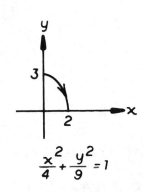

$$\frac{x^2}{4} + \frac{y^2}{9} = 1$$

15.1.13

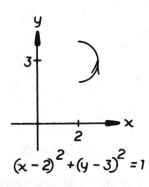

$$(x - 2)^2 + (y - 3)^2 = 1$$

15.1.14

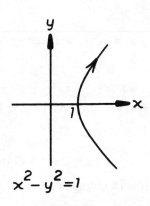

$$x^2 - y^2 = 1$$

15.1.15

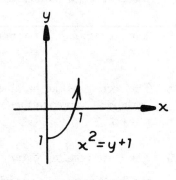

$$x^2 = y + 1$$

15.1.16 The corresponding parametric equations are $x = \cos t$, $y = \sin t$, $z = t$ describes a helix wound around a right circular cylinder of radius 1.

15.1.17 The corresponding parametric equations are $x = 3\cos 2t$, $y = 2\sin 2t$, $z = t$ describes a helix wound around an elliptical cylinder.

15.1.18 The corresponding parametric equations are $x = 2\cos 2t$, $y = 3\sin 2t$, $z = 2$ describes an ellipse $\dfrac{x^2}{4} + \dfrac{y^2}{9} = 1$ in the $z = 2$ plane.

SECTION 15.2

15.2.1 Find $\lim\limits_{t \to 1} \left\langle \ln t, -\sqrt[3]{t}, e^{4t} \right\rangle$.

15.2.2 Find $\mathbf{r}'(t)$ if $\mathbf{r}(t) = e^t\mathbf{i} + e^{2t}\mathbf{j} + \mathbf{k}$.

15.2.3 Find $\mathbf{r}'(t)$ if $\mathbf{r}(t) = \sqrt{t^2 + 2t}\,\mathbf{i} + \ln\sqrt{t^2 + 2t}\,\mathbf{j}$.

15.2.4 Find $\mathbf{r}'(t)$ if $\mathbf{r}(t) = \langle t, \ln\cos 2t, \ln\sin 2t \rangle$.

15.2.5 Find $\mathbf{r}'(t)$ if $\mathbf{r}(t) = \langle \cos t + t\sin t, \sin t - t\cos t \rangle$.

15.2.6 Find $\mathbf{r}'(\pi/3)$ if $\mathbf{r}(t) = t\mathbf{i} + \ln\sin 2t\mathbf{j} + \cos^2 2t\mathbf{k}$.

15.2.7 Find $\mathbf{r}'(t)$ if $\mathbf{r}(t) = \ln t\mathbf{i} - t^{-2}\mathbf{j} + te^{3t}\mathbf{k}$.

15.2.8 Find $\mathbf{r}'(t)$ if $\mathbf{r}(t) = \sin^{-1} 2t\mathbf{i} + \tan^{-1} 2t\mathbf{j}$.

15.2.9 Find the parametric equations of the tangent line to $\mathbf{r}(t) = \langle t, t^2, t^3 \rangle$ at the point where $t = 1$.

15.2.10 Find the parametric equations of the tangent line to $\mathbf{r}(t) = \cos 2t\mathbf{i} + 2\sin 2t\mathbf{j} - 3t\mathbf{k}$ at the point where $t = \pi/6$.

15.2.11 Find the vector equation of the tangent line to $\mathbf{r}(t) = \langle \sec 2t, \cos 2t, 2t \rangle$ at the point where $t = 0$.

15.2.12 Find the vector equation of the tangent line to $\mathbf{r}(t) = \cot^{-1} t\mathbf{i} + \tan^{-1} t\mathbf{j} - 3t\mathbf{k}$ at the point where $t = 1$.

15.2.13 Find the vector equation of the tangent line to $\mathbf{r}(t) = \sin t\mathbf{i} + \sinh 2t\mathbf{j} + \operatorname{sech} 2t\mathbf{k}$ at the point where $t = 0$.

15.2.14 Prove that $\mathbf{r}$ is continuous at $t = \pi/4$ if $\mathbf{r}(t) = \sin 2t\mathbf{i} + \cos 3t\mathbf{j} + \tan t\mathbf{k}$.

15.2.15 Find $\mathbf{r}'(\pi/4)$ if $r(t) = 6\sin 2t\mathbf{i} + 6\cos 2t\mathbf{j}$, then sketch the graph of $\mathbf{r}(t)$ and the tangent vector $\mathbf{r}'(\pi/4)$.

15.2.16 Find $\mathbf{r}'(1)$ if $\mathbf{r}(t) = \langle t^2, 4t^{-2} \rangle$, then sketch the graph of $\mathbf{r}(t)$ and the tangent vector $\mathbf{r}'(1)$.

15.2.17 Find $\mathbf{r}'(1)$ if $\mathbf{r}(t) = t^2\mathbf{i} + 2t\mathbf{j}$, then sketch the graph of $\mathbf{r}(t)$ and the tangent vector $\mathbf{r}'(1)$.

15.2.18 Evaluate $\displaystyle\int_0^{\pi} (t\mathbf{i} + \sin t\mathbf{j} + e^t\mathbf{k})dt$.

15.2.19 Evaluate $\displaystyle\int_0^{\pi/2} \left\langle e^t\sin t, te^t \right\rangle dt$.

15.2.20 Evaluate $\displaystyle\int \left(\sin 2t\mathbf{i} + \cos 3t\mathbf{j} - \frac{1}{\sqrt{1 - 16t^2}}\mathbf{k} \right) dt$.

15.2.21 Find $\mathbf{r}(t)$ if $\mathbf{r}'(t) = \sin 2t\mathbf{i} + \cos 2t\mathbf{j} - t\mathbf{k}$ and $\mathbf{r}(0) = \mathbf{i} + \mathbf{j} + \mathbf{k}$.

15.2.22 Find $\mathbf{r}(t)$ if $\mathbf{r}'(t) = \dfrac{1}{1+t^2}\mathbf{i} + 2\tan 2t\mathbf{j} + e^{-t}\mathbf{k}$ and $\mathbf{r}(0) = \mathbf{i} + \mathbf{j} + \mathbf{k}$.

SOLUTIONS

SECTION 15.2

15.2.1 $\langle 0, 0, 1 \rangle$.

15.2.2 $\mathbf{r}'(t) = e^t \mathbf{i} + 2e^{2t} \mathbf{j}$.

15.2.3 $\mathbf{r}'(t) = \dfrac{t+1}{\sqrt{t^2 + 2t}} i + \dfrac{t+1}{t^2 + 2t} \mathbf{j}$.

15.2.4 $\mathbf{r}'(t) = \langle 1, -2\tan 2t, 2\cot 2t \rangle$.

15.2.5 $\mathbf{r}'(t) = \langle t\cos t, t\sin t \rangle$.

15.2.6 $\mathbf{r}'(t) = \mathbf{i} + 2\cot 2t\mathbf{j} - 4\cos 2t \sin 2t\mathbf{k}; \ f'(\pi/3) = \mathbf{i} - \dfrac{2}{\sqrt{3}}\mathbf{j} + \sqrt{3}\mathbf{k}$.

15.2.7 $\mathbf{r}'(t) = \dfrac{1}{t}\mathbf{i} + \dfrac{2}{t^3}\mathbf{j} + e^{3t}(3t+1)\mathbf{k}$.

15.2.8 $\mathbf{r}'(t) = \dfrac{2}{\sqrt{1 - 4t^2}}\mathbf{i} + \dfrac{2}{1 + 4t^2}\mathbf{j}$.

15.2.9 $\mathbf{r}'(t) = \langle 1, 2t, 3t^2 \rangle; \ x = 1 + t, \ y = 1 + 2t, \ z = 1 + 3t$.

15.2.10 $\mathbf{r}'(t) = \langle -2\sin 2t, 4\cos 2t, -3 \rangle, \ x = \dfrac{1}{2} - \sqrt{3}t, \ y = \sqrt{3} + 2t, \ z = -\dfrac{\pi}{2} - 3t$.

15.2.11 $\mathbf{r}'(t) = \langle 2\sec 2t \tan 2t, -2\sin 2t, 2 \rangle; \ r = \langle 1, 1, 0 \rangle + t\langle 0, 0, 2 \rangle$.

15.2.12 $\mathbf{r}'(t) = \left\langle -\dfrac{1}{1+t^2}, \dfrac{1}{1+t^2}, 3 \right\rangle; \ \mathbf{r} = \left\langle \dfrac{\pi}{4}, \dfrac{\pi}{4}, -3 \right\rangle + t\left\langle -\dfrac{1}{2}, \dfrac{1}{2}, 3 \right\rangle$.

15.2.13 $\mathbf{r}'(t) = \mathbf{i} + 2\mathbf{j}$
$\mathbf{r} = \langle 0, 0, 1 \rangle + t\langle 1, 2, 0 \rangle$.

15.2.14 $\mathbf{r}(\pi/4) = \mathbf{i} - \mathbf{j} + \mathbf{k}, \ \lim\limits_{t \to \pi/4} \mathbf{r}(t) = \mathbf{r}(\pi/4)$.

15.2.15 $\mathbf{r}'(t) = 12\cos 2t\mathbf{i} - 12\sin 2t\mathbf{j}$; $\mathbf{r}'(\pi/4) = -12\mathbf{j}$.

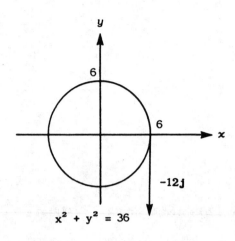

15.2.16 $\mathbf{r}'(t) = 2t\mathbf{i} - 8t^{-3}\mathbf{j}$; $\mathbf{r}'(1) = 2\mathbf{i} - 8\mathbf{j}$.

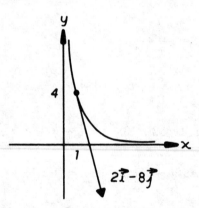

15.2.17 $\mathbf{r}'(t) = 2t\mathbf{i} + 2\mathbf{j}$; $\mathbf{r}'(1) = 2\mathbf{i} + 2\mathbf{j}$.

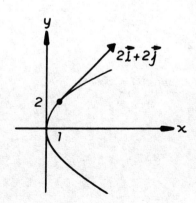

15.2.18 $\left[\dfrac{t^2}{2}\mathbf{i} - \cos t\mathbf{j} + e^t\mathbf{k}\right]_0^\pi = \dfrac{\pi^2}{2}\mathbf{i} + 2\mathbf{j} + (e^\pi - 1)\mathbf{k}.$

15.2.19 $\left[\left\langle \dfrac{e^t}{2}(\sin t - \cos t), e^t(t-1)\right\rangle\right]_0^{\pi/2} = \left\langle \dfrac{e^{\pi/2}}{2} + 1/2, \dfrac{\pi}{2}e^{\pi/2} - e^{\pi/2} + 1\right\rangle.$

15.2.20 $-\dfrac{1}{2}\cos 2t\mathbf{i} + \dfrac{1}{3}\sin 3t\mathbf{j} - \dfrac{1}{4}\sin^{-1} 4t\mathbf{k} + C.$

15.2.21 $\mathbf{r}(t) = -\dfrac{1}{2}\cos 2t\mathbf{i} + \dfrac{1}{2}\sin 2t\mathbf{j} - \dfrac{t^2}{2}\mathbf{k} + C;\ \mathbf{r}(0) = \mathbf{i} + \mathbf{j} + \mathbf{k}$, so

$\mathbf{r}(t) = \left(1 - \dfrac{1}{2}\cos 2t\right)\mathbf{i} + \left(1 + \dfrac{1}{2}\sin 2t\right)\mathbf{j} + \left(1 - \dfrac{t^2}{2}\right)\mathbf{k}.$

15.2.22 $\mathbf{r}(t) = \tan^{-1}\dfrac{3}{2}t\mathbf{i} - \ln\cos 2t\mathbf{j} - e^{-t}\mathbf{k} + C;$

$\mathbf{r}(0) = \mathbf{i} + \mathbf{j} + \mathbf{k},\ \mathbf{r}(t) = \left(1 + \tan^{-1}t\right)\mathbf{i} + (1 - \ln\cos 2t)\mathbf{j} + \left(2 + e^{-t}\right)\mathbf{k}.$

SECTION 15.3

15.3.1   Determine whether $\mathbf{r}$ is a smooth function of the parameter t.

$$\mathbf{r}(t) = t^2\mathbf{i} + (4t^3 - t^2)\mathbf{j} + t^3\mathbf{k}.$$

15.3.2   Determine whether $\mathbf{r}$ is a smooth function of the parameter t.

$$\mathbf{r}(t) = \sin t^2\mathbf{i} - (1 - \ln(t))\mathbf{j} + \cos t^2\mathbf{k}.$$

15.3.3   Calculate $\dfrac{dr}{d\tau}$ by the chain rule for $r = 2t^2 i + t^3 j; t = 2\tau + 5$.

15.3.4   Find the arc length of the curve given by $x = 3\cos t$, $y = 3\sin t$, $z = t$ for $0 \le t \le 2\pi$.

15.3.5   Find the arc length of the curve given by $\mathbf{r}(t) = \langle 6\sin 2t, 6\cos 2t, 5t \rangle$ for $0 \le t \le \pi$.

15.3.6   Find the arc length of the curve given by $\mathbf{r} = 5t\mathbf{i} + 4\sin 3t\mathbf{j} + 4\cos 3t\mathbf{k}$ for $0 \le t \le 2\pi$.

15.3.7   Find the arc length of the curve given by $\mathbf{r} = \cos t\mathbf{i} + \sin t\mathbf{j} + t^{3/2}\mathbf{k}$ for $0 \le t \le \dfrac{20}{3}$.

15.3.8   Find the arc length of the curve given by $\mathbf{r} = \dfrac{t^3}{3}\mathbf{i} + \dfrac{t^2}{\sqrt{2}}\mathbf{j} + t\mathbf{k}$ for $0 \le t \le 3$.

15.3.9   Find parametric equations for $\mathbf{r} = \langle 6\sin 2t, 6\cos 2t \rangle$ using arc length, s, as a parameter. Use the point on the curve where $t = 0$ as the reference point.

15.3.10   Find parametric equations for $\mathbf{r}(t) = 2\cos t\mathbf{i} + 2\sin t\mathbf{j}$ using arc length, s, as a parameter. Use the point on the curve where $t = 0$ as the reference point.

15.3.11   Find parametric equations for $\mathbf{r} = \langle \cos t + t\sin t, \sin t - t\cos t \rangle$ using arc length, s, as a parameter. Use the point on the curve where $t = 0$ as the reference point.

15.3.12   Find parametric equations for $\mathbf{r} = 2t\mathbf{i} - 3\mathbf{j}$ using arc length, s, as a parameter. Use the point on the curve where $t = 0$ as the reference point.

15.3.13   Find parametric equations for $\mathbf{r} = 3t\mathbf{i} + (4 - t)\mathbf{j}$ using arc length, s, as a parameter. Use the point on the curve where $t = 0$ as the reference point.

15.3.14   Find parametric equations for $\mathbf{r} = \langle 2 + \cos 3t, 3 - \sin 3t, 4t \rangle$, $0 \le t \le \dfrac{2\pi}{3}$, using arc length, s, as a parameter. Use the point on the curve where $t = 0$ as the reference point.

15.3.15   Find parametric equations for $\mathbf{r} = \sin^3 2t\mathbf{i} + \cos^3 2t\mathbf{j}$, $0 \le t \le \dfrac{\pi}{4}$, using arc length as a parameter. Use the point on the curve where $t = 0$ as the reference point.

SOLUTIONS

SECTION 15.3

15.3.1 $\mathbf{r}'(t) = 2t\mathbf{i} + (12t^2 - 2t)\mathbf{j} + 3t^2\mathbf{k}$

$\mathbf{r}'(t) = 0$ when $t = 0$. $\mathbf{r}(t)$ is not a smooth function.

15.3.2 $\mathbf{r}'(t) = 2t\cos t^2\mathbf{i} + \dfrac{1}{t}\mathbf{j} - 2t\sin t^2\mathbf{k}$

$\dfrac{1}{t}$ is discontinuous at $t = 0$. $\mathbf{r}(t)$ is not a smooth function.

15.3.3 $\dfrac{dr}{d\tau} = dr/dt \cdot dt/d\tau = (4t\mathbf{i} + 3t^2\mathbf{j})2$

$\qquad = 8t\mathbf{i} + 6t^2\mathbf{j} = 8(2\tau + 5)\mathbf{i} + 6(2\tau + 5)^2\mathbf{j}$

15.3.4 $\left(\dfrac{dx}{dt}\right)^2 + \left(\dfrac{dy}{dt}\right)^2 + \left(\dfrac{dz}{dt}\right)^2 = (-3\sin t)^2 + (3\cos t)^2 + (1)^2$

$\qquad\qquad\qquad\qquad = 9\sin^2 t + 9\cos^2 t + 1 = 10;$

$L = \displaystyle\int_0^{2\pi} \sqrt{10}\,dt = 2\pi\sqrt{10}.$

15.3.5 $\left(\dfrac{dx}{dt}\right)^2 + \left(\dfrac{dy}{dt}\right)^2 + \left(\dfrac{dz}{dt}\right)^2 = (12\cos 2t)^2 + (-12\sin 2t)^2 + (5)^2$

$\qquad\qquad\qquad\qquad = 144\cos^2 t + 144\sin^2 2t + 25 = 169;$

$L = \displaystyle\int_0^{\pi} 13\,dt = 13\pi.$

15.3.6 $\left(\dfrac{dx}{dt}\right)^2 + \left(\dfrac{dy}{dt}\right)^2 + \left(\dfrac{dz}{dt}\right)^2 = (5)^2 + (12\cos 3t)^2 + (-12\sin 3t)^2$

$\qquad\qquad\qquad\qquad = 25 + 144\cos^2 3t + 144\sin^2 3t = 169;$

$L = \displaystyle\int_0^{2\pi} 13\,dt = 26\pi.$

15.3.7 $\left(\dfrac{dx}{dt}\right)^2 + \left(\dfrac{dy}{dt}\right)^2 + \left(\dfrac{dz}{dt}\right)^2 = (-\sin t)^2 + (\cos t)^2 + \left(\dfrac{3}{2}t^{1/2}\right)^2$

$\qquad\qquad\qquad\qquad = \sin^2 t + \cos^2 t + \dfrac{9}{4}t = 1 + \dfrac{9}{4}t;$

$L = \displaystyle\int_0^{20/3} \sqrt{1 + \dfrac{9}{4}t}\,dt = \dfrac{8}{27}\left(1 + \dfrac{9}{4}t\right)^{3/2}\Bigg]_0^{20/3} = \dfrac{56}{3}.$

15.3.8 $\left(\dfrac{dx}{dt}\right)^2 + \left(\dfrac{dy}{dt}\right)^2 + \left(\dfrac{dz}{dt}\right)^2 = \left(t^2\right)^2 + \left(\sqrt{2}t\right)^2 + (1)^2 = t^4 + 2t^2 + 1;\ L = \displaystyle\int_0^3 \left(t^2 + 1\right)dt = 12.$

15.3.9 $x = 6\sin 2u,\ y = 6\cos 2u,$

$$\left(\dfrac{dx}{du}\right)^2 + \left(\dfrac{dy}{du}\right)^2 = (12\cos 2u)^2 + (-12\sin 2u)^2$$

$$= 144\cos^2 2u + 144\sin^2 2u = 144;$$

$$s = \int_0^t 12\,du = 12t,\ t = \frac{s}{12}\ \text{so}\ x = 6\sin\frac{s}{6},\ y = 6\cos\frac{s}{6}.$$

15.3.10 $x = 2\cos u,\ y = 2\sin u,\ \left(\dfrac{dx}{du}\right)^2 + \left(\dfrac{dy}{du}\right)^2 = (-2\sin u)^2 + (2\cos u)^2 = 4\sin^2 u + 4\cos^2 u = 4;$

$$s = \int_0^t 2\,du = 2t,\ t = \frac{s}{2}\ \text{so}\ x = 2\cos\frac{s}{2},\ y = 2\sin\frac{s}{2}.$$

15.3.11 $x = \cos u + u\sin u,\ y = \sin u - u\cos u,$

$\left(\dfrac{dx}{du}\right)^2 + \left(\dfrac{dy}{du}\right)^2 = (u\cos u)^2 + (u\sin u)^2 = u^2\cos^2 u + u^2\sin^2 u = u^2;\ s = \displaystyle\int_0^t u\,du = \frac{t^2}{2},$

$t = \sqrt{2s}\ \text{so}\ x = \cos\sqrt{2s} + \sqrt{2s}\sin\sqrt{2s},\ y = \sin\sqrt{2s} - \sqrt{2s}\cos\sqrt{2s}.$

15.3.12 $x = 2u,\ y = -3,\ \left(\dfrac{dx}{du}\right)^2 + \left(\dfrac{dy}{du}\right)^2 = 4,\ s = \int_0^t 2\,du = 2t,\ t = \dfrac{s}{2}\ \text{so}\ x = s,\ y = -3.$

15.3.13 $x = 3u,\ y = 4 - u,\ \left(\dfrac{dx}{du}\right)^2 + \left(\dfrac{dy}{du}\right)^2 = (3)^2 + (-1)^2 = 10,\ s = \displaystyle\int_0^t \sqrt{10}\,du = \sqrt{10}\,t,\ t = \dfrac{s}{\sqrt{10}}$

so $x = \dfrac{3s}{\sqrt{10}},\ y = 4 - \dfrac{s}{\sqrt{10}}.$

15.3.14 $x = 2 + \cos 3u,\ y = 3 - \sin 3u,\ z = 4u;$

$$\left(\dfrac{dx}{du}\right)^2 + \left(\dfrac{dy}{du}\right)^2 + \left(\dfrac{dz}{du}\right)^2 = (-3\sin 3u)^2 + (-3\cos 3u)^2 + (4)^2$$

$$= 9\sin^2 3u + 9\cos^2 3u + 16 = 25;$$

$$s = \int_0^t 5\,du = 5t,\ t = \frac{s}{5}\ \text{so}\ x = 2 + \cos\frac{3s}{5},\ y = 3 - \sin\frac{3s}{5},\ z = \frac{4s}{5}\ \text{for}\ 0 \le s \le \frac{10\pi}{3}.$$

15.3.15 $x = \sin^3 2u$, $y = \cos^3 2u$,

$$\left(\frac{dx}{du}\right)^2 + \left(\frac{dy}{du}\right)^2 = \left(6\sin^2 2u \cos 2u\right)^2 + \left(-6\cos^2 2u \sin 2u\right)^2$$

$$= 36\sin^2 2u \cos^2 2u;$$

$$s = \int_0^t 6\sin 2u \cos 2u\, du = \frac{3}{2}\sin^2 2t, \quad \sin^2 2t = \frac{2s}{3}, \quad \sin 2t = \left(\frac{2s}{3}\right)^{1/2} \text{ so}$$

$$\cos^2 2t = 1 - \sin^2 2t = 1 - \frac{2s}{3} = \frac{3 - 2s}{3}, \quad \cos 2t = \left(\frac{3 - 2s}{3}\right)^{1/2} \text{ so } x = \left(\frac{2s}{3}\right)^{3/2},$$

$$y = \left(\frac{3 - 2s}{3}\right)^{3/2} \text{ for } 0 \le s \le \frac{3}{2}.$$

SECTION 15.4

15.4.1 Find the unit tangent and unit normal vectors to the curve $\mathbf{r}(t) = t^2\mathbf{i} + 2t\mathbf{j}$ at $t = 1$. Sketch a portion of the curve showing the point of tangency.

15.4.2 Find the unit tangent and unit normal vectors to the curve $\mathbf{r}(t) = 4\cos t\mathbf{i} + \sin t\mathbf{j}$ at $t = \dfrac{\pi}{2}$. Sketch a portion of the curve showing the point of tangency.

15.4.3 Find the unit tangent and unit normal vectors to the curve $\mathbf{r}(t) = \ln t\mathbf{i} + t\mathbf{j}$ at $t = 2$. Sketch a portion of the curve showing the point of tangency.

15.4.4 Find the unit tangent and unit normal vectors to the curve $\mathbf{r}(t) = t^3\mathbf{i} + 2t^2\mathbf{j}$ at $t = 1$. Sketch a portion of the curve showing the point of tangency.

15.4.5 Find the unit tangent and unit normal vectors to $\mathbf{r}(t) = \left\langle t^2 + 1, \dfrac{1}{t} \right\rangle$ at $(2, 1)$.

15.4.6 Find the unit tangent and unit normal vectors to $\mathbf{r}(t) = t\mathbf{i} + \ln\cos t\mathbf{j}$ at $t = \dfrac{\pi}{4}$.

15.4.7 Find the unit tangent and unit normal vectors to the curve $\mathbf{r}(t) = 2\sin t\mathbf{i} + 3\cos t\mathbf{j}$ at $\mathbf{t} = \dfrac{\pi}{6}$. Sketch a portion of the curve showing the point of tangency.

15.4.8 Find the unit tangent and unit normal vectors to $\mathbf{r}(t) = e^t\sin t\mathbf{i} + e^t\cos t\mathbf{j}$ at $t = 0$.

15.4.9 Find the unit tangent and unit normal vectors to $\mathbf{r}(t) = e^{3t}\mathbf{i} + 3e^{2t}\mathbf{j}$ at $t = 0$.

15.4.10 Find the unit tangent and unit normal vectors to $\mathbf{r}(t) = 2\cos t\mathbf{i} + 2\sin t\mathbf{j}$ at $\mathbf{t} = \dfrac{\pi}{4}$. Sketch a portion of the curve showing the point of tangency.

15.4.11 Find the unit tangent and unit normal vectors to $x = 3\cos t$, $y = 3\sin t$, $z = \sqrt{7}t$ at $t = \dfrac{\pi}{2}$.

15.4.12 Find the unit tangent and unit normal vectors to $\mathbf{r}(t) = \langle 6\cos 2t, 6\sin 2t, 5t \rangle$ at $t = \pi$.

15.4.13 Find the unit tangent and unit normal vectors to $\mathbf{r}(t) = 2t\mathbf{i} + 4\sin 3t\mathbf{j} + 4\cos 3t\mathbf{k}$ at $t = \pi/2$.

15.4.14 Find the unit tangent and unit normal vectors to $x = e^t$, $y = e^t\cos t$, $z = e^t\sin t$ at $t = \pi$.

15.4.15   Find the vector equation of the line which is perpendicular to the curve
$\mathbf{r}(t) = \langle 2\cos 3t,\ 2\sin 3t,\ 8t \rangle$ at $t = \dfrac{\pi}{9}$.

15.4.16   Find the parametric equation of the line which is perpendicular to
$\mathbf{r}(t) = \langle 5t,\ 6\sin 2t,\ 6\cos 2t \rangle$ at $t = \dfrac{\pi}{3}$.

15.4.17   Find the direction cosines of the tangent and normal vectors to $x = e^t \sin 2t$, $y = e^t \cos 2t$, $z = 2e^t$ at $t = 0$.

15.4.18   Find the direction cosines of the tangent and normal vectors to $x = 2\cos t$, $y = 2\sin t$, $z = \sqrt{5}$ at $t = \dfrac{\pi}{2}$.

SOLUTIONS

SECTION 15.4

15.4.1 $\mathbf{T}(t) = \dfrac{2t\mathbf{i} + 2\mathbf{j}}{\sqrt{4t^2 + 4}}$,

$\mathbf{T}(1) = \dfrac{1}{\sqrt{2}}\mathbf{i} + \dfrac{1}{\sqrt{2}}\mathbf{j}$,

$\mathbf{N}(1) = -\dfrac{1}{\sqrt{2}}\mathbf{i} + \dfrac{1}{\sqrt{2}}\mathbf{j}$.

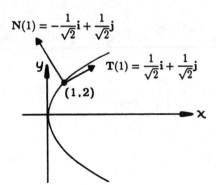

15.4.2 $\mathbf{T}(t) = \dfrac{-4\sin t\mathbf{i} + \cos t\mathbf{j}}{\sqrt{16\sin^2 t + \cos^2 t}}$,

$\mathbf{T}(\pi/2) = -\mathbf{i}$,

$\mathbf{N}(t) = \dfrac{-\cos t\mathbf{i} - 4\sin t\mathbf{j}}{\sqrt{16\sin^2 t + \cos^2 t}}$,

$\mathbf{N}\left(\dfrac{\pi}{2}\right) = -\mathbf{j}$.

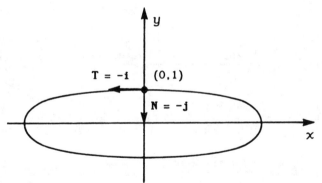

15.4.3 $\mathbf{T}(t) = \dfrac{\dfrac{1}{t}\mathbf{i} + \mathbf{j}}{\sqrt{\dfrac{1}{t^2} + 1}}$,

$\mathbf{T}(2) = \dfrac{1}{\sqrt{5}}\mathbf{i} + \dfrac{2}{\sqrt{5}}\mathbf{j}$,

$\mathbf{N}(2) = -\dfrac{2}{\sqrt{5}}\mathbf{i} + \dfrac{1}{\sqrt{5}}\mathbf{j}$.

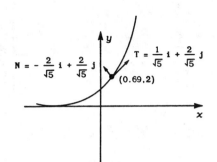

15.4.4 $\mathbf{T}(t) = \dfrac{3t^2\mathbf{i} + 4t\mathbf{j}}{\sqrt{9t^4 + 16t^2}}$, $\mathbf{T}(1) = \dfrac{3}{5}\mathbf{i} + \dfrac{4}{5}\mathbf{j}$, $\mathbf{N}(1) = -\dfrac{4}{5}\mathbf{i} + \dfrac{3}{5}\mathbf{j}$.

15.4.5 $\mathbf{T}(t) = \left\langle \dfrac{2t}{\sqrt{4t^2 + \dfrac{1}{t^4}}}, \dfrac{-1/t^2}{\sqrt{4t^2 + \dfrac{1}{t^4}}} \right\rangle, \mathbf{T}(1) = \left\langle \dfrac{2}{\sqrt{5}}, -\dfrac{1}{\sqrt{5}} \right\rangle, \mathbf{N}(1) = \left\langle \dfrac{1}{\sqrt{5}}, \dfrac{2}{\sqrt{5}} \right\rangle.$

15.4.6 $\mathbf{T}(t) = \dfrac{\mathbf{i} - \tan t\mathbf{j}}{\sec t}; \ \mathbf{T}(\pi/4) = \dfrac{1}{\sqrt{2}}\mathbf{i} - \dfrac{1}{\sqrt{2}}\mathbf{j}; \ \mathbf{N}(\pi/4) = \dfrac{1}{\sqrt{2}}\mathbf{i} + \dfrac{1}{\sqrt{2}}\mathbf{j}.$

15.4.7 $\mathbf{T}(t) = \dfrac{2\cos t\mathbf{i} - 3\sin t\mathbf{j}}{\sqrt{4\cos^2 t + 9\sin^2 t}},$

$\mathbf{T}(\pi/6) = \dfrac{2}{\sqrt{7}}\mathbf{i} - \dfrac{3}{\sqrt{21}}\mathbf{j}$

$\mathbf{N}(\pi/6) = \dfrac{3}{\sqrt{21}}\mathbf{i} + \dfrac{2}{\sqrt{7}}\mathbf{j}.$

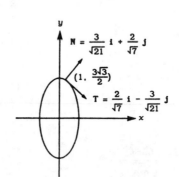

15.4.8 $\mathbf{T}(t) = \dfrac{(e^t \sin t + e^t \cos t)\mathbf{i} + (e^t \cos t - e^t \sin t)\mathbf{j}}{\sqrt{2}}, \ \mathbf{T}(0) = \dfrac{1}{\sqrt{2}}\mathbf{i} + \dfrac{1}{\sqrt{2}}\mathbf{j}; \ \mathbf{N}(0) = -\dfrac{1}{\sqrt{2}}\mathbf{i} + \dfrac{1}{\sqrt{2}}\mathbf{j}.$

15.4.9 $\mathbf{T}(t) = \dfrac{3e^{3t}\mathbf{i} + 6e^{2t}\mathbf{j}}{\sqrt{9e^{6t} + 36e^{4t}}}, \ \mathbf{T}(0) = \dfrac{\mathbf{i}}{\sqrt{5}} + \dfrac{2\mathbf{j}}{\sqrt{5}}; \ \mathbf{N}(0) = \dfrac{-2}{\sqrt{5}}\mathbf{i} + \dfrac{\mathbf{j}}{\sqrt{5}}.$

15.4.10 $\mathbf{T}(t) = \dfrac{-2\sin t\mathbf{i} + 2\cos t\mathbf{j}}{2},$

$\mathbf{T}\left(\dfrac{\pi}{4}\right) = -\dfrac{1}{\sqrt{2}}\mathbf{i} + \dfrac{1}{\sqrt{2}}\mathbf{j},$

$\mathbf{N}\left(\dfrac{\pi}{4}\right) = -\dfrac{1}{\sqrt{2}}\mathbf{i} - \dfrac{1}{\sqrt{2}}\mathbf{j}.$

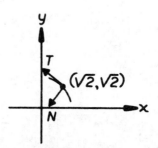

15.4.11 $\mathbf{r}(t) = 3\cos t\mathbf{i} + 3\sin t\mathbf{j} + \sqrt{7}\,t\mathbf{k}; \ \mathbf{r}'(t) = -3\sin t\mathbf{i} + 3\cos t\mathbf{j} + \sqrt{7}\,\mathbf{k}, \ \|\mathbf{r}'(t)\| = 4,$

so $\mathbf{T}(t) = \dfrac{1}{4}\left(-3\sin t\mathbf{i} + 3\cos t\mathbf{j} + \sqrt{7}\,\mathbf{k}\right)$ and $\mathbf{T}(\pi/2) = \dfrac{1}{4}\left(-3\mathbf{i} + \sqrt{7}\,\mathbf{k}\right) = -\dfrac{3}{4}\mathbf{i} - \dfrac{\sqrt{7}}{4}\mathbf{k}$

$\mathbf{T}'(t) = \dfrac{-3}{4}\cos t\mathbf{i} - \dfrac{3}{4}\sin t\mathbf{j}, \ \|\mathbf{T}'(t)\| = 3/4$ and $\mathbf{N}(t) = \dfrac{4}{3}\left(-\dfrac{3}{4}\cos t\mathbf{i} - \dfrac{3}{4}\sin t\mathbf{j}\right)$

$= -\cos t\mathbf{i} - \sin t\mathbf{j};$ thus $\mathbf{N}\left(\dfrac{\pi}{2}\right) = -\mathbf{j}.$

15.4.12 $\mathbf{r}'(t) = -12\sin 2t\mathbf{i} + 12\cos 2t\mathbf{j} + 5\mathbf{k}$, $\|\mathbf{r}'(t)\| = 13$

$$\mathbf{T}(t) = \frac{1}{13}(-12\sin 2t\mathbf{i} + 12\cos 2t\mathbf{j} + 5k)$$

$$= -\frac{12}{13}\sin 2t\mathbf{i} + \frac{12}{13}\cos 2t\mathbf{j} + \frac{5}{13}k \text{ so}, \mathbf{T}(\pi) = \frac{12}{13}\mathbf{j} + \frac{5}{13}\mathbf{k},$$

$$\mathbf{T}'(t) = -\frac{24}{13}\cos 2t\mathbf{i} - \frac{24}{13}\sin 2t\mathbf{j}, \|\mathbf{T}'(t)\| = \frac{24}{13} \text{ and}$$

$$\mathbf{N}(t) = \frac{13}{24}\left(-\frac{24}{13}\cos 2t\mathbf{i} - \frac{24}{13}\sin 2t\mathbf{j}\right) = -\cos 2t\mathbf{i} - \sin 2t\mathbf{j}, \text{ thus}, N(\pi) = -\mathbf{i}.$$

15.4.13 $\mathbf{r}'(t) = 2\mathbf{i} + 12\cos 3t\mathbf{j} - 12\sin 3t\mathbf{k}$, $\|\mathbf{r}(t)\| = 2\sqrt{37}$

$$\mathbf{T}(t) = \frac{1}{2\sqrt{37}}(2\mathbf{i} + 12\cos 3t\mathbf{j} - 12\sin 3t\mathbf{k})$$

$$= \frac{1}{\sqrt{37}}\mathbf{i} + \frac{6}{\sqrt{37}}\cos 3t\mathbf{j} - \frac{6}{\sqrt{37}}\sin 3t\mathbf{k}$$

so $\mathbf{T}(\pi/2) = \frac{1}{\sqrt{37}}\mathbf{i} + \frac{6}{\sqrt{37}}\mathbf{k}$

$$\mathbf{T}'(t) = -\frac{18}{\sqrt{37}}\sin 3t\mathbf{j} - \frac{18}{\sqrt{37}}\cos 3t\mathbf{k}, \|\mathbf{T}'(t)\| = \frac{18}{\sqrt{37}} \text{ and}$$

$$\mathbf{N}(t) = \frac{\sqrt{37}}{18}\left(-\frac{18}{\sqrt{37}}\sin 3t\mathbf{j} - \frac{18}{\sqrt{37}}\cos 3t\mathbf{k}\right) = -\sin 3t\mathbf{j} - \cos 3t\mathbf{k} \text{ thus}, \mathbf{N}(\pi/2) = \mathbf{j}.$$

15.4.14 $\mathbf{r}(t) = e^t\mathbf{i} + e^t\cos t\mathbf{j} + e^t\sin t\mathbf{k}$

$\mathbf{r}'(t) = e^t\mathbf{i} + (e^t\cos t - e^t\sin t)\mathbf{j} + (e^t\sin t + e^t\cos t)\mathbf{k}$; $\|\mathbf{r}(t)\| = \sqrt{3}e^t$

$$\mathbf{T}(t) = \frac{1}{\sqrt{3}e^t}\left[e^t\mathbf{i} + (e^t\cos t - e^t\sin t)\mathbf{j} + (e^t\sin t + e^t\cos t)\mathbf{k}\right] \text{ or}$$

$$\mathbf{T}(t) = \frac{1}{\sqrt{3}}\mathbf{i} + \frac{1}{\sqrt{3}}(\cos t - \sin t)\mathbf{j} + \frac{1}{\sqrt{3}}(\sin t + \cos t)\mathbf{k}; \text{ so } \mathbf{T}(\pi) = \frac{1}{\sqrt{3}}\mathbf{i} - \frac{1}{\sqrt{3}}\mathbf{j} - \frac{1}{\sqrt{3}}\mathbf{k}$$

$$\mathbf{T}'(t) = -\frac{1}{\sqrt{3}}(\sin t + \cos t)\mathbf{j} + \frac{1}{\sqrt{3}}(\cos t - \sin t)\mathbf{k}; \|\mathbf{T}'(t)\| = \sqrt{\frac{2}{3}} \text{ and}$$

$$\mathbf{N}(t) = \sqrt{\frac{3}{2}}\left[-\frac{1}{\sqrt{3}}(\sin t + \cos t)\mathbf{j} + \frac{1}{\sqrt{3}}(\cos t - \sin t)\mathbf{k}\right] \text{ or}$$

$$\mathbf{N}(t) = -\frac{1}{\sqrt{2}}(\sin t + \cos t)\mathbf{j} + \frac{1}{\sqrt{2}}(\cos t - \sin t)\mathbf{k} \text{ thus}, \mathbf{N}(\pi) = \frac{1}{\sqrt{2}}\mathbf{j} - \frac{1}{\sqrt{2}}\mathbf{k}.$$

15.4.15   The line is parallel to the normal cost vector, $\mathbf{N}$, so

$\mathbf{r}(t) = \langle 2\cos 3t, 2\sin 3t, 8t\rangle, \mathbf{r}'(t) = \langle -6\sin 3t, 6\cos 3t, 8\rangle, \|\mathbf{r}'(t)\| = 10$

$$\mathbf{T}(t) = \left\langle -\frac{3}{5}\sin 3t, \frac{3}{5}\cos 3t, \frac{4}{5}\right\rangle$$

$$\mathbf{T}'(t) = \left\langle -\frac{9}{5}\cos 3t, -\frac{9}{5}\sin 3t, 0\right\rangle, \|\mathbf{T}'(t)\| = \frac{9}{5}, \mathbf{N}(t) = \langle -\cos 3t, -\sin 3t, 0\rangle.$$

At $t = \pi/9$, $\mathbf{N}(\pi/9) = \left\langle -\frac{1}{2}, -\frac{\sqrt{3}}{2}, 0 \right\rangle$ thus the line is parallel to $\left\langle -\frac{1}{2}, -\frac{\sqrt{3}}{2}, 0 \right\rangle$ or any

scalar multiple such as $\left\langle 1, \sqrt{3}, 0 \right\rangle$ so the equation of the perpendicular line is

$x = \left\langle 1, \sqrt{3}, 8\pi/9 \right\rangle + t \left\langle 1, \sqrt{3}, 0 \right\rangle.$

15.4.16   The line is parallel to the normal vector, $\mathbf{N}$, so let $\mathbf{r}(t) = \langle 5t, 6\sin 2t, 6\cos 2t \rangle$

$\mathbf{r}'(t) = \langle 5, 12\cos 2t, -12\sin 2t \rangle$, $\|\mathbf{r}'(t)\| = 13$

$\mathbf{T}(t) = \left\langle \frac{5}{13}, \frac{12}{13}\cos 2t, -\frac{12}{13}\sin 2t \right\rangle$

$\mathbf{T}'(t) = \left\langle 0, -\frac{24}{13}\sin 2t, -\frac{24}{13}\cos 2t \right\rangle$, $\|\mathbf{T}(t)\| = \frac{24}{13}$, $\mathbf{N}(t) = \langle 0, -\sin 2t, -\cos 2t \rangle.$

At $t = \pi/3$, $\mathbf{N}(\pi/3) = \left\langle 0, -\frac{\sqrt{3}}{2}, \frac{1}{2} \right\rangle$, thus, the normal line is parallel to $\left\langle 0, -\frac{\sqrt{3}}{2}, \frac{1}{2} \right\rangle$ or

any scalar multiple such as $\left\langle 0, \sqrt{3}, -1 \right\rangle$, so the desired equations are

$x = \frac{5\pi}{3}$, $y = 3\sqrt{3} + \sqrt{3}t$, $z = -3 - t.$

15.4.17   Let $\mathbf{r}(t) = \langle e^t \sin 2t, e^t \cos 2t, 2e^t \rangle$, then

$\mathbf{r}'(t) = \langle e^t \sin 2t + 2e^t \cos 2t, e^t \cos 2t - 2e^t \sin 2t, 2e^t \rangle$, $\|\mathbf{r}'(t)\| = 3e^t$

$\mathbf{T}(t) = \left\langle \frac{\sin 2t + 2\cos 2t}{3}, \frac{\cos 2t - 2\sin 2t}{3}, \frac{2}{3} \right\rangle$

$\mathbf{T}(0) = \left\langle \frac{2}{3}, \frac{1}{3}, \frac{2}{3} \right\rangle$ so the direction cosines of the tangent vector at $t = 0$ are $\cos\alpha = \frac{2}{3}$,

$\cos\beta = \frac{1}{3}$, $\cos\gamma = \frac{2}{3}$; $\mathbf{T}'(t) = \left\langle \frac{2\cos 2t - 4\sin 2t}{3}, \frac{-2\sin 2t - 4\cos 2t}{3}, 0 \right\rangle$;

$\|\mathbf{T}'(t)\| = \frac{2\sqrt{5}}{3}$

$\mathbf{N}(t) = \left\langle \frac{\cos 2t - 2\sin 2t}{\sqrt{5}}, \frac{-\sin 2t - 2\cos 2t}{\sqrt{5}}, 0 \right\rangle$

$\mathbf{N}(0) = \left\langle \frac{1}{\sqrt{5}}, \frac{-2}{\sqrt{5}}, 0 \right\rangle$ so the direction cosines of the normal vector at $t = 0$ are $\cos\alpha = \frac{1}{\sqrt{5}}$,

$\cos\beta = \frac{-2}{\sqrt{5}}$, $\cos\gamma = 0.$

15.4.18   See 15.4.11. So the direction cosines of the tangent vector at $t = \frac{\pi}{2}$ are $\cos\alpha = -\frac{2}{3}$, $\cos\beta = 0$,

$\cos\delta = \sqrt{5}/3$ and the direction cosines of the normal vector at $t = \frac{\pi}{2}$ are $\cos\alpha = 0$,

$\cos\beta = -1$, $\cos\delta = 0.$

SECTION 15.5

15.5.1   Find the curvature for $x = 2\cos t$, $y = \cos 2t$ at $t = \pi/4$.

15.5.2   Find the curvature for $x = t^3$, $y = 2t^2$ at $t = 1$.

15.5.3   Find the curvature for $x = \ln t$, $y = t$ at $t = 2$.

15.5.4   Find the curvature for $x = t^2$, $y = 1/t$ at $t = 1/2$.

15.5.5   Find the curvature for $x = 2e^t$, $y = 2e^{-t}$ at $t = 0$.

15.5.6   Find the curvature for $\mathbf{r}(t) = \langle 6\cos 2t,\ 6\sin 2t,\ 5t \rangle$ at $t = \pi$.

15.5.7   Find the curvature for $\mathbf{r}(t) = \sin t\,\mathbf{i} + \cos t\,\mathbf{j} + \ln \cos t\,\mathbf{k}$ at $t = 0$.

15.5.8   Find the curvature for $x = e^t$, $y = e^t \cos t$, $z = e^t \sin t$ at $t = 0$.

15.5.9   Find the curvature for $\mathbf{r}(t) = \left\langle e^t,\ \sqrt{2}t,\ e^{-t} \right\rangle$ at $t = 0$.

15.5.10   Find the radius of curvature for $x = t^2 + 1$, $y = t - 2$, at $t = 2$.

15.5.11   Find the radius of curvature for $\mathbf{r}(t) = \langle 4\sin t,\ 2t - \sin 2t,\ \cos 2t \rangle$ at $t = \pi/2$.

15.5.12   Sketch $y = \dfrac{x^2}{4}$. Calculate the radius of curvature at $x = 1$ and sketch the oscillating circle.

15.5.13   Sketch $x = 2\cos t$, $y = 3\sin t$ for $0 \le t \le 2\pi$. Calculate the radius of curvature at $t = \pi/2$ and sketch the oscillating circle.

15.5.14   Sketch $y^2 = 4x$. Calculate the radius of curvature at $(1, 2)$ and sketch the oscillating circle.

15.5.15   Sketch $xy = 6$. Calculate the radius of curvature at $x = 2$ and sketch the oscillating circle.

15.5.16   Sketch $y = e^x$. Calculate the radius of curvature at $x = 0$ and sketch the oscillating circle.

15.5.17   Find the curvature for $x^2 + y^2 = 10x$ at $(2, -4)$.

15.5.18   Find the curvature for $y = 3 \cosh \dfrac{x}{3}$ at $x = 0$.

15.5.19   At what point(s) does $x = 2 \cos t$, $\quad y = 3 \sin t$ for $0 \leq t < 2\pi$ have a minimum radius of curvature?

15.5.20   Find the maximum and minimum values of the radius of curvature for the curve $x = \cos t$, $y = \sin t$, $\quad z = \sin t$. $0 \leq t < 2\pi$.

SOLUTIONS

SECTION 15.5

15.5.1 Use the formula of Exercise 24.

$$k(t) = \frac{|(-2\sin t)(-4\cos 2t) - (-2\sin 2t)(-2\cos t)|}{\left(4\sin^2 t + 4\sin^2 2t\right)^{3/2}}; \; k(\pi/4) = \frac{1}{3\sqrt{3}}.$$

15.5.2 $k(t) = \dfrac{\|\langle 3t^2, 4t, 0\rangle \times \langle 6t, 4, 0\rangle\|}{\|\langle 3t^2, 4t, 0\rangle\|^3}; \; k(1) = \dfrac{12}{125}.$

15.5.3 $k(t) = \dfrac{\left\|\left\langle \frac{1}{t}, 1, 0\right\rangle \times \left\langle -\frac{1}{t^2}, 0, 0\right\rangle\right\|}{\left\|\left\langle \frac{1}{t}, 1, 0\right\rangle\right\|^3}; \; k(2) = \dfrac{2}{5\sqrt{5}}.$

15.5.4 $k(t) = \dfrac{\left\|\left\langle 2t, -\frac{1}{t^2}, 0\right\rangle \times \left\langle 2, \frac{2}{t^3}, 0\right\rangle\right\|}{\left\|\left\langle 2t, -\frac{1}{t^2}, 0\right\rangle\right\|^3}; \; k(1/2) = \dfrac{24}{17\sqrt{17}}.$

15.5.5 $k(t) = \dfrac{\|\langle 2e^t, -2e^{-t}, 0\rangle \times \langle 2e^t, 2e^{-t}, 0\rangle\|}{\|\langle 2e^t, -2e^{-t}, 0\rangle\|^3}; \; k(0) = \dfrac{1}{2\sqrt{2}}.$

15.5.6 $k(t) = \dfrac{\|\langle -12\sin 2t, 12\cos 2t, 5\rangle \times \langle -24\cos 2t, -24\sin 2t, 0\rangle\|}{\|\langle -12\sin 2t, 12\cos 2t, 5\rangle\|^3}; \; k(\pi) = \dfrac{24}{169}.$

15.5.7 $k(t) = \dfrac{\|\langle \cos t, -\sin t, -\tan t\rangle \times \langle -\sin t, -\cos t, -\sec^2 t\rangle\|}{\|\langle \cos t, -\sin t, -\tan t\rangle\|^3}; \; k(0) = \sqrt{2}.$

15.5.8 $k(t) = \dfrac{\|\langle e^t, e^t\cos t - e^t\sin t, e^t\sin t + e^t\cos t\rangle \times \langle e^t, -2e^t\sin t, 2e^t\cos t\rangle\|}{\|\langle e^t, e^t\cos t - e^t\sin t, e^t\sin t + e^t\cos t\rangle\|^3} \; k(0) = \dfrac{\sqrt{2}}{3}.$

15.5.9 $k(t) = \dfrac{\|\langle e^t, \sqrt{2}, -e^{-t}\rangle \times \langle e^t, 0, e^{-t}\rangle\|}{\|\langle e^t, \sqrt{2}, -e^{-t}\rangle\|^3}; \; k(0) = \dfrac{\sqrt{2}}{4}.$

15.5.10 $k(t) = \dfrac{\|\langle 2t, 0, 0\rangle \times \langle 2, 0, 0\rangle\|}{\|\langle 2t, 1, 0\rangle\|^3}, \quad k(2) = \dfrac{2}{17\sqrt{17}}, \; \rho(2) = \dfrac{17\sqrt{17}}{2}.$

15.5.11 $k(t) = \dfrac{\| \langle 4\cos t, 2 - 2\cos 2t, -2\sin 2t \rangle \times \langle -4\sin t, 4\sin 2t, -4\cos 2t \rangle \|}{\| \langle 4\cos t, 2 - 2\cos 2t, -2\sin 2t \rangle \|^3}$;

$k\left(\dfrac{\pi}{2}\right) = \dfrac{\sqrt{2}}{4}, \rho\left(\dfrac{\pi}{2}\right) = \dfrac{4}{\sqrt{2}}$

15.5.12   Use the formula for Exercise 15.

$$k(x) = \dfrac{|1/2|}{\left[1 + \left(\dfrac{x}{2}\right)^2\right]^{3/2}}, \; k(1) = \dfrac{4}{5\sqrt{5}}, \; \rho(1) = \dfrac{5\sqrt{5}}{4}.$$

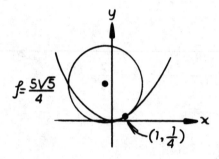

15.5.13   Use the formula of Exercise 24.

$$k(t) = \dfrac{|(-2\sin t)(-3\sin t) - (3\cos t)(-2\cos t)|}{\left(4\sin^2 t + 9\cos^2 t\right)^{3/2}}$$

$$= \dfrac{6}{\left(4\sin^2 t + 9\cos^2 t\right)^{3/2}}; k(\pi/2) = \dfrac{3}{4}, \rho(\pi/2) = 4/3.$$

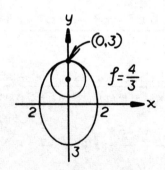

15.5.14   Use implicit differentiation and the formula of Exercise 15.

$$k = \dfrac{\left| -\dfrac{4}{y^3} \right|}{\left[1 + \left(\dfrac{2}{y} \right)^2 \right]^{3/2}}, \quad k \text{ at } (1,2) = \dfrac{1}{4\sqrt{2}} \text{ so } \rho = 4\sqrt{2}.$$

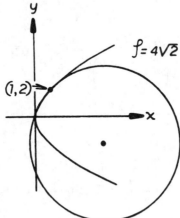

15.5.15   Use the formula of Exercise 15.

$$k(x) = \dfrac{\left| \dfrac{12}{x^3} \right|}{\left[1 + \left(-\dfrac{6}{x^2} \right)^2 \right]^{3/2}}, \quad k(2) = \dfrac{12}{13\sqrt{13}}, \quad \rho(2) = \dfrac{13\sqrt{13}}{12}.$$

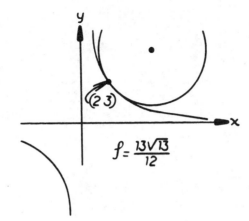

15.5.16   Use the formula of Exercise 15.

$$k(x) = \frac{|e^x|}{[1 + (e^x)^2]^{3/2}}, \ k(0) = \frac{1}{2\sqrt{2}}, \ \rho(0) = 2\sqrt{2}.$$

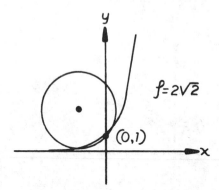

15.5.17   Use implicit differentiation and the formula of Exercise 15.

$$k = \frac{\left| -\dfrac{25}{y^3} \right|}{\left[1 + \left(\dfrac{5-x}{y} \right)^2 \right]^{3/2}}, \text{ at } (2, -4), \ k = \frac{1}{5}.$$

15.5.18   Use the formula of Exercise 15.

$$k(x) = \frac{\left| \dfrac{1}{3} \cosh \dfrac{x}{3} \right|}{\left[1 + \left(\sinh \dfrac{x}{3} \right)^2 \right]^{3/2}}, \ k(0) = \frac{1}{3}.$$

15.5.19 $k(t) = \dfrac{6}{(4\sin^2 t + 9\cos^2 t)^{3/2}}$

$\rho(t) = \dfrac{1}{6}(4\sin^2 t + 9\cos^2 t)^{3/2} = \dfrac{1}{6}(4 + 5\cos^2 t)^{3/2}$

which, by inspection, is minimum when $t = \pi/2$ or $3\pi/2$. The radius of curvature is minimum at $(0,3)$, and $(0,-3)$.

15.5.20 $k(t) = \dfrac{\sqrt{2}}{(\sin^2 t + \cos^2 t + \cos^2 t)^{3/2}} = \dfrac{\sqrt{2}}{(1 + \cos^2 t)^{3/2}}$

$\rho(t) = \dfrac{1}{\sqrt{2}}(1 + \cos^2 t)^{3/2}$

The minimum value of ρ is $\dfrac{1}{\sqrt{2}}$ at $t = \pi/2$ and $t = 3\pi/2$. The maximum value of ρ is 2 at $t = 0$ and $t = \pi$.

SECTION 15.6

15.6.1   Find the velocity, speed, and acceleration of a particle whose position is given by $\mathbf{r}(t) = t^2\mathbf{i} + 2t\mathbf{j}$ at $t = 2$ seconds, then sketch the path of the particle together with the velocity and acceleration vectors at $t = 2$ seconds.

15.6.2   Find the velocity, speed, and acceleration of a particle whose position is given by $\mathbf{r}(t) = \langle 4\cos t,\ \sin t \rangle$ at $t = \pi/2$ seconds, then sketch the path of the particle together with the velocity and acceleration vectors at $t = \pi/2$ seconds.

15.6.3   Find the velocity, speed, and acceleration of a particle whose position is given by $\mathbf{r}(t) = e^t\mathbf{i} + e^{2t}\mathbf{j}$ at $t = 0$ seconds, then sketch the path of the particle together with the velocity and acceleration vectors at $t = 0$ seconds.

15.6.4   Find the velocity, speed, and acceleration of a particle whose position is given by $\mathbf{r}(t) = \langle \cos t, \sin t, t^{3/2} \rangle$ at $t = \pi/2$ seconds.

15.6.5   Find the velocity, speed, and acceleration of a particle whose position is given by $\mathbf{r}(t) = e^t\mathbf{i} + e^t\cos t\mathbf{j} + e^t\sin t\mathbf{k}$ at $t = 0$ seconds,

15.6.6   Find the position and velocity vectors of a particle whose acceleration is given by $a(t) = 2\sin 2t\mathbf{i} + 4\cos 2t\mathbf{j}$ if $\mathbf{v} = (\pi/2) = \mathbf{i}+\mathbf{j}$ and $\mathbf{r}(\pi/2) = \pi\mathbf{i} + \frac{\pi}{2}\mathbf{j}$.

15.6.7   A particle moves through 3-space in such a way that its velocity is $\mathbf{v}(t) = 2\mathbf{i} - 4t^3\mathbf{j} + 6\sqrt{t}\,\mathbf{k}$. Find the coordinates of the particle at $t = 1$ second if the particle was initially at $(1, 5, 3)$ at $t = 0$ seconds.

15.6.8   A particle moves through 3-space in such a way that its velocity is $\mathbf{v}(t) = -\sin\pi t\mathbf{i} + \cos\pi t\mathbf{j}+t\mathbf{k}$. Find the coordinates of the particle at $t = 1$ second if the particle was initially at $\left(\dfrac{1}{\pi}, 0, 9\right)$ at $t = 0$ seconds.

15.6.9   A particle travels along a curve given by $\mathbf{r}(t) = 3\cos 3t\mathbf{i} - 3\sin 3t\mathbf{j} + 2\mathbf{k}$. Find the displacement and distance traveled by the particle during the time interval $0 \le t \le \pi/2$ seconds.

15.6.10   A particle travels along a curve given by $\mathbf{r}(t) = e^t\cos t\mathbf{i} + e^t\sin\mathbf{j} + 4\mathbf{k}$. Find the displacement and distance traveled by the particle during the time interval $0 \le t \le \pi$ seconds.

15.6.11   A particle travels along a curve given by $\mathbf{r}(t) = \langle t^2, t^3 \rangle$. Find the scalar and vector, tangential and normal components of the acceleration and the curvature of the path when $t = 1$ second.

15.6.12   A particle travels along a helical path given by $\mathbf{r}(t) = \cos 2t\mathbf{i} + \sin 2t\mathbf{j} + t\mathbf{k}$. Find the scalar and vector, tangential and normal components of the acceleration and the curvature of the path when $t = \pi/2$ seconds.

15.6.13   A particle travels along a path given by $\mathbf{r}(t) = \langle 1+t^3, 2t^3, 2-t^3 \rangle$. Find the scalar and vector, tangential and normal components of the acceleration and the curvature of the path when $t = 1$ second.

15.6.14   A particle travels along a path given by $\mathbf{r}(t) = \langle \sqrt{2}\,t, e^t, e^t \rangle$. Find the scalar and vector, tangential and normal components of the acceleration and the curvature of the path at the point $(0, 1, 1)$.

15.6.15   Show that the position and velocity vectors of the particle whose position is given by $\mathbf{r}(t) = \sin t \cos t\,\mathbf{i} + \cos^2 t\,\mathbf{j} + \sin t\,\mathbf{k}$ are orthogonal.

15.6.16   A shell is fired from a mortar at ground level with a velocity of 250 meters per second at an elevation of 60°. How far does the shell travel horizontally?

15.6.17   A shell is fired from ground level at an elevation of 60° and strikes a target 6000 meters away. Calculate the muzzle speed of the shell.

15.6.18   A certain calculus text is thrown upward from the roof of a college dormitory 160 feet high with an elevation of 45° with the horizontal. How far from the base of the dormitory will the text strike the ground if its initial speed was 32 feet per second?

SOLUTIONS

SECTION 15.6

15.6.1 $\mathbf{v}(t) = 2t\mathbf{i} + 2\mathbf{j}$,
$\|\mathbf{v}(t)\| = \sqrt{4t^2 + 4}$,
$\mathbf{a}(t) = 2\mathbf{i}$,
$\mathbf{v}(2) = 4\mathbf{i} + 2\mathbf{j}$, $\|\mathbf{v}(2)\| = 2\sqrt{5}$,
$\mathbf{a}(2) = 2\mathbf{i}$.

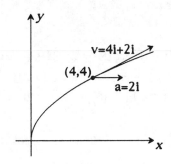

15.6.2 $\mathbf{v}(t) = \langle -4\sin t, \cos t\rangle$
$\|\mathbf{v}(t)\| = \sqrt{16\sin^2 t + \cos^2 t}$
$\mathbf{a}(t) = \langle -4\cos t, -\sin t\rangle$
$\mathbf{v}(\pi/2) = \langle -4, 0\rangle$, $\|\mathbf{v}(\pi/2)\| = 4$
$\|\mathbf{a}(\pi/2)\| = \langle 0, -1\rangle$

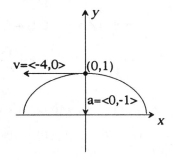

15.6.3 $\mathbf{v}(t) = e^t\mathbf{i} + 2e^{2t}\mathbf{j}$
$\|\mathbf{v}(t)\| = \sqrt{e^{2t} + 4e^{4t}}$
$\mathbf{a}(t) = e^t\mathbf{i} + 4e^{2t}\mathbf{j}$
$\mathbf{v}(0) = \mathbf{i} + 2\mathbf{j}$, $\|\mathbf{v}(0)\| = \sqrt{5}$
$\mathbf{a}(0) = \mathbf{i} + 4\mathbf{j}$

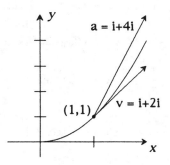

15.6.4 $\mathbf{v}(t) = \left\langle -\sin t, \cos t, \dfrac{3}{2}t^{1/2}\right\rangle$

$\|\mathbf{v}(t)\| = \sqrt{\sin^2 t + \cos^2 t + \dfrac{9}{4}t} = \sqrt{1 + \dfrac{9}{4}t}$

$\mathbf{a}(t) = \left\langle -\cos t, -\sin t, 3/4t^{-1/2}\right\rangle$

$$\mathbf{v}(\pi/2) = \left\langle -1, 0, \frac{3}{2}\sqrt{\pi/2} \right\rangle, \; \|\mathbf{v}(\pi/2)\| = \sqrt{1 + \frac{9\pi}{8}}$$

$$\mathbf{a}(\pi/2) = \left\langle 0, -1, \frac{3}{4}\sqrt{\frac{2}{\pi}} \right\rangle$$

15.6.5 $\mathbf{v}(t) = e^t \mathbf{i} + (e^t \cos t - e^t \sin t)\mathbf{j} + (e^t \sin t + e^t \cos t)\mathbf{k}$,

$\|\mathbf{v}(t)\| = \sqrt{3e^{2t}}$; $\mathbf{a}(t) = e^t \mathbf{i} - 2e^t \sin t\mathbf{j} + 2e^t \cos t\mathbf{k}$;

$\mathbf{v}(0) = \mathbf{i} + \mathbf{j} + \mathbf{k}$; $\|\mathbf{v}(0)\| = \sqrt{3}$; $\mathbf{a}(0) = \mathbf{i} + 2\mathbf{k}$.

15.6.6 $\mathbf{v}(t) = \displaystyle\int \mathbf{a}(t)dt = \int (2\sin 2t\mathbf{i} + 4\cos 2t\mathbf{j})dt + \mathbf{C}_1$,

$\mathbf{v}(t) = -\cos 2t\mathbf{i} + 2\sin 2t\mathbf{j} + \mathbf{C}_1$, $\mathbf{v}(\pi/2) = \mathbf{i} + \mathbf{C}_1 = \mathbf{i} + \mathbf{j}$, so $\mathbf{C}_1 = \mathbf{j}$;

$\mathbf{v}(t) = -\cos 2t\mathbf{i} + (1 + 2\sin 2t)\mathbf{j}$; $\mathbf{r}(t) = \displaystyle\int \mathbf{v}(t)dt = \int [-\cos 2t\mathbf{i} + (1 + 2\sin 2t)\mathbf{j}]dt + \mathbf{C}_1$,

so $\mathbf{r}(t) = -\dfrac{1}{2}\sin 2t\mathbf{i} + (t - \cos 2t)\mathbf{j} + \mathbf{C}_2$; $\mathbf{r}(\pi/2) = (\dfrac{\pi}{2} + 1)\mathbf{j} + \mathbf{C}_2 = \pi\mathbf{i} + \dfrac{\pi}{2}\mathbf{j}$,

so $\mathbf{C}_2 = \pi\mathbf{i} - \mathbf{j}$, $\mathbf{r}(t) = \left(\pi - \dfrac{1}{2}\sin 2t\right)\mathbf{i} + (t - 1 - \cos 2t)\mathbf{j}$

15.6.7 $\mathbf{r}(t) = \displaystyle\int (2\mathbf{i} - 4t^3\mathbf{j} + 6\sqrt{t}\,\mathbf{k})dt = 2t\mathbf{i} - t^4\mathbf{j} + 4t^{3/2}\mathbf{k} + C$, $\mathbf{r}(0) = C = \mathbf{i} + 5\mathbf{j} + 3\mathbf{k}$,

$\mathbf{r}(t) = (1 + 2t)\mathbf{i} + (5 - t^4)\mathbf{j} + (3 + 4t^{3/2})\mathbf{k}$ so $\mathbf{r}(1) = 3\mathbf{i} + 4\mathbf{j} + 7\mathbf{k}$; the particle is located at $(3, 4, 7)$ at $t = 1$ second.

15.6.8 $\mathbf{r}(t) = \displaystyle\int (-\sin \pi t\mathbf{i} + \cos \pi t\mathbf{j} + t\mathbf{k})\,dt$, $\mathbf{r}(t) = \dfrac{1}{\pi}\cos \pi t\mathbf{i} + \dfrac{1}{\pi}\sin \pi t\mathbf{j} + \dfrac{t^2}{2}\mathbf{k} + C$,

$\mathbf{r}(0) = \dfrac{1}{\pi}\mathbf{i} + C = \dfrac{1}{\pi}\mathbf{i} + 9\mathbf{k}$, $C = 9\mathbf{k}$ so $\mathbf{r}(t) = \dfrac{1}{\pi}\cos \pi t\mathbf{i} + \dfrac{1}{\pi}\sin \pi t\mathbf{j} + \left(9 + \dfrac{t^2}{2}\right)\mathbf{k}$ and

$\mathbf{r}(1) = -\dfrac{1}{\pi}\mathbf{i} + \dfrac{19}{2}\mathbf{k}$, the particle is located at $\left(-\dfrac{1}{\pi}, 0, \dfrac{19}{2}\right)$ at $t = 1$ second.

15.6.9 $\mathbf{v}(t) = \dfrac{d\mathbf{r}}{dt} = -9\sin 3t\mathbf{i} - 9\cos 3t\mathbf{j}$, $\|\mathbf{v}(t)\| = \sqrt{81\sin^2 3t + 81\cos^2 3t} = 9$. The displacement

$\mathbf{\Delta r} = \mathbf{r}(\pi/2) - \mathbf{r}(0) = \left(3\cos 3\dfrac{\pi}{2}\mathbf{i} - 3\sin 3\dfrac{\pi}{2}\mathbf{j} + 2\mathbf{k}\right) - (3\cos 0\mathbf{i} - 3\sin 0\mathbf{j} + 2\mathbf{k})$ $\mathbf{\Delta r} = -3\mathbf{i} + 3\mathbf{j}$,

the distance traveled is $L = \displaystyle\int_0^{\pi/2} 9dt = 9\dfrac{\pi}{2}$ units.

15.6.10 $\mathbf{v}(t) = \dfrac{d\mathbf{r}}{dt} = (e^t \cos t - e^t \sin t)\mathbf{i} + (e^t \sin t + e^t \cos t)\mathbf{j}$,

$\|\mathbf{v}(t)\| = \sqrt{(e^t \cos t - e^t \sin t)^2 + (e^t \sin t + e^t \cos t)^2} = \sqrt{2e^{2t}}$. The displacement

$\mathbf{\Delta r} = \mathbf{r}(\pi) - \mathbf{r}(0) = (e^\pi \cos \pi\mathbf{i} - e^\pi \sin \pi\mathbf{j} + 4\mathbf{k}) - (e^0 \cos 0\mathbf{i} + e^0 \sin 0\mathbf{j} + 4\mathbf{k}) = -(e^\pi + 1)\mathbf{i}$,

the distance traveled is $L = \displaystyle\int_0^\pi \sqrt{2e^{2t}}\,dt = \sqrt{2}(e^\pi - 1)$ units.

15.6.11 $\mathbf{v}(t) = \mathbf{r}'(t) = \langle 2t, 3t^2 \rangle$, $\mathbf{a}(t) = \mathbf{v}'(t) = \langle 2, 6t \rangle$,

$$a_T = \frac{\langle 2t, 3t^2 \rangle \cdot \langle 2, 6t \rangle}{\sqrt{(2t)^2 + (3t^2)^2}} = \frac{4t + 18t^3}{\sqrt{4t^2 + 9t^4}},$$

when $t = 1 \sec$, $a_T = \dfrac{22}{\sqrt{13}} \approx 6.10$; $a_N = \dfrac{\| \langle 2t, 3t^2 \rangle \times \langle 2, 6t \rangle \|}{\sqrt{(2t)^2 + (3t^2)^2}}$

$a_N = \dfrac{6t^2}{\sqrt{13}}$, when $t = 1 \sec$, $a_N = \dfrac{6}{\sqrt{13}} \approx 1.66$.

$\mathbf{T}(1) = \dfrac{\mathbf{v}(1)}{\|v(1)\|} = \left\langle \dfrac{2}{\sqrt{13}}, \dfrac{3}{\sqrt{13}} \right\rangle$ so $a_T \mathbf{T}(1) = \dfrac{22}{\sqrt{13}} \left\langle \dfrac{2}{\sqrt{13}}, \dfrac{3}{\sqrt{13}} \right\rangle = \left\langle \dfrac{44}{13}, \dfrac{66}{13} \right\rangle$,

$a_N \mathbf{N}(1) = a(1) - a_T \mathbf{T}(1) = \langle 2, 6 \rangle - \left\langle \dfrac{44}{13}, \dfrac{66}{13} \right\rangle = \left\langle -\dfrac{18}{13}, \dfrac{12}{13} \right\rangle$.

At $t = 1 \sec$, $K = \dfrac{\|\mathbf{v}(1) \times \mathbf{a}(1)\|}{\|\mathbf{v}(1)\|^3} = \dfrac{6}{13\sqrt{13}} \approx 0.13$

15.6.12 $\mathbf{v}(t) = -2 \sin 2t \mathbf{i} + 2 \cos 2t \mathbf{j} + \mathbf{k}$

$\mathbf{a}(t) = -4 \cos 2t \mathbf{i} - 4 \sin 2t \mathbf{j}$

$$a_T = \frac{(-2 \sin 2t \mathbf{i} + 2 \cos 2t \mathbf{j} + \mathbf{k}) \cdot (-4 \cos 2t \mathbf{i} - 4 \sin 2t \mathbf{j})}{\sqrt{(-2 \sin 2t)^2 + (2 \cos 2t)^2 + (1)^2}} = 0$$

$$a_N = \frac{\|(-2 \sin 2t \mathbf{i} + 2 \cos 2t \mathbf{j} + \mathbf{k}) \times (-4 \cos 2t \mathbf{i} - 4 \sin 2t \mathbf{j})\|}{\sqrt{(-2 \sin 2t)^2 + (2 \cos 2t)^2 + (1)^2}}$$

$a_N = \dfrac{\sqrt{80}}{\sqrt{5}} = 4$, when $t = \dfrac{\pi}{2} \sec$, $a_N = 4$,

$T(\pi/2) = \dfrac{V(\pi/2)}{\|V(\pi/2)\|} = -\dfrac{2}{\sqrt{5}}\mathbf{j} + \dfrac{1}{\sqrt{5}}\mathbf{k}$, thus,

$a_T \cdot T(1) = 0$ so $a_N \mathbf{N}(\pi/2) = \mathbf{a}(\pi/2) = 4i$.

At $t = \pi/2$ seconds, $\mathbf{k} = \dfrac{\|\mathbf{v}(\pi/2) \times \mathbf{a}(\pi/2)\|}{\|\mathbf{v}(\pi/2)\|^3} = \dfrac{\sqrt{80}}{5\sqrt{5}} = \dfrac{4}{5}$.

15.6.13 $\mathbf{v}(t) = \mathbf{v}'(t) = \langle 3t^2, 6t^2, -3t^2 \rangle$,

$\mathbf{a}(t) = \mathbf{v}'(t) = \langle 6t, 12t, -6t \rangle$.

$$a_T = \frac{\langle 3t^2, 6t^2, -3t^2 \rangle \cdot \langle 6t, 12t, -6t \rangle}{\sqrt{(3t^2)^2 + (6t^2)^2 + (-3t^2)^2}} = 6\sqrt{6}\, t$$

$$a_N = \frac{\| \langle 3t^2, 6t^2, -3t^2 \rangle \times \langle 6t, 12t, -6t \rangle \|}{\sqrt{(3t^2)^2 + (6t^2)^2 + (-3t^2)^2}} = 0$$

when $t = 1$ sec, $a_T = 6\sqrt{6}$, $\mathbf{T}(1) = \dfrac{V(1)}{\|V(1)\|} = \left\langle \dfrac{\sqrt{6}}{6}, \dfrac{2\sqrt{6}}{6}, \dfrac{-\sqrt{6}}{6} \right\rangle$

$a_T \mathbf{T}(1) = \langle 6, 12, -6 \rangle$

$K = 0$ since $\mathbf{V} \times \mathbf{a} = 0$

15.6.14 $\mathbf{v}(t) = \mathbf{r}'(t) = \left\langle \sqrt{2}, e^t, e^t \right\rangle$, $\mathbf{a}(t) = \mathbf{v}'(t) = \langle 0, e^t, e^t \rangle$.

$$a_T = \frac{\left\langle \sqrt{2}, e^t, e^t \right\rangle \cdot \langle 0, e^t, e^t \rangle}{\sqrt{(\sqrt{2})^2 + (e^t)^2 + (e^t)^2}} = \frac{2e^{2t}}{\sqrt{2 + 2e^{2t}}} \text{ at the point } (0,1,1), \ t = 0 \text{ sec so } a_T = 1.$$

$$a_N = \frac{\| \left\langle \sqrt{2}, e^t, e^t \right\rangle \times \langle 0, e^t, e^t \rangle \|}{\sqrt{(\sqrt{2})^2 + (e^t)^2 + (e^t)^2}} = \frac{\| \langle 0, -\sqrt{2}e^t, \sqrt{2}e^t \rangle \|}{\sqrt{2 + 2e^{2t}}}$$

at $t = 0$ sec, $a_N = 1/2$. $T(0) = \dfrac{V(0)}{\|V(0)\|} = \left\langle \dfrac{\sqrt{2}}{2}, \dfrac{1}{2}, \dfrac{1}{2} \right\rangle$ so $a_T T(0) = \left\langle \dfrac{\sqrt{2}}{2}, \dfrac{1}{2}, \dfrac{1}{2} \right\rangle$

$a_N N(0) = \mathbf{a}(0) - a_T \mathbf{T}(0) = \langle 0, 1, 1 \rangle - \left\langle \dfrac{\sqrt{2}}{2}, \dfrac{1}{2}, \dfrac{1}{2} \right\rangle$ $a_N N(0) = \left\langle \dfrac{-\sqrt{2}}{2}, \dfrac{1}{2}, \dfrac{1}{2} \right\rangle$.

At $t = 0$ sec, $K = \dfrac{\| \langle \sqrt{2}, 1, 1 \rangle \times \langle 0, 1, 1 \rangle \|}{\left(\sqrt{(\sqrt{2})^2 + (1)^2 + (1)^2} \right)^3} = \dfrac{1}{4}$

15.6.15 $\mathbf{r}(t) = (\sin t \cos t\, \mathbf{i} + \cos^2 t\, \mathbf{j} + \sin t\, \mathbf{k}$, $\mathbf{v}(t) = \mathbf{r}'(t) = (\cos^2 t - \sin^2 t)i - 2\cos t \sin t\, j + \cos t\, k$, $\mathbf{r}(t) \cdot \mathbf{v}(t) = [(\cos^2 t - \sin^2 t)\mathbf{i} - 2\cos t \sin t\, \mathbf{j} + \cos t\, \mathbf{k}] \cdot (\sin t \cos t\, \mathbf{i} + \cos^2 t\, \mathbf{j} + \sin t\, \mathbf{k}) = 0$ so the vectors are orthogonal.

15.6.16 $\mathbf{v}_0(t) = 250\cos 60°\mathbf{i} + 250\sin 60°\mathbf{j} = 125\mathbf{i} + 125\sqrt{3}\mathbf{j}$, so $\mathbf{r}(t) = 125t\mathbf{i} + \left(125\sqrt{3} - 4.9t^2 \right)\mathbf{j}$,

thus, $x = 125t$ and $y = 125\sqrt{3} - 4.9t^2$, $y = 0$ when $t = 0$ or $t = \dfrac{125\sqrt{3}}{4.9} \approx 44.2$ sec, thus,

$x = 125 \left(\dfrac{125\sqrt{3}}{4.9} \right) = \dfrac{15625\sqrt{3}}{4.9} \approx 5523$ meters.

15.6.17 Let $v_0 = \|\mathbf{v}_0\|$, then $\mathbf{v}_0 = \dfrac{v_0}{2}\mathbf{i} + \dfrac{v_0\sqrt{3}}{2}\mathbf{j}$. $s(0) = 0$ and $\mathbf{r}(t) = \dfrac{v_0 t}{2}\mathbf{i} + \left(\dfrac{v_0\sqrt{3}t}{2} - 4.9t^2 \right)\mathbf{j}$,

thus, $x(t) = \dfrac{v_0 t}{2}$ and $y(t) = \dfrac{v_0\sqrt{3}}{2}t - 4.9t^2$. $y = 0$ when $t = 0$ or $t = \dfrac{v_0\sqrt{3}}{9.8}$ so that

$x_{\max} = \dfrac{v_0^2\sqrt{3}}{19.6} = 6000$ and $v_0 \approx 261$ m/s.

15.6.18 $\mathbf{v}_0 = 16\sqrt{2}\mathbf{i} + 16\sqrt{2}\mathbf{j}$, $s_0 = 160$ so $\mathbf{r}(t) = 16\sqrt{2}t\mathbf{i} + \left(160 + 16\sqrt{2}t - 16t^2 \right)\mathbf{j}$. $x(t) = 16\sqrt{2}t$

and $y(t) = 160 + 16\sqrt{2}t - 16t^2$. $y = 0$ when $t = \dfrac{\sqrt{2} - \sqrt{42}}{2}$ which is not valid or when

$t = \dfrac{\sqrt{2} + \sqrt{42}}{2} \approx 3.94$ seconds so $x \approx x(3.94) = 89.3$ feet.

SUPPLEMENTARY EXERCISES, CHAPTER 15

In Exercises 1–3,

(a) find $\mathbf{v} = d\mathbf{r}/dt$ and $\mathbf{a} = d^2\mathbf{r}/dt^2$;

(b) sketch the graph of $\mathbf{r}(t)$, showing the direction of increasing t, and find the vectors $\mathbf{r}'(t)$ and $\mathbf{r}''(t)$ at the points corresponding to $t = t_0$ and $t = t_1$.

1. $\mathbf{r}(t) = \sqrt{t+4}\,\mathbf{i} + 2t\mathbf{j}; t_0 = -3, t_1 = 0$.

2. $\mathbf{r}(t) = \langle 2 + \cosh t, 1 - 2\sinh t \rangle; t_0 = 0, t_1 = \ln 2$.

3. $\mathbf{r}(t) = \langle 2t^3 - 1, t^3 + 1 \rangle; t_0 = 0, t_1 = -\frac{1}{2}$.

4. Find the limits.

    (a) $\lim\limits_{t \to e} \langle t + \ln t^2, \ln t + t^2 \rangle$

    (b) $\lim(\cos 2t\mathbf{i} - 3t\mathbf{j})$.

5. Evaluate the integrals.

    (a) $\displaystyle\int (k\mathbf{i} + m\mathbf{j})\,dt$

    (b) $\displaystyle\int_0^{\ln 3} \langle e^{2t}, 2e^t \rangle\,dt$

    (c) $\displaystyle\int_0^2 \| \cos t\mathbf{i} + \sin t\mathbf{j} \|\,dt$

    (d) $\displaystyle\int \frac{d}{dt}[\sqrt{t^2 + 3}\,\mathbf{i} + \ln(\sin t)\mathbf{j}]\,dt$.

In Exercises 6–8, find (a) ds/dt and (b) parametric equations for the curve with arc length s as a parameter, assuming the point corresponding to t_0 is the reference point.

6. $\mathbf{r}(t) = (3e^t + 2)\mathbf{i} + (e^t - 1)\mathbf{j}; t_0 = 0$.

7. $\mathbf{r}(t) = \left\langle \dfrac{t^2 + 1}{t}, \ln t^2 \right\rangle$, where $t > 0; t_0 = 1$.

8. $\mathbf{r}(t) = \langle t^3, t^2 \rangle$, where $t \geq 0; t_0 = 0$.

In Exercises 9 and 10, sketch the graph of the curve, showing the direction of increasing t.

9. $\mathbf{r}(t) = \langle t, t^2 + 1, 1 \rangle$.

10. $x = t, y = t, z = 2\cos(\pi t/2), 0 \leq t \leq 2$.

11. Find the velocity, speed, acceleration, unit tangent vector, unit normal vector, and curvature when $t = 0$ for the motion given by $x = a \sin t, y = a \cos t, z = a \ln(\cos t)$ $(a > 0)$.

12. The position vector of a particle is

$$\mathbf{r}(t) = \sin(2t)\mathbf{i} + \cos(2t)\mathbf{j} + 2e^t\mathbf{k}.$$

(a) Find the velocity, acceleration, and speed as functions of t.

(b) Find the scalar tangential and normal components of acceleration and the curvature when $t = 0$.

In Exercises 13 and 14, find the arc length of the curve.

13. $x = 2t, y = 4 \sin 3t, z = 4 \cos 3t, 0 \le t \le 2\pi$.

14. $\mathbf{r}(t) = \langle e^{-t}, \sqrt{2}t, e^t \rangle, 0 \le t \le \ln 2$.

15. Consider the curve whose position vector is

$$\mathbf{r}(t) = \langle e^{-t}, e^{2t}, t^3 + 1 \rangle$$

Find parametric equations for the tangent line to the curve at the point where $t = 0$.

16. **(a)** Show that the speed of a particle is constant if $\mathbf{r} = 3 \sin 2t\mathbf{i} - 3 \cos 2t\mathbf{j} - 8t\mathbf{k}$.

(b) Show that $\mathbf{v}$ and $\mathbf{a}$ are orthogonal at each point on the path of part (a), and thus verify the result in Exercise 20, Section 15.6.

17. If $\mathbf{u} = \langle 2t, 3, -t^2 \rangle$ and $\mathbf{v} = \langle 0, t^2, t \rangle$, find

(a) $\displaystyle\int_0^3 \mathbf{u}\, dt$ **(b)** $d(\mathbf{u} \times \mathbf{v})/dt$.

18. If $\mathbf{r}(t) = \langle \cos(\pi e^t), \sin(\pi e^t), \pi t \rangle$, find the angle between the acceleration $\mathbf{a}$ and the velocity $\mathbf{v}$ when $t = 0$.

For the curves given in Exercises 19–21, find (a) the unit tangent vector $\mathbf{T}$ at P_0 and (b) the curvature κ at P_0.

19. $\mathbf{r}(t) = (t^2 + 1)\mathbf{i} + (1/t)\mathbf{j}; P_0(2, 1)$.   **20.** $y = \ln x; P_0(1, 0)$.

21. $x = (y - 1)^2; P_0(0, 1)$.

In Exercises 22–25, find the curvature κ of the given curve at P_0.

22. $xy^2 = 1; P_0(1,1)$.

23. $\mathbf{r}(t) = (t + t^3)\mathbf{i} + (t + t^2)\mathbf{j}; P_0(2,2)$.

24. $y = a\cosh(x/a); P_0(a, a\cosh 1)$.

25. $e^x = \sec y; P_0(0,0)$.

26. Find the smallest radius of curvature and the point at which it occurs for $\mathbf{r}(t) = \langle e^{2t}, e^{-2t} \rangle$.

27. Find the equation of the osculating circle for the parabola $y = (x-1)^2$ at the point $(1,0)$. Verify that y' and y'' for the parabola are the same as y' and y'' for the osculating circle at $(1,0)$.

In Exercises 28 and 29, calculate $d\mathbf{r}/du$ by the chain rule, and check the result by first expressing $\mathbf{r}$ in terms of u.

28. $\mathbf{r} = \langle \sin t, 2\cos 2t \rangle; t = e^{u/2}$.

29. $\mathbf{r} = \langle e^t - 1, 2e^{2t} \rangle; t = \ln u$.

In Exercises 30 and 31, find the scalar tangential and normal components of acceleration.

30. $\mathbf{r}(t) = \langle \cosh 2t, \sinh 2t \rangle, t \geq 0$.

31. $\mathbf{r}(t) = \langle \sin t - t\cos t, \cos t + t\sin t \rangle, t \geq 0$.

For the motion described in Exercises 32 and 33,

(a) find $\mathbf{v}, \mathbf{a}$, and ds/dt at P_0;

(b) find κ at P_0;

(c) find a_T and a_N at P_0;

(d) describe the trajectory;

(e) find the center of the osculating circle at P_0.

32. $\mathbf{r}(t) = (1 - t^2)\mathbf{i} + 2t\mathbf{j}; P_0(0,2)$.

33. $\mathbf{r}(t) = \langle e^{-t}, e^t \rangle; P_0(1,1)$.

34. At $t = 0$, a particle of mass m is located at the point $(-2/m, 0)$ and has a velocity of $(2\mathbf{i} - 3\mathbf{j})/m$. Find the position function $\mathbf{r}(t)$ if the particle is acted upon by a force $\mathbf{F} = \langle 2\cos t, 3\sin t \rangle$, for $t \geq 0$.

35. The force acting on a particle of unit mass $(m = 1)$ if $\mathbf{F}(\sin t)\mathbf{i} + (4e^{2t})\mathbf{j}$. If the particle starts at the origin with an initial velocity $\mathbf{v}_0 = \mathbf{i} + 2\mathbf{j}$, find the position function $\mathbf{r}(t)$ at any $t \geq 0$.

36. A curve in a railroad track has the shape of the parabola $x = y^2/100$. If a train is loaded so that its scalar normal component of acceleration cannot exceed 25 units/sec^2, what is its maximum possible speed as it rounds the curve at $(0,0)$?

37. A particle moves along the parabola $y = 2x - x^2$ with a constant x-component of velocity of 4 ft/sec. Find the scalar tangential and normal components of acceleration at the points (a) $(1,1)$ and (b) $(0,0)$.

38. If a particle moves along the curve $y = 2x^2$ with constant speed $ds/dt = 10$, what are a_T and a_N at $P(x, 2x^2)$?

39. A child twirls a weight at the end of a 2-meter string at a rate of 1 revolution/second. Find a_T and a_N for the motion of the weight.

For the motion described in Exercises 40 and 41, find (a) ds/dt and (b) the distance traveled over the interval described.

40. $\mathbf{r}(t) = \langle 2 \sinh t, \sinh^2 t \rangle$, $0 \le t \le 1$.   **41.** $\mathbf{r}(t) = e^t \langle \sin 2t, \cos 2t \rangle$, $0 \le t \le \ln 3$.

SOLUTIONS

SUPPLEMENTARY EXERCISES CHAPTER 15

1. (a) $\mathbf{v} = \dfrac{1}{2}(t+4)^{-1/2}\mathbf{i} + 2\mathbf{j}$, $\mathbf{a} = -\dfrac{1}{4}(t+4)^{-3/2}\mathbf{i}$

   (b) $\mathbf{r}'(-3) = (1/2)\mathbf{i} + 2\mathbf{j}$, $\mathbf{r}''(-3) = -(1/4)\mathbf{i}$
   $\mathbf{r}'(0) = (1/4)\mathbf{i} + 2\mathbf{j}$, $\mathbf{r}''(0) = -(1/32)\mathbf{i}$

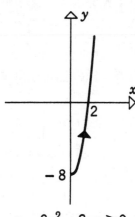

$y = 2x^2 - 8,\; x \geq 0$

2. (a) $\mathbf{v} = \langle \sinh t, -2\cosh t \rangle$, $\mathbf{a} = \langle \cosh t, -2\sinh t \rangle$

   (b) $\mathbf{r}'(0) = \langle 0, -2 \rangle$
   $\mathbf{r}''(0) = \langle 1, 0 \rangle$
   $\mathbf{r}'(\ln 2) = \langle 3/4, -5/2 \rangle$
   $\mathbf{r}''(\ln 2) = \langle 5/4, -3/2 \rangle$

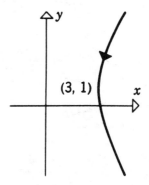

$(3, 1)$

3. (a) $\mathbf{v} = \langle 6t^2, 3t^2 \rangle$, $\mathbf{a} = \langle 12t, 6t \rangle$

   (b) $\mathbf{r}'(0) = \langle 0, 0 \rangle$, $\mathbf{r}''(0) = \langle 0, 0 \rangle$
   $\mathbf{r}'(-1/2) = \langle 3/2, 3/4 \rangle$, $\mathbf{r}''(-1/2) = \langle -6, -3 \rangle$

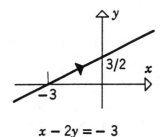

$x - 2y = -3$

4. (a) $\langle e + 2, 1 + e^2 \rangle$ (b) $(1/2)\mathbf{i} - (\pi/2)\mathbf{j}$

5. **(a)** $(k\mathbf{i} + m\mathbf{j})t + \mathbf{C}$

   (b) $\langle e^{2t}/2, 2e^t \rangle \big]_0^{\ln 3} = \langle 9/2, 6 \rangle - \langle 1/2, 2 \rangle = \langle 4, 4 \rangle$

   (c) $\displaystyle\int_0^2 dt = 2$ **(d)** $\sqrt{t^2 + 3}\,\mathbf{i} + \ln(\sin t)\mathbf{j} + \mathbf{C}$

6. **(a)** $ds/dt = \|\mathbf{r}'(t)\| = \|3e^t\mathbf{i} + e^t\mathbf{j}\| = \sqrt{10}\,e^t$

   (b) $s = \displaystyle\int_0^t \sqrt{10}\,e^u du = \sqrt{10}(e^t - 1),\ e^t = 1 + s/\sqrt{10},\ x = 5 + 3s/\sqrt{10},\ y = s/\sqrt{10}$

7. **(a)** $ds/dt = \|\mathbf{r}'(t)\| = \|\langle(t^2 - 1)/t^2, 2/t\rangle\| = 1 + 1/t^2$

   (b) $s = \displaystyle\int_1^t (1 + 1/u^2)du = t - 1/t,\ t^2 - st - 1 = 0,\ t = (s \pm \sqrt{s^2 + 4})/2$, but $t \geq 0$
   so $t = (s + \sqrt{s^2 + 4})/2$ and $x = \sqrt{s^2 + 4},\ y = 2\ln[(s + \sqrt{s^2 + 4})/2]$.

8. **(a)** $ds/dt = \|\mathbf{r}'(t)\| = \|\langle 3t^2, 2t\rangle\| = t\sqrt{9t^2 + 4}$

   (b) $s = \displaystyle\int_0^t u(9u^2 + 4)^{1/2}du = [(9t^2 + 4)^{3/2} - 8]/27,\ t = \dfrac{1}{3}[(27s + 8)^{2/3} - 4]^{1/2}$,
   $x = \dfrac{1}{27}[(27s + 8)^{2/3} - 4]^{3/2},\ y = \dfrac{1}{9}[(27s + 8)^{2/3} - 4]$

9.

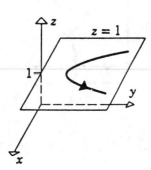

10.

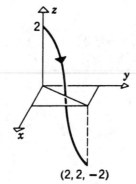

$(2, 2, -2)$

11. $\mathbf{r}(t) = a\sin t\,\mathbf{i} + a\cos t\,\mathbf{j} + a\ln(\cos t)\mathbf{k},\ \mathbf{v} = a\cos t\,\mathbf{i} - a\sin t\,\mathbf{j} - a\tan t\,\mathbf{k}$
   $\|\mathbf{v}\| = a(\cos^2 t + \sin^2 t + \tan^2 t)^{1/2} = a(1 + \tan^2 t)^{1/2} = a\sec t$
   $\mathbf{a} = -a\sin t\,\mathbf{i} - a\cos t\,\mathbf{j} - a\sec^2 t\,\mathbf{k},\ \mathbf{T} = \mathbf{v}/\|\mathbf{v}\| = \cos^2 t\,\mathbf{i} - \sin t\cos t\,\mathbf{j} - \sin t\,\mathbf{k}$
   $d\mathbf{T}/dt = -2\sin t\cos t\,\mathbf{i} - (\cos^2 t - \sin^2 t)\mathbf{j} - \cos t\,\mathbf{k}$; at $t = 0$, $\mathbf{v} = a\mathbf{i}$, $\|\mathbf{v}\| = a$, $\mathbf{a} = -a(\mathbf{j} + \mathbf{k})$,
   $\mathbf{T} = \mathbf{i},\ \mathbf{N} = (d\mathbf{T}/dt)/\|d\mathbf{T}/dt\| = (-\mathbf{j} - \mathbf{k})/\sqrt{2},\ \kappa = \|\mathbf{v} \times \mathbf{a}\|/\|\mathbf{v}\|^3 = \|a^2\mathbf{j} - a^2\mathbf{k}\|/a^3 = \sqrt{2}/a$

12. **(a)** $\mathbf{v} = 2\cos(2t)\mathbf{i} - 2\sin(2t)\mathbf{j} + 2e^t\mathbf{k},\ \|\mathbf{v}\| = 2(1 + e^{2t})^{1/2}$
   $\mathbf{a} = -4\sin(2t)\mathbf{i} - 4\cos(2t)\mathbf{j} + 2e^t\mathbf{k}$

   (b) When $t = 0$, $\mathbf{v} = 2\mathbf{i} + 2\mathbf{k}$, $\mathbf{a} = -4\mathbf{j} + 2\mathbf{k}$, $\|\mathbf{v}\| = 2\sqrt{2}$, $\mathbf{v} \cdot \mathbf{a} = 4$, $\mathbf{v} \times \mathbf{a} = 8\mathbf{i} - 4\mathbf{j} - 8\mathbf{k}$ so
   $a_T = (\mathbf{v} \cdot \mathbf{a})/\|\mathbf{v}\| = \sqrt{2}$, $a_N = \|\mathbf{v} \times \mathbf{a}\|/\|\mathbf{v}\| = 12/(2\sqrt{2}) = 3\sqrt{2}$, and
   $\kappa = \|\mathbf{v} \times \mathbf{a}\|/\|\mathbf{v}\|^3 = 3\sqrt{2}/8$.

13. $(dx/dt)^2 + (dy/dt)^2 + (dz/dt)^2 = 4 + 144\cos^2 3t + 144\sin^2 3t = 148$,

$$L = \int_0^{2\pi} \sqrt{148}\, dt = 2\pi\sqrt{148} = 4\pi\sqrt{37}.$$

14. $\mathbf{r}'(t) = \langle -e^{-t}, \sqrt{2}, e^t \rangle$, $\|\mathbf{r}'(t)\| = (e^{-2t} + 2 + e^{2t})^{1/2} = e^{-t} + e^t$, $L = \int_0^{\ln 2} (e^{-t} + e^t) dt = 3/2$.

15. $\mathbf{r}'(t) = \langle -e^{-t}, 2e^{2t}, 3t^2 \rangle$, $\mathbf{r}'(0) = \langle -1, 2, 0 \rangle$ is parallel to the tangent line to the curve at the tip of $\mathbf{r}(0) = \langle 1, 1, 1 \rangle$ so parametric equations of the tangent line are $x = 1 - t$, $y = 1 + 2t$, $z = 1$.

16. **(a)** $\mathbf{v} = 6\cos 2t\mathbf{i} + 6\sin 2t\mathbf{j} - 8\mathbf{k}$, $\|\mathbf{v}\| = \sqrt{36\cos^2 2t + 36\sin^2 2t + 64} = \sqrt{100} = 10$.
(b) $\mathbf{a} = -12\sin 2t\mathbf{i} + 12\cos 2t\mathbf{j}$, $\mathbf{v} \cdot \mathbf{a} = 0$.

17. **(a)** $\int_0^3 \langle 2t, 3, -t^2 \rangle dt = \langle t^2, 3t, -t^3/3 \rangle \Big]_0^3 = \langle 9, 9, -9 \rangle$
(b) $\mathbf{u} \times \mathbf{v} = \langle 3t + t^4, -2t^2, 2t^3 \rangle$, $d(\mathbf{u} \times \mathbf{v})/dt = \langle 3 + 4t^3, -4t, 6t^2 \rangle$.

18. $\mathbf{v} = \pi\langle -e^t \sin(\pi e^t), e^t \cos(\pi e^t), 1 \rangle$,
$\mathbf{a} = \pi e^t \langle -\pi e^t \cos(\pi e^t) - \sin(\pi e^t), -\pi e^t \sin(\pi e^t) + \cos(\pi e^t), 0 \rangle$; when $t = 0$, $\mathbf{v} = \pi\langle 0, -1, 1 \rangle$,
$\mathbf{a} = \pi\langle \pi, -1, 0 \rangle$, $\cos\theta = (\mathbf{v} \cdot \mathbf{a})/(\|\mathbf{v}\|\,\|\mathbf{a}\|) = 1/\sqrt{2\pi^2 + 2}$, $\theta = \cos^{-1}(1/\sqrt{2\pi^2 + 2}) \approx 78°$.

19. **(a)** $\mathbf{r}'(t) = 2t\mathbf{i} - (1/t^2)\mathbf{j}$, $t = 1$ at P_0 so $\mathbf{T} = \mathbf{r}'(1)/\|\mathbf{r}'(1)\| = (2\mathbf{i} - \mathbf{j})/\sqrt{5}$
(b) $\kappa(t) = \dfrac{6t^4}{(4t^6 + 1)^{3/2}}$ so $\kappa(1) = \dfrac{6}{5^{3/2}}$

20. **(a)** Let $x = t$, then $\mathbf{r}(t) = t\mathbf{i} + \ln t\mathbf{j}$, $\mathbf{r}'(t) = \mathbf{i} + (1/t)\mathbf{j}$,
$t = 1$ at P_0 so $\mathbf{T} = \mathbf{r}'(1)/\|\mathbf{r}'(1)\| = (\mathbf{i} + \mathbf{j})/\sqrt{2}$
(b) $\kappa(t) = \dfrac{t}{(t^2 + 1)^{3/2}}$, $\kappa(1) = \dfrac{1}{2^{3/2}}$

21. **(a)** Let $y = t$, then $\mathbf{r}(t) = (t - 1)^2\mathbf{i} + t\mathbf{j}$,
$\mathbf{r}'(t) = 2(t - 1)\mathbf{i} + \mathbf{j}$, $t = 1$ at P_0 so $\mathbf{T} = \mathbf{r}'(1)/\|\mathbf{r}'(1)\| = \mathbf{j}$
(b) $\kappa(t) = 2/[4(t - 1)^2 + 1]^{3/2}$, $\kappa(1) = 2$

22. Let $y = t$, then $x = 1/t^2$, $\kappa(t) = \dfrac{6|t|^5}{(4 + t^6)^{3/2}}$, $t = 1$ at P_0, $\kappa(1) = \dfrac{6}{5^{3/2}}$

23. $x = t + t^3$, $y = t + t^2$, $\kappa(t) = \dfrac{|2 - 6t - 6t^2|}{[(1 + 3t^2)^2 + (1 + 2t)^2]^{3/2}}$, $t = 1$ at P_0, $\kappa(1) = 2/25$

24. $\kappa(x) = \dfrac{\cosh(x/a)}{a[1 + \sinh^2(x/a)]^{3/2}} = \dfrac{1}{2}\operatorname{sech}^2(x/a)$, $\kappa(a) = \dfrac{1}{a}\operatorname{sech}^2 1$

25. Let $y = t$, then $x = \ln(\sec t)$, $\kappa(t) = |\cos t|$, $t = 0$ at P_0, $\kappa(0) = 1$

26. $\kappa(t) = \dfrac{2}{(e^{4t} + e^{-4t})^{3/2}}$, $\rho(t) = \dfrac{1}{2}(e^{4t} + e^{-4t})^{3/2} = \sqrt{2}(\cosh 4t)^{3/2}$, the smallest radius of curvature is $\rho(0) = \sqrt{2}$, it occurs at the point $(1, 1)$.

27. $\kappa(x) = \dfrac{2}{[1 + 4(x-1)^2]^{3/2}}$, $\kappa(1) = 2$, $\rho = 1/2$. The parabola opens upward and has its vertex at $(1, 0)$ so the center of curvature is at $(1, 1/2)$ and the oscillating circle is $(x-1)^2 + (y - 1/2)^2 = 1/4$. $y' = 0$ and $y'' = 2$ at $(1, 0)$ for both the parabola and the circle.

28. $d\mathbf{r}/du = (d\mathbf{r}/dt)(dt/du) = \dfrac{1}{2}e^{u/2}\langle\cos t, -4\sin 2t\rangle = \langle(1/2)e^{u/2}\cos e^{u/2}, -2e^{u/2}\sin 2e^{u/2}\rangle$

29. $d\mathbf{r}/du = (d\mathbf{r}/dt)(dt/du) = (1/u)\langle e^t, 4e^{2t}\rangle = \langle 1, 4u\rangle$

30. $\mathbf{v} = \langle 2\sinh 2t, 2\cosh 2t\rangle$, $\mathbf{a} = \langle 4\cosh 2t, 4\sinh 2t\rangle$, $\|\mathbf{v}\| = 2\sqrt{\sinh^2 2t + \cosh^2 2t} = 2\sqrt{\cosh 4t}$, $\mathbf{v} \cdot \mathbf{a} = 16\sinh 2t \cosh 2t = 8\sinh 4t$, $\mathbf{v} \times \mathbf{a} = 8\mathbf{k}$ so $a_T = (4\sinh 4t)/\sqrt{\cosh 4t}$, $a_N = 4/\sqrt{\cosh 4t}$.

31. $\mathbf{v} = \langle t\sin t, t\cos t\rangle$, $\mathbf{a} = \langle \sin t + t\cos t, \cos t - t\sin t\rangle$, $\|\mathbf{v}\| = t$, $\mathbf{v} \cdot \mathbf{a} = t$, $\mathbf{v} \times \mathbf{a} = -t^2\mathbf{k}$ so $a_T = 1$, $a_N = t$.

32. (a) $\mathbf{v} = -2t\mathbf{i} + 2\mathbf{j}$, $\mathbf{a} = -2\mathbf{i}$; $t = 1$ at P_0 so $\mathbf{v} = -2\mathbf{i} + 2\mathbf{j}$, $\mathbf{a} = -2\mathbf{i}$, $ds/dt = \|\mathbf{v}\| = 2\sqrt{2}$.

    (b) $\mathbf{v} \times \mathbf{a} = 4\mathbf{k}$, $\kappa = \|\mathbf{v} \times \mathbf{a}\|/\|\mathbf{v}\|^3 = 1/(4\sqrt{2})$.

    (c) $\mathbf{v} \cdot \mathbf{a} = 4$, $a_T = (\mathbf{v} \cdot \mathbf{a})/\|\mathbf{v}\| = \sqrt{2}$, $a_N = \|\mathbf{v} \times \mathbf{a}\|/\|\mathbf{v}\| = \sqrt{2}$.

    (d) The trajectory is the parabola $x = 1 - y^2/4$, traced so that y increases with t.

    (e) From part (b), the radius is $1/\kappa = 4\sqrt{2}$. If the center is (h, k), then $(x - h)^2 + (y - k)^2 = 32$ is an equation of the circle. The circle must be tangent to the curve at P_0 so $h^2 + (2 - k)^2 = 32$ and, equating slopes, $h/(2 - k) = -1$, $h = k - 2$ thus $h^2 + h^2 = 32$, $h^2 = 16$, $h = -4$ (because the center must be to the left of the vertex of the parabola), $k - 2 = -4$, $k = -2$. The center is at $(-4, -2)$.

33. (a) $\mathbf{v} = \langle -e^{-t}, e^t\rangle$, $\mathbf{a} = \langle e^{-t}, e^t\rangle$; $t = 0$ at P_0 so $\mathbf{v} = \langle -1, 1\rangle$, $\mathbf{a} = \langle 1, 1\rangle$, $ds/dt = \|\mathbf{v}\| = \sqrt{2}$

    (b) $\mathbf{v} \times \mathbf{a} = -2\mathbf{k}$, $\kappa = \|\mathbf{v} \times \mathbf{a}\|/\|\mathbf{v}\|^3 = 1/\sqrt{2}$

    (c) $\mathbf{v} \cdot \mathbf{a} = 0$, $a_T = (\mathbf{v} \cdot \mathbf{a})/\|\mathbf{v}\| = 0$, $a_N = \|\mathbf{v} \times \mathbf{a}\|/\|\mathbf{v}\| = \sqrt{2}$

    (d) The trajectory is the branch of the hyperbola $y = 1/x$ in the first quadrant, traced so that y increases with t.

    (e) The radius is $1/\kappa = \sqrt{2}$. If the center is (h, k), then $(x - h)^2 + (y - k)^2 = 2$ is an equation of the circle. The circle must be tangent to the curve at P_0 so $(1 - h)^2 + (1 - k)^2 = 2$ and, equating slopes, $-(1 - h)/(1 - k) = -1$, $1 - h = 1 - k$ thus $(1 - h)^2 = 1$, $(1 - h) = \pm 1$, $h = 0$ (reject, the center must be to the right of $x = 1$) or $h = 2$, $k = h = 2$. The center is at $(2, 2)$.

34. $\mathbf{r}(0) = (-2/m)\mathbf{i}$, $\mathbf{v}(0) = (2\mathbf{i} - 3\mathbf{j})/m$. Use $\mathbf{F} = m\mathbf{a}$ to get $\mathbf{a} = (2\cos t\mathbf{i} + 3\sin t\mathbf{j})/m$,

$\mathbf{v}(t) = \displaystyle\int \mathbf{a}\,dt = (2\sin t\mathbf{i} - 3\cos t\mathbf{j})/m + \mathbf{C}_1$,

$\mathbf{v}(0) = (-3/m)\mathbf{j} + \mathbf{C}_1 = (2\mathbf{i} - 3\mathbf{j})/m$, $\mathbf{C}_1 = (2/m)\mathbf{i}$ so $\mathbf{v}(t) = [(2 + 2\sin t)\mathbf{i} - 3\cos t\mathbf{j}]/m$,

$\mathbf{r}(t) = \displaystyle\int \mathbf{v}\,dt = [(2t - 2\cos t)\mathbf{i} - 3\sin t\mathbf{j}]/m + \mathbf{C}_2$,

$\mathbf{r}(0) = (-2/m)\mathbf{i} + \mathbf{C}_2 = (-2/m)\mathbf{i}$, $\mathbf{C}_2 = \mathbf{0}$ so $\mathbf{r}(t) = [2(t - \cos t)\mathbf{i} - 3\sin t\mathbf{j}]/m$

35. $\mathbf{r}(0) = \mathbf{0}$, $\mathbf{v}(0) = \mathbf{i} + 2\mathbf{j}$. Use $\mathbf{F} = m\mathbf{a}$ with $m = 1$ to get

$\mathbf{a} = \sin t\mathbf{i} + 4e^{2t}\mathbf{j}$, $\mathbf{v}(t) = \displaystyle\int \mathbf{a}\,dt = -\cos t\mathbf{i} + 2e^{2t}\mathbf{j} + \mathbf{C}_1$,

$\mathbf{v}(0) = -\mathbf{i} + 2\mathbf{j} + \mathbf{C}_1 = \mathbf{i} + 2\mathbf{j}$, $\mathbf{C}_1 = 2\mathbf{i}$ so, $\mathbf{v}(t) = (2 - \cos t)\mathbf{i} + 2e^{2t}\mathbf{j}$,

$\mathbf{r}(t) = \displaystyle\int \mathbf{v}\,dt = (2t - \sin t)\mathbf{i} + e^{2t}\mathbf{j} + \mathbf{C}_2$, $\mathbf{r}(0) = \mathbf{j} + \mathbf{C}_2 = \mathbf{0}$ so $\mathbf{C}_2 = -\mathbf{j}$,

$\mathbf{r}(t) = (2t - \sin t)\mathbf{i} + (e^{2t} - 1)\mathbf{j}$

36. $\kappa(y) = \dfrac{|d^2x/dy^2|}{[1 + (dx/dy)^2]^{3/2}} = \dfrac{1/50}{[1 + (y/50)^2]^{3/2}}$, $\kappa(0) = 1/50$, $a_N = \kappa(ds/dt)^2$ so

$ds/dt = (a_N/\kappa)^{1/2} \leq [25/(1/50)]^{1/2} = 25\sqrt{2}$.

37. $dx/dt = 4$, by the chain rule $dy/dt = (2 - 2x)(dx/dt) = 8(1 - x)$,

$ds/dt = [(dx/dt)^2 + (dy/dt)^2]^{1/2} = 4[1 + 4(1 - x)^2]^{1/2}$,

$d^2s/dt^2 = 2[1 + 4(1 - x)^2]^{-1/2}[8(1 - x)](-dx/dt) = -64(1 - x)[1 + 4(1 - x)^2]^{-1/2}$

so $a_T = -64(1 - x)/\sqrt{1 + 4(1 - x)^2}$; $d^2x/dt^2 = 0$, $d^2y/dt^2 = -8dx/dt = -32$,

$\|\mathbf{a}\|^2 = (d^2x/dt^2)^2 + (d^2y/dt^2)^2 = 0 + (-32)^2 = 1024$.

(a) $a_T = 0$, $a_N^2 = \|\mathbf{a}\|^2 - a_T^2 = 1024$, $a_N = 32$

(b) $a_T = -64/\sqrt{5}$, $a_N^2 = 1024 - (64/\sqrt{5})^2 = 1024/5$, $a_N = 32/\sqrt{5}$

38. $a_T = d^2s/dt^2 = 0$, $a_N = \kappa(ds/dt)^2 = \dfrac{4}{(1 + 16x^2)^{3/2}}(10)^2 = 400/(1 + 16x^2)^{3/2}$

39. In one revolution the weight travels a distance that is equal to the circumference of a circle of radius 2 m so $ds/dt = 2\pi(2) = 4\pi$ m/sec, $a_T = d^2s/dt^2 = 0$,

$a_N = \kappa(ds/dt)^2 = (1/2)(4\pi)^2 = 8\pi^2$ m/sec^2.

40. (a) $\mathbf{r}'(t) = \langle 2\cosh t, 2\sinh t\cosh t\rangle = 2\cosh t\langle 1, \sinh t\rangle$

$ds/dt = \|\mathbf{r}'(t)\| = 2\cosh t(1 + \sinh^2 t)^{1/2} = 2\cosh^2 t$

(b) $L = \displaystyle\int_0^1 (ds/dt)dt = \int_0^1 2\cosh^2 t\,dt = \int_0^1 (\cosh 2t + 1)dt = 1 + \frac{1}{2}\sinh 2$

41. **(a)** $\mathbf{r}'(t) = e^t \langle 2\cos 2t, -2\sin 2t \rangle + e^t \langle \sin 2t, \cos 2t \rangle = e^t \langle 2\cos 2t + \sin 2t, \cos 2t - 2\sin 2t \rangle$,

$ds/dt = \|\mathbf{r}'(t)\| = e^t[(2\cos 2t + \sin 2t)^2 + (\cos 2t - 2\sin 2t)^2]^{1/2} = \sqrt{5}\,e^t$

(b) $L = \displaystyle\int_0^{\ln 3} (ds/dt)dt = \int_0^{\ln 3} \sqrt{5}\,e^t dt = 2\sqrt{5}.$

CHAPTER 16
Partial Derivatives

SECTION 16.1

16.1.1   Let $f(x, y, z) = 2\tan^{-1}\dfrac{y}{x} + \ln(x^2 + z^2)$; find $f(1, 1, 1)$ and $f(1, -1, 1)$.

16.1.2   Let $f(x, y, z) = ze^x \sin y$; find $f(\ln 3, \dfrac{\pi}{2}, 1)$.

16.1.3   Let $f(x, y, z) = yze^{\ln(x^2 + y^2)}$; find $f(1, -1, 2)$ and $f(0, 1, 4)$.

16.1.4   Let $f(x, y) = \sin^{-1}\dfrac{x}{y} + \tan^{-1}\left(\dfrac{x}{y}\right)$; find $f(1, 1)$.

16.1.5   Let $f(x, y, z) = xz^2 \cosh(\ln y)$; find $f(2, 2, 1)$.

16.1.6   Sketch the graph of $f(x, y) = 4 - \dfrac{2}{3}x - \dfrac{1}{2}y$ in xyz space and label two points on the surface.

16.1.7   Sketch the graph of $f(x, z) = x^2 + z^2$ in xyz space.

16.1.8   Sketch the graph of $f(x, y) = \sqrt{16 - x^2 - y^2}$ in xyz space.

16.1.9   Sketch the graph of $f(x, y) = \sqrt{16 - x^2 - 2y^2}$ in xyz space.

16.1.10   Let $f(x, y) = 2x^2y + \dfrac{y}{x}$, $x(t) = 2t$, and $y(t) = t^2$; find $f[x(t), y(t)]$ and $f[x(2), y(2)]$.

16.1.11   Let $f(x, y) = \sin(xy) + y\ln(xy) + y$, $x = e^t$, and $y = t^2$; find $f[x(t), y(t)]$ and $f[x(0), y(1)]$.

16.1.12   Describe the family of level curves for $z = x^2 + y^2$, $(z \geq 0)$ and sketch a few of these curves.

16.1.13   Describe the family of level curves for $z = 4x^2 + y^2 (z \geq 0)$ and sketch a few of these curves.

16.1.14   Let $f(x, y) = x^2 + xy + y^2 - 2x - 3y + 1$. Find $f(3, -3)$ and $f(t, s + t)$.

16.1.15   Describe the family of level curves for $z = \sqrt{\dfrac{x + y}{x - y}}$ and sketch a few of these curves.

16.1.16   Sketch the natural domain of $f(x, y) = \sqrt{4 - x^2 - y^2}$. Shade the region included in the natural domain. Use solid lines for the portion of the boundary included in the natural domain.

16.1.17 Sketch $f(x, y) = 1 - y^2$ in xyz space and state its natural domain.

16.1.18 Sketch the level curves for $z = xy$ for $k = 0, 1, -1$, and 4.

16.1.19 Find a parametric representation of the cylinder $x^2 + z^2 = 9$ between the planes $y = 1$ and $y = 4$ in terms of the parameters u and v, where $u = x$ and $v = y$.

16.1.20 Find a parametric representation of $2z - 4x + 3y = 3$ in terms of the parameters u and v, where $u = x$ and $v = y$.

16.1.21 Find a parametric representation of the portion of the sphere $x^2 + y^2 + z^2 = 16$ above the plane $z = 3$ in terms of the parameters r and θ, where (r, θ, z) are cylindrical coordinates of a point on the surface.

16.1.22 Find a parametric representation of $z = \dfrac{1}{4 + x^2 + y^2}$ in terms of parameters r and θ, where (r, θ, z) are cylindrical coordinates of a point of the surface.

16.1.23 Describe $x = 3u + v, y = u - 2v, z = 2v$ for $-\infty < u < +\infty$ and $-\infty < v < +\infty$ by eliminating the parameters to obtain an equation for the surface in rectangular coordinates.

16.1.24 Describe $x = 4 \sin u, y = 3 \cos u, z = 2v$ for $0 \le u \le 2\pi$ and $1 \le v \le 4$ by eliminating the parameters to obtain an equation for the surface in rectangular coordinates.

SOLUTIONS

SECTION 16.1

16.1.1 $f(1,1,1) = \dfrac{\pi}{2} + \ln 2$; $f(1,-1,1) = -\dfrac{\pi}{2} + \ln 2$.

16.1.2 3.

16.1.3 $f(1,-1,2) = -4$; $f(0,1,4) = 4$.

16.1.4 $\dfrac{3\pi}{4}$.

16.1.5 $f(2,2,1) = (2)(1)\left(\dfrac{e^{\ln 2} + e^{-\ln 2}}{2}\right) = \dfrac{5}{2}$.

16.1.6

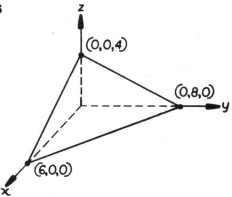

16.1.7

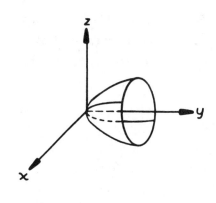

16.1.8

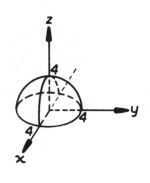

16.1.9

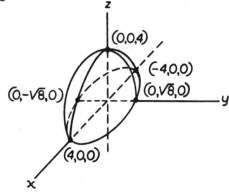

16.1.10 $f[x(t), y(t)] = 8t^4 + \dfrac{t}{2}$; $f[x(2), y(2)] = 129$.

16.1.11 $f[x(t), y(t)] = \sin(t^2 e^t) + t^2 \ln(t^2 e^t) + t^2$ and $f[x(0), y(1)] = \sin 1 + 1$.

16.1.12   The family of level curves
for $z \geq 0$ are families of
concentric circles in the
xy plane.

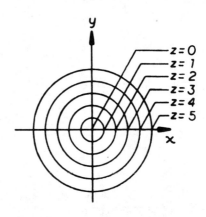

16.1.13   The family of level curves
for $z \geq 0$ are families of
concentric ellipses in the
xy plane.

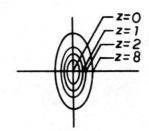

16.1.14 $f(3, -3) = 13$;
$$f(t, s + t) = t^2 + st + t^2 + s^2 + 2st + t^2 - 2t - 3s - 3t + 1$$
$$= 3t^2 + 3st + s^2 - 3s - 5t + 1.$$

16.1.15   The family of level curves   **16.1.16**
are families of straight
lines through the origin.
$y \neq x$

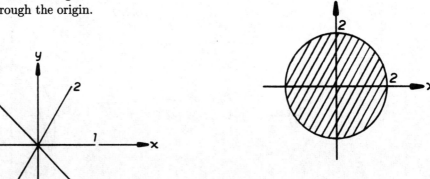

16.1.17   The natural domain includes
all real values of x and y.

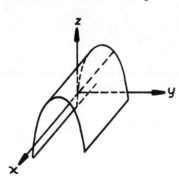

16.1.18

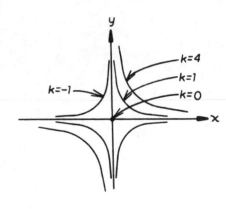

16.1.19   Since $u = x$ and $v = y$, then $z = \sqrt{9 - u^2}$. The parametric representation of the surface is
$x = u$, $y = v$, $z = \sqrt{9 - u^2}$, where $-3 \le u \le 3$ and $1 \le v \le 4$.

16.1.20   Since $u = x$ and $v = y$, then $z = \dfrac{3 + 4u - 3v}{2}$. The parametric representation of this surface
if $x = u$, $y = v$, $z = \dfrac{3 + 4u - 3v}{2}$ where $-\infty < u < +\infty$ and $-\infty < v < +\infty$.

16.1.21   Since $x = r \cos \theta$ and $y = r \sin \theta$, then $z = \sqrt{16 - r^2}$. The parametric representation of this
surface is $x = r \cos \theta$, $y = r \sin \theta$, $z = \sqrt{16 - r^2}$, where $0 \le r < \sqrt{7}$ and $0 \le \theta \le 2\pi$.

16.1.22   Since $x = r \cos \theta$ and $y = r \sin \theta$, then $z = \dfrac{1}{4 + r^2}$. The parametric representation of this
surface is $x = r \cos \theta$, $y = r \sin \theta$, $z = \dfrac{1}{4 + r^2}$, where $0 \le r < +\infty$ and $0 \le \theta \le 2\pi$.

16.1.23   Solve $x = 3u + v$
$y = u - 2v$

to get $v = \dfrac{x - 3y}{14}$. Since $z = 2v$, $z = \dfrac{x - 3y}{7}$ is an equation for the surface in rectangular
coordinates.

16.1.24   Since $x = 4 \sin u$, $y = 3 \cos u$ and $\sin^2 u + \cos^2 u = 1$, then $\dfrac{x^2}{16} + \dfrac{y^2}{9} = 1$. Since $z = 2v$ and
$1 \le v \le 4$, then $2 \le z \le 8$. Thus, $\dfrac{x^2}{16} + \dfrac{y^2}{9} = 1$, $2 \le z \le 8$ is an equation for the surface in
rectangular coordinates.

SECTION 16.2

16.2.1   Evaluate $\lim\limits_{(x,y)\to(1,2)} (x^2 + 3y)$.

16.2.2   Evaluate $\lim\limits_{(x,y)\to(1,1)} \dfrac{4 + x - y}{3 + x - 3y}$.

16.2.3   Evaluate $\lim\limits_{(x,y)\to(1,\pi/2)} x^3 \sin \dfrac{y}{x}$.

16.2.4   Evaluate $\lim\limits_{(x,y)\to(0,0)} \dfrac{2x + y}{x^3 + y^3}$.

16.2.5   Evaluate $\lim\limits_{(x,y)\to(3,-1)} \dfrac{x^2 - 9y^2}{x + 3y}$.

16.2.6   Evaluate $\lim\limits_{(x,y)\to(0,0)} \dfrac{\tan(x^2 + y^2)}{x^2 + y^2}$.

16.2.7   Evaluate $\lim\limits_{(x,y)\to(0,1)} \dfrac{\sin^{-1}(xy - 2)}{\tan^{-1}(3xy - 6)}$.

16.2.8   Evaluate $\lim\limits_{(x,y,z)\to(0,0,0)} \dfrac{xz}{x^2 + 2y^2 + z^2}$.

16.2.9   Sketch and shade the region where $f(x,y) = \dfrac{1}{x^2 + y^2 - 4}$ is continuous.

16.2.10   Sketch and shade the region where $f(x,y) = \ln(2x + 3y)$ is continuous.

16.2.11   Sketch and shade the region where $f(x,y) = \sin^{-1}(2x + 3y)$ is continuous.

16.2.12   Show that $\lim\limits_{(x,y)\to(0,0)} \dfrac{x^2}{x^2 + y^2}$ does not exist.

16.2.13   Show that $f(x,y) = \begin{cases} \dfrac{\tan(x^2 + y^2)}{x^2 + y^2}, & (x,y) \neq (0,0) \\ 1, & (x,y) = (0,0) \end{cases}$ is continuous at $(0,0)$.

16.2.14   Let $f(x,y) = \dfrac{xy}{x^2 + y^2}$. Is it possible to define $f(0,0)$ so that f will be continuous at $(0,0)$?

SOLUTIONS

SECTION 16.2

16.2.1   7.   **16.2.2**   4.   **16.2.3**   1.

16.2.4   The limit does not exist since $\displaystyle\lim_{(x,y)\to(0,0)} \frac{2x+y}{x^3+y^2} = +\infty.$

16.2.5 $\displaystyle\lim_{(x,y)\to(3,-1)} \frac{(x+3y)(x-3y)}{x+3y} = \lim_{(x,y)\to(3,-1)} (x-3y) = 6.$

16.2.6   Let $r^2 = x^2 + y^2$, then $\displaystyle\lim_{(x,y)\to(0,0)} r = 0$, use L'Hôpital's rule to get

$$\lim_{r\to0} \frac{\tan r^2}{r^2} = \lim_{r\to0} \frac{2r\sec^2 r^2}{2r} = \lim_{r\to0} \sec^2 r^2 = \sec^2 0 = 1.$$

16.2.7   Let $z = xy - 2$, then $\displaystyle\lim_{(x,y)\to(2,1)} z = 0$, use L'Hôpital's rule to get

$$\lim_{z\to0} \frac{\sin^{-1} z}{\tan^{-1} 3z} = \lim_{z\to0} \frac{\dfrac{1}{\sqrt{1-z^2}}}{\dfrac{3}{1+9z^2}} = \frac{1}{3}.$$

16.2.8   Along the z axis: $\displaystyle\lim_{(x,y,z)\to(0,0,0)} \frac{0}{z^2} = 0$

Along the line $x = t$, $y = t$, $z = t$, $\displaystyle\lim_{(x,y,z)\to(0,0,0)} t = 0$ so $\displaystyle\lim_{t\to0} \frac{t^2}{t^2+2t^2+t^2} = \lim_{t\to0} \frac{t^2}{4t^2} = \frac{1}{4}$,

thus the lim does not exist.

16.2.9   **16.2.10**

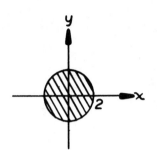

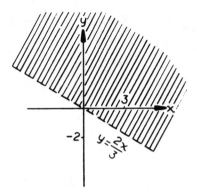

16.2.11 $f(x, y)$ is continuous for $|2x + 3y| \le 1$.

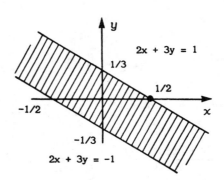

16.2.12 Along $y = 0$, $\displaystyle\lim_{(x,y)\to(0,0)} \frac{x^2}{x^2} = 1$,

along $y = x$, $\displaystyle\lim_{(x,y)\to(0,0)} \frac{x^2}{2x^2} = \frac{1}{2}$,

thus the limit does not exist.

16.2.13 Let $r^2 = x^2 + y^2$, then $\displaystyle\lim_{(x,y)\to(0,0)} r = 0$; use L'Hôpital's rule to get

$\displaystyle\lim_{r\to 0} \frac{\tan r^2}{r^2} = \lim_{r\to 0} \frac{2r \sec^2 r^2}{2r} = \lim_{r\to 0} \sec^2 r^2 = 1 = f(0,0)$ thus, f is continuous at $(0,0)$.

16.2.14 No, $\displaystyle\lim_{(x,y)\to(0,0)} \frac{xy}{x^2 + y^2}$ does not exist

Along $x = 0$: $\displaystyle\lim_{y\to 0} \frac{0}{y^2} = 0$

Along $x = y$: $\displaystyle\lim_{y\to 0} \frac{y^2}{2y^2} = \frac{1}{2}$

SECTION 16.3

16.3.1 Find $\dfrac{\partial z}{\partial x}$ and $\dfrac{\partial z}{\partial y}$ if $z = \dfrac{x}{y} \sin(xy^2)$.

16.3.2 Find $f_x(x, y)$ and $f_y(x, y)$ if $f(x, y) = x^y$.

16.3.3 Find $\dfrac{\partial z}{\partial x}$ and $\dfrac{\partial z}{\partial y}$ if $z = x^3 + xy - y\cos xy$.

16.3.4 Find $\dfrac{\partial z}{\partial x}$ and $\dfrac{\partial z}{\partial y}$ if $z = y^2 e^{-x} + y$.

16.3.5 Find $f_y(1, 1)$ if $f(x, y) = y - e^{xy^2} + \sqrt{x^2 + 1}$.

16.3.6 Find $f_x(4, 2)$ and $f_{xy}(4, 2)$ if $f(x, y) = \ln(xy - 1) + e^y \sqrt{x}$.

16.3.7 Find $f_x(4, 2)$ and $f_{xy}(4, 2)$ if $f(x, y) = y\ln(x + y^2) + y^2 \sqrt{x}$.

16.3.8 Find $f_{xx}(x, y)$ if $f(x, y) = \sqrt{16 - 9x^2 - 4y^2}$.

16.3.9 Find all the second partial derivatives of f if $f(x, y) = \cos(xy^2)$.

16.3.10 Find $f_{yx}(3, 2)$ if $f(x, y) = (1 + x + y^2)^{4/3}$.

16.3.11 Find f_y and f_{yy} if $f(x, y) = (x^2 + xy)^{5/2}$.

16.3.12 Use implicit differentiation to find $\dfrac{\partial z}{\partial x}$ and $\dfrac{\partial z}{\partial y}$ if $x^2 z^2 - 2xyz + y^2 z^3 = 3$.

16.3.13 Use implicit differentiation to evaluate $\dfrac{\partial z}{\partial x}$ at $(1, -2, 1)$ if $x^3 z - 3xy^2 - (yz)^3 = -3$.

16.3.14 Use implicit differentiation to find $\dfrac{\partial z}{\partial x}$ and $\dfrac{\partial z}{\partial y}$ if $x^2 + y^2 + z^2 - 2xyz = 5$.

16.3.15 Use implicit differentiation to find $\dfrac{\partial z}{\partial x}$ and $\dfrac{\partial^2 z}{\partial x^2}$ if $x^{1/3} + y^{1/3} + z^{1/3} = 16$.

16.3.16 Use implicit differentiation to find $\dfrac{\partial z}{\partial x}$ and $\dfrac{\partial z}{\partial y}$ if $(x + y)^2 = (y - z)^3$.

16.3.17 Use implicit differentiation to find $\dfrac{\partial^2 z}{\partial y^2}$ if $\dfrac{x^2}{4} + \dfrac{y^2}{4} - \dfrac{z^2}{9} = 1$.

16.3.18 Find $\dfrac{\partial z}{\partial x}$ and $\dfrac{\partial z}{\partial y}$ if $x^2 z^2 - 2xyz + z^3 y^2 = 2$.

16.3.19 Evaluate $\dfrac{\partial z}{\partial x}$ and $\dfrac{\partial z}{\partial y}$ at $(3, 3, 2)$ if $x^3 + y^3 + z^3 - 3xyz = 8$.

16.3.20 Evaluate $\dfrac{\partial z}{\partial x}$ and $\dfrac{\partial z}{\partial y}$ at $(1, 0, \pi/6)$ if $x^2 \cos^2 z - y^2 \sin z = \sin^2 2z$.

16.3.21 Verify that $x\dfrac{\partial z}{\partial x} + y\dfrac{\partial z}{\partial y} = z$ if $z = x \sin\left(\dfrac{x}{y}\right) + ye^{y/x}$.

16.3.22 Verify that $x\dfrac{\partial z}{\partial x} + y\dfrac{\partial z}{\partial y} = 3z$ if $z = x^3 + 2x^2 y + 3xy^2 + y^3$.

16.3.23 Verify that $4\dfrac{\partial z}{\partial x} - 3\dfrac{\partial z}{\partial y} = 0$ if $z = (3x + 4y)^4$.

SOLUTIONS

SECTION 16.3

16.3.1 $\dfrac{\partial z}{\partial x} = xy\cos(xy^2) + \dfrac{1}{y}\sin(xy^2); \ \dfrac{\partial z}{\partial y} = 2x^2\cos(xy^2) - \dfrac{x}{y^2}\sin(xy^2).$

16.3.2 $f_x(x,y) = yx^{y-1}$; let $z = f(x,y) = x^y$, then $\ln z = y\ln x$ and $\dfrac{1}{z}\dfrac{\partial z}{\partial y} = \ln x$ so

$\dfrac{\partial z}{\partial y} = f_y(x,y) = x^y\ln x.$

16.3.3 $\dfrac{\partial z}{\partial x} = 3x^2 + y + y^2\sin xy; \ \dfrac{\partial z}{\partial y} = x + xy\sin xy - \cos xy.$

16.3.4 $\dfrac{\partial z}{\partial x} = -y^2 e^{-x}; \ \dfrac{\partial z}{\partial y} = 2ye^{-x} + 1.$

16.3.5 $f_y(x,y) = 1 - 2xye^{xy^2}$ so $f_y(1,1) = 1 - 2e.$

16.3.6 $f_x(x,y) = \dfrac{y}{xy-1} + \dfrac{e^y}{2\sqrt{x}}, \ f_x(4,2) = \dfrac{2}{7} + \dfrac{e^2}{4};$

$f_{xy}(x,y) = \dfrac{e^y}{2\sqrt{x}} - \dfrac{1}{(xy-1)^2}, \ f_{xy}(4,2) = \dfrac{e^2}{4} - \dfrac{1}{49}.$

16.3.7 $f_x(x,y) = \dfrac{y}{x+y^2} + \dfrac{y^2}{2\sqrt{x}}, \ f_x(4,2) = \dfrac{5}{4};$

$f_{xy}(x,y) = \dfrac{x-y^2}{(x+y^2)^2} + \dfrac{y}{\sqrt{x}}, \ f_{xy}(4,2) = 1.$

16.3.8 $f_x(x,y) = -\dfrac{9x}{\sqrt{16 - 9x^2 - 4y^2}}; \ f_{xx}(x,y) = \dfrac{36y^2 - 144}{(16 - 9x^2 - 4y^2)^{3/2}}.$

16.3.9 $\dfrac{\partial f}{\partial x} = -y^2\sin(xy^2), \ \dfrac{\partial f}{\partial y} = -2xy\sin(xy^2),$

$\dfrac{\partial^2 f}{\partial x^2} = -y^4\cos(xy^2), \ \dfrac{\partial^2 f}{\partial y\partial x} = -2xy^3\cos(xy^2) - 2y\sin(xy^2),$

$\dfrac{\partial^2 f}{\partial x\partial y} = -2xy^3\cos(xy^2) - 2y\sin(xy^2),$

$\dfrac{\partial^2 f}{\partial y^2} = -4x^2y^2\cos(xy^2) - 2x\sin(xy^2).$

16.3.10 $f_x(x, y) = \dfrac{4}{3}(1 + x + y^2)^{1/3}$, $f_{yx}(x, y) = \dfrac{8y}{9(1 + x + y^2)^{2/3}}$,

$f_{yx}(3, 2) = \dfrac{4}{9}$.

16.3.11 $f_y(x, y) = \dfrac{5}{2}x(x^2 + xy)^{3/2}$, $f_{yy}(x, y) = \dfrac{15}{4}x^2(x^2 + xy)^{1/2}$.

16.3.12 $x^2\left(2z\dfrac{\partial z}{\partial x}\right) + z^2(2x) - 2y\left(x\dfrac{\partial z}{\partial x} + z\right) + y^2\left(3z^2\dfrac{\partial z}{\partial x}\right) = 0$,

$\dfrac{\partial z}{\partial x} = \dfrac{2yz - 2xz^2}{2x^2z - 2xy + 3y^2z^2}$;

$x^2\left(2z\dfrac{\partial z}{\partial y}\right) - 2x\left(y\dfrac{\partial z}{\partial x} + z\right) + y^2\left(3z^2\dfrac{\partial z}{\partial y}\right) + z^3(2y) = 0$,

$\dfrac{\partial z}{\partial y} = \dfrac{2xz - 2yz^3}{2x^2z - 2xy + 3y^2z^2}$.

16.3.13 $x^3\dfrac{\partial z}{\partial x} + z(3x^2) - 3y^2 - 3(yz)^2\left(y\dfrac{\partial z}{\partial x}\right) = 0$,

$\dfrac{\partial z}{\partial x} = \dfrac{3y^2 - 3x^2z}{x^3 - 3y^3z^2}$ so $\dfrac{\partial z}{\partial x}\Big]_{(1,-2,1)} = \dfrac{9}{25}$.

16.3.14 $2x + 2z\dfrac{\partial z}{\partial x} - 2y\left[x\dfrac{\partial z}{\partial x} + z\right] = 0$, $\dfrac{\partial z}{\partial x} = \dfrac{x - yz}{xy - z}$;

$2y + 2z\dfrac{\partial z}{\partial y} - 2x\left[y\dfrac{\partial z}{\partial y} + z\right] = 0$, $\dfrac{\partial z}{\partial y} = \dfrac{xz - y}{z - xy}$.

16.3.15 $\dfrac{1}{3}x^{-2/3} + \dfrac{1}{3}z^{-2/3}\dfrac{\partial z}{\partial x} = 0$, $\dfrac{\partial z}{\partial x} = -\left(\dfrac{z}{x}\right)^{2/3}$;

$\dfrac{\partial^2 z}{\partial x^2} = \dfrac{2z^{1/3}(x^{1/3} + z^{1/3})}{3x^{5/3}}$ or $\dfrac{\partial^2 z}{\partial x^2} = \dfrac{2z^{1/3}(16 - y^{1/3})}{3x^{5/3}}$.

16.3.16 $2(x + y)(1) = 3(y - z)^2\left(-\dfrac{\partial z}{\partial x}\right)$, $\dfrac{\partial z}{\partial x} = -\dfrac{2(x + y)}{3(y - z)^2}$;

$2(x + y)(1) = 3(y - z)^2\left(1 - \dfrac{\partial z}{\partial y}\right)$, $\dfrac{\partial z}{\partial y} = 1 - \dfrac{2(x + y)}{3(y - z)^2}$.

16.3.17 $\dfrac{2y}{4} - \dfrac{2z}{9}\dfrac{\partial z}{\partial y} = 0$, $\dfrac{\partial z}{\partial y} = \dfrac{9y}{4z}$;

$\dfrac{\partial^2 z}{\partial y^2} = \dfrac{9}{4}\left[\dfrac{z - y\dfrac{\partial z}{\partial y}}{z^2}\right] = \dfrac{9}{4}\left[\dfrac{z - y\left(\dfrac{9y}{4z}\right)}{z^2}\right] = \dfrac{9(4z^2 - 9y^2)}{16z^3}$.

16.3.18 $2xz^2 + 2x^2 z \dfrac{\partial z}{\partial x} - 2xy \dfrac{\partial z}{\partial x} - 2yz + 3y^2 z^2 \dfrac{\partial z}{\partial x} = 0,$

$\dfrac{\partial z}{\partial x} = \dfrac{2yz - 2xz^2}{2x^2 z - 2xy + 3y^2 z^2};$

$2x^2 z \dfrac{\partial z}{\partial y} - 2xy \dfrac{\partial z}{\partial y} - 2xz + 2yz^3 + 3y^2 z^2 \dfrac{\partial z}{\partial y} = 0,$

$\dfrac{\partial z}{\partial y} = \dfrac{2xz - 2yz^3}{2x^2 z - 2xy + 3y^2 z}.$

16.3.19 $3x^2 + 3z^2 \dfrac{\partial z}{\partial x} - 3xy \dfrac{\partial z}{\partial x} - 3yz = 0, \quad \dfrac{\partial z}{\partial x} = \dfrac{yz - x^2}{z^2 - xy}, \quad \dfrac{\partial z}{\partial x}\bigg|_{(3,3,2)} = \dfrac{3}{5};$

$3y^2 + 3z^2 \dfrac{\partial z}{\partial y} - 3xy \dfrac{\partial z}{\partial y} - 3xz = 0, \quad \dfrac{\partial z}{\partial y} = \dfrac{xz - y^2}{z^2 - xy}, \quad \dfrac{\partial z}{\partial y}\bigg|_{(3,3,2)} = \dfrac{3}{5}.$

16.3.20 $-2x^2 \cos z \sin z \dfrac{\partial z}{\partial x} + 2x \cos^2 z - y^2 \cos z \dfrac{\partial z}{\partial x} = 4 \sin 2z \cos 2z \dfrac{\partial z}{\partial x},$

$\dfrac{\partial z}{\partial x} = \dfrac{2x \cos^2 z}{4 \cos 2z \sin 2z + 2x^2 \cos z \sin z + y^2 \cos z}, \quad \dfrac{\partial z}{\partial x}\bigg|_{(1,0,\pi/6)} = \dfrac{1}{\sqrt{3}};$

$-2x^2 \cos z \sin z \dfrac{\partial z}{\partial y} - y^2 \cos z \dfrac{\partial z}{\partial y} - 2y \sin z = 4 \sin 2z \cos 2z \dfrac{\partial z}{\partial x},$

$\dfrac{\partial z}{\partial y} = -\dfrac{2y \sin z}{4 \cos 2z \sin 2z + 2x^2 \cos z \sin z + y^2 \cos z}, \quad \dfrac{\partial z}{\partial y}\bigg|_{(1,0,\pi/6)} = 0.$

16.3.21 $x\left(\dfrac{x}{y}\cos\dfrac{x}{y} + \sin\dfrac{x}{y} - \dfrac{y^2}{x^2} e^{y/x}\right) + y\left(-\dfrac{x^2}{y^2}\cos\dfrac{x}{y} + \dfrac{y}{x} e^{y/x} + e^{y/x}\right) = x\sin\dfrac{x}{y} + y e^{y/x} = z.$

16.3.22 $x(3x^2 + 4xy + 3y^2) + y(2x^2 + 6xy + 3y^2) = 3(x^3 + 2x^2 y + 3xy^2 + y^3) = 3z.$

16.3.23 $4[12(3x + 4y)^3] - 3[16(3x + 4y)^3] = 0.$

SECTION 16.4

16.4.1   Verify that $f_{xy} = f_{yx}$ if $f(x, y) = x^2 y^3 + x^4 y^2$.

16.4.2   Verify that $\dfrac{\partial^2 z}{\partial y \partial x} = \dfrac{\partial^2 z}{\partial x \partial y}$ if $z = \tan^{-1}\left(\dfrac{x}{y}\right)$.

16.4.3   Verify that $f_{xy} = f_{yx}$ if $f(x, y) = \sin(3x + 2y) + \ln(3x + 2y)$.

16.4.4   Verify that $f_{xyy} = f_{yxy} = f_{yyx}$ if $f(x, y) = x \sin y$.

16.4.5   Verify that $f_{yxx} = f_{xyx} = f_{xxy}$ if $f(x, y) = e^{2xy} + x \ln y$.

16.4.6   Use the chain rule to evaluate $\dfrac{dz}{dt}$ at $t = 1$ if $z = x^3 y^2$; $x = t^2 + 1$, $y = t^3 + 2$.

16.4.7   Use the chain rule to find $\dfrac{dz}{dt}$ if $z = \sqrt{x^2 + y^2}$; $x = e^t$, $y = \cos t$.

16.4.8   Use the chain rule to find $\dfrac{dz}{dt}$ if $z = y^2 e^x$; $x = \cos t$, $y = t^3$.

16.4.9   Use the chain rule to find $\dfrac{dz}{dx}$ if $z = xy$ and $y = e^x \cos x$.

16.4.10   Use the chain rule to find $\dfrac{dz}{dy}$ if $z = xy$ and $x = y \cos y$.

16.4.11   Use the chain rule to find $\dfrac{\partial z}{\partial s}$ and $\dfrac{\partial z}{\partial t}$ if $z = x \sin y$; $x = se^t$, $y = se^{-t}$.

16.4.12   Use the chain rule to find $\dfrac{\partial z}{\partial u}$ and $\dfrac{\partial z}{\partial v}$ if $z = x^2 \tan y$; $x = u^2 + v^3$, $y = \ln(u^2 + v^2)$.

16.4.13   Use the chain rule to find $\dfrac{\partial z}{\partial r}$ and $\dfrac{\partial z}{\partial \theta}$ if $z = \dfrac{xy}{x^2 + y^2}$; $x = r \cos\theta$, $y = r \sin\theta$.

16.4.14   Use the chain rule to find $\dfrac{\partial z}{\partial u}$ and $\dfrac{\partial z}{\partial v}$ if $z = x \cos y + y \sin x$; $x = uv^2$, $y = u + v$.

16.4.15   Use the chain rule to find $\dfrac{\partial z}{\partial s}$ and $\dfrac{\partial z}{\partial t}$ if $z = x^2 + y^2$; $x = st$, $y = s - t$.

16.4.16 Use the chain rule to find $\dfrac{\partial z}{\partial u}$ and $\dfrac{\partial z}{\partial v}$ if $z = x^3 + xy + y^2$; $x = 2u + v$, $y = u - 2v$.

16.4.17 Verify that $z = f(x^3 - y^2)$ satisfies the equation $2y\dfrac{\partial z}{\partial x} + 3x^2\dfrac{\partial z}{\partial y} = 0$.

16.4.18 Verify that $z = f\left(\dfrac{y}{x}\right)$ satisfies the equation $x\dfrac{\partial z}{\partial x} + y\dfrac{\partial z}{\partial y} = 0$.

SOLUTIONS

SECTION 16.4

16.4.1 $f_y = 3x^2y^2 + 2x^4y$, $f_{xy} = 6xy^2 + 8x^3y$;
$f_x = 2xy^3 + 4x^3y^2$, $f_{yx} = 6xy^2 + 8x^3y$; $f_{xy} = f_{yx}$.

16.4.2 $\dfrac{\partial z}{\partial x} = \dfrac{y}{x^2 + y^2}$, $\dfrac{\partial^2 z}{\partial y \partial x} = \dfrac{x^2 - y^2}{(x^2 + y^2)^2}$;

$\dfrac{\partial z}{\partial y} = -\dfrac{x}{x^2 + y^2}$, $\dfrac{\partial^2 z}{\partial x \partial y} = \dfrac{x^2 - y^2}{(x^2 + y^2)^2}$; $\dfrac{\partial^2 z}{\partial y \partial x} = \dfrac{\partial^2 z}{\partial x \partial y}$.

16.4.3 $f_y = 2\cos(3x + 2y) + \dfrac{2}{3x + 2y}$, $f_{yx} = -6\sin(3x + 2y) - \dfrac{6}{(3x + 2y)^2}$;

$f_x = 3\cos(3x + 2y) + \dfrac{3}{3x + 2y}$, $f_{xy} = -6\sin(3x + 2y) - \dfrac{6}{(3x + 2y)^2}$,

$f_{xy} = f_{yx}$.

16.4.4 $f_y = x\cos y$, $f_{yy} = -x\sin y$, $f_{yyx} = -\sin y$;
$f_y = x\cos y$, $f_{yx} = \cos y$, $f_{yxy} = -\sin y$;
$f_x = \sin y$, $f_{xy} = \cos y$, $f_{xyy} = -\sin y$; $f_{xyy} = f_{yxy} = f_{yyx}$.

16.4.5 $f_x = 2ye^{2xy} + \ln y$, $f_{xx} = 4y^2e^{2xy}$, $f_{xxy} = 8xy^2e^{2xy} + 8ye^{2xy}$;

$f_x = 2ye^{2xy} + \ln y$, $f_{xy} = 4xye^{2xy} + 2e^{2xy} + \dfrac{1}{y}$,

$f_{xyx} = 8xy^2e^{2xy} + 8ye^{2xy}$;

$f_y = 2xe^{2xy} + \dfrac{x}{y}$, $f_{yx} = 4xye^{2xy} + 2e^{2xy} + \dfrac{1}{y}$, $f_{yxx} = 8xy^2e^{2xy} + 8ye^{2xy}$;

$f_{yxx} = f_{xyx} = f_{xxy}$.

16.4.6 $\dfrac{dz}{dt} = 6t(t^2 + 1)^2(t^3 + 2)^2 + 6t^2(t^2 + 1)^3(t^3 + 2)$ so $\dfrac{dz}{dt} = 360$ when $t = 1$.

16.4.7 $\dfrac{dz}{dt} = \dfrac{xe^t}{\sqrt{x^2 + y^2}} - \dfrac{y\sin t}{\sqrt{x^2 + y^2}} = \dfrac{e^{2t} - \sin t \cos t}{\sqrt{e^{2t} + \cos^2 t}}$.

16.4.8 $\dfrac{dz}{dt} = -y^2e^x \sin t + 6yt^2e^x = 6t^5e^{\cos t} - t^6e^{\cos t}\sin t$.

16.4.9 $\dfrac{dz}{dx} = y + x(e^x \cos x - e^x \sin x) = e^x \cos x + xe^x \cos x - xe^x \sin x$.

16.4.10 $\dfrac{dz}{dy} = y(\cos y - y\sin y) + x = -y^2 \sin y + 2y \cos y$.

16.4.11 $\dfrac{\partial z}{\partial s} = (\sin y)e^t + (x\cos y)e^{-t} = e^t \sin(se^{-t}) + s\cos(se^{-t}),$

$\dfrac{\partial z}{\partial t} = (\sin y)(se^t) + (x\cos y)(-se^{-t}) = se^t \sin(se^{-t}) - s^2\cos(se^{-t}).$

16.4.12 $\dfrac{\partial z}{\partial u} = 2x\tan y(2u) + x^2\sec^2 y\left(\dfrac{2u}{u^2+v^2}\right)$

$\qquad = 4u(u^2+v^3)\tan\ln(u^2+v^2) + \dfrac{2u(u^2+v^3)^2\sec^2\ln(u^2+v^2)}{u^2+v^2};$

$\dfrac{\partial z}{\partial v} = 2x\tan y(3v^2) + x^2\sec^2 y\left(\dfrac{2v}{u^2+v^2}\right)$

$\qquad = 6v^2(u^2+v^3)\tan\ln(u^2+v^2) + \dfrac{2v(u^2+v^3)^2\sec^2\ln(u^2+v^2)}{u^2+v^2}.$

16.4.13 $\dfrac{\partial x}{\partial r} = \dfrac{y^3-x^2y}{(x^2+y^2)^2}\cos\theta + \dfrac{x^3-xy^2}{(x^2+y^2)^2}\sin\theta = 0,$

$\dfrac{\partial z}{\partial\theta} = \dfrac{y^3-x^2y}{(x^2+y^2)^2}(-r\sin\theta) + \dfrac{x^3-xy^2}{(x^2+y^2)^2}(r\cos\theta) = \cos^2\theta - \sin^2\theta = \cos 2\theta.$

16.4.14 $\dfrac{\partial z}{\partial u} = (\cos y + y\cos x)v^2 + (-x\sin y + \sin x)(1)$

$\qquad = v^2\cos(u+v) + v^2(u+v)\cos uv^2 - uv^2\sin(u+v) + \sin uv^2,$

$\dfrac{\partial z}{\partial v} = (\cos y + y\cos x)(2uv) + (-x\sin y + \sin x)(1)$

$\qquad = 2uv\cos(u+v) + 2uv(u+v)\cos uv^2 - uv^2\sin(u+v) + \sin uv^2.$

16.4.15 $\dfrac{\partial z}{\partial s} = 2x(t) + 2y(1) = 2st^2 + 2(s-t),$

$\dfrac{\partial z}{\partial t} = 2x(s) + 2y(-1) = 2s^2t - 2(s-t).$

16.4.16 $\dfrac{\partial z}{\partial u} = (3x^2+y)(2) + (x+2y)(1) = 6(2u+v)^2 + 6u - 7v,$

$\dfrac{\partial z}{\partial v} = (3x^2+y)(1) + (x+2y)(-2) = 3(2u+v)^2 - 7u + 4v.$

16.4.17 Let $u = x^3 - y^2$, then $z = f(u)$ and $\dfrac{\partial z}{\partial x} = \dfrac{\partial z}{\partial u}\dfrac{\partial u}{\partial x} = 3x^2\dfrac{\partial z}{\partial u},$

$\dfrac{\partial z}{\partial y} = \dfrac{\partial z}{\partial u}\dfrac{\partial u}{\partial y} = -2y\dfrac{\partial z}{\partial u},$ thus, $2y\left(3x^2\dfrac{\partial z}{\partial u}\right) + 3x^2\left(-2y\dfrac{\partial z}{\partial u}\right) = 0.$

16.4.18 Let $u = \dfrac{y}{x}$, then $z = f(u)$ and $\dfrac{\partial z}{\partial x} = \dfrac{\partial z}{\partial u}\dfrac{\partial u}{\partial x} = -\dfrac{y}{x^2}\dfrac{\partial z}{\partial u},$

$\dfrac{\partial z}{\partial y} = \dfrac{\partial z}{\partial u}\dfrac{\partial u}{\partial y} = \dfrac{1}{x}\dfrac{\partial z}{\partial u},$ thus, $x - \dfrac{y}{x^2}\dfrac{\partial z}{\partial u} + y\left(\dfrac{1}{x}\dfrac{\partial z}{\partial u}\right) = 0.$

SECTION 16.5

16.5.1 Find the equations of the tangent plane and normal line to $4x^2 + 9y^2 + z = 17$ at $(-1, 1, 4)$.

16.5.2 Find the equations of the tangent plane and normal line to $z = e^x \sin \pi y$ at $(2, 1, 0)$.

16.5.3 Find the equations of the tangent plane and normal line to $z = x^2 + y^2$ at $(2, -1, 5)$.

16.5.4 Find the equations of the tangent plane and normal line to $z = xe^{\sin y}$ at $(2, \pi, 2)$.

16.5.5 Find the equations of the tangent plane and normal line to $z = 3x^2 + 2y^2$ at $(2, -1, 14)$.

16.5.6 Find the equations of the tangent plane and normal line to $z = \dfrac{y^2}{3} - x$ at the origin.

16.5.7 Find dz if $z = \ln \sqrt[3]{1 + xy}$.

16.5.8 Find all points on the surface $z = xe^{-y}$ at which the tangent plane is horizontal.

16.5.9 Find a point on the surface $z = 16 - 4x^2 - y^2$ at which the tangent plane is perpendicular to the line $x = 3 + 4t$, $y = 2t$, $z = 2 - t$.

16.5.10 Find dz if $z = x \sin^{-1} y + x^2 y$.

16.5.11 The period, T, of a pendulum is given by $T = 2\pi \sqrt{\dfrac{\ell}{g}}$ where ℓ is the length of the pendulum and g is the acceleration due to gravity. Suppose $\ell = 5.1$ feet with a maximum error of 0.1 feet and $T = 2.5$ seconds with a maximum error of 0.05 seconds. Use differentials to estimate the maximum error in g.

16.5.12 The radius and height of a right-circular cylinder are measured with errors of at most 0.1 inches. If the height and radius are measured to be 10 inches and 2 inches, respectively, use differentials to approximate the maximum possible error in the calculated value of the volume.

16.5.13 The power consumed in an electrical resistor is given by $P = \dfrac{E^2}{R}$ watts. Suppose $E = 200$ volts and $R = 8$ ohms, approximate the change in power if E is decreased by 5 volts and R is decreased by 0.20 ohms.

16.5.14 Let $f(x, y) = \sqrt{x + 2y}$. Use a total differential to approximate the change in $f(x, y)$ as (x, y) varies from $(3, 5)$ to $(2.98, 5.1)$.

16.5.15   The legs of a right triangle are measured to be 6 and 8 inches with a maximum error of 0.10 inches in each measurement. Use differentials to estimate the maximum possible error in the calculated value of the hypotenuse and the area of the triangle.

16.5.16   Find the point on the surface $z = 9 - x^2 - y^2$ at which the tangent plane is parallel to the plane $2x + 3y + 2z = 6$.

16.5.17   The lengths and widths of a rectangle are measured with errors of at most 1%. Use differentials to estimate the maximum percentage error in the calculated area.

16.5.18   Show that the plane $z = 1$ is tangent to the surface $z = \sin xy$ at infinitely many points.

SOLUTIONS

SECTION 16.5

16.5.1 $f(x, y) = 17 - 4x^2 - 9y^2$, $f_x(x, y) = -8x$, $f_x(-1, 1) = 8$, $f_y(x, y) = -18y$, $f_y(-1, 1) = -18$, so the tangent plane at $(-1, 1, 4)$ is $8(x+1) - 18(y-1) - (z-4) = 0$ or $8x - 18y - z + 30 = 0$ and the normal line is $x = -1 + 8t$, $y = 1 - 18t$, $z = 4 - t$.

16.5.2 $f(x, y) = e^x \sin \pi y$, $f_x(x, y) = e^x \sin \pi y$, $f_x(2.1) = 0$, $f_y(x, y) = \pi e^x \cos \pi y$, $f_y(2, 1) = -\pi e^2$, so the equation of the tangent plane at $(2, 1, 0)$ is $\pi e^2(y - 1) + z = 0$ or $\pi e^2 y + z - \pi e^2 = 0$ and the normal line is $x = 2$, $y = 1 - \pi e^2 t$, $z = -t$.

16.5.3 $f(x, y) = x^2 + y^2$, $f_x(x, y) = 2x$, $f_x(2, -1) = 4$, $f_y(xy) = 2y$, $f_y(2, -1) = -2$, so the equation of the tangent plane at $(2, -1, 5)$ is $4(x - 2) - 2(y + 1) - (z - 5) = 0$ or $4x - 2y - z - 5 = 0$ and the normal line is $x = 2 + 4t$, $y = -1 - 2t$, $z = 5 - t$.

16.5.4 $f(x, y) = xe^{\sin y}$, $f_x(x, y) = e^{\sin y}$, $f_x(2, \pi) = 1$, $f_y(x, y) = xe^{\sin y} \cos y$, $f_y(2, \pi) = -2$, so the equation of the tangent plane at $(2, \pi, 2)$ is $1(x-2) - 2(y-\pi) - (z-2) = 0$ or $x - 2y - z + 2\pi = 0$ and the normal line is $x = 2 + t$, $y = \pi - 2t$, $z = 2 - t$.

16.5.5 $f(x, y) = 3x^2 + 2y^2$, $f_x(x, y) = 6x$, $f_x(2, -1) = 12$, $f_y(x, y) = 4y$, $f_y(2, -1) = -4$, so the equation of the tangent plane at $(2, -1, 14)$ is $12(x - 2) - 4(y + 1) - (z - 14) = 0$ or $12x - 4y - z - 14 = 0$ and the normal line is $x = 2 + 12t$, $y = 1 - 4t$, $z = 14 - t$.

16.5.6 $f(x, y) = \dfrac{y^2}{3} - x$, $f_x(x, y) = -1$, $f_x(0, 0) = -1$, $f_y(x, y) = \dfrac{2y}{3}$, $f_y(0, 0) = 0$, so the equation of the tangent plane at the origin is $-1(x - 0) - (z - 0) = 0$ or $x + z = 0$ and the normal line is $x = t$, $z = t$.

16.5.7 $f(x, y) = \ln \sqrt[3]{1 + xy} = \dfrac{1}{3} \ln(1 + xy)$, $f_x(x, y) = \dfrac{y}{3(1 + xy)}$, $f_y(x, y) = \dfrac{x}{3(1 + xy)}$, so $dz = \dfrac{y}{3(1 + xy)} dx + \dfrac{x}{3(1 + xy)} dy$.

16.5.8 There are no points on the surface $z = xe^{-y}$ at which the tangent plane is horizontal because $\dfrac{\partial z}{\partial x} = e^{-y} \neq 0$ for any real number.

16.5.9 $\dfrac{\partial z}{\partial x} = -8x$ and $\dfrac{\partial z}{\partial y} = -2y$, so the equation of the normal to the surface at (x_0, y_0, z_0) is $-8x_0 i - 2y_0 j - k$. A vector parallel to the given line and the normal is $4i + 2j - k$, thus, $-8x_0 = 4$, $x_0 = -1/2$; $-2y_0 = 2$, $y_0 = -1$ and $z = 14$ at $(-1/2, -1)$, so the point on the surface is $(-1/2, -1, 14)$.

16.5.10 $f(x,y) = x\sin^{-1}y + x^2 y$, $f_x(x,y) = \sin^{-1}y + 2xy$, $f_y(x,y) = \dfrac{x}{\sqrt{1-y^2}} + x^2$, so

$$dz = (\sin^{-1}y + 2xy)dx + \left(\frac{x}{\sqrt{1-y^2}} + x^2\right)dy.$$

16.5.11 $g = \dfrac{4\pi^2 \ell}{T^2}$, $dg = \dfrac{4\pi^2}{T^2}d\ell - \dfrac{8\pi^2 \ell}{T^3}dT$, $|d\ell| \leq 0.1$ and $|dT| \leq .05$ so

$$|dg| = \left| \frac{(4\pi^2)(0.1)}{(2.5)^2} - \frac{(8\pi^2)(5.1)(0.05)}{(2.5)^3} \right| \leq 1.93 \text{ ft./sec.}^2$$

Thus the maximum error is approximately 1.93.

16.5.12 $v = \pi r^2 h$, $dv = 2\pi rh\, dr + \pi r^2 dh$. When $h = 10$, $r = 2$, and
$|dh| = |dr| = 0.1$, $|dv| \leq (2\pi)(2)(10)(0.1) + (\pi)(2)^2(0.1) \leq 4.4\pi$.
The maximum error is approximately 4.4π.

16.5.13 The approximate change in power consumption is $dp = \dfrac{2E}{R}dE - \dfrac{E^2}{R^2}dR$, then

$$dP = \frac{(2)(200)}{8} \cdot (-5) - \frac{(200)^2}{(8)^2} \cdot (-0.20) = -125 \text{ watts, so the power consumed is decreased}$$

by 125 watts.

16.5.14 $f_x(x,y) = \dfrac{1}{2\sqrt{x+2y}}$, $f_y(x,y) = \dfrac{1}{\sqrt{x+2y}}$, thus, $df = \dfrac{1}{2\sqrt{x+2y}}dx + \dfrac{1}{\sqrt{x+2y}}dy$, then $x = 3$,
$dx = -0.02$, $y = 5$, and $dy = 0.1$ so $df = -\dfrac{0.01}{\sqrt{13}} + \dfrac{0.1}{\sqrt{13}} = \dfrac{0.09}{\sqrt{13}} \approx 0.025$.

16.5.15 Let $z = \sqrt{x^2 + y^2}$ so $dz = \dfrac{x}{\sqrt{x^2+y^2}}dx + \dfrac{y}{\sqrt{x^2+y^2}}dy$, then $x = 6$, $y = 8$, $|dx| \leq 0.1$,
$|dy| \leq 0.1$, so $|dz| \leq \dfrac{6}{\sqrt{(6)^2+(8)^2}} \cdot (0.1) + \dfrac{8}{\sqrt{(6)^2+(8)^2}} \cdot (0.1) = 0.14$ is the maximum error
in the hypotenuse. Let $A = \dfrac{1}{2}xy$, thus $dA = \dfrac{1}{2}y\,dx + \dfrac{1}{2}x\,dy$ so
$$|dA| \leq \left(\frac{1}{2}\right)(8)(0.1) + \left(\frac{1}{2}\right)(6)(0.1) = 0.7 \text{ is the maximum error in the area.}$$

16.5.16 $\dfrac{\partial z}{\partial x} = -2x$, $\dfrac{\partial z}{\partial y} = -2y$, so the equation of the normal to the surface at (x_0, y_0, z_0) is
$-2x_0 i - 2y_0 j - k$. A vector normal to the given plane is $2i + 3j - k$,
which is also normal to the tangent plane since it is parallel to the given plane, thus,
$-2x_0 = 2$, $x_0 = -1$, $-2y_0 = 3$, $y_0 = -\dfrac{3}{2}$, and $z = \dfrac{23}{4}$ at $(-1, -3/2)$,
so the point on the surface is $\left(-1, -\dfrac{3}{2}, \dfrac{23}{4}\right)$.

16.5.17 $A = xy$, $dA = y\,dx + x\,dy$, $\dfrac{dA}{A} = \dfrac{dx}{x} + \dfrac{dy}{y}$, $|\dfrac{dx}{x}| = 0.01$ and

$|\dfrac{dy}{y}| \le 0.01$, $|\dfrac{dA}{A}| = |\dfrac{dx}{x}| + |\dfrac{dy}{y}| \le 0.02 = 2\%$.

16.5.18 The plane $z = 1$ intersects the surface $z = \sin xy$ when

$xy = (2k+1)\pi/2, k = 0, \pm1, \pm2, \cdots$. At these points $\dfrac{\partial z}{\partial x} = y\cos xy = 0$ and

$\dfrac{\partial z}{\partial y} = x\cos xy = 0$. The tangent plane is $0(x - x_0) + 0(y - y_0) - (z - 1) = 0$.

So $z = 1$ is tangent to the surface $z = \sin xy$ at these infinitely many points.

SECTION 16.6

16.6.1 Find the directional derivative of $f(x,y) = e^x \sin y$ at $(0, \pi/3)$ in the direction of $a = 5\mathbf{i} - 2\mathbf{j}$.

16.6.2 Find the directional derivative of $f(x,y) = \ln \sqrt[3]{x^2 + y^2}$ at $(3,4)$ in the direction of $a = \langle 4, 3 \rangle$.

16.6.3 Find the directional derivative of $f(x,y) = \dfrac{x^2}{16} + \dfrac{y^2}{9}$ at $(4,3)$ in the direction of $a = \mathbf{i} + \mathbf{j}$.

16.6.4 Find the directional derivative of $f(x,y) = e^x \cos y$ at $(2, \pi)$ in the direction of $a = \langle 2, 3 \rangle$.

16.6.5 Find the directional derivative of $f(x,y) = 3xy^2 - 4x^3y$ at $(1,2)$ in the direction of $a = 3\mathbf{i} + 4\mathbf{j}$.

16.6.6 Find the directional derivative of $f(x,y) = e^x \sin \pi y$ at $(0, 1/3)$ in the direction towards $(3, 7/3)$.

16.6.7 Find the directional derivative of $f(x,y) = x\tan^{-1}\dfrac{y}{x}$ at $(1,1)$ in the direction of $a = 2\mathbf{i} - \mathbf{j}$.

16.6.8 Find the rate of change of $f(x,y) = \dfrac{2x}{x-y}$ at $(1,0)$ in the direction of a vector making an angle of $60°$ with the positive x axis.

16.6.9 Find the rate of change of $f(x,y) = \dfrac{x+y}{2x-y}$ at $(1,1)$ in the direction of a vector making an angle of $150°$ with the positive x axis.

16.6.10 Find the rate of change of $f(x,y) = 2xy - \dfrac{y}{x}$ at $(1,2)$ in the direction of a vector making an angle of $120°$ with the positive x axis.

16.6.11 The temperature, T, at a point (x,y) on a semi-circular plate is given by
$T(x,y) = 3x^2y - y^3 + 273$ degrees Celsius.

  (a) Find the temperature at $(1,2)$.
  (b) Find the rate of change of temperature at $(1,2)$ in the direction of $a = \mathbf{i} - 2\mathbf{j}$.
  (c) Find a unit vector in the direction in which the temperature increases most rapidly at $(1,2)$ and find this maximum rate of increase in temperature at $(1,2)$.

16.6.12 The temperature, T, at a point (x,y) in the xy-plane is given by $T(x,y) = x^3y^2$ degrees Celsius. Find a unit vector in the direction in which the temperature decreases most rapidly at $(2,1)$ and find this maximum rate of decrease in temperature at $(2,1)$.

16.6.13   The temperature, T, at a point (x, y) in the xy-plane is given by $T(x, y) = xy - x$. Find a unit vector in the direction in which the temperature increases most rapidly at $(1, 1)$ and find this maximum rate of increase in temperature at $(1, 1)$.

16.6.14   The temperature on the surface of a long, flat plate is given by $T(x, y) = x \sin y$ degrees Celsius. A bug is located at $(1, \pi/2)$.

   (a)   In what direction should the bug move for the most rapid decrease in temperature?
   (b)   If the bug moves in the direction $-2\mathbf{i} + \mathbf{j}$, will the temperature increase or decrease and at what rate?
   (c)   In what direction (there are two) can the bug move so that the temperature remains the same as it is at $(1, \pi/2)$?

16.6.15   At $t = 0$, the position of a particle on a rectangular membrane is given by
$P(x, y) = \sin \dfrac{\pi x}{3} \sin \dfrac{\pi y}{5}$. Find the rate at which P changes if the particle moves from $\left(\dfrac{3}{4}, \dfrac{15}{4} \right)$
in a direction of a vector making an angle $30°$ with the positive x-axis.

16.6.16   Sketch the level curve of $f(x, y) = \dfrac{x}{y^2}$ that passes through $(4, -4)$ and draw the gradient vector at that point.

16.6.17   Sketch the level curve of $f(x, y) = x^2 + y^2$ that passes through $(2, 2)$ and draw the gradient vector at that point.

16.6.18   Sketch the level curve of $f(x, y) = \dfrac{x^2}{4} + \dfrac{y^2}{9}$ that passes through $(2, 3)$ and draw the gradient vector at that point.

SOLUTIONS

SECTION 16.6

16.6.1 $\nabla \mathbf{f}(x,y) = e^x \sin y \mathbf{i} + e^x \cos y \mathbf{j}$, $\nabla \mathbf{f}(0, \pi/3) = \dfrac{\sqrt{3}}{2}\mathbf{i} + \dfrac{1}{2}\mathbf{j}$,

$\mathbf{u} = \dfrac{5}{\sqrt{29}}\mathbf{i} - \dfrac{2}{\sqrt{29}}\mathbf{j}$, $Duf = \nabla \mathbf{f} \cdot \mathbf{u} = \dfrac{5\sqrt{3} - 2}{2\sqrt{29}}$.

16.6.2 $\nabla \mathbf{f}(x,y) = \left\langle \dfrac{2x}{3(x^2 + y^2)}, \dfrac{2y}{3(x^2 + y^2)} \right\rangle$, $\nabla \mathbf{f}(3, 4) = \left\langle \dfrac{2}{25}, \dfrac{8}{75} \right\rangle$,

$\mathbf{u} = \left\langle \dfrac{4}{5}, \dfrac{3}{5} \right\rangle$, $Duf = \nabla \mathbf{f} \cdot \mathbf{u} = \dfrac{16}{125}$.

16.6.3 $\nabla \mathbf{f}(x,y) = \dfrac{x}{8}\mathbf{i} + \dfrac{2y}{9}\mathbf{j}$, $\nabla \mathbf{f}(4, 3) = \dfrac{1}{2}\mathbf{i} + \dfrac{2}{3}\mathbf{j}$,

$\mathbf{u} = \dfrac{1}{\sqrt{2}}\mathbf{i} + \dfrac{1}{\sqrt{2}}\mathbf{j}$, $Duf = \nabla \mathbf{f} \cdot \mathbf{u} = \dfrac{7}{6\sqrt{2}}$.

16.6.4 $\nabla \mathbf{f}(x,y) = \langle e^x \cos y, -e^x \sin y \rangle$, $\nabla \mathbf{f}(2, \pi) = \langle -e^2, 0 \rangle$,

$\mathbf{u} = \left\langle \dfrac{2}{13}, \dfrac{3}{13} \right\rangle$, $Duf = \nabla \mathbf{f} \cdot \mathbf{u} = -\dfrac{2e^2}{13}$.

16.6.5 $\nabla \mathbf{f}(x,y) = (3y^2 - 12x^2 y)\mathbf{i} + (6xy - 4x^3)\mathbf{j}$, $\nabla \mathbf{f}(1, 2) = -12\mathbf{i} + 8\mathbf{j}$,

$\mathbf{u} = \dfrac{3}{5}\mathbf{i} + \dfrac{4}{5}\mathbf{j}$, $Duf = \nabla \mathbf{f} \cdot \mathbf{u} = -\dfrac{4}{5}$.

16.6.6 $\nabla \mathbf{f}(x,y) = e^x \sin \pi y \mathbf{i} + \pi e^x \cos \pi y \mathbf{j}$, $\nabla \mathbf{f}(0, 1/3) = \dfrac{\sqrt{3}}{2}\mathbf{i} + \dfrac{\pi}{2}\mathbf{j}$, designate the given points as

$P_0(0, 1/3)$ and $P_1(3, 7/3)$, then $\mathbf{a} = \overrightarrow{P_0 P_1} = 3\mathbf{i} + 2\mathbf{j}$, $\mathbf{u} = \dfrac{3}{\sqrt{13}}\mathbf{i} + \dfrac{2}{\sqrt{13}}\mathbf{j}$,

$Duf = \nabla \mathbf{f} \cdot \mathbf{u} = \dfrac{3\sqrt{3} + 2\pi}{2\sqrt{13}}$.

16.6.7 $\nabla \mathbf{f}(x,y) = \left(\tan^{-1} \dfrac{y}{x} - \dfrac{xy}{x^2 + y^2} \right)\mathbf{i} + \dfrac{x^2}{x^2 + y^2}\mathbf{j}$,

$\nabla \mathbf{f}(1, 1) = \left(\dfrac{\pi}{4} - \dfrac{1}{2} \right)\mathbf{i} + \dfrac{1}{2}\mathbf{j}$,

$\mathbf{u} = \dfrac{2}{\sqrt{5}}\mathbf{i} - \dfrac{1}{\sqrt{5}}\mathbf{j}$, $Duf = \nabla \mathbf{f} \cdot \mathbf{u} = \dfrac{\pi}{2\sqrt{5}} - \dfrac{3}{2\sqrt{5}}$.

16.6.8 $\nabla \mathbf{f}(x,y) = -\dfrac{2y}{(x - y)^2}\mathbf{i} + \dfrac{2x}{(x - y)^2}\mathbf{j}$, $\nabla \mathbf{f}(1, 0) = 2\mathbf{j}$,

$\mathbf{u} = \cos \theta \mathbf{i} + \sin \theta \mathbf{j} = \dfrac{1}{2}\mathbf{i} + \dfrac{\sqrt{3}}{2}\mathbf{j}$, $Duf = \nabla \mathbf{f} \cdot \mathbf{u} = \sqrt{3}$.

16.6.9 $\nabla \mathbf{f}(x, y) = -\dfrac{3y}{(2x - y)^2}\mathbf{i} + \dfrac{3x}{(2x - y)^2}\mathbf{j},\ \nabla \mathbf{f}(1, 1) = -3\mathbf{i} + 3\mathbf{j},$

$\mathbf{u} = \cos\theta\mathbf{i} + \sin\theta\mathbf{j} = -\dfrac{\sqrt{3}}{2}\mathbf{i} + \dfrac{1}{2}\mathbf{j},\ Duf = \nabla \mathbf{f} \cdot \mathbf{u} = \dfrac{3 + 3\sqrt{3}}{2}.$

16.6.10 $\nabla \mathbf{f}(x, y) = \left(2y + \dfrac{y}{x^2}\right)\mathbf{i} + \left(2x - \dfrac{1}{x}\right)\mathbf{j},\ \nabla \mathbf{f}(1, 2) = 6\mathbf{i} + \mathbf{j},$

$\mathbf{u} = \cos\theta\mathbf{i} + \sin\theta\mathbf{j} = -\dfrac{1}{2}\mathbf{i} + \dfrac{\sqrt{3}}{2}\mathbf{j},\ Duf = \nabla \mathbf{f} \cdot \mathbf{u} = \dfrac{\sqrt{3} - 6}{2}.$

16.6.11   **(a)**   271° C

   (b) $\nabla \mathbf{T}(x, y) = 6xy\mathbf{i} + (3x^2 - 3y^2)\mathbf{j},\ \nabla \mathbf{T}(1, 2) = 12\mathbf{i} - 9\mathbf{j},$

   $\mathbf{u} = \dfrac{1}{\sqrt{5}}\mathbf{i} - \dfrac{2}{\sqrt{5}}\mathbf{j},\ DuT = \nabla \mathbf{T} \cdot \mathbf{u} = 6\sqrt{5}.$

   (c)   The direction of the most rapid increase in temperature is $\nabla \mathbf{T}(1, 2)$. A unit vector in this direction is $\dfrac{\nabla \mathbf{T}(2, 1)}{\|\nabla \mathbf{T}(2, 1)\|} = \dfrac{4}{5}\mathbf{i} - \dfrac{3}{5}\mathbf{j}$ and the most rapid increase in temperature is $\|\nabla \mathbf{T}(1, 2)\| = 15°$ C.

16.6.12   The direction of the most rapid decrease in temperature is $-\nabla \mathbf{T}(2, 1)$.

$\nabla \mathbf{T}(x, y) = 3x^2y^2\mathbf{i} + 2x^3y\mathbf{j},$ so $-\nabla \mathbf{T}(2, 1) = -12\mathbf{i} - 16\mathbf{j}$. A unit vector in this direction is $\dfrac{-\nabla \mathbf{T}(2, 1)}{\|-\nabla \mathbf{T}(2, 1)\|} = -\dfrac{3}{5}\mathbf{i} - \dfrac{4}{5}\mathbf{j}$ and the most rapid decrease in temperature is $\|\nabla \mathbf{T}(2, 1)\| = 20°$ C.

16.6.13   The direction of the most rapid increase in temperature is $\nabla \mathbf{T}(1, 1)$.

$\nabla \mathbf{T}(x, y) = (y - 1)\mathbf{i} + x\mathbf{j}$ so $\nabla \mathbf{T}(1, 1) = \mathbf{j}$ which is also the desired unit vector. The most rapid increase in temperature is thus $\|\nabla \mathbf{T}(1, 1)\| = 1°$ C.

16.6.14   **(a)**   The bug should move in the direction $-\nabla \mathbf{T}(1, \pi/2)$.

   $\nabla \mathbf{T}(x, y) = \sin y\mathbf{i} + x\cos y\mathbf{j},$ so $-\nabla \mathbf{T}(1, \pi/2) = -\mathbf{i}.$

   (b) $\mathbf{u} = -\dfrac{2}{\sqrt{5}}\mathbf{i} + \mathbf{j},$ so $DuT = \nabla \mathbf{T} \cdot \mathbf{u} = -\dfrac{2}{\sqrt{5}}$, thus the temperature is decreasing at the rate of $\dfrac{2\sqrt{5}}{5}°$ C.

   (c)   The bug should move on the isotherm normal to $\nabla \mathbf{T} = \mathbf{i}$ in the direction $+\mathbf{j}$ or $-\mathbf{j}$.

16.6.15 $\nabla \mathbf{P}(x, y) = \dfrac{\pi}{3}\cos\dfrac{\pi x}{3}\sin\dfrac{\pi y}{5}\mathbf{i} + \dfrac{\pi}{5}\sin\dfrac{\pi x}{3}\cos\dfrac{\pi y}{5}\mathbf{j},$

$\nabla \mathbf{P}\left(\dfrac{3}{4}, \dfrac{15}{4}\right) = \dfrac{\pi}{6}\mathbf{i} - \dfrac{\pi}{10}\mathbf{j},\ \mathbf{u} = \cos\theta\mathbf{i} + \sin\theta\mathbf{j} = \dfrac{\sqrt{3}}{2}\mathbf{i} + \dfrac{1}{2}\mathbf{j},$

$DuP = \nabla \mathbf{P} \cdot \mathbf{u} = \dfrac{5\sqrt{3}\pi - 3}{60}.$

16.6.16 $f(4,-4) = \dfrac{1}{4}$ so $y^2 = 4x$ for
$y \neq 0$,
$$\nabla f(x,y) = \frac{1}{y^2}\mathbf{i} - \frac{2x}{y^3}\mathbf{j},$$
$$\nabla f(4,-4) = \frac{1}{16}\mathbf{i} + \frac{1}{8}\mathbf{j}$$

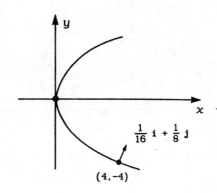

16.6.17 $\nabla f(x,y) = 2x\mathbf{i} + 2y\mathbf{j}$,
$\nabla f(2,2) = 4\mathbf{i} + 4\mathbf{j}$,
$f(2,2) = 8$, so $x^2 + y^2 = 8$
is the level curve.

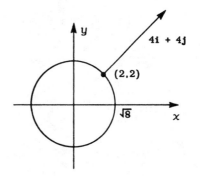

16.6.18 $f(2,3) = 2$ so $\dfrac{x^2}{8} + \dfrac{y^2}{18} = 1$
is the level curve.
$$\nabla f(x,y) = \frac{x}{2}\mathbf{i} + \frac{2y}{3}\mathbf{j},$$
$$\nabla f(2,3) = \mathbf{i} + 2\mathbf{j}.$$

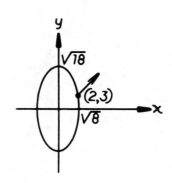

SECTION 16.7

16.7.1 Use the chain rule to find $\dfrac{d\omega}{dt}$ if $\omega = \tan^{-1}(xyz)$ and $x = t^2$, $y = t^3$, $z = t^{-4}$.

16.7.2 Use the chain rule to find $\dfrac{d\omega}{dt}$ if $\omega = \sin xy + y\ln xz + z$ and $x = e^t$, $y = t^2$, $z = 1$.

16.7.3 Find f_{zzy} if $f(yz) = z^4 - 3yz^2 + y\sin z$.

16.7.4 Find $d\omega$ if $\omega = 3x^2 + 2y^2 + z^2 - 2xy + 3xz - 12$.

16.7.5 Find $d\omega$ if $\omega = e^{2z}\sqrt[3]{x^2 + y^2}$.

16.7.6 Find $d\omega$ if $\omega = x^2 + 3xy - 2y^2 + 3xz + z^2$.

16.7.7 Find the equations of the tangent plane and normal line to $x^2z - xy^2 - yz^2 - 18 = 0$ at $(0, -2, 3)$.

16.7.8 Find the equations of the tangent plane and normal line to $xyz + 2x + 3y + 3z - 2 = 0$ at $(1, 2, -2)$.

16.7.9 Find the equations of the tangent plane and normal line to $\dfrac{x^2}{4} + \dfrac{y^2}{9} - \dfrac{z^2}{36} = 1$ at $(2, 3, 6)$.

16.7.10 Find the directional derivative of $f(x, y, z) = y\sin \pi xz + xy + \tan(\pi z)$ at $(1, 2, 1)$ in the direction of $\mathbf{a} = 2\mathbf{i} + 6\mathbf{j} - 9\mathbf{k}$.

16.7.11 Find the directional derivative of $f(x, y, z) = x^2 - 2y^2 + z^2$ at $(3, 3, 1)$ in the direction of $\mathbf{a} = 2\mathbf{i} + \mathbf{j} - \mathbf{k}$.

16.7.12 Find the directional derivative of $f(x, y, z) = x^2y^3 + \sqrt{xz}$ at $(1, -2, 3)$ in the direction of $\mathbf{a} = 5\mathbf{j} + \mathbf{k}$.

16.7.13 Find a unit vector in the direction in which $f(x, y, z) = 4e^{xy}\cos z$ decreases most rapidly at $(0, 1, \pi/4)$ and find the rate of decrease of f in that direction.

16.7.14 Find a unit vector in the direction in which $f(x, y, z) = \ln(1 + x^2 + y^2 - z^2)$ increases most rapidly at $(1, -1, 1)$ and find the rate of increase of f in that direction.

16.7.15 Find the directional derivative of $f(x, y, z) = x^2y + xy^2 + z^2$ if one leaves $(1, 1, 1)$ in the direction of $(3, 1, 2)$.

16.7.16   The temperature distribution of a ball centered at the origin is given by

$$T(x, y, z) = \frac{25}{x^2 + y^2 + z^2 + 1}.$$ Find the maximum rate of increase in temperature at $(3, -1, 2)$
and find a unit vector in that direction.

16.7.17   The temperature of a region in space is given by $T(x, y, z) = x^2 y z^3$. Find the maximum rate
of increase in temperature at $(2, 1, -1)$ and find a unit vector in that direction.

16.7.18   Find the directional derivative of $\phi(x, y, z) = xyz$ at $(1, 1, 1)$ in the direction of the normal
to the surface $x^2 y + y^2 x + yz^2 - 3 = 0$ at $(1, 1, 1)$.

SOLUTIONS

SECTION 16.7

16.7.1 $\dfrac{d\omega}{dt} = \dfrac{1}{1+t^2}$.

16.7.2 $t^2 e^t \cos t^2 e^t + 2te^t \cos t^2 e^t + 3t^2$.

16.7.3 $f_z = 4z^3 - 6yz + y\cos z,\ f_{zz} = 12z^2 - 6y - y\sin z,\ f_{zzy} = -6 - \sin z$.

16.7.4 $d\omega = (6x - 2y + 3z)dx + (4y - 2x)dy + (2z + 3x)dz$.

16.7.5 $d\omega = \dfrac{2xe^{2z}}{3(x^2+y^2)^{2/3}}dx + \dfrac{2ye^{2z}}{3(x^2+y^2)^{2/3}}dy + 2e^{2z}(x^2+y^2)^{1/3}dz$.

16.7.6 $d\omega = (2x + 3y + 3z)dx + (3x - 4y)dy + (3x + 2x)dz$.

16.7.7 $F(x,y,z) = x^2 z - xy^2 - yz^2 - 18$,

$\nabla \mathbf{F}(x,y,z) = (2xz - y^2)\mathbf{i} + (-2xy - z^2)\mathbf{j} + (x^2 - 2yz)\mathbf{k}$,

$\nabla \mathbf{F}(0,-2,3) = -4\mathbf{i} - 9\mathbf{j} + 12\mathbf{k}$; tangent plane $4x + 9y - 12z + 54 = 0$;

normal line $x = -4t,\ y = -2 - 9t,\ z = 3 + 12t$.

16.7.8 $F(x,y,z) = xyz + 2x + 3y + 3z - 2 = 0$

$\nabla \mathbf{F}(x,y,z) = (yz + 2)\mathbf{i} + (xz + 3)\mathbf{j} + (xy + 3)\mathbf{k}$

$\nabla \mathbf{F}(1,2,-2) = -2\mathbf{i} + \mathbf{j} + 5\mathbf{k}$; tangent plane $2x - y - 5z + 10 = 0$;

normal line $x = 1 - 2t,\ y = 2 + t,\ z = -2 + 5t$.

16.7.9 $F(x,y,z) = \dfrac{x^2}{4} + \dfrac{y^2}{9} - \dfrac{z^2}{36} - 1$,

$\nabla \mathbf{F}(x,y,z) = \dfrac{2x}{4}\mathbf{i} + \dfrac{2y}{9}\mathbf{j} - \dfrac{2z}{36}\mathbf{k}$,

$\nabla \mathbf{F}(2,3,6) = \mathbf{i} + \dfrac{2}{3}\mathbf{j} - \dfrac{1}{3}\mathbf{k}$; tangent plane $3x + 2y - z - 6 = 0$;

normal line $x = 2 + t,\ y = 3 + \dfrac{2}{3}t,\ z = 6 - \dfrac{1}{3}t$.

16.7.10 $\nabla \mathbf{f}(x,y,z) = (\pi yz \cos \pi xz + y\tan \pi z)\mathbf{i} + (\sin \pi xz + x\tan \pi z)\mathbf{j} + (\pi xy \cos \pi xz + \pi xy \sec^2 \pi z)\mathbf{k}$,

$\nabla \mathbf{f}(1,2,1) = -2\pi \mathbf{i} + 4\pi \mathbf{k},\ \mathbf{u} = \dfrac{2}{11}\mathbf{i} + \dfrac{6}{11}\mathbf{j} - \dfrac{9}{11}\mathbf{k}$

$Duf = \nabla \mathbf{f} \cdot \mathbf{u} = \dfrac{32\pi}{11}$.

16.7.11 $\nabla\mathbf{f}(x,y,z) = 2x\mathbf{i} - 4y\mathbf{j} + 2z\mathbf{k}$, $\nabla\mathbf{f}(3,3,1) = 6\mathbf{i} - 12\mathbf{j} + 2\mathbf{k}$,

$\mathbf{u} = \dfrac{2}{\sqrt{6}}\mathbf{i} + \dfrac{1}{\sqrt{6}}\mathbf{j} - \dfrac{1}{\sqrt{6}}\mathbf{k}$; $Duf = \nabla\mathbf{f}\cdot\mathbf{u} = -\dfrac{\sqrt{6}}{3}$.

16.7.12 $\nabla\mathbf{f}(x,y,z) = 2xy^3 + \dfrac{1}{2}\sqrt{\dfrac{z}{x}}\mathbf{i} + 3x^2y^2\mathbf{j} + \dfrac{1}{2}\sqrt{\dfrac{x}{z}}\mathbf{k}$,

$\nabla\mathbf{f}(1,-2,3) = \left(\dfrac{\sqrt{3}}{2} - 16\right)\mathbf{i} + 12\mathbf{j} + \dfrac{1}{2\sqrt{3}}\mathbf{k}$, $\mathbf{u} = \dfrac{5}{\sqrt{26}}\mathbf{j} + \dfrac{1}{\sqrt{26}}\mathbf{k}$;

$Duf = \nabla\mathbf{f}\cdot\mathbf{u} = \dfrac{60}{\sqrt{26}} + \dfrac{1}{2\sqrt{78}}$.

16.7.13 $\nabla\mathbf{f}(x,y,z) = 4ye^{xy}\cos z\mathbf{i} + 4xe^{xy}\cos z\mathbf{j} - 4e^{xy}\sin z\mathbf{k}$,

$\nabla\mathbf{f}(0,1,\pi/4) = 2\sqrt{2}\mathbf{i} - 2\sqrt{2}\mathbf{k}$, $\mathbf{u} = (-2\sqrt{2}\mathbf{i} + 2\sqrt{2}\mathbf{k})/4$, $\|\nabla\mathbf{f}(0,1,\pi/4)\| = 4$.

16.7.14 $\nabla\mathbf{f}(x,y,z) = \dfrac{2x}{1+x^2+y^2+z^2}\mathbf{i} + \dfrac{2y}{1+x^2+y^2+z^2}\mathbf{j} - \dfrac{2z}{1+x^2+y^2+z^2}\mathbf{k}$,

$\nabla\mathbf{f}(1,-1,1) = \mathbf{i} - \mathbf{j} - \mathbf{k}$, $\mathbf{u} = (\mathbf{i} - \mathbf{j} - \mathbf{k})/\sqrt{3}$, $\|\nabla\mathbf{f}(1,-1,1)\| = \sqrt{3}$.

16.7.15 $\nabla\mathbf{f}(x,y,z) = (2xy + y^2)\mathbf{i} + (x^2 + 2xy)\mathbf{j} + 2z\mathbf{k}$,

$\nabla\mathbf{f}(1,1,1) = 3\mathbf{i} + 3\mathbf{j} + 2\mathbf{k}$; designate the given points as $P_0(1,1,1)$ and $P_1(3,1,2)$ then
$\mathbf{a} = \overrightarrow{P_0P_1} = 2\mathbf{i} + \mathbf{k}$, $\mathbf{u} = (2\mathbf{i} + \mathbf{k})/\sqrt{5}$; $Duf = \nabla\mathbf{f}\cdot\mathbf{u} = \dfrac{8}{\sqrt{5}}$.

16.7.16 $\nabla\mathbf{T}(x,y,z) = -\dfrac{50x}{225}\mathbf{i} - \dfrac{50y}{225}\mathbf{j} - \dfrac{50z}{225}\mathbf{k}$,

$\nabla\mathbf{T}(3,-1,2) = -\dfrac{6}{9}\mathbf{i} + \dfrac{2}{9}\mathbf{j} - \dfrac{4}{9}\mathbf{k}$, $\mathbf{u} = -\dfrac{3}{\sqrt{14}}\mathbf{i} + \dfrac{1}{\sqrt{14}}\mathbf{j} - \dfrac{2}{\sqrt{14}}\mathbf{k}$,

$\|\nabla\mathbf{T}(3,1,-2)\| = \dfrac{2\sqrt{14}}{9}$.

16.7.17 $\nabla\mathbf{T}(x,y,z) = 2xyz^3\mathbf{i} + x^2z^3\mathbf{j} + 3x^2yz^2\mathbf{k}$,

$\nabla\mathbf{T}(2,1,-1) = -4\mathbf{i} - 4\mathbf{j} + 12\mathbf{k}$, $\mathbf{u} = (-4\mathbf{i} - 4\mathbf{j} + 12\mathbf{k})/\sqrt{176}$,

$\|\nabla\mathbf{T}(2,1,-1)\| = \sqrt{176} = 4\sqrt{11}$.

16.7.18 $\nabla\phi(x,y,z) = yz\mathbf{i} + xz\mathbf{j} + xy\mathbf{k}$, $\nabla\phi(1,1,1) = \mathbf{i} + \mathbf{j} + \mathbf{k}$,

let $\mathbf{F}(x,y,z) = x^2y + y^2x + yz^2 - 3$,

$\nabla\mathbf{F}(x,y,z) = (2xy + y^2)\mathbf{i} + (x^2 + 2xy + z^2)\mathbf{j} + 2yz\mathbf{k}$,

$\nabla\mathbf{F}(1,1,1) = 3\mathbf{i} + 4\mathbf{j} + 2\mathbf{k}$ is normal to the given surface,

$\mathbf{u} = (3\mathbf{i} + 4\mathbf{j} + 2\mathbf{k})/\sqrt{29}$; $Du\phi = \nabla\phi\cdot\mathbf{u} = \dfrac{9}{\sqrt{29}}$.

16.8.1 Use the chain rule to find $\dfrac{\partial w}{\partial r}$ and $\dfrac{\partial w}{\partial s}$ if $w = \ln(x^2 + y^2 + 2z)$, $x = r + s$, $y = r - s$, $z = 2rs$.

16.8.2 Use the chain rule to find $\dfrac{\partial w}{\partial r}$, $\dfrac{\partial w}{\partial \theta}$, and $\dfrac{\partial w}{\partial z}$ if $w = xy + yz$, $x = r \cos \theta$, $y = r \sin \theta$, $z = z$.

16.8.3 Use the chain rule to find $\dfrac{\partial w}{\partial r}$ and $\dfrac{\partial w}{\partial s}$ if $w = \ln(x^2 + y^2 + z^2)$, $x = e^r \cos s$, $y = e^r \sin s$, $z = e^s$.

16.8.4 Use the chain rule to find $\dfrac{dw}{dt}$ if $w = x^2 + y^2 + z^2$, $x = e^t \cos t$, $y = e^t \sin t$, $z = e^t$.

16.8.5 Use the chain rule to find $\dfrac{\partial w}{\partial t}$, $\dfrac{\partial w}{\partial u}$, and $\dfrac{\partial w}{\partial v}$ if $w = 3x - 2y + z$, $x = t^2 - u^2$, $y = u^2 + v^2$, $z = v^2 - t^2$.

16.8.6 Use the chain rule to find $\dfrac{\partial w}{\partial u}$ and $\dfrac{\partial w}{\partial v}$ if $w = 4x - y + 2z$, $x = u \sin v$, $y = v \sin u$, $z = \sin u \sin v$.

16.8.7 Use the chain rule to find $\dfrac{\partial w}{\partial r}$ and $\dfrac{\partial w}{\partial s}$ if $w = \sqrt{x^2 + y^2 + z^2}$, $x = r \cos s$, $y = r \sin s$, $z = r \tan s$.

16.8.8 Use the chain rule to find $\dfrac{\partial w}{\partial r}$ and $\dfrac{\partial w}{\partial s}$ if $w = x^2 + y^2 + z^2$, $x = r \cos s$, $y = r \sin s$, $z = rs$.

16.8.9 The length, width and height of a rectangular box are increasing at rates of 1 cm/sec, 2 cm/sec, and 2 cm/sec, respectively. At what rate is the volume increasing when the length is 3 cm, the width is 5 cm, and the height is 7 cm?

16.8.10 Evaluate $\dfrac{\partial z}{\partial x}$ and $\dfrac{\partial z}{\partial y}$ at $(1/2, -1, 2)$ if $e^{xz} + \ln(yz + 4) = y + 1 + e$ and z is a differentiable function of x and y.

16.8.11 Evaluate $\dfrac{\partial w}{\partial \phi}$ at the point whose spherical coordinates are $(4, \pi/3, \pi/6)$ if $w = (x^2 - 2y + z)^2$ and $x = \rho \sin \phi \cos \theta$, $y = \rho \sin \phi \sin \theta$, $z = \rho \cos \phi$.

16.8.12 Use the chain rule to find $\dfrac{\partial z}{\partial u}$ and $\dfrac{\partial z}{\partial v}$ if $z = x^2 y^3 + x \sin y$, $x = u^2$, $y = uv$.

16.8.13 Show that $u = z \tan^{-1} \dfrac{y}{x}$ satisfies $\dfrac{\partial^2 u}{\partial x^2} + \dfrac{\partial^2 u}{\partial y^2} + \dfrac{\partial^2 u}{\partial z^2} = 0$.

16.8.14   Show that $u = \dfrac{1}{\sqrt{x^2 + y^2 + z^2}}$ satisfies $\dfrac{\partial^2 u}{\partial x^2} + \dfrac{\partial^2 u}{\partial y^2} + \dfrac{\partial^2 u}{\partial z^2} = 0$.

16.8.15   Show that if $z = x + f(x, y)$ that $x\dfrac{\partial z}{\partial x} - y\dfrac{\partial z}{\partial y} = x$.

16.8.16   Let $z = f(x, y)$ with $x = r\cos\theta$ and $y = r\sin\theta$. Show that
$$\left(\frac{\partial z}{\partial r}\right)^2 + \frac{1}{r^2}\left(\frac{\partial z}{\partial \theta}\right)^2 = \left(\frac{\partial z}{\partial x}\right)^2 + \left(\frac{\partial z}{\partial y}\right)^2.$$

SOLUTIONS

SECTION 16.8

16.8.1 $\dfrac{\partial \omega}{\partial r} = \dfrac{2}{r+s}$; $\dfrac{\partial \omega}{\partial s} = \dfrac{2}{r+s}$.

16.8.2 $\dfrac{\partial \omega}{\partial r} = r \sin 2\theta + z \sin \theta$; $\dfrac{\partial \omega}{\partial \theta} = r^2 \cos 2\theta + rz \cos \theta$; $\dfrac{\partial \omega}{\partial z} = r \sin \theta$.

16.8.3 $\dfrac{\partial \omega}{\partial r} = \dfrac{2e^{2r}}{e^{2r} + e^{2s}}$; $\dfrac{\partial \omega}{\partial s} = \dfrac{2e^{2s}}{e^{2r} + e^{2s}}$.

16.8.4 $4e^{2t}$.

16.8.5 $\dfrac{\partial \omega}{\partial t} = 4t$, $\dfrac{\partial \omega}{\partial u} = -10u$, $\dfrac{\partial \omega}{\partial v} = -2v$.

16.8.6 $\dfrac{\partial \omega}{\partial u} = 4 \sin v - v \cos u + 2 \cos u \sin v$; $\dfrac{\partial \omega}{\partial v} = 4u \cos v - \sin u + 2 \sin u \cos v$.

16.8.7 $\dfrac{\partial \omega}{\partial r} = \sec s$; $\dfrac{\partial \omega}{\partial s} = r \tan s \sec s$.

16.8.8 $\dfrac{\partial \omega}{\partial r} = 2r + 2rs^2$; $\dfrac{\partial \omega}{\partial s} = 2r^2 s$.

16.8.9 $v = lwh, \dfrac{dv}{dt} = wh \dfrac{dl}{dt} + lh \dfrac{dw}{dt} + lw \dfrac{dh}{dt}$

$\qquad = (5)(7)(1) + (3)(7)(2) + (3)(5)(2)$

$\qquad = 107 \text{cm}^3/sec.$

16.8.10 $xe^{xz} \dfrac{\partial z}{\partial x} + ze^{xz} + \dfrac{y}{yz+4} \dfrac{\partial z}{\partial x} = 0$, $\dfrac{\partial z}{\partial x} = -\dfrac{ze^{xz}}{xe^{xz} + \dfrac{y}{yz+4}}$,

$\qquad \left. \dfrac{\partial z}{\partial x} \right|_{(1/2,-1,2)} = \dfrac{4e}{1-e}$; $xe^{xz} \dfrac{\partial z}{\partial y} + \dfrac{y}{yz+4} \dfrac{\partial z}{\partial y} + \dfrac{z}{yz+4} = 1$,

$\qquad \dfrac{\partial z}{\partial y} = \dfrac{1 - \dfrac{z}{yz+4}}{xe^{xz} + \dfrac{y}{yz+4}}$, $\left. \dfrac{\partial z}{\partial y} \right|_{(1/2,-1,2)} = 0.$

16.8.11 $\dfrac{\partial \omega}{\partial \phi} = 2(x^2 - 2y + z)(2x)(\rho \cos \phi \cos \theta) + 2(x^2 - 2y + z)(-2)(\rho \cos \phi \sin \theta)$
$\qquad\qquad + 2(x^2 - 2y + z)(1)(-\rho \sin \phi);$

$\qquad \left. \dfrac{\partial \omega}{\partial \phi} \right|_{(4,\pi/3,\pi/6)} = 4\sqrt{3} - 16.$

16.8.12 $\dfrac{\partial z}{\partial u} = (2xy^3 + \sin y)(2u) + (3x^2y^2 + x\cos y)v = 7u^6v^3 + 2u\sin uv + u^2v\cos uv,$

$\qquad\quad \dfrac{\partial z}{\partial v} = (2xy^3 + \sin y)(0) + (3x^2y^2 + x\cos y)u = 3u^7v^2 + u^3\cos uv.$

16.8.13 $\dfrac{\partial u}{\partial x} = -\dfrac{yz}{x^2 + y^2}, \dfrac{\partial^2 u}{\partial x^2} = \dfrac{2xyz}{(x^2 + y^2)^2}; \dfrac{\partial u}{\partial y} = \dfrac{xz}{x^2 + y^2}, \dfrac{\partial^2 u}{\partial x^2} = \dfrac{-2xyz}{(x^2 + y^2)^2};$

$\qquad\quad \dfrac{\partial u}{\partial z} = \tan^{-1}\dfrac{y}{x}, \dfrac{\partial^2 u}{\partial z^2} = 0;$ so $\dfrac{2xyz}{(x^2 + y^2)^2} - \dfrac{2xyz}{(x^2 + y^2)^2} + 0 = 0.$

16.8.14 $\dfrac{\partial u}{\partial x} = \dfrac{x}{(x^2 + y^2 + z^2)^{3/2}}, \dfrac{\partial^2 u}{\partial x^2} = \dfrac{y^2 + z^2 - 2x^2}{(x^2 + y^2 + z^2)^{5/2}};$

$\qquad\quad \dfrac{\partial u}{\partial y} = \dfrac{y}{(x^2 + y^2 + z^2)^{3/2}}, \dfrac{\partial^2 u}{\partial y^2} = \dfrac{x^2 + z^2 - 2y^2}{(x^2 + y^2 + z^2)^{5/2}};$

$\qquad\quad \dfrac{\partial u}{\partial z} = \dfrac{z}{(x^2 + y^2 + z^2)^{3/2}}, \dfrac{\partial^2 u}{\partial z^2} = \dfrac{x^2 + y^2 - 2z^2}{(x^2 + y^2 + z^2)^{5/2}};$

$\qquad\quad$ so $\dfrac{y^2 + z^2 - 2x^2}{(x^2 + y^2 + z^2)^{5/2}} + \dfrac{x^2 + z^2 - 2y^2}{(x^2 + y^2 + z^2)^{5/2}} + \dfrac{x^2 + y^2 - 2z^2}{(x^2 + y^2 + z^2)^{5/2}} = 0.$

16.8.15 Let $u = xy$, then $z = x + f(u)$, $\dfrac{\partial z}{\partial x} = 1 + y\dfrac{\partial u}{\partial x}$, $\dfrac{\partial z}{\partial y} = x\dfrac{\partial u}{\partial y}$, so

$\qquad\quad x\left(1 + y\dfrac{\partial u}{\partial x}\right) - y\left(x\dfrac{\partial u}{\partial y}\right) = x.$

16.8.16 $\dfrac{\partial z}{\partial r} = \dfrac{\partial z}{\partial x}\cos\theta + \dfrac{\partial z}{\partial y}\sin\theta, \dfrac{\partial z}{\partial\theta} = -r\sin\theta\dfrac{\partial z}{\partial x} + r\cos\theta\dfrac{\partial z}{\partial y},$

$\qquad\quad$ so $\left(\dfrac{\partial z}{\partial x}\cos\theta + \dfrac{\partial z}{\partial y}\sin\theta\right)^2 + \dfrac{1}{r^2}\left(-r\sin\theta\dfrac{\partial z}{\partial x} + r\cos\theta\dfrac{\partial z}{\partial y}\right)^2 = \left(\dfrac{\partial z}{\partial x}\right)^2 + \left(\dfrac{\partial z}{\partial y}\right)^2.$

SECTION 16.9

16.9.1 Locate all relative maxima, relative minima, and saddle points for
$f(x,y) = 5xy - 7x^2 - y^2 - 3x - 6y + 2.$

16.9.2 Locate all relative maxima, relative minima, and saddle points for
$f(x,y) = x^3 + y^2 - 12x + 6y - 7.$

16.9.3 Locate all relative maxima, relative minima, and saddle points for
$f(x,y) = x^2 + 3xy + 3y^2 - 6x + 3y - 6.$

16.9.4 Locate all relative maxima, relative minima, and saddle points for
$f(x,y) = x^2 - xy + y^2 + 2x + 2y - 4.$

16.9.5 Locate all relative maxima, relative minima, and saddle points for
$f(x,y) = x^2 - 2y^2 - 6x + 8y + 3.$

16.9.6 Locate all relative maxima, relative minima, and saddle points for
$f(x,y) = x^2 + 3xy + y^2 - 10x - 10y.$

16.9.7 Locate all relative maxima, relative minima, and saddle points for
$f(x,y) = 2x^2 + y^2 - 4x - 6y.$

16.9.8 Locate all relative maxima, relative minima, and saddle points for $f(x,y) = x^3 - 9xy + y^3$.

16.9.9 Locate all relative maxima, relative minima, and saddle points for
$f(x,y) = x^2 + \frac{1}{3}y^3 - 2xy - 3y.$

16.9.10 A rectangular box, open at the top, is to contain 256 cubic inches. Find the dimensions of the box for which the surface area is a minimum.

16.9.11 Find the point on the plane $2x - 3y + z = 19$ that is closest to $(1, 1, 0)$.

16.9.12 Find the shortest distance to $2x + y - z = 5$ from $(1, 1, 1)$.

16.9.13 An open rectangular box containing 18 cubic inches is to be constructed so that the base material costs 3 cents per square inch, the front face costs 2 cents per square inch, and the sides and back each cost 1 cent per square inch. Find the dimensions of the box for which the cost of construction will be a minimum.

16.9.14   Find the points on $z^2 = x^2 + y^2$ that are closest to $(2, 2, 0)$.

16.9.15   Find the point on $x + 2y + z = 1$ that is closest to the origin.

16.9.16   Find the maximum product of x, y, and z where x, y, and z are positive numbers such that $4x + 3y + z = 108$.

16.9.17   Find the minimum sum of $9x + 5y + 3z$ if x, y, and z are positive numbers such that $xyz = 25$.

16.9.18   Find the maximum product of x^2yz if x, y, and z are positive numbers such that $3x + 2y + z = 24$.

SOLUTIONS

SECTION 16.9

16.9.1 $f_x = -14x + 5y - 3$, $f_y = 5x - 2y - 6$, critical point at $(-12, -33)$;
$f_{xx} = -14$, $f_{yy} = -2$, $f_{xy} = 5$. $D > 0$ and $f_{xx} < 0$ at $(-12, -33)$;
relative maximum.

16.9.2 $f_x = 3x^2 - 12$, $f_y = 2y + 6$, critical points at $(-2, -3)$ and $(2, -3)$;
$f_{xx} = 6x$, $f_{yy} = 2$, $f_{xy} = 0$. $D < 0$ at $(-2, -3)$ so saddle point at $(-2, -3)$;
$D > 0$ and $f_{xx} > 0$ at $(2, -3)$; relative minimum.

16.9.3 $f_x = 2x + 3y - 6$, $f_y = 3x + 6y + 3$, critical point at $(15, -8)$;
$f_{xx} = 2$, $f_{yy} = 6$, $f_{xy} = 3$. $D > 0$ and $f_{xx} > 0$ at $(15, -8)$; relative minimum.

16.9.4 $f_x = 2x - y + 2$, $f_y = -x + 2y + 2$, critical point at $(-2, -2)$; $f_{xx} = 2$, $f_{yy} = 2$, $f_{xy} = -1$.
$D > 0$ and $f_{xx} > 0$ at $(-2, -2)$; relative minimum.

16.9.5 $f_x = 2x - 6$, $f_y = -4y + 8$, critical point at $(3, 2)$; $f_{xx} = 2$, $f_{yy} = -4$, $f_{xy} = 0$. $D < 0$ at
$(3, 2)$; saddle point.

16.9.6 $f_x = 2x + 3y - 10$, $f_y = 3x + 2y - 10$, critical point at $(2, 2)$; $f_{xx} = 2$, $f_{yy} = 2$, $f_{xy} = 3$.
$D < 0$ at $(2, 2)$; saddle point.

16.9.7 $f_x = 4x - 4$, $f_y = 2y - 6$, critical point at $(1, 3)$; $f_{xx} = 4$, $f_{yy} = 2$, $f_{xy} = 0$. $D > 0$ and
$f_{xx} > 0$ at $(1, 3)$; relative minimum.

16.9.8 $f_x = 3x^2 - 9y$, $f_y = -9x + 3y^2$, critical points at $(0, 0)$ and $(3, 3)$; $f_{xx} = 6x$, $f_{yy} = 6y$,
$f_{xy} = -9$. $D < 0$ at $(0, 0)$, saddle point; $D > 0$ and $f_{xx} > 0$ at $(3, 3)$; relative minimum.

16.9.9 $f_x = 2x - 2y$, $f_y = y^2 - 2x - 3$, critical points at $(-1, -1)$ and $(3, 3)$; $f_{xx} = 2$, $f_{yy} = 2y$,
$f_{xy} = -2$. $D < 0$ at $(-1, -1)$, saddle point; $D > 0$ and $f_{xx} > 0$ at $(3, 3)$; relative minimum.

16.9.10 Minimize $S = xy + 2yz + 2xz$ subject to $xyz = 256$, $x > 0$, $y > 0$, $z > 0$, thus, $z = \dfrac{256}{xy}$

and $S = xy + \dfrac{512}{x} + \dfrac{512}{y}$; $S_x = y - \dfrac{512}{x^2}$, $S_y = x - \dfrac{512}{y^2}$; critical point at $(8, 8)$; $S_{xx} = \dfrac{1024}{x^3}$,
$S_{yy} = \dfrac{1024}{y^3}$, $S_{xy} = 1$ so $S_{xx}S_{yy} - (S_{xy})^2 > 0$ and $S_{xx} > 0$ at $(8, 8)$, thus, the minimum
surface area occurs when $x = 8$, $y = 8$, and $z = 4$.

16.9.11 Minimize $W = D^2 = (x-1)^2 + (y-1)^2 + (z-0)^2$ subject to $2x - 3y + z = 19$, thus,
$z = 19 - 2x + 3y$ and $W = (x-1)^2 + (y-1)^2 + (19 - 2x + 3y)^2$;
$W_x = 2(x-1) + 2(19 - 2x + 3y)(-2)$, $W_y = 2(y-1) + 2(19 - 2x + 3y)(3)$, critical point
at $\left(\dfrac{27}{7}, -\dfrac{23}{7}\right)$; $W_{xx} = 10$, $W_{yy} = 20$, $W_{xy} = -12$; $W_{xx}W_{yy} - (W_{xy})^2 > 0$ and $W_{xx} > 0$ at
$\left(\dfrac{27}{7}, -\dfrac{23}{7}\right)$, so $\left(\dfrac{27}{7}, -\dfrac{23}{7}, \dfrac{10}{7}\right)$ is the closest point on $2x - 3y + z = 19$ to $(1,1,0)$.

16.9.12 Find the point on $2x + y - z = 5$ that is closest to $(1,1,1)$. Minimize
$W = D^2 = (x-1)^2 + (y-1)^2 + (z-1)^2$ subject to $2x + y - z = 5$, thus, $z = 2x + y - 5$
and $W = (x-1)^2 + (y-1)^2 + (2x + y - 5 - 1)^2$; $W_x = 2(x-1) + 2(2x + y - 6)(2)$,
$W_y = 2(y-1) + 2(2x + y - 6)(1)$; critical point at $\left(2, \dfrac{3}{2}\right)$; $W_{xx} = 10$, $W_{yy} = 4$, $W_{xy} = 4$ so
$W_{xx}W_{yy} - (W_{xy})^2 > 0$, $W_{xx} > 0$ at $\left(2, \dfrac{3}{2}\right)$ so the closest point on $2x + y - z = 5$ to $(1,1,1)$
is $\left(2, \dfrac{3}{2}, \dfrac{1}{2}\right)$ and thus, the shortest distance is $\sqrt{\dfrac{3}{2}}$ or $\dfrac{\sqrt{6}}{2}$.

16.9.13 Let x, y, z be, respectively, the length, width, and height of the box. Minimize the cost,
$C = 3xy + 3xz + 2yz$ subject to $v = xyz = 18$, $x > 0$, $y > 0$, $z > 0$, thus, $z = \dfrac{18}{xy}$ and
$C = 3xy + \dfrac{54}{y} + \dfrac{36}{x}$; $C_x = 3y - \dfrac{36}{x^2}$, $C_y = 3x - \dfrac{54}{y^2}$, critical point at $(2,3)$; $C_{xx} = \dfrac{72}{x^3}$,
$C_{yy} = \dfrac{108}{y^3}$, $C_{xy} = 3$; $C_{xx}C_{yy} - (C_{xy})^2 > 0$ and $C_{xx} > 0$ at $(2,3)$ so the cost is a minimum
when $x = 2$, $y = 3$, and $z = 3$.

16.9.14 Minimize $W = D^2 = (x-2)^2 + (y-2)^2 + (z-0)^2$ subject to $z^2 = x^2 + y^2$, then,
$W = (x-2)^2 + (y-2)^2 + x^2 + y^2$; $W_x = 2(x-2) + 2x$, $W_y = 2(y-2) + 2y$, critical
point at $(1,1)$; $W_{xx} = 4$, $W_{yy} = 4$, $W_{xy} = 0$; $W_{xx}W_{yy} - (W_{xy})^2 > 0$, $W_{xx} > 0$ at $(1,1)$ so
relative minima occur when $x = 1$, $y = 1$, and $z = \pm\sqrt{2}$. The closest points are $(1, 1, \sqrt{2})$
and $(1, 1, -\sqrt{2})$.

16.9.15 Minimize $W = D^2 = x^2 + y^2 + z^2$ subject to $x + 2y + z = 1$, thus, $z = 1 - x - 2y$ and
$W = x^2 + y^2 + (1 - x - 2y)^2$; $W_x = 2x + 2(1 - x - 2y)(-1)$, $W_y = 2y + 2(1 - x - 2y)(-2)$,
critical point at $(1/6, 1/3)$; $W_{xx} = 4$, $W_{yy} = 10$, $W_{xy} = 4$; $W_{xx}W_{yy} - (W_{xy})^2 > 0$ and
$W_{xx} > 0$ thus, $(1/6, 1/3, 1/6)$ is the closest point on $x + 2y + z = 1$ to the origin.

16.9.16 Maximize $P = xyz$ subject to $4x + 3y + z = 108$, $x > 0$, $y > 0$, and $z > 0$, thus,
$z = 108 - 4x - 3y$ and $P = xy(108 - 4x - 3y)$; $P_x = 108y - 8xy - 3y^2$, $P_y = 108x - 4x^2 - 6xy$;
critical point at $(9,12)$; $P_{xx} = -8y$, $P_{yy} = -6x$, $P_{xy} = 108 - 8x - 6y$; $P_{xx}P_{yy} - (P_{xy})^2 > 0$
and $P_{xx} < 0$ so a relative maximum occurs at $(9, 12, 36)$ and the maximum product of xyz
is $(9)(12)(36) = 3888$.

16.9.17   Minimize $S = 9x + 5y + 3z$ subject to $xyz = 25$, thus, $z = \dfrac{25}{xy}$ and $S = 9x + 5y + \dfrac{75}{xy}$,

$S_x = 9 - \dfrac{75}{x^2y}$, $S_y = 5 - \dfrac{75}{xy^2}$; critical point at $(5/3, 3)$; $S_{xx} = \dfrac{150}{x^3y}$, $S_{yy} = \dfrac{150}{xy^3}$, $S_{yx} = \dfrac{75}{x^2y^2}$,

$S_{xx}S_{yy} - (S_{xy})^2 > 0$ and $S_{xx} > 0$ at $(5/3, 3)$ so a relative minimum occurs at $(5/3, 3, 5)$ and

the minimum sum is $9\left(\dfrac{5}{3}\right) + 5(3) + 3(5) = 45$.

16.9.18   Maximize $P = x^2yz$ subject to $3x + 2y + z = 24$, $x > 0$, $y > 0$, $z > 0$, thus, $z = 24 - 3x - 2y$
and $P = x^2y(24 - 3x - 2y)$; $P_x = 48xy - 9x^2y - 4xy^2$, $P_y = 24x^2 - 3x^3 - 4x^2y$; critical point
at $(4, 3)$; $P_{xx} = 48y - 18xy - 4y^2$, $P_{yy} = -4x^2$, $P_{xy} = 48x - 9x^2 - 8xy$; $P_{xx}P_{yy} - (P_{xy})^2 > 0$,
$P_{xx} < 0$ at $(4, 3)$ so a relative maximum occurs at $(4, 3, 6)$ and the maximum product of x^2yz
is $(4)^2(3)(6) = 288$.

SECTION 16.10

16.10.1   Use the Lagrange multiplier method to find the point on the surface $z = xy + 1$ that is closest to the origin.

16.10.2   Use the Lagrange multiplier method to find the point on the plane $x + 2y + z = 1$ that is closest to the $(1, 1, 0)$.

16.10.3   Use the Lagrange multiplier method to find three positive numbers whose sum is 12 and whose product, x^2yz is a maximum.

16.10.4   Use the Lagrange multiplier method to find the maximum sum of $x^2 + y^2 + z^2$ if
$$x + 2y + 2z = 12.$$

16.10.5   An open rectangular box is to contain 256 cubic inches. Use the Lagrange multiplier method to find the dimensions of the box which uses the least amount of material.

16.10.6   An open rectangular box containing 18 cubic inches is constructed of material costing 3 cents per square inch for the base, 2 cents per square inch for the front face, and 1 cent per square inch for the sides and back. Use the Lagrange multiplier method to find the dimensions of the box for which the cost of construction is a minimum.

16.10.7   Use the Lagrange multiplier method to find the volume of the largest rectangular box that can be inscribed in the ellipsoid $2x^2 + 3y^2 + 4z^2 = 12$.

16.10.8   Use the Lagrange multiplier method to find the volume of the largest rectangular box that can be inscribed in the ellipsoid $2x^2 + 3y^2 + 6z^2 = 18$.

16.10.9   Use the Lagrange multiplier method to find the point on the plane $2x - 3y + z = 19$ that is closest to $(1, 1, 0)$.

16.10.10   Use the Lagrange multiplier method to find the shortest distance from $(1, 1, 1)$ to
$$2x + y - z = 5.$$

16.10.11   Use the Lagrange multiplier method to find the points on $z^2 = x^2 + y^2$ that are closest to $(2, 2, 0)$.

16.10.12   Use the Lagrange multiplier method to find three positive numbers whose sum is 36 and whose product is as large as possible.

16.10.13   Use the Lagrange multiplier method to find three positive numbers whose product is 64 and whose sum is as small as possible.

16.10.14   Use the Lagrange multiplier method to find three positive numbers whose product is as large as possible if their sum is given by $2x + 2y + z = 84$.

16.10.15   The base of a rectangular box costs three times as much per square foot as do the sides and top. Use the Lagrange multiplier method to find the dimensions of the box with least cost if the box contains 54 cubic feet.

16.10.16   The top and sides of a rectangular display case cost 5 times as much per square foot as the base. Use the Lagrange multiplier method to find the dimensions of the case with least cost if its case holds 12 cubic feet.

16.10.17   The temperature, T, at a point in space is given by $T(x, y, z) = 400xyz^2$ degrees Celsius. Use the Lagrange multiplier method to find the highest temperature on the sphere $x^2 + y^2 + z^2 = 1$.

SOLUTIONS

SECTION 16.10

16.10.1 $f(x, y, z) = D^2 = x^2 + y^2 + z^2$, $g(x, y, z) = z - xy - 1$;

$\nabla f = 2x\mathbf{i} + 2y\mathbf{j} + 2z\mathbf{k}$, $\nabla g = -x\mathbf{i} - y\mathbf{j} + \mathbf{k}$; equate $\nabla f = \lambda \nabla g$;

$2x = -\lambda y$, $2y = -\lambda x$, $2z = \lambda$; thus $-\dfrac{2x}{y} = -\dfrac{2y}{x}$ so $y = \pm x$, then $z = -\dfrac{x}{y}$ so $z = \pm 1$;

substitute into $z = xy + 1$ to get the critical points $(0, 0, 1)$, $(\sqrt{2}, -\sqrt{2}, -1)$, $(-\sqrt{2}, \sqrt{2}, 1)$; $(0, 0, 1)$ is the closest point to the origin.

16.10.2 $f(x, y, z) = D^2 = (x - 1)^2 + (y - 1)^2 + z^2$, $g(x, y, z) = x + 2y + z - 1$;

$\nabla f = 2(x - 1)\mathbf{i} + 2(y - 1)\mathbf{j} + 2z\mathbf{k}$, $\nabla g = \mathbf{i} + 2\mathbf{j} + \mathbf{k}$; equate $\nabla f = \lambda \nabla g$; $2(x - 1) = \lambda$, $2(y - 1) = 2\lambda$, $2z = \lambda$; thus $2(x - 1) = y - 1$ so $y = 2x - 1$, then $2(x - 1) = 2z$, thus, $z = x - 1$; substitute into $x + 2y + z = 1$ to get the $x = 2/3$, thus the closest point to $(1, 1, 0)$ is $(2/3, 1/3, -1/3)$.

16.10.3 $f(x, y, z) = P = x^2yz$, $g(x, y, z) = x + y + z - 10$, where x, y, and z are the three positive numbers; $\nabla f = 2xyz\mathbf{i} + x^2z\mathbf{j} + x^2y\mathbf{k}$, $\nabla g = \mathbf{i} + \mathbf{j} + \mathbf{k}$; equate $\nabla f = \lambda \nabla g$; $2xyz = \lambda$, $x^2z = \lambda$, $x^2y = \lambda$, thus, $2xyz = x^2z$ so $y = \dfrac{x}{2}$, then, $2xyz = x^2y$ so $z = \dfrac{x}{2}$; substitute into $x + y + z = 12$ to get $x = 6$, thus, the three positive numbers are $x = 6$, $y = 3$, and $z = 3$ and the maximum product is 324.

16.10.4 $f(x, y, z) = S = x^2 + y^2 + z^2$, $g(x, y, z) = x + 2y + 2z - 12$; $\nabla f = 2x\mathbf{i} + 2y\mathbf{j} + 2z\mathbf{k}$, $\nabla g = \mathbf{i} + 2\mathbf{j} + 2\mathbf{k}$; equate $\nabla f = \lambda \nabla g$; $2x = \lambda$, $2y = 2\lambda$, $2z = 2\lambda$, thus, $y = 2x$, and $z = 2x$; substitute into $x + 2y + 2z = 12$ to get $x = \dfrac{4}{3}$, thus, $y = \dfrac{8}{3}$ and $z = \dfrac{8}{3}$, the maximum sum is $\left(\dfrac{4}{3}\right)^2 + \left(\dfrac{8}{3}\right)^2 + \left(\dfrac{8}{3}\right)^2 = 16$.

16.10.5 Let x, y, and z be, respectively, the length, width, and height of the box.

$f(x, y, z) = xy + 2xz + 2yz$, $g(x, y, z) = xyz - 256$; $\nabla f = (y + 2z)\mathbf{i} + (x + 2z)\mathbf{j} + (2x + 2y)\mathbf{k}$, $\nabla g = yz\mathbf{i} + xz\mathbf{j} + xy\mathbf{k}$; equate $\nabla f = \lambda \nabla g$; $y + 2z = \lambda yz$, $x + 2z = \lambda xz$, $2x + 2y = \lambda xy$; thus, $\dfrac{y + 2z}{yz} = \dfrac{x + 2z}{xz}$ so $y = x(z \neq 0)$, then $\dfrac{y + 2z}{yz} = \dfrac{2x + 2y}{xy}$ so $z = \dfrac{x}{2}$ $(y \neq 0)$; substitute into $xyz = 256$ to get $x = 8$, thus, $x = 8$, $y = 8$, and $z = 4$.

16.10.6 Let x, y, and z be, respectively, the length, width, and height of the box and let the cost be $f(x, y, z) = 3xy + 3xz + 2yz$ subject to $g(x, y, z) = xyz - 18$;

$\nabla f = (3y + 3z)\mathbf{i} + (3x + 2z)\mathbf{j} + (3x + 2y)\mathbf{k}$, $\nabla g = yz\mathbf{i} + xz\mathbf{j} + xy\mathbf{k}$; equate $\nabla f = \lambda \nabla g$; $3y + 3z = \lambda yz$, $3x + 2z = \lambda xz$, $3x + 2y = \lambda xy$; $\dfrac{3y + 3z}{yz} = \dfrac{3x + 2z}{xz}$ so $y = \dfrac{3x}{2}(z \neq 0)$, $\dfrac{3y + 3z}{yz} = \dfrac{3x + 2y}{xy}$ so $z = \dfrac{3x}{2}(y \neq 0)$; substitute into $xyz = 18$ to get $x = 2$ so the required dimensions are $x = 2$, $y = 3$, and $z = 3$.

16.10.7 Let (x, y, z) be a point on the portion of the ellipsoid that lies in the first octant, thus, $V = (2x)(2y)(2z) = 8xyz$. Let $f(x, y, z) = 8xyz$, $g(x, y, z) = 2x^2 + 3y^2 + 4z^2 - 12$; $\nabla f = 8yz\mathbf{i} + 8xz\mathbf{j} + 8xy\mathbf{k}$, $\nabla g = 4x\mathbf{i} + 6y\mathbf{j} + 8z\mathbf{k}$; equate $\nabla f = \lambda\nabla g$ thus, $8yz = 4\lambda x$, $8xz = 6\lambda y$, $8xy = 8\lambda z$; thus, $\dfrac{8yz}{4x} = \dfrac{8xz}{6y}$ so $y = \pm\sqrt{\dfrac{2}{3}}x$, $\dfrac{8yz}{4x} = \dfrac{8xy}{8z}$ so $z = \pm\dfrac{x}{\sqrt{2}}$; substitute into $2x^2 + 3y^2 + 4z^2 = 12$ to get $x = \pm\sqrt{2}$, so, for (x, y, z) in the first octant, $x = \sqrt{2}$, $y = \dfrac{2}{\sqrt{3}}$, and $z = 1$, thus, the maximum volume is ' $8\left(\sqrt{2}\right)\left(\dfrac{2}{\sqrt{3}}\right)(1) = \dfrac{16\sqrt{6}}{3}$.

16.10.8 Let (x, y, z) be a point on the portion of the ellipsoid that lies in the first octant, thus, $V = (2x)(2y)(2z) = 8xyz$. Let $f(x, y, z) = 8xyz$, $g(x, y, z) = 3x^2 + 2y^2 + 6z^2 - 18$; $\nabla f = 8yz\mathbf{i} + 8xz\mathbf{j} + 8xy\mathbf{k}$, $\nabla g = 6x\mathbf{i} + 4y\mathbf{j} + 12z\mathbf{k}$; equate $\nabla f = \lambda\nabla g$ thus, $8yz = 6\lambda x$, $8xz = 4\lambda y$, $8xy = 12\lambda z$; thus, $\dfrac{8yz}{6x} = \dfrac{8xz}{4y}$ so $y = \pm\sqrt{\dfrac{3}{2}}x$, $\dfrac{8yz}{6x} = \dfrac{8xy}{12z}$ so $z = \pm\sqrt{\dfrac{1}{2}}x$; substitute into $3x^2 + 2y^2 + 6z^2 = 18$ to get $x = \pm\sqrt{2}$, so, for x, y, and z in the first octant, $x = \sqrt{2}$, $y = \sqrt{3}$, and $z = 1$, thus, the maximum volume is $8(\sqrt{2})(\sqrt{3})(1) = 8\sqrt{6}$.

16.10.9 Let $f(x, y, z) = D^2 = (x - 1)^2 + (y - 1)^2 + z^2$, $g(x, y, z) = 2x - 3y + z - 19$, $\nabla f = 2(x - 1)\mathbf{i} + 2(y - 1)\mathbf{j} + 2z\mathbf{k}$, $\nabla g = 2\mathbf{i} - 3\mathbf{j} + \mathbf{k}$, equate $\nabla f = \lambda\nabla g$, thus, $2(x - 1) = 2\lambda$, $2(y - 1) = -3\lambda$, $2z = \lambda$; thus, $x = \lambda + 1$, $y = \dfrac{2 - 3\lambda}{2}$, and $z = \dfrac{\lambda}{2}$; substitute into $2x - 3y + z = 19$ to get $\lambda = \dfrac{20}{7}$, then $x = \dfrac{27}{7}$, $y = -\dfrac{23}{7}$, and $z = \dfrac{10}{7}$ so the closest point is $\left(\dfrac{27}{7}, -\dfrac{23}{7}, \dfrac{10}{7}\right)$.

16.10.10 Find the point on the plane $2x + y - z = 5$ that is closest to $(1, 1, 1)$, thus, $f(x, y, z) = D^2 = (x - 1)^2 + (y - 1)^2 + (z - 1)^2$, $g(x, y, z) = 2x + y - z - 5$, $\nabla f = 2(x - 1)\mathbf{i} + 2(y - 1)\mathbf{j} + 2(z - 1)\mathbf{k}$, $\nabla g = 2\mathbf{i} + \mathbf{j} - \mathbf{k}$; equate $\nabla f = \lambda\nabla g$, thus, $2(x - 1) = 2\lambda$, $2(y - 1) = \lambda$, $2(z - 1) = -\lambda$; thus, $x = \lambda + 1$, $y = \dfrac{\lambda}{2} + 1$, $z = 1 - \dfrac{\lambda}{2}$, substitute into $2x + y - z = 5$ to get $\lambda = 1$, then, $x = 2$, $y = 3/2$, and $z = 1/2$, so the closest point is $(2, 3/2, 1/2)$ and the shortest distance is $\sqrt{(2 - 1)^2 + \left(\dfrac{3}{2} - 1\right)^2 + (1/2 - 1)^2} = \dfrac{\sqrt{6}}{2}$.

16.10.11 Let $f(x, y, z) = D^2 = (x - 2)^2 + (y - 2)^2 + z^2$, $g(x, y, z) = x^2 + y^2 - z^2$, $\nabla f = 2(x - 2)\mathbf{i} + 2(y - 2)\mathbf{j} + 2z\mathbf{k}$, $\nabla g = 2x\mathbf{i} + 2y\mathbf{j} - 2z\mathbf{k}$; equate $\nabla f = \lambda\nabla g$, thus, $2(x - 2) = 2\lambda x$, $2(y - 2) = 2\lambda y$, $2z = -2\lambda z$; if $z \neq 0$, $\lambda = -1$ then $x = y = 1$, substitute into $z^2 = x^2 + y^2$ to get $z = \pm\sqrt{2}$ yielding the critical points $(1, 1, \sqrt{2})$ and $(1, 1, -\sqrt{2})$; if $z = 0$, then $x = y = 0$ and $(0, 0, 0)$ is a critical point. Test $(1, 1, \sqrt{2})$ and $(1, 1, -\sqrt{2})$ to show that they are the closest points to $(2, 2, 0)$.

16.10.12 Let x, y, and z be the three numbers and let their product be xyz, thus, $f(x,y,z) = xyz$, $g(x,y,z) = x+y+z-36$, $\nabla f = yz\mathbf{i} + xz\mathbf{j} + xy\mathbf{k}$, $\nabla g = \mathbf{i}+\mathbf{j}+\mathbf{k}$; equate $\nabla f = \lambda\nabla g$, thus, $yz = \lambda$, $xz = \lambda$, $xy = \lambda$, then, $yz = xz$, so $y = x(z \neq 0)$; $yz = xy$, so $z = x$ $(y \neq 0)$; substitute into $x+y+z = 36$ to get $x = 12$, so the three positive numbers are $x = 12$, $y = 12$, $z = 12$ and their maximum product is $(12)(12)(12) = 1728$.

16.10.13 Let x, y, and z be the three positive numbers and let their sum be $x+y+z$, thus, $f(x,y,z) = x+y+z$, $g(x,y,z) = xyz - 64$; $\nabla f = \mathbf{i}+\mathbf{j}+\mathbf{k}$, $\nabla g = yz\mathbf{i} + xz\mathbf{j} + xy\mathbf{k}$; equate $\nabla f = \lambda\nabla g$, thus, $1 = \lambda yz$, $1 = \lambda xz$, $1 = \lambda xy$, then $\dfrac{1}{yz} = \dfrac{1}{xz}$ so $y = x$; $\dfrac{1}{yz} = \dfrac{1}{xy}$ so $z = x$; substitute into $xyz = 64$ to get $x = 4$, so the three positive numbers are $x = 4$, $y = 4$, $z = 4$ and their minimum sum is $4+4+4 = 12$.

16.10.14 Let x, y, and z be the three positive numbers and let their product be xyz, thus, $f(x,y,z) = xyz$, $g(x,y,z) = 2x+2y+z-84$, $\nabla f = yz\mathbf{i}+xz\mathbf{j}+xy\mathbf{k}$, $\nabla g = 2\mathbf{i}+2\mathbf{j}+\mathbf{k}$; equate $\nabla f = \lambda\nabla g$, thus, $yz = 2\lambda$, $xz = 2\lambda$, $xy = \lambda$, thus, $\dfrac{yz}{2} = \dfrac{xz}{2}$ so $y = x(z \neq 0)$; $\dfrac{yz}{2} = xy$ so $z = 2x(y \neq 0)$; substitute into $2x+2y+z = 84$ to get $x = 14$, so the three positive numbers are $x = 14$, $y = 14$, $z = 28$ and their maximum product is $(14)(14)(28) = 5488$.

16.10.15 Let $p = $ cost/square inch of material and let x, y, and z be the length, width, and height of the box, then $f(x,y,z) = 4pxy+2pxz+2pyz$ is the cost of the box subject to $xyz = 54$ thus, $g(x,y,z) = xyz - 54$; $\nabla f = (4py+2pz)\mathbf{i} + (4px+2pz)\mathbf{j} + (2px+2py)\mathbf{k}$, $\nabla g = yz\mathbf{i}+xz\mathbf{j}+xy\mathbf{k}$; equate $\nabla f = \lambda\nabla g$ to get $4py+2pz = \lambda yz$, $4px+2pz = \lambda xz$, $2px+2py = \lambda xy$ and $xyz = 54$; thus, $\dfrac{4py+2pz}{yz} = \dfrac{4px+2pz}{xz}$, $y = x(z \neq 0)$; $\dfrac{4py+2pz}{yz} = \dfrac{2px+2py}{xy}$, $z = 2x(y \neq 0)$; substitute into $xyz = 54$ to get $x = 3$, so the dimensions of the box with least cost is $x = 3$, $y = 3$, and $z = 6$.

16.10.16 Let x, y, and z be the length, width, and height of the case, then, $f(x,y,z) = 6xy + 10xz + 10yz$ is the cost of the case subject to $xyz = 12$ so let $g(x,y,z) = xyz - 12$; $\nabla f = (6y+10z)\mathbf{i} + (6x+10z)\mathbf{j} + (10x+10y)\mathbf{k}$, $\nabla g = yz\mathbf{i}+xz\mathbf{j}+xy\mathbf{k}$; equate $\nabla f = \lambda\nabla g$ thus, $6y+10z = \lambda yz$, $6x+10z = \lambda xz$, $10x+10y = \lambda xy$, thus, $\dfrac{6y+10z}{yz} = \dfrac{6x+10z}{xz}$ so $y = x(z \neq 0)$; $\dfrac{6y+10z}{yz} = \dfrac{10x+10y}{xy}$ so $z = \dfrac{3}{5}x(y \neq 0)$; substitute into $xyz = 12$ to get $x = \sqrt[3]{20}$, thus the dimensions of the case with least cost is $x = \sqrt[3]{20}$, $y = \sqrt[3]{20}$, and $z = \dfrac{3\sqrt[3]{20}}{5}$.

16.10.17 $f(x, y, z) = T(x, y, z) = 400xyz^2$, $g(x, y, z) = x^2 + y^2 + z^2 - 1$,

$\nabla f = 400yz^2\mathbf{i} + 400xz^2\mathbf{j} + 800xyz\mathbf{k}$, $\nabla g = 2x\mathbf{i} + 2y\mathbf{j} + 2z\mathbf{k}$; equate $\nabla f = \lambda\nabla g$, thus,

$400yz^2 = 2\lambda x$, $400xz^2 = 2\lambda y$, $800xyz = 2\lambda z$ then, $\dfrac{400yz^2}{2x} = \dfrac{400xz^2}{2y}$ so $y = \pm x (z \neq 0)$,

$\dfrac{400yz^2}{2x} = \dfrac{800xzy}{2z}$ so $z = \pm\sqrt{2}x$ $(y \neq 0)$; substitute into $x^2 + y^2 + z^2 = 1$ to get $x = \pm\dfrac{1}{2}$,

thus, $y = \pm\dfrac{1}{2}$ and $z = \pm\dfrac{\sqrt{2}}{2}$. The maximum temperature of $50°$ occurs on the sphere whenever x and y have the same sign.

SUPPLEMENTARY EXERCISES CHAPTER 16

1. Let $f(x, y) = e^x \ln y$. Find
   (a) $f(0, e)$
   (b) $f(\ln y, e^x)$
   (c) $f(r + s, rs)$.

2. Sketch the domain of f using solid lines for portions of the boundary included in the domain and dashed lines for portions not included.
   (a) $f(x, y) = \sqrt{x - y}/(2x - y)$ (b) $f(x, y) = \ln(xy - 1)$ (c) $f(x, y) = (\sin^{-1} x)/e^y$.

3. Describe the graph of f.
   (a) $f(x, y) = \sqrt{x^2 + 4y^2}$
   (b) $f(x, y) = 1 - x/a - y/b$.

4. Find f_x, f_y, and f_z if $f(x, y, z) = x^2/(y^2 + z^2)$.

5. Find $\partial w/\partial r$ if $w = \ln(xy)/\sin yz$, $x = r + s, y = s, z = 3r - s$.

6. Find $\partial f/\partial x$, $\partial f/\partial y$, and $\partial f/\partial u$ if $f(x, y, z, u) = (e^{yz}/x) + \ln(u - x)$.

7. Find f_x, f_y, f_{xy}, and f_{yzx} at $(0, \pi/2, 1)$ if $f(x, y, z) = e^{xy} \sin yz$.

8. Find $g_{xy}(0, 3)$ and $g_{yy}(2, 0)$ if $g(x, y) = \sin(xy) + xe^y$.

9. Find $\partial w/\partial \theta|_{r=1, \theta=0}$ if $w = \ln(x^2 + y^2)$, $x = re^\theta$, $y = \tan(r\theta)$.

10. Find $\partial w/\partial r$ if $w = x \cos y + y \sin x$, $x = rs^2$, $y = r + s$.

11. Find $\partial w/\partial s$ if $w = \ln(x^2 + y^2 + 2z)$, $x = r + s, y = r - s, z = 2rs$.

In Exercises 12-15, verify the assertion.

12. If $w = \tan(x^2 + y^2) + x\sqrt{y}$, then $w_{xy} = w_{yx}$.

13. If $w = \ln(3x - 3y) + \cos(x + y)$, then $\partial^2 w/\partial x^2 = \partial^2 w/\partial y^2$.

14. If $F(x, y, z) = 2z^3 - 3(x^2 + y^2)z$, then $F_{xx} + F_{yy} + F_{zz} = 0$.

15. If $f(x, y, z) = xyz + x^2 + \ln(y/z)$, then $f_{xyzx} = f_{zxxy}$.

16. Find the slope of the tangent line at $(1, -2, -3)$ to the curve of intersection of the surface $z = 5 - 4x^2 - y^2$ with
    (a) the plane $x = 1$
    (b) the plane $y = -2$.

795

17. The pressure in newtons/meter^2 of a gas in a cylinder is given by $P = 10T/V$ with T in kelvins (K) and V in meters^3.

    (a) If T is increasing at a rate of 3 K/min with V held fixed at 2.5 m^3, find the rate at which the pressure is changing when $T = 50$ K.

    (b) If T is held fixed at 50 K while V is decreasing at the rate of 3 m^3/min, find the rate at which the pressure is changing when $V = 2.5$ m^3.

In Exercises 18 and 19,

    (a) find the limit of $f(x,y)$ as $(x,y) \to (0,0)$ if it exists;

    (b) determine whether f is continuous at $(0,0)$.

18. $f(x,y) = \dfrac{x^4 - x + y - x^3 y}{x - y}$.

19. $f(x,y) = \begin{cases} \dfrac{x^4 - y^4}{x^2 + y^2} & \text{if } (x,y) \neq (0,0) \\ 0 & \text{if } (x,y) = (0,0). \end{cases}$

20. Find dw/dt using the chain rule.

    (a) $w = \sin xy + y \ln xz + z$, $x = e^t$, $y = t^2$, $z = 1$

    (b) $w = \sqrt{xy - e^z}$, $x = \sin t$, $y = 3t$, $z = \cos t$.

21. Use Formula (17) of Section 16.4 to find dy/dx.

    (a) $3x^2 - 5xy + \tan xy = 0$ **(b)** $x \ln y + \sin(x - y) = \pi$.

22. If $F(x,y) = 0$, find a formula for d^2y/dx^2 in terms of partial derivatives of F. [*Hint:* Use formula (17) of Section 16.4.]

23. The voltage V across a fixed resistance R in series with a variable resistance r is $V = RE/(r+R)$, where E is the source voltage. Express dV/dt in terms of dE/dt and dr/dt.

24. Let $f(x,y,z) = 1/(z - x^2 - 4y^2)$.

    (a) Describe the domain of f.

    (b) Describe the level surface $f(x,y,z) = 2$.

    (c) Find $f(3t, uv, e^{3t})$.

In Exercises 25–29, find

    (a) the gradient of f at P_0;

    (b) the directional derivative at P_0 in the indicated direction.

25. $f(x,y) = x^2 y^5$, $P_0(3,1)$; from P_0 toward $P_1(4,-3)$.

26. $f(x,y,z) = ye^x \sin z$, $P_0(\ln 2, 2, \pi/4)$; in the direction of $\mathbf{a} = \langle 1, -2, 2 \rangle$.

27. $f(x, y, z) = \ln(xyz)$, $P_0(3, 2, 6)$; $\mathbf{u} = \langle -1, 1, 1 \rangle / \sqrt{3}$.

28. $f(x, y) = x^2 y + 2xy^2$, $P_0(1, 2)$; $\mathbf{u}$ makes an angle of $60°$ with the positive x-axis.

29. $f(x, y, z) = xy + yz + zx$, $P_0(1, -1, 2)$; from P_0 toward $P_1(11, 10, 0)$.

30.   Let $f(x, y, z) = (x + y)^2 + (y + z)^2 + (z + x)^2$. Find the maximum rate of decrease of f at $P_0(2, -1, 2)$ and the direction in which this rate of decrease occurs.

31.   Find all unit vectors $\mathbf{u}$ such that $D_{\mathbf{u}}f = 0$ at P_0.
   (a) $f(x, y) = x^3 y^3 - xy$, $P_0(1, -1)$ **(b)** $f(x, y) = xe^y$, $P_0(-2, 0)$.

32.   The directional derivative $D_{\mathbf{u}}f$ at (x_0, y_0) is known to be 2 when $\mathbf{u}$ makes an angle of $30°$ with the positive x-axis, and 8 when this angle is $150°$. Find $D_{\mathbf{u}}f(x_0, y_0)$ in the direction of the vector $\sqrt{3}\mathbf{i} + 2\mathbf{j}$.

33.   At the point $(1, 2)$, the directional derivative $D_{\mathbf{u}}f$ is $2\sqrt{2}$ toward $P_1(2, 3)$ and -3 toward $P_2(1, 0)$. Find $D_{\mathbf{u}}f(1, 2)$ toward the origin.

In Exercises 34 and 35,
   (a)   find a normal vector $\mathbf{N}$ at $P_0(x_0, y_0, f(x_0, y_0))$;
   (b)   find an equation for the tangent plane at P_0.

34. $f(x, y) = 4x^2 + y^2 + 1$; $P_0(1, 2, 9)$.

35. $f(x, y) = 2\sqrt{x^2 + y^2}$; $P_0(4, -3, 10)$.

36.   Find equations for the tangent plane and normal line to the given surface at P_0.
   (a) $z = x^2 e^{2y}$; $P_0(1, \ln 2, 4)$ **(b)** $x^2 y^3 z^4 + xyz = 2$; $P_0(2, 1, -1)$.

37.   Find all points P_0 on the surface $z = 2 - xy$ at which the normal line passes through the origin.

38.   Show that for all tangent planes to the surface $x^{2/3} + y^{2/3} + z^{2/3} = 1$, the sum of the squares of the x-, y-, and z-intercepts is 1.

39.   Find all points on the elliptic paraboloid $z = 9x^2 + 4y^2$ at which the normal line is parallel to the line through the points $P(4, -2, 5)$ and $Q(-2, -6, 4)$.

40.   If $w = x^2 y - 2xy + y^2 x$, find the increment Δw and the differential dw if (x, y) varies from $(1, 0)$ to $(1.1, -0.1)$

41. Use differentials to approximate the change in the volume $V = \frac{1}{3}x^2h$ of a pyramid with a square base when its height h is increased from 2 to 2.2 m, while its base dimension x is decreased from 1 to 0.9 m. Compare this to ΔV.

42. If $f(x, y, z) = x^2y^4/(1+z^2)$, use differentials to approximate $f(4.996, 1.003, 1.995)$.

In Exercises 43–45, locate all relative minima, relative maxima, and saddle points.

43. $f(x, y) = x^2 + 3xy + 3y^2 - 6x + 3y$ **44.** $f(x, y) = x^2y - 6y^2 - 3x^2$.

45. $f(x, y) = x^3 - 3xy + \frac{1}{2}y^2$

Solve Exercises 46 and 47 two ways:
    (a) Use the constraint to eliminate a variable.
    (b) Use Lagrange multipliers.

46. Find all relative extrema of x^2y^2 subject to the constraint $4x^2 + y^2 = 8$

47. Find the dimensions of the rectangular box of maximum volume that can be inscribed in the ellipsoid $(x/a)^2 + (y/b)^2 + (z/c)^2 = 1$.

In Exercises 48 and 49, use Lagrange multipliers.

48. Find the points on the curve $5x^2 - 6xy + 5y^2 = 8$ whose distance from the origin is (i) minimum and (ii) maximum.

49. A current I branches into currents I_1, I_2, and I_3 throught resistors with resistances R_1, R_2, and R_3 (see figure) in such a way that the total energy to the three resistors is a minimum. If the energy delivered to R_i is $I_i^2R_i (i = 1, 2, 3)$, find the ratios $I_1:I_2:I_3$.

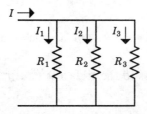

SOLUTIONS

SUPPLEMENTARY EXERCISES CHAPTER 16

1. **(a)** 1   **(b)** xy **(c)** $e^{r+s}\ln rs$

2. **(a)**   **(b)**   **(c)**

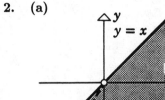
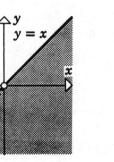
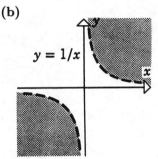
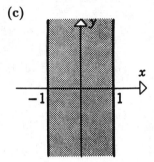

3. **(a)** upper half of the elliptic cone $z^2 = x^2 + 4y^2$

     (b) the plane with x, y, and z intercepts of 1, a, and b.

4. $f_x = 2x/\left(y^2 + z^2\right)$, $f_y = -2x^2 y/\left(y^2 + z^2\right)^2$, $f_z = -2x^2 z/\left(y^2 + z^2\right)^2$

5. $1/(x\sin yz) - 3y(\csc yz \cot yz)\ln xy$

6. $\partial f/\partial x = -e^{yz}/x^2 - 1/(u-x)$, $\partial f/\partial y = ze^{yz}/x$, $\partial f/\partial u = 1/(u-x)$

7. $\pi/2$, 0, 1, $-\pi^2/4$ **8.** $1 + e^3$, 2

9. 2   **10.** $s^2(\cos y + y\cos x) - x\sin y + \sin x$

11. $(2x - 2y + 4r)/\left(x^2 + y^2 + 2z\right) = \dfrac{2}{r+s}$

12. $w_{xy} = w_{yx} = 8xy\sec^2\left(x^2 + y^2\right)\tan\left(x^2 + y^2\right) + (1/2)y^{-1/2}$

13. $\partial^2 w/\partial x^2 = \partial^2 w/\partial y^2 = -(x-y)^{-2} - \cos(x+y)$

14. $F_{xx} = F_{yy} = -6z$, $F_{zz} = 12z$ **15.** $f_{xyzx} = f_{zxxy} = 0$

16. **(a)** $\partial z/\partial y = -2y$, slope $= -2(-2) = 4$ **(b)** $\partial z/\partial x = -8x$, slope $= -8(1) = -8$

17. **(a)** $dP/dt = (\partial P/\partial T)(dT/dt) = (10/V)(dT/dt) = (10/2.5)(3) = 12$ newtons/m^2/min

(b) $dP/dt = (\partial P/\partial V)(dV/dt) = -\left(10T/V^2\right)(dV/dt)$

$$= -(500/6.25)(-3) = 240 \text{ newtons/m}^2/\text{min}$$

18. **(a)** $\displaystyle\lim_{(x,y)\to(0,0)} \frac{(x-y)\left(x^3-1\right)}{(x-y)} = \lim_{(x,y)\to(0,0)} \left(x^3-1\right) = -1$

(b) not continuous at $(0,0)$ because $f(0,0)$ is not defined

19. **(a)** $\displaystyle\lim_{(x,y)\to(0,0)} \frac{\left(x^2-y^2\right)\left(x^2+y^2\right)}{x^2+y^2} = \lim_{(x,y)\to(0,0)} \left(x^2-y^2\right) = 0$

(b) continuous at $(0,0)$ because $\displaystyle\lim_{(x,y)\to(0,0)} f(x,y) = f(0,0)$

20. **(a)** $(y\cos xy + y/x)e^t + 2t(x\cos xy + \ln xz)$ **(b)** $(1/2)(y\cos t + 3x + e^z \sin t)/\sqrt{xy - e^z}$

21. **(a)** $-(6x - 5y + y\sec^2 xy)/\left(-5x + x\sec^2 xy\right)$ **(b)** $-[\ln y + \cos(x-y)]/[x/y - \cos(x-y)]$

22. $dy/dx = -F_x/F_y; \; d^2y/dx^2 = -(F_y dF_x/dx - F_x dF_y/dx)/F_y^2$

$$= -(F_y[F_{xx} + F_{xy}dy/dx] - F_x[F_{yx} + F_{yy}dy/dx])/F_y^2,$$

replace dy/dx by $-F_x/F_y$ and assume that $F_{xy} = F_{yx}$ to get

$d^2y/dx^2 = -\left(F_y^2 F_{xx} - 2F_x F_y F_{xy} + F_x^2 F_{yy}\right)/F_y^3$

23. $dV/dt = (\partial V/\partial E)(dE/dt) + (\partial V/\partial r)(dr/dt) = \dfrac{R}{r+R}\dfrac{dE}{dt} - \dfrac{RE}{(r+R)^2}\dfrac{dr}{dt}$

24. **(a)** all (x,y,z) not on the elliptic paraboloid $z = x^2 + 4y^2$

(b) $1/(z - x^2 - 4y^2) = 2, \; z - x^2 - 4y^2 = 1/2, \; z = 1/2 + x^2 + 4y^2$ which is an elliptic paraboloid

(c) $1/\left(e^{3t} - 9t^2 - 4u^2 v^2\right)$

25. **(a)** $\nabla f(3,1) = \langle 6, 45 \rangle$

(b) $\overrightarrow{P_0 P_1} = \langle 1, -4 \rangle, \; \mathbf{u} = \langle 1, -4 \rangle/\sqrt{17}, \; D_{\mathbf{u}}f = -174/\sqrt{17}$

26. **(a)** $\nabla f(\ln 2, 2, \pi/4) = \sqrt{2}\,\langle 2, 1, 2 \rangle$ **(b)** $\mathbf{u} = \langle 1, -2, 2 \rangle/3, \; D_{\mathbf{u}}f = 4\sqrt{2}/3$

27. **(a)** $\nabla f(3, 2, 6) = \langle 1/3, 1/2, 1/6 \rangle$ **(b)** $D_{\mathbf{u}}f = \sqrt{3}/9$

28. **(a)** $\nabla f(1, 2) = \langle 12, 9 \rangle$ **(b)** $\mathbf{u} = \langle 1/2, \sqrt{3}/2 \rangle, \; D_{\mathbf{u}}f = 6 + 9\sqrt{3}/2$

29. **(a)** $\nabla f(1, -1, 2) = \langle 1, 3, 0 \rangle$

(b) $\overrightarrow{P_0 P_1} = \langle 10, 11, -2 \rangle, \; \mathbf{u} = \langle 10, 11, -2 \rangle/15, \; D_{\mathbf{u}}f = 43/15$

30. $\nabla f(2,-1,2) = 2\langle 5,2,5\rangle, \|\nabla f(2,-1,2)\| = 6\sqrt{6}$, the maximum rate of decrease is $6\sqrt{6}$ in the direction of $-\langle 5,2,5\rangle$.

31.   **(a)** $\nabla f(1,-1) = 2(-\mathbf{i}+\mathbf{j})$, $D_{\mathbf{u}}f = 0$ if $\mathbf{u}$ is normal to ∇f so $\mathbf{u} = \pm(\mathbf{i}+\mathbf{j})/\sqrt{2}$

     (b) $\nabla f(-2,0) = \mathbf{i} - 2\mathbf{j}$, $\mathbf{u} = \pm(2\mathbf{i}+\mathbf{j})/\sqrt{5}$

32. $\nabla f(x_0, y_0) = a\mathbf{i}+b\mathbf{j}$. $D_{\mathbf{u}}f = 2$ when $\mathbf{u} = \left(\sqrt{3}\mathbf{i}+\mathbf{j}\right)/2$ and $D_{\mathbf{u}}f = 8$ when $\mathbf{u} = \left(-\sqrt{3}\mathbf{i}+\mathbf{j}\right)/2$ so $\left(\sqrt{3}a+b\right)/2 = 2$ and $\left(-\sqrt{3}a+b\right)/2 = 8$, solve for a and b to get $a = -2\sqrt{3}$, $b = 10$. If $\mathbf{u} = \left(\sqrt{3}\mathbf{i}+2\mathbf{j}\right)/\sqrt{7}$ then $D_{\mathbf{u}}f = \left(-2\sqrt{3}\mathbf{i}+10\mathbf{j}\right) \cdot \left(\sqrt{3}\mathbf{i}+2\mathbf{j}\right)/\sqrt{7} = 2\sqrt{7}$.

33. $\nabla f(1,2) = a\mathbf{i}+b\mathbf{j}$; $D_{\mathbf{u}}f = 2\sqrt{2}$ when $\mathbf{u} = (\mathbf{i}+\mathbf{j})/\sqrt{2}$ and $D_{\mathbf{u}}f = -3$ when $\mathbf{u} = -\mathbf{j}$ so $(a+b)/\sqrt{2} = 2\sqrt{2}$ and $-b = -3$, $a = 1$, $b = 3$. If $\mathbf{u} = -(\mathbf{i}+2\mathbf{j})/\sqrt{5}$ then $D_{\mathbf{u}}f = -7/\sqrt{5}$.

34.   **(a)** $\langle f_x(1,2), f_y(1,2), -1\rangle = \langle 8,4,-1\rangle = \mathbf{N}$ **(b)** $8x + 4y - z = 7$

35.   **(a)** $\langle f_x(4,-3), f_y(4,-3), -1\rangle = \langle 8/5, -6/5, -1\rangle$, let $\mathbf{N} = \langle 8,-6,-5\rangle$

     (b) $8x - 6y - 5z = 0$

36.   **(a)** $f(x,y,z) = z - x^2 e^{2y}$, $\nabla f(1, \ln 2, 4) = \langle -8,-8,1\rangle$, $\mathbf{n} = \langle 8,8,-1\rangle$; tangent plane
        $8x + 8y - z = 4 + 8\ln 2$; normal line $x = 1 + 8t$, $y = \ln 2 + 8t$, $z = 4 - t$.

     (b) $f(x,y,z) = x^2 y^3 z^4 + xyz$, $\nabla f(2,1,-1) = \langle 3,10,-14\rangle = \mathbf{n}$; tangent plane
        $3x + 10y - 14z = 30$; normal line $x = 2 + 3t$, $y = 1 + 10t$, $z = -1 - 14t$.

37.   Let $f(x,y,z) = z + xy$; $\nabla f(x_0, y_0, z_0) = \langle y_0, x_0, 1\rangle$ is normal to the surface at $P_0(x_0, y_0, z_0)$. The normal line passes through the origin when $\langle x_0, y_0, z_0\rangle$ and $\langle y_0, x_0, 1\rangle$ are parallel so $\langle x_0, y_0, z_0\rangle = k\langle y_0, x_0, 1\rangle = \langle ky_0, kx_0, k\rangle$ for some value of k. Equate the third component of these vectors to find that $k = z_0$ so $x_0 = y_0 z_0$ and $y_0 = x_0 z_0$, eliminate y_0 to get $x_0 = x_0 z_0^2$, $x_0\left(1 - z_0^2\right) = 0$, $x_0 = 0$ or $z_0 = \pm 1$. If $x_0 = 0$ then $y_0 = (0)z_0 = 0$ and, from the equation of the surface, $z_0 = 2 - (0)(0) = 2$ so $(0,0,2)$ is one of the points. If $z_0 = 1$ then $y_0 = x_0$ so $1 = 2 - x_0^2$, $x_0^2 = 1$, $x_0 = \pm 1$ so $(1,1,1)$ and $(-1,-1,1)$ are also points where the normal line passes through the origin. If $z_0 = -1$ then $y_0 = -x_0$ so $-1 = 2 + x_0^2$, $x_0^2 = -3$ which has no real solution.

38.   Let $P_0(x_0, y_0, z_0)$ be a point on the surface then if $x_0 \neq 0$, $y_0 \neq 0$, and $z_0 \neq 0$ the vector $\langle x_0^{-1/3}, y_0^{-1/3}, z_0^{-1/3}\rangle$ is normal to the surface at P_0 and the tangent plane is
$x_0^{-1/3}x + y_0^{-1/3}y + z_0^{-1/3}z = x_0^{2/3} + y_0^{2/3} + z_0^{2/3} = 1$, the x, y and z intercepts are $x_0^{1/3}$, $y_0^{1/3}$, $z_0^{1/3}$, the sum of the squares is $x_0^{2/3} + y_0^{2/3} + z_0^{2/3} = 1$ because (x_0, y_0, z_0) is on the surface.

39. $\langle 18x_0, 8y_0, -1\rangle$ is normal to the surface at a point (x_0, y_0, z_0). $\overrightarrow{PQ} = \langle -6, -4, -1\rangle$ so the normal line is parallel to $\overrightarrow{PQ}$ if $\langle 18x_0, 8y_0, -1\rangle = k\langle -6, -4, -1\rangle$ for some value of k. By inspection $k = 1$ so $18x_0 = -6$ and $8y_0 = -4$, $x_0 = -1/3$ and $y_0 = -1/2$ thus $z_0 = 2$. The only point is $(-1/3, -1/2, 2)$.

40. Let $f(x,y) = w$ then $\Delta w = f(1.1, -0.1) - f(1,0) = 0.11$,

$$dw = \left(2xy - 2y + y^2\right)dx + \left(x^2 - 2x + 2yx\right)dy = (0)(0.1) + (-1)(-0.1) = 0.1$$

41. $dV = (2/3)xhdx + (1/3)x^2dh = (2/3)(1)(2)(-0.1) + (1/3)(1)^2(0.2) = -0.2/3 \approx -0.067 \text{ m}^3$
$\Delta V = (1/3)(0.9)^2(2.2) - (1/3)(1)^2(2) = -0.218/3 \approx -0.073 \text{ m}^3$

42. $df = \left[2xy^4/\left(1+z^2\right)\right]dx + \left[4x^2y^3/\left(1+z^2\right)\right]dy - \left[2x^2y^4z/\left(1+z^2\right)^2\right]dz$
$\quad = (10/5)(-0.004) + (100/5)(0.003) - (100/25)(-0.005) = 0.072$

$f(4.996, 1.003, 1.995) \approx f(5,1,2) + df = 5 + 0.072 = 5.072$

43. $f_x = 2x + 3y - 6 = 0$, $f_y = 3x + 6y + 3 = 0$; critical point $(15, -8)$; $f_{xx}f_{yy} - f_{xy}^2 > 0$ and $f_{xx} > 0$ at $(15, -8)$, relative minimum.

44. $f_x = 2xy - 6x = 0$, $f_y = x^2 - 12y = 0$; critical points $(0,0)$ and $(\pm 6, 3)$; $D > 0$ and $f_{xx} < 0$ at $(0,0)$, relative maximum; $D < 0$ at $(\pm 6, 3)$, saddle points.

45. $f_x = 3x^2 - 3y = 0$, $f_y = -3x + y = 0$; critical points $(0,0)$ and $(3,9)$; $D < 0$ at $(0,0)$, saddle point; $D > 0$ and $f_{xx} > 0$ at $(3,9)$, relative minimum.

46. **(a)** $w = x^2y^2$; $y^2 = 8 - 4x^2$ so $w = 8x^2 - 4x^4$ for $-\sqrt{2} \le x \le \sqrt{2}$. $dw/dx = 16x\left(1 - x^2\right) = 0$ if $x = 0, \pm 1$. If $x = 0$ then $y = \pm 2\sqrt{2}$ and $d^2w/dx^2 > 0$ so relative minima occur at $(0, \pm 2\sqrt{2})$. If $x = -1$ or 1 then $y = \pm 2$ and $d^2w/dx^2 < 0$ so relative maxima occur at $(-1, \pm 2)$ and $(1, \pm 2)$. At the endpoints $x = \pm\sqrt{2}$ we find that $y = 0$ thus $w = \left(\pm\sqrt{2}\right)(0) = 0$ so relative minima occur at $(\pm\sqrt{2}, 0)$ because $w = x^2y^2 \ge 0$ everywhere.

(b) $2xy^2 = 8x\lambda$, $2x^2y = 2y\lambda$. If $x \ne 0$ then $\lambda = y^2/4$ and thus $2x^2y = y^3/2$, $4x^2y - y^3 = 0$, $y\left(4x^2 - y^2\right) = 0$, $y = 0$ or $y^2 = 4x^2$; if $y = 0$ then $4x^2 + (0)^2 = 8$ so $x = \pm\sqrt{2}$, if $y^2 = 4x^2$ then $4x^2 + 4x^2 = 8$ so $x = \pm 1$. If $x = 0$ then $4(0)^2 + y^2 = 8$ so $y = \pm 2\sqrt{2}$. Test $(\pm\sqrt{2}, 0)$, $(1, \pm 2)$, $(-1, \pm 2)$ and $(0, \pm 2\sqrt{2})$. $w = 0$ at $(\pm\sqrt{2}, 0)$ and $(0, \pm 2\sqrt{2})$, $w = 4$ at $(1, \pm 2)$ and $(-1, \pm 2)$. The maximum value occurs at $(1, \pm 2)$ and $(-1, \pm 2)$, the minimum value at $(\pm\sqrt{2}, 0)$ and $(0, \pm 2\sqrt{2})$.

47. **(a)** Let (x, y, z) be a point on the portion of the ellipsoid that is in the first octant then $V = (2x)(2y)(2z) = 8xyz$. For convenience introduce the new variables $u = x/a$, $v = y/b$, and $w = z/c$ so $V = (8abc)uvw$ where $u^2 + v^2 + w^2 = 1$. Also for convenience we will maximize $S = u^2v^2w^2$ instead of V. $w^2 = 1 - u^2 - v^2$ so $S = u^2v^2 - u^4v^2 - u^2v^4$, $S_u = 2uv^2(1 - 2u^2 - v^2) = 0$, $S_v = 2vu^2\left(1 - u^2 - 2v^2\right) = 0$; critical point $(1/\sqrt{3}, 1/\sqrt{3})$; $S_{uu}S_{vv} - S_{uv}^2 > 0$ and $S_{uu} < 0$ at this point so a relative maximum occurs there. If $u = v = 1/\sqrt{3}$ then $w = 1/\sqrt{3}$ so $x = a/\sqrt{3}$, $y = b/\sqrt{3}$, and $z = c/\sqrt{3}$. The dimensions of the box are $2a/\sqrt{3}$, $2b/\sqrt{3}$, and $2c/\sqrt{3}$.

(b) $f(x, y, z) = 8xyz$, $(x/a)^2 + (y/b)^2 + (z/c)^2 = 1$; $8yz = (2x/a^2)\lambda$, $8xz = (2y/b^2)\lambda$, $8xy = (2z/c^2)\lambda$; $4a^2yz/x = 4b^2xz/y = 4c^2xy/z$, $y^2/b^2 = x^2/a^2$ and $z^2/c^2 = x^2/a^2$ so $3\left(x^2/a^2\right) = 1$, $x = a/\sqrt{3}$ and therefore $y = b/\sqrt{3}$ and $z = c/\sqrt{3}$. The dimensions agree with those in part (a).

48. $f(x, y) = x^2 + y^2$; $2x = (10x - 6y)\lambda$; $2y = (-6x + 10y)\lambda$. If $10x - 6y \neq 0$ and $-6x + 10y \neq 0$ then $x/(5x - 3y) = y/(-3x + 5y)$, $y^2 = x^2$, $y = \pm x$; if $y = x$ then $5x^2 - 6x^2 + 5x^2 = 8$ so $x = \pm\sqrt{2}$, if $y = -x$ then $5x^2 + 6x^2 + 5x^2 = 8$ so $x = \pm 1/\sqrt{2}$. If $10x - 6y = 0$ or $-6x + 10y = 0$ then $x = y = 0$, which does not satisfy the equation of the curve. The test points are $(\sqrt{2}, \sqrt{2})$, $(-\sqrt{2}, -\sqrt{2})$, $(1/\sqrt{2}, -1/\sqrt{2})$, and $(-1/\sqrt{2}, 1/\sqrt{2})$. $f\left(\sqrt{2}, \sqrt{2}\right) = f\left(-\sqrt{2}, -\sqrt{2}\right) = 4$, $f\left(1/\sqrt{2}, -1/\sqrt{2}\right) = f\left(-1/\sqrt{2}, 1/\sqrt{2}\right) = 1$ so the distance from the origin is minimum at $\left(1/\sqrt{2}, -1/\sqrt{2}\right)$ and $\left(-1/\sqrt{2}, 1/\sqrt{2}\right)$, maximum at $\left(\sqrt{2}, \sqrt{2}\right)$ and $\left(-\sqrt{2}, -\sqrt{2}\right)$.

49. $f(I_1, I_2, I_3) = I_1^2 R_1 + I_2^2 R_2 + I_3^2 R_3$, $I_1 + I_2 + I_3 = I$. $2I_1 R_1 = \lambda$, $2I_2 R_2 = \lambda$, $2I_3 R_3 = \lambda$; $2I_1 R_1 = 2I_2 R_2 = 2I_3 R_3$, $I_1/I_2 = R_2/R_1 = R_1^{-1}/R_2^{-1}$ and $I_2/I_3 = R_2^{-1}/R_3^{-1}$ so $I_1 : I_2 : I_3 = R_1^{-1} : R_2^{-1} : R_3^{-1}$.

CHAPTER 17

Multiple Integrals

SECTION 17.1

17.1.1 Evaluate $\int_0^\pi \int_0^1 y \cos xy \, dx \, dy$.

17.1.2 Evaluate $\int_0^1 \int_0^1 (x^2 + y^2) dy \, dx$.

17.1.3 Evaluate $\int_0^3 \int_{-2}^0 \left(\frac{1}{2} x^2 y - xy \right) dy \, dx$.

17.1.4 Evaluate $\int_2^4 \int_0^3 (3 - y) x^2 dy \, dx$.

17.1.5 Evaluate $\int_0^1 \int_0^2 (x + 2) dy \, dx$.

17.1.6 Evaluate the double integral $\iint_R (2xy - x^2) dA$ where R is the rectangle bounded by $-1 \le x \le 2$ and $0 \le y \le 4$.

17.1.7 Evaluate $\int_1^4 \int_{-1}^2 (x + 3x^2 y) dy \, dx$.

17.1.8 Evaluate $\int_1^2 \int_0^1 y \, dy \, dx$.

17.1.9 Evaluate the double integral $\iint_R x^2 y \, dA$ where R is the rectangular region bounded by the lines $x = -1$, $x = 2$, $y = 0$, and $y = 2$.

17.1.10 Evaluate $\int_0^1 \int_0^1 e^{x+y} dy \, dx$.

17.1.11 Evaluate $\int_1^e \int_1^{\ln y} e^x dx \, dy$.

17.1.12 Find the volume under the surface $z = x\sqrt{x^2 + y}$ and over the rectangle $R = \{(x, y) : 0 \le x \le 1, 0 \le y \le 3\}$.

17.1.13 Find the volume under the plane $z = \frac{x}{2} + y$ and over the rectangle $R = \{(x, y) : 1 \le x \le 3, 2 \le y \le 7\}$.

17.1.14 Evaluate $\int_1^4 \int_1^3 (x^2 - y) dx \, dy$.

17.1.15 Evaluate $\int_0^{1/2} \int_0^2 \frac{1}{\sqrt{1 - x^2}} dy \, dx$.

17.1.16 Evaluate $\int_0^4 \int_0^1 \frac{1}{1 + y^2} dy \, dx$.

17.1.17 Evaluate $\int_0^2 \int_0^1 \frac{1}{\sqrt{4 - x^2}} dx \, dy$.

17.1.18 Evaluate $\int_0^{\pi/4} \int_0^1 x \cos y \, dx \, dy$.

SOLUTIONS

SECTION 17.1

17.1.1 $\quad \int_0^{\pi} \int_0^1 y \cos xy \, dx \, dy = \int_0^{\pi} \sin y \, dy = 2.$

17.1.2 $\quad \int_0^1 \int_0^1 (x^2 + y^2) dy \, dx = \int_0^1 \left(x^2 + \frac{1}{3} \right) dx = \frac{2}{3}.$

17.1.3 $\quad \int_0^3 \int_{-2}^0 (1/2x^2 y - xy) dy \, dx = \int_0^3 (2x - x^2) dx = 0.$

17.1.4 $\quad \int_2^4 \int_0^3 (3 - y)x^2 dy \, dx = \int_2^4 \frac{9}{2} x^2 dx = 84.$

17.1.5 $\quad \int_0^1 \int_0^2 (x + 2) dy \, dx = \int_0^1 2(x + 2) dx = 5.$

17.1.6 $\quad \int_{-1}^2 \int_0^4 (2xy - x^2) dy \, dx = \int_{-1}^2 (16x - 4x^2) dx = 12.$

17.1.7 $\quad \int_1^4 \int_{-1}^2 (x + 3x^2 y) dy \, dx = \int_1^4 (3x + \frac{9}{2} x^2) dx = 117.$

17.1.8 $\quad \int_1^2 \int_0^1 y \, dy \, dx = \int_1^2 \frac{1}{2} dx = \frac{1}{2}.$

17.1.9 $\quad \int_{-1}^2 \int_0^2 x^2 y \, dy \, dx = \int_{-1}^2 2x^2 dx = 6.$

17.1.10 $\quad \int_0^1 \int_0^1 e^{x+y} dy \, dx = \int_0^1 (e^{x+1} - e^x) dx = e^2 - 2e + 1.$

17.1.11 $\quad \int_1^e \int_1^{\ln y} e^x dx \, dy = \int_1^e (y - e) dy = -\frac{e^2}{2} + e - \frac{1}{2}.$

17.1.12 $\quad V = \int_0^3 \int_0^1 x\sqrt{x^2 + y} \, dx \, dy = \int_0^3 \frac{1}{3} [(1 + y)^{3/2} - y^{3/2}] dy = \frac{62}{15} - \frac{6\sqrt{3}}{5}.$

17.1.13 $\quad V = \int_1^3 \int_2^7 \left(\frac{x}{2} + y \right) dy \, dx = \int_1^3 \left(\frac{5x}{2} + \frac{45}{2} \right) dx = 55.$

17.1.14 $\displaystyle\int_1^4 \int_1^3 (x^2 - y)dx\,dy = \int_1^4 \left(\frac{26}{3} - 2y\right)dy = 11.$

17.1.15 $\displaystyle\int_0^{1/2} \int_0^2 \frac{1}{\sqrt{1-x^2}}dy\,dx = \int_0^{1/2} \frac{2}{\sqrt{1-x^2}}dx = 2\sin^{-1} x \Big]_0^{1/2} = \frac{\pi}{3}.$

17.1.16 $\displaystyle\int_0^4 \int_0^1 \frac{1}{1+y^2}dy\,dx = \int_0^4 \frac{\pi}{4}dx = \pi.$

17.1.17 $\displaystyle\int_0^2 \int_0^1 \frac{1}{\sqrt{4-x^2}}dx\,dy = \int_0^2 \frac{\pi}{6}dy = \frac{\pi}{3}.$

17.1.18 $\displaystyle\int_0^{\pi/4} \int_0^1 x\cos y\,dx\,dy = \int_0^{\pi/4} \frac{1}{2}\cos y\,dy = \frac{\sqrt{2}}{4}.$

SECTION 17.2

17.2.1 Evaluate $\int_0^1 \int_y^1 e^{x^2} \, dx \, dy$ by first sketching R then reversing the order of integration.

17.2.2 Evaluate $\int_0^1 \int_{x^2}^x (x^2 + y^2) \, dy \, dx$ by first sketching R then reversing the order of integration.

17.2.3 Evauate $\int_0^1 \int_0^{2\sqrt{1-y^2}} x \, dx \, dy$ by first sketching R then reversing the order of integration.

17.2.4 Evaluate $\int_1^2 \int_0^{\sqrt{x}} y \ln x^2 \, dy \, dx$.

17.2.5 Evaluate $\int_0^1 \int_{2y}^2 \cos(x^2) \, dx \, dy$ by expressing it as an equivalent double integral with order of integration reversed.

17.2.6 Evaluate $\int_0^1 \int_0^x y\sqrt{x^2 + y^2} \, dy \, dx$.

17.2.7 Sketch R and express $\int_0^{\pi/4} \int_{\sin x}^{\cos x} f(x,y) \, dy \, dx$ as an equivalent double integral with order of integration reversed.

17.2.8 Sketch R and express $\int_0^1 \int_{1-y}^{2-y} f(x,y) \, dx \, dy$ as an equivalent double integral with order of integration reversed.

17.2.9 Use a double integral to find the area enclosed by $y = x^2$ and $y = \sqrt{x}$.

17.2.10 Use a double integral to find the area enclosed by $x = y - y^2$ and $x + y = 0$.

17.2.11 Find the volume of the solid enclosed by $y = x^2 - x$, $y = x$, and $z = x + 1$.

17.2.12 Find the volume of the solid in the first octant enclosed by $y = x^2/4$, $z = 0$, $y = 4$, $x = 0$, and $x - y + 2z = 2$.

17.2.13 Find the volume of the solid enclosed by $x = 0$, $z = 0$, $z = 4 - x^2$, $y = 2x$, and $y = 4$.

17.2.14 Find the volume of the solid enclosed by $y = x^2 - x + 1$, $y = x + 1$, and $z = x + 1$.

17.2.15 Find the volume of the solid that is enclosed by $z = x^2 + y^2$, $y = 2x$, $y = x^2$, and $z = 0$.

17.2.16 Find the volume of the solid in the first octant enclosed by $z = 4 - y^2$, $z = 0$, $x = 0$, $y = x$, and $y = 2$.

17.2.17 Find the volume of the solid in the first octant enclosed by $x^2 + y^2 = 4$, $y = z$, and $z = 0$.

17.2.18 Find the volume enclosed by $x^2 + y^2 = 1$ and $y^2 + z^2 = 1$.

SOLUTIONS

SECTION 17.2

17.2.1 $\displaystyle\int_0^1 \int_y^1 e^{x^2}\, dx\, dy$

$$= \int_0^1 \int_0^x e^{x^2}\, dy\, dx$$

$$= \int_0^1 x e^{x^2}\, dx = \frac{1}{2}(e-1).$$

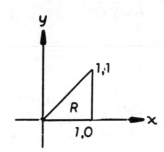

17.2.2 $\displaystyle\int_0^1 \int_{x^2}^x (x^2+y^2)\, dy\, dx$

$$= \int_0^1 \int_y^{\sqrt{y}} (x^2+y^2)\, dx\, dy$$

$$= \int_0^1 \left(\frac{y^{3/2}}{3} + y^{5/2} - \frac{4y^3}{3} \right) dy$$

$$= \frac{3}{35}.$$

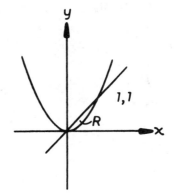

17.2.3 $\displaystyle\int_0^1 \int_0^{2\sqrt{1-y^2}} x\, dx\, dy$

$$= \int_0^2 \int_0^{\sqrt{1-\frac{x^2}{4}}} x\, dy\, dx$$

$$= \int_0^2 x\sqrt{1 - \frac{x^2}{4}}\, dx = \frac{4}{3}.$$

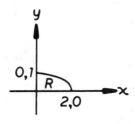

17.2.4 $\displaystyle\int_1^2 \int_0^{\sqrt{x}} y \ln x^2\, dx = \int_1^2 x \ln x\, dx = 2\ln 2 - \frac{3}{4}.$

17.2.5 $\displaystyle\int_0^1 \int_{2y}^2 \cos(x^2)\, dx\, dy = \int_0^2 \int_0^{x/2} \cos(x^2)\, dy\, dx = \frac{1}{2}\int_0^2 x\cos(x^2)\, dx = \frac{1}{4}\sin 4.$

17.2.6 $\displaystyle\int_0^1\int_0^x y\sqrt{x^2+y^2}\,dy\,dx = \int_0^1 \frac{2\sqrt{2}-1}{3}x^3dx = \frac{2\sqrt{2}-1}{12}.$

17.2.7 $\displaystyle\int_0^{\pi/4}\int_{\sin x}^{\cos x} f(x,y)dy\,dx$

$\displaystyle = \int_0^{\frac{\sqrt{2}}{2}}\int_0^{\sin^{-1}y} f(x,y)dx\,dy$

$\displaystyle + \int_{\sqrt{2}/2}^1\int_0^{\cos^{-1}y} f(x,y)dx\,dy.$

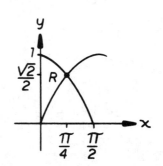

17.2.8 $\displaystyle\int_0^1\int_{1-y}^{2-y} f(x,y)dx\,dy$

$\displaystyle = \int_0^1\int_{1-x}^1 f(x,y)dy\,dx$

$\displaystyle + \int_1^2\int_0^{2-x} f(x,y)dy\,dx.$

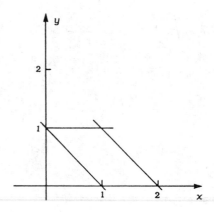

17.2.9 $\displaystyle A = \int_0^1\int_{x^2}^{\sqrt{x}} dy\,dx = \int_0^1 (\sqrt{x}-x^2)dx = \frac{1}{3}.$

17.2.10 $\displaystyle A = \int_0^2\int_{-y}^{y-y^2} dx\,dy = \int_0^2 (2y-y^2)dy = \frac{4}{3}.$

17.2.11 $\displaystyle V = \int_0^2\int_{x^2-x}^{x} (x+1)dy\,dx = \int_0^2 (2x+x^2-x^3)dx = \frac{8}{3}.$

17.2.12 $\displaystyle V = \int_0^4\int_0^{\sqrt{4y}} \frac{1}{2}(2-x+y)dx\,dy = \frac{1}{2}\int_0^4 (2y^{3/2}-2y+4y^{1/2})dy = \frac{232}{15}.$

17.2.13 $\displaystyle V = \int_0^4\int_0^{y/2} (4-x^2)dx\,dy = \int_0^4 \left(2y-\frac{y^3}{24}\right)dy = \frac{40}{3}.$

17.2.14 $\quad V = \displaystyle\int_0^2 \int_{x^2-x+1}^{x+1} (x+1)dy\,dx = \int_0^2 (2x + x^2 - x^3)dx = \dfrac{8}{3}.$

17.2.15 $\quad V = \displaystyle\int_0^2 \int_{x^2}^{2x} (x^2 + y^2)dy\,dx = \int_0^2 \left(\dfrac{14}{3}x^3 - x^4 - \dfrac{1}{3}x^6\right)dx = \dfrac{216}{35}.$

17.2.16 $\quad V = \displaystyle\int_0^2 \int_0^y (4 - y^2)dx\,dy = \int_0^2 (4y - y^3)dy = 4.$

17.2.17 $\quad V = \displaystyle\int_0^2 \int_0^{\sqrt{4-x^2}} y\,dy\,dx = \dfrac{1}{2}\int_0^2 (4 - x^2)dx = \dfrac{8}{3}.$

17.2.18 $\quad V = 8\displaystyle\int_0^1 \int_0^{\sqrt{1-y^2}} \sqrt{1-y^2}\,dx\,dy = 8\int_0^1 (1 - y^2)dy = \dfrac{16}{3}.$

SECTION 17.3

17.3.1   Use a double integral in polar coordinates to find the volume in the first octant enclosed by $x = 0$, $y = 0$, $z = 0$, the plane $z + y = 3$, and the cylinder $x^2 + y^2 = 4$.

17.3.2   Use polar coordinates to evaluate $\iint\limits_{R} 2(x + y)dA$ where R is the region enclosed by $x^2 + y^2 = 9$ and $x \geq 0$.

17.3.3   Use a double integral in polar coordinates to find the volume in the first octant of the solid enclosed by $x^2 + y^2 = 4$, $y = z$, and $z = 0$.

17.3.4   Use a double integral in polar coordinates to find the volume of the solid enclosed by $x^2 + y^2 = 10 - z$ and $z = 1$.

17.3.5   Use a double integral in polar coordinates to find the volume of the solid enclosed by the sphere $x^2 + y^2 + z^2 = 9$ and the cylinder $x^2 + y^2 = 1$.

17.3.6   Use a double integral in polar coordinates to find the volume of the solid enclosed by the paraboloid $z = 4 - x^2 - y^2$ and $z = 0$.

17.3.7   Evaluate $\displaystyle\int_{-2}^{2} \int_{0}^{\sqrt{4-x^2}} e^{-(x^2+y^2)}dy\,dx$ by converting to an equivalent integral in polar coordinates. Sketch R.

17.3.8   Evaluate $\displaystyle\int_{0}^{3} \int_{-\sqrt{9-y^2}}^{\sqrt{9-y^2}} \frac{1}{\sqrt{x^2 + y^2}}dx\,dy$ by converting to an equivalent integral in polar coordinates. Sketch R.

17.3.9   Evaluate $\displaystyle\int_{-2}^{2} \int_{0}^{\sqrt{4-x^2}} y\,dy\,dx$ by converting to an equivalent integral in polar coordinates.

17.3.10   Evaluate $\displaystyle\int_{0}^{2} \int_{0}^{\sqrt{4-x^2}} (x^2 + y^2)dy\,dx$ by converting to an equivalent double integral in polar coordinates. Sketch R.

17.3.11   Use a double integral in polar coordinates to find the volume of the solid in the first octant enclosed by the ellipsoid $9x^2 + 9y^2 + 4z^2 = 36$ and the planes $x = \sqrt{3}y$, $x = 0$, and $z = 0$.

17.3.12   Use a double integral in polar coordinates to find the volume enclosed by the sphere $x^2 + y^2 + z^2 = 16$ and the cylinder $(x - 2)^2 + y^2 = 4$.

17.3.13 Use a double integral in polar coordinates to find the volume enclosed by $z = 0$, $x + 2y - z = -4$, and the cylinder $x^2 + y^2 = 1$.

17.3.14 Use a double integral in polar coordinates to find the volume that is inside the sphere $x^2 + y^2 + z^2 = 9$, outside the cylinder $x^2 + y^2 = 4$ and above $z = 0$.

17.3.15 Use a double integral in polar coordinates to find the area enclosed by the limacon $r = 6 + \sin\theta$.

17.3.16 Use a double integral in polar coordinates to find the area that is inside $r = 1 + \cos\theta$ and outside $r = 1$.

17.3.17 Use a double integral in polar coordinates to find the area enclosed by $r = 3\sin 3\theta$.

SOLUTIONS

SECTION 17.3

17.3.1 $V = \int_0^{\pi/2} \int_0^2 (3 - r\sin\theta)r\,dr\,d\theta = \int_0^{\pi/2} \left(6 - \frac{8}{3}\sin\theta\right)d\theta = 3\pi - \frac{8}{3}.$

17.3.2 $\int_{-\pi/2}^{\pi/2} \int_0^3 2(r\cos\theta + r\sin\theta)r\,dr\,d\theta = 18\int_{-\pi/2}^{\pi/2} (\cos\theta + \sin\theta)d\theta = 36.$

17.3.3 $V = \int_0^{\pi/2} \int_0^2 (r\sin\theta)r\,dr\,d\theta = \int_0^{\pi/2} \frac{8}{3}\sin\theta\,d\theta = \frac{8}{3}.$

17.3.4 $V = \int_0^{2\pi} \int_0^3 [(10 - r^2) - 1]r\,dr\,d\theta = \int_0^{2\pi} \frac{81}{4}d\theta = \frac{81}{2}\pi.$

17.3.5 $2\int_0^{2\pi} \int_0^1 \sqrt{9 - r^2}\,r\,dr\,d\theta = \frac{2}{3}(27 - 8\sqrt{8})\int_0^{2\pi} d\theta = \frac{4\pi}{3}(27 - 16\sqrt{2}).$

17.3.6 $V = \int_0^{2\pi} \int_0^2 (4 - r^2)r\,dr\,d\theta = \int_0^{2\pi} 4d\theta = 8\pi.$

17.3.7 $\int_0^{\pi} \int_0^2 e^{-r^2}r\,dr\,d\theta$

$= \frac{1}{2}(1 - e^{-4})\int_0^{\pi} d\theta$

$= \frac{\pi}{2}(1 - e^{-4}).$

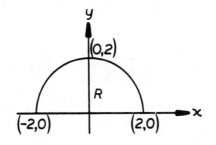

17.3.8 $\int_0^{\pi} \int_0^3 \frac{1}{r} \cdot r\,dr\,d\theta$

$= \int_0^{\pi} 3d\theta = 3\pi.$

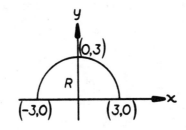

17.3.9 $\quad \displaystyle\int_{-2}^{2}\int_{0}^{\sqrt{4-x^2}} y\,dy\,dx = \int_{0}^{\pi}\int_{0}^{2}(r\sin\theta)r\,dr\,d\theta = \frac{8}{3}\int_{0}^{\pi}\sin\theta\,d\theta = \frac{16}{3}.$

17.3.10 $\quad \displaystyle\int_{0}^{2}\int_{0}^{\sqrt{4-x^2}}(x^2+y^2)dy\,dx$

$$= \int_{0}^{\pi/2}\int_{0}^{2}(r^2)r\,dr\,d\theta$$

$$= \int_{0}^{\pi/2} 4d\theta = 2\pi.$$

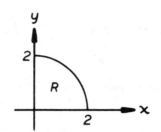

17.3.11 $\quad V = \displaystyle\int_{0}^{\pi/3}\int_{0}^{2}\frac{\sqrt{36-9r^2}}{2}r\,dr\,d\theta = \int_{0}^{\pi/3} 4d\theta = \frac{4\pi}{3}.$

17.3.12 $\quad V = 4\displaystyle\int_{0}^{\pi/2}\int_{0}^{4\cos\theta}\sqrt{16-r^2}\,r\,dr\,d\theta = \frac{256}{3}\int_{0}^{\pi/2}(1-\sin^3\theta)d\theta = \frac{128}{9}(3\pi-4).$

17.3.13 $\quad V = \displaystyle\int_{0}^{2\pi}\int_{0}^{1}(4+r\cos\theta+2r\sin\theta)r\,dr\,d\theta = \int_{0}^{2\pi}\left(2+\frac{1}{3}\cos\theta+\frac{2}{3}\sin\theta\right)d\theta = 4\pi.$

17.3.14 $\quad V = \displaystyle\int_{0}^{2\pi}\int_{2}^{3}\sqrt{9-r^2}\,r\,dr\,d\theta = \int_{0}^{2\pi}\frac{5\sqrt{5}}{3}d\theta = \frac{10\sqrt{5}}{3}\pi.$

17.3.15 $\quad A = \displaystyle\int_{0}^{2\pi}\int_{0}^{6+\sin\theta} r\,dr\,d\theta = \frac{1}{2}\int_{0}^{2\pi}(36+12\sin\theta+\sin^2\theta)d\theta = 73\pi.$

17.3.16 $\quad A = \displaystyle\int_{-\pi/2}^{\pi/2}\int_{1}^{1+\cos\theta} r\,dr\,d\theta = \frac{1}{2}\int_{-\pi/2}^{\pi/2}(\cos^2\theta+2\cos\theta)d\theta = 2+\frac{\pi}{4}.$

17.3.17 $\quad A = 3\displaystyle\int_{0}^{\pi/3}\int_{0}^{3\sin 3\theta} r\,dr\,d\theta = \frac{27}{2}\int_{0}^{\pi/3}\sin^2 3\theta\,d\theta = \frac{9\pi}{4}.$

SECTION 17.4

17.4.1 Find the surface area cut from the plane $z = 4x + 3$ by the cylinder $x^2 + y^2 = 25$.

17.4.2 Find the surface area of that portion of the paraboloid $z = x^2 + y^2$ which lies below the plane $z = 1$.

17.4.3 Find the surface area cut from the plane $2x - y - z = 0$ by the cylinder $x^2 + y^2 = 4$.

17.4.4 Find the surface area of that portion of the plane $3x + 4y + 6z = 12$ that lies in the first octant.

17.4.5 Find the surface area of that portion of the paraboloid $z = 25 - x^2 - y^2$ for which $z \geq 0$.

17.4.6 Find the surface area of that portion of the sphere $x^2 + y^2 + z^2 = 4$ that lies inside the cylinder $x^2 + y^2 = 2x$ and above the xy-plane.

17.4.7 Find the surface area of that portion of the paraboloid $z = 25 - x^2 - y^2$ that lies inside the cylinder $x^2 + y^2 = 9$ and above the xy-plane.

17.4.8 Find the surface area of the surface $z = \dfrac{1}{a}(y^2 - x^2)$ cut by the cylinder $x^2 + y^2 = a^2$ that lies above the xy-plane.

17.4.9 Find the surface area of the portion of the cone $z^2 = x^2 + y^2$ that is above the region in the first quadrant bounded by $y = x$ and the parabola $y = x^2$.

17.4.10 Find the surface area of that portion of the cylinder $x^2 + z^2 = 25$ that lies inside the cylinder $x^2 + y^2 = 25$.

17.4.11 Find the surface area of the portion of the sphere $x^2 + y^2 + z^2 = 18$ that is cut out by the cone $z = \sqrt{x^2 + y^2}$.

17.4.12 Find the surface area of that portion of the plane $z = x + y$ in the first octant which lies inside the cylinder $4x^2 + 9y^2 = 36$.

17.4.13 Find the surface area of that portion of the cylinder $y^2 + z^2 = 4$ which lies above the region in the xy-plane enclosed by the lines $y = \sqrt{3} - x, x = 0, y = 0$.

17.4.14 Find the surface area of that portion of the cylinder $z = y^2$ which lies above the triangular region with vertices at $(0,0,0)$, $(0,1,0)$, and $(1,1,0)$.

17.4.15 Find the surface area of that portion of the cylinder $x^2 = 1 - z$ which lies above the triangular region with vertices at $(0,0,0)$, $(1,0,0)$, and $(1,1,0)$.

17.4.16   Find the surface area of that portion of the sphere $x^2 + y^2 + z^2 = 4$ that lies inside the cylinder $x^2 + y^2 = 2y$.

17.4.17   Find the surface area of that portion of the sphere $x^2 + y^2 + z^2 = 4$ that lies in the first octant between the planes $y = 0$ and $y = x$.

17.4.18   Find the surface area of that portion of the sphere $x^2 + y^2 + z^2 = 4$ that lies in the first octant between the planes $y = 0$ and $y = \sqrt{3}x$.

17.4.19   Find the surface area of the sphere $r(u, v) = 3 \sin u \cos v i + 3 \sin u \sin v j + 3 \cos u k$ for which $0 \le u \le \pi/2, 0 \le v \le 2\pi$.

17.4.20   Find the surface area of the portion of the paraboloid $r(u, v) = u \cos v i + u \sin v j + u^2 k$ for which $0 \le u \le 1, 0 \le v \le \pi$.

SOLUTIONS

SECTION 17.4

17.4.1 $\dfrac{\partial z}{\partial x} = 4$, $\dfrac{\partial z}{\partial y} = 0$,

$$S = \iint\limits_{R} \sqrt{17}\, dA = \int_0^{2\pi} \int_0^5 \sqrt{17}\, r\, dr\, d\theta = \frac{25\sqrt{17}}{2} \int_0^{2\pi} d\theta = 25\sqrt{17}\pi.$$

17.4.2 $\dfrac{\partial z}{\partial x} = 2x$, $\dfrac{\partial z}{\partial y} = 2y$,

$$S = \iint\limits_{R} \sqrt{4x^2 + 4y^2 + 1}\, dA = \int_0^{2\pi} \int_0^1 \sqrt{4r^2 + 1}\, r\, dr\, d\theta$$

$$= \frac{1}{12}(5\sqrt{5} - 1) \int_0^{2\pi} d\theta = \frac{\pi}{6}(5\sqrt{5} - 1).$$

17.4.3 $\dfrac{\partial z}{\partial x} = 2$, $\dfrac{\partial z}{\partial y} = -1$,

$$S = \iint\limits_{R} \sqrt{6}\, dA = \int_0^{2\pi} \int_0^2 \sqrt{6}\, r\, dr\, d\theta = 2\sqrt{6} \int_0^{2\pi} d\theta = 4\sqrt{6}\,\pi.$$

17.4.4 $\dfrac{\partial z}{\partial x} = -\dfrac{1}{2}$, $\dfrac{\partial z}{\partial y} = -\dfrac{2}{3}$,

$$S = \iint\limits_{R} \sqrt{\frac{1}{4} + \frac{4}{9} + 1}\, dA = \int_0^4 \int_0^{\frac{12-3x}{4}} \sqrt{\frac{1}{4} + \frac{4}{9} + 1}\, dy\, dx$$

$$= \frac{\sqrt{61}}{6} \int_0^4 \left(\frac{12 - 3x}{4} \right) dx = \sqrt{61}.$$

17.4.5 $\dfrac{\partial z}{\partial x} = -2x$, $\dfrac{\partial z}{\partial y} = -2y$,

$$S = \iint\limits_{R} \sqrt{4x^2 + 4y^2 + 1}\, dA = \int_0^{2\pi} \int_0^5 \sqrt{4r^2 + 1}\, r\, dr\, d\theta$$

$$= \frac{1}{12}(101\sqrt{101} - 1) \int_0^{2\pi} d\theta = \frac{\pi}{6}(101\sqrt{101} - 1).$$

17.4.6 $\dfrac{\partial z}{\partial x} = -\dfrac{x}{z}$, $\dfrac{\partial z}{\partial y} = -\dfrac{y}{z}$, $\sqrt{\dfrac{x^2}{z^2} + \dfrac{y^2}{z^2} + 1} = \sqrt{\dfrac{4}{4 - x^2 - y^2}}$,

$$S = \iint\limits_{R} \sqrt{\frac{4}{4 - x^2 - y^2}}\, dA = \int_{-\pi/2}^{\pi/2} \int_0^{2\cos\theta} \frac{2}{\sqrt{4 - r^2}}\, r\, dr\, d\theta = \int_{-\pi/2}^{\pi/2} 4(1 - \sin\theta)d\theta = 4\pi.$$

17.4.7 $\dfrac{\partial z}{\partial x} = -2x,\ \dfrac{\partial z}{\partial y} = -2y,$

$$S = \iint\limits_{R} \sqrt{4x^2 + 4y^2 + 1}\, dA = \int_0^{2\pi}\int_0^3 \sqrt{4r^2+1}\, r\, dr\, d\theta$$

$$= \frac{37\sqrt{37}-1}{12}\int_0^{2\pi} d\theta = \frac{\pi}{6}(37\sqrt{37}-1).$$

17.4.8 $\dfrac{\partial z}{\partial x} = -\dfrac{2x}{a},\ \dfrac{\partial z}{\partial y} = \dfrac{2y}{a},$

$$S = \iint\limits_{R} \sqrt{\frac{4x^2}{a^2} + \frac{4y^2}{a^2} + 1}\, dA = \int_0^{2\pi}\int_0^a \sqrt{\frac{4r^2}{a^2}+1}\, r\, dr\, d\theta$$

$$= \frac{a^2}{12}(5\sqrt{5}-1)\int_0^{2\pi} d\theta = \frac{\pi a^2}{6}(5\sqrt{5}-1).$$

17.4.9 $\dfrac{\partial z}{\partial x} = \dfrac{x}{z},\ \dfrac{\partial z}{\partial y} = \dfrac{x}{z}$

$$S = \iint\limits_{R} \sqrt{\frac{x^2}{z^2} + \frac{y^2}{z^2} + 1}\, dA = \int_0^1\int_{x^2}^x \sqrt{2}\, dy\, dx$$

$$= \sqrt{2}\int_0^1 (x - x^2)dx = \frac{\sqrt{2}}{6}.$$

17.4.10 $\dfrac{\partial z}{\partial x} = -\dfrac{x}{z},\ \dfrac{\partial z}{\partial y} = 0,\ \sqrt{\dfrac{x^2}{z^2}+1} = \sqrt{\dfrac{25}{25-x^2}},\ S = \iint\limits_{R}\sqrt{\dfrac{25}{25-x^2}}\, dA$

$$\int_{-5}^{5}\int_{-\sqrt{25-x^2}}^{\sqrt{25-x^2}}\sqrt{\frac{25}{25-x^2}}\, dy\, dx = \int_{-5}^5 10dx = 100.$$

17.4.11 $\dfrac{\partial z}{\partial x} = -\dfrac{x}{z},\ \dfrac{\partial z}{\partial y} = -\dfrac{y}{z}$

$$S = \iint\limits_{R} \sqrt{\frac{x^2}{z^2} + \frac{y^2}{z^2} + 1}\, dA = \iint\limits_{R}\sqrt{\frac{18}{18 - x^2 - y^2}}\, dA$$

$$= \int_0^{2\pi}\int_0^3 \sqrt{\frac{18}{18-r^2}}\, r\, dr\, d\theta = \left(18 - 9\sqrt{2}\right)\int_0^{2\pi} d\theta$$

$$= 18(2 - \sqrt{2})\pi.$$

17.4.12 $\dfrac{\partial z}{\partial x} = 1,\ \dfrac{\partial z}{\partial y} = 1,$

$$S = \iint\limits_{R} \sqrt{3}\, dA = \int_0^3 \int_0^{\sqrt{\frac{36-4x^2}{9}}} \sqrt{3}\, dy\, dx = \sqrt{3}\int_0^3 \sqrt{\frac{36-4x^2}{9}}\, dx = \frac{3\sqrt{3}\pi}{2}.$$

17.4.13 $\dfrac{\partial z}{\partial x} = 0,\ \dfrac{\partial z}{\partial y} = -\dfrac{y}{z},\ \sqrt{\dfrac{y^2}{z^2}+1} = \sqrt{\dfrac{4}{4-y^2}},$

$$S = \iint\limits_{R} \sqrt{\frac{4}{4-y^2}}\, dA = \int_0^{\sqrt{3}} \int_0^{\sqrt{3}-y} \sqrt{\frac{4}{4-y^2}}\, dx\, dy$$

$$= 2\int_0^{\sqrt{3}} \left(\frac{\sqrt{3}}{\sqrt{4-y^2}} - \frac{y}{\sqrt{4-y^2}}\right) dy$$

$$= 2\left(\frac{\sqrt{3}\pi}{3} - 1\right).$$

17.4.14 $\dfrac{\partial z}{\partial x} = 0,\ \dfrac{\partial z}{\partial y} = 2y,$

$$S = \iint\limits_{R} \sqrt{4y^2+1}\, dA = \int_0^1 \int_0^y \sqrt{4y^2+1}\, dy = \int_0^1 y\sqrt{4y^2+1}\, dy = \frac{1}{12}(5\sqrt{5}-1).$$

17.4.15 $\dfrac{\partial z}{\partial x} = -2x,\ \dfrac{\partial z}{\partial y} = 0,$

$$S = \iint\limits_{R} \sqrt{4x^2+1}\, dA = \int_0^1 \int_0^x \sqrt{4x^2+1}\, dy\, dx = \int_0^1 x\sqrt{4x^2+1}\, dx = \frac{1}{12}(5\sqrt{5}-1).$$

17.4.16 $\dfrac{\partial z}{\partial x} = -\dfrac{x}{z},\ \dfrac{\partial z}{\partial y} = -\dfrac{y}{z},\ \sqrt{\dfrac{x^2}{z^2}+\dfrac{y^2}{z^2}+1} = \sqrt{\dfrac{4}{4-x^2-y^2}},$

$$S = \iint\limits_{R} \sqrt{\frac{4}{4-x^2-y^2}}\, dA = 4\int_0^{\pi/2} \int_0^{2\sin\theta} \sqrt{\frac{4}{4-r^2}}\, r\, dr\, d\theta$$

$$= 8\int_0^{\pi/2} (2 - 2\cos\theta)d\theta = 8\pi - 16.$$

17.4.17 $\dfrac{\partial z}{\partial x} = -\dfrac{x}{z},\ \dfrac{\partial z}{\partial y} = -\dfrac{y}{z},\ \sqrt{\dfrac{x^2}{z^2}+\dfrac{y^2}{z^2}+1} = \sqrt{\dfrac{4}{4-x^2-y^2}},$

$$S = \iint\limits_{R} \sqrt{\frac{4}{4-x^2-y^2}}\, dA = \int_0^{\pi/4} \int_0^2 \sqrt{\frac{4}{4-r^2}}\, r\, dr\, d\theta = 4\int_0^{\pi/4} d\theta = \pi.$$

17.4.18 $\dfrac{\partial z}{\partial x} = -\dfrac{x}{z}, \dfrac{\partial z}{\partial y} = -\dfrac{y}{z}, \sqrt{\dfrac{x^2}{z^2} + \dfrac{y^2}{z^2} + 1} = \sqrt{\dfrac{4}{4 - x^2 - y^2}},$

$$S = \iint\limits_{R} \sqrt{\dfrac{4}{4 - x^2 - y^2}}\, dA = \int_0^{\pi/3} \int_0^2 \sqrt{\dfrac{4}{4 - r^2}}\, r\, dr\, d\theta = 4 \int_0^{\pi/3} d\theta = \dfrac{4\pi}{3}.$$

17.4.19 $\dfrac{\partial r}{\partial u} = 3\cos u \cos v\, i + 3\cos u \sin v\, j - 3\sin u\, k$

$\dfrac{\partial r}{\partial v} = -3\sin u \sin v\, i + 3\sin u \cos v\, j$

$$\dfrac{\partial r}{\partial u} \times \dfrac{\partial r}{\partial v} = \begin{vmatrix} i & j & k \\ 3\cos u \cos v & 3\cos u \sin v & -3\sin u \\ -3\sin u \sin v & 3\sin u \cos v & 0 \end{vmatrix}$$

$$S = \iint\limits_{R} \left\| \dfrac{\partial r}{\partial u} \times \dfrac{\partial r}{\partial v} \right\| dA = \int_0^{2\pi} \int_0^{\pi/2} 9\sin u\, du\, dv$$

$$= \int_0^{2\pi} 9\, dv = 18\pi$$

17.4.20 $\dfrac{\partial r}{\partial u} = \cos v\, i + \sin v\, j + 2u\, k$

$\dfrac{\partial r}{\partial v} = -u\sin v\, i + u\cos v\, j$

$$\dfrac{\partial r}{\partial u} \times \dfrac{\partial r}{\partial v} = \begin{vmatrix} i & j & k \\ \cos v & \sin v & 2u \\ -u\sin v & u\cos v & 0 \end{vmatrix}$$

$$S = \iint\limits_{R} \left\| \dfrac{\partial r}{\partial u} \times \dfrac{\partial r}{\partial v} \right\| dA = \int_0^{\pi} \int_0^1 u\sqrt{4u^2 + 1}\, du\, dv$$

$$= \dfrac{5^{3/2} - 1}{12}\pi$$

17.5.1 Use a triple integral to find the volume of the solid enclosed by $z = 0$, $y = x^2 - x$, $y = x$, and $z = x + 1$.

17.5.2 Use a triple integral to find the volume of the solid enclosed by $x^2 = 4y$, $y + z = 1$, and $z = 0$.

17.5.3 Use a triple integral to find the volume of the solid enclosed by $y^2 = 4x$, $z = 0$, $z = x$, and $x = 4$.

17.5.4 Use a triple integral to find the volume of the tetrahedron enclosed by $2x + 2y + z = 6$ and the coordinate planes.

17.5.5 Use a triple integral to find the volume of the solid in the first octant enclosed by $z = y$, $y^2 = x$, and $x = 1$.

17.5.6 Use a triple integral to find the volume of the solid in the first octant enclosed by the cylinder $x = 4 - y^2$ and the planes $z = y$, $x = 0$, and $z = 0$.

17.5.7 Use a triple integral to find the volume of the solid in the first octant enclosed by $z = x^2 + y^2$, $y = x$, and $x = 1$.

17.5.8 Use a triple integral to find the volume of the solid in the first octant enclosed by the cylinder $z = 4 - y^2$ and the planes $y = x$, $z = 0$, $x = 0$, and $y = 2$.

17.5.9 Use a triple integral to find the volume of the tetrahedron enclosed by the plane $3x + 6y + 4z = 12$ and the coordinate planes.

17.5.10 Use a triple integral to find the volume of the solid whose base is the region in the xy-plane enclosed by $y = x^2 - x + 1$, $y = x + 1$, and $z = x + 1$.

17.5.11 Use a triple integral to find the volume of the solid enclosed by $z = x^2 + y^2$, $y = x^2$, $z = 0$, and $y = x$.

17.5.12 Use a triple integral to find the volume of the solid enclosed by $y = x^2$, $x = y^2$, $z = 0$, and $z = 3$.

17.5.13 Use a triple integral to find the volume of the solid enclosed by $z = \dfrac{4}{y^2 + 1}$, $z = 0$, $y = x$, $y = 3$, and $x = 0$.

17.5.14 Evaluate $\iiint\limits_{G} x\, dv$ where G is the solid in the first octant enclosed by $x + y + z = 3$ and the coordinate planes.

17.5.15 Evaluate $\displaystyle\int_0^1 \int_0^z \int_0^{\sqrt{yz}} x\, dx\, dy\, dz$.

17.5.16 Use a triple integral to find the volume of the solid enclosed by $z = 0$, $y = 4 - x^2$, $y = 3x$, and $z = x + 4$.

17.5.17 Evaluate $\displaystyle\iiint_G y\, z\, dv$ where G is the solid in the first octant enclosed by $y = 0$, $y = \sqrt{1 - x^2}$, and $z = x$.

17.5.18 Evaluate $\displaystyle\iiint_G y\, dv$ where G is the solid in the first octant enclosed by $y = 1$, $y = x$, $z = x+1$, and the coordinate planes.

SOLUTIONS

SECTION 17.5

17.5.1 $v = \int_0^2 \int_{x^2-x}^{x} \int_0^{x+1} dz\, dy\, dx = \int_0^2 \int_{x^2-x}^{x} (x+1) dy\, dx = \int_0^2 (2x + x^2 - x^3) dx = \frac{8}{3}$.

17.5.2 $v = \int_{-2}^2 \int_{\frac{x^2}{4}}^1 \int_0^{1-y} dz\, dy\, dx = \int_{-2}^2 \int_{\frac{x^2}{4}}^1 (1-y) dy\, dx = \int_{-2}^2 \left(\frac{1}{2} - \frac{x^2}{4} + \frac{x^4}{32} \right) dx = \frac{16}{15}$.

17.5.3 $v = \int_{-4}^4 \int_{\frac{y^2}{4}}^4 \int_0^x dz\, dx\, dy = \int_{-4}^4 \int_{\frac{y^2}{4}}^4 x\, dx\, dy = \int_{-4}^4 \frac{1}{2} \left(16 - \frac{y^4}{16} \right) dy = \frac{256}{5}$.

17.5.4 $v = \int_0^3 \int_0^{3-y} \int_0^{6-2x-2y} dz\, dx\, dy = \int_0^3 \int_0^{3-y} (6 - 2x - 2y) dx\, dy = \int_0^3 (3-y)^2 dy = 9$.

17.5.5 $v = \int_0^1 \int_0^{\sqrt{x}} \int_0^y dz\, dy\, dx = \int_0^1 \int_0^{\sqrt{x}} y\, dy\, dx = \int_0^1 \frac{x}{2} dx = \frac{1}{4}$.

17.5.6 $v = \int_0^4 \int_0^{\sqrt{4-x}} \int_0^y dz\, dy\, dx = \int_0^4 \int_0^{\sqrt{4-x}} y\, dy\, dx = \int_0^4 \left(\frac{4-x}{2} \right) dx = 4$.

17.5.7 $v = \int_0^1 \int_0^x \int_0^{x^2+y^2} dz\, dy\, dx = \int_0^1 \int_0^x (x^2 + y^2) dy\, dx = \int_0^1 \frac{4x^3}{3} dx = \frac{1}{3}$.

17.5.8 $v = \int_0^2 \int_0^y \int_0^{4-y^2} dz\, dx\, dy = \int_0^2 \int_0^y (4 - y^2) dx\, dy = \int_0^2 (4y - y^3) dy = 4$.

17.5.9 $v = \int_0^4 \int_0^{\frac{12-3x}{6}} \int_0^{\frac{12-3x-6y}{4}} dz\, dy\, dx = \int_0^4 \int_0^{\frac{12-3x}{6}} \left(\frac{12 - 3x - 6y}{4} \right) dy\, dx$

$= \int_0^4 \frac{3}{16}(4-x)^2 dx = 4$.

17.5.10 $v = \int_0^2 \int_{x^2-x+1}^{x+1} \int_0^{x+1} dz\, dy\, dx = \int_0^2 \int_{x^2-x+1}^{x+1} (x+1) dy\, dx = \int_0^2 (2x + x^2 - x^3) dx = \frac{8}{3}$.

17.5.11 $v = \int_0^1 \int_{x^2}^x \int_0^{x^2+y^2} dz\, dy\, dx = \int_0^1 \int_{x^2}^x (x^2 + y^2) dy\, dx = \int_0^1 \left(\frac{4x^3}{3} - x^4 - \frac{x^6}{3} \right) dx = \frac{3}{35}$.

17.5.12 $v = \int_0^1 \int_{x^2}^{\sqrt{x}} \int_0^3 dz\, dy\, dx = \int_0^1 \int_{x^2}^{\sqrt{x}} 3\, dy\, dx = \int_0^1 3(\sqrt{x} - x^2) dx = 1$.

17.5.13 $v = \int_0^3 \int_0^y \int_0^{\frac{4}{y^2+1}} dz\,dx\,dy = \int_0^3 \int_0^y \frac{4}{y^2+1}\,dx\,dy = \int_0^3 \frac{4y}{y^2+1}\,dy = 2\ln 10.$

17.5.14 $\int_0^3 \int_0^{3-x} \int_0^{3-x-y} x\,dz\,dy\,dx = \int_0^3 \int_0^{3-x} (3x - x^2 - xy)dy\,dx$

$$= \int_0^3 \frac{1}{2}(9x - 6x^2 + x^3)dx = \frac{27}{8}.$$

17.5.15 $\int_0^1 \int_0^z \int_0^{\sqrt{yz}} x\,dx\,dy\,dz = \int_0^1 \int_0^z \frac{yz}{2}\,dy\,dz = \int_0^1 \frac{z^3}{4}\,dz = \frac{1}{16}.$

17.5.16 $v = \int_{-4}^1 \int_{3x}^{4-x^2} \int_0^{x+4} dz\,dy\,dx = \int_{-4}^1 \int_{3x}^{4-x^2} (x+4)dy\,dx$

$$= \int_{-4}^1 (16 - 8x - 7x^2 - x^3)dx = \frac{625}{12}.$$

17.5.17 $\int_0^1 \int_0^{\sqrt{1-x^2}} \int_0^x yz\,dz\,dy\,dx = \int_0^1 \int_0^{\sqrt{1-x^2}} \frac{x^2 y}{2}\,dy\,dx = \int_0^1 \frac{1}{4}(x^2 - x^4)dx = \frac{1}{30}.$

17.5.18 $\int_0^1 \int_0^y \int_0^{x+1} y\,dz\,dx\,dy = \int_0^1 \int_0^y (xy + y)dx\,dy = \int_0^1 \left(\frac{y^3}{2} + y^2\right)dy = \frac{11}{24}.$

SECTION 17.6

17.6.1   Find the center of gravity of the lamina enclosed by $x = 0$, $x = 4$, $y = 0$, and $y = 3$ if its density is given by $\delta(x, y) = k(x + y^2)$.

17.6.2   Find the centroid of the lamina enclosed by $x = 4y - y^2$ and the y-axis.

17.6.3   Find the centroid of the lamina enclosed by $y = 4 - x$, $x = 0$, and $y = 0$.

17.6.4   Find the centroid of the lamina enclosed by $y = x^2$ and the line $y = 4$.

17.6.5   Find the centroid of the lamina enclosed by $y = x^3$, $x = 2$, and $y = 0$.

17.6.6   Find the centroid of the lamina enclosed by $y = x^2 - 2x$ and $y = 0$.

17.6.7   Find the centroid of the lamina enclosed by $x^2 = 8y$, $y = 0$, and $y = 4$.

17.6.8   Find the center of gravity of the lamina enclosed by $y^2 = 4x$, $x = 4$, and $y = 0$ if its density is given by $\delta(x, y) = ky$.

17.6.9   Find the center of gravity of the lamina enclosed by $x = 0$, $x = 4$, $y = 0$, and $y = 3$ if its density is given by $\delta(x, y) = kx^2y$.

17.6.10   Find the center of gravity of the lamina enclosed by $y = \sin x$, $y = 0$, $0 \leq x \leq \pi$ if its density is proportional to the distance from the x-axis.

17.6.11   Find the center of gravity of the lamina enclosed by $r = a \cos \theta$, $0 \leq \theta \leq \pi/2$, if its density is proportional to the distance from the origin.

17.6.12   Find the centroid of the lamina enclosed by $y = \sqrt{4 - x^2}$ and $y = 0$.

17.6.13   Find the mass of the lamina in the first quadrant that is inside $r = 8 \cos \theta$ and outside $r = 4$ if the density of the region is given by $\delta(r, \theta) = \sin \theta$.

17.6.14   Find the mass of the lamina cut from the circle $x^2 + y^2 = 36$ by the line $x = 3$ if its density is given by $\delta(x, y) = \dfrac{x^2}{x^2 + y^2}$.

17.6.15   Find the centroid of the solid in the first octant enclosed by $x^2 + z^2 = 1$ and the plane $y = 3$.

17.6.16   Find the center of gravity of the solid enclosed by $-1 \leq x \leq 1$, $-1 \leq y \leq 1$, $-1 \leq z \leq 1$ if its density is given by $\delta(x, y, z) = x^2y^2z^2$.

17.6.17   Find the mass of the tetrahedron in the first octant enclosed by the coordinate planes and the plane $x + y + z = 1$ if its density is given by $\delta(x, y, z) = xy$.

17.6.18   Use the theorem of Pappas to find the volume of the solid generated when the region enclosed by $y = x^2$ and $y = 8 - x^2$ is revolved about the line $y = -2$. [Hint: Obtain the centroid by symmetry.]

SOLUTIONS

SECTION 17.6

17.6.1 $M = \int_0^4 \int_0^3 k(x + y^2) dy\, dx = 60k$, $M_x = \int_0^4 \int_0^3 ky(x + y^2) dy\, dx = 117k$,

$M_y = \int_0^4 \int_0^3 kx(x + y^2) dy\, dx = 136k$; $\bar{x} = \dfrac{M_y}{M} = \dfrac{34}{15}$, $\bar{y} = \dfrac{M_x}{M} = \dfrac{39}{20}$;

center of graavity $\left(\dfrac{34}{15}, \dfrac{39}{20}\right)$.

17.6.2 $A = \int_0^4 \int_0^{4y-y^2} dx\, dy = \dfrac{32}{3}$, $\iint_R x\, dA = \int_0^4 \int_0^{4y-y^2} x\, dx\, dy = \dfrac{256}{15}$,

$\iint_R y\, dA = \int_0^4 \int_0^{4y-y^2} y\, dx\, dy = \dfrac{64}{3}$; centroid $\left(\dfrac{8}{5}, 2\right)$.

17.6.3 $A = \int_0^4 \int_0^{4-x} dy\, dx = 8$, $\iint_R x\, dA = \int_0^4 \int_0^{4-x} x\, dy\, dx = \dfrac{32}{3}$,

$\iint_R y\, dA = \int_0^4 \int_0^{4-x} y\, dy\, dx = \dfrac{32}{3}$; centroid $\left(\dfrac{4}{3}, \dfrac{4}{3}\right)$.

17.6.4 $A = \int_{-2}^2 \int_{x^2}^4 dy\, dx = \dfrac{32}{3}$, $\iint_R y\, dA = \int_{-2}^2 \int_{x^2}^4 y\, dy\, dx = \dfrac{128}{5}$,

$\bar{x} = 0$ by symmetry of the region; centroid $\left(0, \dfrac{12}{5}\right)$.

17.6.5 $A = \int_0^2 \int_0^{x^3} dy\, dx = 4$, $\iint_R x\, dA = \int_0^2 \int_0^{x^3} x\, dy\, dx = \dfrac{32}{5}$,

$\iint_R y\, dA = \int_0^2 \int_0^{x^3} y\, dy\, dx = \dfrac{64}{7}$; centroid $\left(\dfrac{8}{5}, \dfrac{16}{7}\right)$.

17.6.6 $A = -\int_0^2 \int_0^{x^2-2x} dy\, dx = \dfrac{4}{3}$, $\iint_R y\, dA = -\int_0^2 \int_0^{x^2-2x} y\, dy\, dx = -\dfrac{8}{15}$,

$\bar{x} = 1$ by symmetry of the region; centroid $\left(1, -\dfrac{2}{5}\right)$.

17.6.7 $A = \int_0^4 \int_0^{\frac{x^2}{8}} dy\, dx = \dfrac{8}{3}$, $\displaystyle\iint_R x\, dA = \int_0^4 \int_0^{\frac{x^2}{8}} x\, dy\, dx = 8$,

$\displaystyle\iint_R y\, dA = \int_0^4 \int_0^{\frac{x^2}{8}} y\, dy\, dx = \dfrac{8}{5}$; centroid $(3, 3/5)$.

17.6.8 $M = \int_0^4 \int_0^{\sqrt{4x}} ky\, dy\, dx = 16k$, $M_x = \int_0^4 \int_0^{\sqrt{4x}} k\, y^2 dy\, dx = \dfrac{512k}{15}$,

$M_y = \int_0^4 \int_0^{\sqrt{4x}} k\, xy\, dy\, dx = \dfrac{128k}{3}$; $\bar{x} = \dfrac{M_y}{M} = \dfrac{8}{3}$, $\bar{y} = \dfrac{M_x}{M} = \dfrac{32}{15}$;

center of gravity $\left(\dfrac{8}{3}, \dfrac{32}{15}\right)$.

17.6.9 $M = \int_0^4 \int_0^3 kx^2 y\, dy\, dx = 96k$, $M_x = \int_0^4 \int_0^3 k\, x^2 y^2 dy\, dx = 192k$,

$M_y = \int_0^4 \int_0^3 k\, x^3 y\, dy\, dx = 288k$; $\bar{x} = \dfrac{M_y}{M} = 3$, $\bar{y} = \dfrac{M_x}{M} = 2$;

center of gravity $(3, 2)$.

17.6.10 $M = \int_0^\pi \int_0^{\sin x} ky\, dy\, dx = \dfrac{k\pi}{4}$, $M_x = \int_0^\pi \int_0^{\sin x} k\, y^2 dy\, dx = \dfrac{4k}{9}$;

$\bar{x} = \dfrac{\pi}{2}$ by symmetry of the region, $\bar{y} = \dfrac{M_x}{M} = \dfrac{16}{9\pi}$;

center of gravity $\left(\dfrac{\pi}{2}, \dfrac{16}{9\pi}\right)$.

17.6.11 $M = \int_0^{\pi/2} \int_0^{a\cos\theta} k\, r^2 dr\, d\theta = \dfrac{2ka^3}{9}$,

$M_x = \int_0^{\pi/2} \int_0^{a\cos\theta} k\, r^3 \sin\theta\, dr\, d\theta = \dfrac{ka^4}{20}$,

$M_y = \int_0^{\pi/2} \int_0^{a\cos\theta} k\, r^3 \cos\theta\, dr\, d\theta = \dfrac{2k/a^4}{15}$; $\bar{x} = \dfrac{M_y}{M} = \dfrac{3a}{5}$, $\bar{y} = \dfrac{M_x}{M} = \dfrac{9a}{40}$;

center of gravity $\left(\dfrac{3a}{5}, \dfrac{9a}{40}\right)$.

17.6.12 Change to polar coordinates, then $A = \int_0^\pi \int_0^2 r\, dr\, d\theta = 2\pi$,

$$\iint_R y\, dA = \int_0^\pi \int_0^2 r^2 \sin\theta\, dr\, d\theta = \frac{16}{3}; \ \bar{x} = 0 \text{ by symmetry of the region,}$$

$$\bar{y} = \frac{M_x}{M} = \frac{8}{3\pi}; \text{ center of gravity } \left(0, \frac{8}{3\pi}\right).$$

17.6.13 $M = \int_0^{\pi/3} \int_4^{8\cos\theta} r\sin\theta\, dr\, d\theta = \frac{16}{3}.$

17.6.14 Change to polar coordinates, then

$$M = \int_{-\pi/3}^{\pi/3} \int_{\frac{3}{\cos\theta}}^{6} r\cos^2\theta\, dr\, d\theta = 3\pi + \frac{9\sqrt{3}}{2}.$$

17.6.15 $V = \int_0^1 \int_0^3 \int_0^{\sqrt{1-x^2}} dz\, dy\, dx = \frac{3\pi}{4}, \ \iiint_G x\, dv = \int_0^1 \int_0^3 \int_0^{\sqrt{1-x^2}} x\, dz\, dy\, dx = 1,$

$$\iiint_G z\, dv = \int_0^1 \int_0^3 \int_0^{\sqrt{1-x^2}} z\, dz\, dy\, dx = 1; \ \bar{x} = \frac{4}{3\pi}, \ \bar{y} = \frac{3}{2} \text{ by symmetry of the solid,}$$

$$\bar{z} = \frac{4}{3\pi}; \text{ centroid } \left(\frac{4}{3\pi}, \frac{3}{2}, \frac{4}{3\pi}\right).$$

17.6.16 The center of gravity is located at $(0,0,0)$ by symmetry of the solid.

17.6.17 $M = \int_0^1 \int_0^{1-x} \int_0^{1-x-y} xy\, dz\, dy\, dx = \frac{1}{120}.$

17.6.18 $A = \int_{-2}^2 \int_{x^2}^{8-x^2} dy\, dx = \frac{64}{3} \text{ so } V = \left(\frac{64}{3}\right)(12\pi) = 256\pi.$

SECTION 17.7

17.7.1   Evaluate $\displaystyle\int_0^{\frac{\pi}{4}} \int_1^{\cos\theta} \int_1^r \frac{1}{r^2 z^2}\, dz\, dr\, d\theta.$

17.7.2   Evaluate $\displaystyle\int_0^{\pi} \int_0^{\pi/4} \int_0^{\cos\theta} \rho^2 \sin\theta\, d\rho\, d\phi\, d\theta.$

17.7.3   Evaluate $\displaystyle\int_0^{2\pi} \int_1^2 \int_0^5 e^z r\, dz\, dr\, d\theta.$

17.7.4   Evaluate $\displaystyle\int_0^{2\pi} \int_0^{\pi} \int_1^2 \rho^4 \cos^2\phi \sin\phi\, d\rho\, d\phi\, d\theta.$

17.7.5   Use cylindrical coordinates to find the volume of the solid in the first octant enclosed by the coordinate planes, the cylinder $x^2 + y^2 = 4$, and the plane $z + y = 3$.

17.7.6   Use cylindrical coordinates to find the volume and centroid of the cylinder enclosed by $x^2 + y^2 = 4$, $z = 0$, and $z = 4$.

17.7.7   Use cylindrical coordinates to find the volume inside $x^2 + y^2 = 4x$, above $z = 0$, and below $x^2 + y^2 = 4z$.

17.7.8   Use spherical coordinates to find the volume of the solid enclosed by $x^2 + y^2 + z^2 = 2$ and $x^2 + y^2 + z^2 = 1$.

17.7.9   Use cylindrical coordinates to evaluate $\displaystyle\iiint_G \sqrt{x^2 + y^2}\, dV$ where G is the solid enclosed by $z = x^2 + y^2$ and $z = 8 - x^2 - y^2$.

17.7.10   Use cylindrical coordinates to find the volume and centroid of the solid enclosed by the paraboloid $z = x^2 + y^2$ and the plane $z = 4$.

17.7.11   Use spherical coordinates to find the mass of the sphere $x^2 + y^2 + z^2 = 9$ if its density is given by $\delta(x, y, z) = \dfrac{z^2}{x^2 + y^2 + z^2}.$

17.7.12   Use cylindrical coordinates to find the volume and centroid of the solid enclosed by $z = \sqrt{x^2 + y^2}$ and the plane $z = 1$.

17.7.13   Use spherical coordinates to find the volume of the sphere $x^2 + y^2 + z^2 = 2z$.

17.7.14   Use spherical coordinates to find the mass and center of gravity of the sphere $x^2 + y^2 + z^2 = 2z$ if its density is given by $\delta(x, y, z) = \sqrt{x^2 + y^2 + z^2}$.

17.7.15   Use spherical coordinates to find the mass of the sphere $x^2 + y^2 + z^2 = 4$ if its density is given by $\delta(x, y, z) = x^2 + y^2$.

17.7.16   Use cylindrical coordinates to find the mass of the ellipsoid $x^2 + y^2 + \dfrac{z^2}{4} = 1$ if its density is given by $\delta(x, y, z) = z$.

17.7.17   Use cylindrical coordinates to find the volume of the solid enclosed by the sphere $x^2 + y^2 + z^2 = 9$, $z = 0$, and the cylinder $x^2 + y^2 = 3y$.

17.7.18   Use spherical coordinates to find the mass of a sphere of radius 4 if its density is proportional to the distance from its center.

17.7.19   Use sperical coordinates to find the mass of the solid bounded below by the cone $z = \sqrt{x^2 + y^2}$ and above by the sphere $x^2 + y^2 + z^2 = 9$ if its density is given by $\delta(x, y, z) = x^2 + y^2 + z^2$.

SOLUTIONS

SECTION 17.7

17.7.1 $\displaystyle\int_0^{\pi/4}\int_1^{\cos\theta}\int_1^r \frac{1}{r^2z^2}\,dz\,dr\,d\theta = \int_0^{\pi/4}\int_1^{\cos\theta}\left(\frac{1}{r^2}-\frac{1}{r^3}\right)dr\,d\theta$

$$= \int_0^{\pi/4}\left(-\sec\theta+\frac{1}{2}\sec^2\theta+\frac{1}{2}\right)d\theta$$

$$= \frac{1}{2}+\frac{\pi}{8}-\ln(\sqrt{2}+1).$$

17.7.2 $\displaystyle\int_0^{\pi}\int_0^{\pi/4}\int_0^{\cos\theta} \rho^2\sin\theta\,d\rho\,d\phi\,d\theta$

$$= \frac{1}{3}\int_0^{\pi}\int_0^{\pi/4}\cos^3\theta\sin\theta\,d\phi\,d\theta$$

$$= \frac{\pi}{12}\int_0^{\pi}\cos^3\theta\sin\theta\,d\theta = 0.$$

17.7.3 $\displaystyle\int_0^{2\pi}\int_1^2\int_0^5 e^z r\,dz\,dr\,d\theta = \int_0^{2\pi}\int_1^2 (e^5-1)r\,dr\,d\theta$

$$= \int_0^{2\pi}\frac{3}{2}(e^5-1)d\theta = 3\pi(e^5-1).$$

17.7.4 $\displaystyle\int_0^{2\pi}\int_0^{\pi}\int_1^2 \rho^4\cos^2\phi\sin\phi\,d\rho\,d\phi\,d\theta = \int_0^{2\pi}\int_0^{\pi}\frac{31}{5}\cos^2\phi\sin\phi\,d\phi\,d\theta$

$$= \int_0^{2\pi}\frac{62}{15}\,d\theta = \frac{124\pi}{15}.$$

17.7.5 $\displaystyle V = \int_0^{\pi/2}\int_0^2\int_0^{3-r\sin\theta} r\,dz\,dr\,d\theta = \int_0^{\pi/2}\int_0^2 (3r - r^2\sin\theta)dr\,d\theta$

$$= \int_0^{\pi/2}\left(6-\frac{8}{3}\sin\theta\right)d\theta = 3\pi - \frac{8}{3}.$$

17.7.6 $\displaystyle V = \int_0^{2\pi}\int_0^2\int_0^4 r\,dz\,dr\,d\theta = 16\pi,\ \bar{x}=\bar{y}=0$ by symmetry of the region,

$\displaystyle\bar{z} = \frac{1}{V}\int_0^{2\pi}\int_0^2\int_0^4 rz\,dz\,dr\,d\theta = 2.$ The volume is 16π and the centroid is located at $(0,0,2)$.

This problem could have been done without the use of calculus.

17.7.7 $V = \int_{-\frac{\pi}{2}}^{\frac{\pi}{2}} \int_0^{4\cos\theta} \int_0^{\frac{r^2}{4}} r\,dz\,dr\,d\theta = \int_{-\frac{\pi}{2}}^{\frac{\pi}{2}} \int_0^{4\cos\theta} \frac{r^3}{4}\,dr\,d\theta = \int_{-\frac{\pi}{2}}^{\frac{\pi}{2}} 16\cos^4\theta\,d\theta = 6\pi.$

17.7.8 $V = 2 \int_0^{2\pi} \int_0^{\pi/2} \int_1^{\sqrt{2}} \rho^2 \sin\phi\,d\rho\,d\phi\,d\theta$

$\qquad = \dfrac{2}{3} \int_0^{2\pi} \int_0^{\pi/2} (2\sqrt{2}-1)\sin\phi\,d\phi\,d\theta$

$\qquad = \dfrac{2}{3} \int_0^{2\pi} (2\sqrt{2}-1)\,d\theta = \dfrac{4\pi}{3}(2\sqrt{2}-1)$

17.7.9 The intersection of $z = x^2 + y^2$ and $z = 8 - x^2 - y^2$ is the circle $x^2 + y^2 = 4$, so

$\int_0^{2\pi} \int_0^2 \int_{r^2}^{8-r^2} r^2\,dz\,dr\,d\theta = 2 \int_0^{2\pi} \int_0^2 (4r^2 - r^4)\,dr\,d\theta = \int_0^{2\pi} \dfrac{128}{15}\,d\theta = \dfrac{256\pi}{15}.$

17.7.10 $V = \int_0^{2\pi} \int_0^2 \int_{r^2}^4 r\,dz\,dr\,d\theta = 8\pi$, $\bar{x} = \bar{y} = 0$ by symmetry of the region,

$\bar{z} = \dfrac{1}{V} \int_0^{2\pi} \int_0^2 \int_{r^2}^4 rz\,dz\,dr\,d\theta = \dfrac{8}{3}$ so, the volume is 8π and the centroid is located at $\left(0, 0, \dfrac{8}{3}\right)$.

17.7.11 $M = \int_0^{2\pi} \int_0^\pi \int_0^3 \rho^2 \cos^2\phi \sin\phi\,d\rho\,d\phi\,d\theta$

$\qquad = \int_0^{2\pi} \int_0^\pi 9\cos^2\phi\sin\phi\,d\phi\,d\theta = \int_0^{2\pi} 6\,d\theta = 12\pi.$

17.7.12 $V = \int_0^{2\pi} \int_0^1 \int_r^1 r\,dz\,dr\,d\theta = \dfrac{\pi}{3}$, $\bar{x} = \bar{y} = 0$ by symmetry of the region,

$\bar{z} = \dfrac{1}{V} \int_0^{2\pi} \int_0^1 \int_r^1 rz\,dz\,dr\,d\theta = \dfrac{3}{4}$. The volume is $\dfrac{\pi}{3}$ and the centroid is located at $\left(0, 0, \dfrac{3}{4}\right)$.

17.7.13 $V = \int_0^{2\pi} \int_0^{\pi/2} \int_0^{2\cos\phi} \rho^2 \sin\phi\,d\rho\,d\phi\,d\theta = \int_0^{2\pi} \int_0^{\pi/2} \dfrac{8}{3}\cos^3\phi\sin\phi\,d\phi = \int_0^{2\pi} \dfrac{2}{3}\,d\theta = \dfrac{4\pi}{3}.$

17.7.14 $M = \int_0^{2\pi} \int_0^{\pi/2} \int_0^{2\cos\phi} \rho^3 \sin\phi\,d\rho\,d\phi\,d\theta = \dfrac{8\pi}{5}$, $\bar{x} = \bar{y} = 0$ by symmetry of the region,

$\bar{z} = \dfrac{1}{M} \int_0^{2\pi} \int_0^{\pi/2} \int_0^{2\cos\phi} \rho^4 \cos\phi\sin\phi\,d\rho\,d\phi\,d\theta = \dfrac{8}{7}$. The mass is $\dfrac{8\pi}{5}$ and the center of gravity is located at $\left(0, 0, \dfrac{8}{7}\right)$.

17.7.15 $M = \int_0^{2\pi} \int_0^{\pi} \int_0^{2} \rho^4 \sin^3\phi\, d\rho\, d\phi\, d\theta = \int_0^{2\pi} \int_0^{\pi} \frac{32}{5} \sin^3\phi\, d\phi\, d\theta = \int_0^{2\pi} \frac{128}{15} d\theta = \frac{256\pi}{15}.$

17.7.16 $M = \int_0^{2\pi} \int_0^{1} \int_0^{2\sqrt{1-r^2}} rz\, dz\, dr\, d\theta = \int_0^{2\pi} \int_0^{1} (2r - 2r^3)dr\, d\theta = \int_0^{2\pi} \frac{1}{2} d\theta = \pi.$

17.7.17 $V = 2 \int_0^{\frac{\pi}{2}} \int_0^{3\sin\theta} \int_0^{\sqrt{9-r^2}} r\, dz\, dr\, d\theta = 2 \int_0^{\frac{\pi}{2}} \int_0^{3\sin\theta} r\sqrt{9-r^2}\, dr\, d\theta$

$= 2 \int_0^{\frac{\pi}{2}} \frac{1}{3}(27 - 27\cos^3\theta)d\theta = 9\pi - 12.$

17.7.18 $M = \int_0^{2\pi} \int_0^{\pi} \int_0^{4} k\rho^3 \sin\phi\, d\rho\, d\phi\, d\theta = k \int_0^{2\pi} \int_0^{\pi} 64\sin\phi\, d\phi\, d\theta$

$= k \int_0^{2\pi} 128 d\theta = 256\pi k.$

17.7.19 $M = \int_0^{2\pi} \int_0^{\pi/4} \int_0^{3} \delta(x,y,z)\rho^2 \sin\phi\, d\rho\, d\phi\, d\theta$

$= \int_0^{2\pi} \int_0^{\pi/4} \int_0^{3} (\rho^2)\, \rho^2 \sin\phi\, d\rho\, d\phi\, d\theta$

$= \int_0^{2\pi} \int_0^{\pi/4} \int_0^{3} \rho^4 \sin\phi\, d\rho\, d\phi\, d\theta$

$= \frac{243}{5} \int_0^{2\pi} \int_0^{\pi/4} \sin\phi\, d\phi\, d\theta$

$= -\frac{243}{5} \int_0^{2\pi} \cos\phi \Big|_0^{\pi/4} d\theta$

$= -\frac{243\pi\,(\sqrt{2}-2)}{5}$

SECTION 17.8

17.8.1 Find the Jacobian $\dfrac{\partial(x,y)}{\partial(u,v)}$ if $x = u^2 - v^2$ and $y = u^2 - 2v^2$.

17.8.2 Solve for x and y in terms of U and V to find the Jacobian $\dfrac{\partial(x,y)}{\partial(u,v)}$ if $U = 3x - 2y$ and $V = x + y$.

17.8.3 Find the Jacobian $\dfrac{\partial(x,y,z)}{\partial(u,v,w)}$ if $x = 3\cosh u \cos v$, $y = 3\sinh u \sin v$, and $z = w$.

17.8.4 Use the transformation $U = 2x - y, V = x + 2y$ to evaluate $\displaystyle\iint\limits_{R} \dfrac{2x-y}{x+2y}\, dA$ where R is the rectangular region enclosed by $2x - y = 4$, $2x - y = 8$, $x + 2y = 3$, and $x + 2y = 6$.

17.8.5 Use the transformation $U = x + y$ and $V = 2x - 3y$ to evaluate $\displaystyle\iint\limits_{R} x\, dA$ where R is the region bounded by $x + y = 1$, $x + y = 2$, $2x - 3y = 2$, and $2x - 3y = 5$.

17.8.6 Use an appropriate transform to evaluate $\displaystyle\iint\limits_{R} xy\, dA$ where R is the region enclosed by $y = \dfrac{1}{4}x, y = 2x, y = \dfrac{2}{x}$ and $y = \dfrac{4}{x}$.

17.8.7 Use an appropriate transform to find the area of the region in the first quadrant enclosed by $x + y = 1$, $x + y = 2$, $3x - 2y = 2$, and $3x - 2y = 5$.

17.8.8 Use an appropriate transform to evaluate $\displaystyle\iint\limits_{R} (x+y)\, dA$ where R is the region enclosed by the rectangle whose vertices are $(0,0)$, $(2,2)$, $(6,-2)$, and $(4,-4)$.

SOLUTIONS

SECTION 17.8

17.8.1 $J(x, y) = \dfrac{\partial(x, y)}{\partial(u, v)} = \begin{vmatrix} 2u & -2v \\ 2u & -4v \end{vmatrix} = -4uv.$

17.8.2 Solve $u = 3x - 2y$ and $v = x + y$ for x and y to get $x = \dfrac{1}{5}(u + 2v)$ and $y = \dfrac{1}{5}(3v - u)$, then,

$$\dfrac{\partial(x, y)}{\partial(u, v)} = \begin{vmatrix} 1/5 & 2/5 \\ -1/5 & 3/5 \end{vmatrix} = \dfrac{1}{5}.$$

17.8.3 $\dfrac{\partial(x, y, z)}{\partial(u, v, w)} = \begin{vmatrix} 3\sinh u \cos v & -3\cosh u \sin v & 0 \\ 3\cosh u \sin v & 3\sinh u \cos v & 0 \\ 0 & 0 & 1 \end{vmatrix} = 9\sinh^2 u \cos^2 v + 9\cosh^2 u \sin^2 v.$

17.8.4 Solve $u = 2x - y$ and $v = x + 2y$ in terms of x and y to get $x = \dfrac{1}{5}(2u + v)$ and $y = \dfrac{1}{5}(2v - u)$.

S is the region enclosed by $4 \le u \le 8$ and $3 \le v \le 6$, then, $\dfrac{\partial(x, y)}{\partial(u, v)} = \begin{vmatrix} 2/5 & 1/5 \\ -1/5 & 2/5 \end{vmatrix} = \dfrac{1}{5}$,

thus $\displaystyle\iint\limits_{R} \dfrac{2x - y}{x + 2y}\, dA = \int_3^6 \int_4^8 \dfrac{u}{v}\left|\dfrac{1}{5}\right| du\, dv = 24\int_3^6 \dfrac{1}{v}\, dv = 24\ln 2.$

17.8.5 Solve $u = x + y$ and $v = 2x - 3y$ in terms of x and y to get $x = \dfrac{1}{5}(3u + v)$ and $y = \dfrac{1}{5}(2u - v).$

S is the region enclosed by $1 \le u \le 2$ and $2 \le v \le 5$, then $\dfrac{\partial(x, y)}{\partial(u, v)} = \begin{vmatrix} 3/5 & 1/5 \\ 2/5 & -1/5 \end{vmatrix} = -1/5,$

thus, $\displaystyle\int_2^5 \int_1^2 \dfrac{1}{5}(3u + v)\left|-\dfrac{1}{5}\right| du\, dv = \dfrac{1}{25}\int_2^5 \left(\dfrac{9}{2} + v\right) du = \dfrac{24}{25}.$

17.8.6 Let $u = \dfrac{y}{x}$ and $v = xy$ be an appropriate transform. Solve u and v in terms of x and y to get

$x = \sqrt{\dfrac{v}{u}}$ and $y = \sqrt{uv}$. S is the region enclosed by $\dfrac{1}{4} \le u \le 2$ and $2 \le v \le 4$, then, $\dfrac{\partial(x, y)}{\partial(u, v)} =$

$\begin{vmatrix} -\dfrac{1}{2u}\sqrt{\dfrac{v}{u}} & \dfrac{1}{2\sqrt{uv}} \\ \dfrac{1}{2}\sqrt{\dfrac{v}{u}} & \dfrac{1}{2}\sqrt{\dfrac{u}{v}} \end{vmatrix} = -\dfrac{1}{2u}$, thus, $\displaystyle\int_2^4 \int_{1/4}^2 v\left|-\dfrac{1}{2u}\right| du\, dv = \dfrac{3}{2}\ln 2\int_2^4 v\, du = 9\ln 2.$

17.8.7 Let $U = x + y$ and $V = 3x - 2y$ be an appropriate transform. Solve U and V in terms of x and y to get $x = \frac{1}{5}(2u + v)$ and $y = \frac{1}{5}(3u - v)$. S is the region enclosed by $1 \le u \le 2$ and $2 \le v \le 5$, then, $\dfrac{\partial(x, y)}{\partial(u, v)} = \begin{vmatrix} 2/5 & 1/5 \\ 3/5 & -1/5 \end{vmatrix} = -1/5$, thus, $\displaystyle\int_2^5 \int_1^2 \left| -\frac{1}{5} \right| \, du \, dv = \frac{3}{5}$.

17.8.8 The region R is enclosed by the lines $y = -x$, $x - y = 8$, $x + y = 4$, and $y = x$. Let $u = x + y$ and $v = x - y$ be an appropriate transform. Solve U and V in terms of x and y to get $x = \frac{1}{2}(u + v)$ and $y = \frac{1}{2}(u - v)$. S is the region enclosed by $0 \le u \le 4$ and $0 \le v \le 8$, then, $\dfrac{\partial(x, y)}{\partial(u, v)} = \begin{vmatrix} 1/2 & 1/2 \\ 1/2 & -1/2 \end{vmatrix} = -1/2$, thus, $\displaystyle\int_0^8 \int_0^4 u \left| -\frac{1}{2} \right| \, du \, dv = 4 \int_0^8 dv = 32$.

SUPPLEMENTARY EXERCISES, CHAPTER 17

In Exercises 1–4, evaluate the iterated integrals.

1. $\displaystyle\int_{1/2}^{1}\int_{0}^{2x}\cos(\pi x^2)\,dy\,dx.$

2. $\displaystyle\int_{0}^{2}\int_{-y}^{2y}xe^{y^3}\,dx\,dy.$

3. $\displaystyle\int_{-1}^{0}\int_{0}^{y^2}\int_{xy}^{1}2y\,dz\,dx\,dy.$

4. $\displaystyle\int_{0}^{1}\int_{0}^{z}\int_{0}^{\sqrt{yz}}x\,dx\,dy\,dz.$

In Exercises 5 and 6, express the iterated integral as an equivalent integral with the order of integration reversed.

5. $\displaystyle\int_{0}^{2}\int_{0}^{x/2}e^{x}e^{y}\,dy\,dx.$

6. $\displaystyle\int_{0}^{\pi}\int_{y}^{\pi}\frac{\sin x}{x}\,dx\,dy.$

7. Use a double integral to find the area of the region bounded by $y = 2x^3$, $2x + y = 4$, and the x-axis.

8. Use a double integral to find the area of the region bounded by $x = y^2$ and $x = 4y - y^2$.

9. Sketch the region R whose area is given by the iterated integral.

(a) $\displaystyle\int_{0}^{\pi/2}\int_{\tan(x/2)}^{\sin x}dy\,dx$

(b) $\displaystyle\int_{\pi/6}^{\pi/2}\int_{a}^{a(1+\cos\theta)}r\,dr\,d\theta \quad (a > 0).$

In Exercises 10–12, evaluate the double integral over R using either rectangular or polar coordinates.

10. $\displaystyle\iint_{R}xy\,dA$; R is the region bounded by $y = \sqrt{x}$, $y = 2 - \sqrt{x}$, and the y-axis.

11. $\displaystyle\iint_{R}x^2\sin y^2\,dA$; R is the region bounded by $y = x^3$, $y = -x^3$, and the $y = 8$.

12. $\displaystyle\iint_{R}(4 - x^2 - y^2)\,dA$; R is the sector in the first quadrant bounded by the circle $x^2 + y^2 = 4$ and the coordinate axes.

In Exercises 13–15, use a double integral in rectangular or polar coordinates to find the volume of the solid.

13. The solid in the first octant bounded by the coordinate planes and the plane $3x + 2y + z = 6$.

14. The solid enclosed by the cylinders $y = 3x + 4$ and $y = x^2$, and such that $0 \le z \le \sqrt{y}$.

15. The solid $G = \{(x, y, z) : 2 \le x^2 + y^2 \le 4 \text{ and } 0 \le z \le 1/(x^2 + y^2)^2\}$.

16. Convert to polar coordinates and evaluate:

$$\int_0^{\sqrt{2}} \int_x^{\sqrt{4-x^2}} 4xy \, dy \, dx$$

17. Convert to rectangular coordinates and evaluate:

$$\int_0^{\pi/2} \int_0^{2a \sin \theta} r \sin 2\theta \, dr \, d\theta \quad (a > 0)$$

In Exercises 18–20, find the area of the region using a double integral in polar coordinates.

18. The region outside the circle $r = \sqrt{2}a$ and inside the lemniscate $r^2 = 4a^2 \cos 2\theta$.

19. The region enclosed by the rose $r = \cos 3\theta$.

20. The region inside the circle $r = 2\sqrt{3} \sin \theta$ and outside the circle $r = 3$.

In Exercises 21–23, find the area of the surface described.

21. The part of the paraboloid $z = 3x^2 + 3y^2 - 3$ below the xy-plane.

22. The part of the plane $2x + 2y + z = 7$ in the first octant.

23. The part of the cone $z^2 = x^2 + y^2$ between the planes $z = 1$ and $z = 4$.

24. Evaluate $\iiint\limits_G x^2 yz \, dV$, where G is the set of points satisfying the inequalities $0 \le x \le 2$, $-x \le y \le x^2$, and $0 \le z \le x + y$.

25. Evaluate $\iiint\limits_G \sqrt{x^2 + y^2} \, dV$, where G is the set of points satisfying $x^2 + y^2 \le 16$, $0 \le z \le 4 - y$.

26. If $G = \{(x, y, z) : x^2 + y^2 \le z \le 4x\}$, express the volume of G as a triple integral in (a) rectangular coordinates and (b) cylindrical coordinates.

27. In each part find an equivalent integral of the form $\iiint\limits_{G}(\ )\,dx\,dz\,dy$.

(a) $\displaystyle\int_0^1\int_0^{(1-x)/2}\int_0^{1-x-2y} z\,dz\,dy\,dx$

(b) $\displaystyle\int_0^2\int_{x^2}^4\int_0^{4-y} 3\,dz\,dy\,dx$.

28. (a) Change to cylindrical coordinates and then evaluate:

$$\int_{-2}^2\int_{-\sqrt{4-x^2}}^{\sqrt{4-x^2}}\int_{(x^2+y^2)^2}^{16} x^2\,dz\,dy\,dx$$

(b) Change to spherical coordinates and evaluate:

$$\int_0^1\int_0^{\sqrt{1-x^2}}\int_0^{\sqrt{1-x^2-y^2}}\frac{1}{1+x^2+y^2+z^2}\,dz\,dy\,dx$$

29. If G is the region bounded above by the sphere $\rho=a$ and bounded below by the cone $\phi=\pi/3$, express $\iiint_G(x^2+y^2)\,dV$ as an iterated integral in (a) spherical coordinates, (b) cylindrical coordinates, and (c) Cartesian coordinates.

In Exercises 30–33, find the volume of G.

30. $G=\{(r,\theta,z):0\le r\le 2\sin\theta,0\le z\le r\sin\theta\}$.

31. G is the solid enclosed by the "inverted apple" $\rho=a(1+\cos\phi)$.

32. G is the solid that is enclosed between the surfaces $x=y^2+z^2$ and $x=1-y^2$.

33. G is the solid bounded below by the upper nappe of the cone $\phi=\pi/6$ and above the plane $z=a$.

In Exercises 34–36, find the centroid $(\overline{x},\overline{y})$ of the plane region R.

34. The region R is the upper half of the ellipse $(x/a)^2+(y/b)^2=1$.

35. The region R is enclosed by the cardioid $r=a(1+\sin\theta)$.

36. The region R is bounded by $y^2=4x$ and $y^2=8(x-2)$.

In Exercises 37 and 38, find the center of gravity of the lamina with density δ.

37. The triangular lamina with vertices $(a,0),(-a,0)$, and $(0,b)$, where $a>0$ and $b>0$; and $\delta(x,y)$ is proportional to the distance from (x,y) to the y-axis.

38. The lamina enclosed by the circle $r = 3\cos\theta$, but outside the cardioid $r = 1 + \cos\theta$, and with $\delta(r,\theta)$ proportional to the distance from (r,θ) to the x-axis.

In Exercises 39 and 40, find the mass of the solid G if its density is δ.

39. The solid G is the part of the first octant under the plane $x/a + y/b + z/c = 1$, where a, b, c are positive, and $\delta(x, y, z) = kz$.

40. The spherical solid G is bounded by $\rho = a$ and $\delta(x, y, z)$ is twice the distance from (x, y, z) to the origin.

In Exercises 41–43, find the centroid of G.

41. The solid G bounded by $y = x^2, y = 4, z = 0$, and $y + z = 4$.

42. The solid G is the part of the sphere $\rho \leq a$ lying within the cone $\phi \leq \phi_0$, where $\phi_0 \leq \pi/2$.

43. The solid G is bounded by the cone with vertex $(0, 0, h)$ and base $x^2 + y^2 \leq R^2$ in the xy-plane.

SOLUTIONS

SUPPLEMENTARY EXERCISES CHAPTER 17

1. $\displaystyle\int_{1/2}^{1}\int_{0}^{2x}\cos(\pi x^2)\,dy\,dx = \int_{1/2}^{1} 2x\cos(\pi x^2)\,dx = -1/(\sqrt{2}\pi)$

2. $\displaystyle\int_{0}^{2}\int_{-y}^{2y}xe^{y^3}\,dx\,dy = \int_{0}^{2}\frac{3}{2}y^2 e^{y^3}\,dy = (e^8-1)/2$

3. $\displaystyle\int_{-1}^{0}\int_{0}^{y^2}\int_{xy}^{1}2y\,dz\,dx\,dy = \int_{-1}^{0}\int_{0}^{y^2}(2y-2xy^2)\,dx\,dy = \int_{-1}^{0}(2y^3-y^6)\,dy = -9/14$

4. $\displaystyle\int_{0}^{1}\int_{0}^{z}\int_{0}^{\sqrt{yz}}x\,dx\,dy\,dz = \int_{0}^{1}\int_{0}^{z}\frac{1}{2}yz\,dy\,dz = \int_{0}^{1}\frac{1}{4}z^3\,dz = 1/16$

5. $\displaystyle\int_{0}^{1}\int_{2y}^{2}e^x e^y\,dx\,dy$ **6.** $\displaystyle\int_{0}^{\pi}\int_{0}^{x}\frac{\sin x}{x}\,dy\,dx$

7. $\displaystyle A = \int_{0}^{1}\int_{0}^{2x^3}dy\,dx + \int_{1}^{2}\int_{0}^{4-2x}dy\,dx = 1/2+1 = 3/2$

8. $\displaystyle A = \int_{0}^{2}\int_{y^2}^{4y-y^2}dx\,dy = 8/3$

9. **(a)** 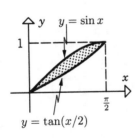 **(b)**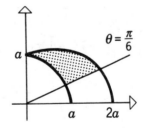

10. $\displaystyle\int_{0}^{1}\int_{\sqrt{x}}^{2-\sqrt{x}}xy\,dy\,dx = \int_{0}^{1}x(2-2\sqrt{x})\,dx = 1/5$

11. $\displaystyle\int_{0}^{8}\int_{-\sqrt[3]{y}}^{\sqrt[3]{y}}x^2\sin(y^2)\,dx\,dy = \int_{0}^{8}\frac{2}{3}y\sin(y^2)\,dy = (1-\cos 64)/3$

12. $\displaystyle\int_{0}^{\pi/2}\int_{0}^{2}(4-r^2)r\,dr\,d\theta = 2\pi$

13. $V = \int_0^2 \int_0^{(6-3x)/2} (6 - 3x - 2y)dy\, dx = \int_0^2 \frac{1}{4}(6 - 3x)^2 dx = 6$

14. $V = \int_{-1}^4 \int_{x^2}^{3x+4} \sqrt{y}\, dy\, dx = \int_{-1}^4 \frac{2}{3}[(3x + 4)^{3/2} - x^3]dx = 1453/30$

15. $V = \int_0^{2\pi} \int_{\sqrt{2}}^2 \frac{1}{r^3}dr\, d\theta = \int_0^{2\pi} \frac{1}{8}d\theta = \pi/4$

16. $\int_{\pi/4}^{\pi/2} \int_0^2 4r^3 \cos\theta \sin\theta\, dr\, d\theta = 4$

17. $\int_0^{2a} \int_0^{\sqrt{2ay-y^2}} \frac{2xy}{x^2 + y^2}dx\, dy = \int_0^{2a} (y\ln 2a - y\ln y)dy = a^2$

18. $A = 4\int_0^{\pi/6} \int_{\sqrt{2a}}^{2a\sqrt{\cos 2\theta}} r\, dr\, d\theta = 4a^2 \int_0^{\pi/6} (2\cos 2\theta - 1)d\theta = \frac{2}{3}(3\sqrt{3} - \pi)a^2$

19. $A = 6\int_0^{\pi/6} \int_0^{\cos 3\theta} r\, dr\, d\theta = 3\int_0^{\pi/6} \cos^2 3\theta\, d\theta = \pi/4$

20. $A = 2\int_{\pi/3}^{\pi/2} \int_3^{2\sqrt{3}\sin\theta} r\, dr\, d\theta = \int_{\pi/3}^{\pi/2} (12\sin^2\theta - 9)d\theta = (3\sqrt{3} - \pi)/2$

21. $z_x = 6x,\ z_y = 6y,\ z_x^2 + z_y^2 + 1 = 36(x^2 + y^2) + 1;$

$S = \int_0^{2\pi} \int_0^1 r\sqrt{36r^2 + 1}dr\, d\theta = (37\sqrt{37} - 1)\pi/54$

22. $z_x^2 + z_y^2 + 1 = 9;\ S = \int_0^{7/2} \int_0^{7/2-x} 3\, dy\, dx = 147/8$

23. $z_x = x/z,\ z_y = y/z,\ z_x^2 + z_y^2 + 1 = 2;\ S = \int_0^{2\pi} \int_1^4 \sqrt{2}\, r\, dr\, d\theta = 15\sqrt{2}\pi$

24. $\int_0^2 \int_{-x}^{x^2} \int_0^{x+y} x^2 yz\, dz\, dy\, dx = \int_0^2 \int_{-x}^{x^2} \frac{1}{2}x^2(x^2 y + 2xy^2 + y^3)dy\, dx$

$= \int_0^2 \left(\frac{1}{4}x^8 + \frac{1}{3}x^9 + \frac{1}{8}x^{10} - \frac{1}{24}x^6\right)dx = \frac{245,552}{3465}$

25. $\displaystyle\int_0^{2\pi}\int_0^4\int_0^{4-r\sin\theta} r^2\,dz\,dr\,d\theta = \int_0^{2\pi}\int_0^4 (4r^2 - r^3\sin\theta)\,dr\,d\theta$

$$= \int_0^{2\pi} \frac{64}{3}(4 - 3\sin\theta)\,d\theta = 512\pi/3$$

26. $z = x^2 + y^2$ and $z = 4x$ intersect in the curve whose projection onto the xy-plane is $x^2 + y^2 = 4x$ or, in polar coordinates, $r = 4\cos\theta$.

(a) $\displaystyle\int_0^4\int_{-\sqrt{4x-x^2}}^{\sqrt{4x-x^2}}\int_{x^2+y^2}^{4x} dz\,dy\,dx$ (b) $\displaystyle\int_{-\pi/2}^{\pi/2}\int_0^{4\cos\theta}\int_{r^2}^{4r\cos\theta} r\,dz\,dr\,d\theta$

27. (a) G is the region in the first octant bounded by the coordinate planes and the plane $z = 1 - x - 2y$; the integral is $\displaystyle\int_0^{1/2}\int_0^{1-2y}\int_0^{1-2y-z} z\,dx\,dz\,dy$

(b) G is the region in the first octant bounded by the planes $x = 0$, $z = 0$, $z = 4 - y$, and the parabolic cylinder $y = x^2$; the integral is $\displaystyle\int_0^4\int_0^{4-y}\int_0^{\sqrt{y}} 3\,dx\,dz\,dy$

28. (a) $\displaystyle\int_0^{2\pi}\int_0^2\int_{r^4}^{16} r^3\cos^2\theta\,dz\,dr\,d\theta = \int_0^{2\pi}\int_0^2 (16r^3 - r^7)\cos^2\theta\,dr\,d\theta = 32\int_0^{2\pi}\cos^2\theta\,d\theta = 32\pi$

(b) $\displaystyle\int_0^{\pi/2}\int_0^{\pi/2}\int_0^1 \frac{\rho^2\sin\phi}{1+\rho^2}\,d\rho\,d\phi\,d\theta = (1 - \pi/4)\int_0^{\pi/2}\int_0^{\pi/2}\sin\phi\,d\phi\,d\theta = \pi(4-\pi)/8$

29. (a) $\displaystyle\int_0^{2\pi}\int_0^{\pi/3}\int_0^a \rho^4\sin^3\phi\,d\rho\,d\phi\,d\theta$ (b) $\displaystyle\int_0^{2\pi}\int_0^{\sqrt{3}a/2}\int_{r/\sqrt{3}}^{\sqrt{a^2-r^2}} r^3\,dz\,dr\,d\theta$

(c) $\displaystyle\int_{-\sqrt{3}a/2}^{\sqrt{3}a/2}\int_{-\sqrt{3a^2/4-x^2}}^{\sqrt{3a^2/4-x^2}}\int_{\sqrt{x^2+y^2}/\sqrt{3}}^{\sqrt{a^2-x^2-y^2}} (x^2+y^2)\,dz\,dy\,dx$

30. $V = \displaystyle\int_0^{\pi}\int_0^{2\sin\theta}\int_0^{r\sin\theta} r\,dz\,dr\,d\theta = \int_0^{\pi}\int_0^{2\sin\theta} r^2\sin\theta\,dr\,d\theta = \frac{8}{3}\int_0^{\pi}\sin^4\theta\,d\theta = \pi$

31. $V = \displaystyle\int_0^{2\pi}\int_0^{\pi}\int_0^{a(1+\cos\phi)} \rho^2\sin\phi\,d\rho\,d\phi\,d\theta = \int_0^{2\pi}\int_0^{\pi} \frac{1}{3}a^3(1+\cos\phi)^3\sin\phi\,d\phi\,d\theta$

$$= \frac{4}{3}a^3\int_0^{2\pi} d\theta = 8\pi a^3/3$$

32. $x = y^2 + z^2$ and $x = 1 - y^2$ intersect in the curve whose projection onto the yz-plane is the ellipse $2y^2 + z^2 = 1$,

$$V = 4 \int_0^{1/\sqrt{2}} \int_0^{\sqrt{1-2y^2}} \int_{y^2+z^2}^{1-y^2} dx \, dz \, dy$$

$$= 4 \int_0^{1/\sqrt{2}} \int_0^{\sqrt{1-2y^2}} (1 - 2y^2 - z^2) dz \, dy = \frac{8}{3} \int_0^{1/\sqrt{2}} (1 - 2y^2)^{3/2} dy = \sqrt{2}\pi/4$$

33. $V = \int_0^{2\pi} \int_0^{a/\sqrt{3}} \int_{\sqrt{3}r}^{a} r \, dz \, dr \, d\theta = \int_0^{2\pi} \int_0^{a/\sqrt{3}} (ar - \sqrt{3}r^2) dr \, d\theta = \frac{1}{18} a^3 \int_0^{2\pi} d\theta = \pi a^3/9$

34. $\bar{x} = 0$ from the symmetry of the region, $A = \pi ab/2$,

$$\bar{y} = \frac{1}{A} \int_{-a}^{a} \int_0^{b\sqrt{1-(x/a)^2}} y \, dy \, dx = \frac{2}{\pi ab} (2ab^2/3) = 4b/(3\pi); \text{ centroid } \left(0, \frac{4b}{3\pi}\right)$$

35. $\bar{x} = 0$ from the symmetry of the region,

$$A = \int_0^{2\pi} \int_0^{a(1+\sin\theta)} r \, dr \, d\theta = 3\pi a^2/2,$$

$$\bar{y} = \frac{1}{A} \int_0^{2\pi} \int_0^{a(1+\sin\theta)} r^2 \sin\theta \, dr \, d\theta = \frac{1}{A} \int_0^{2\pi} \frac{1}{3} a^3 (1 + \sin\theta)^3 \sin\theta \, d\theta \text{ but}$$

$$(1 + \sin\theta)^3 \sin\theta = \sin\theta + 3\sin^2\theta + 3\sin^3\theta + \sin^4\theta \text{ and } \int_0^{2\pi} \sin\theta \, d\theta = \int_0^{2\pi} 3\sin^3\theta \, d\theta = 0$$

$$\text{so } \bar{y} = \frac{1}{A} \int_0^{2\pi} \frac{1}{3} a^3 (3\sin^2\theta + \sin^4\theta) d\theta - \frac{2}{3\pi a^2} (5\pi a^3/4) = 5a/6; \text{ centroid } (0, 5a/6)$$

36. $\bar{y} = 0$ from the symmetry of the region, $A = 2 \int_0^4 \int_{y^2/4}^{y^2/8+2} dx \, dy = 32/3$,

$$\bar{x} = \frac{2}{A} \int_0^4 \int_{y^2/4}^{y^2/8+2} x \, dx \, dy = (3/16)(128/15) = 8/5; \text{ centroid } (8/5, 0)$$

37. $\bar{x} = 0$ from the symmetry of density and region, $M = 2 \int_0^a \int_0^{b(1-x/a)} kx \, dy \, dx = ka^2 b/3$,

$$\bar{y} = \frac{2}{M} \int_0^a \int_0^{b(1-x/a)} k \, xy \, dy \, dx = \frac{6}{ka^2 b} (ka^2 b^2/24) = b/4; \text{ center of gravity } (0, b/4)$$

38. $\bar{y} = 0$ from the symmetry of density and region,

$$M = 2 \int_0^{\pi/3} \int_{1+\cos\theta}^{3\cos\theta} kr^2 \sin\theta \, dr \, d\theta = 115k/48,$$

$$\bar{x} = \frac{2}{M} \int_0^{\pi/3} \int_{1+\cos\theta}^{3\cos\theta} kr^3 \cos\theta \sin\theta \, dr \, d\theta = \frac{2}{M} \int_0^{\pi/3} \frac{k}{4}[81\cos^4\theta - (1+\cos\theta)^4]\cos\theta \sin\theta \, d\theta$$

$$= \frac{k}{2M}\left[\int_0^{\pi/3} 81\cos^5\theta \sin\theta \, d\theta - \int_0^{\pi/3}(1+\cos\theta)^4 \cos\theta \sin\theta \, d\theta\right]$$

and with $u = \cos\theta$, $v = 1 + \cos\theta$

$$\bar{x} = \frac{k}{2M}\left[-\int_1^{1/2} 81u^5 du + \int_2^{3/2} v^4(v-1)dv\right]$$

$$= (24/115)(1701/128 - 7463/1920) = 4513/2300; \text{ center of gravity } (4513/2300, 0)$$

39. $M = \displaystyle\int_0^a \int_0^{b(1-x/a)} \int_0^{c(1-x/a-y/b)} kz \, dz \, dy \, dx = \frac{1}{2}kc^2 \int_0^a \int_0^{b(1-x/a)}(1 - x/a - y/b)^2 dy \, dx$

$$= \frac{1}{6}kbc^2 \int_0^a (1 - x/a)^3 dx = \frac{1}{24}kabc^2$$

40. $M = \displaystyle\int_0^{2\pi} \int_0^{\pi} \int_0^a 2\rho^3 \sin\phi \, d\rho \, d\phi \, d\theta = 2\pi a^4$

41. $\bar{x} = 0$ from the symmetry of the region, $V = 2\displaystyle\int_0^2 \int_{x^2}^4 \int_0^{4-y} dz \, dy \, dx = 256/15,$

$$\bar{y} = \frac{2}{V}\int_0^2 \int_{x^2}^4 \int_0^{4-y} y \, dz \, dy \, dx = (15/128)(512/35) = 12/7,$$

$$\bar{z} = \frac{2}{V}\int_0^2 \int_{x^2}^4 \int_0^{4-y} z \, dz \, dy \, dx = (15/128)(1024/105) = 8/7; \text{ centroid } (0, 12/7, 8/7)$$

42. $\bar{x} = \bar{y} = 0$ from the symmetry of the region,

$$V = \int_0^{2\pi} \int_0^{\phi_0} \int_0^a \rho^2 \sin\phi \, d\rho \, d\phi \, d\theta = \frac{2}{3}\pi a^3(1 - \cos\phi_0),$$

$$\bar{z} = \frac{1}{V}\int_0^{2\pi} \int_0^{\phi_0} \int_0^a \rho^3 \cos\phi \sin\phi \, d\rho \, d\phi \, d\theta = \frac{\pi a^4(1 - \cos^2\phi_0)/4}{2\pi a^3(1 - \cos\phi_0)/3} = \frac{3}{8}a(1 + \cos\phi_0);$$

centroid $\left(0, 0, \dfrac{3}{8}a(1 + \cos\phi_0)\right)$

43. $\bar{x} = \bar{y} = 0$ from the symmetry of the region, $V = \pi R^2 h/3,$

$$\bar{z} = \frac{1}{V}\int_0^{2\pi} \int_0^R \int_0^{h(1-r/R)} zr \, dz \, dr \, d\theta = \frac{3}{\pi R^2 h}(\pi R^2 h^2/12) = h/4; \text{ centroid } (0, 0, h/4)$$

CHAPTER 18

Topics in Vector Calculus

SECTION 18.1

18.1.1 Sketch the vector field, $\mathbf{F}(x,y) = i - 2j$, by drawing some typical non-intersecting vectors. The vectors need not be drawn to the same scale as the coordinate axes, but they should be in the correct proportions relative to each other.

18.1.2 Sketch the vector field, $\mathbf{F}(x,y) = -xj$, by drawing some typical non-intersecting vectors. The vectors need not be drawn to the same scale as the coordinate axes, but they should be in the correct proportions relative to each other.

18.1.3 Sketch the vector field, $\mathbf{F}(x,y) = xi - yj$, by drawing some typical non-intersecting vectors. The vectors need not be drawn to the same scale as the coordinate axes, but they should be in the correct proportions relative to each other.

18.1.4 Sketch the vector field, $\mathbf{F}(x,y) = 2xj$ by drawing some typical non-intersecting vectors. The vectors need not be drawn to the same scale as the coordinate axes, but they should be in the correct proportions relative to each other.

18.1.5 Sketch the vector field, $\mathbf{F}(x,y) = \sqrt{x}\,i$ by drawing some typical non-intersecting vectors. The vectors need not be drawn to the same scale as the coordinate axes, but they should be in the correct proportions relative to each other.

18.1.6 Sketch the vector field, $\mathbf{F}(x,y) = \sqrt{x}\,j$ by drawing some typical non-intersecting vectors. The vectors need not be drawn to the same scale as the coordinate axes, but they should be in the correct proportions relative to each other.

18.1.7 Find div $\mathbf{F}$ and curl $\mathbf{F}$ of $\mathbf{F}(x,y,z) = x^2 y\mathbf{i} + xy^2\mathbf{j} + (xyz)\mathbf{k}$.

18.1.8 Find div $\mathbf{F}$ and curl $\mathbf{F}$ of $\mathbf{F}(x,y,z) = \cosh x\,i + \sinh y\mathbf{j} + \ln(xy)\mathbf{k}$

18.1.9 Find div $\mathbf{F}$ and curl $\mathbf{F}$ of $\mathbf{F}(x,y,z) = e^x \cos y\,i + e^x \sin y\mathbf{j} + z\mathbf{k}$

18.1.10 Find div $\mathbf{F}$ and curl $\mathbf{F}$ of $\mathbf{F}(x,y,z) = x^3 y\mathbf{i} + xy^3\mathbf{j} + 2\mathbf{k}$

18.1.11 Find div $\mathbf{F}$ and curl $\mathbf{F}$ of $\mathbf{F}(x,y,z) = \sin x\mathbf{i} + \cos y\mathbf{j} + xyz\mathbf{k}$

18.1.12 Find div $\mathbf{F}$ and curl $\mathbf{F}$ of $\mathbf{F}(x,y,z) = ye^{x^2}i + ze^{y^2}j + xe^{z^2}k$

18.1.13   Sketch the gradient field of $\phi(x, y) = -x - y$

18.1.14   Sketch the gradient field of $\phi(x, y) = -x + y$

18.1.15   Sketch the gradient field of $\phi(x, y) = x + 2y$

18.1.16   Sketch the gradient field of $\phi(x, y) = 2x - y$

18.1.17   Sketch the gradient field of $\phi(x, y) = xy + x$

18.1.18   Sketch the gradient field of $\phi(x, y) = y + xy$

SOLUTIONS

SECTION 18.1

18.1.1

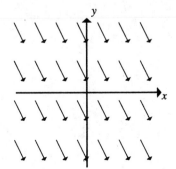

18.1.2

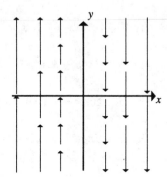

18.1.3

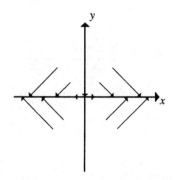

18.1.4

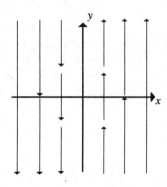

18.1.5

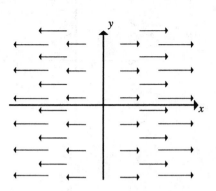

18.1.6

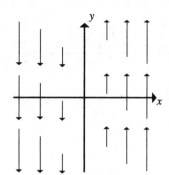

18.1.7 div $\mathbf{F} = 2xy + 2xy + xy = 5xy$. Curl $\mathbf{F} = xz\,\mathbf{i} - yz\,\mathbf{j} + (y^2 - x^2)\mathbf{k}$.

18.1.8 div $\mathbf{F} = \sinh x + \cosh y$. Curl $\mathbf{F} = \dfrac{1}{y}\mathbf{i} - \dfrac{1}{x}\mathbf{j}$.

18.1.9 div $\mathbf{F} = 2e^x \cos y + 1$. Curl $\mathbf{F} = 2e^x \sin y\mathbf{k}$.

18.1.10 Div $\mathbf{F} = 3x^2y + 3xy^2$. Curl $\mathbf{F} = (y^3 - x^3)\mathbf{k}$.

18.1.11 Div $\mathbf{F} = \cos x - \sin y + xy$. Curl $\mathbf{F} = xz\mathbf{i} - yz\mathbf{j}$

18.1.12 Div $\mathbf{F} = 2xye^{x^2} + 2yze^{y^2} + 2xze^{z^2}$. Curl $\mathbf{F} = -e^{y^2}\mathbf{i} - e^{z^2}\mathbf{j} - e^{x^2}\mathbf{k}$

18.1.13

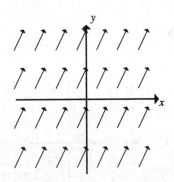

18.1.14

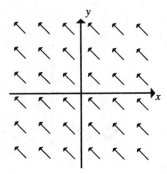

18.1.15

18.1.16

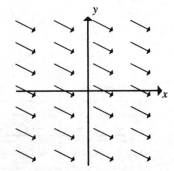

18.1.17

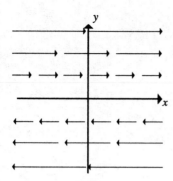

18.1.18

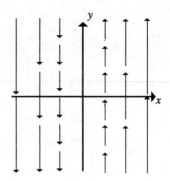

SECTION 18.2

18.2.1 Find the area of the surface extending upward from the parabola $y = x^2 (0 \le x \le 3)$ to the plane $z = 4x$.

18.2.2 Find the area of the surface extending upward from the semicircle $y = \sqrt{25 - x^2}$ to the surface $z = xy$.

18.2.3 Evaluate the line integral $\int_c \dfrac{1}{1 + x^2} ds$, where C is the curve $x = t$, $y = \dfrac{t^2}{2}$, $0 \le t \le 2$.

18.2.4 Evaluate the line integral $\int_c \dfrac{ze^{(z^2+2)}}{x^2 + y^2} ds$, where C is the helix $x = \cos t$, $y = \sin t$, $z = t$, $0 \le t \le 2\pi$.

18.2.5 Evaluate $\int_c 2xy \, dx + (e^x + x^2) dy$ where C is the line segment from $(0,0)$ to $(1,1)$.

18.2.6 Evaluate $\int_c y^2 dx - x^2 dy$ where C is the line segment from $(0,1)$ to $(1,0)$.

18.2.7 Evaluate $\int_c xy \, dx - y^2 dy$ where C is the line segment from $(0,0)$ to $(2,1)$.

18.2.8 Evaluate $\int_c (x^2 - y^2) dx - 2xy \, dy$ where C is the parabola $y = 2x^2$ from $(0,0)$ to $(1,2)$.

18.2.9 Evaluate $\int_c (3x^2 + y) dx + 4xy \, dy$ where C is the path from $(0,0)$ to $(2,0)$ to $(0,4)$ to $(0,0)$.

18.2.10 Evaluate $\int_c (e^x - 3y) dx + (e^y + 6x) dy$ where C is the path from $(0,0)$ to $(1,0)$ to $(0,2)$ to $(0,0)$.

18.2.11 Evaluate $\int_c \mathbf{F} \cdot d\mathbf{r}$ where $\mathbf{F}(x,y,z) = -yz\mathbf{i} - xz\mathbf{j} + (1 + xy)\mathbf{k}$ and C is the circular helix $\mathbf{r}(t) = 2\cos t\mathbf{i} + 2\sin t\mathbf{j} + 3t\mathbf{k}$ from $(2,0,0)$ to $(2,0,6\pi)$.

18.2.12 Evaluate $\int_c \mathbf{F} \cdot d\mathbf{r}$ where $\mathbf{F}(x,y) = x^2y\mathbf{i} + 4\mathbf{j}$ and C is the curve $\mathbf{r}(t) = e^t\mathbf{i} + e^{-t}\mathbf{j}$ for $0 \le t \le 1$.

18.2.13   Evaluate $\int_c \mathbf{F} \cdot d\mathbf{r}$ where $\mathbf{F}(x, y, z) = z\mathbf{i} + x\mathbf{j} + y\mathbf{k}$ and C is the helix

$\mathbf{r}(t) = \sin t\mathbf{i} + 3\sin t\mathbf{j} + \sin^2 t\mathbf{k}$ for $0 \le t \le \dfrac{\pi}{2}$.

18.2.14   Find the work done by the force $\mathbf{F}(x, y, z) = 2x\mathbf{i} + 3xy\mathbf{j} + 4z\mathbf{k}$ acting on a particle that moves along the curve $\mathbf{r}(t) = t\mathbf{i} + 2t^2\mathbf{j} + 3t^3\mathbf{k}$ from the origin to $(1, 2, 3)$.

18.2.15   Find the work done by the force $\mathbf{F}(x, y) = y^2\mathbf{i} - 2x^2\mathbf{j}$ acting on a particle that moves:

    (a)   along the line segment from $(0, 2)$ to $(1, 1)$.
    (b)   along the circle $x = \cos t$, $y = \sin t$, from $(1, 0)$ to $(0, 1)$.

18.2.16   Find the work done by the force $\mathbf{F}(x, y) = 2\left[\dfrac{-y\mathbf{i} + x\mathbf{j}}{x^2 + y^2}\right]$ acting on a particle that moves along the curve $\mathbf{r}(t) = 3\cos t\mathbf{i} + 3\sin t\mathbf{j}$ for $0 \le t \le 2\pi$.

18.2.17   Evaluate $\int_c \mathbf{F} \cdot d\mathbf{r}$ where $\mathbf{F}(x, y, z) = -y\mathbf{i} + x\mathbf{j} + z\mathbf{k}$ and C is the circle $x = \cos t$, $y = \sin t$ for $0 \le t \le 2\pi$.

18.2.18   Find the work done by the force $\mathbf{F}(x, y, z) = z\mathbf{i} + y\mathbf{j} + x\mathbf{k}$ acting on a particle that moves along the curve $\mathbf{r}(t) = \mathbf{i} + \sin t\mathbf{j} + \cos t\mathbf{k}$ for $0 \le t \le \dfrac{\pi}{3}$.

18.2.19   Find the work done by the force $\mathbf{F}(x, y, z) = y\mathbf{i} + x^2\mathbf{j} + z^2\mathbf{k}$ acting on a particle that moves along the curve $\mathbf{r}(t) = \sqrt{t}\,\mathbf{i} + t\mathbf{j} + \pi t\mathbf{k}$ for $1 \le t \le 2$.

18.2.20   Evaluate $\int_c \mathbf{F} \cdot d\mathbf{r}$ where $\mathbf{F}(x, y, z) = 4xy\mathbf{i} - 8y\mathbf{j} + 3\mathbf{k}$ and C is the curve given by $y = 2x$, $z = 3$ from $(0, 0, 3)$ to $(3, 6, 3)$.

18.2.21   Find the work done by the force $\mathbf{F}(x, y, z) = (x^2 - y)\mathbf{i} + (y^2 - z)\mathbf{j} + (z^2 - x)\mathbf{k}$ acting on a particle that moves along the curve given by $y = x^2$, $z = x^3$ from $(0, 0, 0)$ to $(1, 1, 1)$.

18.2.22   Evaluate $\int_c \mathbf{F} \cdot d\mathbf{r}$ where $\mathbf{F}(x, y, z) = x\sin y\mathbf{i} + \cos y\mathbf{j} + (x + y)\mathbf{k}$ and C is the straight line $x = y = z$ from $(0, 0, 0)$ to $(1, 1, 1)$.

18.2.23   Find the work done by the force $\mathbf{F}(x, y, z) = z\sin x\mathbf{i} + y\sin x\mathbf{j} + yz\cos x\mathbf{k}$ acting on a particle that moves along the straight line $x = y = z$ from $(0, 0, 0)$ to $(1, 1, 1)$.

18.2.24   Find the mass of a thin wire shaped in the form of the circular arc $y = \sqrt{4 - x^2}$ $(0 \le x \le 2)$ if the density function is $f(x, y) = kxy^{3/2}$ $(k > 0)$.

SOLUTIONS

SECTION 18.2

18.2.1 Denote the parabola by C and represent it as $\mathbf{r}(t) = t\mathbf{i} + t^2\mathbf{j}$ $(0 \le t \le 3)$.

$$A = \int_c 4x\, ds = \int_0^3 (4t)\sqrt{(1)^2 + (2t)^2}\, dt = 4\int_0^3 t\sqrt{1 + 4t^2}\, dt = \frac{37^{3/2} - 1}{3}.$$

18.2.2 Denote the semicircle by C and represent it as $r(t) = 5\cos t\, i + 5\sin t\, j$ $(0 \le t \le \pi)$.

$$A = \int_c xy\, ds = 2\int_0^{\pi/2} (5\cos t)(5\sin t)\sqrt{(-5\sin t)^2 + (5\cos t)^2}\, dt$$

$$= 250\int_0^{\pi/2} \cos t \sin t\, dt = 125.$$

18.2.3 $ds = \sqrt{1 + t^2}\, dt$

$$\int_c \frac{1}{1 + x^2}\, ds = \int_0^2 \frac{1}{1 + t^2}\sqrt{1 + t^2}\, dt = \int_0^2 \frac{1}{\sqrt{1 + t^2}}\, dt = \ln(\sqrt{5} + 2).$$

18.2.4 $ds = \sqrt{(-\sin t)^2 + (\cos t)^2 + (1)^2} = \sqrt{2}$

$$\int_c \frac{ze^{(z^2 + 2)}}{x^2 + y^2}\, ds = \sqrt{2}\int_0^{2\pi} te^{(t^2 + 2)}\, dt = \frac{\sqrt{2}}{2}\left[e^{(4\pi^2 + 2)} - e^2\right].$$

18.2.5 $x = y = t$; $dx = dy = dt$; $\displaystyle\int_0^1 (3t^2 + e^t)\, dt = e$.

18.2.6 $x = t$, $y = 1 - t$; $dx = dt$, $dy = -dt$; $\displaystyle\int_0^1 (1 - 2t + 2t^2)\, dt = \frac{2}{3}$.

18.2.7 $x = 2t$, $y = t$; $dx = 2dt$, $dy = dt$; $\displaystyle\int_0^1 3t^2\, dt = 1$.

18.2.8 $x = t$, $y = 2t^2$; $dx = dt$, $dy = 4t\, dt$; $\displaystyle\int_0^1 (t^2 - 20t^4)\, dt = -\frac{11}{3}$.

18.2.9 $C_1 : x = 2t$, $y = 0$; $dx = 2dt$, $dy = 0$; $0 \le t \le 1$;

$C_2 : x = 2 - 2t$, $y = 4t$; $dx = -2dt$, $dy = 4dt$; $0 \le t \le 1$;

$C_3 : x = 0$, $y = 4 - 4t$; $dx = 0$, $dy = -4dt$; $0 \le t \le 1$;

$$\int_0^1 24t^2\, dt + \int_0^1 (-24 + 168t - 152t^2)\, dt + \int_0^1 0\, dt = \frac{52}{3}.$$

18.2.10 $C_1 : x = t,\ y = 0;\ dx = dt,\ dy = 0;\ 0 \le t \le 1;$
  $C_2 : x = 1 - t,\ y = 2t;\ dx = -dt,\ dy = 2dt;\ 0 \le t \le 1;$
  $C_3 : x = 0,\ y = 2 - t,\ dx = 0,\ dy = -dt,\ 0 \le t \le 2;$

$$\int_0^1 e^t dt + \int_0^1 (-6t - e^{1-t} + 2e^{2t} + 12)dt + \int_0^2 (-e^{2-t})dt = 9.$$

18.2.11 $\displaystyle\int_0^{2\pi} (6\sin 2t - 12t\cos 2t + 3)dt = 6\pi.$

18.2.12 $\displaystyle\int_0^1 (e^{2t} - 4e^{-t})dt = \frac{e^2}{2} + \frac{4}{e} - \frac{9}{2}.$

18.2.13 $\displaystyle\int_0^{\pi/2} (7\sin^2 t \cos t + 3\sin t \cos t)dt = \frac{23}{6}.$

18.2.14 $\displaystyle\int_0^1 (2t + 24t^4 + 108t^5)dt = \frac{119}{5}.$

18.2.15   **(a)** $x = t,\ y = 2 - t;\ dx = dt,\ dy = -dt,\ \mathbf{r}(t) = t\mathbf{i} + (2 - t)\mathbf{j};$

$$\int_c \mathbf{F} \cdot d\mathbf{r} = \int_0^1 (3t^2 - 4t + 4)dt = 3.$$

  (b) $\displaystyle\int_c \mathbf{F} \cdot d\mathbf{r} = \int_0^{\pi/2} (-2\cos^3 t - \sin^3 t)dt = -2.$

18.2.16 $\displaystyle\int_0^{2\pi} 2dt = 4\pi.$

18.2.17 $\mathbf{r}(t) = \cos t\mathbf{i} + \sin t\mathbf{j};\ \displaystyle\int_0^{2\pi} dt = 2\pi.$

18.2.18 $\displaystyle\int_0^{\pi/3} (\cos t \sin t - \sin t)dt = -\frac{1}{8}.$

18.2.19 $\displaystyle\int_1^2 \left(\frac{\sqrt{t}}{2} + t + \pi^3 t^2\right)dt = \frac{2\sqrt{2}}{3} + \frac{7}{6} - \frac{7\pi^3}{3}.$

18.2.20 $\mathbf{r}(t) = t\mathbf{i} + 2t\mathbf{j} + 3\mathbf{k};\ \displaystyle\int_0^3 (8t^2 - 32t)dt = -72.$

18.2.21 $\mathbf{r}(t) = t\mathbf{i} + t^2\mathbf{j} + t^3\mathbf{k};\ \displaystyle\int_0^1 (3t^8 + 2t^5 - 2t^4 - 3t^3)dt = -\frac{29}{60}.$

18.2.22 $\displaystyle\int_0^1 (t\sin t + \cos t + 2t)dt = 2\sin 1 - \cos 1 + 1.$

18.2.23 $\displaystyle\int_0^1 (2t\sin t + t^2\cos t)dt = \sin 1.$

18.2.24 Represent the circular arc as $r(t) = 2\cos ti + 2\sin tj, 0 \le t \le \pi/2.$

$ds = \sqrt{(-2\sin t)^2 + (2\cos t)^2}\, dt = 2\,dt.$

$$M = \int_c \delta(x,y)ds = \int_c kxy^{3/2}ds = \int_0^{\pi/2} k\cos t(\sin t)^{3/2}2\,dt = 16\sqrt{2}/5k$$

SECTION 18.3

18.3.1 Determine whether $\mathbf{F}(x, y) = 2xy\mathbf{i} + (x^2 + 2y)\mathbf{j}$ is conservative. If it is, find a potential function for it.

18.3.2 Determine whether $\mathbf{F}(x, y) = (1 + \sqrt{y})\mathbf{i} + \dfrac{x}{2\sqrt{y}}\mathbf{j}$ is conservative. If it is, find a potential function for it.

18.3.3 Determine whether $\mathbf{F}(x, y) = 3x^2 y\mathbf{i} + (x^3 + 3y^2)\mathbf{j}$ is conservative. If it is, find a potential function for it.

18.3.4 Show that $\displaystyle\int_{(0,0)}^{(2,2)} 2xy\, dx + (x^2 + 1)dy$ is independent of path and evaluate.

18.3.5 Show that $\displaystyle\int_{(0,0)}^{(1,1)} 2xy\, dx + (x^2 + 2y)dy$ is independent of path and evaluate.

18.3.6 Determine whether $\mathbf{F}(x, y) = (2x + y^3)\mathbf{i} + (3xy^2 - e^{-2y})\mathbf{j}$ is conservative. If it is, find a potential function for it.

18.3.7 Show that $\displaystyle\int_{(0,0)}^{(1,\pi/2)} (\sin y + y\sin x)dx + (x\cos y - \cos x)dy$ is independent of path and evaluate.

18.3.8 Show that $\displaystyle\int_{(1,0)}^{(2,1)} (2x^4 + 2xy^3)dx + (3y^2 x^2 + 3y^4)dy$ is independent of path and evaluate.

18.3.9 Determine whether $\mathbf{F}(x, y) = (3\cos y + 2\sin x)\mathbf{i} + (3y^2 - 3x\sin y)\mathbf{j}$ is conservative. If it is, find a potential function for it.

18.3.10 Find the work done by the conservative force
$\mathbf{F}(x, y) = (y\sec^2 x + \sec x \tan x)\mathbf{i} + (\tan x + 2y)\mathbf{j}$ as it acts on a particle moving from
$P(0, 0)$ to $Q\left(\dfrac{\pi}{4}, 1\right)$.

18.3.11 Determine whether $\mathbf{F}(x, y) = (y^2 - 2\sin y)\mathbf{i} + (2xy - 2x\cos y)\mathbf{j}$ is conservative. If it is, find a potential function for it.

18.3.12 Find $\displaystyle\int \mathbf{F} \cdot d\mathbf{r}$ where $\mathbf{F}(x, y) = (2x + 3y)\mathbf{i} + (3x - 2y)\mathbf{j}$ and C is the curve $\mathbf{r}(t) = \sin t\mathbf{i} + \cos t\sin^2 t\mathbf{j}$; $0 \le t \le \pi/2$.

18.3.13 Find $\int_C \mathbf{F} \cdot d\mathbf{r}$ where $\mathbf{F}(x, y) = (2xy^2 + 1)\mathbf{i} + 2x^2 y\mathbf{j}$ and C is the curve $\mathbf{r}(t) = e^t \sin t\,\mathbf{i} + e^t \cos t\,\mathbf{j}, 0 \leq t \leq \pi/2$.

18.3.14 Find $\int_C \mathbf{F} \cdot d\mathbf{r}$ where $\mathbf{F}(x, y) = (2x^4 + 2xy^3)\mathbf{i} + (3x^2 y^2 + 3y^4)\mathbf{j}$ and C is the curve $\mathbf{r}(t) = te^t\mathbf{i} + (1 + t)\mathbf{j}, 0 \leq t \leq 1$.

18.3.15 Find $\int_C \mathbf{F} \cdot d\mathbf{r}$ where $F(x, y) = (y + 2xe^y)i + (x + x^2 e^y)i$ and C is the curve $\mathbf{r}(t) = \sqrt{t}\mathbf{i} + \ln t\,\mathbf{j}$ for $1 \leq t \leq 4$.

18.3.16 Find the work done by the conservative force $\mathbf{F}(x, y) = -\dfrac{y}{x^2} \sinh \dfrac{y}{x}\mathbf{i} + \dfrac{1}{x} \sinh \dfrac{y}{x}\mathbf{j}$ as it acts on a particle moving from $P(1, 1)$ to $Q(2, 2)$.

18.3.17 Find the work done by the conservative force $\mathbf{F}(x, y) = -\dfrac{y}{x^2 + y^2}\mathbf{i} + \dfrac{x}{x^2 + y^2}\mathbf{j}$ as it acts on a particle moving from $P(0, 1)$ to $Q(1, 1)$.

18.3.18 Find the work done by the conservative force $\mathbf{F}(x, y) = \dfrac{2x}{x^2 + y^2}\mathbf{i} + \dfrac{2y}{x^2 + y^2}\mathbf{j}$ as it acts on a particle moving from $P(1, 0)$ to $Q(2, 3)$.

SOLUTIONS

SECTION 18.3

18.3.1 $\frac{\partial}{\partial y}(2xy) = 2x = \frac{\partial}{\partial x}(x^2 + 2y)$, conservative, so $\frac{\partial \phi}{\partial x} = 2xy$ and $\frac{\partial \phi}{\partial y} = x^2 + 2y$;

$\phi = x^2 y + k(y)$, $x^2 + k'(y) = x^2 + 2y$, $k'(y) = 2y$, $k(y) = y^2 + K$ and $\phi = x^2 y + y^2 + K$.

18.3.2 $\frac{\partial}{\partial y}(1 + \sqrt{y}) = \frac{1}{2\sqrt{y}} = \frac{\partial}{\partial x}\left(\frac{x}{2\sqrt{y}}\right)$, conservative, so $\frac{\partial \phi}{\partial x} = 1 + \sqrt{y}$ and $\frac{\partial \phi}{\partial y} = \frac{x}{2\sqrt{y}}$;

$\phi = x + x\sqrt{y} + k(y)$, $\frac{x}{2\sqrt{y}} + k'(y) = \frac{x}{2\sqrt{y}}$, $k'(y) = 0$, $k(y) = K$ and $\phi = x + x\sqrt{y} + K$.

18.3.3 $\frac{\partial(3x^2 y)}{\partial y} = 3x^2 = \frac{\partial(x^3 + 3y^2)}{\partial x}$, conservative, so $\frac{\partial \phi}{\partial x} = 3x^2 y$ and $\frac{\partial \phi}{\partial y} = x^3 + 3y^2$;

$\phi = x^3 y + k(y)$, $x^3 + k'(y) = x^3 + 3y^2$, $k'(y) = 3y^2$, $k(y) = y^3 + K$ and

$\phi = x^3 y + y^3 + K$.

18.3.4 $\frac{\partial}{\partial y}(2xy) = 2x = \frac{\partial}{\partial x}(x^2 + 1)$, $\phi = x^2 y + y$, $\phi(2, 2) - \phi(0, 0) = 10$.

18.3.5 $\frac{\partial}{\partial y}(2xy) = 2x = \frac{\partial}{\partial x}(x^2 + 2y)$, $\phi = x^2 y + y^2$, $\phi(1, 1) - \phi(0, 0) = 2$.

18.3.6 $\frac{\partial}{\partial y}(2x + y^3) = 3y^2 = \frac{\partial}{\partial x}(3xy^2 - e^{-2y})$, conservative, so $\frac{\partial \phi}{\partial x} = 2x + y^3$ and

$\frac{\partial \phi}{\partial y} = 3xy^2 - e^{-2y}$, $\phi = x^2 + xy^3 + k(y)$, $3xy^2 + k'(y) = 3xy^2 - e^{-2y}$, $k'(y) = -e^{-2y}$,

$k(y) = \frac{1}{2}e^{-2y} + K$ and $\phi = x^2 + xy^3 + \frac{1}{2}e^{-2y} + K$.

18.3.7 $\frac{\partial}{\partial y}(\sin y + y \sin x) = \cos y + \sin x = \frac{\partial}{\partial x}(x \cos y - \cos x)$, $\phi = x \sin y - y \cos x$,

$\phi(1, \pi/2) - \phi(0, 0) = 1 - \frac{\pi}{2}\cos 1$.

18.3.8 $\frac{\partial}{\partial y}(2x^4 + 2xy^3) = 6xy^2 = \frac{\partial}{\partial x}(3y^2 x^2 + 3y^4)$, $\phi = \frac{2}{5}x^5 + x^2 y^3 + \frac{3}{5}y^5$, $\phi(2, 1) - \phi(1, 0) = 17$.

18.3.9 $\frac{\partial(3\cos y + 2\sin x)}{\partial y} = -3\sin y = \frac{\partial(3y^2 - 3x\sin y)}{\partial x}$, conservative, so

$\frac{\partial \phi}{\partial x} = 3\cos y + 2\sin x$ and $\frac{\partial \phi}{\partial y} = 3y^2 - 3x\sin y$;

$\phi = 3x\cos y - 2\cos x + k(y)$, $-3x\sin y + k'(y) = -3x\sin y + 3y^2$, $k'(y) = 3y^2$, $k(y) = y^3 + K$,

and $\phi = 3x\cos y - 2\cos x + y^3 + K$.

18.3.10 $\phi = y \tan x + \sec x + y^2$, $\phi\left(\dfrac{\pi}{4}, 1\right) - \phi(0,0) = 1 + \sqrt{2}$.

18.3.11 $\dfrac{\partial(y^2 - 2\sin y)}{\partial y} = 2y - 2\cos y = \dfrac{\partial(2xy - 2x\cos y)}{\partial x}$, conservative, so $\dfrac{\partial \phi}{\partial x} = y^2 - 2\sin y$ and

$\dfrac{\partial \phi}{\partial y} = 2xy - 2x\cos y$; $\phi = xy^2 - 2x\sin y + k(y)$, $2xy + 2x\cos y + k'(y) = 2xy - 2x\cos y$,

$k'(y) = 0$, $k(y) = K$, and $\phi = xy^2 - 2x\sin y + K$.

18.3.12 $\dfrac{\partial}{\partial y}(2x + 3y) = 3 = \dfrac{\partial}{\partial x}(3x - 2y)$ so $\mathbf{F}$ is conservative, then,

$\phi = \displaystyle\int (2x + 3y)dx + K(y) = x^2 + 3xy + k(y)$; $\dfrac{\partial \phi}{\partial y} = 3x + k'(y) = 3x - 2y$,

$k'(y) = -2y$, $k(y) = -y^2 + K$ thus, $\phi = x^2 + 3xy - y^2 + K$. $\tilde{r}(0) = \langle 0, 0 \rangle$,

$\tilde{r}(\pi/2) = \langle 1, 0 \rangle$ so $\displaystyle\int \mathbf{F} \cdot d\mathbf{r} = \phi(1, 0) - \phi(0,0) = 1 - 0 = 1$.

18.3.13 $\dfrac{\partial}{\partial y}(2xy^2 + 1) = 4xy = \dfrac{\partial}{\partial x}(2x^2 y)$ so $\mathbf{F}$ is conservative, then,

$\phi = \displaystyle\int (2xy^2 + 1)dx + k(y) = x^2 y^2 + x + k(y)$; $\dfrac{\partial \phi}{\partial y} = 2x^2 y + k'(y) = 2x^2 y$

$k'(y) = 0$, $k(y) = K$, thus $\phi = x^2 y^2 + x + K$. $\mathbf{r}(0) = \langle 0, 1 \rangle$, $\mathbf{r}(\pi/2) = \langle e^{\pi/2}, 0 \rangle$

so $\displaystyle\int \mathbf{F} \cdot d\mathbf{r} = \phi(e^{\pi/2}, 0) - \phi(0, 1) = e^{\pi/2}$.

18.3.14 $\dfrac{\partial}{\partial y}(2x^4 + 2xy^3) = 6xy^2 = \dfrac{\partial}{\partial x}(3x^2 y^2 + 3y^4)$ so $\mathbf{F}$ is conservative, then,

$\phi = \displaystyle\int (2x^4 + 2xy^3)dx + k(y) = \dfrac{2}{5}x^5 + x^2 y^3 + k(y)$; $\dfrac{\partial \phi}{\partial y} = 3x^2 y^2 + k'(y) = 3x^2 y^2 + 3y^4$,

$k'(y) = 3y^4$, $k(y) = \dfrac{3}{5}y^5 + K$, thus, $\phi = \dfrac{2}{5}x^5 + x^2 y^3 + \dfrac{3}{5}y^5 + K$. $\mathbf{r}(0) = \langle 0, 1 \rangle$, $\mathbf{r}(1) = \langle e, 2 \rangle$ so

$\displaystyle\int \mathbf{F} \cdot d\mathbf{r} = \phi(e, 2) - \phi(0, 1) = \dfrac{2}{5}e^5 + 8e^2 + \dfrac{93}{5}$.

18.3.15 $\dfrac{\partial}{\partial y}(y + 2xe^y) = 1 + 2xe^y = \dfrac{\partial}{\partial x}(x + x^2 e^y)$ so $\mathbf{F}$ is conservative, then,

$\phi \displaystyle\int (y + 2xe^y)dx + k(y) = xy + x^2 e^y + k(y)$;

$\dfrac{\partial \phi}{\partial y} = x + x^2 e^y + k'(y) = x + x^2 e^y$, $k'(y) = 0$, $k(y) = K$, thus, $\phi = xy + x^2 e^y + K$.

$\mathbf{r}(1) = \langle 1, 0 \rangle$, $\mathbf{r}(4) = \langle 2, \ln 4 \rangle$ so

$\displaystyle\int \mathbf{F} \cdot d\mathbf{r} = \phi(2, \ln 4) - \phi(1, 0) = 15 + 4\ln 2$.

18.3.16 $\phi = \cosh \dfrac{y}{x}$, $w = \phi(2,2) - \phi(1,1) = \cosh 1 - \cosh 1 = 0$.

18.3.17 $\phi = -\tan^{-1} \dfrac{x}{y}$, $w = \phi(1,1) - \phi(0,1) = -\dfrac{\pi}{4}$.

18.3.18 $\phi = \ln(x^2 + y^2)$, $w = \phi(2,3) - \phi(1,0) = \ln 13$.

18.4.1   Use Green's Theorem to evaluate $\int_c (3x^2 + y)dx + 4xy\,dy$ where C is the triangular region with vertices $(0,0)$, $(2,0)$ and $(0,4)$. Assume that the curve is traversed in a counterclockwise manner.

18.4.2   Use Green's Theorem to evaluate $\int_c (2xy - y^2)dx + (x^2 - y^2)dy$ where C is the boundary of the region enclosed by $y = x$ and $y = x^2$. Assume that the curve c is traversed in a counterclockwise manner.

18.4.3   Use Green's Theorem to evaluate $\int_c (3x^2 + y)dx + 4y^2 dy$ where C is the boundary of the region enclosed by $x = y^2$ and $y = \dfrac{x}{2}$ traversed in a counterclockwise manner.

18.4.4   Use Green's Theorem to evaluate $\int_c (y - \sin x)dx + \cos x dy$ where C is the boundary of the region with vertices $(0,0)$, $\left(\dfrac{\pi}{2},0\right)$, and $\left(\dfrac{\pi}{2},1\right)$ traversed in a counterclockwise manner.

18.4.5   Use Green's Theorem to evaluate $\int (3x^2 + y)dx + 2xy^3 dy$ where C is the rectangle bounded by $x = -1$, $x = 3$, $y = 0$, and $y = 2$.

18.4.6   Use Green's Theorem to evaluate $\int_c (2xy - y^2)dx + x^2 dy$ where C is the boundary of the region enclosed by $y = x + 1$ and $y = x^2 + 1$, traversed in a counterclockwise manner.

18.4.7   Use Green's Theorem to evaluate $\int_c (x^3 - 3y)dx + (x + \sin y)dy$ where C is the boundary of the triangular region with vertices $(0,0)$, $(1,0)$, and $(0,2)$ traversed in a counterclockwise manner.

18.4.8   Use Green's Theorem to evaluate $\int_c (x^2 - \cosh y)dx + (y + \sin x)dy$ where C is the boundary of the region enclosed by $0 \le x \le \pi$, and $0 \le y \le 1$, traversed in a counterclockwise manner.

18.4.9   Use Green's Theorem to evaluate $\int_c -xy^2 dx + x^2 y\,dy$ where C is the boundary of the region in the first quadrant enclosed by $y = 1 - x^2$ traversed in a counterclockwise manner.

18.4.10   Use Green's Theorem to evaluate $\int_c y^3 dx + (x^3 + 3xy^2)dy$ where C is the boundary of the region enclosed by $y = x^2$ and $y = x$ traversed in a counterclockwise manner.

18.4.11 Use Green's Theorem to evaluate $\int_c -x^2y\,dx + xy^2\,dy$ where C is the boundary of the circle $x^2 + y^2 = 16$ traversed in a counterclockwise manner.

18.4.12 Use Green's Theorem to evaluate $\int 3x^2y\,dx + (e^{2x} + x^3)\,dy$ where C is the boundary of the triangular region with vertices $(0,0)$, $(a,0)$, and (a,a) traversed in a counterclockwise manner, and a is positive.

18.4.13 Use a line integral to find the area of the region in the first quadrant enclosed by $y = x$ and $y = x^3$.

18.4.14 Use a line integral to find the area of the region enclosed by $y = 1 - x^4$ and $y = 0$.

18.4.15 Use Green's Theorem to evaluate $\int_C 2\tan^{-1}\frac{y}{x}\,dx + \ln(x^2 + y^2)\,dy$ where C is the boundary of the circle $(x-2)^2 + y^2 = 1$ traversed in a counterclockwise manner.

18.4.16 Use a line integral to find the area of the region enclosed by $x^2 + 4y^2 = 4$.

18.4.17 Use a line integral to find the area of the region enclosed by $y = x$ and $y = x^2$.

18.4.18 Use a line integral to find the area of the region enclosed by $y = \sin x$, $y = \cos x$, and $x = 0$.

18.4.19 A particle, starting at $(1,0)$, traverses the upper semicircle $x^2 + y^2 = 1$ and returns to its starting point along the x-axis. Use Green's Theorem to find the work done on the particle by a force $\mathbf{F}(x,y) = xy^2\mathbf{i} + \left(\frac{1}{3}x^3 + x^2y\right)\mathbf{j}$.

SOLUTIONS

SECTION 18.4

18.4.1 $\frac{\partial}{\partial x}(4xy) = 4y$, $\frac{\partial}{\partial y}(3x^2 + y) = 1$, $\int_0^2 \int_0^{-2x+4} (4y - 1)dy\,dx = \frac{52}{3}$.

18.4.2 $\frac{\partial}{\partial x}(x^2 - y^2) = 2x$, $\frac{\partial}{\partial y}(2xy - y^2) = 2x - 2y$, $\int_0^1 \int_{x^2}^{x} 2y\,dy\,dx = \frac{2}{15}$.

18.4.3 $\frac{\partial}{\partial x}(4y^2) = 0$, $\frac{\partial}{\partial y}(3x^2 + y) = 1$, $\int_0^4 \int_{\frac{x}{2}}^{\sqrt{x}} dy\,dx = -\frac{4}{3}$.

18.4.4 $\frac{\partial}{\partial x}(\cos x) = -\sin x$, $\frac{\partial}{\partial y}(y - \sin x) = 1$, $\int_0^{\frac{\pi}{2}} \int_0^{\frac{2x}{\pi}} (-\sin x - 1)dy\,dx = -\frac{2}{\pi} - \frac{\pi}{4}$.

18.4.5 $\frac{\partial}{\partial x}(2xy^3) = 2y^3$, $\frac{\partial}{\partial y}(3x^2 + y) = 1$, $\int_{-1}^3 \int_0^2 (2y^3 - 1)dy\,dx = 24$.

18.4.6 $\frac{\partial}{\partial x}(x^2) = 2x$, $\frac{\partial}{\partial y}(2xy - y^2) = 2x - 2y$, $\int_0^1 \int_{x^2+1}^{x+1} 2y\,dy\,dx = \frac{7}{15}$.

18.4.7 $\frac{\partial}{\partial x}(x + \sin y) = 1$, $\frac{\partial}{\partial y}(x^3 - 3y) = -3$,

$\iint\limits_R 4dA = 4[\text{area of triangle}] = 4\left[\frac{1}{2}(1)(2)\right] = 4.$

18.4.8 $\frac{\partial}{\partial x}(y + \sin x) = \cos x$, $\frac{\partial}{\partial y}(x^2 - \cosh y) = -\sinh y$,

$\int_0^{\pi} \int_0^1 (\cos x + \sinh y)dy\,dx = \pi(\cosh 1 - 1).$

18.4.9 $\frac{\partial}{\partial x}(x^2 y) = 2xy$, $\frac{\partial}{\partial y}(-xy^2) = -2xy$, $\int_0^1 \int_0^{1-x^2} 4xy\,dy\,dx = \frac{1}{3}$.

18.4.10 $\frac{\partial}{\partial x}(x^3 + 3xy^2) = 3x^2 + 3y^2$, $\frac{\partial}{\partial y}(y^3) = 3y^2$, $\int_0^1 \int_{x^2}^{x} 3x^2 dy\,dx = \frac{3}{20}$.

18.4.11 $\frac{\partial}{\partial x}(xy^2) = y^2$, $\frac{\partial}{\partial y}(-x^2 y) = -x^2$, $\iint\limits_R (y^2 + x^2)dA = \int_0^{2\pi} \int_0^4 r^3 dr\,d\theta = 128\pi$.

18.4.12 $\dfrac{\partial}{\partial x}(e^{2x} + x^3) = 2e^{2x} + 3x^2,\ \dfrac{\partial}{\partial y}(3x^2 y) = 3x^2,\ \displaystyle\int_0^a \int_0^x 2e^{2x}\,dy\,dx = \dfrac{1}{2}\left(e^{2a}(2a-1)+1\right)$

18.4.13 Take C_1 along $y = x^3$, $(0,0)$ to $(1,1)$, $x = t$, $y = t^3$, $0 \leq t \leq 1$ and C_2 along $y = x$, $(1,1)$ to
$(0,0)$, $x = 1-t$, $y = 1-t$, $0 \leq t \leq 1$. $A = \dfrac{1}{2}\displaystyle\int_0^1 2t^3\,dt + \dfrac{1}{2}\int_0^1 0\,dt = \dfrac{1}{4}$.

18.4.14 Take C_1 along $y = 0$, $(-1,0)$ to $(1,0)$, $x = t$, $y = 0$, $-1 \leq t \leq 1$ and C_2 along $y = 1 - x^4$,
$(1,0)$ to $(-1,0)$, $x = -t$, $y = 1 - t^4$, $-1 \leq t \leq 1$. $A = \displaystyle\int_c -y\,dx = 0 + \int_{-1}^1 (1-t^4)\,dt = \dfrac{8}{5}$.

18.4.15 $\dfrac{\partial}{\partial x}[\ln(x^2 + y^2)] = \dfrac{2x}{x^2 + y^2},\ \dfrac{\partial}{\partial y}\left[2\tan^{-1}\dfrac{y}{x}\right] = \dfrac{2x}{x^2 + y^2},$

$\displaystyle\iint_R \left(\dfrac{2x}{x^2 + y^2} - \dfrac{2x}{x^2 + y^2}\right) dA = 0.$

18.4.16 Let $x = 2\cos t$, $y = \sin t$, for $0 \leq t \leq 2\pi$, $A = \dfrac{1}{2}\displaystyle\int -y\,dx + x\,dy$, $A = \dfrac{1}{2}\displaystyle\int_0^{2\pi} 2\,dt = 2\pi$.

18.4.17 Take C_1 along $y = x^2$, $(0,0)$ to $(1,1)$, $x = t$, $y = t^2$, $0 \leq t \leq 1$ and $y = x$, $(1,1)$ to $(0,0)$,
$x = 1-t$, $y = 1-t$, $0 \leq t \leq 1$,

$A = \dfrac{1}{2}\displaystyle\int -y\,dx + x\,dy = \dfrac{1}{2}\int_0^1 t^2\,dt + \dfrac{1}{2}\int_0^1 0\,dt = \dfrac{1}{6}$.

18.4.18 $C_1 : y = \sin x : (0,0)$ to $\left(\dfrac{\pi}{4}, \dfrac{\sqrt{2}}{2}\right)$, $x = t$, $y = \sin t$, $0 \leq t \leq \pi/4$

$C_2 : y = \cos x : \left(\dfrac{\pi}{4}, \dfrac{\sqrt{2}}{2}\right)$ to $(0,1)$, $x = \dfrac{\pi}{4} - t$, $y = \cos\left(\dfrac{\pi}{4} - t\right)$, $0 \leq t \leq \dfrac{\pi}{4}$

$C_3 : x = 0 : (0,1)$ to $(0,0)$, $y = 1-t$, $0 \leq t \leq 1$

$A = \displaystyle\int -y\,dx = \int_0^{\pi/4} -\sin t\,dt + \int_0^{\pi/4} \cos\left(\dfrac{\pi}{4} - t\right)dt + \int_0^1 0\,dt = \sqrt{2} - 1$.

18.4.19 $\dfrac{\partial g}{\partial x} = x^2 + 2xy,\ \dfrac{\partial f}{\partial y} = 2xy$

$W = \displaystyle\iint_R x^2\,dA = \int_0^\pi \int_0^1 r^3 \cos^2 \theta\,dr\,d\theta$

$= \dfrac{1}{4}\displaystyle\int_0^\pi \cos^2 \theta\,d\theta = \dfrac{\pi}{8}$.

SECTION 18.5

18.5.1   Evaluate the surface integral $\iint_\sigma (x^2 + y^2) dS$ where σ is the portion of the cone

$z = \sqrt{3(x^2 + y^2)}$ for $0 \le z \le 3$.

18.5.2   Evaluate the surface integral $\iint_\sigma 8x\, dS$ where σ is the surface enclosed by $z = x^2$,

$0 \le x \le 2$, and $-1 \le y \le 2$.

18.5.3   Evaluate the surface integral $\iint_\sigma 3x^3 \sin y\, dS$ where σ is the surface enclosed by $z = x^3$,

$0 \le x \le 2$, and $0 \le y \le \pi$.

18.5.4   Evaluate the surface integral $\iint_\sigma (\cos x + \sin y) dS$ where σ is that portion of $x + y + z = 1$

which lies in the first octant.

18.5.5   Evaluate the surface integral $\iint_\sigma \tan^{-1}\frac{y}{x} dS$ where σ is that portion of the paraboloid

$z = x^2 + y^2$ enclosed by $1 \le z \le 9$.

18.5.6   Evaluate the surface integral $\iint_\sigma x\, dS$ where σ is that portion of the plane $x + 2y + 3z = 6$

which lies in the first octant.

18.5.7   Evaluate the surface integral $\iint_\sigma (x^2 + y^2) dS$ where σ is that portion of the plane

$z = 4x + 20$ cut by the cylinder $x^2 + y^2 = 9$.

18.5.8   Evaluate the surface integral $\iint_\sigma y\, dS$ where σ is that portion of the plane $z = x + y$ inside

the elliptic cylinder $4x^2 + 9y^2 = 36$ which lies in the first octant.

18.5.9   Evaluate the surface integral $\iint_\sigma y\, dS$ where σ is that portion of the cylinder $y^2 + z^2 = 4$

which lies above the region in the xy-plane enclosed by the lines $x + y = 1$, $x = 0$, and $y = 0$.

18.5.10   Evaluate the surface integral $\iint_\sigma y^4 dS$ where σ is that portion of the surface $z = y^4$ which

lies above the triangle in the xy-plane with vertices $(0,0)$, $(0,1)$, and $(1,1)$.

18.5.11   Evaluate the surface integral $\iint\limits_{\sigma} x^2 dS$ where σ is that portion of the surface $z = x^3$ which lies above the triangle in the xy-plane with vertices $(0,0)$, $(1,0)$, and $(1,1)$.

18.5.12   Evaluate the surface integral $\iint\limits_{\sigma} x^2 dS$ where σ is that portion of the surface $x + y + z = 1$ which lies inside the cylinder $x^2 + y^2 = 1$.

18.5.13   Evaluate the surface integral $\iint\limits_{\sigma} y^2 dS$ where σ is that portion of the plane $x + y + z = 1$ that lies in the first octant.

18.5.14   Evaluate the surface integral $\iint\limits_{\sigma} y^2 dS$ where σ is that portion of the cylinder $y^2 + z^2 = 1$ that lies above the xy-plane enclosed by $0 \le x \le 5$ and $-1 \le y \le 1$.

18.5.15   Evaluate the surface integral $\iint\limits_{\sigma} (x^2 + y^2) dS$ where σ is that portion of the cylinder $x^2 + z^2 = 1$ that lies above the xy-plane enclosed by $0 \le y \le 5$.

18.5.16   Evaluate the surface integral $\iint\limits_{\sigma} (y^2 + z^2) dS$ where σ is the portion of the cone $x = \sqrt{3(y^2 + z^2)}$ for $0 \le x \le 3$.

18.5.17   Evaluate the surface integral $\iint\limits_{\sigma} 8x \, dS$ where σ is the surface enclosed by $y = x^2$, $0 \le x \le 2$, and $-1 \le z \le 2$.

18.5.18   Evaluate the surface integral $\iint\limits_{\sigma} (\sin y + \cos z) dS$ where σ is that portion of the plane $x + y + z = 1$ which lies in the first octant.

18.5.19   Evaluate the surface integral $\iint\limits_{\sigma} xy^3 z \, dS$, where σ is the portion of the cone $\mathbf{r}(u,v) = u \cos v \mathbf{i} + u \sin v \mathbf{j} + u \mathbf{k}$ for which $1 \le u \le 2$, $0 \le v \le \pi/2$.

18.5.20   Evaluate the surface integral $\iint\limits_{\sigma} \dfrac{x^2 + z^2}{y^3} dS$ where σ is the portion of the cylinder $\mathbf{r}(u,v) = 5 \cos v \mathbf{i} + u \mathbf{j} + 5 \sin v \mathbf{k}$ for which $1 \le u \le 3$, $0 \le v \le 2\pi$.

SOLUTIONS

SECTION 18.5

18.5.1 R is the circular region enclosed by $x^2 + y^2 = 3$

$$\iint_\sigma (x^2 + y^2)dS = \iint_R (x^2 + y^2)\sqrt{\frac{9x^2}{3(x^2 + y^2)} + \frac{9y^2}{3(x^2 + y^2)} + 1}\, dA$$

$$= 2\iint_R (x^2 + y^2)dA = 2\int_0^{2\pi}\int_0^{\sqrt{3}} r^3 dr\, d\theta = 9\pi.$$

18.5.2 R is the rectangular region in the plane $z = 0$ enclosed by $0 \le x \le 2$ and $-1 \le y \le 2$.

$$\iint_\sigma 8x\, dS = \iint_R 8x\sqrt{4x^2 + 1}\, dA = \int_0^2\int_{-1}^2 8x\sqrt{4x^2 + 1}\, dy\, dx = 2(17\sqrt{17} - 1).$$

18.5.3 R is the rectangular region in the plane $z = 0$ enclosed by $0 \le x \le 2$ and $0 \le y \le \pi$.

$$\iint_\sigma 3x^3 \sin y\, dS = \iint_R 3x^3 \sin y\sqrt{9x^4 + 1}\, dA$$

$$= \int_0^2\int_0^\pi 3x^3 \sin y\sqrt{9x^4 + 1}\, dy\, dx = \frac{1}{9}(145\sqrt{145} - 1).$$

18.5.4 R is the region in the first quadrant enclosed by the coordinate axes and $x + y = 1$.

$$\iint_\sigma (\cos x + \sin y)dS = \iint_R (\cos x + \sin y)\sqrt{1 + 1 + 1}\, dA$$

$$= \sqrt{3}\int_0^1\int_0^{1-x} (\cos x + \sin y)dy\, dx = \sqrt{3}(2 - \cos 1 - \sin 1).$$

18.5.5 R is the annulus enclosed between $x^2 + y^2 = 1$ and $x^2 + y^2 = 9$.

$$\iint_\sigma \tan^{-1}\frac{y}{x}dS = \iint_R \tan^{-1}\frac{y}{x}\sqrt{4x^2 + 4y^2 + 1}\, dA$$

$$= \int_0^{2\pi}\int_1^3 \theta r\sqrt{4r^2 + 1}\, dr\, d\theta = \frac{\pi^2}{6}(37\sqrt{37} - 5\sqrt{5}).$$

18.5.6 R is the region in the first quadrant enclosed by the coordinate axes and the line $x + 2y = 6$.

$$\iint_\sigma x\, dS = \iint_R \frac{\sqrt{14}}{3}x\, dA = \int_0^3\int_0^{6-2y} \frac{\sqrt{14}}{3}x\, dx\, dy = 6\sqrt{14}.$$

18.5.7 R is the circular region enclosed by $x^2 + y^2 = 9$.

$$\iint_\sigma (x^2 + y^2)dS = \iint_R (x^2 + y^2)\sqrt{17}\, dA = \int_0^{2\pi}\int_0^3 \sqrt{17}\, r^3 dr\, d\theta = \frac{81\sqrt{17}\pi}{2}.$$

18.5.8 R is the elliptical region enclosed by $4x^2 + 9y^2 = 36$ or $0 \le x \le 3$ and $0 \le y \le 2$.

$$\iint\limits_{\sigma} y \, dS = \iint\limits_{R} \sqrt{3} y \, dA = \int_0^3 \int_0^{\sqrt{\frac{36-4x^2}{9}}} \sqrt{3} \, y \, dy \, dx = 4\sqrt{3}.$$

18.5.9 R is the region in the first quadrant enclosed by $x + y = 1$, $x = 0$, and $y = 0$.

$$\iint\limits_{\sigma} y \, dS = \iint\limits_{R} y \sqrt{\frac{4}{4-y^2}} \, dA = \int_0^1 \int_0^{1-x} \frac{2y}{\sqrt{4-y^2}} \, dy \, dx = 4 - \sqrt{3} - \frac{2\pi}{3}.$$

18.5.10 R is the triangular region in the xy-plane enclosed by the lines $x = 0$, $y = 1$, $y = x$.

$$\iint\limits_{\sigma} y^4 dS = \iint\limits_{R} y^4 \sqrt{16y^6 + 1} \, dA = \int_0^1 \int_0^y y^4 \sqrt{16y^6 + 1} \, dx \, dy = \frac{1}{144}(17\sqrt{17} - 1).$$

18.5.11 R is the triangular region in the xy-plane enclosed by the lines $x = 1$, $y = 0$, and $y = x$.

$$\iint\limits_{\sigma} x^2 dS = \iint\limits_{R} x^2 \sqrt{9x^4 + 1} \, dA = \int_0^1 \int_0^x x^2 \sqrt{9x^4 + 1} \, dy \, dx = \frac{1}{54}(10\sqrt{10} - 1).$$

18.5.12 R is the circular region enclosed by $x^2 + y^2 = 1$ in the xy-plane.

$$\iint\limits_{\sigma} x^2 dS = \iint\limits_{R} x^2 \sqrt{3} \, dA = \int_0^{2\pi} \int_0^1 \sqrt{3} \, r^3 \cos^2 \theta \, dr \, d\theta = \frac{\sqrt{3} \, \pi}{4}.$$

18.5.13 R is the triangular region in the first quadrant enclosed by $y = 0$, $x = 0$, and $y = 1 - x$.

$$\iint\limits_{\sigma} y^2 dS = \iint\limits_{R} \sqrt{3} y^2 dA = \int_0^1 \int_0^{1-x} \sqrt{3} \, y^2 dy \, dx = \frac{\sqrt{3}}{12}.$$

18.5.14 R is the rectangular region enclosed by $-1 \le y \le 1$ and $0 \le x \le 5$.

$$\iint\limits_{\sigma} y^2 dS = \iint\limits_{R} \frac{y^2}{\sqrt{1-y^2}} \, dA.$$ By symmetry of the region and the fact that the inner integral

is improper, $\displaystyle \iint\limits_{R} \frac{y^2}{\sqrt{1-y^2}} dA = \lim_{y_0 \to 1} 2 \int_0^5 \int_0^{y_0} \frac{y^2}{\sqrt{1-y^2}} dy \, dx = \frac{5\pi}{2}.$

18.5.15 R is the rectangular region enclosed by $-1 \le x \le 1$ and $0 \le y \le 5$.

$$\iint\limits_{\sigma} (x^2 + y^2) dS = \iint\limits_{R} \frac{x^2 + y^2}{\sqrt{1-x^2}} dA.$$ By symmetry of the region and the fact that the inner

integral is improper, $\displaystyle \iint\limits_{R} \frac{x^2 + y^2}{\sqrt{1-x^2}} dA = \lim_{x_0 \to 1} 2 \int_0^5 \int_0^{x_0} \frac{x^2 + y^2}{\sqrt{1-x^2}} dy \, dx = \frac{265\pi}{6}.$

18.5.16 R is the circular region in the yz-plane enclosed by $y^2 + z^2 = 3$.

$$\iint\limits_{\sigma} (y^2 + z^2)dS = \iint\limits_{R} (y^2 + z^2)\sqrt{\frac{9y^2}{3(y^2 + z^2)} + \frac{9z^2}{3(y^2 + z^2)} + 1}\, dA$$

$$= \iint\limits_{R} 2(y^2 + z^2)dA = 2\int_0^{2\pi}\int_0^{\sqrt{3}} r^3 dr\, d\theta = 9\pi.$$

18.5.17 R is the rectangular region in the xz-plane enclosed by $0 \le x \le 2$, $-1 \le z \le 2$.

$$\iint\limits_{\sigma} 8x\, dS = \iint\limits_{R} 8x\sqrt{4x^2 + 1}\, dA = \int_0^2\int_{-1}^2 8x\sqrt{4x^2 + 1}\, dz\, dx = 2(17\sqrt{17} - 1).$$

18.5.18 R is the region in the yz-plane enclosed by the coordinate axes and $y + z = 1$.

$$\iint\limits_{\sigma} (\sin y + \cos z)dS = \iint\limits_{R} (\sin y + \cos z)\sqrt{3}dA$$

$$= \sqrt{3}\int_0^1\int_0^{1-z} (\sin y + \cos z)dy\, dz = \sqrt{3}(2 - \cos 1 - \sin 1).$$

18.5.19 $\dfrac{\partial r}{\partial u} = \cos v\mathbf{i} + \sin v\mathbf{j} + \mathbf{k}$

$\dfrac{\partial r}{\partial v} = -u\sin v\mathbf{i} + u\cos v\mathbf{j}$

$\dfrac{\partial r}{\partial u} \times \dfrac{\partial r}{\partial v} = (-u\cos v)\mathbf{i} + (-u\sin v)\mathbf{j} + u\mathbf{k}$

$\left\| \dfrac{\partial r}{\partial u} \times \dfrac{\partial r}{\partial v} \right\| = \sqrt{(-u\cos v)^2 + (-u\sin v)^2 + u^2}$

$\qquad\qquad = \sqrt{2}u$

$$\iint\limits_{\sigma} xy^3 z\, dS = \int_0^{\pi/2}\int_1^2 (u\cos v)(u\sin v)^3(u)(\sqrt{2}u)du\, dv$$

$$= \sqrt{2}\int_0^{\pi/2}\int_1^2 u^6 \cos v\sin^3 v\, du\, dv = \frac{127\sqrt{2}}{28}$$

18.5.20 $\dfrac{\partial r}{\partial u} = \mathbf{j}$ $\qquad \dfrac{\partial r}{\partial v} = -5\sin v\mathbf{i} + 5\cos v\mathbf{k}$

$\dfrac{\partial r}{\partial u} \times \dfrac{\partial r}{\partial u} = (5\cos v)\mathbf{i} + (5\sin v)\mathbf{k}$

$\left\| \dfrac{\partial r}{\partial u} \times \dfrac{\partial r}{\partial v} \right\| = \sqrt{(5\cos v)^2 + (0)^2 + (5\sin v)^2} = 5$

$$\iint\limits_{\sigma} \frac{x^2 + z^2}{y^3}dS = \int_0^{2\pi}\int_1^3 \frac{(5\cos v)^2 + (5\sin v)^2}{u^3} 5du\, dv$$

$$= 125\int_0^{2\pi}\int_1^3 u^{-3}du\, dv = \frac{1000\pi}{9}$$

SECTION 18.6

18.6.1 Evaluate $\iint\limits_{\sigma} \mathbf{F} \cdot \mathbf{n}\, dS$ where $\mathbf{F}(x, y, z) = y\mathbf{i} - x\mathbf{j} + 8\mathbf{k}$ and σ is that portion of the paraboloid $z = x^2 + y^2$ that lies below the plane $z = 4$ and is oriented by downward unit normals.

18.6.2 Evaluate $\iint\limits_{\sigma} \mathbf{F} \cdot \mathbf{n}\, dS$ where $\mathbf{F}(x, y, z) = y\mathbf{i} - x\mathbf{j} + 9\mathbf{k}$ and σ is that portion of the paraboloid $z = 4 - x^2 - y^2$ that lies above $z = 0$ and is oriented by upward unit normals.

18.6.3 Evaluate $\iint\limits_{\sigma} \mathbf{F} \cdot \mathbf{n}\, dS$ where $\mathbf{F}(x, y, z) = x\mathbf{i} + y\mathbf{j} + z\mathbf{k}$ and σ is that portion of the plane $2x + 3y + 4z = 12$ which lies in the first octant and is oriented by upward unit normals.

18.6.4 Evaluate $\iint\limits_{\sigma} \mathbf{F} \cdot \mathbf{n}\, dS$ where $\mathbf{F}(x, y, z) = x\mathbf{i} + y\mathbf{j} + 2z\mathbf{k}$ and σ is that portion of the surface $z = 4 - x^2 - y^2$ above the xy-plane oriented by upward unit normals.

18.6.5 Evaluate $\iint\limits_{\sigma} \mathbf{F} \cdot \mathbf{n}\, dS$ where $\mathbf{F}(x, y, z) = y\mathbf{i} - x\mathbf{j} - 4z^2\mathbf{k}$ and σ is that portion of the cone $z = \sqrt{x^2 + y^2}$ which lies above the square in the xy-plane with vertices $(0, 0)$, $(1, 0)$, $(1, 1)$, and $(0, 1)$, and oriented by downward unit normals.

18.6.6 Evaluate $\iint\limits_{\sigma} \mathbf{F} \cdot \mathbf{n}\, dS$ where $\mathbf{F}(x, y, z) = y\mathbf{i} - x\mathbf{j} - \mathbf{k}$ and σ is that portion of the hemisphere $z = -\sqrt{4 - x^2 - y^2}$ which lies below the plane $z = 0$ and is oriented by downward unit normals.

18.6.7 Evaluate $\iint\limits_{\sigma} \mathbf{F} \cdot \mathbf{n}\, dS$ where $\mathbf{F}(x, y, z) = z\mathbf{i} + x\mathbf{j} + y\mathbf{k}$ and σ is that portion of the cylinder $x^2 + y^2 = 4$ in the first octant between $z = 0$ and $z = 4$. The surface is oriented by right unit normals.

18.6.8 Evaluate $\iint\limits_{\sigma} \mathbf{F} \cdot \mathbf{n}\, dS$ where $\mathbf{F}(x, y, z) = x\mathbf{i} + y\mathbf{j} - z\mathbf{k}$ and σ is that portion of the cone $z = \sqrt{x^2 + y^2}$ which lies in the first octant between $z = 1$ and $z = 2$. The surface is oriented by downward unit normals.

18.6.9 Evaluate $\iint\limits_{\sigma} \mathbf{F} \cdot \mathbf{n}\, dS$ where $\mathbf{F}(x, y, z) = -xy^2\mathbf{i} + z\mathbf{j} + xz\mathbf{k}$ and σ is that portion of the surface $z = xy$ bounded by $0 \leq x \leq 3$ and $0 \leq y \leq 2$. The surface is oriented by upward unit normals.

18.6.10 Evaluate $\iint\limits_{\sigma} \mathbf{F} \cdot \mathbf{n}\, dS$ where $\mathbf{F}(x, y, z) = y\mathbf{i} + 2x\mathbf{j} + xy\mathbf{k}$ and σ is that portion of the cylinder $x^2 + y^2 = 9$ in the first octant between $z = 1$ and $z = 4$. The surface is oriented by right unit normals.

18.6.11 Evaluate $\iint\limits_{\sigma} \mathbf{F} \cdot \mathbf{n}\, dS$ where $\mathbf{F}(x, y, z) = y\mathbf{i} + z\mathbf{j} + y\mathbf{k}$ and σ is that portion of the cone $x = \sqrt{y^2 + z^2}$ which lies in the first octant between $x = 1$ and $x = 3$. The surface is oriented by forward unit normals.

18.6.12 Evaluate $\iint\limits_{\sigma} \mathbf{F} \cdot \mathbf{n}\, dS$ where $\mathbf{F}(x, y, z) = x\mathbf{i} + y\mathbf{j} - 2z\mathbf{k}$ and σ is that portion of the sphere $x^2 + y^2 + z^2 = 9$ which lies above the xy-plane and is oriented by upward unit normals.

18.6.13 Evaluate $\iint\limits_{\sigma} \mathbf{F} \cdot \mathbf{n}\, dS$ where $\mathbf{F}(x, y, z) = x\mathbf{i} + 4\mathbf{j} + 2x^2\mathbf{k}$ and σ is that portion of the paraboloid $z = x^2 + y^2$ which lies above the xy-plane enclosed by the parabolas $y = 1 - x^2$ and $y = x^2 - 1$. The surface is oriented by downward unit normals.

18.6.14 Evaluate $\iint\limits_{\sigma} \mathbf{F} \cdot \mathbf{n}\, dS$ where $\mathbf{F}(x, y, z) = 2\mathbf{i} - z\mathbf{j} + y\mathbf{k}$ and σ is that portion of the paraboloid $x = y^2 + z^2$ between $x = 0$ and $x = 4$. The surface is oriented by forward unit normals.

18.6.15 Evaluate $\iint\limits_{\sigma} \mathbf{F} \cdot \mathbf{n}\, dS$ where $\mathbf{F}(x, y, z) = 9\mathbf{i} - z\mathbf{j} + y\mathbf{k}$ and σ is that portion of the paraboloid $x = 4 - y^2 - z^2$ to the right of $x = 0$ oriented by forward unit normals.

18.6.16 Evaluate $\iint\limits_{\sigma} \mathbf{F} \cdot \mathbf{n}\, dS$ where $\mathbf{F}(x, y, z) = -x\mathbf{i} - 2x\mathbf{j} + (z - 1)\mathbf{k}$ and σ is the surface enclosed by that portion of the paraboloid $z = 4 - y^2$ which lies in the first octant and is bounded by the coordinate planes and the plane $y = x$. The surface is oriented by upward unit normals.

18.6.17 Evaluate $\iint\limits_{\sigma} \mathbf{F} \cdot \mathbf{n}\, dS$ where $F(x, y, z) = j$ and σ is the portion of the plane $6x + 3y + z = 12$ in the first octant oriented by upward unit normals.

18.6.18 Evaluate $\iint\limits_{\sigma} \mathbf{F} \cdot \mathbf{n}\, dS$ where $\mathbf{F}(x, y, z) = z^2\mathbf{k}$ and σ is the upper hemisphere $z = \sqrt{4 - x^2 - y^2}$ oriented by upward unit normals.

SOLUTIONS

SECTION 18.6

18.6.1 R is the circular region enclosed by $x^2 + y^2 = 4$.

$$\iint_\sigma \mathbf{F} \cdot \mathbf{n}\, dS = \iint_R -8 dA = -8(\text{area of circle}) = -8[\pi(2)^2] = -32\pi.$$

18.6.2 R is the circular region enclosed by $x^2 + y^2 = 4$.

$$\iint_\sigma \mathbf{F} \cdot \mathbf{n}\, dS = \iint_R 9 dA = 9(\text{area of circle}) = 9[\pi(2)^2] = 36\pi.$$

18.6.3 R is the triangular region in the xy-plane enclosed by $x = 0$, $y = 0$, and $2x + 3y = 12$.

$$\iint_\sigma \mathbf{F} \cdot \mathbf{n}\, dS = \iint_R 3 dA = 3(\text{area of rt triangle}) = 3\left[\frac{1}{2}(6)(4)\right] = 36.$$

18.6.4 R is the circular region enclosed by $x^2 + y^2 = 4$.

$$\iint_\sigma \mathbf{F} \cdot \mathbf{n}\, dS = \iint_R 8\, dA = 8(\text{area of circle}) = 8\pi(2)^2 = 32\pi$$

18.6.5 R is the square in the xy-plane with vertices $(0,0)$, $(1,0)$, $(1,1)$, and $(0,1)$.

$$\iint_\sigma \mathbf{F} \cdot \mathbf{n}\, dS = \iint_R 4z^2 dA = \int_0^1 \int_0^1 4(x^2 + y^2) dy\, dx = \frac{8}{3}.$$

18.6.6 R is the circular region in the plane $z = 0$ enclosed by $x^2 + y^2 = 4$.

$$\iint_\sigma \mathbf{F} \cdot \mathbf{n}\, dS = \iint_R dA = \text{ area of circle } = \pi(2)^2 = 4\pi.$$

18.6.7 R is the region in the xz-plane for $0 \le x \le 2$ and $0 \le z \le 4$.

$$\iint_\sigma \mathbf{F} \cdot \mathbf{n}\, dS = \iint_R \left(\frac{xz}{y} + x\right) dA = \lim_{x_0 \to 2^-} \int_0^{x_0} \int_0^4 \left(\frac{xz}{\sqrt{4 - x^2}} + x\right) dz\, dx = 24.$$

18.6.8 R is the annular region enclosed between $x^2 + y^2 = 1$ and $x^2 + y^2 = 2$.

$$\iint_\sigma \mathbf{F} \cdot \mathbf{n}\, dS = \iint_R 2\sqrt{x^2 + y^2}\, dA = \int_0^{\pi/2} \int_1^2 2r^2 dr\, d\theta = \frac{7\pi}{3}.$$

18.6.9 R is the rectangular region in the xy-plane enclosed by $0 \le x \le 3$ and $0 \le y \le 2$.

$$\iint_\sigma \mathbf{F} \cdot \mathbf{n}\, dS = \iint_R xy^3 dA = \int_0^3 \int_0^2 xy^3 dy\, dx = 18.$$

18.6.10 R is the rectangular region in the xz-plane for $0 \le x \le 3$ and $1 \le z \le 4$.

$$\iint_\sigma \mathbf{F} \cdot \mathbf{n}\, dS = \iint_R 3x\, dA = \int_1^4 \int_0^3 3x\, dx\, dz = \frac{81}{2}.$$

18.6.11 R is the annular region enclosed between $y^2 + z^2 = 1$ and $y^2 + z^2 = 9$ in the yz-plane.

$$\iint_\sigma \mathbf{F} \cdot \mathbf{n}\, dS = \iint_R \left(y - \frac{2yz}{\sqrt{y^2 + z^2}} \right) dA$$

$$= \int_0^{\pi/2} \int_1^3 (r^2 \sin\theta - 2r^2 \cos\theta \sin\theta)\, dr\, d\theta = 0.$$

18.6.12 R is the circular region in the xy-plane enclosed by $x^2 + y^2 = 9$.

$$\iint_\sigma \mathbf{F} \cdot \mathbf{n}\, dS = \iint_R \frac{3(x^2 + y^2) - 18}{\sqrt{9 - x^2 - y^2}}\, dA = \lim_{r_0 \to 3^-} \int_0^{2\pi} \int_0^{r_0} \frac{3r^3 - 18r}{\sqrt{9 - r^2}}\, dr\, d\theta = 0.$$

18.6.13 R is the region in the xy-plane enclosed by the parabolas $y = 1 - x^2$ and $y = x^2 - 1$.

$$\iint_\sigma \mathbf{F} \cdot \mathbf{n}\, dS = \iint_R 8y\, dA = \int_{-1}^1 \int_{x^2-1}^{1-x^2} 8y\, dy\, dx = 0.$$

18.6.14 R is the circular region in the yz-plane enclosed by $y^2 + z^2 = 4$.

$$\iint_\sigma \mathbf{F} \cdot \mathbf{n}\, dS = \iint_R 2\, dA = 2[\text{area of circle}] = 2[\pi(2)^2] = 8\pi.$$

18.6.15 R is the circular region in the yz-plane enclosed by $y^2 + z^2 = 4$.

$$\iint_\sigma \mathbf{F} \cdot \mathbf{n}\, dS = \iint_R 9\, dA = 9[\text{area of circle}] = 9[\pi(2)^2] = 36\pi.$$

18.6.16 R is the circular region in the xy-plane enclosed by $y = x$, $x = 0$, and $y = 2$.

$$\iint_\sigma \mathbf{F} \cdot \mathbf{n}\, dS = \int_0^2 \int_0^y (3 - 4xy - y^2)\, dx\, dy = -6.$$

18.6.17 R is the triangular region enclosed by $6x + 3y = 12$, $x = 0$, $y = 0$

$$\iint_\sigma \mathbf{F} \cdot \mathbf{n}\, dS = \iint_R 3\, dA = 3\,(\text{area of triangle}) = 3\left[\frac{1}{2}(4)(2)\right] = 12$$

18.6.18 R is the circular region enclosed by $x^2 + y^2 = 4$

$$\iint_\sigma \mathbf{F} \cdot \mathbf{n}\, dS = \iint_R z^2\, dA = \int_0^{2\pi} \int_0^2 (4r - r^3)\, dr\, d\theta = 8\pi$$

SECTION 18.7

18.7.1   Use the divergence theorem to evaluate $\iint\limits_{\sigma} \mathbf{F} \cdot \mathbf{n}\, dS$ where

$\mathbf{F}(x, y, z) = (x + \cos z)\mathbf{i} + (2y + \sin z)\mathbf{k} + (z + e^x)\mathbf{k}$, $\mathbf{n}$ is the outer unit normal to σ, and σ is the surface of the paraboloid $z = x^2 + y^2$ which is inside the cylinder $x^2 + y^2 = 1$.

18.7.2   Use the divergence theorem to evaluate $\iint\limits_{\sigma} \mathbf{F} \cdot \mathbf{n}\, dS$ where $\mathbf{F}(x, y, z) = x\mathbf{i} + y\mathbf{j} + z\mathbf{k}$, $\mathbf{n}$ is the outer unit normal to σ, and σ is the surface of the cube enclosed by the planes $-1 \le x \le 1$, $-1 \le y \le 1$, and $-1 \le z \le 1$.

18.7.3   Use the divergence theorem to evaluate $\iint\limits_{\sigma} \mathbf{F} \cdot \mathbf{n}\, dS$ where $\mathbf{F}(x, y, z) = x\mathbf{i} + y\mathbf{j} - z\mathbf{k}$, $\mathbf{n}$ is the outer unit normal to σ, and σ is the surface formed by the intersection of the two paraboloids, $z = x^2 + y^2$ and $z = 4 - (x^2 + y^2)$.

18.7.4   Use the divergence theorem to evaluate $\iint\limits_{\sigma} \mathbf{F} \cdot \mathbf{n}\, dS$ where

$\mathbf{F}(x, y, z) = (2x + z)\mathbf{i} + y\mathbf{j} - (2z + \sin x)\mathbf{k}$, $\mathbf{n}$ is the outer unit normal to σ, and σ is the surface of the cylinder $x^2 + y^2 = 4$ enclosed between the planes $z = 0$ and $z = 4$.

18.7.5   Use the divergence theorem to evaluate $\iint\limits_{\sigma} \mathbf{F} \cdot \mathbf{n}\, dS$ where $\mathbf{F}(x, y, z) = \dfrac{x^3}{3}\mathbf{i} + \dfrac{y^3}{3}\mathbf{j} + \dfrac{z^3}{3}\mathbf{k}$, $\mathbf{n}$ is the outer unit normal to σ, and σ is the surface of the cylinder $x^2 + y^2 = 1$ enclosed between the planes $z = 0$ and $z = 1$.

18.7.6   Use the divergence theorem to evaluate $\iint\limits_{\sigma} \mathbf{F} \cdot \mathbf{n}\, dS$ where $\mathbf{F}(x, y, z) = x\mathbf{i} + y\mathbf{j} + z\mathbf{k}$, $\mathbf{n}$ is the outer unit normal to σ, and σ is the surface bounded by $x + y + z = 1$, $x = 0$, $y = 0$, and $z = 0$.

18.7.7   Use the divergence theorem to evaluate $\iint\limits_{\sigma} \mathbf{F} \cdot \mathbf{n}\, dS$ where $\mathbf{F}(x, y, z) = (x^3 + 3xy^2)\mathbf{i} + z^3\mathbf{k}$, $\mathbf{n}$ is the outer unit normal to σ, and σ is the surface of the sphere of radius a centered at the origin.

18.7.8   Use the divergence theorem to evaluate $\iint\limits_{\sigma} \mathbf{F} \cdot \mathbf{n}\, dS$ where $\mathbf{F}(x, y, z) = e^x\mathbf{i} - ye^x\mathbf{j} + 4x^2 z\mathbf{k}$, $\mathbf{n}$ is the outer unit normal to σ, and σ is the surface of the solid enclosed by $x^2 + y^2 = 4$ and the planes $z = 0$ and $z = 9$.

18.7.9   Use the divergence theorem to evaluate $\iint\limits_{\sigma} \mathbf{F} \cdot \mathbf{n} \, dS$ where $\mathbf{F}(x,y,z) = e^x\mathbf{i} - ye^x\mathbf{j} + 3z\mathbf{k}$, $\mathbf{n}$ is the outer unit normal to σ, and σ is the surface of the sphere by $x^2 + y^2 + z^2 = 9$.

18.7.10 Use the divergence theorem to evaluate $\iint\limits_{\sigma} \mathbf{F} \cdot \mathbf{n} \, dS$ where $\mathbf{F}(x,y,z) = x^2\mathbf{i} + y^2\mathbf{j} + z^2\mathbf{k}$, $\mathbf{n}$ is the outer unit normal to σ, and σ is the surface of the cube enclosed by the planes $0 \leq x \leq 1$, $0 \leq y \leq 1$, and $0 \leq z \leq 1$.

18.7.11 Use the divergence theorem to evaluate $\iint\limits_{\sigma} \mathbf{F} \cdot \mathbf{n} \, dS$ where $\mathbf{F}(x,y,z) = x^3\mathbf{i} + x^2y\mathbf{j} + x^2z\mathbf{k}$, $\mathbf{n}$ is the outer unit normal to σ, and σ is the surface enclosed by the cylinder $x^2 + y^2 = 2$ and the planes $z = 0$ and $z = 2$.

18.7.12 Use the divergence theorem to evaluate $\iint\limits_{\sigma} \mathbf{F} \cdot \mathbf{n} \, dS$ where

$\mathbf{F}(x,y,z) = z^2(x + 5y + z^2)\mathbf{i} + x^2(x^3 + y^+e^z)\mathbf{j} + y^2(x + y + z)\mathbf{k}$, $\mathbf{n}$ is the outer unit normal to σ, and σ is the surface enclosed by $x^2 + y^2 + z^2 = 16$.

18.7.13 Use the divergence theorem to evaluate $\iint\limits_{\sigma} \mathbf{F} \cdot \mathbf{n} \, dS$ where $\mathbf{F}(x,y,z) = x^3\mathbf{i} + x^2y\mathbf{j} - x^2z\mathbf{k}$, $\mathbf{n}$ is the outer unit normal to σ, and σ is the surface enclosed by the hemisphere $z = \sqrt{4 - x^2 - y^2}$ and the xy-plane.

18.7.14 Use the divergence theorem to evaluate $\iint\limits_{\sigma} \mathbf{F} \cdot \mathbf{n} \, dS$ where $\mathbf{F}(x,y,z) = x^2\mathbf{i} + y^2\mathbf{j} + z^2\mathbf{k}$, $\mathbf{n}$ is the outer unit normal to σ, and σ is the surface enclosed by the cylinder $x^2 + y^2 = 4$ and the planes $z = 0$ and $z = 5$.

18.7.15 Use the divergence theorem to evaluate $\iint\limits_{\sigma} \mathbf{F} \cdot \mathbf{n} \, dS$ where $\mathbf{F}(x,y,z) = yz\mathbf{i} + xy\mathbf{j} + xz\mathbf{k}$, $\mathbf{n}$ is the outer unit normal to σ, and σ is the surface enclosed by the cylinder $x^2 + z^2 = 1$ and the planes $y = -1$ and $y = 1$.

18.7.16 Use the divergence theorem to evaluate $\iint\limits_{\sigma} \mathbf{F} \cdot \mathbf{n} \, dS$ where $\mathbf{F}(x,y,z) = y^2x\mathbf{i} + yz^2\mathbf{j} + x^2y^2\mathbf{k}$, $\mathbf{n}$ is the outer unit normal to σ, and σ is the sphere $x^2 + y^2 + z^2 = 4$.

18.7.17 Determine whether the flow field $\mathbf{F}(x,y,z) = (x + z)\mathbf{i} + (y + z)\mathbf{j} - (2z - xy)\mathbf{k}$ is free of all sources and sinks. If it is not, find the location of all sources and sinks.

18.7.18   Determine whether the flow field $\mathbf{F}(x,y,z) = x^2\mathbf{i} + y^2\mathbf{j} + x^2\mathbf{k}$ is free of all sources and sinks. If it is not, find the location of all sources and sinks.

18.7.19   Determine whether the flow field $\mathbf{F}(x,y,z) = 2x^3\mathbf{i} + 2y^3\mathbf{j} + 2z^3\mathbf{k}$ is free of all sources and sinks. If it is not, find the location of all sources and sinks.

SOLUTIONS

SECTION 18.7

18.7.1 G is the solid bounded by $z = x^2 + y^2$ and $z = 1$.

$$\iiint_G \operatorname{div} \mathbf{F} \, dv = \iiint_G 4 dv = 4 \int_0^{2\pi} \int_0^1 \int_{r^2}^1 r \, dz \, dr \, d\theta = 2\pi$$

18.7.2 G is the cube enclosed by σ.

$$\iiint_G \operatorname{div} \mathbf{F} \, dv = \iiint_G 3 dv = 3(\text{volume of cube}) = 3(2^3) = 24.$$

18.7.3 G is the solid enclosed by σ.

$$\iiint_G \operatorname{div} \mathbf{F} \, dv = \iiint_G dv = \int_0^{2\pi} \int_0^{\sqrt{2}} \int_{r^2}^{4-r^2} r \, dz \, dr \, d\theta = 4\pi.$$

18.7.4 G is the solid enclosed by σ.

$$\iiint_G \operatorname{div} \mathbf{F} \, dv = \iiint_G dv = \text{volume of cylinder} = \pi(2)^2(4) = 16\pi.$$

18.7.5 G is the solid enclosed by σ.

$$\iiint_G \operatorname{div} \mathbf{F} \, dv = \iiint_G (x^2 + y^2 + z^2) dv$$

$$= \int_0^{2\pi} \int_0^1 \int_0^1 (r^2 + z^2) r \, dz \, dr \, d\theta = \frac{5\pi}{6}.$$

18.7.6 G is the solid enclosed by σ.

$$\iiint_G \operatorname{div} \mathbf{F} \, dv = \iiint_G 3 dv = \int_0^1 \int_0^{1-x} \int_0^{1-x-y} 3 \, dx \, dy \, dx = \frac{1}{2}.$$

18.7.7 G is the spherical solid.

$$\iiint_G \operatorname{div} \mathbf{F} \, dv = \iiint_G 3(x^2 + y^2 + z^2) dv$$

$$= 3a^2 (\text{volume of sphere}) = 3a^2 \left(\frac{4}{3}\pi a^3 \right) = 4\pi a^5.$$

18.7.8 G is the cylindrical solid.

$$\iiint_G \operatorname{div} \mathbf{F} \, dv = \iiint_G 4x^2 dv = \int_0^{2\pi} \int_0^2 \int_0^9 4r^3 \cos^2 \theta \, dz \, dr \, d\theta = 144\pi.$$

18.7.9 G is the spherical solid.

$$\iiint_G \text{div } \mathbf{F}\, dv = \iiint_G 3\, dv = 3(\text{volume of sphere}) = 3\left[\frac{4}{3}\pi(3)^3\right] = 108\pi.$$

18.7.10 G is the cube.

$$\iiint_G \text{div } \mathbf{F}\, dv = \iiint_G (2x + 2y + 2z)\, dv = \int_0^1 \int_0^1 \int_0^1 2(x + y + z)\, dz\, dy\, dx = 3.$$

18.7.11 G is the cylindrical solid.

$$\iiint_G \text{div } \mathbf{F}\, dv = \iiint_G 5x^2\, dv = \int_0^{2\pi} \int_0^{\sqrt{2}} \int_0^2 5r^3 \cos^2\theta\, dz\, dr\, d\theta = 10\pi.$$

18.7.12 G is the spherical solid.

$$\iiint_G \text{div } \mathbf{F}\, dv = \iiint_G (x^2 + y^2 + z^2)\, dv$$

$$= 16(\text{volume of sphere}) = 16\left[\frac{4}{3}\pi(4)^3\right] = \frac{\pi}{3}(4)^6.$$

18.7.13 G is the solid enclosed by σ.

$$\iiint_G \text{div } \mathbf{F}\, dv = \iiint_G 3x^2\, dv = \int_0^{2\pi} \int_0^{\pi/2} \int_0^2 3\rho^4 \sin^3\phi \cos^2\theta\, d\rho\, d\phi\, d\theta = \frac{64\pi}{5}.$$

18.7.14 G is the solid enclosed by σ.

$$\iiint_G \text{div } \mathbf{F}\, dv = \iiint_G (2x + 2y + 2z)\, dv$$

$$= \int_0^{2\pi} \int_0^2 \int_0^5 2[r^2(\cos\theta + \sin\theta) + zr]\, dz\, dr\, d\theta = 100\pi.$$

18.7.15 G is the solid enclosed by σ and using polar coordinates in the xz-plane.

$$\iiint_G \text{div } \mathbf{F}\, dv = \iiint_G 2x\, dv = \int_0^{2\pi} \int_0^1 \int_{-1}^1 2r^2 \cos\theta\, dy\, dr\, d\theta = 0.$$

18.7.16 G is the solid enclosed by σ.

$$\iiint_G \text{div } \mathbf{F}\, dv = \iiint_G (y^2 + z^2)\, dv$$

$$= \int_0^{2\pi} \int_0^\pi \int_0^2 (\rho^2 \sin^2\phi \sin\theta + \rho^2 \cos^2\phi)\rho^2 \sin\phi\, d\rho\, d\phi\, d\theta = \frac{128\pi}{15}.$$

18.7.17   div $\mathbf{F} = 0$, no sources or sinks

18.7.18   div $\mathbf{F} = x + y$; sources where $y > -x$, sinks where $y < -x$

18.7.19   div $\mathbf{F} = 6x^2 + 6y^2 + 6z^2$; sources at all points except the origin; no sinks

SECTION 18.8

18.8.1   Verify Stokes' Theorem if σ is the portion of the sphere $x^2 + y^2 + z^2 = 1$ for which $z \geq 0$ and $\mathbf{F}(x, y, z) = (2x - y)\mathbf{i} - yz^2\mathbf{j} - y^2 z\mathbf{k}$.

18.8.2   Use Stokes' Theorem to evaluate $\displaystyle\int_c (z - y)dx + (x - z)dy + (y - x)dz$ where c is the boundary, in the xy plane, of the surface σ given by $z = 4 - (x^2 + y^2)$, $z \geq 0$.

18.8.3   Use Stokes' Theorem to evaluate $\displaystyle\int_c y^2 dx + x^2 dy - (x + z)dz$ where c is a triangle in the xy plane with vertices $(0, 0, 0)$, $(1, 0, 0)$, and $(1, 1, 0)$ with a counterclockwise orientation looking down the positive z axis.

18.8.4   Use Stokes' Theorem to evaluate $\displaystyle\int_c -3y\,dx + 3x\,dy + z\,dz$ over the circle $x^2 + y^2 = 1$, $z = 1$ traversed counterclockwise.

18.8.5   Use Stokes' Theorem to evaluate $\displaystyle\int_c z\,dx + x\,dy + y\,dz$ over the triangle with vertices $(1, 0, 0)$, $(0, 1, 0)$, and $(0, 0, 1)$ traversed in a counterclockwise manner.

18.8.6   Use Stokes' Theorem to evaluate $\displaystyle\iint_\sigma (\text{curl } \mathbf{F}) \cdot \mathbf{n}\,dS$ where $\mathbf{F}(x, y, z) = x^2\mathbf{i} + z^2\mathbf{j} - y^2\mathbf{k}$ and σ is that portion of the paraboloid $z = 4 - x^2 - y^2$ for which $z \geq 0$.

18.8.7   Use Stokes' Theorem to evaluate $\displaystyle\iint_\sigma (\text{curl } \mathbf{F}) \cdot \mathbf{n}\,dS$ where

$\mathbf{F}(x, y, z) = (z - y)\mathbf{i} + (z^2 + x)\mathbf{j} + (x^2 - y^2)\mathbf{k}$ and σ is that portion of the sphere $x^2 + y^2 + z^2 = 4$ for which $z \geq 0$.

18.8.8   Use Stokes' Theorem to evaluate $\displaystyle\iint_\sigma (\text{curl } \mathbf{F}) \cdot \mathbf{n}\,dS$ where $\mathbf{F}(x, y, z) = y\mathbf{k}$ and σ is that portion of the ellipsoid $4x^2 + 4y^2 + z^2 = 4$ for which $z \geq 0$.

18.8.9   Use Stokes' Theorem to evaluate $\displaystyle\int_c \sin z\,dx - \cos x\,dy + \sin y\,dz$ over the rectangle $0 \leq x \leq \pi$, $0 \leq y \leq 1$, and $z = 2$ traversed in a counterclockwise manner.

18.8.10   Use Stokes' Theorem to evaluate $\displaystyle\int_c (x + y)dx + (2x - 3)dy + (y + z)dz$ over the boundary of the triangle with vertices $(2, 0, 0)$, $(0, 3, 0)$, and $(0, 0, 6)$ traversed in a counterclockwise manner.

18.8.11 Use Stokes' Theorem to evaluate $\int_C 4z\,dx - 2x\,dy + 2x\,dz$ where C is the intersection of the cylinder $x^2 + y^2 = 1$ and the plane $z = y + 1$.

18.8.12 Use Stokes' Theorem to evaluate $\int_C -yz\,dx + xz\,dy + xy\,dz$ where C is the circle $x^2 + y^2 = 2$, $z = 1$.

18.8.13 Use Stokes' Theorem to evaluate $\int_C (4x - 2y)dx - yz^2\,dy - y^2 z\,dz$ where C is the circular region enclosed by $x^2 + y^2 = 4$, $z = 2$.

18.8.14 Use Stokes' Theorem to evaluate $\int_C \left(e^{-x^2} - yz\right)dx + \left(e^{-y^2} + xz + 2x\right)dy + e^{-z^2}\,dz$ over the circle $x^2 + y^2 = 1$, $z = 1$.

18.8.15 Use Stokes' Theorem to evaluate $\int_C xz\,dx + y^2\,dy + x^2\,dz$ where C is the intersection of the plane $x + y + z = 5$ and the cylinder $x^2 + \dfrac{y^2}{4} = 1$.

18.8.16 Use Stokes' Theorem to find the circulation around the triangle with vertices $(0,0,0)$, $(1,0,0)$, and $(1,1,0)$ traversed in a counterclockwise manner looking down the positive z-axis if the flow field is given by $\mathbf{F}(x,y,z) = -y^3\mathbf{i} + x^3\mathbf{j} - (x+z)\mathbf{k}$.

SOLUTIONS

SECTION 18.8

18.8.1 If σ is oriented by upward normals, then C is the intersection of the sphere $x^2 + y^2 + z^2 = 1$ with $z = 0$, thus, C is the circle $x^2 + y^2 = 1$ which can be parametrized as

$$r(t) = \cos t\mathbf{i} + \sin t\mathbf{j} \text{ for } 0 \leq t \leq 2\pi, \text{ so, } \int_c \mathbf{F} \cdot d\mathbf{r} = \int_0^{2\pi} (\sin^2 t - 2\cos t \sin t)dt = \pi;$$

Curl $\mathbf{F} = \mathbf{k}$, $\mathbf{n} = -\dfrac{x}{\sqrt{1-x^2-y^2}}\mathbf{i} - \dfrac{y}{\sqrt{1-x^2-y^2}}\mathbf{j} + \mathbf{k}$, and R is the circular region in the

xy plane enclosed by C, so $\displaystyle\iint_\sigma (\text{curl } \mathbf{F}) \cdot \mathbf{n}\,dS = \iint_R dA = $ area of a circle of radius

$1 = \pi(1)^2 = \pi$.

18.8.2 $\mathbf{F} = (z-y)\mathbf{i} + (x-z)\mathbf{j} + (y-x)\mathbf{k}$, curl $\mathbf{F} = 2\mathbf{i} + 2\mathbf{j} + 2\mathbf{k}$, $\mathbf{n} = \dfrac{2x\mathbf{i} + 2y\mathbf{j} + \mathbf{k}}{\sqrt{4x^2 + 4y^2 + 1}}$, and R is the

circular region in the xy plane enclosed by $x^2 + y^2 = 4$, so,

$$\iint_\sigma (\text{curl } \mathbf{F}) \cdot \mathbf{n}\,dS = \iint_R (4x + 4y + 2)dA$$

$$= \int_0^{2\pi} \int_0^2 (4r^2 \cos\theta + 4r^2 \sin\theta + 2r)dr\,d\theta = 8\pi.$$

18.8.3 Let σ be the portion of the plane $z = 0$, oriented with upward normals for which curl $\mathbf{F} = \mathbf{j} + (2x - 2y)\mathbf{k}$, $\mathbf{n} = \mathbf{k}$, and R is the triangular region in the xy plane enclosed by C, then

$$\int_c \mathbf{F} \cdot d\mathbf{r} = \iint_\sigma (\text{curl } \mathbf{F}) \cdot \mathbf{n}\,dS = \iint_R (2x - 2y)dA = \int_0^1 \int_0^x (2x - 2y)\,dy\,dx = \frac{1}{3}.$$

18.8.4 Let σ be the portion of the plane $z = 1$, oriented with upward normals for which curl $\mathbf{F} = 6\mathbf{k}$, $\mathbf{n} = \mathbf{k}$, and R is the circular region in the xy plane enclosed by $x^2 + y^2 = 1$, then

$$\int_c \mathbf{F} \cdot d\mathbf{r} = \iint_\sigma (\text{curl } \mathbf{F}) \cdot \mathbf{n}\,dS = \iint_\sigma 6ds = \iint_R 6dA = 6(\text{area of circle}) = 6[\pi(1)^2] = 6\pi.$$

18.8.5 Let σ be the portion of the plane $z = 1 - x - y$, oriented with upward normals for which curl $\mathbf{F} = \mathbf{i} + \mathbf{j} + \mathbf{k}$, $\mathbf{n} = \dfrac{\mathbf{i} + \mathbf{j} + \mathbf{k}}{\sqrt{3}}$, and R is the triangular region in the xy plane enclosed by $x + y = 1$, $x = 0$, and $y = 0$, thus

$$\int_c \mathbf{F} \cdot d\mathbf{r} = \iint_\sigma (\text{curl } \mathbf{F}) \cdot \mathbf{n}\,dS = \iint_R 3dA = 3(\text{area of triangle}) = 3\left[\frac{1}{2}(1)(1)\right] = \frac{3}{2}.$$

18.8.6 Let σ be oriented with upward normals and C be the circle $x^2 + y^2 = 4$ which can be parametrized as $\mathbf{r}(t) = 2\cos t\mathbf{i} + 2\sin t\mathbf{j}$ for $0 \le t \le 2\pi$, then,

$$\iint_\sigma (\text{curl } \mathbf{F}) \cdot \mathbf{n}\, dS = \int_c \mathbf{F} \cdot d\mathbf{r} = \int_0^{2\pi} -8\cos^2 t \sin t\, dt = 0.$$

18.8.7 Let σ be oriented with upward normals and C be the circle $x^2 + y^2 = 4$ which can be parametrized as $r(t) = 2\cos t\mathbf{i} + 2\sin t\mathbf{j}$ for $0 \le t \le 2\pi$, then,

$$\iint_\sigma (\text{curl } \mathbf{F}) \cdot \mathbf{n}\, dS = \int_c \mathbf{F} \cdot d\mathbf{r} = \int_0^{2\pi} 4\, dt = 8\pi.$$

18.8.8 Let σ be oriented with upward normals and C be the circle $4x^2 + 4y^2 = 4$ or $x^2 + y^2 = 1$ which can be parametrized as $r(t) = \cos t\mathbf{i} + \sin t\mathbf{j}$ for $0 \le t \le 2\pi$, then,

$$\iint_\sigma (\text{curl } \mathbf{F}) \cdot \mathbf{n}\, dS = \int_c \mathbf{F} \cdot d\mathbf{r} = \int_c 0 = 0.$$

18.8.9 Let σ be that portion of the plane $z = 2$ oriented with upward normals for which curl $\mathbf{F} = \cos y\mathbf{i} + \cos z\mathbf{j} + \sin x\mathbf{k}$, $\mathbf{n} = \mathbf{k}$, and R is the rectangular region in the xy plane enclosed by $0 \le x \le \pi$ and $0 \le y \le 1$, then

$$\iint_c \mathbf{F} \cdot d\mathbf{r} = \iint_\sigma (\text{curl } \mathbf{F}) \cdot \mathbf{n}\, dS = \iint_R \sin x\, dA = \int_0^\pi \int_0^1 \sin x\, dy\, dx = 2.$$

18.8.10 Let σ be part of the plane $z = 6 - 3x - 2y$ oriented with upward normals for which curl $\mathbf{F} = \mathbf{i} + \mathbf{k}$, $n = \dfrac{3i + 2j + k}{\sqrt{14}}$, and R is the triangular region in the xy plane enclosed by $3x + 2y = 6$, $x = 0$, and $y = 0$, thus,

$$\iint_\sigma (\text{curl } \mathbf{F}) \cdot \mathbf{n}\, dS = \iint_R 4\, dA = 4(\text{area of triangle}) = 4\left[\left(\frac{1}{2}\right)(2)(3)\right] = 12.$$

18.8.11 Let σ be the portion of the plane $z = y + 1$ oriented with upward normals for which curl $\mathbf{F} = 2\mathbf{j} - 2\mathbf{k}$, $n = \dfrac{-j + k}{\sqrt{2}}$, and R is the circular region in the xy plane enclosed by $x^2 + y^2 = 1$, then

$$\int_c \mathbf{F} \cdot d\mathbf{r} = \iint_\sigma (\text{curl } \mathbf{F}) \cdot \mathbf{n}\, dS = \iint_R -4\, dA = -4(\text{area of circle}) = -4[\pi(1)^2] = -4\pi.$$

18.8.12 Let σ be the portion of the plane $z = 1$ oriented with upward normals for which curl $\mathbf{F} = -2y\mathbf{j} + 2z\mathbf{k}$, $\mathbf{n} = \mathbf{k}$, and R is the circular region in the xy plane enclosed by $x^2 + y^2 = 2$, then

$$\int_c \mathbf{F} \cdot d\mathbf{r} = \iint_\sigma (\text{curl } \mathbf{F}) \cdot \mathbf{n}\, dS = \iint_R 2z\, dA = 2\iint_R dA$$

$$= 2(\text{area of circle}) = 2[\pi(\sqrt{2})^2] = 4\pi.$$

18.8.13 Let σ be that portion of the plane $z = 2$ oriented with upward normals for which curl $\mathbf{F} = 2\mathbf{k}$, $\mathbf{n} = \mathbf{k}$, and R is the circular region in the xy plane enclosed by $x^2 + y^2 = 4$, then

$$\int_c \mathbf{F} \cdot d\mathbf{r} = \iint_\sigma (\text{curl } \mathbf{F}) \cdot \mathbf{n}\, dS = \iint_R 2\, dA = 2(\text{area of circle}) = 2[\pi(2)^2] = 8\pi.$$

18.8.14 Let σ be the portion of the plane $z = 1$ oriented with upward normals for which curl $\mathbf{F} = -x\mathbf{i} + y\mathbf{j} + (2 + 2z)\mathbf{k}$, $\mathbf{n} = \mathbf{k}$, and R is the circular region in the xy plane enclosed by $x^2 + y^2 = 1$, then

$$\int_c \mathbf{F} \cdot d\mathbf{r} = \iint_\sigma (\text{curl } \mathbf{F}) \cdot \mathbf{n}\, dS = \iint_R (2 + 2z)\, dA = \iint_R 4\, dA$$
$$= 4(\text{area of circle}) = 4[\pi(1)^2] = 4\pi.$$

18.8.15 Let z be the portion of the plane $z = 5 - x - y$, oriented with upward normals for which curl $\mathbf{F} = -x\mathbf{j}$, $\mathbf{n} = \mathbf{i} + \mathbf{j} + \mathbf{k}$, and R is the elliptical region in the xy plane enclosed by $x^2 + \dfrac{y^2}{4} = 1$, then

$$\int_c \mathbf{F} \cdot d\mathbf{r} = \iint_\sigma (\text{curl } \mathbf{F}) \cdot \mathbf{n}\, dS = \iint_R -x\, dA = \int_{-2}^{2} \int_{-\sqrt{1-\frac{y^2}{4}}}^{\sqrt{1-\frac{y^2}{4}}} -x\, dx\, dy.$$

18.8.16 Let σ be part of the plane $z = 0$, oriented with upward normals for which curl $\mathbf{F} = \mathbf{j} + 3(x^2 + y^2)\mathbf{k}$ and $\mathbf{n} = \mathbf{k}$, then

$$\iint_\sigma (\text{curl } \mathbf{F}) \cdot \mathbf{n}\, dS = \iint_\sigma 3(x^2 + y^2)\, dS = \int_0^1 \int_0^x 3(x^2 + y^2)\, dy\, dx = 1.$$

SUPPLEMENTARY EXERCISES, CHAPTER 18

In Exercises 1–6, evaluate the line integral by any method.

1. $\displaystyle\int_C 2y\,dx + 3\,dy$ along $y = \sin x$ from the point $(0,0)$ to the point $(\pi, 0)$.

2. $\displaystyle\int_C x^5\,dy$ along the curve C given by $x = 1/t$, $y = 4t^2$, $1 \le t \le 2$.

3. $\displaystyle\int_C \langle 2e^y, -x \rangle \cdot d\mathbf{r}$ along $y = \ln x$ from the point $(1,0)$ to the point $(3, \ln 3)$.

4. $\displaystyle\int_C (y\mathbf{i} + z\mathbf{j} + x\mathbf{k}) \cdot d\mathbf{r}$ along the curve C given by $\mathbf{r} = \langle t^2 - 2t, -2t, t - 2 \rangle$, $0 \le t \le 2$.

5. $\displaystyle\int_C z\,dx + y\,dy - x\,dz$ along the line segment from $(0,0,0)$ to $(1,2,3)$.

6. $\displaystyle\int_C x \sin xy\,dx - y \sin xy\,dy$ along the line segment from $(0,0)$ to $(1, \pi)$.

In Exercises 7–9, determine whether $\mathbf{F}$ is conservative. If it is, find a potential function for it.

7. $\mathbf{F}(x,y) = y \sin xy\,\mathbf{i} - x \cos xy\,\mathbf{j}$.

8. $\mathbf{F}(x,y) = 2x(\ln y - 1)\mathbf{i} + \left(\dfrac{x^2}{y} - 3y^2\right)\mathbf{j}$.

9. $\mathbf{F}(x,y) = \left(3x^2 - \dfrac{y^2}{x^2}\right)\mathbf{i} + \left(\dfrac{2y}{x} + 4y\right)\mathbf{j}$.

SOLUTIONS

SUPPLEMENTARY EXERCISES CHAPTER 18

1. $x = t$, $y = \sin t$, $0 \le t \le \pi$; $\displaystyle\int_0^\pi (2\sin t + 3\cos t)dt = 4$

2. $\displaystyle\int_1^2 8/t^4 dt = 7/3$

3. $x = t$, $y = \ln t$, $1 \le t \le 3$; $\displaystyle\int_1^3 (2t - 1)dt = 6$

4. $\displaystyle\int_0^2 (4 - 3t^2)dt = 0$

5. $x = t$, $y = 2t$, $z = 3t$, $0 \le t \le 1$; $\displaystyle\int_0^1 4t\, dt = 2$

6. $x = t$, $y = \pi t$, $0 \le t \le 1$; $\displaystyle\int_0^1 (1 - \pi^2)t\sin(\pi t^2)dt = (1 - \pi^2)/\pi$

7. $\partial(y\sin xy)/\partial y = xy\cos xy + \sin xy$, $\partial(-x\cos xy)/\partial x = xy\sin xy - \cos xy$, not conservative.

8. $\partial[2x(\ln y - 1)]/\partial y = 2x/y = \partial(x^2/y - 3y^2)/\partial x$, conservative so $\partial\phi/\partial x = 2x(\ln y - 1)$ and $\partial\phi/\partial y = x^2/y - 3y^2$, $\phi = x^2(\ln y - 1) + k(y)$, $x^2/y + k'(y) = x^2/y - 3y^2$, $k'(y) = -3y^2$, $k(y) = -y^3 + K$, $\phi = x^2(\ln y - 1) - y^3 + K$.

9. $\partial(3x^2 - y^2/x^2)/\partial y = -2y/x^2 = \partial(2y/x + 4y)/\partial x$, conservative so $\partial\phi/\partial x = 3x^2 - y^2/x^2$ and $\partial\phi/\partial y = 2y/x + 4y$, $\phi = x^3 + y^2/x + k(y)$, $2y/x + k'(y) = 2y/x + 4y$, $k'(y) = 4y$, $k(y) = 2y^2 + K$, $\phi = x^3 + y^2/x + 2y^2 + K$.

10. $h(x)F(x,y)$ is conservative if $\partial[yh(x)]/\partial y = \partial[-2xh(x)]/\partial x$, $h(x) = -2xh'(x) - 2h(x)$, $2xh'(x) + 3h(x) = 0$ which is both separable and first-order-linear; the solution is $h(x) = C|x|^{-3/2}$. $g(y)F(x,y)$ is conservative if $\partial[yg(y)]/\partial y = \partial[-2xg(y)]/\partial x$, $yg'(y) + g(y) = -2g(y)$, $yg'(y) + 3g(y) = 0$ so $g(y) = Cy^{-3}$.

11. $\partial(\cos 2y - 3x^2y^2)/\partial y = -2\sin 2y - 6x^2y$ and $\partial(\cos 2y - 2x\sin 2y - 2x^3y)/\partial x = -2\sin 2y - 6x^2y$ so it is independent of path. The line segment from $(1, \pi/4)$ to $(2, \pi/4)$ is $x = 1 + t$, $y = \pi/4$, $0 \le t \le 1$; the line integral along this path is
$$-\int_0^1 3(\pi/4)^2 (1 + t)^3 dt = -7\pi^2/16.$$

12. $\partial(x^2y^4)/\partial y = 4x^2y^3$, $\partial(y^2x^4)/\partial x = 4y^2x^3$, not independent of path

903

13. $\partial(1/y)/\partial y = -1/y^2 = \partial(-x/y^2)/\partial x$, independent of path;
$\phi = x/y$, $\phi(2,1) - \phi(1,2) = 2 - 1/2 = 3/2$

14. $\partial(ye^{xy} - 1)/\partial y = xye^{xy} + e^{xy} = \partial(xe^{xy})/\partial x$, independent of path;
$\phi = e^{xy} - x$, $\phi(1,0) - \phi(0,1) = 0 - 1 = -1$

15. $\int_0^2 \int_0^{2x} (y^2 - 2x)dy\, dx = 0$

16. $\iint_R -7dA = -7(\pi) = -7\pi$

17. $\int_{-2}^2 \int_0^{4-y^2} 3\, dx\, dy = 32$

18. $\int_0^2 \int_{x^2}^{2x} (3x^2 - 5x)dy\, dx = -28/15$

19. $\iint_R (-3x^2 - 3y^2)dA = -3\int_0^\pi \int_1^2 r^3 dr\, d\theta = -45\pi/4$

20. $x = r\cos\theta = 2\cos^2\theta = 1 + \cos 2\theta$, $y = r\sin\theta = 2\sin\theta\cos\theta = \sin 2\theta$;
$A = \frac{1}{2}\int_C x\, dy - y\, dx = \int_0^\pi (\cos 2\theta + 1)d\theta = \pi$

21. $C: x = t$, $y = t^2$, $0 \le t \le 1$; $W = \int_C \mathbf{F} \cdot d\mathbf{r} = \int_0^1 (2t + 3t^2 + 2t^4)dt = 12/5$

22. $\partial(3x^2y^3)/\partial y = 9x^2y^2 = \partial(3x^3y^2)/\partial x$ so $\mathbf{F}$ is conservative and $W = \int_C \mathbf{F} \cdot d\mathbf{r} = 0$ because C is closed

23. $C: x = t$, $y = 2t$, $z = 3t$, $0 \le t \le 1$; $W = \int_C \mathbf{F} \cdot d\mathbf{r} = \int_0^1 4t\, dt = 2$

24. $\nabla f = \frac{x}{x^2 + y^2 + z^2}\mathbf{i} + \frac{y}{x^2 + y^2 + z^2}\mathbf{j} + \frac{z}{x^2 + y^2 + z^2}\mathbf{k}$; on σ, $\nabla f = x\mathbf{i} + y\mathbf{j} + z\mathbf{k}$ because
$x^2 + y^2 + z^2 = 1$. $D_\mathbf{n}f = \nabla f \cdot \mathbf{n}$ so
$$\iint_\sigma D_\mathbf{n}f\, dS = \iint_\sigma \nabla f \cdot \mathbf{n}\, dS = \iint_R \frac{1}{\sqrt{1 - x^2 - y^2}}dA$$
$$= \lim_{r_0 \to 1^-} \int_0^{\pi/2} \int_0^{r_0} \frac{r}{\sqrt{1 - r^2}}dr\, d\theta = \lim_{r_0 \to 1^-} \frac{\pi}{2}(1 - \sqrt{1 - r_0^2}) = \frac{\pi}{2}.$$

25. $D_\mathbf{n}f = \nabla f \cdot \mathbf{n}$ so $\iint_\sigma D_\mathbf{n}f\, dS = \iint_\sigma \nabla f \cdot \mathbf{n}\, dS = -\iiint_G \text{div}\,(\nabla f)\, dV$ by the Divergence
Theorem. $\nabla f = 2x\mathbf{i} + 2y\mathbf{j} + 2z\mathbf{k}$; div $(\nabla f) = 6$ so
$$\iint_\sigma D_\mathbf{n}f\, dS = -6\iiint_G dV = -6\left[\frac{4}{3}\pi(1)^3\right] = -8\pi.$$

26. $D_{\mathbf{n}}\phi = \nabla\phi \cdot \mathbf{n}$ so $\iint\limits_{\sigma} D_{\mathbf{n}}\phi \, dS = \iint\limits_{\sigma} \nabla\phi \cdot \mathbf{n} \, dS = \iiint\limits_{G} \text{div} \, (\nabla\phi) dV$ by the Divergence

Theorem. $\nabla\phi = \dfrac{\partial\phi}{\partial x}\mathbf{i} + \dfrac{\partial\phi}{\partial y}\mathbf{j} + \dfrac{\partial\phi}{\partial z}\mathbf{k}$ so $\text{div} \, (\nabla\phi) = \dfrac{\partial^2\phi}{\partial x^2} + \dfrac{\partial^2\phi}{\partial y^2} + \dfrac{\partial^2\phi}{\partial z^2}$ and

$$\iint\limits_{\sigma} D_{\mathbf{n}}\phi \, dS = \iiint\limits_{G} \left(\frac{\partial^2\phi}{\partial x^2} + \frac{\partial^2\phi}{\partial y^2} + \frac{\partial^2\phi}{\partial z^2} \right) dV.$$

CHAPTER 19

Second-Order Differential Equations

SECTION 19.1

19.1.1 Verify that e^x and xe^x are both solutions to $y'' - 2y' + y = 0$.

19.1.2 Find the general solution of $y'' - y' - 2y = 0$.

19.1.3 Find the general solution of $y'' + 3y' - y = 0$.

19.1.4 Find the general solution of $y'' + 3y' - 4y = 0$.

19.1.5 Find the general solution of $4y'' - 4y' + y = 0$.

19.1.6 Find the general solution of $y'' - 10y' + 25y = 0$.

19.1.7 Find the general solution of $2y'' + y' - y = 0$.

19.1.8 Find the general solution of $2y'' - 5y' - 3y = 0$.

19.1.9 Find the general solution of $y'' - 2y' + 4y = 0$.

19.1.10 Find the general solution of $4y'' + 2y' + y = 0$.

19.1.11 Find the general solution of $y'' + 6y' + 13y = 0$.

19.1.12 Solve the initial value problem $y'' - y' - 6y = 0$; $y(0) = 2$, $y'(0) = 1$.

19.1.13 Find a second order linear homogeneous differential equation with constant coefficients that has $y_1 = e^{x/2}$ and $y_2 = xe^{x/2}$ as solutions.

19.1.14 Find a second order linear homogeneous differential equation with constant coefficients that has $y_1 = e^x \cos \dfrac{x}{2}$ and $y_2 = e^x \sin \dfrac{x}{2}$ as solutions.

19.1.15 Find a second order linear homogeneous differential equation with constant coefficients that has $y = e^{-x} \cos 2x$ and $y = e^{-x} \sin 2x$ as solutions.

19.1.16 Solve the initial value problem $y'' - 6y' + 10y = 0$; $y(0) = 1$, $y'(0) = 4$.

19.1.17 Solve the initial value problem $y'' + 4y' + 8y = 0$; $y(0) = 0$, $y'(0) = 8$.

19.1.18 Find the general solution of $y'' - 9y = 0$ and express your answer in terms of hyperbolic functions.

19.1.19 Find the general solution of $y'' - 36y = 0$ and express your answer in terms of hyperbolic functions.

SOLUTIONS

SECTION 19.1

19.1.1 Let $y_1 = e^x$ and $y_2 = xe^x$, substitute into the given differential equation to get
$e^x - 2e^x + e^x = 0$ for y_1 and $(xe^x + 2e^x) - 2(xe^x + e^x) + xe^x = 0$ for y_2.

19.1.2 The auxiliary equation, $m^2 - m - 2 = 0$ or $(m+1)(m-2) = 0$ has real roots $m = -1$, $m = 2$, so the general solution is $y = c_1 e^{-x} + c_2 e^{2x}$.

19.1.3 The auxiliary equation, $m^2 + 3m - 1 = 0$ has real roots $m = \dfrac{-3 + \sqrt{13}}{2}$, $m = \dfrac{-3 - \sqrt{13}}{2}$ so
the general solution is $y = c_1 e^{\left(\frac{-3 + \sqrt{13}}{2}\right)x} + c_2 e^{\left(\frac{-3 - \sqrt{13}}{2}\right)x}$.

19.1.4 The auxiliary equation, $m^2 + 3m - 4 = 0$ or $(m+4)(m-1) = 0$ has real roots $m = -4$, $m = 1$ so the general solution is $y = c_1 e^{-4x} + c_2 e^x$.

19.1.5 The auxiliary equation $4m^2 - 4m + 1 = 0$ or $(2m-1)^2 = 0$ has one real root $m = \dfrac{1}{2}$ so the general solution is $y = c_1 e^{x/2} + c_2 x e^{x/2}$.

19.1.6 The auxiliary equation $m^2 - 5m + 25 = 0$ or $(m-5)^2 = 0$ has one real root $m = 5$ so the general solution is $y = c_1 e^{5x} + c_2 x e^{5x}$.

19.1.7 The auxiliary equation $2m^2 + m - 1 = 0$ or $(m+1)(2m-1) = 0$ has real roots $m = -1$, $m = 1/2$ so the general solution is $y = c_1 e^{-x} + c_2 e^{x/2}$.

19.1.8 The auxiliary equation $2m^2 - 5m - 3 = 0$ or $(2m+1)(m-3) = 0$ has real roots $m = -1/2$, $m = 3$ so the general solution is $y = c_1 e^{-x/2} + c_2 e^{3x}$.

19.1.9 The auxiliary equation $m^2 - 2m + 4 = 0$ has complex roots $m = 1 + \sqrt{3}i$, $m = 1 - \sqrt{3}i$ so the general solution is $y = e^x \left(c_1 \cos \sqrt{3}x + c_2 \sin \sqrt{3}x \right)$.

19.1.10 The auxiliary equation $4m^2 + 2m + 1 = 0$ has complex roots $m = \dfrac{1}{4} + \dfrac{\sqrt{3}}{4}i$, $m = \dfrac{1}{4} - \dfrac{\sqrt{3}}{4}i$
so the general solution is $y = e^{-x/4} \left(c_1 \cos \dfrac{\sqrt{3}}{4}x + c_2 \sin \dfrac{\sqrt{3}}{4}x \right)$.

19.1.11 The auxiliary equation $m^2 + 6m + 13 = 0$ has complex roots $m = -3 + 2i$, $m = -3 - 2i$ so the general solution is $y = e^{-3x} (c_1 \cos 2x + c_2 \sin 2x)$.

19.1.12   The auxiliary equation $m^2 - m - 6 = 0$ or $(m+2)(m-3) = 0$ has real roots $m = -2$, $m = 3$ so the general solution is $y(x) = c_1 e^{-2x} + c_2 e^{3x}$ and whose derivative is

$y'(x) = -2c_1 e^{-2x} + 3c_2 e^{3x}$. Substitution of $x = 0$ into both equations and applying the initial conditions yields

$$\begin{aligned} c_1 + c_2 &= 2 \\ -2c_1 + 3c_2 &= 1. \end{aligned}$$

Solve for c_1 and c_2 to get $c_1 = 1$, $c_2 = 1$, thus $y(x) = e^{-2x} + e^{3x}$ is the solution to the initial value problem.

19.1.13   The auxiliary equation has one root $m = 1/2$ so $(m - 1/2)^2 = 0$ or $m^2 - m + \dfrac{1}{4} = 0$ which can be written as $4m^2 - 4m + 1 = 0$, thus $4y'' - 4y' + y = 0$ is the differential equation.

19.1.14   The auxiliary equation has roots $m = 1 + \dfrac{1}{2}i$, $m = 1 - \dfrac{1}{2}i$ so

$\left(m - 1 - \dfrac{1}{2}i\right)\left(m - 1 + \dfrac{1}{2}i\right) = 0$ or $m^2 - 2m + \dfrac{5}{4} = 0$ which can be written as

$4m^2 - 8m + 5 = 0$, thus $4y'' - 8y' + 5y = 0$ is the differential equation.

19.1.15   The auxiliary equation has roots $m = -1 + 2i$, $m = -1 - 2i$ so $(m + 1 - 2i)(m + 1 - 2i) = 0$ or $m^2 + 2m + 5 = 0$, thus, $y'' + 2y' + 5y = 0$ is the differential equation.

19.1.16   The auxiliary equation $m^2 - 6m + 10 = 0$ has complex roots $m = 3 + i$, $m = 3 - i$ so the general solution is $y(x) = e^{3x}(c_1 \cos x + c_2 \sin x)$ and whose derivative is

$y'(x) = e^{3x}(-c_1 \sin x + c_2 \cos x) + 3e^{3x}(c_1 \cos x + c_2 \sin x)$. Substitution of $x = 0$ into both equations and applying the initial conditions yields

$$\begin{aligned} c_1 &= 1 \\ 3c_1 + c_2 &= 4. \end{aligned}$$

Solve for c_1 and c_2 to get $c_1 = 1$, $c_2 = 1$ thus $y(x) = e^{3x}(\cos x + \sin x)$ is the solution to the initial value problem.

19.1.17   The auxiliary equation $m^2 + 4m + 8 = 0$ has complex roots $m = -2 + 2i$, $m = -2 - 2i$ so the general solution is $y(x) = e^{-2x}(c_1 \cos 2x + c_2 \sin 2x)$ and whose derivative is

$y'(x) = e^{-2x}(-2c_1 \sin 2x + 2c_2 \cos 2x) - 2e^{-2x}(c_1 \cos 2x + c_2 \sin 2x)$. Substitution of $x = 0$ into both equations and applying the initial conditions yields

$$\begin{aligned} c_1 &= 0 \\ -2c_1 + 2c_2 &= 8. \end{aligned}$$

Solve for c_1 and c_2 to get $c_1 = 0$, $c_2 = 4$ thus $y(x) = 4e^{-2x} \sin 2x$ is the solution to the initial value problem.

19.1.18   The auxiliary equation $m^2 - 9 = 0$ or $(m+3)(m-3) = 0$ has real roots $m = -3$, $m = 3$ so the general solution is $y(x) = c_1 e^{-3x} + c_2 e^{3x}$. Let $c_1 = c_2 = 1/2$ to get

$$y_1(x) = \frac{e^{-3x} + e^{3x}}{2} = \cosh 3x \text{ and let } c_1 = 1/2, c_2 = -1/2 \text{ to get}$$

$$y_2(x) = \frac{e^{-3x} - e^{-3x}}{2} = \sinh 3x \text{ thus, the general solution can be written as}$$

$y(x) = c_1 \cosh 3x + c_2 \sinh 3x$ (see Exercise 24).

19.1.19   The auxiliary equation $m^2 - 36 = 0$ or $(m+6)(m-6) = 0$ has real roots $m = -6$, $m = 6$ so the general solution is $y = c_1 e^{-6x} + c_2 e^{6x}$. Let $c_1 = c_2 = 1/2$ to get

$$y_1 = \frac{e^{-6x} + e^{6x}}{2} = \cosh 6x \text{ and let } c_1 = 1/2, c_2 = -1/2 \text{ to get}$$

$$y_2 = \frac{e^{-6x} - e^{-6x}}{2} = \sinh 6x \text{ thus, the general solution can be written as}$$

$y(x) = c_1 \cosh 6x + c_2 \sinh 6x$ (see Exercise 24).

SECTION 19.2

19.2.1   Use the method of undetermined coefficients to find the general solution of
$$y'' - 3y' + 2y = e^{-4x}.$$

19.2.2   Use the method of undetermined coefficients to find the general solution of
$$y'' - 2y' = 8e^{3x} + 2\sin x.$$

19.2.3   Use the method of undetermined coefficients to find the general solution of
$$y'' - 3y' + 2y = e^x.$$

19.2.4   Use the method of undetermined coefficients to find the general solution of
$$y'' + 5y' + 4y = 3 - 2x.$$

19.2.5   Use the method of undetermined coefficients to find the general solution of
$$y'' - 3y' = 6e^{3x} + 4\cos x.$$

19.2.6   Use the method of undetermined coefficients to find the general solution of
$$y'' - y' - 20y = e^{5x}.$$

19.2.7   Use the method of undetermined coefficients to find the general solution of
$$y'' - 5y' - 6y = 2e^{-x} + 3.$$

19.2.8   Use the method of undetermined coefficients to find the general solution of
$$y'' + 9y = 2\sin x - 3\cos x.$$

19.2.9   Use the method of undetermined coefficients to find the general solution to
$$y'' + 16y = -4\sin 4x.$$

19.2.10   Use the method of undetermined coefficients to find the general solution of
$$y'' - y' - 12y = e^{9x} - 2e^{-x} + 1.$$

19.2.11   Use the method of undetermined coefficients to find the general solution of
$$y'' + 4y = 2\sin x + \cos 2x.$$

19.2.12   Use the method of undetermined coefficients to find the general solution of
$$2y'' - 5y' - 3y = \sqrt{e^x} - 4 + \sqrt{e^{-x}}.$$

19.2.13   Use the method of undetermined coefficients to find the general solution of
$$y'' - 2y' - 8y = 5e^{-x}\cos 2x.$$

19.2.14   Use the method of undetermined coefficients to find the general solution of
$y'' - y' - 2y = 6x + 6e^{-x}$.

19.2.15   Use the method of undetermined coefficients to solve the initial value problem
$y'' - 3y' - 4y = 8x^2 + 4;\ y(0) = -\dfrac{17}{4},\ y'(0) = 4$.

19.2.16   Use the method of undetermined coefficients to solve the initial value problem
$y'' + 4y' + 13y = 5\sin 2x;\ y(0) = 0,\ y'(0) = 2/29$.

19.2.17   Use the method of undetermined coefficients to solve the initial value problem
$y'' + 4y = 2\cos 2x;\ y(0) = 0,\ y'(0) = 2$.

SOLUTIONS

SECTION 19.2

19.2.1 The complementary solution is $y_c(x) = c_1 e^x + c_2 e^{2x}$ so the particular solution is of the form $y_p(x) = Ae^{-4x}$. Substitute y_p into the original differential equation to get $A = 1/30$, thus the general solution is

$$y(x) = c_1 e^x + c_2 e^{2x} + 1/30 e^{-4x}.$$

19.2.2 The complementary solution is $y_c(x) = c_1 + c_2 e^{2x}$ so the particular solution is of the form $y_p(x) = Ae^{3x} + B\sin x + C\cos x$. Substitute y_p into the original differential equation to get $A = 8/3$, $B = -2/5$, and $C = 4/5$, thus the general solution is

$$y(x) = c_1 + c_2 e^{2x} + \frac{8}{3}e^{3x} - \frac{2}{5}\sin x + \frac{4}{5}\cos x.$$

19.2.3 The complementary solution is $y_c(x) = c_1 e^x + c_2 e^{2x}$ so the particular solution is of the form $y_p(x) = Axe^x$. Substitute y_p into the original differential equation to get $A = -1$, thus the general solution is

$$y(x) = c_1 e^x + c_2 e^{2x} - xe^x.$$

19.2.4 The complementary solution is $y_c(x) = c_1 e^{-4x} + c_2 e^{-x}$ so the particular solution is of the form $y_p(x) = A + Bx$. Substitute y_p into the original differential equation to get $A = 11/8$ and $B = -1/2$, thus the general solution is

$$y(x) = c_1 e^{-4x} + c_2 e^{-x} + \frac{11}{8} - \frac{1}{2}x.$$

19.2.5 The complementary solution is $y_c(x) = c_1 + c_2 e^{3x}$ so the particular solution is of the form $y_p(x) = Axe^{3x} + B\cos x + C\sin x$. Substitute y_p into the original differential equation to get $A = 2$, $B = -2/5$, and $C = -6/5$, thus the general solution is

$$y(x) = c_1 + c_2 e^{3x} + 2xe^{3x} - \frac{2}{5}\cos x - \frac{6}{5}\sin x.$$

19.2.6 The complementary solution is $y_c(x) = c_1 e^{-4x} + c_2 e^{5x}$ so the particular solution is of the form $y_p(x) = Axe^{5x}$. Substitute y_p into the original differential equation to get $A = \frac{1}{9}$, thus the general solution is

$$y(x) = c_1 e^{-4x} + c_2 e^{5x} + \frac{1}{9}xe^{5x}.$$

19.2.7 The complementary solution is $y_c(x) = c_1 e^{-x} + c_2 e^{6x}$ so the particular solution is of the form $y_p(x) = Axe^{-x} + B$. Substitute y_p into the original differential equation to get $A = -\frac{2}{7}$ and $B = -1/2$, thus the general solution is

$$y(x) = c_1 e^{-x} + c_2 e^{6x} - \frac{2}{7}xe^{-x} - \frac{1}{2}.$$

19.2.8 The complementary solution is $y_c(x) = c_1 \cos 3x + c_2 \sin 3x$ so the particular solution is of the form $y_p(x) = A \sin x + B \cos x$. Substitute y_p into the original differential equation to get $A = -1/4$ and $B = 3/8$, thus the general solution is

$$y(x) = c_1 \cos 3x + c_2 \sin 3x + 1/4 \sin x - 3/8 \cos x.$$

19.2.9 The complementary solution is $y_c(x) = c_1 \cos 4x + c_2 \sin 4x$ so the particular solution is of the form $y_p(x) = Ax \sin 4x + Bx \cos 4x$. Substitute y_p into the original differential equation to get $A = 0$ and $B = 1/2$, thus the general solution is

$$y(x) = c_1 \cos 4x + c_2 \sin 4x + \frac{1}{2} x \cos 4x.$$

19.2.10 The complementary solution is $y_c(x) = c_1 e^{-3x} + c_2 e^{4x}$ so the particular solution is of the form $y_p(x) = Ae^{9x} + Be^{-x} + C$. Substitute y_p into the original differential equation to get $A = \frac{1}{60}$, $B = \frac{1}{5}$ and $C = -\frac{1}{12}$, thus the general solution is

$$y(x) = c_1 e^{-3x} + c_2 e^{4x} + \frac{1}{60} e^{9x} + \frac{1}{5} e^{-x} - \frac{1}{12}.$$

19.2.11 The complementary solution is $y_c(x) = c_1 \cos 2x + c_2 \sin 2x$ so the particular solution is of the form $y_p(x) = A \sin x + B \cos x + Cx \cos 2x + Dx \sin 2x$. Substitute y_p into the original differential equation to get $A = \frac{2}{3}$, $B = C = 0$, and $D = 1/4$, thus the general solution is

$$y(x) = c_1 \cos 2x + c_2 \sin 2x + \frac{2}{3} \sin x + \frac{1}{4} x \sin 2x.$$

19.2.12 The complementary solution is $y_c(x) = c_1 e^{-x/2} + c_2 e^{3x}$ so the particular solution is of the form $y_p(x) = Ae^{x/2} + B + Cxe^{-x/2}$. Substitute y_p into the original differential equation to get $A = -1/5$, $B = 4/3$, and $C = -1/7$, thus the general solution is

$$y(x) = c_1 e^{-x/2} + c_2 e^{3x} - \frac{1}{5} e^{x/2} + \frac{4}{3} - \frac{1}{7} xe^{-x/2}.$$

19.2.13 The complementary solution is $y_c(x) = c_1 e^{-2x} + c_2 e^{4x}$ so the particular solution is of the form $y_p(x) = Ae^{-x} \cos 2x + Be^{-x} \sin 2x$. Substitute y_p into the original differential equation to get $A = -9/29$ and $B = -8/29$, thus the general solution is

$$y(x) = c_1 e^{-2x} + c_2 e^{4x} - \frac{9}{29} e^{-x} \cos 2x - \frac{8}{29} e^{-x} \sin 2x.$$

19.2.14 The complementary solution is $y_c(x) = c_1 e^{-x} + c_2 e^{2x}$ so the particular solution is of the form $y_p(x) = A + Bx + Cxe^{-x}$. Substitute y_p into the original differential equation to get $A = 3/2$, $B = -3$, and $C = -2$, thus the general solution is

$$y(x) = c_1 e^{-x} + c_2 e^{2x} + \frac{3}{2} - 3x - 2xe^{-x}.$$

19.2.15 The complementary solution is $y_c(x) = c_1 e^{-x} + c_2 e^{4x}$ so the particular solution is of the form $y_p(x) = A + Bx + Cx^2$. Substitute y_p into the original differential equation to get $A = 1/4$, $B = 3$, and $C = -2$, thus the general solution is

$$y(x) = c_1 e^{-x} + c_2 e^{4x} - \frac{17}{4} + 3x - 2x^2.$$

Differentiate $y(x)$ to get $y'(x) = -c_1 e^{-x} + 4c_2 e^{4x} + 3 - 4x$ and apply the initial conditions yielding:

$$\left. \begin{array}{r} c_1 + c_2 = 0 \\ -c_1 + 4c_2 = 1 \end{array} \right\}$$

from which $c_1 = -\dfrac{1}{5}$ and $c_2 = \dfrac{1}{5}$ so the solution to the initial value problem is

$$y(x) = -\frac{1}{5}e^{-x} + \frac{1}{5}e^{4x} - \frac{17}{4} + 3x - 2x^2.$$

19.2.16 The complementary solution is $y_c(x) = e^{-2x}(c_1 \cos 3x + c_2 \sin 3x)$ so the particular solution is of the form $y_p(x) = A \sin 2x + B \cos 2x$. Substitute y_p into the original differential equation to get $A = 9/29$ and $B = -8/29$, thus the general solution is

$$y(x) = e^{-2x}(c_1 \cos 3x + c_2 \sin 3x) + \frac{9}{29}\sin 2x - \frac{8}{29}\cos 2x.$$

Differentiate y to get

$$y' = e^{-2x}(-3c_1 \sin 3x + 3c_2 \cos 3x) - 2e^{-2x}(c_1 \cos 3x + c_2 \sin 3x) + \frac{18}{29}\cos 2x + \frac{16}{29}\sin 2x \text{ and}$$

apply the initial conditions to yield $c_1 = 8/29$ and $c_2 = 0$, so the solution to the initial value problem is

$$y(x) = \frac{8}{29}e^{-2x}\cos 3x + \frac{9}{29}\sin 2x - \frac{8}{29}\cos 2x.$$

19.2.17 The complementary solution is $y_c(x) = c_1 \cos 2x + c_2 \sin 2x$ so the particular solution is of the form $y_p(x) = Ax \cos 2x + Bx \sin 2x$. Substitute y_p into the original differential equation to get $A = 0$ and $B = 1/2$, thus the general solution is

$$y(x) = c_1 \cos 2x + c_2 \sin 2x + \frac{1}{2}x \sin 2x.$$

Differentiate y to get $y'(x) = -2c_1 \sin 2x + 2c_2 \cos 2x + x \cos 2x + \dfrac{1}{2}\sin 2x$ and apply the initial conditions to yield $c_1 = 0$ and $c_2 = 1$, so the solution to the initial value problem is $y(x) = \sin 2x + \dfrac{1}{2}x \sin 2x.$

SECTION 19.3

19.3.1 Use variation of parameters to find the general solution of $y'' + 4y = \tan 2x$.

19.3.2 Use variation of parameters to find the general solution of $y'' + 9y = \csc 3x$.

19.3.3 Use variation of parameters to find the general solution of $y'' + 4y = \cot 2x$.

19.3.4 Use variation of parameters to find the general solution of $9y'' + y = \sec \dfrac{x}{3}$.

19.3.5 Use variation of parameters to find the general solution of $y'' + y = \cos^2 x$.

19.3.6 Use variation of parameters to find the general solution of $y'' + y = \sec^2 x$.

19.3.7 Use variation of parameters to find the general solution of $y'' + 6y' + 9y = x^{-3}e^{-3x}$.

19.3.8 Use variation of parameters to find the general solution of $y'' + 3y' + 2y = \dfrac{1}{1 + e^x}$.

19.3.9 Use variation of parameters to find the general solution of $y'' + y = \dfrac{1}{1 + \sin x}$.

19.3.10 Use variation of parameters to find the general solution of $y'' + y = \dfrac{1}{1 + \cos x}$.

19.3.11 Use variation of parameters to find the general solution of $y'' - y = \cosh x$.

19.3.12 Use variation of parameters to find the general solution of $y'' - y = \sinh x$.

19.3.13 Use variation of parameters to solve $y'' + 2y' - 3y = 10e^x$.

19.3.14 Use variation of parameters to solve $y'' + y = \tan x$.

19.3.15 Use variation of parameters to solve $y'' + 2y' + y = e^{-x}$.

SOLUTIONS

SECTION 19.3

19.3.1   The complementary solution is $y_c = c_1 \cos 2x + c_2 \sin 2x$ so the particular solution is of the form $y_p = u \cos 2x + v \sin 2x$. Substitute into (9), solve and get $u' = -\dfrac{1}{2}(\sec 2x - \cos 2x)$ and $v' = \dfrac{1}{2}\sin 2x$, then $u = -\dfrac{1}{4}\ln|\sec 2x + \tan 2x| + \dfrac{1}{4}\sin 2x$ and $v = -\dfrac{1}{4}\cos 2x$ so the general solution is

$$y(x) = c_1 \cos 2x + c_2 \sin 2x - \frac{1}{4}\cos 2x \ln|\sec 2x + \tan 2x| - \frac{1}{4}\sin 2x \cos 2x.$$

19.3.2   The complementary solution is $y_c = c_1 \cos 3x + c_2 \sin 3x$ so the particular solution is of the form $y_p = u \cos 3x + v \sin 3x$. Substitute into (9), solve and get $u' = -\dfrac{1}{3}$ and $v' = \dfrac{1}{3}\cot 3x$, then $u = -\dfrac{x}{3}$ and $v = \dfrac{1}{9}\ln|\sin 3x|$ so the general solution is

$$y(x) = c_1 \cos 3x + c_2 \sin 3x - \frac{x}{3}\cos 3x + \frac{1}{9}\sin 3x \ln|\sin 3x|.$$

19.3.3   The complementary solution is $y_c = c_1 \cos 2x + c_2 \sin 2x$ so the particular solution is of the form $y_p = u \cos 2x + v \sin 2x$. Substitute into (9), solve and get $u' = -\dfrac{1}{2}\cos 2x$ and $v' = \dfrac{1}{2}(\csc 2x - \sin 2x)$, then $u = -\dfrac{1}{4}\sin 2x$ and $v = \dfrac{1}{4}\ln|\csc 2x - \cot 2x| + \dfrac{1}{4}\cos 2x$ so the general solution is

$$y(x) = c_1 \cos 2x + c_2 \sin 2x + \frac{1}{4}\sin 2x \ln|\csc 2x - \cot 2x|.$$

19.3.4   The complementary solution is $y_c = c_1 \cos \dfrac{x}{3} + c_2 \sin \dfrac{x}{3}$ so the particular solution is of the form $y_p = u \cos \dfrac{x}{3} + v \sin \dfrac{x}{3}$. Substitute into (9), solve and get $u' = -3\tan \dfrac{x}{3}$ and $v' = 3$ then $u = 9 \ln \left| \cos \dfrac{x}{3} \right|$ and $v = 3x$ so the general solution is

$$y(x) = c_1 \cos \frac{x}{3} + c_2 \sin \frac{x}{3} + 9 \cos \frac{x}{3} \ln \left| \cos \frac{x}{3} \right| + 3x \sin \frac{x}{3}.$$

19.3.5   The complementary solution is $y_c = c_1 \cos x + c_2 \sin x$ so the particular solution is of the form $y_p = u \cos x + v \sin x$. Substitute into (9), solve and get $u' = -\cos^2 x \sin x$ and $v' = \cos^3 x$, then $u = \dfrac{1}{3}\cos^3 x$ and $v = \sin x - \dfrac{1}{3}\sin^3 x$ so the general solution is

$$y(x) = c_1 \cos x + c_2 \sin x + \frac{1}{3}\cos^4 x + \sin^2 x - \frac{1}{3}\sin^4 x$$

or

$$y(x) = c_1 \cos x + c_2 \sin x + \frac{1}{3} + \frac{1}{3}\sin^2 x.$$

19.3.6   The complementary solution is $y_c = c_1 \cos x + c_2 \sin x$ so the particular solution is of the form $y_p = u \cos x + v \sin x$. Substitute into (9), solve and get $u' = -\sec^2 x \sin x$ and $v' = \cos x \sec^2 x$, then $u = -\sec x$ and $v = \ln|\sec x + \tan x|$ so the general solution is

$$y(x) = c_1 \cos x + c_2 \sin x + \sin x \ln|\sec x + \tan x| - 1.$$

19.3.7   The complementary solution is $y_c = c_1 e^{-3x} + c_2 x e^{-3x}$ so the particular solution is of the form $y_p = u e^{-3x} + v x e^{-3x}$. Substitute into (9), solve and get $u' = -x^{-2}$ and $v' = x^{-3}$ so the general solution is

$$y(x) = c_1 e^{-3x} + c_2 x e^{-3x} + \frac{1}{2x} e^{-3x}.$$

19.3.8   The complementary solution is $y_c = c_1 e^{-2x} + c_2 e^{-x}$ so the particular solution is of the form $y_p = u e^{-2x} + v e^{-x}$. Substitute into (9), solve and get $u' = -\dfrac{e^{2x}}{1 + e^x}$ and $v' = \dfrac{e^x}{1 + e^x}$, then, $u = -(1 + e^x) + \ln(1 + e^x)$ and $v = \ln(1 + e^x)$ so the general solution is

$$y(x) = c_1 e^{-2x} + c_2 e^{-x} + \left(e^{-2x} + e^{-x}\right) \ln(1 + e^x).$$

19.3.9   The complementary solution is $y_c = c_1 \cos x + c_2 \sin x$ so the particular solution is of the form $y_p = u \cos x + v \sin x$. Substitute into (9), solve and get $u' = -\dfrac{\sin x}{1 + \sin x}$ and $v' = \dfrac{\cos x}{1 + \sin x}$ then $u = -\sec x + \tan x - x$ and $v = \ln(1 + \sin x)$ so the general solution is

$$y(x) = c_1 \cos x + c_2 \sin x - x \cos x + \sin x \ln(1 + \sin x) - 1.$$

19.3.10   The complementary solution is $y_c = c_1 \cos x + c_2 \sin x$ so the particular solution is of the form $y_p = u \cos x + v \sin x$. Substitute into (9), solve and get $u' = -\dfrac{\sin x}{1 + \cos x}$ and $v' = \dfrac{\cos x}{1 + \cos x}$ then $u = \ln(1 + \cos x)$ and $v = -\csc x + \cot x - x$ so the general solution is

$$y(x) = c_1 \cos x + c_2 \sin x + \cos x \ln(1 + \cos x) + x \sin x - 1.$$

19.3.11   The complementary solution is $y_c = c_1 e^{-x} + c_2 e^x$ so the particular solution is of the form $y_p = u e^{-x} + v e^x$. Substitute into (9), solve and get $u' = -\dfrac{e^x \cosh x}{2} = -\dfrac{1}{4}\left(e^{2x} + 1\right)$, $v' = \dfrac{e^{-x} \cosh x}{2} = \dfrac{1}{4}\left(1 + e^{-2x}\right)$, then $u = -\dfrac{e^{2x}}{8} - \dfrac{x}{4}$ and $v = \dfrac{x}{4} - \dfrac{e^{-2x}}{8}$ so the general solution is

$$y(x) = c_1 e^{-x} + c_2 e^x + \frac{x}{2} \sinh x.$$

19.3.12   The complementary solution is $y_c = c_1 e^{-x} + c_2 e^x$ so the particular solution is of the form $y_p = u e^{-x} + v e^x$. Substitute into (9), solve and get $u' = -\dfrac{e^x \sinh x}{2} = -\dfrac{1}{4}\left(e^{2x} - 1\right)$,

$$v' = \frac{e^{-x}\sinh x}{2} = \frac{1}{4}\left(1 - e^{-2x}\right), \text{ then } u = -\frac{e^{2x}}{8} - \frac{x}{4} \text{ and } v = \frac{x}{4} + \frac{e^{-2x}}{8} \text{ so the general}$$
solution is

$$y(x) = c_1 e^{-x} + c_2 e^{x} + \frac{x}{2}\cosh x.$$

19.3.13 The complementary solution is $y_c = c_1 e^{x} + c_2 e^{-3x}$ so the particular solution is of the form $y_p = ue^{x} + ve^{-3x}$. Substitute into (9), solve and get $u' = \frac{5}{2}$ and $v' = -\frac{5}{2}e^{4x}$, then $u = \frac{5}{2}x$ and $v = -\frac{5}{8}e^{4x}$ so the general solution is

$$y(x) = c_1 e^{x} + c_2 e^{-3x} + \frac{5}{2}xe^{x}$$

19.3.14 The complementary solution is $y_c = c_1 \cos x + c_2 \sin x$ so the particular solution is of the form $y_p = u\cos x + v\sin x$. Substitute into (9), solve and get $u' = -\sin x \tan x$ and $v' = \sin x$, then $u = \ln|\sec x + \tan x| + \sin x$ and $v = -\cos x$ so the general solution is

$$y(x) = c_1 \cos x + c_2 \sin x + (\ln|\sec x + \tan x| + \sin x)\cos x + \sin^2 x.$$

19.3.15 The complementary solution is $y_c = c_1 e^{-x} + c_2 x e^{-x}$ so the particular solution is

$ue^{-x} + vxe^{-x}$. Substitute into (9), solve and get $u' = 1$ and $v' = x$, then $u = x$ and $v = \frac{x^2}{2}$ so the general solution is

$$y(x) = c_1 e^{-x} + c_2 x e^{-x} + \frac{x^2}{2}e^{-x}$$

SECTION 19.4

19.4.1   A weight of 12 pounds is attached to a vertical spring with spring constant $k = 96$ pounds per foot. The weight is pulled down 2 inches below its equilibrium position and released. Assume the absence of a damping force.

     (a)   Find an initial value problem whose solution, $y(t)$ is the position function of the weight.

     (b)   Solve the initial value problem.

19.4.2   A weight of 4 pounds is attached to a vertical spring with spring constant $k = 8$ pounds per foot and comes to rest in its equilibrium position. The weight is the given a downward velocity of 2 feet per second. Assume the absence of a damping force.

     (a)   Find an initial value problem whose solution, $y(t)$ is the position function of the weight.

     (b)   Solve the initial value problem and find the amplitude, period, and frequency of the vibration.

19.4.3   A vertical spring with weight attached is vibrating with a period $\pi/4$ seconds. Find the numerical value of the weight if the spring constant is $k = 16$ pounds per foot.

19.4.4   A weight of 24 pounds is attached to a vertical spring with spring constant $k = 72$ pounds per foot. The weight is released 3 inches above its equilibrium position. Assume the absence of a damping force.

     (a)   Find an initial value problem whose solution $y(t)$ is the position function of the weight.

     (b)   Solve the initial value problem and find the amplitude, period, and frequency of the vibration.

19.4.5   A weight of 64 pounds is attached to a vertical spring with spring constant $k = 32$ pounds per foot. The weight is then pulled 1 foot below its equilibrium position and released. Assume the absence of a damping force.

     (a)   Find an initial value problem whose solution, $y(t)$ is the position function of the weight.

     (b)   Solve the initial value problem.

     (c)   What is the position of the weight at $t = \dfrac{5\pi}{12}$ seconds? How fast and which way is it moving at that time?

19.4.6 A mass of 50 kilograms is attached to a vertical spring with spring constant $k = 200$ newtons per meter and comes to rest in its equilibrium position. The mass is then given an upward velocity of 10 meters per second. Assume the absence of a damping force.

(a) Find an initial value problem whose solution, $y(t)$ is the position function of the mass.
(b) Solve the initial value problem.
(c) What is the position of the mass at $t = \dfrac{5\pi}{12}$ seconds? How fast and which way is it moving at that time?

19.4.7 A vertical spring with mass attached is vibrating with a period of $\pi/2$ seconds. Find the weight of the mass if the spring constant is $k = 8$ pounds per foot.

19.4.8 A weight of 32 pounds is attached to a vertical spring with spring constant $k = 16$ pounds per foot and comes to rest in its equilibrium position. The weight is pulled down 6 inches below its equilibrium position and released. The surrounding medium has a damping constant $c = 4$.

(a) Find an initial value problem whose solution, $y(t)$ is the position function of the weight.
(b) Solve the initial value problem.

19.4.9 A weight of 4 pounds is attached to a vertical spring with spring constant $k = 2$ pounds per foot. The weight is released 1 foot above its equilibrium position with a downward velocity of 8 feet per second. The surrounding medium has a damping constant $c = 1$.

(a) Find an initial value problem whose solution, $y(t)$ is the position function of the weight.
(b) Solve the initial value problem.
(c) Find the time and position for which the weight attains its maximum displacement from the equilibrium position.

19.4.10 A mass of 40 grams is attached to a vertical spring with spring constant $k = 3920$ dynes per centimeter. The mass is released from its equilibrium position with a downward velocity of 2 centimeters per second. The surrounding medium has a damping constant $c = 560$.

(a) Find an initial value problem whose solution, $y(t)$ is the position function of the mass.
(b) Solve the initial value problem.
(c) Find the period and frequency of the vibration.

SOLUTIONS

SECTION 19.4

19.4.1 **(a)** $m = \dfrac{12}{32} = \dfrac{3}{8}$, so $y'' + 256y = 0$, $y(0) = -1/6$, $y'(0) = 0$.

    (b) $y(t) = c_1 \cos 16t + c_2 \sin 16t$; differentiate y to get $y'(t) = -16c_1 \sin 16t + 16c_2 \cos 16t$ and apply the initial conditions, thus $c_1 = -1/6$ and $c_2 = 0$, so $y(t) = -\dfrac{1}{6} \cos 16t$ is the solution to the initial value problem.

19.4.2 **(a)** $m = \dfrac{4}{32} = \dfrac{1}{8}$ so $y'' + 64y = 0$, $y(0) = 0$, $y'(0) = -2$.

    (b) $y(t) = c_1 \cos 8t + c_2 \sin 8t$, differentiate y to get $y'(t) = -8c_1 \sin 8t + 8c_2 \cos 8t$ and apply the initial conditions, thus $c_1 = 0$ and $c_2 = -1/4$, so $y(t) = -\dfrac{1}{4} \sin 8t$ is the solution to the initial value problem; amplitude $= |-1/4| = 1/4$, $T = 2\pi\sqrt{\dfrac{1/8}{8}} = \dfrac{\pi}{4}$ seconds, $f = \dfrac{4}{\pi}$ cycle per second.

19.4.3 $T = 2\pi\sqrt{\dfrac{m}{k}}$ so $m = \dfrac{kT^2}{4\pi^2} = \dfrac{(16)\left(\frac{\pi}{4}\right)^2}{4\pi^2} = \dfrac{1}{4}$, so $W = mg = \left(\frac{1}{4}\right)(32) = 8$ pounds is the weight.

19.4.4 **(a)** $m = \dfrac{24}{32} = \dfrac{3}{4}$ so $y'' + 96y = 0$, $y(0) = 1/4$, $y'(0) = 0$.

    (b) $y(t) = c_1 \cos 4\sqrt{6}t + c_2 \sin 4\sqrt{6}t$, differentiate y to get $y'(t) = -4\sqrt{6}c_1 \sin 4\sqrt{6}t + 4\sqrt{6}c_2 \cos 4\sqrt{6}t$ and apply the initial conditions, thus $c_1 = 1/4$ and $c_2 = 0$ so $y(t) = \dfrac{1}{4} \cos 4\sqrt{6}t$; amplitude $= 1/4$, $T = 2\pi\sqrt{\dfrac{3/4}{72}} = \dfrac{\pi}{2\sqrt{6}}$ seconds and $f = \dfrac{2\sqrt{6}}{\pi}$ cycles per second.

19.4.5 **(a)** $m = \dfrac{64}{32} = 2$ so $y'' + 16y = 0$, $y(0) = -1$, $y'(0) = 0$.

    (b) $y(t) = c_1 \cos 4t + c_2 \sin 4t$, differentiate y to get $y'(t) = -4c_1 \sin 4t + 4c_2 \cos 3t$ and apply the initial conditions, thus $c_1 = -1$ and $c_2 = 0$ so $y(t) = -\cos 4t$ is the solution to the initial value problem.

    (c) $y(5\pi/12) = -\cos\dfrac{5\pi}{3} = -\dfrac{1}{2}$ so the weight is positioned 6 inches below the equilibrium position; $v(t) = y'(t) = 4\sin 4t$, $y'\left(\dfrac{5\pi}{12}\right) = 4\sin\dfrac{5\pi}{3} = -2\sqrt{3}$ feet per second thus the weight is moving with a downward velocity of $2\sqrt{3}$ feet per second.

19.4.6 **(a)** $y'' + 4y = 0$; $y(0) = 0$, $y'(0) = 10$.

(b) $y(t) = c_1 \cos 2t + c_2 \sin 2t$, differentiate y to get $y'(t) = -2c_1 \sin 2t + 2c_2 \cos 2t$ and apply the initial conditions, thus $c_1 = 0$ and $c_2 = 5$ so $y(t) = 5 \sin 2t$ is the solution to the initial value problem.

(c) $y(5\pi/12) = 5 \sin \dfrac{5\pi}{6} = \dfrac{5}{2}$ so the mass is 2.5 meters above the equilibrium position; $v(t) = y'(t) = 10 \cos 2t$ so $y'\left(\dfrac{5\pi}{12}\right) = 10 \cos \dfrac{5\pi}{6} = -5\sqrt{3}$ meters per second moving downward.

19.4.7 $T = 2\pi \sqrt{\dfrac{m}{k}}$ so $m = \dfrac{kT^2}{4\pi^2} = \dfrac{(8)\left(\dfrac{\pi}{2}\right)^2}{4\pi^2} = \dfrac{1}{2}$ slug so the weight, $w = mg = \left(\dfrac{1}{2}\right)(32) = 16$ pounds.

19.4.8 **(a)** $m = \dfrac{32}{32} = 1$, so $y'' + 4y' + 16y = 0$; $y(0) = -1/2$, $y'(0) = 0$.

(b) $y(t) = e^{-2t}(c_1 \cos 2\sqrt{3}t + c_2 \sin 2\sqrt{3}t)$, differentiate y to get $y(t) = e^{-2t}\left(-2\sqrt{3}c_1 \sin 2\sqrt{3}t + 2\sqrt{3}c_2 \cos 2\sqrt{3}t\right) - 2e^{-2t}\left(c_1 \cos 2\sqrt{3}t + c_2 \sin 2\sqrt{3}t\right)$ and apply the initial conditions, thus $c_1 = \dfrac{-1}{2}$ and $c_2 = \dfrac{-1}{2\sqrt{3}}$ so $y(t) = e^{-2t}\left(1/2 \cos 2\sqrt{3}t + \dfrac{1}{2\sqrt{3}} \sin 2\sqrt{3}t\right)$.

19.4.9 **(a)** $m = \dfrac{4}{32} = \dfrac{1}{8}$ so $y'' + 8y' + 16y = 0$; $y(0) = 1$, $y'(0) = -8$.

(b) $y(t) = c_1 e^{-4t} + c_2 t e^{-4t}$, differentiate y to get $y'(t) = -4c_1 e^{-4t} + c_2\left(-4te^{-4t} + e^{-4t}\right)$ and apply the initial conditions, thus $c_1 = 1$ and $c_2 = -4$ so $y(t) = e^{-4t}(1 - 4t)$.

(c) When $y'(t) = 0$, $t = 1/2$ seconds and the maximum displacement is $y = -e^{-2}$ feet or e^{-2} feet below the equilibrium position.

19.4.10 **(a)** $y'' + 14y' + 98y = 0$; $y(0) = 0$, $y'(0) = -2$.

(b) $y(t) = e^{-7t}(c_1 \cos 7t + c_2 \sin 7t)$, differentiate y to get $y'(t) = e^{-7t}\left(-7c_1 \sin 7t + 7c_2 \cos 7t\right) - 7e^{-7t}\left(c_1 \cos 7t + c_2 \sin 7t\right)$ and apply the initial conditions, thus $c_1 = 0$ and $c_2 = -\dfrac{2}{7}$ so $y(t) = -\dfrac{2}{7}e^{-7t} \sin 7t$.

(c) $T = \dfrac{(4)(40\pi)}{\sqrt{(4)(40)(3920) - (560)^2}} = \dfrac{2\pi}{7}$ seconds so $f = \dfrac{7}{2\pi}$ cycles per second.